3

Fatal Army Air Forces Aviation Accidents
in the United States, 1941–1945

Fatal Army Air Forces Aviation Accidents in the United States 1941–1945

ANTHONY J. MIRELES

Volume 3 : August 1944–December 1945,
Appendices, Indexes

McFarland & Company, Inc., Publishers
Jefferson, North Carolina, and London

Volume 3

LIBRARY OF CONGRESS CATALOGUING-IN-PUBLICATION DATA

Mireles, Anthony J., 1963–
Fatal Army Air Forces aviation accidents in the
United States, 1941–1945 / Anthony J. Mireles.
p.   cm.
Includes bibliographical references and index.

ISBN 0-7864-2106-1 (3 volume set : softcover : 50# alkaline paper) ∞
   ISBN 0-7864-2788-4 (v. 1 : softcover : 50# alkaline paper)
   ISBN 0-7864-2789-2 (v. 2 : softcover : 50# alkaline paper)
   ISBN 0-7864-2790-6 (v. 3 : softcover : 50# alkaline paper)

1. Aircraft accidents — United States — History — 20th century.
2.   United States. Army Air Forces — History.
I. Title.
TL553.52.M57 2006   358.4'134 — dc22   2006008342

British Library cataloguing data are available

On the front cover: A Stearman PT-13B (AAF serial number 40-1643),
Cope, South Carolina, March 6, 1945, United States Army Air
Forces *(courtesy Craig Fuller, Aviation Archeology Investigation
and Research);* fire and smoke ©2006 PhotoSpin

Manufactured in the United States of America

*McFarland & Company, Inc., Publishers
Box 611, Jefferson, North Carolina 28640
www.mcfarlandpub.com*

Contents

Abbreviations

1Lt.	First Lieutenant
2Lt.	Second Lieutenant
AAB	Army Air Base
AAC	Army Air Corps
AAF	Army Air Forces
A/C	Aviation Cadet
AFCE	Automatic Flight Control Equipment
agl	above ground level
A/S	Aviation Student
Aspt.	Aspirant
CAA	Civil Aeronautic Administration
CAP	Civil Air Patrol
Capt.	Captain
Col.	Colonel
Contact	Visual flying conditions
Cpl.	Corporal
CWT	Central War Time
Ens.	Ensign
EWT	Eastern War Time
F/C	Flying Cadet
F/O	Flying Officer
GMT	Greenwich Mean Time
Hg	mercury (in inches)
IFR	Instrument Flight Rules
Lt.	Lieutenant
Lt. Col.	Lieutenant Colonel
Lt. (j.g.)	Lieutenant (junior grade)
Maj.	Major
MSgt.	Master Sergeant
msl	mean sea level
MWT	Mountain War Time
Pfc.	Private First Class
Pvt.	Private
PWT	Pacific War Time
RAF	Royal Air Force
RCAF	Royal Canadian Air Force
Sgt.	Sergeant
SSgt.	Staff Sergeant
Sub-Lt.	Sub-Lieutenant
TSgt.	Technical Sergeant
USMA	U.S. Military Academy
VFR	Visual Flight Rules
WASP	Women's Air Force Service Pilots
W/O	Warrant Officer
Zulu Time	Greenwich Mean Time

A Note about Dates, Times, and Entry Numbers

All times are local unless otherwise indicated and are based on World War II era time zones. Beginning February 9, 1942, most regions in the continental United States moved clocks one hour ahead to "War Time," which ended in October of 1945.

Dates serve as entry numbers and are listed in the Month/Day/Year format. Example: the entry 1-2-41 represents January 2, 1941. When more than one accident occurred on a given day, alphabetical indicators have been attached to the date entries to show the order in which the accidents were recorded and to help the reader distinguish among them. Example: the entry 7-19-43D represents the fourth fatal accident recorded for July 19, 1943.

(1944 continued)

August

8-1-44A. Death Valley, California. At 0815 PWT, a Consolidated B-24D and a B-24J collided in mid-air and crashed 10 miles south of Furnace Creek, Death Valley, California, killing 17 fliers aboard both airplanes. Top turret gunnery student Pvt. Newton J. Steven was hurled from B-24J #42-78522 as it broke apart and he was able to parachute to safety, receiving only minor injuries. The airplanes were part of a flight of six B-24s that had taken off from the Army Air Field at Muroc, California, on a gunnery and formation training mission. The formation, consisting of two, three-ship elements in V-flights, had assembled at 9,000 feet indicated altitude over the AAF Bombing Range near Muroc. The formation then climbed to altitude in preparation for the gunnery exercise over Muroc. The formation flew to the northeast for a short time and then to the east, making a 180-degree left turn over Death Valley. The formation was beginning to level out of the turn and was flying west at 20,000 feet when the collision occurred. Other pilots in the formation reported that during the flight they had observed B-24J #42-78522, which was the lead airplane of the second element, flying too close vertically to the first element. Several minutes before the collision, B-24D #42-72862, flying in the number-two position of the first element, had reported to the flight leader that it was suffering a problem with the number-three turbo-supercharger. Other pilots in the flight stated that the B-24D had trouble holding formation, lagging slightly behind the lead element for much of the flight. Apparently trying to leave the formation — or perhaps continuing in the turn that the formation had completed — the pilot of the B-24D was seen executing a gentle turn to the left, causing the pilot of the B-24J, who was flying a tight formation on the first element, to attempt the same gentle left turn in an attempt to parallel the course of the B-24D and avoid a collision. The B-24D apparently became caught in the propeller turbulence of the leading airplanes, causing its port wing to dip further down in the turn. Unable to follow the maneuver and anticipating the collision, the pilot of the B-24J attempted a banking maneuver to the right as the B-24D continued to bank to the left and into the path of his airplane. The B-24J was attempting to roll to the right and under the B-24D but the number-one engine and propeller collided with the B-24D fuselage from underneath near the waist windows, severing the tail of the B-24D and sending it plummeting to earth where it exploded into flames upon impact. B-24s flying in the second element had to maneuver to avoid striking flaming debris and the severed tail section of the B-24D that were hurtling through the formation. The B-24J appeared to "hover" in mid-air momentarily as the two port engines and pieces of the flight deck and fuselage peeled away before the bomber spun to earth and exploded into flames. Pvt. Steven was thrown clear at this time. The wreckage of B-24J #42-78522 was scattered over two square miles on the talus slope near Devil's Golf Course and B-24D #42-72862 slammed to earth on the salt marsh south of Devil's Golf Course. Investigation revealed that this was the first high-altitude formation flight for all four of the pilots killed in the collision. Investigators recommended that all crewmembers wear their parachute harness at all times while airborne and that crew wear their parachutes in flight whenever possible. Killed in the crash of B-24J #42-78522 were: 2Lt. Ernest J. Chapman, pilot; 2Lt. William Johansen, co-pilot; F/O Eugene E. Hechtman, navigator; F/O John H. Tilson, bombardier; Cpl. Arlington S. Leininger, engineer; Cpl. Richard A. Lira, radio operator; Cpl. Louis C. Bartlett, gunner; Pfc. Ted W. Srygley, gunner; Pfc. Donald C. Becker, gunner. Killed in the crash of B-24D #42-72862 were: 2Lt. Sam B. Johnson, pilot; 2Lt. Rex Phillips, co-pilot; Cpl. Edgar A. Peloquin, engineer; Pfc. Robert T. Thomas, engineer; Cpl. Kenneth D. Towns, radio operator; Cpl. Carroll B. Ball, gunner; Pfc. Ellis B. Crowley, gunner; Pfc. Donald F. Hickok, gunner.

8-1-44B. Fort Myers, Florida. At 1359 EWT, a Curtiss P-40N suffering an in-flight fire crashed one mile southeast of Page Field, Fort Myers, Florida, killing pilot 2Lt. Robert N. Winslow, Jr. The airplane had taken off at 1325 EWT from Page Field on an engineering mission. The airplane was observed at 10,000 feet with thick white smoke trailing. Two pilots observed the P-40 spinning at about 1,500 agl. They observed that there was a fire in the cockpit and the cockpit canopy was in the closed position. The P-40 spun to the ground and exploded into flames. It was speculated that the engine threw a rod.

8-1-44C. Liberal, Kansas. At 1426 CWT, a Consolidated RB-24E crashed one-half mile north of the Army Air Field at Liberal, Kansas, killing three crewmembers and injuring another. The airplane was practicing simulated two-engine landings at the field and was in the pattern at an altitude of 600 feet agl. The airplane entered a right-hand turn from the base leg to final. As the pilot attempted to roll out of the turn, the airplane entered a side-slip to the right; the starboard wing was low and the airplane was losing altitude rapidly. The pilot leveled out and then the nose came up. The starboard wing dropped again and struck an airport obstruction light at an elevation of 25 feet.

The airplane then cartwheeled into the ground 50 yards further on, bursting into flames as it careened for another 150 yards before flipping to an inverted position and coming to rest. B-24 instructor pilot 2Lt. Martin J. Hanson, co-pilot/B-24 student 2Lt. John W. Scott, Jr. and engineer Pvt. Leonard Satzman were killed in the crash. Pilot/B-24 student 1Lt. John V. Schuster received very serious injuries.

8-1-44D. Amarillo, Texas. At 1656, a Bell P-39Q crashed while attempting a landing at Amarillo Army Air Field, Amarillo, Texas, killing pilot 2Lt. Leslie E. Cannon. Investigation revealed that the airplane was turning from the base leg to final approach when it stalled and spun to the ground. The airplane smashed to the ground and burst into flames one mile from the end of the runway.

8-1-44E. Aberdeen Gunnery Range, Idaho. At 1658 MWT, a Republic P-47D crashed at the AAF Aberdeen Gunnery Range 15 miles northwest of Pocatello, Idaho, killing pilot 2Lt. Robert W. Froelich. The airplane was one of a two-ship flight that had taken off from Pocatello Army Air Field on a routine gunnery mission. The airplanes had completed four passes at the targets. After completing a fifth firing pass, the subject airplane attempted a steep right turn at very low altitude, causing the starboard wing to strike the ground. The airplane went out of control and cartwheeled into the ground, exploding violently into flames. The pilot was killed instantly.

8-2-44A. Grayling, Michigan. At 1115 EWT, a Republic P-47D struck and killed an elderly civilian woman who was fishing on a rowboat on Lake Margrethe near Grayling, Michigan. The airplane was part of three-ship flight that had taken off from Oscoda Army Air Field, Oscoda, Michigan, on a high-altitude formation mission. The flight was led by AAF fighter instructor 2Lt. Kenneth W. Wassing, Minneapolis, Minnesota. Free French Air Force pilots Sgt. Antoine P. Fabby and Sgt. Andre C. Erard were flying the wing aircraft. The flight had completed the high-altitude exercise and had descended to buzz the lake. The flight buzzed several boats on the lake prior to the accident, coming in so low that the boaters had to take cover. The flight leader decided to make one last low-level buzz before returning to Oscoda Field. One of the wing airplanes struck the woman on the last pass and the pilot was apparently unaware that a person had been struck. The flight then climbed to 1,500 feet and flew back to the field, landing at 1145. Mrs. Mary Meyer, 72, Madeira, Ohio, was killed when the propeller of the low-flying P-47D piloted by Sgt. Fabby struck her across the back as she was crouched low in the boat in an attempt to avoid the airplane. The propeller made a five-inch deep gash horizontally across the middle of her back, killing her instantly. There were no other marks or signs of injury on her body. Her son, Walter G. Meyer, Plainville, Ohio, who was also on board the boat, was able to avoid injury. It was several hours before officers at Oscoda Army Air Field were aware that an accident had taken place and it was soon determined that the airplanes in question had taken off from there. The three P-47s showed no signs of propeller damage and forensic investigation was unable to find blood evidence on any of the airplanes or propellers. After a thorough investigation, investigators were able to determine that the airplane piloted by Sgt. Fabby actually struck the woman. Lt. Wassing and the two Free French pilots were charged with her death by a General Courts Martial at Selfridge Field, Michigan. Published accounts reported that Lt. Wassing, who was leading the flight, was found guilty and sentenced to 18 months confinement at hard labor and immediately dismissed from the service.

8-2-44B. Foster Field, Texas. At 1400, a Curtiss P-40R dove into the ground and exploded seven miles north of Foster Field, Victoria, Texas, killing pilot 2Lt. Donald E. Stevens. The airplane had taken off from Foster Field on a cross-country navigation flight and apparently was returning to the field for a landing. The airplane was observed at an altitude of approximately 6,000 feet agl in a diving spin to the right. The airplane continued in this attitude until it struck the ground and exploded. Investigators later speculated that the pilot had blacked out while performing aerobatic maneuvers. It was noted that the pilot was not to exceed 2,000 feet agl on the mission.

8-2-44C. Seguin, Texas. At 1810, a Vultee BT-13A crashed six miles south of Seguin, Texas, killing pilots 2Lt. Duane W. Stevens and 2Lt. Raymond A. Tilford, Jr. The airplane had taken off from Randolph Field, San Antonio, Texas, on an authorized aerobatic flight. The airplane was observed buzzing the area before the accident. Investigation revealed that the airplane stalled out of a low-altitude turn, entering a spin at an altitude too low to allow recovery. The airplane struck the ground in a nose down attitude, killing both pilots instantly.

8-3-44A. Miami, Florida. At 0631 EWT, a North American B-25C suffering an engine fire crash landed after taking off from the 36th Street Airport, Miami, Florida, killing two passengers and injuring six passengers and crew, five seriously. The B-25 was returning to the United States from an active war theater and had arrived at Miami the day before, flying in from Borinquen Field, Puerto Rico. The war-weary airplane took off at 0629 EWT on a ferry flight to Brooks Field, San Antonio, Texas. The B-25 became airborne and the pilot retracted the landing gear and began to climb out when the port engine burst into flames. The pilot then started to make a 180-degree turn to the left—into the bad engine—in an attempt to return to the airport. The pilot was unable to hold altitude and the airplane smashed into the ground about one and a half miles southeast of the end of the runway where it burst into flames. Fliers seriously injured were: Capt. Roy F. Menning, pilot; 1Lt. Richard A. Irby, bombardier-rated passenger; Sgt. Thomas R. Smith, TSgt. Edwin N. Beckylhymer, SSgt. Clarence Haymes. Engineer TSgt. John J. Taylor sustained minor injuries. Passenger TSgt. Leslie H. Wolfe, Jr. and U.S. Navy passenger Radioman 2C Marvin Lewis Duncan were killed in the crash.

8-3-44B. DeRidder, Louisiana. At 1115, a Martin B-26B crashed while attempting an emergency forced landing 20 miles south of DeRidder, Louisiana, killing four crewmembers and injuring four others. The airplane had taken off from Lake Charles Army Air Field, Louisiana, on a cross-country navigation flight via DeRidder, Louisiana. The airplane successfully navigated to DeRidder and had taken off at 1100 CWT on the return leg to Lake Charles. Investigators stated, "When approximately 15 minutes out of DeRidder, some sort of malfunction was experienced and the airplane began a surging flight, going up and then losing altitude. Within a very short while, the crash alarm bell was rung and the pilot attempted immediately to make a wheels-up forced landing in a stump field. The airplane landed satisfactorily and slid some distance, but ran squarely into a large stump, demolishing the aircraft. Killed in the crash-landing were: 1Lt. Fred M. Miller, B-26 instructor pilot; 2Lt. Noah A. Lunsford, co-pilot/B-26 student; 2Lt. Billie E. Isgrigg, co-pilot/B-26 student; 2Lt. Abraham Spero, bombardier. Crewmembers injured in the crash-landing were: Cpl. George G. Geisler, engineer; Cpl. Franklin H. Babcock, radio operator instructor; Cpl. Albert H. Hiebert, radio operator; Cpl. Burl E. Pritt, gunner.

8-3-44C. Wright Field, Ohio. At 1525 EWT, a North American B-25C crashed while attempting an emergency landing three miles east of Wright Field, Dayton, Ohio, killing five crewmembers. The airplane took off at 1512 from Wright Field on a routine maintenance test flight. At 1518, the airplane was observed approaching the field trailing thick black smoke from the port engine. The pilot apparently overshot the runway and the airplane passed over the field with landing gear extended and the port propeller feathered. The airplane turned to the left in an effort to enter the normal traffic pattern and come back around for another attempt to land. As the airplane turned into the dead engine it began to loose altitude rapidly. The starboard engine sounded as if it were running at full power. The airplane began slipping toward the ground on the left wing. The airplane then smashed into the ground and burst into flames. Killed in the crash were: 1Lt. Roland E. Dufrene, pilot; 2Lt. John S. Stewart, co-pilot; 1Lt. William L. Crawford, passenger; 2Lt. Stanley W. Penna, passenger; Cpl. Frank R. Mann, passenger.

8-3-44D. Neuse, North Carolina. At 1614, a Vultee-Stinson L-5 collided with a power line and crashed near U.S. Highway 1 at Neuse, North Carolina, killing SSgt. Kenneth E. Williams. The airplane had taken off from Raleigh-Durham Army Air Field, North Carolina, with another L-5 on a low-level contour-flying mission. The airplane was apparently flying at very low altitude when the port landing gear struck the power line, which was strung across the road at an elevation of about 30 feet agl. The airplane did not snap the power line but dragged it off of its pole. The power line pulled taut and the airplane flipped over on its back and smashed to the ground. Investigation revealed that the pilot was flying low in the area prior to the accident. Witnesses on the ground stated that they had observed the pilot waving from the cockpit as he flew over at low altitude.

8-3-44E. Mather Field, California. At 1825, a North American RB-25D crashed nine miles south of Mather Field, Sacramento, California, killing pilot 2Lt. Russell Paine and co-pilot 2Lt. John R. Penttinen. The airplane had completed two supervised low-altitude missions and returned to Mather Field. Investigators stated, "At the conclusion of the second cross-country mission, [the instructor] sent his two students, Lt. Paine and Lt. Penttinen, on a local solo mission to complete the flying period. They took off from Mather Field at 1722 PWT with Lt. Penttinen as pilot. At 1805 PWT, a landing was made at Mather Field, and the pilots changed seats. Lt. Paine moving into the pilot's seat and Lt. Penttinen, flying as co-pilot. Take-off was made from Mather Field at 1813 PWT. At approximately 1820 PWT, witnesses on the ground say they saw the aircraft heading in a generally northwesterly direction at an altitude of about 300 feet [agl]. The aircraft made a turn to the left of 165 degrees, and at an altitude estimated to be 50 feet, stalled out and crashed. Contact with the ground was followed immediately by explosion and fire." Investigation revealed that the port engine had failed because of a ruptured fuel diaphragm and the students did not trim the airplane for single-engine flight.

8-3-44F. Naper, Nebraska. At 2030 CWT, a Douglas C-47A flying in severe thunderstorms broke up in flight and crashed three miles west of Naper, Nebraska, killing four crewmembers and 24 pilot-rated passengers. The airplane took off at 1907 CWT from Bruning Army Air Field, Nebraska, on a personnel ferry flight to Pierre Army Air Base, Pierre, South Dakota. The 24 fighter pilots were being transported to Pierre for gunnery training. The airplane had been flying for one hour and eighteen minutes when it encountered a wall of thunderstorms. Investigators noted that the thunderstorms were reported to contain moderate rain, hail, strong winds, cloud to ground lighting and severe turbulence. Investigators speculated that the C-47 had entered the thunderstorm and was soon thrown out of control by either a lighting strike or powerful turbulence. Witnesses reported that a loud thunderclap was heard and then the airplane's engines apparently stopped. The airplane was seen hurtling out of the overcast out of control minus one wing. The airplane rolled inverted and additional pieces began to fly off. The airplane then smashed into the ground in a slight ravine in an inverted 45-degree dive, exploding violently into flames upon impact. Wreckage and bodies were strewn over a large area. The starboard wing was found over a mile from the main wreckage. Investigation of the wreckage revealed that the starboard wing was stressed beyond its design limits, causing it to fail and separate. Investigators speculated that the airplane was out of control and in an inverted position when it was buffeted by strong turbulence. The pilot apparently overstressed the wing while attempting to recover the airplane. Investigators noted that all passengers were wearing parachutes at the time of the crash. Crewmembers killed in the crash were: Capt. Stanley J. Meadows, pilot; Capt. Robert K. Bohle, co-pilot; Capt. Leslie B. Roberts, navigator; Sgt. Orson H. Hutslar, engineer. Pilot-rated passengers killed in the crash were: 2Lt. William F. Acree, F/O John F. Albert, 2Lt. William C. Armstrong, 2Lt. Millard F. Arnett, Jr., 2Lt. Herbert A. Blakeslee, 2Lt. George E. Boeckman, 2Lt. Jack L. Brown, 2Lt. Richard E. Brown, 2Lt. James C. Burke, 2Lt. Donald J. Clarkson, 1Lt. Lloyd L. Hemphill, 2Lt. Arthur Johnson, 1Lt. Clayton R. Jolley, 1Lt. Leonard C. Jolley, 2Lt. Gerald C. Keller, 2Lt. Jack E. Lytle, 2Lt. Robert E. Nesbitt, Jr., 2Lt. Bernard W. O'Malley, 2Lt. Anthony J. Paladino, 2Lt. Bruce S. Patterson, 2Lt. Leland A. Pope, 2Lt. Charles V. Porter, 2Lt. Pat N. Roberts, 2Lt. Levon Sehorn.

8-4-44A. Hunter Field, Georgia. At 1623 EWT, a Martin B-26G crashed while attempting a take-off at Hunter Field, Savannah, Georgia, killing the crew of six. Investigators stated, "[The pilot] taxied to take-off position at the northwest corner of the field at a point designated as Runway 15R. He was observed to spend a considerable amount of time at this position with his engines running between an estimated 800 to 1,000 RPM. Upon being cleared by the Control Tower for take-off he proceeded to take off towards the southeast. [The airplane] was observed by [another pilot] at location 15R awaiting take-off instructions and that the engines on B-26G #403 emitted a considerable amount of black smoke. According to the statements of witnesses ... the ship gained speed very slowly and did not become airborne until after it had passed the end of the runway, at which point the pilot pulled the ship off. The airplane proceeded in a straight direction for approximately one-half mile, gained altitude of approximately 150 feet, made a sharp roll to the left and crashed and burned. It was observed ... that both engines functioned until the ship crashed. It was also observed ... that both engines emitted black smoke until the airplane crashed. The direct cause of this accident cannot be determined by this board. It appears that there was considerable loss of power on each engine which probably was caused by the spark plugs being fouled before the take-off." The airplane was taking off on an instrument calibration flight. Killed in the crash were: F/O James R. Cleland, pilot; F/O Gerald M. O'Connell, Jr., co-pilot; 2Lt. Raymond S. Liptak, navigator; Cpl. Edward A. Varble, engineer; Cpl.

Peter F. Yagodzinski, gunner; Cpl. Erwin Wong, radio operator.

8-4-44B. Iowa, Louisiana. At 1705, a Martin B-26B crashed four miles north of Iowa, Louisiana, killing the crew of three. The airplane took off from Lake Charles Army Air Field, Lake Charles, Louisiana, at 1629 on a local transition flight. The airplane apparently went into a spin at an altitude of approximately 1,500 feet agl. The pilots were unable to recover the airplane and it spun to the ground in a vertical position, exploding into flames upon impact. The crew was killed instantly. Investigators were not able to determine the cause of the accident. Pilot 2Lt. Charles J. Conn, co-pilot 2Lt. Raymond L. Roelle and engineer Cpl. Bennie D. Hudson were killed in the crash.

8-4-44C. Morris Field, North Carolina. At 2350, a Douglas A-20J crashed one and one-half miles northeast of Morris Field, Charlotte, North Carolina, killing pilot 2Lt. Herman Boyd. The airplane took off from Morris Field on a routine transition flight. Investigators stated, "[The pilot] took off at 2238 and did not make any landings until approximately 2345 when he called for landing instructions and acknowledged 'Wheels down and locked. Gas checked.' After turning on his final approach and lining up with the runway, the aircraft disappeared from sight of the tower. Witnesses stated that the engines were normal and then began to sputter and quit. The aircraft nosed over and plunged straight into the ground from an altitude of about 200 feet."

8-5-44A. Columbia River, Washington. At 0708, a Bell P-39K crashed into the Columbia River ten miles northwest of Peach, Washington, killing pilot F/O Robert L. Parcher. The P-39 was part of a flight of five that took off at 0635 from Ephrata Army Air Field, Washington, on a formation flight. The formation was flying at 7,500 feet msl in one three-ship flight and one two-ship flight. The subject airplane was flying in the number-two position in the lead three-ship element. The formation leader led the flight in a turn to the left. The subject airplane apparently stalled and fell off to the right as the formation turned to the left. Every pilot in the flight lost sight of the subject airplane. Moments later a pilot in the flight observed the airplane dive into the Columbia River about 50 feet from the shore. Investigators could not determine the cause of the accident. It was later speculated that the airplane entered a flat spin and that the pilot was unable to fully recover before diving into the river.

8-5-44B. Silver Lake, California. At 0835, a Douglas P-70B suffered the structural failure of the port horizontal stabilizer and port elevator and crashed 40 miles northeast of Silver Lake, California, killing pilot 1Lt. Stuart B. Swaggerty and observer 2Lt. Henry W. Seitz, Jr. Investigators stated, "Lt. Swaggerty was the flight leader of two P-70 airplanes that were firing on aerial gunnery target towed by an RA-20 airplane. When the ammunition was expended, Lt. Swaggerty informed the tow ship and prepared to return to the home field. As the wingman started to close in to join formation he saw Lt. Swaggerty's airplane start a dive and then pull up. He saw something fly off the airplane as it started into a spin to the right. After about three turns it recovered and went into another spin to the left. It recovered again and started a third spin and crashed, bursting into flames and burning. Investigation of the wreckage showed that the airplane was in a flat position turning slightly to the left when it struck the ground. Both propellers were feathered indicating the pilot was attempting to abandon the airplane but lacked sufficient altitude. The safety belts of both pilot and observer were still buckled. The airplane had practically no forward momentum at the point of impact." The port horizontal stabilizer and elevator were found a few hundred yards from the main wreckage. Examination of the port horizontal stabilizer revealed that it had failed at the point where it joined the fuselage.

8-5-44C. Dixon, Nebraska. At 1145, two Boeing B-17G airplanes collided in mid-air and crashed three miles west of Dixon, Nebraska, about 35 miles west of the Army Air Base at Sioux City, Iowa, killing 17 fliers aboard both airplanes. The pilot of B-17G #42-107144, 2Lt. Charles A. Van Pelt, was able to parachute to safety, sustaining only minor injuries. The Aircraft Accident Classification Committee stated, "Crews departed from the home field at approximately 0715 CWT for the purpose of accomplishing a high-altitude, simulated interception-attack, formation practice mission. Crew had flown for approximately four hours in a three-squadron, 18-ship, 'Combat Training Formation' consisting of high (right), lead (center), and low (left) squadrons, each made up of two elements of three ships each. At a point approximately 35 miles from the home field [at an altitude of approximately 9,000 feet msl], the formation executed a normal 180-degree turn to the left. During the turn, the high squadron moved over to the left and above the lead squadron. The turn had been completed, but the high squadron had not as yet begun to move back to its normal position. The pilot of the ship flying in the number-two position of the first element of the high squadron (B-17G #42-107144) reports that at this time he was flying 'pretty close formation;' that at certain times his wing had overlapped the wing of the [high squadron] lead airplane, and that at certain times he had been holding his nose closer than 50 feet from his lead ship. After the formation's 180-degree turn to the left had been completed, but before the high squadron had moved back to its normal position, the pilot flying the number-two position of the first element of the high squadron throttled back, and in doing so, dropped back and over behind his lead ship. In that position he encountered the [propeller turbulence] of his lead ship, and as a result, his left wing dropped. In spite of the pilot's effort to lift his left wing and leave the formation to the right, he continued downward and to

the left. In this manner, and while he was unable to control the movement of his ship, the airplane came down onto the left wing of the ship flying in the number-two position of the lead element of the lead squadron [B-17G #42-107157]. As a result of the impact, both aircraft went out of control and crashed. While one aircraft (B-17G #42-107144) fell earthward, the pilot reports he operated the alarm bell switch and directed the co-pilot to clear the crew from the ship. Soon thereafter the airplane apparently broke apart and the pilot was thrown free; he pulled his ripcord and made a successful descent." Killed in the crash of B-17G #42-107144 were: 2Lt. George M. Nelson, Jr., co-pilot; F/O William Kavner, bombardier; Cpl. Ben E. Wall, engineer; Cpl. Richard L. Davis, radio operator; Pfc. Val J. McClellan, gunner; Pfc. Jack C. Ince, gunner; Cpl. George H. Grada, gunner; Cpl. Walter L. Jacques, gunner. Killed in the crash of B-17G #42-107157 were: 2Lt. Joseph E. Mead, pilot; 2Lt. Archie E. Moran, Jr., co-pilot; 2Lt. Kenneth M. Dewey, bombardier; Cpl. Nevin B. Mathews, radio operator; Sgt. Chester A. Jurkowski, engineer; Cpl. Bernard E. Ryder, gunner; Pfc. Wilbur J. Utter, gunner; Pfc. Dean W. Still, gunner; Pfc. Paul J. Beruman, gunner.

8-5-44D. Woodson, Texas. At 1250, a Consolidated B-24D crashed-landed one and a half miles southeast of Woodson, Texas, killing aerial engineer SSgt. Conley J. Cornelius. B-24 instructor pilot 1Lt. Joseph Kwasnik, B-24 student 2Lt. Richard B. Cozens and B-24 student 2Lt. Robert C. Craig were seriously injured in the crash landing. The airplane had taken off from the Army Air Field at Fort Worth, Texas, on a routine training flight. Lt. Cozen was flying the airplane from the pilot's seat and the instructor was in the co-pilot's seat. Investigators stated, "Students were practicing flying on one engine at the time of the accident at an approximate altitude of 8,000 feet indicated. Upon losing 2,500 feet altitude (indicated altitude of 5,500 feet) the instructor pilot attempted to bring in engines number-two and number-four, which were feathered. Although the propellers unfeathered the engines failed to start. Number-one and number-three engines would not supply sufficient power to remain airborne and a forced landing was made into a group of trees."

8-5-44E. Stout Field, Indiana. At 1415 CWT, a Beech AT-10BH flying in poor weather suffered a catastrophic structural failure of the starboard wing and crashed seven miles southeast of Stout Field, Indianapolis, Indiana, killing two students. A/C Lawrence P. Nelson and A/C Joseph F. O'Brien, Jr. fell to their deaths in an unsuccessful parachute jump. The AT-10BH had taken off from Freeman Field, Seymour, Indiana, on a routine training flight. The airplane apparently encountered a line of thunderstorms. The students entered the thunderstorms, which included strong turbulence, powerful winds and violent updrafts. It was speculated that the airplane went out of control and was stressed beyond its structural limitations when the students attempted to regain level flight. The students bailed out at an altitude too low to allow the parachutes to deploy.

8-6-44. Daggett, California. At 1258, a Martin AT-23A (a version of the B-26 used for advanced target towing) suffering an in-flight engine fire crash landed ten miles northeast of the Army Air Field at Daggett, California, seriously injuring pilot 2Lt. Joseph W. McCloskey. Aerial engineer TSgt. Maurice J. Blanton was killed in an unsuccessful parachute jump. Target tow-reel operator Sgt. James L. Curtis and radio operator Cpl. Darrell E. Shryock parachuted to safety and were uninjured. Investigators stated, "The AT-23A ... took off about 1240 from Daggett Field. Lt. McCloskey circled the field once, climbing to 4,000 feet when the left engine began backfiring intermittently and cutting out. The pilot had just called the tower stating he was returning to the field when the tow-reel operator saw flames coming from the left engine and notified the pilot. By that time altitude had been lost down to approximately 1,000 feet above the ground, so the pilot ordered the crew to bail out through the bomb bay. By the time the third man was out there was insufficient altitude to jump, so the pilot headed for an open spot and made a crash landing. The pilot then crawled through the burning wreckage to safety. However, he is unable as yet to give a clear picture of the accident. The first man to jump was successful. The engineer who was second failed to open his parachute soon enough and he struck the ground before the canopy fully opened. The third man out was successful and stated that when he jumped the pilot had moved to the co-pilot's seat away from the fire and was standing up."

8-7-44A. Sutter, California. At 0740, two Bell P-63A airplanes collided in mid-air and crashed two miles west of Sutter, California, killing 2Lt. John Oetgen aboard P-63A #42-69130. The pilot of P-63A #42-68897, 2Lt. Leroy C. Porter, parachuted to safety and was uninjured. The airplanes were part of a five-ship flight that took off from the Army Air Field at Chico, California, on a camera-gunnery mission and "operational ceiling mission." Investigators stated, "The flight had been at approximately 20,000 feet firing camera-gunnery and were coming down in string in preparation for returning to the home field. Lt. Oetgen, who had been flying in the number-four position, pulled out of the string at about 15,000 feet and accordingly Lt. Porter, who had been flying in the number-five position, moved into the number-four position. The flight leader, noticing that Lt. Oetgen had dropped back, called him on the radio to ascertain his position. Lt. Oetgen answered the flight leader that he was now in the number-five position and catching up. Lt. Porter, now flying in the number-four position, looked back and saw Lt. Oetgen approximately one-quarter mile behind. At about 4,000 feet, in order to close up his formation, the flight leader started doing a series of

180-degree turns. The first four ships had closed up sufficiently to start to join up in a normal two-ship formation and were making a turn to the left when Lt. Porter stated he felt a sharp blow and heard grinding metal. Lt. Porter stated he was in a steep bank to the left and that the blow caused the nose of his airplane to be knocked to the left and downward, and that he himself was thrown against the right window. His airplane appears to have gone into a spin and ... at approximately 4,000 feet ... [he] bailed out."

8-7-44B. Van Nuys, California. At 0829, a Lockheed P-38G crashed one mile southeast of the municipal airport at Van Nuys, California, killing Capt. Herman G. Humphreys. The airplane was the leadship of a four-ship flight that had taken off from Van Nuys Municipal Airport on a formation training flight. The flight approached the airport and the pilot radioed the control tower that he was going to make a single-engine landing, but he allowed the three wing P-38s to land ahead of him. The subject airplane was seen approaching the airport with the port propeller feathered. The pilot made a "good" single engine approach with landing gear extended and flaps down, but when he was about one-third down the runway at an altitude of about 30 feet he decided to go around. He retracted the landing gear and advanced the throttle on the good engine. The airplane began to turn to the left and then mush toward the ground. The pilot attempted to bring up the port wing but the airplane fell off on the port wing and smashed into the ground in a vertical dive, bursting into flames upon impact. The pilot was killed instantly. The P-38 smashed to the ground near a small house, which sustained minor damage. Investigators noted that the airplane had been written up for "high oil temperature" in the port engine on the previous two flights.

8-7-44C. Great Falls, Montana. At 1200, a Douglas A-20H crashed 13 miles northwest of Great Falls, Montana, killing 2Lt. Lance E. Hull, Jr. The airplane had taken off from Gore Field, Great Falls, Montana, on a routine training flight. Investigators stated, "At about 1200 the subject aircraft was noticed by a farmer ... to have gone into a spin from about 2,000 feet, and crash behind some slopes. The airplane did not burn but it was badly broken up and the injuries to the pilot were fatal. From all indications ... an almost partial recovery of the spin was made by the pilot and that the airplane stalled into the ground. The airplane did not skid over the ground and the engines and nose section were only embedded about a foot into the ground." Investigators noted that this was the pilot's first solo flight in an A-20 type airplane.

8-7-44D. Rome, New York. At 1427, a de Havilland F-8 (the photo-reconnaissance version of the Mosquito fighter) crashed three miles southeast of the Army Air Field at Rome, New York, killing pilot 1Lt. William F. Cass and navigator 2Lt. Orlando Del Vecchio. The airplane was being ferried to Bangor, Maine.

A stream of white smoke was seen trailing the starboard engine while on the take-off run. The airplane became airborne and climbed to about 300 feet when the starboard engine burst into flames. The airplane was seen to lose altitude rapidly and then crash into the ground, exploding into flames upon impact. Investigation revealed that the engines were overheated because of excessive ground operation. A wooden plug was also found in the Automatic Manifold Pressure Regulator, preventing it from working correctly. A representative of the de Havilland Aircraft Manufacturing Company stated, "This has been the practice of the RAF Ferrying Command because of the inability to detect ice in the carburetor when the automatic regulator is working." Investigators stated, "With the wooden plug in this regulator, it is possible to obtain almost unlimited manifold pressure by using full throttle as compared to twelve pounds, or approximately 54 inches, when the automatic regulator is not blocked by this plug."

8-7-44E. Location Unknown. At an unknown time between 1526 and 1926 EWT, a Vultee BT-13B disappeared with its pilot 2Lt. Donald R. Smith. The airplane had made the daily mail run from Punta Gorda, Florida, to Drew Field, Tampa, Florida. The airplane took off at 1526 EWT from Drew Field for the return flight to Punta Gorda. The airplane failed to arrive at Punta Gorda and at 1926 — the estimated time that the fuel was exhausted — the BT-13 was declared missing. Over 100 AAF airplanes participated in a massive search for the airplane and pilot. No trace of either was ever found. It was later speculated that the airplane had probably crashed into a swamp, the Gulf of Mexico or some other body of water. Weather was not thought to be a factor in the disappearance of the airplane.

8-7-44F. Abilene, Texas. At 1600, a Vultee BT-13B crashed 15 miles northeast of the Army Air Field at Abilene, Texas, killing pilot 2Lt. Gerald L. Cline. 2Lt. Robert L. Senn suffered very serious injuries when he bailed out at very low altitude. The airplane had taken off at 1330 CWT from Abilene Army Air Field on a routine instrument training flight. Lt. Senn was the safety pilot in the front cockpit and Lt. Cline was flying as the instrument pilot in the rear cockpit. The pilots had practiced instrument flight in the area north of Abilene all afternoon and at about 1530 endeavored to return to the field. The pilots encountered a line of thunderstorms and attempted to get around them. Lt. Senn, who was flying the airplane, lost control of the airplane when he inadvertently entered instrument conditions. Lt. Senn ordered Lt. Cline to bail out three times. Lt. Senn bailed out at an altitude of 300 feet agl and was seriously injured. It was not known why Lt. Cline did not parachute to safety.

8-7-44G. Camilla, Georgia. At 1730 EWT, a North American AT-6A spun to the ground five miles southeast of Camilla, Georgia, killing A/C Daniel R. Lamberson. The airplane was part of a three-ship flight that had taken off from Spence Field, Moultrie, Georgia, on a formation training flight. Investigators stated, "After approximately one hour of three-ship formation flying, the instructor put his wing men into a trail formation and started a Lufbery circle between 2,000 and 3,000 feet in which he cut A/C Lamberson off and got fairly close behind him before breaking out of the Lufbery and signaling for [them to] rejoin into a three-ship V. The next time [the instructor] looked back his number-three ship, A/C Lamberson, was in a slow spin. The ship made a complete spin to the left, appeared to pause in the spin, and then immediately started to spin again, crashing into the ground and bursting into flames."

8-7-44H. Great Salt Lake, Utah. At 1730 MWT, a Republic P-47D crashed into the Great Salt Lake near Antelope Island, 15 miles west of Hill Field, Ogden, Utah, killing pilot 2Lt. Clarence W. Kusky. The P-47 had taken off from Wendover Field, Utah, at 1620 MWT on a low altitude navigation mission. The aircraft was observed flying at an altitude of 30 feet as it approached Antelope Island. The aircraft was then observed to gradually descend into the water. The P-47 bounded back into the air and climbed a short distance before going out of control and crashing into the lake.

8-8-44A. Sioux City, Iowa. At 1100, two Boeing B-17G airplanes collided in mid-air 13 miles north of the Army Air Base at Sioux City, Iowa, killing nine flyers aboard B-17G #42-102477. Radio operator Pfc. Scott Hays was able to parachute to safety but suffered serious injuries. Six crewmembers parachuted to safety from B-17G #42-102763 and were uninjured. Three crewmembers were injured parachuting to safety, one seriously. Pilot 2Lt. Perry A. Olsen and co-pilot 2Lt. Stanley C. Lochrie, Jr. were uninjured and were able to safely land B-17G #42-102763 at Sioux City Army Air Base. The Accident Classification Committee stated, "Crew of B-17s departed from their home field on a high-altitude formation practice mission. The formation had flown approximately four hours in a two-squadron, 12-ship, 'Combat Training Formation' consisting of a lead (center) and low (left) squadron, each squadron was made up of two elements of three ships each. Formation was returning to the home field at an altitude of about 15,000 feet. According to testimony of witnesses, the second element of the lead squadron was in a slight turn to the left. Pilot and co-pilot of ship #42-102763, flying in the number-three position of the second element in the low squadron, [testified] that during the turn their ship crossed over and above number-one and number-two ships of this element. At this time and while the pilot of #763 was attempting to regain his position, and in order to avoid coming any closer to these ships, pilot pulled up and slightly to the left, catching the [propeller turbulence] of [B-17G #876], which was flying number-three position of the lead element of the low squadron. On encountering the [propeller turbulence], the pilot reports that he lost control of his ship and settled down from the right and onto the empennage of [B-17G #42-102477], flying the lead position of the second element of the low squadron, knocking off the vertical stabilizer and tail assembly off of #477 which went out of control and crashed, one man being able to parachute from this airplane successfully. Ship #763 had received a hole in the nose section, [the number-two] engine was damaged and the left forequarter of the airplane sustained major damage. Pilot of this ship gave the alarm to bail out and all of the crew parachuted, except the pilot and co-pilot, who regained control of the ship, feathering the [number-two propeller], and flew ship #763 back to the home field where a landing was successfully accomplished." Crewmembers killed in the crash of B-17G #42-102477 were: Capt. Alan G. Van Sant, B-17 instructor pilot; 2Lt. Donald G. Shanley, pilot/B-17 student; 2Lt. Wallace Crowier, co-pilot/B-17 student; 2Lt. Donald C. Younger, bombardier; Cpl. Howard L. Clark, engineer; Pfc. Howard J. Machia, Jr., gunner; Cpl. Russell J. Pellegrind, gunner; Pfc. Bernard E. Majewski, gunner; TSgt. Roberto Gonzalez, aerial engineer instructor. Gunner Pfc. Joseph Kozdron received serious injuries and gunners Pfc. Dominic F. Berta and Pfc. Clyde E. Adams sustained minor injuries parachuting to safety from B-17G #763. Crewmembers uninjured parachuting to safety from B-17G #763 were: F/O Chester J. Teich, bombardier; Cpl. Delmar N. Mayfield, engineer; Cpl. Clarence D. Anderson, radio operator; Cpl. Richard C. Harsin, gunner.

8-8-44B. Shickley, Nebraska. At 1425 CWT, a Republic P-47D crashed two miles southeast of Shickley, Nebraska, killing pilot 2Lt. Robert G. Dickey. The airplane had taken off from Bruning Army Air Field, Bruning, Nebraska, on a routine training flight. Investigators stated, "Pilot stalled ship after making a pull-up.... Investigation and evidence shows that the aircraft spun from an altitude of approximately 9,000 feet, going into a flat spin before hitting the ground. The position of the body of the pilot shows that he was clear of the airplane before it hit the ground. It is possible that the pilot was hit by the airplane when bailing out, however, the ripcord was not pulled and the witnesses who saw the crash did not see the pilot leave the airplane. In the opinion of the accident committee the pilot could not recover from the spin and waited until the ship was too low to bail out."

8-9-44A. Soap Lake, Washington. At 0620, a Bell P-39N and a P-39Q collided in mid-air and crashed 12 miles north of Soap Lake, Washington, killing P-39N pilot 2Lt. Eroy W. Dewberry. P-39Q pilot 2Lt. Robert E. Crowell suffered minor injuries parachuting to safety. The airplanes were part of a flight of five that had taken off from Ephrata Army Air Field, Ephrata, Washington, on a dive-bombing and camera-gunnery mission. The number-three airplane took off late and was unable to join the formation. The flight circled the dive-bombing range in a left turn at an indicated altitude of 9,500 feet. The flight leader then signaled for a right echelon formation. Lt. Crowell, flying in the number-three position as the second element leader, crossed under to the right in an effort to achieve his position. The flight leader then signaled the flight to peel-off to make the dive-bombing pass. Lt. Dewberry, flying in the number-two position of the lead element, had fallen back and was endeavoring to regain his position when he collided with Lt. Crowell, who was crossing over to the right in an effort to take the number-two position. Lt. Dewberry's airplane collided with Lt. Crowell's from the left and below. Both airplanes went out of control and spun to the ground where they exploded into flames upon impact. It was later speculated that Lt. Dewberry was killed or seriously injured in the collision and was unable to parachute to safety. Investigators noted that the flight leader did nor order Lt. Crowell to take the number-two position.

8-9-44B. Windsor Locks, Connecticut. At 1000, a Republic P-47D crashed while attempting a landing at Bradley Field, Windsor Locks, Connecticut, killing pilot 2Lt. Francis R. Betz. The airplane made an approach but had come in too high and too fast and had to go around. The airplane was seen to make a steep left turn from base leg to final at about 1,000 feet agl, entering a stall. The nose came up in the turn and the airplane fell off on the left wing, recovered slightly and then fell off to the right. The pilot leveled the wings an instant before the airplane smashed into the ground vertically and exploded about one mile south of the runway.

8-9-44C. Oklahoma City, Oklahoma. At 1018, a Cessna AT-17 and a Lockheed F-5E collided in mid-air and crashed five miles south of Will Rogers Field, Oklahoma City, Oklahoma, killing A/C James P. Olney and A/C Stanley R. Olsen aboard the AT-17. The pilot of the F-5E, 2Lt. George W. Morgan, was able to parachute to safety. He was uninjured. Investigators stated, "[Lt. Morgan] took off from Will Rogers Field at approximately 1015 on a mission to photograph Amarillo at 25,000 feet. He climbed out of the field slowly and made a slow turn to the left, which is normal for leaving the traffic pattern at this base. Then he immediately started a slow turn to the right, climbing at 500 to 700 feet a minute to pick up a heading of 257 degrees to reach his first objective. Pilot stated that he had just leveled out of this turn on his heading when he suddenly perceived an [AT-17] about 300 feet away coming in at [a heading] approximately 30 degrees on a collision course. The pilot immediately banked again to the right and pulled up in a climb to avoid collision with the other aircraft. At this point the pilot states he was at approximately 1,500 to 1,800 feet of altitude above the terrain. The [AT-17] was on an instrument flight plan to cruise at 5,000 feet msl. The

[AT-17] was equipped with the green glass and red goggles for instrument flying and it is believed that they were being used. The [AT-17] took no evasive action, indicating that they did not see the [F-5E]. It is the belief of this committee that the [AT-17] was in a very dangerous position, being at such a low altitude close to the field, directly in line with take-off, where very high performance aircraft such as the P-38 are operating. A witness stated that prior to the collision the [AT-17] was maneuvering considerably in an attempt to bracket the beam or pick up a heading. The position of the wing in a climbing turn on the P-38 obstructs the pilot's vision from the side on which the [AT-17] was traveling. Examination of the wreckage reveals that the tail assembly and the right engine were cut off of the [AT-17] in such a manner to indicate that the pilot of the [AT-17] had made a last minute attempt to turn to his left and away from the [F-5E], but the maneuver was made too late."

8-9-44D. Abilene, Texas. At 1208, a Republic P-47D crashed at the Army Air Base at Abilene, Texas, killing pilot 2Lt. Chester A. Stelmach. The airplane had taken off from the base at 1155 on a routine training flight. Soon after take off, the airplane began to experience propeller trouble. The pilot returned to the field, radioed the control tower and asked for landing instructions. The control tower operator asked the pilot if it was an emergency landing and the pilot responded that it was not. The pilot entered the traffic pattern and on the final approach the control tower operator noticed that the landing gear was not completely extended. The pilot applied throttle and began to go around. The airplane was about 500 feet above the departure end of Runway 17 when it was seen to enter a slow flat turn to the left with the flaps fully extended. When about 45 degrees through this turn the nose dropped about 45 degrees. The airplane remained in a nose down attitude until about 100 feet above the ground. The nose came up about five degrees as the airplane dropped to the ground, exploding into flames upon impact.

8-9-44E. Shickley, Nebraska. At 1410 CWT, a Republic P-47D crashed three miles east of Shickley, Nebraska, killing 2Lt. John A. Wagner. The P-47 was part of a two-ship flight that had taken off from the Army Air Field at Bruning, Nebraska, on a high-altitude training flight. The airplanes were flying at about 19,000 feet indicated altitude when Lt. Wagner, who was leading the flight, entered a wingover maneuver. The number-two ship followed and pulled out of the dive at about 9,000 feet at a speed of about 350 mph. The subject airplane was observed in a dive toward the ground at an angle of approximately 60 degrees. The airplane continued in a dive and struck the ground at a steep angle and at very high speed, exploding violently into flames upon impact.

8-9-44F. Kiowa, Colorado. At 1725, a Consolidated B-24J suffering an in-flight fire crashed five miles southeast of Kiowa, Colorado, killing the crew

of ten. The airplane was part of six-ship flight that had taken off from the Army Air Base at Pueblo, Colorado, on a camera-bombing mission. The formation was flying at an indicated altitude of 20,000 feet when a pilot in formation noticed that the subject airplane's number-four engine was on fire. The observing pilot radioed the subject airplane a warning and was informed by the subject pilot that the number-four propeller was running away. The subject B-24 left the formation to return to Pueblo Army Air Base. The airplane was next observed at an altitude of about 3,000 feet agl with the number-four engine and nacelle in flames. The starboard wing and the number-four engine then separated from the aircraft, causing it to go into a slow spin to the right. The airplane then smashed into the ground on the right side and exploded into flames. The starboard wing was found about one mile from the main wreckage. Investigators speculated that the wing was weakened by the intense fire, causing it to separate from the aircraft. Killed in the crash were: 2Lt. Charles W. Swallow, pilot; 2Lt. Vincent J. Mannix, co-pilot; 2Lt. Martin F. Affeldt, bombardier; Cpl. Bill Fowler, engineer; Cpl. George A. Ritzel, radio operator; TSgt. Harold E. Humphrey, radio operator instructor; Cpl. Elmer M. Mihalik, gunner; Pfc. James V. Lenavitt, gunner; Pfc. John A. Hoffman, gunner; Cpl. Edward S. Meszaros, gunner.

8-10-44A. Barksdale Field, Louisiana. At 0750, Free French Air Force pilot trainee Sgt. Roger Lamic was killed when he walked into the rotating port propeller of a Martin B-26C preparing to take-off from Barksdale Field, Shreveport, Louisiana. The B-26 was piloted by Free French Air Force pilots Sgt. Gilbert R. Crausaz and Sgt. Jean Carton, who were uninjured. AAF instructor pilot Capt. Harry J. Hildeburn was observing the pilots. Investigation revealed that the instructor had decided that seven occupants were too many for the scheduled training flight and he ordered three Free French pilot-rated passengers off of the aircraft just before it was to taxi out to the take-off position. Sgt. Lamic exited the B-26, walked near the port engine nacelle and then forward into the rotating port propeller, killing him instantly. The French pilots were scheduled to practice single-engine operation.

8-10-44B. Ferguson, Missouri. At 0925, a Lockheed AT-18A crashed near Ferguson, Missouri, killing the crew of six. Killed in the crash were: 2Lt. William E. Barrett, pilot; Cpl. Johnny E. Sharp, engineer; Pfc. Ervin Keefer, engineer; Sgt. Albert B. Corsini, passenger; Pfc. Andrew Polinsky, passenger; Pvt. Harold E. Cooper, passenger. The airplane had taken off from Scott Field, Illinois, on a maintenance test flight to check the landing gear operation. The pilot apparently stalled the airplane and it went into a spin. The pilot attempted to recover but the airplane snapped over to an inverted position. The pilot was unable to recover and the airplane smashed into the ground where it exploded into flames. Investigation revealed that both engines were operating and producing power at the time of impact.

8-10-44C. Rocky Ford, Colorado. At 1424, a North American AT-24A (a version of the B-25 used for advanced twin-engine pilot training) crashed one mile southeast of the Auxiliary Airfield at Rocky Ford, Colorado, killing an AAF instructor and two students. Instructor 2Lt. Alan W. Hobbs and students A/C Robert C. Rood and A/C Edwin A. Russell were killed in the crash. Investigators stated, "[The instructor] was giving his two students their first demonstration ride in an AT-24. The airplane took off to the north at 1418 from La Junta Army Air Field and crashed ten miles west of the field at 1424. The instructor was flying in the pilot's seat and it is assumed that he was probably climbing at the time of the trouble. According to eyewitnesses, the airplane was flying south at an altitude of about 3,000 to 4,000 feet when it nosed down and started to spin to the right. At about 200 to 300 feet above the ground, the airplane seemed to momentarily come out of the spin as though a recovery were being attempted. The airplane then dived into the ground at a very steep angle, probably making a half turn to the right. It was possible that a secondary stall occurred when the airplane momentarily came out of the spin." Investigators speculated that the airplane entered a spin because of a possible runaway propeller in a climbing turn, an engine failure in a climbing turn, or the malfunction or failure of the flight controls.

8-10-44D. Ballinger, Texas. At 1700, a Fairchild PT-19B crashed seven miles west of Bruce Field, Ballinger, Texas, killing A/C Ralph A. Bolda. Investigation revealed that the student attempted a low speed turn and stalled the airplane at an altitude too low to allow recovery. The airplane entered a spin and smashed into the ground in a nose down attitude, killing the student instantly.

8-10-44E. Walterboro, South Carolina. At 1815, a Republic P-47D crashed three miles west of the Army Air Field at Walterboro, South Carolina, killing pilot 2Lt. Richard H. Bell. The airplane had taken off from Walterboro Army Air Field on a four-ship formation flight. The formation was flying at 20,000 feet when the flight leader's airplane began to experience engine trouble and he alerted the flight that he was returning to the field. The flight leader ordered the number-three and number-four airplanes to complete the formation mission. He ordered the subject airplane, which was the number-two P-47D, to fly individually for the remainder of the flight period. The airplane was seen flying at high altitude and then a short time later it was seen in a vertical dive at about 5,000 feet. Pieces of the airplane were seen falling away as it dove to the ground. The airplane smashed into the ground at a steep angle and at high speed, exploding violently into flames. Pieces of the starboard horizontal stabilizer and elevator, the vertical fin and rudder and the starboard wing tip and aileron were found about a mile

from the main wreckage. The cause of the accident could not be determined.

8-11-44A. Moses Lake, Washington. At 0640, a Bell P-39Q crashed approximately 20 miles SSW of Moses Lake, Washington, killing pilot 2Lt. Leonard A. Fehrenbach. The airplane was part of a three-ship flight that had taken off from Moses Lake on a low-altitude navigation flight and a dive-bombing mission. After successfully completing the navigation flight the formation flew to AAF Dive-bombing Range #3. The flight dove on the target from an altitude of 6,500 feet, the leader pulling out at about 5,000 feet. The flight leader looked back and observed the subject airplane in a spin to the right. The airplane described three turns of spin before reversing direction and entering a spin to the left. The airplane stopped spinning momentarily before entering a flat spin to the right. The pilot never recovered the airplane and it spun to the ground, exploding into flames upon impact.

8-11-44B. Pinellas Army Air Field, Florida. At 1045 EWT, a North American P-51C crashed while attempting a landing at Pinellas Army Air Field, St. Petersburg, Florida, killing 2Lt. Seymour Segal. The airplane was returning to the field after successfully completing a low-altitude navigation mission. The airplane was coming in for a landing on Runway 9 when the pilot decided to go around. The airplane was about three-quarters of the way down the runway and the landing gear was retracting when the engine failed at an altitude of 500 feet agl. The airplane entered a stall, fell off on the port wing and then smashed into the ground just north of Runway 9. The pilot was killed instantly. Investigation revealed that carburetor discharge nozzle was obstructed, causing the engine to stop.

8-11-44C. Wendover Field, Utah. At 0858 MWT, a Republic P-47D crashed near Wendover Field, Utah, killing pilot 2Lt. William A. Dougherty. The airplane was part of a flight that had taken off from Wendover Field on a routine training mission. The flight returned to the field and the flight leader signaled the formation to peel-off for landing. The subject airplane peeled off and entered the traffic pattern. While making the turn from base to final approach, the airplane was seen to make a steep turn to the right. The airplane was seen to straighten out and then suddenly go into a steep bank to the left, losing altitude rapidly. The pilot attempted to recover by applying full throttle but the airplane continued into the ground and crashed. The pilot was killed upon impact.

8-11-44D. Passage Key, Florida. At 1300 EWT, a Curtiss P-40N crashed into the Gulf of Mexico 500 yards east of Passage Key, Florida, killing pilot 1Lt. John Simenitzky. The airplane had taken off from Sarasota Army Air Field, Florida, on a low-level bombing mission and then an aerial gunnery mission, which was to be conducted at an altitude of 5,000 feet. Investigators stated, "Lt. Simenitzky was on a skip bombing and aerial gunnery mission, flying number three in a four-ship flight. Lt. Simenitzky was in normal position as the flight, in string, made its run over the target, and his bomb was seen to hit the island. The flight pulled up from the target in a climbing left turn and started to join the formation. Lt. Simenitzky's airplane reached an altitude of about 500 feet, leveled off for a few seconds, and then started a shallow diving turn in the same general direction as the formation was headed. The airplane continued to lose altitude in the same rate of turn and dive and finally struck the water with the left wing. The airplane exploded on contact with the water." Investigation revealed that the pilot had unfastened his shoulder harness at some time during the flight. He was killed when his head collided with the interior of the aircraft when it collided with the sea. An AAF crash boat recovered the body immediately after the crash.

8-11-44E. Richmond, Virginia. At 1615, a taxiing Vultee BT-13B struck and killed pilot 2Lt. Robert W. Small, who was walking near the taxiway at the Army Air Base at Richmond, Virginia. Investigators stated, "[At 1615, the] subject aircraft was cleared by the Control Tower to taxi from the transient aircraft area to the end of Runway 20. A 'Follow Me' jeep, used at this base to direct transient aircraft, preceded the airplane and led it to the taxiway paralleling Runway 2-20 and then both proceeded north. As Hangar #1, located to the west of the taxiway, was neared a few men crossed the taxiway going from the hangar to the physical training area east of the taxiway. Having made an [S-turn] to the right and noticed all was clear ahead beyond the hangar, the pilot made a gradual turn to the left to see which taxiway the jeep would take to the runway. As he did so, [2Lt. Small] came out of the hangar and started walking north three to five feet off the taxiway on the west side. At a point about 100 feet north of the hangar, the left wing of the airplane struck [Lt. Small]. According to witnesses, Lt. Small gave no indication whatsoever that he realized an airplane was approaching. The pilot brought his airplane to a stop and immediately called the Control Tower for an ambulance." The pilot of the BT-13B, 2Lt. Ira C. Akin, was uninjured.

8-11-44F. Riverside, California. At 1615, a Lockheed P-38F suffering a fire in the port engine was abandoned in flight, crashing five miles north of March Field, Riverside, California, killing a civilian technician. Robert C. Oelschlaeger did not parachute to safety and went in with the airplane, a modified "piggyback" version of the P-38F. Pilot 1Lt. Thomas C. Taylor parachuted to safety and was uninjured. The P-38 had taken off from March Field on a maintenance test flight. Another pilot had reported that there was some trouble with the port engine on a flight the previous day. The necessary adjustments were made to the engine and the flight was scheduled to check the engine's performance. Mr. Oelschlaeger had installed the port engine and he asked to come along on the flight to observe the performance of the port engine. Mr. Oelschlaeger was allowed to fly and was issued a chest parachute pack. The pilot took off and climbed the airplane to about 10,000 feet msl. After cruising for a short time at this altitude smoke was noticed in the cockpit. The pilot opened the side window and smoke filled the cockpit. The pilot jettisoned the canopy and dove the airplane slightly. He ordered the technician to follow him out of the cockpit. The pilot leveled the airplane and stood up and was immediately sucked out of the cockpit. The airplane crashed into the ground near a small building and exploded into flames. Mr. Oelschlaeger, of Fort Wayne, Indiana, was found in the wreckage of the airplane.

8-11-44G. Carmel, California. At 1740, a Douglas P-70A crashed ten miles south of Carmel, California, killing pilot 1Lt. Frank R. Pate, 25, Wilson, North Carolina. The airplane took off from Salinas Army Air Field, California, on a pilot proficiency-/high-altitude flight. A civilian hunter observed the airplane in a vertical dive with the starboard propeller stopped and apparently feathered. The pilot pulled out of the dive, stressing the airplane beyond its design limits. Pieces of the airplane were seen to fall away as the pilot pulled out of the dive. The airplane came out of the dive, flying "wobbly" for a few moments before it went into a steep left turn, crashing into the side of a hill and exploding into flames. The exploding airplane caused a forest fire that burned for nearly two weeks. Investigation revealed that the port wing had failed and separated just prior to the airplane diving into the hill. The outer port wing was found about one-quarter mile from the main wreckage.

8-11-44H. Willacoochee, Georgia. At 2327 EWT, a North American AT-6C flying in poor weather crashed four miles southwest of Willacoochee, Georgia, killing A/C William R. Faris. The airplane had taken off from Spence Field, Moultrie, Georgia, on a navigation flight to McRae, Georgia, to Alma, Georgia, and return to Spence Field. The airplane had checked in at Alma and was on course for Spence Field. Farmers in the area reported that it was raining hard and the wind was strong when the airplane was heard to crash into a wooded area. Investigation revealed that the airplane hit the ground in a 70-degree dive, turning slightly to the left at the time of impact. It was later speculated that the student lost control of the airplane while flying in instrument conditions.

8-12-44A. Stuttgart, Arkansas. At 0005, two Beech AT-10 airplanes collided in mid-air three miles northeast of the Army Air Field at Stuttgart, Arkansas, killing A/C Matthew L. Bergin, Jr. and A/C Waldo F. Belden aboard AT-10 #42-2385. A/C Orson F. Poley and instructor 2Lt. Allan Gardner escaped injury and were able to safely land AT-10 #42-34856. Investigators stated, "A normal traffic pattern was being flown with [A/C Poley] at the controls. The turn onto the base leg was about half completed when both the instructor and the student momentarily saw another airplane off of their left wing, apparently below and slightly to the rear, and at the same instant heard a noise described by the instructor to indicate that the two airplanes had brushed together. The instructor immediately took over the controls, maneuvering the ship up and to the right, and causing the instructor and student to lose sight of the airplane, which had collided with them. The instructor then radioed the occurrence to the tower and landed after receiving landing instructions." Investigators faulted all of the pilots for failing to see and avoid the lighted aircraft in the traffic pattern.

8-12-44B. Richmond, Virginia. At 0703, a Republic P-47C crashed while attempting a landing at the Army Air Base at Richmond, Virginia, killing pilot 2Lt. Louis G. Cox. The P-47C was the number-four airplane of a four-ship flight that took off from Richmond AAB on a high-altitude formation flight. A few minutes after take-off the pilot radioed the flight leader that he was experiencing propeller trouble and was returning to the base. The pilot received landing instructions and entered the traffic pattern. The pilot made an approach but came in too high. The control officer ordered him to go around. The pilot advanced the throttle for the go-around but failed to retract the landing gear. The airplane was flying at about 150 feet agl above the end of the runway when the pilot began a turn to the left. The airplane stalled, snapped left to an inverted attitude and then rolled through and into a spin to the left. The pilot was unable to recover the airplane and it smashed into the ground in a diving spin to the left, exploding violently into flames upon impact. The flap lever was found in the "up" position and the landing gear was in the extended position.

8-12-44C. Memphis, Tennessee. At 1256 CWT, a Douglas C-47A crashed while attempting a take-off at the Memphis Municipal Airport, Memphis, Tennessee, killing pilot 2Lt. Romaine N. Yates and seriously injuring co-pilot F/O Bernard J. Reed and engineer Sgt. Eugene W. Wymola. Passenger 2Lt. Henrietta L. St. Clair sustained minor injuries escaping the burning aircraft. Twenty-three other passengers were uninjured. The airplane was taking off on a "Hospital Evacuation" flight to Chickasha, Oklahoma. The Accident Classification Committee stated, "Everything appeared normal until the ship was about one-quarter of the way down the runway, the left wing was seen to dip, a loud report was heard, and soon after the ship careened off of the runway to the right with flames emitting from the cockpit and tail surfaces. After the ship had started its take-off run, the engineer released the spring latch lock. When the ship had attained approximately 60 mph airspeed, the co-pilot either mistook some movement of the pilot's hand or thought that he had heard the pilot say, 'Gear up,' and pulled the latch into the unlatched position and the gear handle to the

up position. The engineer immediately returned the landing gear handle to the neutral position, but the locks had been pulled and the left landing gear started to retract which caused the left propeller to strike the runway a number of times for a distance of approximately 25 feet. The reduction gear case broke, the propeller flew loose, traveled forward and then inboard, slashed the cockpit in a number of places, broke the instrument panel, the control columns, numerous hydraulic lines, [punctured the] de-icer tank, and incapacitated the pilot. However, it is believed that when the pilot realized that the left landing gear was collapsing, he attempted to pull the ship off of the runway and undoubtedly applied right rudder and right aileron causing the ship to go to the right hand side of the runway. That, and the possibility that the co-pilot may have applied right rudder and brake are the only explanation of the ship's turning to the right following a left landing gear collapse and the loss of the left propeller. Two large holes in the runway were made by the propeller when it came out of the cockpit and struck the runway. It then became lodged against the left wing and left landing gear and was dragged off the runway to within a few feet of where the ship finally stopped. The fire, which was noticed before the ship swerved from the runway, was caused by a combination of shorted electrical circuits, hydraulic fluid, and a demolished de-icer tank. The flaming liquid was blown back and ignited the tail surfaces."

8-13-44A. Hunter Field, Georgia. At 0846 EWT, a North American B-25H exploded in mid-air and crashed four miles southwest of Hunter Field, Savannah, Georgia, killing the crew of five. Killed in the crash were: 2Lt. Dale T. Veghte, pilot; 2Lt. Robert W. Madden, navigator; Cpl. Carl H. Dilyeu, engineer; Cpl. Nicholas C. Martinelli, radio operator; Cpl. Don P. Moldenhauer, gunner. The B-25H took off at 0808 EWT from Hunter Field on an instrument calibration test. Witnesses stated that the B-25 exploded at an altitude of approximately 2,000 feet agl and began losing altitude immediately. The airplane dropped about 1,000 feet and then a second explosion was observed. The entire forward section of the B-25 was blown off in the blast. The airplane broke up, scattering wreckage and bodies over an area of about a mile. Investigators speculated that gasoline fumes had collected in the port wing and were ignited by an unknown source, causing the initial explosion. The blast had caused fuel lines in the bomb bay area of the fuselage to rupture or become dislodged, spilling raw gasoline and allowing fumes to accumulate and cause a second explosion. It was speculated that electrical sparks or the fire in the port wing had ignited gasoline fumes in the fuselage, causing the second explosion. The port wing was found about a mile from the main wreckage.

8-13-44B. Gulf of Mexico, Texas. At 1150, a Republic P-47D collided with an aerial target and crashed into the Gulf of Mexico 60 miles east of Palacios Army

Air Field, Texas, killing pilot F/O Vance Whitten. The P-47 had taken off from Palacios on a low-altitude aerial gunnery mission. The airplane made a pass from the right and had shot the target off of the target tow ship. The target broke free and fell into the path of the P-47, fouling the propeller and impairing the pilot's ability to control the ship. The P-47 rolled out of control and crashed into the Gulf of Mexico half a mile off shore.

8-14-44A. Punta Gorda, Florida. At 1038 EWT, two Curtiss P-40N airplanes collided in mid-air and crashed 18 miles west of Punta Gorda, Florida, killing 1Lt. Robert M. Day aboard P-40N #43-23221. The pilot of P-40N #42-106013, 2Lt. Harold E. Hanson, was able to parachute to safety but suffered serious injuries. The airplanes had taken off from Punta Gorda Army Air Field, Florida, on an over-water aerial gunnery mission. Investigators stated, "Lt. Hanson, tow pilot, was circling to the left at 9,000 feet waiting for his flight to fire. When he spotted his flight, they were below him in a climbing turn to the right. Thinking Lt. Day, the flight leader, saw him, Lt. Hanson started out into the Gulf. He was in this turn when his ship and Lt. Day's ship collided. It is believed that Lt. Day's ship exploded [in the collision]. Lt. Hanson bailed out and received major injuries. Lt. Day was killed. Flying debris from the explosion struck F/O Ralph Mann's aircraft [P-40N 42-105741], resulting in major damage."

8-14-44B. Bartow, Florida. At 1048 EWT, a North American P-51B crashed while attempting an emergency landing at the Army Air Field at Bartow, Florida, killing pilot 2Lt. John R. Sealy. The airplane took off from Bartow on a camera-gunnery mission. When the flight had reached an altitude of 8,000 feet, the pilot radioed the flight leader that his engine was running rough and "would not take throttle." The pilot was ordered to leave the formation and head back to the field for an emergency landing. Tower personnel observed the airplane trailing thick smoke as it was approaching the field parallel to Runway 14, immediately clearing the airplane for an emergency landing on any runway. The pilot radioed that his landing gear would not extend and that the airplane was on fire. The pilot then made a 270-degree turn for an approach on Runway 23. As the airplane made the approach, the smoke stopped and then a large sheet of yellow flame was seen engulfing the cockpit. The P-51 rolled inverted and dove into the ground just short of the runway where it exploded into flames. Investigators speculated that smoke and flames incapacitated the pilot, causing him to lose control of the airplane. Investigation revealed that the engine threw two rods through the engine block, causing the engine to catch fire.

8-14-44C. Leesburg, Florida. At 1103 EWT, a North American P-51B crashed one and a half miles NNE of Leesburg, Florida, killing pilot 2Lt. Byron E. Dillant. The airplane had taken off from Bartow

Army Air Field, Bartow, Florida, on a low altitude navigation mission. The airplane was flying at about 400 feet agl when it suddenly climbed to about 800 feet, rolled to an inverted attitude and then dove into the ground. The airplane exploded violently into flames upon impact, killing the pilot instantly. Investigators could not determine the cause of the crash.

8-14-44D. Middletown, Pennsylvania. At 1444, a North American AT-6A crashed while attempting to make a landing at Olmsted Field, Middletown, Pennsylvania, fatally injuring pilot 1Lt. Philip Stewart and seriously injuring civilian passenger Arthur J. Dettmers. The airplane made a landing on Runway 11 with a slight crosswind from the south. The airplane began to ground loop toward the left, the port wing almost scraping the terrain as the airplane turned 60 degrees from the runway heading. The pilot opened the throttle and attempted to take-off. The airplane, attempting to take off downwind, failed to gain enough speed to become airborne and collided with a wooden cafeteria building. The airplane did not burn. Rescue personnel experienced great difficulty trying to free the two occupants. The pilot died of his injuries at about 1700 EWT.

8-14-44E. Sherman, Texas. At 1530, a Vultee BT-13A crashed 15 miles north of Perrin Field, Sherman, Texas, killing WASP pilot Mary P. Hartson, Portland, Oregon, and passenger SSgt. Orville H. Eitzen, Coin, Iowa. The airplane had taken off from Perrin Field on an "engineering and radio check flight." Investigation revealed that the engine apparently failed and the pilot attempted to make an emergency forced landing. The airplane stalled and fell off into a spin to the right. The airplane spun to the ground and crashed in a nose-down attitude, bursting into flames upon impact.

8-14-44F. Merced, California. At 1430, a Vultee BT-13A crashed six miles southwest of the Army Air Field at Merced, California, killing instructor 2Lt. John T. Smith and student A/C John A. Armstrong. The instructor and student were apparently practicing simulated forced landings in the area prior to the crash. Soon afterward the engine failed and the instructor took over the controls to attempt an actual emergency forced landing. The instructor apparently stretched the glide in an attempt to clear an irrigation ditch. The airplane stalled and spun to the ground from an altitude of 50 feet agl.

8-14-44G. Perry, Georgia. At 1800 EWT, a Vultee BT-13A crashed three and a half miles northeast of Perry, Georgia, killing student pilot SSgt. O.B. Vickery. The airplane was spotted in a spin to the left at an altitude of 5,000 feet. The airplane was seen to come out of the spin momentarily before entering a spin to the right. The airplane crashed and burst into flames upon impact. The airplane had taken off from Cochran Field, Macon, Georgia, on a routine student solo flight.

8-14-44H. Clayton, Georgia. At 2350 EWT, a North American B-25J flying in scattered thunderstorms collided with rising terrain seven and one-half miles northeast of Clayton, Georgia, killing the crew of six. Investigators stated, "The pilot was on an authorized [navigation-training] flight from MacDill Field, Tampa, Florida, to Greenville Army Air Base, South Carolina. The aircraft departed MacDill Field at 2050 EWT and at 2220 EWT the pilot contacted the radio ground station at Greenville AAB and reported that he was in the vicinity of Waycross, Georgia. The next contact was made at 2340 EWT. The pilot was contacted on VHF by Greenville AAB Control Tower after [the] Control Tower Officer heard the pilot of the airplane trying to contact the control tower of a field he was circling in order to make a landing. (This field was later identified as Le Tourneau Field, Toccoa, Georgia) Instructions were given the pilot by the Greenville AAB Tower Officer and the pilot acknowledged and answered that he would follow out the instructions. The aircraft left the Toccoa Airport and headed northward and crashed into Flat Top Mountain at approximately 2350 EWT, ten minutes after leaving the Toccoa Airport." The elevation of the crash site was estimated to be about 3,000 feet. Killed in the crash were: 2Lt. Robert E. Austin, pilot; 2Lt. Charles M. Case, co-pilot; 2Lt. Calbin J. Eaby, navigator; Cpl. James P. Keck, engineer; Cpl. Gerald E. Doyle, radio operator; Cpl. Marvin L. Barr, gunner.

8-15-44A. Tuscaloosa, Alabama. At 1107, a Stearman PT-17 crashed eight miles northwest of Van De Graaff Airport, Tuscaloosa, Alabama, killing Free French Aviation Student Robert Chigaeff de Procofieff. The airplane was observed performing steep turns at an altitude of 200 feet agl above very hilly terrain and the student was observed waving to people on the ground. The airplane lost altitude as the student performed the steep turns while approaching an area of high trees. The student attempted to pull up and the airplane entered a stall. The PT-17 fell off and dove into the ground at an angle of 70 degrees, fatally injuring the student. He died a short time after the crash.

8-15-44B. Edison Junction, Florida. At 1600 EWT, a North American P-51B flying in a severe thunderstorms crashed at Edison Junction, Florida, killing pilot 2Lt. Howard G. Cropsey, who died in an unsuccessful parachute jump. The airplane was part of a two-ship flight that had taken off from Bartow Army Air Field, Florida, on an individual combat mission. The two airplanes became separated and the subject airplane inadvertently flew into a severe thunderstorm. Lt. Cropsey lost control of the airplane and he attempted to parachute to safety while the airplane was flying at very high speed. The pilot either pulled the ripcord too early or his parachute seat pack was struck by part of the airplane, causing damage the pack and the parachute and possibly causing an inadvertent deployment. Witnesses on the ground stated that the pilot

appeared to be "lifeless" as he descended rapidly in the damaged parachute. He struck the ground hard; investigators speculated that the pilot was killed when he struck the ground. His body was found a short distance from the spot where the airplane smashed into the ground.

8-15-44C. Chandler, Arizona. At 1500 MWT, a North American AT-6C crashed three miles northeast of Santan Peak near Chandler, Arizona, killing instructor 2Lt. Jack W. Stewart and student A/C William R. Germain. Investigators stated, "Lt. Stewart and A/C Germain were on a local transition flight and were supposed to go to an auxiliary airfield to shoot landings and possibly solo A/C Germain, after doing some air work. When they did not show up at the auxiliary airfield, an investigation was started, and the ship was found approximately 12 miles southeast of Williams Field. Upon investigation [the members of the Accident Classification Committee] were unable to find any eyewitness of the accident, and the only conclusion that can be drawn is an assumption that they had stalled on the last turn into the auxiliary airfield. The airplane was not damaged sufficiently to indicate that they had gone in from a very high altitude, which leads us to assume that it was a simulated forced landing."

8-15-44D. Gardner Field, California. At 1500, two Vultee BT-13A type airplanes collided in mid-air 15 miles ENE of Gardner Field, Taft, California, killing four fliers. Student A/C Courtney G. Taylor was killed in the crash of BT-13A #41-21725 and instructor 2Lt. Stephen J. Ginter fell to his death in an unsuccessful parachute attempt. Instructor 2Lt. Dick L. Eriksen and student pilot 2Lt. David W. Gemmill were killed in the crash of BT-13A #41-21757. Investigators stated, "Airplane #41-21757 was in a spin, and airplane #41-21725 was apparently doing either a power-on stall to the left or a maximum climbing turn to the left at approximately 6,000 feet indicated altitude. The engine of airplane #41-21757 struck the center of the right outer wing panel of airplane #41-21725. Airplane #41-21757 then stopped its spinning motion and seemed to be under control for a short time but almost immediately fell off into a vertical dive and hit the ground with terrific force. It is believed that the occupants of airplane #41-21757 were injured at the time of the crash in the air. With the right wing cut off, airplane #41-21725 fell off into a high-speed spin and subsequently made contact with the ground at terrific speed. Lt. Ginter apparently attempted to parachute to safety at an altitude of approximately 50 to 100 feet above the ground, his body was found 200 feet from the wreckage and the ripcord was pulled. The pilot chute was out of the pack, but the rest of the parachute was still folded in the pack. The body of A/C Taylor was found in the wreckage, which was completely buried in very soft ground. No attempt was made by the occupants of the airplane #41-21757 to parachute to safety."

8-15-44E. Atlanta, Georgia. At 1900 EWT, a North American B-25J flying in severe thunderstorms suffered a catastrophic structural failure and crashed six miles east of the Army Air Base at Atlanta, Georgia, killing pilot 2Lt. Robert B. Deyerle and co-pilot F/O Tedd R. Levy. The airplane had taken off from Fairfax Field, Kansas City, Kansas, on a ferry flight to the Atlanta Army Air Base. Witnesses on the ground observed the airplane falling out of control from a large thunderstorm cloud. The airplane was seen breaking up as it fell to the ground. Investigators speculated that the pilots had inadvertently flown into a severe thunderstorm cloud and had encountered powerful turbulence, causing the airplane to break up in mid-air.

8-15-44F. Buffalo, New York. At 2200 EWT, Frank Montville, a civilian employee of Curtiss-Wright Aircraft, was fatally injured when he walked into the rotating port propeller of a Curtiss C-46A at Buffalo Municipal Airport, Buffalo, New York. Nelson Davis and George Rauch, both civilian mechanics for Curtiss-Wright Aircraft, were running up the port engine when the incident occurred. The starboard engine was not running at the time. An employee on the ground anticipated the accident and yelled a caution to Mr. Montville, but because of aircraft engine noise was unsuccessful.

8-15-44G. Laughlin Field, Texas. At 2156, a Martin AT-23 (a version of the B-26 used for advanced target towing) crashed while attempting a take-off at Laughlin Field, Del Rio, Texas, killing two fliers and seriously injuring two others. The airplane was taking off on a routine night training mission. The take-off was normal and the pilots retracted the landing gear. The pilot evidently lowered the nose too soon to gain airspeed and flew into the ground. The airplane broke up and burst into flames. Pilot F/O James H. Carmack and co-pilot 2Lt. Stuart R. Knox, Jr. received serious injuries in the crash and aerial engineers Sgt. Frank M. Kollar and SSgt. Charles F. Leyde were unable to escape the burning airplane and were killed. Rescue personnel from Laughlin Field were on the scene immediately and were able to extinguish the flames in a short period.

8-16-44A. Fort Worth, Texas. At 0115, engineer Pfc. Edward F. Donovan was killed when he walked into the rotating number-three propeller of a Consolidated B-24E at the Army Air Field at Fort Worth, Texas. B-24 Pilot 2Lt. John M. McCall and co-pilot 2Lt. John C. McCoy were uninjured. The airplane had just returned to the field and was taxied up to the rear of the flight line. The engineer exited the airplane to help guide the pilots to the proper parking space. The engineer exited through the open bomb bay and walked forward into the number-three propeller.

8-16-44B. Yuma, Arizona. At an unknown time between 0715 and 0915 MWT, a Bell P-39K crashed 25 miles northeast of the Army Air Field at Yuma, Arizona, killing pilot 2Lt. William L. Galbraith.

The airplane had taken off from Yuma at 0715 on a routine training flight and failed to return. The airplane was located on 8-19-44 at 1715 MWT. Investigation revealed that the airplane had struck the ground in a spin.

8-16-44C. Oplin, Texas. At 1030, a Republic P-47C and a P-47D collided in mid-air and crashed two miles northwest of Oplin, Texas, killing P-47C pilot 2Lt. Russell R. Bedwell, Jr. P-47D pilot 2Lt. Ralph N. Costilow parachuted to safety, receiving minor injuries. The airplanes were part of a six-ship flight that had taken off from Abilene Army Air Field, Texas, on a tactical formation training flight. Lt. Bedwell was leading the second element and Lt. Costilow was his wingman. The formation was flying 25,000 feet msl. The flight leader began a gentle turn to the left. After about 45 degrees of turn, Lt. Bedwell's airplane skidded slightly behind the flight leader. As he attempted to recover his position he overshot and then overcorrected. Lt. Costilow attempted to follow the maneuver but was unable to do so. Lt. Bedwell's airplane skidded underneath Lt. Costilow' airplane. Lt. Bedwell, apparently unaware that he was underneath Lt. Costilow, pulled up and into the P-47D. The P-47D propeller chopped into the P-47C at the cockpit area, killing Lt. Bedwell instantly and severing the tail of the P-47C. Lt. Costilow's airplane then went out of control and he bailed out, landing a short distance from the area where the two airplanes slammed into the ground and exploded.

8-16-44D. Patterson Field, Ohio. At 1530 EWT, a Taylorcraft L-2A crashed four miles southeast of Patterson Field, Ohio, killing Maj. Thomas F. Vickers and civilian passenger Oscar L. Goldfarb. The airplane had taken off from Patterson Field at 1450 on a maintenance test flight after being taken out of temporary storage and reassembled. Investigation revealed that the engine had failed and the pilot had stalled the airplane while attempting an emergency forced landing. The airplane fell off into a spin and was turning slightly to the right as it plummeted to the ground in a nose-down attitude.

8-16-44E. Walterboro, South Carolina. At 2026, a Republic P-47D performing a landing overran a P-47G on the runway at Walterboro Army Air Field, South Carolina, killing P-47G pilot 2Lt. George E. Cisco. P-47D pilot F/O Ralph L. Turner was uninjured. Lt. Cisco had taken off with another pilot on an instrument flight and formation flight. F/O Turner was on routine solo transition flight. He had originally been scheduled for an instrument flight but the wing ship failed to take off because of mechanical difficulties. After flying around solo for a short time, F/O Turner joined on the formation of Lt. Cisco and his wingman. After flying around the local area for a short time, the flight came in to land. F/O Turner was unable to transmit with the tower for landing instructions. Lt. Cisco landed on Runway 16 and was rolling about three-quarters of the way down the runway when F/O Turner landed behind and rolled into the P-47G, killing Lt. Cisco instantly. Lt. Cisco had radioed the tower and had asked if the runway was clear ahead of him. The tower operator, thinking that he was talking to F/O Turner, instructed the pilot to "slow it down." Lt. Cisco, thinking the tower was talking to him and the runway ahead was not clear, slowed down his landing roll. Moments later, F/O Turner collided with the P-47G. Investigators faulted both pilots and the control tower officer for the accident. The P-47G tail was severed in the collision. Both airplanes stuck together as they slid for over a 100 yards. Investigators noted that F/O Tuner was not authorized to fly formation with the Lt. Cisco's flight.

8-17-44A. Pierre, South Dakota. At 1004 MWT, a Boeing B-17G suffering an in-flight fire and attempting an emergency forced landing crashed nine miles southeast of the Army Air Base at Pierre, South Dakota, killing six crewmembers. Gunners Cpl. George D. Honey, Cpl. Harry F. Mroch and Cpl. Sidney Nylaan parachuted to safety, receiving minor injuries. Killed in the crash were: 2Lt. Arnold P. Sparman, pilot; 2Lt. Robert M. Biggers, co-pilot; 2Lt. Arthur M. Lippman, bombardier; 2Lt. John F. Hanzes, navigator; Sgt. Samuel J. Lamazzo, engineer; Cpl. Frederick J. Bruening, gunner. "In the opinion of the Aircraft Accident [Classification] Committee, the following are the series of events: At 14,000 feet, after being notified by his crew that the number-four engine was throwing oil [and smoking], the pilot tried to feather the engine and did not have pressure enough to accomplish same. He kept circling at this altitude trying to accomplish the feathering, then he started letting down. He decided to make a landing at Pierre Army Air Base, which was only a few miles away, so at 10,000 feet over the Air Base he informed the crew to stand by for an emergency landing and let his landing gear down. The more he thought of the situation probably the more frantic he became as is indicated by his excited calling over the interphone, trying to contact the Army Tower, until he was told to change [over to the command set]. He circled the field from 10,000 feet down to traffic altitude and made a downwind leg. He flew out this leg approximately three miles when his number-four engine started trailing flames. He gave a signal to bail out and decided to land in the wheat field, so he turned the ship to the right and attempted to land upwind. In attempting to level off, his left wing went down at low speed, hit the ground, [dug in] for 50 feet, and then aircraft exploded."

8-17-44B. Moses Lake, Washington. At 1030, a Bell P-39Q crashed approximately 18 miles southwest of Moses Lake, Washington, killing pilot 2Lt. Charles A. Weatherford. The P-39 was the number-two airplane in a four-ship flight that had taken off from Moses Lake at 1015 on a dive-bombing and simulated gunnery mission. The flight was in an echelon formation to the right when the dive on the target was commenced at about 6,000 feet msl. The flight leader pulled off of the target at about 3,000 feet. The subject airplane followed and dropped its bomb. The flight leader observed the subject airplane in a spin to the left at about 2,500 feet, describing one turn. The airplane then entered a flat spin and continued in a flat spin for two more turns until it struck the ground and exploded. The pilot made no attempt to leave the airplane and it was later speculated that the force of the spin prevented the pilot from bailing out.

8-17-44C. Hemet, California. At 1410, a Stearman PT-27 crashed seven and one-half miles southwest of the Army Air Field at Hemet, California, killing student A/C William S. Barstow. The airplane had taken off from Hemet Army Air Field on a routine student solo training flight. The airplane was apparently stalled and had entered a spin at an altitude too low to allow recovery. The airplane was seen to describe one turn of spin before it smashed into the ground. The student had unfastened his safety belt and was hurled from the wreckage upon impact.

8-18-44A. Nashville, Tennessee. At 0958 CWT, a Martin B-26G suffering the failure of the port engine immediately after take off crashed three miles north of Nashville Municipal Airport, Nashville, Tennessee, killing pilot 1Lt. George W. Blancard and co-pilot F/O Harold D. Strickland. The airplane was taking off on a ferry mission to an active war theater via West Palm Beach, Florida. The airplane used up the entire 5,500-foot runway and only reluctantly became airborne. The B-26 never climbed above 200 feet agl and was observed flying very slow, trailing black smoke from the port engine. The port wing was seen to drop and then the airplane dove into the ground at a steep angle. Investigators could not determine the cause of the apparent diminished engine performance or the subsequent failure of the port engine. Investigators noted that the pilot was very experienced and had completed a combat duty tour in B-26 aircraft.

8-18-44B. Key Field, Mississippi. At 1320, a Vultee BT-13B crashed two and a half miles east of Key Field, Meridian, Mississippi, killing pilots 2Lt. Lawrence P. Delaney and 1Lt. Joseph O. Hall. The airplane had taken off from Key Field on a scheduled instrument training flight. The airplane was observed flying normally and then was seen to gradually lose altitude until it crashed into the side of a small hill in a flying attitude. Investigators speculated that the instrument pilot in the rear cockpit, Lt. Hall, thought that the safety pilot, Lt. Delaney, was flying the airplane and had let go of the controls. The BT-13 was returning to Key Field so that the pilots could change seats.

8-18-44C. Commerce, Texas. At 1400, a Vultee BT-13A crashed one and a half miles south of Commerce, Texas, killing instructor pilot 2Lt. Manley T. Olson and student pilot Sgt. Robert G. Green. The airplane was observed flying east at an altitude of 1,000 feet agl and then as seen to make a 180-degree turn to the west. The BT-13 fell off into a spin and dove into the ground at angle of 60 degrees, bursting into flames upon impact. Investigation of the wreckage revealed that the propeller was in low pitch, the mixture control was full rich, the flaps were at 20 degrees, and the throttle wide open at the time of the crash. Investigators speculated that the airplane was performing a simulated forced landing when it entered a stall at an altitude too low to allow recovery.

8-18-44D. Daykin, Nebraska. At 1650 CWT, a Republic P-47D crashed while attempting an emergency forced landing three miles east of Daykin, Nebraska, fatally injuring pilot F/O Logan D. Massey. Investigators stated, "F/O Massey was on an elementary formation training mission and attempted an emergency landing probably due to power failure. According to statements of witnesses, the airplane started circling and descending rapidly, and on his last turn, the pilot came in at approximately a 30-degree angle with the left wing low. The airplane's left wing caught on the ground then a cartwheeling action occurred with the propeller and engine striking the ground [then] the right wing contacting, and then the vertical fin and stabilizer. From the position of the pilot's body, relative to the airplane, he in all probability was thrown clear of the airplane by the shock of the impact. F/O Massey suffered major injuries and died in the hospital approximately one hour after the time of accident. It is the opinion of the investigating committee that the pilot's safety belt was not fastened at the time of impact with the ground.... The cause of the accident ... was due to power failure."

8-18-44E. Scott, Mississippi. At 1830 CWT, a Cessna UC-78B crashed seven miles southwest of Scott, Mississippi, killing instructor pilot 2Lt. John W. Graybill and student A/C Joseph J. Corbisello. The Accident Classification Committee stated, "Investigation of the scene revealed that both engines were not running at the time of the crash. No flaps were extended and all controls were in normal position for normal level flight with the exception of the gas tank selector [lever], which in the 'off' position. The left gas tank was completely dry and not damaged by the crash. The right gas tank lacked a few gallons of being full. It is believed that the pilot flew on the left tank until it went dry and that when the engines stopped from lack of gas the pilot attempted to switch to the right tank but accidentally turned it to the 'off' position, when already at too low an altitude [to abandon] the aircraft or maneuver to a safe forced landing."

8-18-44F. Pacific Ocean, California. At 1655, a Lockheed P-38J was accidentally shot down by a P-38L over the Pacific Ocean one and a half miles off shore and due west of the mouth of the Russian River, California. P-38L pilot 1Lt. Raymond G. Farkas was able to successfully parachute to safety but drowned before he could be rescued from the sea. Lt. Farkas was leading a flight of four P-38 airplanes that had taken off from Santa Rosa Army Air Field, California, on a high-altitude gunnery mission. The flight climbed to an indicated altitude of 20,000 feet and then headed out to sea at the mouth of the Russian River. As the flight headed out to sea, the lead airplane of the second element moved his flight directly behind the lead element. During this maneuver, the number-four pilot, 2Lt. Marvin E. Thomas, decided to make a "sighter burst" (with his camera) on the port engine of the lead ship. Lt. Thomas, thinking that his gun switch was on the "camera" position, pressed the trigger and a stream of fifty-caliber bullets penetrated the port engine of the lead airplane. The port engine began streaming thick black smoke and Lt. Thomas radioed the lead airplane, alerting the flight leader of what had happened. As Lt. Farkas acknowledged the message the port engine erupted into flames. Lt. Farkas began maneuvering the airplane in an attempt to extinguish the flames but was unsuccessful. Lt. Farkas then bailed out and was seen to descend toward the sea in his fully deployed parachute. The flight of three ships descended and circled the area where the pilot hit the water, but they were unable to locate him. Lt. Farkas' body was found two hours after the accident. His Mae West had been activated and he was free of his parachute. An autopsy revealed that the pilot had drowned. Investigators speculated that the pilot had probably unbuckled his harness and jumped free of his harness just above the surface. It was speculated that the pilot had struck the water with sufficient enough force to render him unconscious, causing him to drown.

8-18-44G. Mountain Home, Idaho. At 1846 MWT, a Martin AT-23A (a version of the B-26 used for advanced target towing) and a Consolidated B-24J collided in mid-air ten miles southeast of the Army Air Field at Mountain Home, Idaho, killing two fliers aboard the AT-23 and seven aboard the B-24. Investigation revealed that the AT-23 was making a simulated fighter pass at a ten-ship formation of B-24 aircraft. The subject B-24 was the lead airplane of the three-ship element flying in the high position. The AT-23 was concentrating on making passes on this high flight. The AT-23 made several passes at the B-24 flight from the two o'clock and eight o'clock positions in a figure-eight flight path. Sometimes the AT-23 would pass above and sometimes below the high-position flight as it made its passes. The B-24 flight leader then turned the formation toward the local AAF bombing range. The pilot of the subject B-24 radioed the B-24 flight leader and asked for permission to leave the formation because they had no bombardiers on board. The B-24 flight leader gave them permission to leave the formation. The high flight then began a turn to the right to leave the formation, losing altitude as it turned. As the flight turned away from the formation, the AT-23 made a pass from approximately the two o'clock position, expecting to pass underneath. The pilots of both airplanes noticed that a collision was imminent and attempted to maneuver out of danger by turning sharply. The starboard wing of the AT-23 struck the B-24J outboard of the number-one engine as the AT-23 passed underneath, severing the AT-23's wing and the B-24's port wing. The AT-23 tail section struck the B-24's number-one engine, causing the engine to separate and the AT-23 to break in two pieces. Both airplanes burst into flames in the collision. The AT-23 went out of control and spun crazily to the ground, exploding violently into flames upon impact. The B-24 rolled to the left and then went straight down, exploding into flames upon impact. The occupants of the AT-23 were unable to escape and went in with the airplane, suffering "complete disintegration." Killed in the crash of the B-24 were: 1Lt. Marion P. Guthrie, pilot; 2Lt. Dorise C. Nickeil, co-pilot; Cpl. Howard F. Gallagher, engineer; Cpl. Harold F. Skotteslad, gunner; Pfc. Fay R. Windle, gunner; Cpl. Eugene F. Bingham, gunner; Pfc. Milton G. Dilling, gunner. Pilot 2Lt. Glen S. Admeyers and engineer Sgt. Archie M. King, Jr. were killed in the crash of the AT-23.

8-18-44H. Famoso, California. At 2130, a Cessna UC-78B crashed three miles southwest of AAF Auxiliary Airfield at Famoso, California, killing A/C Lawrence J. Crudder, Jr. and A/C Aress Emrie. The airplane had taken off from Minter Field, Bakersfield, California, on a routine student night flight. The students entered the traffic pattern for the auxiliary airfield and were going to shoot a landing. The airplane was observed turning from the downwind leg to the base leg. The airplane continued in the turn until it flew into the ground and burst into flames. There was no evidence of engine failure or structural failure. Investigators could not determine what caused the airplane to fly into the ground. It was later speculated that the students probably lost the horizon in the turn.

8-19-44A. Bradley Field, Connecticut. At 0730, a Republic P-47D suffering an engine failure crashed while attempting a landing at Bradley Field, Windsor Locks, Connecticut, killing pilot 1Lt. Walter L. Gibson, Jr. The airplane was suffering diminished engine performance and the pilot returned to the field for an emergency landing. The airplane entered the pattern and while turning left from the base leg to final approach, the engine failed completely. The airplane stalled and then fell off on the port wing. The pilot partially recovered but then the airplane flipped over to an inverted attitude and then dove into a wooded area where it exploded into flames upon impact.

8-19-44B. Walla Walla, Washington. At 0505 PWT, gunner Pfc. George B. Zallaha was killed when he inadvertently walked into the rotating number-two propeller of a Consolidated B-24J operating at Walla Walla Army Air Field, Washington. The airplane was undergoing a pre-flight engine test before the flight

crew was allowed on board. Pfc. Zallaha was at the front end of the B-24 working on the nose turret while the ground crew ran up the engines. A gunnery instructor was seen to walk up to Pfc. Zallaha, speak some words to him and then the gunnery instructor walked around the end of the port wing and then aft to the rear of the airplane. Pfc. Zallaha then picked up an ammunition box and then began walking aft between the fuselage and the number-two engine. He walked directly into the rotating number-two propeller, which severed the right half of his head, killing him instantly. Investigators noted that all personnel had been briefed about avoiding walking through propellers even with engines stopped.

8-19-44C. Ramsey, New Jersey. At 0915, a North American AT-6D suffered a catastrophic structural failure of the starboard wing and crashed at Ramsey, New Jersey, killing A/C J. Lee Forney and injuring A/C Jesse H. Johnson, who was able to parachute to safety. Both were cadets at the U.S. Military Academy at West Point, New York. The airplane had taken off from Stewart Field, New York, on a routine instrument team ride. Investigators stated, "[The cadets] took off at 0900 and climbed to 5,000 feet. Cadet Johnson stated that he then did a barrel roll. Cadet Forney took over the controls and climbed to 10,000 feet and did a loop. After that they flew straight and level for a short time. Johnson then asked Forney if they were not getting too far south. Forney said 'yes' and started a normal turn to the right. Cadet Johnson stated that Cadet Forney had been flying the airplane since they left the 5,000-foot level. After they had turned about 70 or 80 degrees, the ship went into a spiral and straight down [into a dive]. Cadet Johnson saw the airspeed indicator go beyond the 250 mph mark. He then called Forney on the interphone and asked him what he was doing, at the same time cutting the throttle. He did not get an answer. By this time, the airspeed exceeded 300 mph. Cadet Johnson took the controls and attempted to pull the ship out of the dive. At the moment, the airplane started to break up and he thinks he heard something hit the rear part of the canopy. Cadet Johnson bailed out. The airplane hit in a thicket.... It was completely demolished but did not burn.... [A/C] Forney had made no effort to get out." Investigators noted that aerobatic flight was not authorized on the scheduled instrument flight and the airplane was more than 20 miles outside the authorized flying area. It was also noted that the cadets were specifically instructed by their flight leader not to perform aerobatics. Investigation revealed that Cadet Forney never went under the hood to fly on instruments.

8-19-44D. Tonopah, Nevada. At 0800 PWT, a Consolidated B-24J crashed while attempting a landing at the Army Air Field at Tonopah, Nevada, killing the crew of nine. Investigators stated, "Airplane 42-50834 with trainee crew and instructor pilot aboard were performing a transition flight, part of which normally includes an emergency procedure demonstration and/or check. Pilot called into the Tower and requested clearance to make a simulated two-engine landing wherein two engines would be reduced in power to simulate feathered conditions, i.e. 12" Hg manifold pressure and constant RPM. Pilot was making a normal traffic pattern approach to the field, when on the base leg at 1,500 and 2,000 feet estimated, the airplane appeared to stall and drop right and left wings several times, with a nose high attitude, before a final stall during which it crashed into the earth." The B-24 exploded into flames upon impact, killing the crew instantly. Killed in the crash were: 1Lt. Charles H. Enoch, B-24 instructor; F/O John E. Davis, pilot/B-24 student; F/O Johnnie W. Mundell, co-pilot/B-24 student; SSgt. Thomas D. Brown, Engineer instructor; Cpl. Donnie J. Riordan, engineer student; Cpl. Richard N. Stringer, gunner/assistant engineer; Pfc. Walter E. Von Schenk, gunner; Pfc. Junnal E. Tank, gunner; Pfc. Robert E. Pariseau, gunner.

8-19-44E. Pacific Ocean, California. At 0835, a Lockheed P-38G crashed into the Pacific Ocean 26 miles due west of Point Reyes, California, killing pilot 2Lt. Richard K. Holzapfel. The subject airplane was the number-two airplane of a six-ship flight that had taken off from the Army Air Field at Santa Rosa, California, on a high-altitude aerial gunnery mission and a dive-bombing mission. The flight flew 25 miles out to sea to the bombing range. While flying at about 8,000 feet, the flight leader ordered the flight into a string formation. The flight leader then peeled off for his bomb run and the remaining airplanes in the flight followed. The flight leader released his bomb and pulled out at 3,500 feet. The subject airplane continued in the dive until it crashed into the sea and exploded. The pilot's body was never recovered. Investigators could not determine why the airplane dove into the sea.

8-19-44F. Tonopah, Nevada. At 0900 PWT, a Consolidated B-24E suffered a catastrophic structural failure and crashed 45 miles southeast of the Army Air Field at Tonopah, Nevada, killing the crew of nine. The airplane had taken off from Tonopah Army Air Field on a routine training flight and failed to return. It was soon discovered that the airplane had crashed. There were no known witnesses. Investigation revealed the starboard vertical stabilizer and rudder failed and separated, causing the B-24 to go out of control and crash into the desert. Investigators speculated that the pilots had inadvertently allowed the airplane to enter an unusual attitude, which resulted in the airplane being stressed beyond its design limits. The starboard vertical stabilizer and rudder were found about a quarter of a mile from the main wreckage. Investigators noted that the rudders had just been changed due to excessive play in their operation. The airplane and the rudders performed satisfactorily on a maintenance test flight the night before. Investigators did not believe

that the rudder maintenance was a factor in the accident. Killed in the accident were: Capt. Robert E. Sweet, B-24 instructor; 2Lt. Robert L. Pyle, pilot/B-24 student; 2Lt. Gerald C. Wangler, co-pilot/B-24 student; TSgt. Walter J. Johnson, Engineer instructor; Cpl. Frank M. Medrano, engineer student; Pfc. Raymond B. Hoffman, engineer student; Pfc. Phillip A. Knight, gunner; Cpl. Robert J. Gruss, gunner; Pfc. Richard E. Jeffery, gunner.

8-19-44G. Gonzales, Texas. At 1215, a Curtiss P-40N crashed seven miles north of Gonzales, Texas, killing pilot 2Lt. Ward A. Chapman. The airplane had taken off at 1145 CWT from Aloe Field, Victoria, Texas, on a routine training flight. The P-40N dove into a sheep pen on the Joseph Pavlica Farm, exploding violently into flames upon impact. The airplane apparently struck the ground at high speed and at an angle exceeding 45 degrees. Investigators could not determine the cause of the crash.

8-19-44H. Easton, Maryland. At 1415, a Republic P-47G crashed three miles northeast of Easton, Maryland, killing pilot 2Lt. Robert J. Sineath. The airplane had taken off at 1255 from Camp Springs Army Air Field, Washington, D.C., on a routine navigation mission. The airplane was seen in a steep dive, traveling at high speed. The airplane continued in the dive until it smashed into a small, shallow river and exploded. The pilot was killed instantly. Investigators noted that the testimony of several of the civilian witnesses was contradictory and unreliable.

8-19-44I. Imlay City, Michigan. At 1640 EWT, a Consolidated B-24J suffered a catastrophic structural failure of the elevators and crashed two miles ESE of Imlay City, Michigan, killing two army pilots and two civilian engineers. Killed in the crash were: 1Lt. John Kenneth Howmiller, 25, Lansing, Illinois, pilot; Capt. Thomas W. Vaughn, 27, Elyria, Ohio, co-pilot; Harvey D. Jenkins, 26, Ann Arbor, Michigan, civilian aerial engineer, Ford Motor Company; C.R. Womack, 35, Dearborn, Michigan, civilian aerial engineer, Ford Motor Company. The airplane was selected as representative of a production lot of B-24J airplanes manufactured by the Ford Motor Company at Willow Run, Ypsilanti, Michigan, and was scheduled to perform a high-altitude flight test and bomb dropping exercise. The B-24J took off from the factory airport at Willow Run at 1227 EWT and climbed to 20,000 feet msl to perform the bomb dropping exercise on the AAF Bombing Range on the shore of Lake Huron. After the bombs were dropped the airplane flew inland to test the bomb door operation at high speed and in a dive. The airplane descended to 12,000 feet in a power dive while the bomb door tests were being conducted. The B-24 was observed attempting to pull up from a diving attitude at about 8,500 feet when numerous small pieces of "what appeared to be paper" were seen trailing the airplane. The B-24 then went into a steep "power dive" until it struck the ground and violently exploded into flames. Investigation revealed that the elevators had failed and disintegrated, causing the airplane to go out of control and smash to earth in a high-speed dive. Wreckage was scattered over a wide area. Published accounts of the accident reported that Capt. Vaughn had completed a combat tour in B-24 aircraft in the Mediterranean Theater. Lt. Howmiller was assigned as a test pilot and was stationed at Willow Run.

8-19-44J. Luke Field, Arizona. At 1505 MWT, a North American AT-6C and a North American TB-25C collided in mid-air and crashed 30 miles north of Luke Field, Phoenix, Arizona, killing four army fliers. Killed in the crash of the TB-25C were: 1Lt. George Hunter, pilot; 2Lt. Patrick D. Holland, co-pilot; Sgt. James A. Ramey, passenger. 2Lt. Orland L. Luhr was killed in the crash of the AT-6C. Investigators stated, "[TB-25C #42-32383 and AT-6C #41-32833] were flying in two separate formations for the motion picture *God is My Co-Pilot* [Warner Brothers 1945]. The formation of B-25s was heading north at approximately 10,000 feet indicated. The formation of AT-6s' was flying at approximately 12,000 feet, heading south. The AT-6s,' which were depicting [Japanese] Zeros in the picture, were to dive down ahead of the B-25 formation and pull up, making a pass at the bombers from ahead and below. The pilots had been thoroughly briefed on the exact positions — both the bombers and the AT-6s'—and at no time were to come closer than 300 feet vertically. Lt. Luhr was flying on the left wing of the second element of the AT-6s.' As this element made its pass on the B-25s, Lt. Luhr failed to follow his element leader and passed on into the formation of B-25s, striking the right wing of airplane number 42-32383 with the right wing of his airplane [41-32833]. The collision occurred at approximately 9,500 feet indicated. The AT-6 rolled [to the] left and down into a slow spiral then headed almost straight for the ground. The pilot in one of the other AT-6s' followed the airplane down and noticed that the canopy was open approximately four inches but there was no attempt made by Lt. Luhr to bail out. After the collision, B-25 number 42-32383 kept going in almost level flight for a few seconds and then stalled off on the right wing, heading straight down and exploding upon contact with the ground."

8-19-44K. Wilmington, North Carolina. At 1840, a Republic P-47D crashed 15 miles east of Bluethenthal Field, Wilmington, North Carolina, killing 2Lt. Richard I. Mulroney. The airplane had taken off from Bluethenthal Field on a routine training flight and soon after was flying with another P-47 in formation. After a short time of flying together, the subject airplane peeled off of the formation and was last seen heading for Bluethenthal Field. Witnesses on the ground stated that the airplanes were chasing each other and executing tight turns close to the ground. Investigation revealed that the subject airplane stalled at an altitude too low to allow recovery, smashed to the ground and burned.

8-20-44A. Bergstrom Field, Texas. At 1625 CWT, engineer Cpl. Ubaldo Polidore was killed when he was struck by the starboard propeller of a North American B-25D at Bergstrom Field, Austin, Texas. The airplane landed and shut down its engines at 1613 CWT. Cpl. Polidore climbed on the starboard wing to assist in servicing operations. The engineer walked out on to the starboard engine nacelle and cowling and then slid down to the ground on the starboard propeller. The act of sliding down the propeller caused a slight rotation, which "probably caused a cylinder to receive a fuel charge which was ignited by either glowing carbon or an extremely hot plug." The engine fired and the propeller rotated slightly, striking Cpl. Polidore in the head with enough force to cause a fatal injury.

8-20-44B. Location Unknown. At an unknown time after 1900 CWT, a Curtiss P-40N disappeared and is presumed to have crashed into the Gulf of Mexico near Galveston, Texas. The pilot, 2Lt. Robert W. Brown, was declared missing and is presumed to have been killed. The airplane was part of a flight of five P-40s that were to fly on a scheduled aerial gunnery mission. The first four airplanes had taken off in pairs from Galveston Army Air Field and the subject airplane, which was suffering apparent radio problems, took off last at 1758 CWT. The airplane never rendezvoused with its flight for the gunnery mission. An AAF aircraft spotter observed the subject airplane flying low over the beach near Galveston at approximately 1900. The enlisted man stated that the engine sounded as if it were cutting out and that the airplane appeared to be losing altitude as it made a wide turn over the Galveston Hotel. The airplane turned toward the sea and flew away. Despite a thorough search lasting two weeks, no trace of the pilot or the P-40 could be found.

8-20-44C. Rosecrans Field, Missouri. At about 2259 CWT, a North American B-25C crashed six and one-half miles NNW of Rosecrans Field, St. Joseph, Missouri, killing pilot Capt. Lyle R. Torbenson and co-pilot F/O Samuel J. Albin. The B-25 took off from Rosecrans Field at 2252 CWT on a routine transition mission. Several minutes later the airplane smashed through a small stand of trees on the top of a small bluff at an altitude of approximately 10 feet agl. Twelve feet of each wing tip and both ailerons were sheared off in the collision with the trees. The heavily damaged airplane bounded back into the air for another half mile before smashing into the ground in an upright position on a small slope. The airplane burst into flames as it landed on its belly, smashing down two trees and flipping over on its back before sliding to a stop at the edge of a ravine with a 25-foot drop. Investigation revealed that both propellers were not turning and not feathered at the time of the impact with the small bluff.

8-21-44A. Model, Colorado. At 1009, two Consolidated B-24J airplanes collided in mid-air and crashed two miles east of Model, Colorado, killing 18 fliers. Bombardier 2Lt. Robert S. Clark was able to parachute to safety from B-24J #42-51517 and was un-injured. Killed in the crash of B-24J #42-51517 were: 2Lt. Horace M. Jenkins, pilot; 2Lt. Clair A. Clark, co-pilot; Capt. Royal J. Brock, B-24 instructor; Cpl. Francis J. O'Donahue, engineer; Cpl. Herbert J. Rogove, radio operator; Cpl. William A. Powell, Jr., gunner; Pfc. Daniel J. Reccio, gunner; Pfc. Luther C. Killinger, gunner; Pfc. William C. Joyce, Jr., gunner. Killed in the crash of B-24J #42-99946 were: 2Lt. Frank W. Adams, Jr., pilot; F/O William J. Robbins, co-pilot; 2Lt. Claire E. Gates, bombardier; TSgt. Albert J. Sermon, engineer; Cpl. Boyce E. Forester, radio operator; Cpl. Joseph L. Hartman, gunner; Pfc. Malcolm R. Crawford, gunner; Pfc. Morton Kaplan, gunner; Cpl. Carl E. Miller, gunner. B-24 #42-51517 was flying in the number-one (lead) position and B-24 #42-99946 in the number-three position of the high squadron of an eighteen-ship formation that had taken off from Pueblo Army Air Base, Pueblo, Colorado, on a routine high-altitude formation mission. The formation had climbed to an altitude of about 20,000 feet msl approximately 15 miles northeast of Trinidad, Colorado; the B-24 crews were waiting for a target-towing aircraft for a scheduled gunnery exercise. While waiting for the target tow-ship, the formation leader began executing a series of shallow turns and mild evasive action maneuvers. A shallow turn to the right was commenced. B-24J #42-99946 (number-three) overran his position in the turn and ended up flying abreast of B-24J #42-51517 (number-one). Moments later, B-24J #42-99946 turned to the right in a steep (near vertical) bank and pulled up into B-24J #42-51517. The port wing of B-24J #42-51517 cut through the fuselage of B-24J #42-99946 at about the camera hatch, severing 42-99946's tail. B-24J #42-51517's port wing continued through 42-99946's fuselage and into the center section. The port wing folded back toward the fuselage of 42-51517. The two airplanes then separated and 42-99946 exploded into flames, breaking into about five large pieces and several hundred smaller bits. The crewmembers of 42-99946 were ejected from the airplane as it broke up and cascaded to the ground. Many were not wearing parachutes. After great difficulty, Lt. Clark parachuted to safety through the nose wheel well as B-24J #42-51517 plunged to the ground in a diving slip to the left, exploding violently into flames upon impact. The bombardier stated that the airspeed indicator in the nose section indicated approximately 350 mph when he exited the airplane.

8-21-44B. Alamogordo, New Mexico. At 1100, a Boeing YB-29 crashed 25 miles northeast of the Army Air Base at Alamogordo, New Mexico, killing ten crewmembers. The airplane had taken off from Alamogordo AAB on a routine instrument training flight, which included hooded instrument let-downs. An old man and a young boy observed the airplane flying at an altitude of approximately 3,000 feet agl with smoke and flames trailing. The airplane then nosed over, went out of control and then slammed into the ground in an inverted 60-degree dive, exploding violently into flames upon impact. Investigation revealed that the horizontal stabilizers, elevators and various other pieces separated from the airplane prior to impact. Investigators could not determine the cause of the accident. It was later speculated that the B-29 suffered an in-flight fire in one or two engines, causing the flight crew to lose control of the airplane. Killed in the crash were: Capt. Byron L. Bostick, pilot; 2Lt. Jerone F. Muller, co-pilot; 2Lt. Meyer S. Hodiss, pilot-rated passenger; 2Lt. James J. Higgins, flight engineer; Sgt. John R. Uczen, aerial engineer; SSgt. Thomas A. Boyle, radio operator; Pfc. Benjamin E. Smith, gunner; Pfc. Conda M. Smith, gunner; Pvt. Irwin Handelman, assistant radio operator; Pfc. Camillo P. Parenteau, gunner.

8-21-44C. Paine Field, Washington. At 1445, a Bell P-39Q suffering an apparent engine failure just after take-off crashed one mile northwest of Paine Field, Everett, Washington, killing pilot 1Lt. Charles E.R. Blair. The airplane was taking off on a "local test flight." The airplane was trailing black smoke and backfiring as it climbed to about 300 feet agl. The engine apparently quit moments later. The pilot made no attempt to turn and set up a glide straight ahead for an emergency forced landing. The pilot attempted to reach Puget Sound for his forced landing and he released the starboard door to facilitate his escape in the event he made it to the water. The airplane lost altitude rapidly and the pilot realized he was not going to make it to the water. The airplane descended into a large tree about 150 yards from the shoreline. The pilot evidently did not see the tree in the blind spot of his wings as the airplane descended. The airplane broke up and burst into flames. The pilot was killed instantly.

8-21-44D. Ellington Field, Texas. At 1836, a Fairchild AT-21FB (similar to the AT-13 bombardier trainer) crashed one mile south of Ellington Field, Houston, Texas, killing radio operator Cpl. Hyman Thaler. Pilot 2Lt. Herman B. Landry and engineer Pfc. G.E. Pounds. The airplane was being evacuated from Ellington Field to the Army Air Field at Greenville, Texas, in anticipation of a hurricane forecasted to hit the Texas Gulf coast. The airplane took off and the pilot lowered the nose to pick up airspeed before climbing out. The pilot neglected to retract the landing gear. Moments after becoming airborne, the starboard engine began suffering diminished performance. The AT-21 began rolling to the right and the pilot retarded the throttles and straightened the airplane out. The airplane began to fall off to the right and the pilot leveled out at about 75 feet agl and established a glide and attempted to retract the landing gear. The airplane descended into the ground on the extended landing gear and then bounced back into the air. The airplane then slammed back to the ground and rolled about 500 feet before the nose wheel collapsed. The airplane careened for a short distance before flipping over on its back. Cpl. Thayer was killed when the airplane flipped over. He was trapped in the wreckage and his body had to be cut free. The pilot was also pinned in the wreckage and he needed assistance to free himself from the cockpit.

8-21-44E. Pooler, Georgia. At 2024 EWT, a Consolidated B-24E flying in instrument conditions crashed at Pooler, Georgia, killing the crew of seven. The airplane had taken off from the Army Air Base at Liberal, Kansas, on an instrument navigation flight to Chatham Field, Georgia. The airplane approached the area of the field in instrument conditions. The pilot radioed the control tower at Chatham and requested landing instructions. The pilot was cleared on the airway that was being used and was told to descend until visual contact was made. The airplane broke through the overcast at an altitude of about 670 feet with the field in sight. The pilot then entered the traffic pattern for a landing to the east on the main East-West Runway. Control tower personnel had visual contact with the airplane while it flew the pattern. The pilot flew too far on the downwind leg and visual contact was lost. The pilot radioed the tower and advised that he was turning to the base leg. The airplane made no other radio contact with the control tower. After a few minutes, the control tower twice attempted to contact the airplane but to no avail. Control tower personnel noticed a column of black smoke rising west of the East-West Runway and a crash truck was sent to investigate. The tower received a telephone call from Pooler confirming that the B-24 had crashed. Investigation revealed that the airplane flew into the ground in a wooded area and exploded into flames. The ceiling at the wreck site was estimated to be about 200 feet. Killed in the crash were: 1Lt. Frank S. Cash, B-24 instructor; 2Lt. George J. Copa, pilot/B-24 student; 2Lt. Richard G. Devoe, pilot/B-24 student; F/O Louis W. Frangos, co-pilot/B-24 student.

8-22-44A. Santa Maria, California. At an unknown time between 1530 and 1700 PWT, a Cessna UC-78B crashed in a farm field 11 miles west of the Army Air Field at Santa Maria, California, killing the crew of three. A witness on the ground stated that at approximately 1530 the airplane dove out of the clouds and then pulled up sharply and climbed back into the clouds. A short time later he found the airplane wreckage on his property. The airplane had smashed into the ground in a steep dive and was destroyed utterly. Investigators could not determine the cause of the accident. Killed in the crash were: 1Lt. Wayne M. Manlove, 25, Milton, Indiana, pilot; 2Lt. Otto O. Uotila, 21, West Islip, New York, co-pilot; 2Lt. Wilber G. Vinson, 20, Minneapolis, Minnesota, pilot-rated passenger. Investigators noted that Lt. Manlove had completed a combat

tour overseas in P-38 type airplanes and was highly decorated.

8-22-44B. Location Unknown. At an unknown time after 2034 EWT, a Beech AT-10 and two students disappeared on a routine night navigation flight. The airplane had taken off from Moody Field, Valdosta, Georgia, at 2034 EWT and was never heard from again. Missing and presumed killed were A/C Oma G. Capps and A/C Herbert R. Beebe. The airplane was to fly to Atlanta, Georgia, then to McRae, Georgia, and then return to land at Moody Field. Investigators noted that there were scattered thunderstorms on the airplane's intended route. Despite a massive search effort, the airplane and students were never found.

8-23-44A. Williams Field, Arizona. At 0200 MWT, a North American AT-6A crashed seven miles southeast of Williams Field, Chandler, Arizona, killing student A/C William P. Morgan. The airplane had taken off from Williams Field on a routine student night cross-country flight. The airplane was observed flying at an altitude of approximately 8,000 feet msl when it suddenly rolled to an inverted attitude and then dove into the ground. Investigators speculated that the student had either lost his horizon or had lost control of the airplane while performing an unauthorized aerobatic maneuver.

8-23-44B. Midlothian, Texas. At 1030 CWT, a North American P-51D crashed seven miles south of Midlothian, Texas, killing pilot 1Lt. Jospeh W. Garard. The P-51 took off from the North American Aircraft Company facilities at Hensley Field, Grand Prairie, Texas, on a factory test flight. The airplane had flown only two times before and on both occasions the test pilots noted "excessive heat" in the cockpit. Witnesses on the ground observed the airplane flying at an altitude of about 3,000 feet agl in a slight dive. The airplane was seen to enter a slow roll to the right. The airplane then entered a series of rolls to the right, rolling three or four times in a 45-degree dive. The airplane slammed into the ground at high speed in an inverted attitude, exploding violently into flames upon impact. Only small portions of the pilot's body were found.

8-23-44C. Marana, Arizona. At 0945 MWT, a Vultee BT-13A crashed 19 miles northwest of Marana, Arizona, killing A/C Delos G. Wager. The airplane took off at 0815 from Marana Army Air Field on a routine student solo flight. Another student noticed the subject airplane in a spin at about 3,000 feet agl. The airplane continued in a spin until it struck the ground and burst into flames.

8-23-44D. Key Field, Mississippi. At 1256, a Curtiss P-40N crashed while attempting a take-off at Key Field, Meridian, Mississippi, killing 2Lt. John C. Murray, Jr. The P-40 used up the entire runway while taking off on a routine formation flight. The airplane was seen to trailing black smoke as it passed by the runway control vehicle, turning to the left as it climbed slightly just after take-off. The P-40 stalled out of the turn at an altitude of about 175 feet agl, crashing into a heavily wooded area and bursting into flames. Investigation revealed that the spark plugs were loaded with lead deposits and it was speculated that the engine was improperly used on a previous flight. Investigators noted that the airplane had been serviced with 91-octane gasoline. Investigators stated that all pilots had been briefed on the proper power settings while using this fuel. It was suggested that all high-performance pursuit airplanes operating at Key Field be serviced with 100-octane fuel only. It was also noted that this was the third such accident at Key Field. Lt. Murray had received his wings and his commission on 8-12-44. He was unable to bail out because of the low altitude.

8-23-44E. Angelica, New York. At an unknown time after 2130 EWT, a Beech AT-10BH flying in instrument conditions crashed into rising terrain four miles north of Angelica, New York, killing pilot Capt. Benton L. Lewis. The airplane had taken off from the Army Air Field at Columbus, Ohio, on a personal proficiency cross-country flight to Rochester, New York. The airplane failed to arrive at Rochester and was soon declared missing. The airplane was discovered on 9-16-44. Investigators speculated that the pilot was letting down through overcast conditions and while attempting to establish visual contact collided with the hill.

8-23-44F. Lancaster, California. At 2234, a Consolidated B-24J suffering an in-flight fire was abandoned in flight and crashed ten miles northeast of Lancaster, California, killing radio operator Pfc. James W. Phillips, who fell to his death in an unsuccessful parachute jump. Ten crewmembers were injured parachuting to safety. The airplane had taken off from Muroc Army Air Field, California, on a high-altitude bombing mission to be conducted at the AAF Bombing Range near Muroc. After climbing to 20,000 feet, the number-four propeller governor malfunctioned, preventing the pilot from controlling rpm. The manifold pressure became erratic and the pilot elected to return to Muroc for a landing. The B-24 began descending and at about 13,000 feet the engineer took off his oxygen mask and immediately smelled something burning. The waist gunner removed his oxygen mask and also smelled something burning. The engineer examined some wiring on the port side of the bomb bay compartment and noticed that it was arcing. The arcing wires soon ignited gasoline fumes that were present in the bomb bay and the port side of the bomb bay erupted in flames. The engineer received burns while attempting to put out the fire. The flames soon filled the bomb bay section. After all attempts to extinguish the fire had failed, the pilot ordered the crew to bail out. Pfc. Phillips was either hurled from the airplane or had jumped without his parachute. His body,

with his parachute harness attached but with no parachute pack, was found about 300 yards from the main wreckage. Navigator 2Lt. Daniel Prica lost his shoe while bailing out and landed hard, breaking his ankle. Crewmembers injured parachuting to safety were: 2Lt. Verne K. Priester, pilot; 2Lt. Jack D. Rhodes, co-pilot; 2Lt. Donald G. Deininger, bombardier; Cpl. Charles K. York, engineer; Cpl. John J. DeRocili, assistant engineer; Cpl. Robert J. Haney, radio operator; Sgt. Ralph R. Cisneros, gunner; Cpl. Dwight E. Jarvis, gunner; Cpl. James F. Flathers, gunner.

8-23-44G. Muroc, California. At 2255, a Consolidated B-24D suffering an in-flight fire crashed 18 miles northwest of the Army Air Field at Muroc, California, killing five crewmembers. Four crewmembers were injured parachuting to safety, one seriously; gunner Cpl. Alvin J. Fish was uninjured. Killed in the crash were: 2Lt. John L. Graves, pilot; F/O Clifton C. Watts, navigator; Cpl. Bernard D. Fogel, engineer; Pfc. James W. Flitcroft, assistant radio operator; Cpl. Robert F. Nieman, gunner. Crewmembers injured parachuting to safety were: 2Lt. George D. Troen, bombardier; Cpl. Scotty C. Reynolds, radio operator; Cpl. Theodore Chzescikanek, gunner; Co-pilot F/O James L. Redd received serious injuries. The airplane had taken off from Muroc Army Air Field on a high-altitude bombing mission at the AAF Bombing Range near Muroc. Soon after the airplane climbed to 20,000 feet the number-three propeller ran away. Seconds later, the number-three engine erupted into flames and the airplane dropped out of the formation. The fire burned through the wing section, causing the starboard fuel tanks to explode. Cpl. Fish jettisoned the waist section escape hatch and jumped out. The starboard wing separated and the airplane went out of control. The remaining surviving crewmembers were hurled from the airplane as it broke up in flight. Wreckage and bodies were scattered over a wide area. Investigators speculated that the excessive vibration caused by the malfunctioning propeller allowed fuel lines to separate or rupture, causing a massive fire to erupt. Investigators recommended that all crewmembers wear their parachutes at all times during flight and that B-24 pilots be issued "back-type" parachutes so that they can be worn while flying.

8-24-44A. Americus, Georgia. At 1340 CWT, two Stearman PT-17 airplanes collided in mid-air and crashed at Americus, Georgia, killing student A/C James E. Cooper, Jr., Pawtucket, Rhode Island, aboard PT-17 #42-16651. Civilian instructor James J. Mulva, Oshkosh, Wisconsin parachuted to safety and was uninjured. A/C John Edwards, Hammond, Indiana, parachuted to safety from PT-17 #41-8389 and was uninjured. Investigation revealed that PT-17 #41-8389 was performing a power-on stall at about 3,000 feet agl and apparently fell into the path of PT-17 #42-16651. Investigators speculated that A/C Cooper was killed in the collision and was unable to bail out.

8-24-44B. Carlsbad, New Mexico. At 2144, a Beech AT-11 flew into rising terrain seven miles south of the Army Air Field at Carlsbad, New Mexico, killing the crew of five. The airplane had taken off from Carlsbad from on routine bombardier-training mission. The airplane was maneuvering near the field when it smashed into the mountain at a speed described as "faster than normal." The airplane exploded into flames upon impact and all occupants were killed instantly. Wreckage was scattered over an area of 150 yards. The crash was observed from the control tower at Carlsbad. Killed in the crash were: 2Lt. Robert N. Spring, pilot; 2Lt. Raymond F. Chloupek, Bombardier instructor; A/C Billy B. Lane, student bombardier; A/C Vernon C. Johnson, student bombardier; Casimer A. Zielinski, student bombardier.

8-25-44A. Halletsville, Texas. At 1725, two North American AT-6B airplanes collided in mid-air and crashed six miles south of Halletsville, Texas, killing A/C James Q. Tingle aboard AT-6B #41-17110. Instructor pilot 2Lt. John E. Lau received minor injuries parachuting to safety from AT-6B #41-17092. The airplanes had taken off at 1630 CWT from Foster Field, Victoria, Texas, on a three-ship aerobatic flight. Lt. Lau was leading the flight in a trail formation at 10,000 feet. The flight performed various aerobatic maneuvers. The flight broke up in a "rat race" for a short time. The flight leader attempted to reform the flight and was flying to the right of A/C Tingle's airplane at an altitude of approximately 7,500 feet. A short time later, Lt. Lau felt something strike his airplane. Lt. Lau then noticed A/C Tingle's airplane in a dive toward the ground. Lt. Lau could not maintain flight with his heavily damaged airplane and bailed out. Investigation revealed that A/C Tingle's port wing had contacted Lt. Lau's rudder and horizontal stabilizers. A/C Tingle was attempting to regain his number-two position in the flight while in a steep bank to the left, slipping into the leader's tail section. A/C Tingle's port wing was severed about halfway and his airplane went out of control. Both airplanes exploded into flames upon impact with the ground. A/C Tingle was found in the wreckage of his airplane.

8-25-44B. Smyrna, Tennessee. At 2024 CWT, two Consolidated B-24D airplanes collided in mid-air and crashed ten miles southwest of the Army Air Field at Smyrna, Tennessee, killing eight fliers. Killed aboard B-24D #42-40601 were: 2Lt. Graham K. Hobbs, B-24 instructor; 2Lt. Edward A. Kennedy, pilot/B-24 student; 2Lt. Dudley J. Luntsford, Jr., co-pilot/B-24 student; SSgt. Jasper Van Niel, engineer. Killed in the crash of B-24D #42-40708 were: 1Lt. Paul B. DeWitt, B-24 instructor; 2Lt. John S. Schultz, pilot/B-24 student; 2Lt. James A. Grumbles, co-pilot/B-24 student; Sgt. Joseph J. Froelich, engineer. Investigation revealed that B-24D #42-40708 had taken off from Smyrna at 1935 CWT on a scheduled local instrument flight and B-24D #42-40601 took off from Smyrna at 2014 CWT on a "round robin" navigation flight to Atlanta, Georgia, to Birmingham, Alabama, and return to Smyrna. A short time later the airplanes collided at an altitude of approximately 3,500 feet. Investigators stated, "From the examination of the wreckage the left wing of 42-40601 hit the middle of the left vertical stabilizer on 42-40708, and sheared off the left vertical stabilizer to the trailing edge and across the horizontal stabilizer at an angle of about 30 degrees from the perpendicular. The altitude of the airplanes can only be assumed, but must have been between 3,000 and 4,000 feet indicated. The force of the impact caused the fuselage of 42-40708 to break in two at approximately three feet to the rear of the waist gun windows. The horizontal stabilizer, right vertical stabilizer and rudder, elevators, and rear section of the fuselage of 42-40708 fell away separately. [B-24D #42-40708] fell to the ground, landing inverted in a level attitude on the down-slope of a hill and burst into flames.... The left wing of [B-24D #42-40601] from the wing tip to a point eight feet outboard of the number-one engine was torn off by the impact. B-24D #42-40601 was observed by ground witnesses to descend in a spin, but leveled off before it hit the ground. It landed in a level attitude and immediately burst into flames." Both airplanes smashed into the ground within five seconds of each other and in close proximity. Control tower personnel at Smyrna observed the flames of the burning wreckage.

8-26-44A. Omaha, Nebraska. At 0410 CWT, two crewmembers were killed when they bailed out of a Boeing B-29 near Omaha Municipal Airport, Omaha, Nebraska. Three other crewmembers parachuted to safety. Seven crewmembers stayed with the ship and the pilots were able to make a successful emergency landing at Omaha. The airplane had taken off at 1230 CWT (8-25-44) from Smoky Hill Army Air Field, Salina, Kansas, on a bombing mission to the Batista Bombing Range, Cuba, and then return for a landing at Walker Army Air Field, Walker, Kansas. On the return leg the airplane let down near Bruning, Nebraska, through an overcast that topped out at 7,000 feet. The airplane came out of the overcast at an altitude of 800 feet agl and moments later the number-four engine quit. The pilot increased rpm on the remaining engines. After a short time, the remaining engines began to run rough. The number-four engine was trailing fire from the exhaust and the number-two engine began backfiring. Moments later the number-two engine began trailing flames and sparks. The pilot then ordered the men in the aft compartment to bail out. Radio operator Cpl. Harold E. Woodruff parachuted into Lake Carter and drowned before he could be rescued. Gunner Cpl. Kenneth C. Clawson was apparently struck in the head by the aircraft and apparently killed instantly. His parachute was not deployed. Radar operator Sgt. Peter D. Karagannis, gunner Cpl. James W. Winslow and gunner David E. Willis parachuted to safety and received minor injuries. The engine fires went out and the pilots were able to make an emergency landing at Omaha at 0440 CWT. Crewmembers uninjured in the emergency landing were: Lt. Col. John E. Dougherty, pilot; 2Lt. Willard F. Shorey, co-pilot; Col. Claude E. Putnam, co-pilot; Capt. Prescott L. Martin, flight engineer; 1Lt. Arthur H. Rand, flight engineer; Capt. Richard E. Hale, bombardier/navigator; Capt. Berry R. Thompson, navigator.

8-26-44B. Farmington, New Mexico. At 0915, a Boeing B-17F suffering an in-flight fire crashed 12 miles northwest of Farmington, New Mexico, killing eight crewmembers. Three crewmembers were injured parachuting to safety and two other crewmembers were uninjured. The airplane had taken off from Fairmont Army Air Field, Nebraska, on a navigation-training mission to Mines Field, Los Angeles, California. Investigators stated, "When approximately one hour out of Fairmont Army Air Field ... a slight oil leak was noticed on the number-two engine. The oil leak did not appear to be excessive and, since all instrument readings were normal, no attempt was made to feather the [propeller]. Somewhat later, the oil leak appeared to have increased, and the pilot and co-pilot agreed to land at the first suitable field to investigate the oil leak. After crossing the highest range of mountains, and after consulting with the navigator, the pilot and co-pilot selected the airfield at Farmington, New Mexico, in which to land and make their investigation. The Farmington airfield was estimated to be 20 to 30 miles away, and a descent was made to attain a normal traffic altitude. When nearing the traffic altitude, the co-pilot, who was in the navigator's compartment at the time, noticed flames and smoke coming from the number-two engine. The co-pilot directed the bombardier and the navigator to attach their parachutes in preparation of abandoning the airplane. The bombardier was directed to notify all other crewmembers. The co-pilot then returned to the cockpit and notified the pilot of the fire. Attempts were made to slip the airplane to extinguish the flames, but when these were unsuccessful the signal was given to abandon ship. All survivors have stated that the airplane must have been very low when they jumped, because they struck the ground very soon after leaving the airplane. It is the opinion of the investigating officers that those persons whose parachutes were used unsuccessfully received fatal injuries in their jump because of the low altitude of the airplane. Those persons who were fatally injured without using their parachutes apparently had too little time to abandon the airplane." Because of the complete destruction of the airplane and engines, investigators were not able to determine the cause of the engine fire. Co-pilot 1Lt. James C. Boozer and gunner Pfc. Paul E. Butcher were able to parachute to safety and were uninjured. Engineer TSgt. James E. Rodriguez, gunner Pfc. Richard P. Gooden and passenger Cpl. Vincent P. Kiniry received minor injuries parachuting to safety.

Bombardier 2Lt. Roy F. Weiss, engineer Sgt. William P. Steinberger and gunner Cpl. Norbert L. Blue fell to their deaths in unsuccessful parachute attempts. Killed in the crash of the aircraft were: Capt. Robert F. Forbes, pilot; 1Lt. Timothy J. Cavanaugh, navigator; SSgt. Joseph A. Chase, radio operator; Pfc. Vincent A. Alessi, gunner; Cpl. Jose R. Flores, passenger.

8-26-44C. Ephrata, Washington. At 1040, a Bell P-63A crashed ten miles east of the Army Air Base at Ephrata, Washington, killing Capt. John W. Power. The airplane had taken off at 1015 from Spokane Army Air Field, Washington, on a ferry flight to Ephrata AAB. At 1040, an officer in the control tower at Ephrata AAB noticed an explosion and fire east of the field. Investigation revealed that the airplane had struck the ground in a spin. Investigators could not determine how the airplane entered the fatal spin.

8-26-44D. Dixon, California. At 1045, a North American TB-25D crashed 12 miles southeast of Dixon, California, killing students A/C Ellis C. Lassiter and A/C Donald J. Langlais. The airplane had taken off at 1031 from Mather Field, Sacramento, California, on a routine instrument flight. A B-24 co-pilot observed the subject airplane flying at 4,000 feet indicated altitude. The B-25 was seen to suddenly roll to an inverted attitude and then go into a vertical dive. The students made an abrupt recovery but the airplane rolled over into an inverted spin, describing two turns before entering a flat spin. The airplane remained in a flat spin until it struck the ground and exploded into flames. Investigation revealed that the port wing had failed and separated. The port wing was found a mile and a half from the main wreckage. The port engine nacelle, port engine, port aileron, and port landing gear doors were found scattered over a wide area.

8-27-44A. Vinton, Louisiana. At 1130, a Republic P-47D flying in poor weather crashed two miles northwest of Vinton, Louisiana, killing pilot 1Lt. William A. Kerr, Jr. The pilot took off from Harding Field, Baton Rouge, Louisiana, on a medium-altitude cross-country navigation flight with four other P-47s. The formation encountered thunderstorms near Lake Charles, Louisiana, and Lt. Kerr ordered the flight to fly to Alexandria, Louisiana, and wait for him to arrive. He radioed that he would meet them after he headed off an errant aircraft that was part of his flight. The flight leader never arrived at Alexandria and the flight returned to Harding Field. The errant aircraft, which was suffering radio problems and was unable to receive, inadvertently entered the front but the pilot immediately performed a 180-degree turn and was able to get out and head back to Alexandria. Lt. Kerr, apparently not realizing that the errant aircraft had turned out of the storm, entered the thunderstorm in an attempt to intercept the errant airplane on the other side of the front, which contained very turbulent air, lightning and heavy rain. Visibility was zero. Investigators speculated that the pilot lost control of the airplane in the heavy turbulence. Investigators stated that the airplane dove into the ground at high speed from an altitude of approximately 14,000 feet msl.

8-27-44B. Bay St. Louis, Mississippi. At 2006, a Boeing B-17G flying in poor weather crashed 20 miles north of Bay St. Louis, Mississippi, killing the crew of ten. Killed in the crash were: 2Lt. Harold S. Cline, pilot; 2Lt. Francis C. Feasel, co-pilot; 2Lt. David Grolnic, bombardier; 2Lt. Paul B. Lyles, navigator; Cpl. Robert R. Cole, engineer; Cpl. Norman N. Rocheleau, radio operator; Cpl. Raymond T. Samode, engineer; Cpl. James M. Hummel, gunner; Cpl. Charles W. Marvin, gunner; Cpl. Harold R. Van Marter, gunner. Investigators stated, "2Lt. Harold S. Cline took off to fly a Local Celestial navigation mission and was assigned an altitude of 3,500 feet, and was to fly from Gulfport [Mississippi] to Bogalusa [Louisiana] and return. After clearing the tower, the aircraft proceeded on course to Bogalusa and after ten minutes on course ran into a thunderstorm at which time the pilot lost control and crashed. By the weather reports, there were severe thunderstorms on the outside of the local flying area." Investigation revealed that the airplane had slammed into the ground at a steep angle in a swampy area with few trees, exploding violently into flames upon impact.

8-28-44A. Quarry Peak, California. At 0230, a Douglas P-70 crashed into Quarry Peak 15 miles west of Mono Lake, California, killing pilot 2Lt. Eugene K. Wicks and radar operator 2Lt. Richard O. Davis. The airplane had taken off from Hammer Field, Fresno, California, on a high-altitude navigation flight. The airplane had passed over higher peaks just prior to colliding with 10,000-foot Quarry Peak several feet below the summit, indicating to investigators that the airplane was losing altitude at the time of the crash. Flaming wreckage was hurled over the peak strewn down the side of the mountain. Investigators noted that had the airplane missed colliding with Quarry Peak, it would have certainly collided with higher peaks in its flight path. Investigators could not determine why the aircraft was losing altitude.

8-28-44B. Moapa, Nevada. At 1114, a Bell P-63A crashed ten miles NNW of Moapa, Nevada, killing pilot 2Lt. Lewis A. Shaffer. The airplane had taken off from the Army Air Field at Las Vegas, Nevada, on a gun-camera mission. Witnesses on the ground observed the airplane trailing smoke as it flew over. The engine apparently quit and the airplane entered a spin. The pilot was able to recover momentarily before entering another spin. The airplane spun to the ground in a flat spin and exploded into flames. Investigators speculated that the pilot might have been incapacitated by smoke in the cockpit.

8-28-44C. Ajo, Arizona. At 1555 MWT, a North American AT-6C crashed at the AAF Gunnery Range #1 near Ajo, Arizona, killing pilot 1Lt. Neil M. Jelley. Investigators stated, "Lt. Jelley was on a ground

gunnery mission and while making his last turn onto the approach to the target he evidently allowed his airspeed to drop too low by applying back pressure upon the stick. This caused the airplane to do a partial snap over on its back and it spun to the ground."

8-29-44A. Columbus, Ohio. At 1015 EWT, a Vultee BT-13B crashed while attempting a landing at Don Scott Field, Columbus, Ohio, killing pilot 2Lt. Jack K. Eickhoff and passenger 2Lt. Robert E. Sheep. The airplane had taken off from Lockbourne Army Air Field, Columbus, Ohio, on a routine training flight to Don Scott Field, Ohio, to practice take-offs and landings. The airplane had made several successful landings and take-offs prior to the crash. There were no witnesses to the accident and investigators could not determine its cause.

8-29-44B. Goodfellow Field, Texas. At 1000, a Fairchild PT-19A crashed eight miles east of Goodfellow Field, San Angelo, Texas, killing instructor 2Lt. Thomas R. Henderson and A/C Porter F. Collis. Investigators speculated that the student and instructor were practicing simulated forced landings when they stalled the airplane at an altitude too low to allow recovery. The airplane struck the ground in a mushing attitude.

8-29-44C. Douglas, Georgia. At 1230 EWT, a Stearman PT-17 crashed ten miles northwest of Douglas, Georgia, killing A/C Donald W. Chupp. Witnesses on the ground observed the airplane stall out of a slow roll and enter a very tight spin at about 2,000 feet. The student partially recovered at an altitude of about 300 feet agl. The airplane then dove straight into the ground and burst into flames.

8-29-44D. Gordon, Alabama. At 1330, two North American AT-6A airplanes collided in mid-air and crashed five miles south of Gordon, Alabama, killing A/C Cadwalder H. Koons, Jr., who fell to his death attempting to parachute to safety from AT-6A #41-16625. Instructor 2Lt. Chester V. Bogle and student A/C Alan Joseph Kuntze parachuted to safety from AT-6A #41-17016. The airplanes were part of a three-ship flight that had taken off from Napier Field, Dothan, Alabama, on a formation training flight. It was noted that this was the first formation training flight for the students flying in the number-two and number-three positions. Lt. Bogle was flying the lead airplane. The formation executed several maneuvers before entering a Lufbery circle. Investigation revealed that A/C Koons had fallen out of the Lufbery circle and was flying about 500 feet below the altitude of the circle. A/C Koons apparently pulled up and into the lead airplane while trying to rejoin the circle, severing Lt. Bogle's starboard wing. Both airplanes went out of control and plunged to the ground. A/C Koons bailed out an altitude too low to allow his parachute to deploy.

8-29-44E. Wendover Field, Utah. At 1425 MWT, a Republic P-47D crashed 25 miles ESE of Wendover Field, Utah, killing pilot 2Lt. Robert M. Kellerman. The airplane was part of a flight of five that had taken off from Wendover Field on a high-altitude aerobatic mission. The subject airplane performed a slow roll and then completed a complete loop. The pilot then attempted an Immelmann turn. The airplane rolled out on top of the Immelmann in a nose-high attitude with low airspeed. The airplane stalled and entered a spin. The pilot stopped the spin and the airplane entered a vertical dive from which he could not recover. The pilot exited the airplane and was struck on the right leg with the starboard horizontal stabilizer, severing the leg. The pilot was also struck on the head by the vertical stabilizer, killing him instantly and causing the parachute pack to rupture. The parachute was damaged and it did not fully deploy. The pilot's body plunged to the ground near the spot where the airplane smashed into the ground and exploded.

8-29-44F. Homestead, Florida. At 1842 EWT, a Consolidated B-24D crashed while attempting a simulated two-engine landing at Homestead Army Air Field, Homestead, Florida, killing assistant engineer Cpl. George R. Corey. Pilot/B-24 student 1Lt. James O. Shaddy and co-pilot/B-24 instructor 1Lt. Elmer E. Emigh were uninjured in the crash. Engineers SSgt. Clyde R. Rice and SSgt. William R. Standlee were also uninjured. Investigators stated, "On a local training flight the instructor was giving the student his final landing at the end of a flight period. On the downwind leg, the airplane and all four engines were functioning normally. Student called for landing gear down and ran through the normal landing check. On the base leg the instructor cut the number-four engine by retarding the throttle. He then reduced the rpm to low rpm or high pitch position. Student went through the normal emergency procedure [and then] called for increased power and [retracted] the landing gear. The student started his normal three-engine approach and while turning from the base leg to the final approach, the instructor cut the number-three engine by retarding the throttle, he also reduced the rpm to low rpm or high pitch position. When at a distance of approximately two miles and at an altitude of 600 feet the instructor was attempting to increase the rpm on the number-three engine for use in case the necessity became apparent. The student called for the lowering of the landing gear. As the landing gear was being lowered, Cpl. George E. Corey descended into the nose section to check the correct position of the nose landing gear. SSgt. Standlee went to the rear of the ship to check the main landing gear. The ship continued its approach on two engines and when it became apparent that they were undershooting, more power was applied to the engines in an attempt to raise the landing gear. Power on engines number one and number two took hold, but number three and four did not respond. The torque created by the two left engines pulled the ship off to the right and it struck the ground approximately 300 yards inside the boundary of the field, 175 yards off to the right of the runway." The damage to

the airplane might not have been great had the B-24 not struck a two-foot high tree stump in the landing area.

8-29-44G. Quincy, Washington. At 2015, a Bell P-63A crashed eight miles southeast of Quincy, Washington, killing pilot 2Lt. Gordon G. Anderson. The airplane had taken off from Ephrata Army Air Field, Washington, on a low-level navigation mission. The airplane was flying in the number-four position of a four-ship flight. After flying at low altitude for about a half hour, the flight leader climbed the formation up to about 12,000 feet. He then led the flight through a steep dive into a chandelle maneuver to the left. The subject airplane apparently stalled out of the maneuver and was seen to enter an inverted spin. The airplane described about five or six turns of spin when the flight leader radioed the subject pilot to bail out. Almost immediately the starboard door was jettisoned and the pilot was seen to leave the airplane; he apparently opened his parachute immediately. The airplane then spun around and struck the partially deployed parachute, collapsing it. The pilot struck the ground near where the airplane smashed to the ground and exploded.

8-30-44A. Atlantic Ocean, Georgia. At 0815 EWT, a Curtiss P-40N crashed into the Atlantic Ocean five miles due east of Sapelo Sound, Georgia, killing F/O James E. Beard. The subject airplane was the number-two airplane in a three-ship flight that had taken off from Harris Neck Army Air Field, Georgia, on a skip-bombing mission at the AAF South Bombing Range. On the first bombing run, F/O Beard was seen to drop his bomb and then enter a steep climbing turn. The airplane stalled and winged over and then it dove into the sea. Investigators speculated that the pilot had used excessive backpressure on the stick while recovering from his dive, entering a stalling position. The pilot added full throttle to recover, causing the P-40 to enter a climbing left turn and wingover because of the excessive torque.

8-30-44B. Panama City, Florida. At 1059 CWT, a Martin TB-26B operating at Tyndall Field, Panama City, Florida, struck an airplane tug tractor, fatally injuring tug driver Pfc. Salvatore Di Marco. Investigators stated, "Pilot was taxiing in normal fashion with engineer observing from top escape hatch. Driver of the tug was approaching the airplane from the pilot's left side at an angle that would intersect the path of the airplane approximately 40 feet ahead. Driver of the tug had an unobstructed view of the airplane at all times during the last 100 feet of his path, and did not make any effort to avoid the airplane until within approximately 15 feet of the nose of the airplane. When the pilot first saw the driver of the tug, the driver appeared to be looking directly at the airplane; however, the driver made no effort to slow down, so the pilot immediately applied full brakes and the tug continued on his original path until within approximately

15 feet of the nose of the airplane when he turned to the left about 15 degrees, which put him on a path 90 degrees to the path of the airplane. The tug struck the nose of the airplane on the bottom and rear section of the plexi-glass nose. Airplane then rolled the tug over about one and one-half turns, causing the nose wheel unit to collapse, allowing the full weight of the nose section to rest on the tug. Driver of the tug received injuries that proved to be fatal 16 hours later." The B-26 was piloted by 2Lt. LeRoy M. Brown and co-pilot 2Lt. Charles E. Martin. They were uninjured.

8-30-44C. Jet, Oklahoma. At 1145 CWT, a Curtiss P-40N crashed at the Great Salt Lake Gunnery Range near Jet, Oklahoma, killing pilot 2Lt. Melvin R. Wardyn. The airplane was flying in the number-six position of a six-ship flight that had taken off from Strother Field, Kansas, on a dive-bombing mission. Witnesses stated that the airplane had made a normal bombing run but performed an abrupt pull-up. After the pull-up the airplane snapped over into a half-roll, nosed down and then dove into the ground in an inverted position. The airplane exploded into flames upon impact. Investigators speculated that the pilot might have blacked out during the abrupt pullout.

8-30-44D. Lebanon Junction, Kentucky. At 1610 CWT, a Martin B-26G crashed four miles northeast of Lebanon Junction, Kentucky, killing pilot 2Lt. William K. Neal and non-rated co-pilot Pvt. John E. Price. The airplane had taken off from Baltimore, Maryland, on a ferry flight to Nashville, Tennessee. Witnesses on the ground observed the airplane fly over at an altitude estimated to be about 6,000 feet. Moments later, one engine began to make "unusual noises." The engine was heard to stop and the airplane was then seen to enter a spin, continuing in a spin until it struck the ground at a steep angle. The B-26 exploded violently into flames upon impact. It was later speculated that the airplane suffered the failure of one engine and the pilot was unable to feather the propeller and trim the aircraft for single-engine flight in time to prevent losing control.

8-30-44E. Bainbridge, Georgia. At 1920 CWT, a Vultee BT-13A crashed at Vada Auxiliary Airfield near Bainbridge, Georgia, killing A/C Morgan E. Hein. The Accident Classification Committee stated, "At approximately 1920 CWT, A/C Hein, after completing a verbal cockpit check, was cleared for take-off by the tower, taxied out on the runway, advanced the throttle and took off. The first portion of the take-off appeared to be normal. However, when the aircraft attained an altitude of approximately 30 feet it was seen to start a roll to the left with the nose high. After continuing the roll to almost an inverted position, the airplane dived into the ground, shearing the engine from its mounts and coming to rest on its nose approximately 100 feet to the left of the runway, completely demolishing the aircraft and causing fatal injuries to the pilot."

8-30-44F. Victorville, California. At 1815, a Bell TP-39N crashed one-half mile northwest of Victorville, California, killing pilot 2Lt. James D. Laur. Investigation revealed that the pilot was in the traffic pattern and was turning from the base leg onto final approach when the ship seemed to snap roll about a quarter of a turn to the right, then rolled back about the same amount to the left. The airplane then rolled to the right again. The pilot leveled the airplane out just before it slammed into the ground and exploded. The control tower operator witnessed the accident. Investigators speculated that the pilot realized that the airplane had lost altitude in the turn and had pulled up sharply, causing the airplane to stall.

8-31-44. St. Louis, Missouri. At 0906 CWT, a North American AT-6A crashed in a vacant lot at 3100 Easton Street, St. Louis, Missouri, killing pilot 1Lt. James W. Bethka and passenger 1Lt. Vivian A. Lynn. The airplane took off from Scott Field, Bellville, Illinois, on an administrative flight to Lambert Field, St. Louis. Witnesses observed the airplane falling out of control at an altitude of about 2,000 feet agl. Investigation revealed that the airplane had stalled out of a steep turn, snapping over to an inverted position and then entering a split-S type maneuver. The airplane then entered an 80-degree dive toward the ground, continuing in this attitude until it smashed into the ground and burst into flames. Investigators could not determine the cause of the accident.

September

9-1-44A. Muroc, California. At 0718, a Consolidated B-24J crashed five miles southeast of Muroc Army Air Field, California, killing the crew of ten. The airplane took off from Muroc Army Air Field on a high-altitude formation and camera mission. Soon after take-off and climbing out of the area, the number-four engine began "acting up." The pilot elected to return to the field and radioed Muroc Tower that he was coming in. Moments later, the number-three engine began losing rpm. At 0717, the pilot radioed that he was not going to make the field and he was going to attempt an emergency forced landing on Muroc Dry Lake. The pilot then radioed that he was not going to make the dry lakebed. Personnel in the control tower spotted the B-24 as it attempted to shoot its emergency forced landing. The B-24 struck the ground on its landing gear at a slight glide angle, bouncing back into the air for about 150 yards. The airplane then struck the ground hard and began skidding sideways. The airplane broke up and burst into flames as it skidded for about 150 yards, coming to rest approximately 300 yards from the edge of the dry lakebed. Investigation revealed that all B-24 pilots who had flown the subject airplane complained that the airplane could not climb satisfactorily or operate well on three engines. Investigators noted that the B-24 pilots at Muroc had similar performance problems with all B-24J model airplanes in the 42-73377 — 42-73399 block. Killed in the crash were: 2Lt. Lawrence D. Lampton, pilot; F/O William L. O'Brien, co-pilot; F/O William F. Wray, navigator; F/O Howard Beck, bombardier; Cpl. James P. Davis, engineer; Cpl. Robert L. Metcalf, radio operator; Cpl. Boyd H. Jones, gunner; Pfc. Robert W. Breedlove, assistant radio operator; Cpl. Wayne E. Brown, gunner; Pvt. John E. Krstich, gunner.

9-1-44B. Bullrun, Oregon. At 0745, a Bell P-63A crashed ten miles east of Bullrun, Oregon, killing pilot 2Lt. Ernest N. Dorazio. The airplane was part of a five-ship flight that took off from the Army Air Base at Portland, Oregon, on an interception-training mission. The airplane was flying in the number-two position of the formation, which was flying at about 10,500 feet indicated altitude. After completing the interception mission, the formation descended and encountered a broken deck of clouds at about 8,500 feet. The flight leader thought that he saw a hole in the clouds and led the flight through. The formation encountered instrument conditions and broke up. All of the pilots had lost contact with each other. Several minutes later, witnesses on the ground observed the subject airplane dive out of the overcast near rising terrain in the area near the Zig Zag Ranger station. The airplane was diving at the mountain so the pilot pulled up abruptly, inadvertently climbing back into the clouds. Moments later the subject airplane was seen falling out of the clouds, spinning out of control. The airplane smashed into the mountain and exploded into flames, killing the pilot instantly.

9-1-44C. Vada Auxiliary Airfield, Georgia. At 1410 CWT, a Vultee BT-13A crashed while attempting a take-off at Vada Auxiliary Airfield, Georgia, killing student A/C Kenneth L. Sowers (BT-13A #42-42668). The airplane was part of a flight of three that had taken off from the Army Air Field at Bainbridge, Georgia, on a formation training flight. Investigators stated, "At approximately 1300 [A/C Mike Tisik (BT-13A #42-41357)] took off from Bainbridge Army Air Field and joined the left wing of a formation with Lt. William Lockett [BT-13A #41-22363] in the lead airplane and A/C Kenneth L. Sowers on the right wing. After practicing routine air work the formation proceeded to Vada Auxiliary Airfield and satisfactorily completed three landings. When taxiing back to take off for the fourth landing, A/C Tisik was assigned the right wing position and A/C Sowers the left wing position. Just after the formation became airborne, A/C Tisik lost control of his aircraft, veered toward the lead airplane in such a manner that his left wing tip [was caught] in the [propeller turbulence of the lead aircraft], [which] caused loss of control and the wing tip dragged on the ground. As the airplane moved forward, the propeller struck the right horizontal stabilizer of the lead aircraft. A/C Tisik's airplane then cartwheeled on its left wing

and nose, sheared the engine from the airplane, turned 180 degrees and came to rest, completely wrecked. Meanwhile, A/C Sowers on the left wing apparently saw A/C Tisik's airplane out of control veering toward him. In attempting to pull up out of the formation to avoid the impending collision, A/C Sowers stalled his airplane, whereupon it completed a partial snap roll and crashed into the ground, shearing the engine and coming to rest on its nose, fatally injuring the pilot. Lt. Lockett, the formation leader, executed an emergency landing on the field."

9-1-44D. Abilene, Texas. At 1435, two Republic P-47D airplanes dove into the ground 24 miles west of the Army Air Base at Abilene, Texas, killing 2Lt. Edward Lemay and 2Lt. James B. Lesher. The airplanes had taken off from Abilene Army Air Base on an operational ceiling mission. Investigators stated, "Lt. Lemay was No. 3 man in a 5-ship formation which had gone to 30,000 feet msl. At 30,000 feet, the leader of the flight put the formation in string and started losing altitude with lazy-eights and turns. When the flight reached approximately 22,000 feet the leader pulled back up to approximately 25,000 in a straight climb. There he circled once to allow the flight to catch up with him and then he did a steep wing-over, pulling out of the resulting dive at around 18,000 feet. The four airplanes behind the leader lost the leader after the wing-over and the No. 2 man recovered from his dive at approximately 9,000 feet. During this dive, the No. 2 man stated he was indicating over 400 miles an hour airspeed. The man in the No. 4 position was following Lt. Lemay and stated that both of them [Lt. Lemay and 2Lt. James B. Lesher] were going straight down and that his indicated airspeed was between 500 and 600 miles an hour. The No. 4 man decided to pull out eventually and completed his pull out at approximately 4,000 feet above the terrain. Just before he lost sight of Lt. Lemay he stated that Lt. Lemay was still going straight toward the ground and there was no indication that he was trying to pull out. In the wreckage found later it was clearly demonstrated that Lt. Lemay's airplane had gone in to the ground at approximately a 90-degree angle.... [The] No. 5 man, Lt. Lesher, [was killed when he] failed to pull out of the dive and crashed perpendicular to the ground. There were no witnesses of the course of Lt. Lesher's flight after the wing-over since he was the last man in the formation, but since the No. 3 man of the formation seemed to be diving straight down and since Lt. Lesher's airplane wreckage was less than a mile from the wreckage of the No. 3 airplane it is logical to assume that the accidents were similar in cause. See 9-1-44E.

9-1-44E. Abilene, Texas. At 1435, a Republic P-47D dove into the ground 24 miles west of the Army Air Base at Abilene, Texas, killing pilot 2Lt. James B. Lesher. The airplane was part of a five-ship flight that had taken off from Abilene Army Air Base on an operational ceiling mission. See 9-1-44D for details.

9-1-44F. Orland, California. At 1845, a Lockheed P-38L crashed one mile east of Orland, California, killing pilot 2Lt. Leo L. Morel. The airplane had taken off from the Army Air Field at Chico, California, flying in the number-two position of a three-ship flight on a supervised aerobatic mission. The airplane completed three aileron rolls and two Immelmann turns. The airplane was then seen to leave the formation, which was flying at an altitude of 9,500 feet indicated. Witnesses on the ground, who had seen the three airplanes maneuvering, observed the subject airplane in a shallow dive and fully expected the P-38 to pull out at any moment. The airplane continued in the dive until it smashed into the ground and exploded into flames. Investigators could not determine the cause of the accident.

9-3-44A. Foster Field, Texas. At 0900, a North American AT-6D crashed seven miles south of Foster Field, Victoria, Texas, killing pilots 2Lt. Rudolph L. Bullington and F/O David V. James. The airplane took off from Aloe Field, Texas, at 0820 on an instrument team ride. Lt. Bullington was flying in the front cockpit as the safety pilot and F/O David V. James was the instrument pilot flying under the hood in the rear cockpit. The airplane was next observed in a diving turn to the right with the engine running at high rpm. Condensation trails were seen trailing the wingtips just prior to the airplane smashing into the ground at high speed. The airplane burst into flames upon impact and the pilots were killed instantly.

9-3-44B. Niagara Falls, New York. At 1402 EWT, a Martin TB-26C crashed after taking off from the municipal airport at Niagara Falls, New York, killing nine crew and passengers. The B-26 took off at 1359 on an administrative flight to Newark, New Jersey. Investigators stated, "After performing a normal take-off to the northwest the airplane executed a steep turn to the left and began a tight circle around the airport, turning inside the traffic pattern to pass almost directly over the control tower and administrative building where [the pilot's] parents were waiting to watch the ship's departure. A maximum altitude of approximately 500 feet was obtained west of the field whereupon a gradual descent was started and the approach toward the tower made at approximately 400 feet with the ship still descending. As the airplane passed by the tower it started what appeared to be, from witnesses' statements, an abrupt and very steep climbing turn to the left. However, this maneuver was not completed as the airplane continued rolling to the left to an inverted position and crashed on the southeast end of the NW-SE runway, bursting into flames immediately upon contact with the ground." Killed in the crash were: Maj. Willard L. Redans, pilot; 2Lt. William E. Lund, co-pilot; Sgt. Eugene E. Swicker, engineer; Lt. Col. Jay P. Rousek, pilot-rated passenger; Maj. Harry M. Devers, pilot-rated passenger; Capt. William J. Hayes, passenger; 1Lt. Martin Steinman, passenger; 2Lt. Frederick Charifson, passenger; Dr. W. D. Knott, civilian medical doctor.

9-3-44C. Panama City, Florida. At 1533 CWT, a Vultee BT-13A crashed at Tyndall Field, Panama City, Florida, killing pilot 2Lt. Gerald F. O'Shea and passenger Pvt. William E. Martin. The airplane was taking off on Runway 27 on a routine weather check flight. The airplane climbed to about 150 feet and then suffered an engine failure. The wings were seen to wobble and then the airplane started a turn to the left. Investigators speculated that the pilot was trying to make an emergency landing on Runway 13. As the turn progressed, the turn became steeper and the nose pitched up. As the airplane turned through 90 degrees of the turn, it snapped to an inverted position and then described about one and a half turns of spin to the left before smashing into the ground at the intersection of Runway 27 and Runway 13. The airplane burst into flames upon impact and the occupants killed instantly. Investigators noted that a safe emergency landing could have been made straight ahead.

9-4-44. Dodge City, Kansas. At 0935 CWT, a Martin TB-26C exploded in mid-air and crashed 13 miles south of the Army Air Field at Dodge City, Kansas, killing the crew of four. The airplane had taken off on a routine training mission. Numerous witnesses on the ground observed or heard the airplane explode as it passed overhead. Investigation revealed that the airplane exploded at an altitude of about 800 feet and smashed to the ground out of control. Wreckage was scattered over a wide area. Investigators could not determine the cause of the explosion. Killed in the crash were: 2Lt. Robert R. Hawkins, pilot; F/O John L. Hammill, co-pilot; Pfc. James M. Staton, engineer; 2Lt. Willett C. Kanehl, pilot-rated passenger.

9-5-44A. Shaw Field, South Carolina. At 1235 EWT, a Vultee BT-13A crashed northeast of Shaw Field, Sumter, South Carolina, killing French Aviation Student Lucian E. Berain, who fell to his death in an unsuccessful parachute jump. The airplane had taken off from Shaw Field on a scheduled aerobatic flight. The airplane was seen to enter a slow roll. The airplane did not complete the roll and performed a split-S maneuver. The student over controlled while attempting to re-establish level flight and the airplane entered a spin from which the student could not recover. The student was killed when he bailed out at an altitude too low to allow his parachute to deploy.

9-5-44B. Greenville, Texas. At 1420, two Vultee BT-13A airplanes collided in mid-air ten miles east of Majors Field, Greenville, Texas, killing A/C Thomas W. Dorsey, who fell to his death in an unsuccessful parachute attempt from BT-13A #41-11025. A/C George J. Croft was uninjured and was able to safely crash land BT-13A #42-43224, minus the port main wheel, at Majors Field. The collision occurred at an altitude of about 5,000 feet. A/C Dorsey's port wing struck A/C Croft's port main wheel, severing the wing.

A/C Dorsey's airplane went out of control and entered a spin. A/C Dorsey was killed when he bailed out at an altitude too low to allow his parachute to deploy.

9-5-44C. Muroc, California. At 1800, a Lockheed P-38J dove into the ground at the AAF Bombing and Gunnery Range at Muroc, California, killing pilot Capt. Kenneth C. Sparks. The airplane was part of a four-ship flight that had taken off from the Army Air Field at Ontario, California, for a dive-bombing mission at the Muroc Range. The number-four airplane also crashed and its pilot killed. Investigators stated, "Captain Sparks was leader of the flight composed of 2Lt. William H. Stiles, Jr., 2Lt. Francis D. Sullivan, and 2Lt. Sidney A. Story. After approximately 45 minutes of flight, Capt. Sparks led his flight over the target at 26,000 feet. Capt. Sparks then half-rolled and dove vertically on the target. Capt. Sparks dropped his bombs and soon thereafter parts began coming off of his airplane, the airplane went into a violent spin and crashed at approximately 1800 PWT. 2Lt. Sidney A. Story followed Capt. Sparks in the vertical dive and also crashed. Lt. Stiles had one engine [that] was not operating properly so he half-rolled but immediately pulled out and spiraled down. Lt. Sullivan half-rolled after his leader but, because his wheel warning lights came on, also pulled out and spiraled down." The subject airplane exploded into flames upon impact. It is likely that the pilot could not recover from the dive because of the tendency of the P-38 type airplane to suffer from severe control surface buffeting caused by compressibility. See 9-5-44D.

9-5-44D. Muroc, California. At 1800, a Lockheed P-38F crashed at the AAF Bombing and Gunnery Range at Muroc, California, killing pilot 1Lt. Sidney A. Story. The airplane was part of a four-ship flight that had taken off from the Army Air Field at Ontario, California, on a dive-bombing mission at the Muroc Range. Lt. Story followed his flight leader in a vertical dive on the target and was unable to recover. The airplane dove into the ground at high speed, exploding violently into flames upon impact and killing the pilot instantly. It is likely that the pilot could not recover from the dive because of the tendency of the P-38 airplane to suffer severe control surface buffeting caused by compressibility. Flight leader Capt. Kenneth C. Sparks also crashed and was killed. See 9-5-44C for details.

9-5-44E. Gowen Field, Idaho. At 2146 MWT, bombardier 2Lt. Carl F. Fossen was fatally injured when he walked into the rotating number-three propeller of a Consolidated B-24J operating at Gowen Field, Boise, Idaho. The crew had successfully completed an air-to-air gunnery mission; the pilots landed the B-24, taxied in and parked in the proper spot. The pilots were shutting down the engines when the bombardier exited the airplane through the open bomb bay doors and walked into the arc of the rotating propeller. He was struck on the right side of his head and died about nine hours later at the station hospital. The B-24 was being operated by pilot 2Lt. Earl L. Nelson, co-pilot F/O Jack F. Bartuska and B-24 instructor 1Lt. Walter H. Euston. The three pilots and eight other crewmembers were uninjured.

9-6-44A. Douglas, Arizona. At 0830 MWT, a Vultee BT-13A crashed in rugged terrain 65 miles north of the Army Air Field at Douglas, Arizona, killing Service Pilot F/O Stanley T. Ambro. The airplane took off from Coolidge, Arizona, on a flight to El Paso, Texas. The airplane was found to have crashed in a rugged area between Jesse James Canyon and Pinary Canyon. Investigators could not determine the cause of the crash. It was later speculated that the airplane might have encountered an area of severe downdrafts and was forced into the terrain.

9-6-44B. Moses Lake, Washington. At 0949, a Lockheed P-38L attempting a landing collided with vehicles near Runway 3 at Moses Lake Army Air Field, Moses Lake, Washington, killing the driver of the Airfield Control Truck, Cpl. Manfor L. Apodaca. Pilot 2Lt. Richard C. Livingston received minor injuries. Investigators stated, "Pilot was number-three man of a four-ship flight returning from an engineering mission. The flight made a normal approach. Lt. Livingston apparently hit the [propeller turbulence of the preceding airplanes] just before landing as his left wing dropped, then his right wing and as he corrected, the airplane veered off to the right side of the runway, crashing into the control truck and other vehicles parked there." The airplane damaged or destroyed the control truck, a Chevrolet airfield crash truck, a Dodge WC-54 ambulance and a GMC 2½ ton truck.

9-6-44C. Hanford, California. At 1100, a Douglas A-24B crashed ten miles southeast of Hanford, California, killing pilots Maj. Robert E. Selby and 1Lt. Homer B. Meyers. The airplane took off from Visalia, California, on an instrument training flight. Maj. Selby was in the rear cockpit as the instrument pilot and Lt. Meyers was in the front cockpit as the safety pilot. Witnesses on the ground observed the airplane enter an uncontrolled dive until it struck the ground and burst into flames. Lt. Meyers was found a short distance from the wreckage, indicating that he had attempted to parachute to safety at an extremely low altitude. Maj. Selby's safety belt was found to be unfastened. Investigators could not determine what caused the airplane to dive into the ground.

9-6-44D. Victorville, California. At 1110, a Bell TP-39N spun to the ground 20 miles southwest of Victorville, California, killing pilot 2Lt. Pat Montgomery. The airplane had taken off from Victorville Army Air Field on a routine training mission. A witness in the air observed the airplane performing an Immelmann turn at an altitude of 12,000 feet. The subject airplane rolled to the right at the top of the maneuver and then stalled, entering a spin to the right. The airplane continued to spin to the right until it struck the ground and exploded.

9-6-44E. Austin, Texas. At 1622, a Beech AT-7 crashed three and a half miles northeast of Robert Mueller Airport, Austin, Texas, killing four fliers. The airplane had taken off from Hondo Army Air Field, Hondo, Texas, on a flight to ferry personnel and cargo to Texarkana, Texas. The airplane had been flying 500 feet above the overcast for most of the flight. The airplane was seen flying at extremely low altitude prior to the crash, apparently attempting to stay under the low-hanging overcast. The ceiling was estimated to be about 300 feet with moderate to severe rain showers. The airplane went into a steep left bank to avoid striking a utility pole carrying two cable power lines. The port wing struck the ground and the starboard wing struck and severed the lower power line. The airplane then cartwheeled into the terrain and burst into flames. Killed in the crash were: 2Lt. John D. Lawson, pilot; Cpl. Arthur J. English, engineer; Pvt. Jesus U. Martinez, engineer; Pvt. Duncan M. Sheats, radio operator.

9-6-44F. Barksdale Field, Louisiana. At 2034, a Martin B-26C (AAF #41-35123) collided with three other B-26 airplanes while attempting a take-off at Barksdale Field, Shreveport, Louisiana, killing one flier and injuring eight others, one seriously. Eleven fliers escaped injury. The Accident Classification Committee stated, "B-26C No. 41-35123 piloted by 2Lt. Stuart L. Main in the process of taking off crashed into B-26C No. 41-35269 piloted by F/O Richard E. Davis, B-26B No. 41-32021 piloted by 1Lt. Roland A. Ball, and B-26B No. 41-31792 piloted by 2Lt. Edgar L. Johnson, during his take-off run on the center runway (32-14) at Barksdale Field.... [The] subject airplanes were parked on the west side of the center runway (32-14) in preparation for their routine night mission take-off.... After the subject pilot, Lt. Main, had finished his pre-flight check he called the tower for permission to take off on the center runway (32-14). After the tower cleared Lt. Main, he proceeded to start his take-off. Due to the fact that another aircraft had been cleared to take off on the extreme right hand side of the runway, just before Lt. Main had been cleared for take-off. Lt. Main started to take-off further to the left than is normally done, using this type of take-off procedure. Subject pilot being unaware of the parked airplanes to his left and in front of him, gave his aircraft full power for take-off, after reaching a speed of approximately 70 miles per hour, he saw a green light and immediately felt a sharp jar which threw his aircraft to the left and temporarily out of control. He immediately proceeded to attempt to straighten subject aircraft, but not in time to keep from colliding with the second aircraft, which by this time rendered Lt. Mains' aircraft completely out of control causing him to collide with the third aircraft. The first aircraft Lt. Main hit was 41-35269, which was damaged on its right wing, and then being temporarily out of control this caused him to hit 41-32021 with his left engine causing 41-32021 to catch on fire and completely burn. The third aircraft that was hit by Lt. Main was 41-31792; this aircraft was completely cut in half directly behind the wings. It was hit by the right engine and wing of Lt. Main's aircraft. After colliding with these three aircraft Lt. Main's aircraft came to rest on the left side of Runway 32 and completely burned. This accident resulted in fatal injury to armorer gunner Cpl. Thomas L. Langford, who was riding in aircraft No. 41-35123, all other crewmembers involved escaped with slight burns and bruises." Injured in the crash of B-26C #41-35123 were: 2Lt. Stuart L. Main, pilot; 2Lt. Raymond H. Pigman, co-pilot; F/O Edwin E. Clark, bombardier/navigator; Cpl. Paul D. Eagins, engineer; Cpl. Wallace J. McNamara, radio operator. Crew uninjured aboard B-26C #41-35269 were: F/O Richard E. Davis, pilot; F/O Thomas L. Moffitt, co-pilot; F/O James B. Turner, bombardier/navigator; Cpl. Gerald C. Watkins, engineer; Cpl. Carl E. Jones, radio operator; Cpl. James W. Calloway, gunner. The pilot of B-26B #41-32021, 1Lt. Roland H. Bell, was seriously injured. Co-pilot 2Lt. Glen J. Van Wye and engineer Thomas D. Gleason sustained minor injuries. Crew uninjured aboard B-26B #41-31792 were: 2Lt. Edgar L. Johnson, pilot; 2Lt. Kendall B. Keary, co-pilot; 2Lt. Kenneth D. Tewell, bombardier/navigator; Cpl. Warren H. Knouse, engineer; Cpl. James A. Sowers, radio operator.

9-7-44A. Presque Isle, Maine. At 1024, a Douglas C-54A crashed two and a half miles south of the Army Air Field at Presque Isle, Maine, killing three fliers. The airplane had been cleared to take-off on a routine local flight but had to return to the line because of a "pass valve was stuck." The valve became unstuck and the airplane was taxied back out to Runway 19 for take-off. The airplane made what appeared to be a normal take-off. The pilots retracted the landing gear and climbed to about 1,000 feet agl. The airplane deviated from its flight path in a 30-degree turn to the left. After about 30 seconds, the airplane began to lose altitude. The C-54 was next seen to fall into a diving turn to the left and smash into the ground where it exploded into flames immediately upon impact. Pilot Maj. George H. Shafer, co-pilot Capt. Knute Nordahl and engineer MSgt. T.W. Marshall were killed in the crash. Investigators speculated that the elevator trim tab controls might have been improperly rigged.

9-7-44B. Bartow, Florida. At 1135 EWT, a North American P-51C suffered a catastrophic structural failure and crashed eight miles ESE of the Army Air Field at Bartow, Florida, killing 2Lt. Rocco Foglia. The airplane was part of a flight of P-51 fighters making simulated attacks on a formation of Boeing B-17s. The subject airplane dove on a B-17, and after achieving a speed estimated to be about 400 mph the pilot attempted to pull out of the dive. Investigation of the wreckage indicated that the starboard landing gear door separated and the landing gear dropped down. The extreme drag at high speed caused the starboard wing to fail and separate. The airplane rolled violently to the right and then totally disintegrated. The engine, port wing, and tail section all separated from the aircraft as it fell out of control. The main wreckage smashed to the ground and exploded; the wings, tail section and small bits of wreckage were scattered over a wide area.

9-7-44C. Moultrie, Georgia. At 1605 EWT, a North American AT-6D crashed while attempting an instrument take-off at the Spence Field, Moultrie, Georgia, killing instructor 1Lt. Gaines C. King and seriously injuring student A/C W.W. Fowler, Jr. Investigators stated, "As the airplane left the ground it was seen to start a turn to the right in a nose high attitude continuing into a vertical bank in which it apparently stalled. As the airplane stalled it continued the roll to the right and crashed back to the ground in an inverted position nose first. The airplane bounced and came to rest approximately 20 feet from where it first hit the ground. The airplane was a complete wreck."

9-7-44D. Winston-Salem, North Carolina. At 1740, a North American B-25J suffering an in flight fire crashed 22 miles north of Winston-Salem, North Carolina, killing the crew of five. The airplane was part of a six-ship flight that had taken off from the Army Air Base at Greenville, South Carolina, on a formation/bombing mission. "It is the opinion of this Committee that the fire started in the left engine section due to a break in the gasoline lines of that engine. Statements from members of the formation lead the committee to believe that the fire started in the engine section and led to the fuel cells. The opening of the bomb bay doors while still in formation and at an altitude of 8,000 feet leads the committee to believe that there was evidence of gasoline fumes prior to the fire in the left nacelle. It is further believed that the instructor pilot thought he could bring the fire under control by the use of the Lux system and maintain flight on a single engine and, therefore, did not order the crew to abandon the aircraft while there were still sufficient time and altitude. It is further believed that the pop heard by people on the ground was cause by the blowing out of the tire in the left engine nacelle, which caused the opening of the left engine nacelle doors. It is further believed by the Committee that the pilot did not open the bomb bay doors for the purpose of [bomb salvo] as the aircraft was in approximately a 110-degree bank and that the bomb bay door lever was pulled in error for the emergency bomb release handle. Killed in the crash were: 1Lt. Walter A. Knecum, pilot; 2Lt. James G. King, Jr., co-pilot; 2Lt. William H. Barnes, Jr., bombardier- navigator; Cpl. Leonard J. Colburn, engineer; Cpl. Irving Raber, radio operator.

9-8-44A. Tonopah, Nevada. At 0025, a Consolidated B-24J flew into the ground five miles northeast of the Army Air Field at Tonopah, Nevada, killing eight fliers and seriously injuring gunner Cpl. Caius M. Carpenter. Killed in the crash were: 2Lt. George B. Moss, pilot; 2Lt. Forrest L. Carl, co-pilot; 2Lt. Henry J. Koller, bombardier; Cpl. Raymond H. Connor, engineer; Cpl. Arnold D. Branson, radio operator; Pfc. Eugene M. Kruitsick, radio operator; Pfc. Lincoln C. Smith, engineer; Pfc. John R. Campbell, gunner. The airplane had taken off from Tonopah Army Air Field at 2357 PWT (9-7-44) on a local night transition flight. At 0003, the pilot radioed the control tower that the number-two engine was on fire and requested permission for an emergency landing. A short time later, the airplane was observed flying higher than traffic altitude and inside of the normal traffic pattern. No fire was observed. Witnesses observed the airplane north of the field moments before it crashed into the ground and exploded into flames. Rescue personnel did not reach the crash site until 0415. Investigation of the number-two engine revealed no evidence of in-flight fire. Investigators speculated that the engine was probably experiencing a phenomenon known as "torching." The pilots apparently were trouble shooting the engine and had inadvertently allowed the airplane to fly into the ground.

9-8-44B. Marana, Arizona. At 0945 MWT, a Vultee BT-13 crashed ten miles SSE of the Army Air Field at Marana, Arizona, killing A/C Gerald A. Speetzen, who fell to his death in an unsuccessful parachute jump. The student had been practicing landings at Avra Auxiliary Airfield. The airplane was observed to enter a spin at an altitude of 4,000 feet agl. At about 2,000 feet, the rate of spin seemed to slow down momentarily before returning to the initial rate of rotation. The student could not recover and jumped out of the airplane at an altitude of about 300 feet agl, too low to allow his parachute to deploy.

9-8-44C. Bruning, Nebraska. At 1540 CWT, a Boeing B-17G and Republic P-47D collided in midair and crashed five miles northeast of the Army Air Field at Bruning, Nebraska, killing seven fliers. Four other fliers were injured, one seriously, parachuting to safety. 2Lt. John T. McCarthy was killed in the crash of the P-47D. Killed in the crash of the B-17G were: 2Lt. William F. Washburn, pilot; 2Lt. Bernard I. Hall, co-pilot; F/O George A. Budovsky, navigator; 2Lt. Lyle C. Baxman, bombardier; Cpl. John E. Tuchols, engineer; Pvt. Henry c. Sedberry, gunner. Gunner Pfc. Rueben L. Larson, gunner Pvt. Albert L. Mikels and radio operator Cpl. LeNoir A. Greer sustained light injuries parachuting to safety and gunner Cpl. Walter A. Divan was seriously injured. All four were blown from the B-17 when it exploded and broke up in the collision. The B-17 was part of an 18-ship formation that had taken off from the Sioux City Army Air Base, Iowa, on a formation training flight to Bruning Army Air Field where they were to be intercepted by P-47 airplanes stationed there. The P-47 flight was airborne and successfully intercepted the B-17s. The attacking P-47s were supervised by AAF fighter pilot instructors flying above the formations. The subject P-47 made an attack on the subject B-17 from high on the port side. The P-47 began to pull out of the diving attack about 300 yards from the B-17. The P-47 mushed into the B-17's port wing near the number-one engine while attempting to pull up and away from the bomber. Both airplanes exploded into flames and began to break up as they fell. The P-47 entered a very tight spiral and remained in this attitude until it struck the ground and exploded. The B-17 broke up into several pieces as it fell, the main section exploding into flames upon impact with the ground.

9-9-44A. Pueblo, Colorado. At 0848, a Consolidated B-24J crashed while attempting an emergency forced landing three-quarters of a mile ESE of the East-West Runway at Pueblo Army Air Base, Pueblo, Colorado, killing a Chinese Air Force bombardier and gunner. Six other Chinese Air Force fliers and an AAF gunnery instructor were injured. Two other fliers escaped injury. The airplane had taken off from Pueblo AAB on a high-altitude formation and air-to-air gunnery mission. The airplane was flying at about 6,000 feet approximately ten miles WSW of the base when the number-one engine began losing oil pressure. The pilot elected to return to the base. Investigators stated, "When approaching the field the landing gear was lowered but the nose wheel failed to lock and the gear was then retracted and lowered three times in an effort to lock the nose wheel, which was finally accomplished by the engineer holding the gear handle in the down position. While operating the landing gear, the pilot had entered the traffic pattern and turned on his downwind leg, at which time number-one engine caught on fire but was extinguished by feathering the propeller and closing the cowl flaps. In turning on the base leg and final approach sufficient altitude was lost and the pattern extended to such a distance from the field that the pilot was unable to reach the runway. In the pilot's attempt to land, the left landing gear struck the Air Base ammunition revetment, throwing the airplane in such an attitude that number-one and number-two engines struck the concrete retaining wall of the revetment, shearing off number-one and number-two superchargers and turning the airplane 180 degrees either before or after striking the ground beyond, causing complete destruction of the airplane by the crash and fire which later developed." Investigators noted that the most of the crewmembers had not assumed safe crash positions. The bombardier, Capt. Pie-Chih Chen, had been standing on the bomb bay catwalk aft of the flight deck, which was crowded with crewmembers, contributing to his death in the crash-landing. It was also noted that the gunner, Sgt. Ta-Fu Chang, mistakenly braced himself in front of the ball turret rather than aft, greatly contributing to his death in the crash landing. Co-pilot Capt. Yen-Chu Chang, navigator Capt. Kwok-Leung Li and gunner Sgt. Hing-Fa Yen were seriously injured. Crew injured in the crash

were: Capt. Taih-Cheng Wang, flight engineer; 1Lt. Poi-Hung Lin, radio operator; Sgt. Sun-Ping Chiao, gunner; Pfc. Ronald S. Flood, AAF gunnery instructor. Pilot Capt. Tsu-Lun Chien and gunner Kit-Han Chang escaped injury.

9-9-44B. Santa Maria, California. At 0932, a Lockheed P-38L crashed three miles south of the Army Air Field at Santa Maria, California, killing pilot 2Lt. William L. Harris. The airplane was part of a four-ship flight that had taken off at 0730 from Santa Maria Army Air Field on a high-altitude formation mission. The flight was returning to the field after successfully completing the mission, but had to circle the field twice due to heavy traffic at the field. As the flight maneuvered to get back in the traffic pattern, the subject pilot radioed the flight leader that his port engine was running out of gas. The port engine quit and the pilot peeled off of the formation. The flight leader ordered the pilot to switch fuel tanks, which he did. The engine restarted and the pilot had control of the airplane. Two minutes later, a black puff of smoke was seen and the port engine quit. Moments later, a black puff of smoke was seen and the starboard engine quit and then the pilot lost control; the airplane stalled and entered a spin to the left. The pilot was unable to recover and the airplane dove into the ground, killing him instantly.

9-9-44C. Strother Field, Kansas. At 1655 CWT, a Curtiss P-40N crashed one mile north of Strother Field, Winfield, Kansas, killing pilot 2Lt. William P. Johnson, Jr. The airplane was part of a six-ship flight that had taken off from Strother Field on a low-altitude navigation and ground gunnery/bombing mission. The pilot was returning to the field after losing his generator during the mission. The airplane, still carrying bombs, returned to the field and entered the traffic pattern. The airplane was seen turning from the base leg to final approach, continuing in the turn for 270 degrees in a nose-high attitude. The airplane stalled and snapped over into a spin to the left, describing one and a half turns of spin before smashing into the ground and exploding into flames.

9-10-44. DeRidder, Louisiana. At 1724, a Martin B-26B crashed while attempting an emergency forced landing three miles northeast of the Army Air Base at DeRidder, Louisiana, killing one crewmember and injuring five others, one seriously. The airplane took off from DeRidder Army Air Base and climbed to about 300 feet agl when the port engine quit and the starboard engine apparently lost power. The pilot attempted a shallow turn to the right in an attempt to crash-land in a field. The pilot was able to get the airplane to the field and attempted to shoot the emergency landing. The engineer, Cpl. Camillo R. Schiappa, was thrown out of the airplane as it crashed and broke up, killing him instantly. Co-pilot 2Lt. Elmer L. Riber, bombardier F/O Joseph L. Savoni and radio operator Cpl. Norman G. Sambuchi received minor injuries in the crash. Pilot 2Lt. Kay A. Spooner sustained serious injuries. Investigation of the

909 9-12-44D

wreckage revealed that the number-fourteen cylinder of the port engine had failed due to damage to the spark plug threads in the cylinder head.

9-11-44A. American Falls, Idaho. At 1515 MWT, a Grumman OA-14 seaplane crashed while operating at the American Falls Reservoir, American Falls, Idaho, killing pilot Maj. William H. Clark and passenger MSgt. Tandy B. Belew. Passenger MSgt. Alyn E. Trego received serious injuries. The airplane had taken off from Pocatello Army Air Field, Pocatello, Idaho. The Accident Classification Committee stated, "Major Clark was on an authorized air search and rescue training flight. The crash boat had been ordered to stand by and the pilot waited for the boat to get into position before attempting his landing. He made his approach at apparently too high an airspeed and on striking the water the airplane's nose section was broken off; the airplane cartwheeled and was broken up. The landing approach appeared normal except for excess speed and the pilot failed to break his glide or level off.... Pilot approached for landing at too high an airspeed probably because he was unable to judge his altitude above the water, which was smooth, glassy and clear. Inexperience in water landings was a contributing factor." It was noted that the pilot had received his original rating on 2-25-18.

9-11-44B. Choctawhatchee Bay, Florida. At 2100 CWT, a North American B-25H crashed into Choctawhatchee Bay about five miles southeast of Eglin Field, Valparaiso, Florida, killing three fliers and seriously injuring three others. The airplane was returning to Eglin Field after completing a successful routine navigation training flight. The pilot contacted the Eglin Field control tower and requested landing instructions. The pilot entered the traffic pattern. The airplane was turning from the base leg to final approach when the port engine failed. The pilot could not see the runway and could not maintain altitude. The pilot turned toward the bay and pointed the nose down in an attempt to make an emergency forced landing on the bay. The airplane, with the landing gear in the extended position, descended into the bay. The airplane went out of control and broke up on the water. A passing civilian fishing boat rescued the surviving crewmembers from the water. AAF crash boats were on the scene shortly after the airplane crashed and took on the survivors from the civilian boat. The B-25 had taken off from Camp Springs, Washington, D.C., and had been airborne for over four and a half hours. Investigators noted that the floodlights at Eglin Field were not illuminated and the pilot had not requested them. It was speculated that had the floodlights been on the pilot would have set up a better approach and would have been in a better position to make the field when the port engine failed. Engineer SSgt. James J. Connolly, passenger Sgt. John Kowalewski and SSgt. Charles P. Zeltman were killed in the crash. Pilot Maj. Daniel D. Champlain, co-pilot 2Lt. Merritt E. London and passenger SSgt. John M. Spielman sustained serious injuries.

9-12-44A. Merryville Bombing Range, Louisiana. At 1329, a Curtiss P-40N flew into the ground at the Merryville Bombing Range, Louisiana, killing pilot 2Lt. Jesse H. Hudson. The P-40 was part of a flight that had taken off from DeRidder Army Air Base, Louisiana, on a gunnery mission. On the seventh strafing pass the subject airplane came in too low and mushed into the ground while attempting to pull out. The P-40 burst into flames, bounded back into the air and then smashed back into the ground and exploded, scattering wreckage over an area of a mile.

9-12-44B. Valparaiso, Florida. At 1627 CWT, an Aeronca L-3B crashed at Auxiliary Airfield #3 at Eglin Field Military Reservation, Valparaiso, Florida, killing pilot 1Lt. Edgar B. Worley, Jr. and pilot-rated passenger 2Lt. William K. Crawford. The airplane was ferrying personnel from Auxiliary Airfield #5 to Auxiliary Airfield #3. The pilot approached the airfield from the south at an altitude of 50 feet agl. The pilot pulled up in a climbing turn and stalled the airplane at an altitude too low to allow recovery. The airplane spun to the ground, killing the pilots instantly.

9-12-44C. Moses Lake, Washington. At 1606, a Lockheed P-38L crashed 35 miles south of the Army Air Field at Moses Lake, Washington, killing pilot 2Lt. Robert Q. Highland. The airplane was part of a flight of four that had taken off from Moses Lake on a routine formation flight. The flight, flying at about 2,000 feet ten miles SSW of Othello, encountered two U.S. Navy Grumman F6F fighters. The navy fighters were on a simulated escort mission with a Lockheed-Vega B-34 type airplane. The navy fliers stated that the P-38 flight made a wide turn and got on the tail of the B-34. The navy airplanes intercepted the P-38s. The navy fighters made a pass and then broke away in a chandelle to the left. The P-38s followed and the subject airplane, flying in the number-four position, stalled out of the turn and fell into a spin. The subject P-38 described one-half a turn of spin and then dove into the ground, exploding violently into flames upon impact. The pilot was killed instantly. The P-38 flight leader stated that he had turned 90 degrees to the left, then 180 degrees to the right, and then 90 degrees to the left to avoid the navy airplanes and return to base. The AAF flight leader, contrary to testimony made by the navy pilots, denied that the P-38 flight purposely engaged the navy airplanes.

9-12-44D. Ora Grande, California. At 1715, a Bell TP-39Q crashed two and one-half miles north of Ora Grande, California, killing pilot 2Lt. William L. Doose. The airplane had taken off from the Army Air Field at Victorville, California, on a routine formation flight. The airplane left the formation with engine trouble and radioed the Victorville control tower that he wanted a straight-in approach. The tower officer gave the pilot permission to land and moments later he observed the airplane spin to the ground from an altitude of 1,000 feet agl. The airplane burst into flames upon impact, killing the pilot instantly. Investigators could not determine the cause of the engine trouble.

9-12-44E. Walla Walla, Washington. At 2110, a Consolidated B-24E crashed five miles southwest of the Army Air Field at Walla Walla, Washington, killing the crew of six. The airplane had taken off from Walla Walla at 2108 on a routine instrument cross-country mission. The airplane apparently flew into the ground approximately two minutes after take-off and exploded into flames. Investigation of the wreckage revealed that all four engines were operating at the time of the accident. Investigators speculated that the pilot lost the horizon soon after take-off and inadvertently flew the airplane into the ground at East Alder Street. Investigators speculated that poor visibility was a contributing factor. Killed in the crash were: 2Lt. Byford E. Rains, pilot; 2Lt. Cano L. Hobgood, co-pilot; 2Lt. Harold H. Jackson, navigator; 2Lt. Carlos A. Edwards, bombardier; Cpl. John P. Schneider, engineer; Cpl. Allen L. Wade, radio operator.

9-12-44F. Idalou Field, Texas. At 2345, a Douglas C-47B attempting to tow a Waco CG-4A glider aloft from Idalou Field, Idalou, Texas, mistakenly was attached to a line that was connected to a small truck, killing Glider Tow Rope Inspector Pvt. Elvin R. McKenzie. Investigation revealed that a tow-rope, which was being positioned for future use, was being laid out on the ground by airmen manning the tow-rope truck. The rear tow-rope handlers attached a tow-rope on a CG-4 and the forward rope handlers mistakenly attached the tow-rope that was still connected to the truck to the C-47 tow ship. Pvt. McKenzie, the tow-rope inspector, failed to notice the mistake. The flagman gave the C-47 pilot take-off clearance and the C-47 began its take-off. The tow-rope connected to the truck pulled tight and forced the truck to turn over. The tow-rope snapped and struck Pvt. McKenzie, fatally injuring him. The C-47 was piloted by 1Lt. Raymond J. Quartemont and F/O William L. McCutcheon.

9-13-44A. Peterson Field, Colorado. At 0757 MWT, a Consolidated B-24J and a Curtiss P-40N collided in mid-air near Peterson Field, Colorado Springs, Colorado, killing P-40N pilot 2Lt. Edward R. Schwarz. The B-24 was part of a 15-ship formation on a routine camera-bombing mission with interception. The P-40 was part of a flight of four that was scheduled to intercept the bombers. The fighters intercepted the bombers at an altitude of approximately 10,500 feet and made several passes on the B-24s from various angles, closing in to a distance of about 300 yards. As the simulated attacks continued, the fighters began breaking away at distances of 300 to 100 feet or less. B-24 crewmembers stated that the close-in passes became "dangerous." The subject P-40N made a pass from approximately the 2-3 o'clock high position. The P-40

half-rolled out during the pull-out and collided with the bomber, which was flying in the high squadron. The starboard wing of the P-40N struck the starboard wing of the B-24J. The P-40, with a portion of its starboard wing severed, "cartwheeled" across the top of the B-24, damaging the top turret, fuselage and wings. The P-40N entered a spin, smashing to the ground and exploding about 200 feet southeast of Runway 21 at Peterson Field. The P-40N pilot made no apparent attempt to parachute to safety. The B-24 was able to land safely at Pueblo Army Air Base, Pueblo, Colorado. Bombardier 2Lt. John A. Whittaker, who was manning the top turret, received minor injuries. B-24 crewmembers were: F/O Walter L Stiner, B-24 instructor pilot; 2Lt. Raymond H. DuFlon, pilot/B-24 student; 2Lt. Roberto Rulz, co-pilot/B-24 student; 2Lt. Alexander S. Sidie, navigator; Cpl. Bryant C. Edgerton, engineer; Cpl. Mathew A. Cregg, radio operator; Cpl. Albert F. Giardina, gunner; Cpl. Patrick G. Waddell, gunner; Pfc. John D. Sherman, gunner; Pfc. Milton Genes, gunner.

9-13-44B. Victoria, Texas. At 1000, a North American AT-6C crashed near Auxiliary Airfield #2, Victoria, Texas, killing instructor pilot 2Lt. James E. Eastmoore. The airplane had taken off from Foster Field, Victoria, to practice landings at the auxiliary airfield. The instructor was demonstrating "hurdle landings" to his students. The AT-6 had made one successful landing at the field and was demonstrating another when the accident occurred. The airplane, with flaps down and landing gear extended, was flying in the pattern at 700 feet agl behind two other airplanes. The subject airplane apparently slowed down to avoid overtaking the two airplanes ahead. The AT-6 made a tight right turn from the base leg to the approach leg and was observed to suddenly snap over into a spin to the right and smash into the ground where it burst into flames.

9-13-44C. Mountain Home, Idaho. At 1040 MWT, a Consolidated RB-24E attempting an emergency forced landing crashed 55 miles southwest of the Army Air Field at Mountain Home, Idaho, killing nine fliers. The airplane took off from Mountain Home at 0959 on a routine instrument flight. The pilots were cleared to fly at 9,500 feet with one pilot flying under the hood. Investigators stated, "Aircraft must have had engine trouble and [the pilot] tried for a crash-landing with the wheels down. [The pilot] overshot the flat space intended for landing and hit on the other side of the valley. Aircraft turned over on its back and burned.... [The pilot] was flying on a magnetic course of approximately 300 degrees and then, seeing that he could not land on the intended creek valley, attempted to turn right up a small valley running approximately north. The airplane hit the ground right wing first as it mushed into the rocky ridge." Killed in the crash were: 2Lt. Joseph C. Leutze, Jr., pilot; 2Lt. Harry N. Jones, co-pilot; F/O Richard A. Zager, bombardier;

Cpl. Charles Kelley, engineer; Cpl. John D. Weaver, radio operator; Cpl. Billie Davis, gunner; Pfc. James T. Wiley, gunner; Pfc. Lawrence E. Kronberg, gunner; Pfc. Clyde D. Mahon, gunner.

9-13-44D. Clovis, New Mexico. At 1048 MWT, a Boeing B-29 crash-landed one mile south of the Army Air Field at Clovis, New Mexico, killing B-29 instructor pilot Capt. Jack F. Clark. Twelve other crewmembers were injured, six seriously. Investigators stated, "B-29 entered the traffic pattern [at 2,000 feet agl] at a 45-degree angle on the upwind leg. Because the aircraft was higher than it should have been at this point, the pilot let down his wheels and retarded the throttles in order to nose down to the proper traffic pattern altitude, which is 1,500 feet above the surface. The aircraft then turned into the crosswind leg where upon it began to loose altitude rapidly in a very steep nose high attitude. The co-pilot started the landing gear up and full power was applied to all four engines. Stall recovery was not effected and the aircraft crashed. The cause of the accident was the result of a stall attributed to the failure of the elevator trim tab. The flattener hinge pin had worked out and into a position where it caused a restriction of proper elevator travel. The wire that holds the two halves of hinge together probably worked itself out to where it caught on the fuselage, jamming the controls." Fliers seriously injured in the crash were: 2Lt. Gordon M. Kogl, pilot; 2Lt. James C. Phillips, Jr., co-pilot; 2Lt. O.D. Crane, navigator; 2Lt. Joseph J. Scudder, flight engineer instructor; SSgt. Elijah C. Hester, flight engineer; Pfc. John S. Stenger, radio operator. Fliers injured in the crash were: Pfc. Ben F. Jones, central fire control instructor; Cpl. Karl F. Weimer, central fire control; Cpl. Walter R. Couch, starboard gunner; Cpl. George C. Coleman, port gunner; Cpl. Frank Corradini, tail gunner; Cpl. Wilbur J. Curd, passenger.

9-13-44E. Port Isabel, Texas. At 1220, a North American AT-6A and an AT-6C collided in mid-air at Port Isabel, Texas, killing WASP pilot Alice E. Lovejoy in the crash of the AT-6A. 1Lt. Carl F. Brady was able to parachute to safety from the AT-6A and was uninjured. 1Lt. Gustave J. Vetter and F/O Elmer V. Zoon were able to safely land the damaged AT-6C and were uninjured. The airplanes were part of a six-ship flight that had taken off from Brownsville Municipal Airport, Brownsville, Texas, on a formation-training flight. Investigators stated, "After practicing formation for approximately 25 minutes, one of the airplanes dropped out of the formation. At this time they were flying a V-formation. Shortly after the number-six airplane had broken away from the formation, the leader gave a signal for an echelon to the right. At this time Lt. Carl F. Brady was flying in the number-three position and Lt. G.J. Vetter was flying the number-four position. Shortly after the echelon to the right was formed the leader gave a signal for a turn to the left. After investigation ... it was not definitely determined

whether the number-four and number-five ships had established their positions in the echelon to the right.... Shortly after the number-three ship ... established his bank there was a violent jolt. At this time Lt. Brady noticed that he had lost his right aileron, which later proved his airplane and the number-four airplane collided. No one in the formation witnessed the actual impact of the two ships."

9-13-44F. Peterson Field, Colorado. At 1434, civilian Caterpillar tractor operator Howard G. Lawrence was killed when a Curtiss P-40N operating at Peterson Field, Colorado Springs, Colorado, taxied into the tractor. P-40N pilot F/O James P. Ames was uninjured. The P-40N and the Caterpillar tractor were seriously damaged. The airplane was returning from a scheduled skip bombing and strafing mission, landing on Runway 12. The pilot turned onto the taxiway and began S-turning to the parking area. The tractor maneuvered onto the taxiway and the P-40 pilot failed to see it in time to avoid the accident.

9-14-44A. Warrensburg, Missouri. At 0100 CWT, two Douglas C-47A airplanes collided in mid-air near Sedalia Army Air Field, Warrensburg, Missouri, killing four fliers. Pilot 2Lt. Keith W. Blackwelder and co-pilot 2Lt. John M.V. Boomgaard were killed in the crash of C-47A #42-24321; pilot 2Lt. James J. Cram and co-pilot 2Lt. David A. Flack were killed in the crash of C-47A #42-24317. Investigators stated, "The two aircraft were members of a nine-ship V of V's formation having taken off from Sedalia Army Air Field, flying to Vichy Army Air Field, Vichy, Missouri, where a landing was made. The formation then took off and flew back to Sedalia via Springfield, Missouri. Lt. Boomgaard was flying in the second element on the right wing of the element leader, and Lt. Flack was in the same element flying the left wing of the element leader. The formation leader received clearance from the tower for landing instructions, and proceeded to bring the formation down [for a landing on Runway 31]. The lead element peeled off and the second element leader peeled off when he was clear. Next, Lt. Boomgaard's aircraft peeled off, colliding with Lt. Flack's aircraft. Lt. Boomgaard's aircraft crashed immediately after the collision just west of the field. Lt. Flack's aircraft continued on, crashing on the other side of the field, or just east of the field, approximately two and one-half miles from the point of collision."

9-14-44B. Warden, Washington. At 0910, a Lockheed P-38L crashed five miles west of Warden, Washington, killing pilot 2Lt. Glenn W. Ingersoll. The airplane was flying in the number-three position of a four-ship flight that had taken off from Moses Lake Army Air Field, Moses Lake, Washington, on a high-altitude oxygen mission. The flight had climbed to 25,000 feet and the flight leader ordered the three wing airplanes to break formation and spread out. The pilots were then ordered to "familiarize" themselves with the performance and feel of the airplane at that high altitude. The formation broke up and the pilots began maneuvering. The subject airplane was then seen to enter a nose-down attitude. The airplane continued in a dive until it struck the ground and exploded. The airplane was seen breaking up at an altitude of approximately 4,000 feet. It is likely that the pilot put the airplane in a diving maneuver from which he could not recover because of the tendency of the P-38 type airplane to suffer from severe control surface buffeting caused by compressibility.

9-14-44C. Cordova, Nebraska. At 1215 CWT, a Republic P-47D crashed at Cordova, Nebraska, killing pilot F/O George F. Robbins. The airplane was flying in the number-two position of a five-ship flight that had taken off from Bruning Army Air Field, Bruning, Nebraska, on a high-altitude aerobatic flight. The formation was at an altitude of 20,000 feet in an echelon to the right. The flight leader peeled off and then performed a dive, pulling up into an Immelmann turn. The four wing aircraft attempted to follow. The pilot of the number-three airplane reported that he had blacked out while attempting to pull up into the maneuver, reaching an indicated airspeed of approximately 400 mph. After recovering from the blackout and regaining control of his airplane, the number-three pilot observed the subject airplane in a steep spiral at about 10,000 feet. The subject airplane continued in this attitude until it struck the ground and exploded. The starboard aileron separated from the subject airplane at approximately 4,000 feet. A short time later, the number-five airplane crashed and the pilot killed. See 9-14-44E.

9-14-44D. Meacham, Oregon. At 1030, a Beech AT-7C flying in instrument conditions crashed eight miles west of Meacham, Oregon, killing the crew of five. The airplane had taken off from the Army Air Field at Pocatello, Idaho, on a routine training flight to Portland, Oregon. A rancher working in a nearby field observed the subject airplane dive out of the overcast and then pull up steeply back into the clouds. Moments later the airplane fell out of the overcast and smashed to the ground where it exploded into flames. Killed in the crash were: 1Lt. Robert C. Horner, Jr., instructor pilot; 2Lt. Harold E. Knaus, pilot; F/O Mark A. Meyer, co-pilot; Seaman 2nd class M.A. Roberts, U.S. Navy passenger; Seaman 1st class Leslie L. Martin, U.S. Navy passenger.

9-14-44E. Bruning, Nebraska. At 1240 CWT, a Republic P-47D crashed near the Army Air Field at Bruning, Nebraska, killing pilot 2Lt. Albert G. Leoni. The airplane was flying in the number-five position of a five-ship flight that had taken off from the field on a high-altitude aerobatic mission. The formation was flying at about 20,000 feet in an echelon to the right when the flight leader peeled off into a dive and then pulled up into an Immelmann turn. The formation followed the leader into the maneuver. The number-two airplane entered a steep spiral and crashed, killing the pilot. The number-three pilot blacked out during the

steep pull up. He recovered from the blackout and was able to recover control of his airplane. After the crash of the number-two airplane, the flight leader ordered the number-five airplane to take over the number-two position. The pilot assumed the number-two position and the flight then returned to Bruning Army Air Field. When the formation arrived at Bruning, the flight leader and the remaining ships peeled off and entered the pattern. The flight leader was too close to another airplane in the traffic pattern and the subject flight was ordered to go around. The subject airplane was seen to enter a double snap roll. The pilot recovered at an altitude of about 300 feet agl, but the airplane stalled again and rolled to an inverted position, diving into the ground and exploding into flames. See 9-14-44C.

9-14-44F. Indian Pass, Florida. At 1600 CWT, a Consolidated B-24H suffering an engine fire crashed at Indian Pass, killing four crewmembers. Another crewmember fell to his death in an unsuccessful parachute jump. Eight crewmembers were able to parachute to safety and were uninjured; one crewmember was injured parachuting to safety. The airplane had taken off from Tyndall Field and was on a routine gunnery mission. The airplane was flying operating at the gunnery range at an altitude of 5,000 feet about five miles off shore when fire was noticed coming from the number-four engine. The pilot turned the airplane toward land and ordered the crew to attach their parachutes and to stand by to abandon ship. The B-24, with the number-four engine on fire and the number-two engine smoking, was steadily losing altitude and when the airplane reached land the pilot ordered the crew to bail out. The crew bailed out in an "orderly fashion" at an altitude of approximately 2,000 to 1,500 feet and all landed within an area of one mile. The airplane continued flying in a wings-level attitude but was still losing altitude rapidly. The B-24 continued losing altitude and as it approached the ground it went out of control and made a complete roll to the right. The B-24 then dove into the ground at a very steep angle, exploding into flames upon impact about 35 miles southeast of Tyndall Field, Panama City, Florida. Killed in the crash were: 2Lt. William J. Monohan, pilot; F/O Anthony T. Fusco, co-pilot; Cpl. John C. Murphy, engineer instructor; Cpl. Raymond B. Wise, engineer instructor. Student gunner A/C Nicholas A. Ravella was fell to his death attempting to parachute to safety. Crewmembers parachuting to safety were: 2Lt. Joseph L. Louther, instructor; Sgt. Rae Elliot, instructor; Pvt. Thomas L. Jordan, student gunner; Cpl. Sidney Kowarsky, student gunner; A/C Stanley M. Nuttall, student gunner; Pvt. Cletus A. Keller, student gunner; Pvt. Marvin L. Johnson, student gunner; Pfc. Lowell E. Matlock, student gunner. Student gunner A/C Paul Simcich was injured parachuting to safety.

9-14-44G. Peterson Field, Colorado. At 1616, a Consolidated B-24J crashed five miles northeast of Peterson Field, Colorado Springs, Colorado, killing nine fliers. Killed in the crash were: 2Lt. Wilbur H. Steinkamp, pilot; 2Lt. Harold R. Shaffer, co-pilot; 2Lt. William A. Price, navigator; F/O Dewayne N. Eyler, bombardier; Cpl. Francis E. Gaffey, engineer; Cpl. Carl W. Vest, radio operator; Cpl. Dwight F. Marquiss, gunner; Pfc. Pedro Morales, gunner; Pfc. Paul E. Weissman, gunner. The airplane was part of an 18-ship formation that had taken off from the Army Air Base at Pueblo, Colorado, on a high-altitude formation and interception flight. The formation was flying at 10,000 feet indicated altitude and the subject airplane was attempting to take the number-four position of the high squadron. As the pilot maneuvered the B-24, he struck the propeller turbulence of the lead element of the high squadron. The airplane banked sharply to the left and then entered a stall, flipping over to an inverted position. The pilot apparently tried to recover with a split-S type maneuver but the airplane entered a flat spin to the left, continuing in this attitude until it struck the ground and exploded into flames.

9-14-44H. Shreveport, Louisiana. At 1740, a Martin B-26B crashed seven miles northwest of Shreveport, Louisiana, killing the crew of seven. The airplane had taken off from Barksdale Field, Louisiana, on an instrument training flight. The pilot was to fly under the hood as the instrument pilot and the co-pilot was to act as the safety pilot. Investigators stated, "The subject pilot was flying instruments at an estimated altitude of 4,500 to 5,000 feet and over banked the subject aircraft, causing a wing tip stall, which threw the aircraft over on its back and into a tight spiral to the right. Being under the hood and the rapidity of the movement of the subject aircraft, caused the pilot to lose control and go into a spin. He apparently gained partial control again after 4 to 6 turns but immediately lost control and fell into a dive from approximately 500 feet. The airplane crashed into the ground and burned." French Air Force crewmembers killed in the crash were: SSgt. Jacques E. Moricut, pilot; SSgt. Alfred M. Duclaux, co-pilot; SSgt. Jacques Florin, bombardier/navigator; SSgt. Marcel Barthelemy, engineer; SSgt. Roger Pericaud, radio operator; Cpl. Andre Melin, gunner; SSgt. Paul Grimbert, radio operator.

9-14-44I. Columbia, South Carolina. At 2053, a North American B-25J crashed near the Army Air Base at Columbia, South Carolina, killing four crewmembers and seriously injuring two others. The airplane was cleared to take-off on a night navigation flight and smashed into a hill one-half mile from the end of the runway soon after take-off. Investigators stated, "It is apparent that the aircraft failed to make a normal climb after take-off. [The aircraft] sheared off the top of adjacent trees before striking the brow of a hill. The clean cut of the trees is some indication of the speed of the aircraft. After striking about 200 feet (about 12 feet in elevation above the runway) down the slope, the aircraft continued on over the hill for about 200 yards scattering parts as it went, ending in a pile,

which immediately caught fire. The pilots were thrown clear along with their respective seats." Pilot 2Lt. Jack K. Pound and co-pilot F/O Norman J. Seher sustained very serious injuries. Crewmembers killed in the crash were: Cpl. David B. Mermell, engineer; Cpl. Arthur H. Stedman, radio operator; Cpl. Paul C. Smith, gunner; 2Lt. Yaroslav Naveschuk, bombardier/navigator.

9-14-44J. Tucson, Arizona. At 1925 MWT, a Consolidated B-24D crashed while attempting an emergency forced landing eight miles southeast of Davis-Monthan Field, Tucson, Arizona, killing three fliers and seriously injuring another. Investigators stated, "According to the statement of the only survivor, just after the airplane passed over the field preparatory to entering the traffic pattern, there was a 'terrific bang' after which the pilots seemed to lose control of the airplane and it crashed shortly thereafter. It is assumed that the master switch was involuntarily placed in the off position causing [all four] engines to backfire; and that the pilots, not realizing what had happened, were unable to find the cause of the loss of the four engines." The airplane went into a steep descending glide and smashed into the ground with terrific force. Pilot 2Lt. Chester L. Welenc, co-pilot F/O Stephen P. Pedone and engineer SSgt. Nick H. Outten were killed in the crash. Gunner Pfc. Arland R. Smith, Jr. sustained very serious injuries.

9-15-44A. Flagstaff, Arizona. At 0317 MWT, a Consolidated TB-24J crashed into rising terrain in the San Francisco Mountains ten miles north of Flagstaff, Arizona, killing eight crewmembers. The airplane had taken off from the Army Air Field at Bakersfield, California, on a cross-country navigation flight to Kirtland Field, Albuquerque, New Mexico. At 0302 MWT, the airplane reported in over Ashfork, Arizona. At the time of the accident the airplane was 15 miles off course and on the wrong side of the CFR airway that the crew had proposed to follow in the flight plan that was filed. The B-24 smashed into the western slope of Kachina Peak at an elevation of approximately 12,000 feet, exploding into flames upon impact. Investigation revealed that all four engines were operating at cruise rpm at the time of the crash. Killed in the crash were: 2Lt. Warren E. Crowther, B-24 instructor; F/O Ray P. Shipley, pilot/B-24 student; F/O Clyce H. McCelvey, B-24 student; F/O Charles W. McDonald, B-24 student; F/O Patrick E. Pertuset, B-24 student; SSgt. James R. Hartzog, engineer; Cpl. John A. Franke, Jr., radio operator; Pfc. Hugh H. Brown, student engineer.

9-15-44B. Craig Field, Alabama. At 0945, a North American AT-6C crashed one mile northwest of Craig Field, Selma, Alabama, killing French Air Force instructor Claude H. Gillet and French Air Force student pilot Sgt. Joseph C. Lancon. The airplane had taken off at 0735 on a routine student navigation mission. After partially completing the student navigation flight, the airplane returned to the field and entered the traffic pattern. The airplane stalled while making the turn from the base leg to final approach. The AT-6 entered a spin to the left, describing one-half a turn before the instructor stopped the spin. The airplane remained in a stalling attitude so a burst of throttle was added, but the instructor was unable to recover. The nose dropped and the AT-6C dove into the ground from an altitude of 300 feet agl, bursting into flames upon impact and killing the occupants instantly.

9-15-44C. Patterson Field, Ohio. At 1130 EWT, a Curtiss P-40N and a Beech UC-45F collided on the runway at Patterson Field, Ohio, killing three fliers and injuring two others aboard the UC-45. The P-40N pilot, WASP Helen A. Fremd, was uninjured. The P-40 was taking off on a ferry flight to Atlanta, Georgia, and the UC-45 was taking off on an administrative flight to Nashville, Tennessee. Traffic at Patterson Field was heavy because of a large number of aircraft taking off for the east coast. Many of these airplanes had been evacuated from the coastal regions because of a hurricane threat. The tower was allowing some airplanes with instrument clearances to take off and receive their clearances in the air. A number of P-40s were waiting to take off on a ferry mission. A Consolidated B-24 that took off at 1121 returned to the field and made an emergency landing on Runway 21 at 1129. The control tower cleared the UC-45 for take-off on Runway 21 as soon as the B-24 was safely off of the runway. The UC-45 taxied to the intersection of Runway 21 and Runway 25 and swung around into take-off position. Moments later, the P-40N, which was not cleared to take-off, collided with the UC-45 from the rear, between the port wing and empennage. The P-40's port landing gear and wheel struck the UC-45 on the tail section, severing it. The P-40's port wing cut into the cockpit area of the UC-45, killing three instantly. The P-40's propeller chewed into the UC-45's starboard wing, causing it to burst into flames. Investigation revealed that the subject P-40 was the second to be cleared for take-off on Runway 25. Pilot Fremd had to refuse the clearance because the oil temperature was not at the proper level. When the oil temperature reached operating temperature, the subject P-40 was cleared to the take-off runway but not cleared to take-off. It is possible that Pilot Fremd misunderstood a take-off clearance given to a P-47 about 50 seconds later, thinking it was for her. Pilot Fremd then began her take-off roll and collided with the UC-45 in the port rear quarter at the runway intersection. The P-40 skidded to a halt a short distance away. Tower personnel testified that the subject P-40 was not cleared for take-off. Pilot 1Lt. Jack Vance, engineer Sgt. Verl R. Singer and passenger CWO Alexander I. Murin were killed in the UC-45F. Flight engineer 1Lt. Henry M. Saffren sustained serious injuries and passenger Pvt. Roger B. Godfrey received minor injuries.

9-15-44D. Orangeburg, South Carolina. At 1300, two Stearman PT-17 airplanes collided in mid-air five miles west of Hawthorne Field, Orangeburg, South Carolina, killing French Air Force student Sgt. Robert Tine aboard PT-17 #42-16402. French Air Force student Sgt. Guy Huc was uninjured and was able to safely land PT-17 #42-16403 at Hawthorne Field. The airplanes were flying in a large echelon formation on a ferry flight from Americus, Georgia, to Orangeburg after being evacuated because of a hurricane threat. The flight leader signaled the formation to peel off for landing at Hawthorne Field. The collision occurred moments later as the airplanes attempted to leave the formation.

9-16-44A. MacDill Field, Florida. At 0630 EWT, pilot 2Lt. Preston G. McCarson was fatally injured when he walked into the rotating number-two propeller of a Boeing B-17G operating at MacDill Field, Tampa, Florida. The B-17, with all four engines running and the entire crew aboard, was stopped with wheel chocks in place and was waiting to taxi out to the runway to take-off on a routine gunnery mission. The Expediting Officer was observed to drive up to the front of the airplane and begin talking with the maintenance crew chief. For an unknown reason the pilot exited the B-17 by the front escape hatch. After he exited the airplane he was struck by the rotating number-two propeller. Because of miscommunication between the crewmembers, 20 minutes had passed before the control tower was notified of the incident. Soon after the tower was notified, the pilot was picked up by an ambulance and transported to the station hospital where died of his injuries at 1230 EWT. Co-pilot 2Lt. Ronald D. Dowden and six other fliers were uninjured.

9-16-44B. Chesapeake Bay, Maryland. At 1000, a Republic P-47C crashed into Chesapeake Bay east of Bay Ridge, Maryland, killing F/O Henry F. Marad. The airplane was part of a four-ship flight that had taken off from Camp Springs Army Air Field, Washington, D.C., on a high-altitude formation flight. The formation climbed to 20,000 feet and began practicing formation "join-ups." After a short time, the subject airplane was found to be missing. It was later discovered that the airplane had crashed into Chesapeake Bay halfway between Bay Ridge and Kent Island. Investigators could not determine the cause of the accident.

9-16-44C. New Orleans, Louisiana. At 1145, a Martin B-26C crashed while attempting an emergency forced landing at the municipal airport at New Orleans, Louisiana, killing two crewmembers and injuring four others, two seriously. The B-26C had taken off from the Army Air Field at Lake Charles, Louisiana, on a routine over-water navigation flight. During the flight the aircraft suffered the failure of the starboard engine. The pilot radioed New Orleans Municipal Airport's control tower, requesting permission for an emergency landing. The pilot approached the runway high and at 160 mph and appeared to dive the airplane toward the runway. The airplane touched down and the co-pilot pulled the emergency brake system. After skidding about

800 feet, the starboard tire blew and the airplane careened off the left side of the runway. The port brake remained locked and the airplane continued to skid out of control toward a sea wall on the three wheels. The nose landing gear did not collapse. The B-26 smashed into the seawall and the fuselage broke into two pieces just aft of the trailing edge, the forward portion of the shattered airplane catapulting over the seawall and coming to rest in the surf. Pilot 2Lt. Ernest R. Pruitt, Jr. and gunner Cpl. James E. Gross were killed. Co-pilot 2Lt. Robert G. Clark and engineer Cpl. Francis G. Padalick sustained serious injuries. Bombardier 2Lt. Gerald W. Vath and radio operator Cpl. George R. Roemlein received minor injuries.

9-16-44D. Mill Creek Canyon, California. At 1830, a Taylorcraft L-2B collided with rising terrain at Mill Creek Canyon, 25 miles east of San Bernardino, California, fatally injuring pilot F/O Walter S. Jake. The airplane had taken off from Palmdale, California, on a ferry flight to Palm Springs, California. The L-2 collided with a tall tree with the starboard wing, causing the airplane to go out of control and crash into the mountainside. The airplane came to rest on its right side. The pilot was found in the cockpit, alive and still strapped to his seat with a map spread out over his lap. Civilians in the area immediately rescued the pilot, but he died a short time later. Investigators noted that the pilot was about six miles north of his proposed course.

9-17-44A. Westover Field, Massachusetts. At 0302, a Consolidated B-24J crashed near Westover Field, Chicopee, Massachusetts, killing the crew of seven. Killed in the crash were: 2Lt. Gene R. Asay, pilot; 2Lt. John W. Woodrow, co-pilot; Sgt. Neal W. Johnson, engineer; Pfc. Jack W. Hairston, assistant engineer; Cpl. John A. Perry, radio operator; Pfc. Clifford K. Nordby, assistant radio operator; Sgt. William D. Haynes, gunner. The airplane had taken off from Westover Field on a local night flight at 2208 (9-16-44) and had returned to the area of the field at about 0245. The B-24 entered the traffic pattern and the pilot had considerable difficulty in lining the airplane up with the runway on the final approach. At an altitude of approximately 400 feet, the pilot radioed the tower that he was aborting the landing and going around. The airplane continued in a descending glide until reaching an altitude of about ten feet above the runway. The pilot advanced the throttles and began the go-around. The airplane, with landing gear in the extended position and flaps fully retracted, never gained more than 100 feet. The B-24 descended into trees about two miles from the end of the runway about 15 degrees to the right of the runway heading, immediately bursting into flames.

9-17-44B. Bangor, Maine. At 1158, a Douglas A-26B crashed three and one-half miles northeast of Dow Field, Bangor, Maine, killing pilot 1Lt. Jack W. Williams and navigator 2Lt. Albert L. Keener. The airplane had taken off from Dow Field with two other A-26 airplanes at 1154 EWT on a ferry flight to the European Theater of Operations via Goose Bay, Labrador. The airplane climbed out for a short time and then was seen to enter a nose-down attitude until it struck a wooded ridge and exploded into flames. Witnesses stated that the starboard engine appeared to smoking as the airplane began climbing. Investigation revealed that the airplane struck the ground in a 30-degree dive in a wings-level attitude. Both propellers appeared to have been turning at low rpm. Investigators speculated that the airplane might have suffered a fuel system failure.

9-17-44C. Lake Huron, Michigan. At 1710 EWT, a Republic P-47D crashed into Lake Huron three miles ENE of Sturgeon Point, Michigan, killing French Air Force pilot Sgt. Francois Messinger. The airplane was in the number-three position of a four-ship flight that had taken off from Oscoda Army Air Field at 1605 on a high-altitude formation mission. The formation climbed to 8,000 feet, then to 10,000 feet, and then to 19,000 feet, making oxygen system checks at each level. The formation then climbed to 25,000 feet and flew at this altitude for several minutes. The flight leader then began a slight turn to the left. The subject airplane continued almost straight ahead, flying to the right of the turning first element. The number-four airplane stayed with the subject airplane. The subject airplane entered a nose-down attitude, entering a diving spiral to the right. The wingman called the number-three airplane on the radio but received no response. As the dive approached 80 degrees from the horizontal, the wingman broke off and leveled out. The number-four pilot then observed the subject airplane dive into the lake. The subject pilot made no attempt to recover or parachute to safety and did not answer repeated radio calls. It was later speculated that the pilot might have suffered an oxygen system failure.

9-17-44D. Fairfield, California. At 1430, a Consolidated B-24J crashed seven miles south of the Fairfield-Suisun Army Air Base, Fairfield, California, killing four fliers and injuring three others, one seriously. The airplane had taken off from Hamilton Field, San Rafael, California, on a flight to test the fuel transfer system prior to the airplane being ferried overseas. The airplane was airborne about an hour and a half when the pilots began testing the fuel transfer system. The fuel system functioned properly for about 25 minutes then the number-three and number-four engines quit. The co-pilot feathered the number-three propeller and was attempting to feather the number-four propeller when the airplane snapped violently to the right and entered a dive at an indicated altitude of 2,800 feet. The pilot added power to the port engines and was able to level the airplane just above the terrain, but the airplane stalled again and smashed into the ground from very low altitude. The airplane did not burn immediately, but caught on fire many minutes later as rescue personnel attempted to cut two dead bodies free from the wreckage. Killed in the crash were: 2Lt. Joseph B. Weller, pilot; 1Lt. George P. Buffington, navigator; 1Lt. Julius J. Szemetko, pilot-rated passenger; Pfc. Edward R. Puck, radio operator. Engineer SSgt. Burton R. Robbins and passenger Sgt. Robert T. Kincaid were injured and co-pilot 1Lt. Lawrence H. Piper sustained serious injuries.

9-17-44E. Copeland, Kansas. At approximately 2146 CWT, a Boeing B-29 crashed into a farm house one mile south of Copeland, Kansas, killing ten crewmembers and two civilians on the ground. Two civilians were seriously injured. The airplane had taken off from Walker Army Air Field, Walker, Kansas, on a gunnery mission. The subject airplane had successfully completed the gunnery mission and had returned and landed at Walker Army Air Field. A navigator that had been on board had become sick and was let off the airplane. The crew was then assigned to fly a formation training flight with another B-29, but the other B-29 failed to take-off on the mission because of mechanical difficulties. The subject B-29 took off on an instrument-training mission instead. At approximately 2145, the subject airplane flew over Copeland from the northeast to the southwest at an altitude of about 200 feet agl, making a low-altitude pass. The B-29, flying at very low altitude with its landing lights illuminated, began a shallow turn to the right. The airplane lost altitude in the turn and descended into the ground, smashing into an occupied farmhouse and a barn. The airplane exploded into flames upon impact. Farmer O.H. Hatfield and his 14-month-old grandson were killed. Mrs. O.H. Hatfield and her adult daughter sustained serious injuries. The official accident report did not list the names of the adult daughter or her child. Investigation revealed that the co-pilot had worked for four summers at Copeland and had lived with two of his uncles during that time. It was discovered that the subject B-29 had buzzed the uncle's house just prior to the accident. Crewmembers killed in the crash were: 2Lt. Wayne B. Cooper, pilot; 2Lt. Lawrence H. Eslinger, co-pilot; 2Lt. William P. Kane, flight engineer; 2Lt. Kenneth O. West, navigator; Sgt. Bruno Gutt, radio operator; Sgt. George V. Johnson, radar operator; Cpl. Robert L. Cooper, gunner; Cpl. John F. Kelly, gunner; Cpl. Gerald J. Murphy, gunner; SSgt. Clarence A. Boston, gunner.

9-17-44F. Flagstaff, Arizona. At 2105 PWT, a Boeing B-17G suffering an in-flight fire crashed in the San Francisco Mountains 15 miles northeast of Flagstaff, Arizona, killing the crew of four. The airplane had taken off from Kingman Army Air Field, Kingman, Arizona, on a navigation flight to Winslow, Arizona, to Tucson, Arizona, and return to Kingman. The airplane was seen to make a 360-degree turn over the mountains north of Flagstaff. Witnesses said that the engines sounded like they were running at high rpm and the airspeed appeared to be slow and that the airplane

appeared to be climbing. At about 13,000 feet two fires were seen to break out on the port wing. The B-17 then entered a descending glide toward the mountains, striking trees on the peak and severing the tail section. The airplane smashed onto the mountain and exploded into flames. Investigators were unable to determine the cause of the apparent engine fires. Killed in the crash were: 1Lt. Richard V. Howell, B-17 instructor; 2Lt. Richard A. Davis, co-pilot/B-17 student; F/O Lafayette R. Brown, co-pilot/B-17 student; Sgt. Edmund M. Sikora, engineer.

9-17-44G. Kingman, Arizona. At 2325 PWT, a North American AT-6A crashed while attempting a take-off at Kingman Army Air Field, Kingman, Arizona, killing non-rated pilot Pfc. Roy L. Duncan. Investigation revealed that the non-pilot enlisted man had taken the airplane without permission and was attempting to take-off on an unauthorized flight. Investigators stated, "Control Tower Operators ... noticed an AT-6 type aircraft with navigation lights burning, pull out into take-off position on Runway 21. Having received no word of an impending clearance, they checked with the Base Operations Dispatcher and found no clearance had been authorized. At this moment, the aircraft started its take-off roll, and it was evident to them that the pilot had little control over the aircraft as it was seen to swerve from side to side on the runway. They immediately turned in the crash alarm and then watched the airplane leave the ground. When the aircraft reached an altitude of approximately 100 feet, the pilot was seen to lose all control of the aircraft and crash in the center of Runway 7-25, directly in front of the Sub-Depot hangar. The crash truck had begun to move before the aircraft struck the ground. Upon arrival, the Crash Crew found Pfc. Duncan severely injured but still alive. He was removed as quickly as possible to the Base Hospital but died shortly thereafter."

9-18-44A. Belleville, Illinois. At 0820, a Martin B-26F crashed while attempting a take-off at Scott Field, Belleville, Illinois, killing five fliers and injuring another. The airplane was taking off on a routine navigation flight to its station at Lake Charles Army Air Field, Lake Charles, Louisiana. After a normal take-off run on Runway 18, the B-26 became airborne and at an altitude of about five feet agl the port engine began losing power. The airplane dropped back down to the runway left wheel first. The emergency brakes were applied and the airplane began veering to the left at an angle of 40 degrees to the take-off heading. After careening out of control for about 1,000 feet, the B-26 nose wheel collapsed and the airplane collided with a small tractor. After smashing into the tractor the airplane skidded over a set of railroad tracks, flipping over and coming to rest upside down. The airplane did not burn but was destroyed utterly. Killed in the crash were: 2Lt. Bruce R. Bell, pilot; 2Lt. Bernard J. Fein, co-pilot; 2Lt. Dudley T. Mabee, bombardier/navigator; Cpl.

Weston L. Lugg, engineer; Cpl. Marvink Benson, radio operator. Gunner Samuel C. Berlotti sustained serious injuries.

9-18-44B. Tucumcari, New Mexico. At 1150 MWT, a Curtiss P-40N crashed 20 miles SSW of Tucumcari, New Mexico, killing pilot 2Lt. Franck Beall, Jr. The subject airplane was part of a four-ship formation that had taken off from the Army Air Field at Ft. Sumner, New Mexico, on a simulated individual combat flight at medium altitude. Lt. Beall was leading and the three other P-40s were following in a string formation. The subject airplane led the flight into a shallow dive. The subject airplane leveled out at an altitude of about 50 feet agl and then performed a slow roll. As the pilot completed the roll, the airplane mushed into the ground right side up. The airplane skidded along the ground for about 200 feet, bursting into flames as it came to rest.

9-18-44C. Fannin, Texas. At 1825, a North American AT-6D crashed two miles south of Fannin, Texas, killing instructor 2Lt. James W. Marshall and student A/C William G. Molitor. The airplane had taken off from Aloe Field, Texas, at 1715 on a routine training flight. Investigation of the wreckage revealed that the airplane suffered an apparent engine failure. The airplane was stalled and it entered a spin at an altitude too low to allow recovery. Both occupants were killed instantly upon impact.

9-19-44A. Boyce, Louisiana. At 0045, a Boeing B-17G collided with trees and crashed two miles north of Boyce, Louisiana, killing five fliers and injuring three others, two seriously. The airplane had taken off from Alexandria Army Air Field, Alexandria, Louisiana, on a high-altitude bombing mission. The crew had successfully completed the bombing mission and was returning to base. The surviving pilot stated that he was letting down from about 8,000 feet with the intention of entering the traffic pattern for Alexandria Army Air Field when the crash occurred. The pilot stated that he could not remember the airplane descending into trees and crashing into the ground. Killed in the crash were: 2Lt. Stemati Savas, co-pilot; 2Lt. Chris Bartley, navigator; F/O Charles W. Nick, bombardier; Cpl. Charles W. Jennings, radio operator; Cpl. William T. Hancock, gunner. Pilot 2Lt. Maxon W. Eggleston and engineer Cpl. Carl O. Fite sustained serious injuries. Cpl. Fite was in a coma for a few weeks following the crash. Radio operator Cpl. John P. Penn received minor injuries.

9-19-44B. Mountain Home, Idaho. At 1115 MWT, gunner Cpl. Harry J. Troy was killed when he was asphyxiated after being crushed in a nose turret he was operating on board a Consolidated B-24J that was flying on a gunnery exercise at the AAF Gunnery Range near Mountain Home Army Air Field, Idaho. "Investigation revealed that the gunner must have leaned forward to fix his guns and failed to turn off the valve that stops the turret from operating. He must

have pushed the [gun elevation] control forward as he leaned over causing the sighting cradle to come down on his head and neck." The B-24J was piloted by 2Lt. Dan M. Wilson and 2Lt. Louis B. Higgins.

9-19-44C. Camp Rucker, Alabama. At 1555, two Curtiss P-40N airplanes collided in mid-air and crashed near Camp Rucker, Alabama, killing 2Lt. William M. Kemp aboard P-40N #43-23991. F/O James R. Kesling received minor injuries parachuting to safety from P-40N #42-105365. The airplanes were part of a five-ship flight that was maneuvering in a trail formation. The two subject airplanes broke off from the formation and began a series of "dives and zooms." F/O Keeling dived below the base of a cloud, which was at an altitude of about 4,500 feet msl. He then pulled up and around it with Lt. Kemp following in trail. F/O Kesling's airplane approached stalling speed and he pushed the nose down to gain airspeed. Lt. Kemp's airplane over-ran F/O Kesling's airplane from below and behind and they collided as F/O Kesling's airplane descended and Lt. Kemp's airplane climbed. F/O Kesling stated that he did not see the other airplane at any time. Lt. Kemp was not able to parachute to safety and it was later speculated that he was killed in the collision.

9-19-44D. Speculator, New York. At about 2300 EWT, a Curtiss C-46A crashed into the northeast slope of the southern ridge of Blue Ridge Mountain 12 miles NNW of Speculator, New York, killing the crew of three. The C-46 had taken off from the Army Air Base at Syracuse, New York, at 1932 EWT on a routine navigation training flight to Waterfront, New York, to Dansville, New York, to Norwich, New York, and return to Syracuse AAB. The airplane failed to return to Syracuse AAB and was declared missing, finally being discovered on 8-4-45 by two CAP pilots that were flying on a search mission for a missing civilian airplane. It was discovered that the C-46 had smashed into Blue Ridge Mountain at an elevation of approximately 3,200 feet, about 400 feet from the top. Investigators stated that they did not consider weather to be a factor in the crash. Investigation revealed that several civilian witnesses heard the airplane fly over Lewey Lake House, followed by a muffled explosion. The civilians stated that they did not report their observations to authorities because they did not realize that an airplane had crashed and were not aware that an airplane was missing. The AAF had embarked on a highly publicized search for the C-46, which included requests for information from the public via newspaper and radio. Pilot 2Lt. William R. Barohn, co-pilot 2Lt. Charles G. Pate and engineer TSgt. Edward V. Poska were killed in the crash.

9-20-44A. Redmond, Oregon. At 1030, a Beech UC-45 crashed just after take-off from Redmond Army Air Field, Redmond, Oregon, killing five fliers and injuring another. The airplane was taking off on an administrative flight from Redmond to Chico,

California. The airplane climbed to an altitude of about 800 feet agl when both engines suddenly quit. The pilot attempted to make an emergency forced landing but the airplane struck a large tree and several large boulders. The airplane broke up and burst into flames. Killed in the crash were: Lt. Col. Ernest C. Young, pilot; Col. Earl L. Naiden, co-pilot; Maj. F.A. Zamboni, passenger; Lt. Col. William T. Moody, passenger; W/O Clair K. Benzer, passenger. Engineer Sgt. James F. Barton received minor injuries. Investigators speculated that the double engine failure was caused by a short circuit in the master switch or the master switch was inadvertently turned to the off position (possibly by the co-pilot's knee).

9-20-44B. Mt. Harvard, California. At 1533, a Lockheed P-38L collided with Mt. Harvard, Angeles Forest, California, killing Lockheed test pilot Roy Carl Cameron. The airplane had taken off from the Lockheed Air Terminal, Burbank, California, on a high-altitude production test flight. At 1428, the pilot reported that he was at 29,000 feet msl over Bakersfield, California. Radio engineers at Mt. Harvard observed the airplane flying just below an overcast that filled the canyon at the base of the mountains. Seconds later the P-38 collided with the peak of Mt. Harvard, exploding into flames upon impact.

9-20-44C. Long Island Sound, New York. At 2348, a North American AT-6C flying in poor weather was abandoned in flight and crashed into Long Island Sound, New York. Pilot-rated passenger 1Lt. Louis L. Mathews parachuted to safety but drowned before he could be rescued. Pilot Lt. Col. Malcolm G. Spooner was able to parachute to safety. He was immediately rescued from the sound by a passing boat. The airplane took off from Bolling Field, Washington, D.C., on a routine flight to LaGuardia Field, New York. The airplane arrived in the New York City area at about 2110 EWT, but the pilot was not able to find the field. The pilot flew around for two hours under a 400-foot ceiling trying to find the field, but he was unsuccessful. The airplane was low on fuel and a decision to bail out over the sound was made.

9-21-44A. Malden, Missouri. At 0036 CWT, a Douglas C-47A collided with a church steeple and crashed one mile north of the Army Air Field at Malden, Missouri, killing two fliers and injuring six others. Co-pilot 2Lt. Keith H. Taylor and radio operator Pvt. Walter D. Campana were killed. Fliers seriously injured in the crash were: F/O Gene R. Tippy, pilot; Pvt. James N. Fisher, engineer; 2Lt. Glenn J. Schwandt, pilot-rated passenger; Pvt. William A. Fargnoli, engineer-rated passenger; Pvt. Charles Burris, radio operator-rated passenger. Pilot-rated passenger F/O Alvin M. Weiss sustained minor injuries. The airplane had taken off from Malden Army Air Field on a cross-country navigation flight. The airplane carried two separate four-man crews; one crew flew the outbound leg and one crew flew the inbound leg. The

airplane was returning to the field following the successful navigation flight and was cleared to land. The airplane was on the final approach to land when it collided with a church steeple that was approximately 50 feet tall. The port engine and propeller struck the steeple, causing the port propeller to separate from the engine. It was found 117 feet from the church steeple. The C-47, with landing gear extended and locked, swerved to the left and then smashed into a field, bursting into flames upon impact. AAF rescue personnel were on the scene immediately and the fire was quickly extinguished.

9-21-44B. Greenville, South Carolina. At 0901, a North American B-25J attempting an emergency forced landing crashed eight miles ENE of the Army Air Base at Greenville, South Carolina, killing three fliers and seriously injuring another. B-25 instructor Capt. Bertel H. Carlson, pilot/B-25 student F/O Eugene A. Nivens and engineer Sgt. Stephen Mentes were killed in the crash; bombardier F/O Windsor B. Wade received serious injuries. Investigators stated, "Pilot was flying in the local flying area and instructing a trainee pilot in single-engine operation. Capt. Carlson, the instructor pilot, was in the right seat of the aircraft. Approximately 30 minutes after take-off, the pilot attempted to feather the left [propeller], but without success, and all efforts to again start the engine failed. The aircraft lost altitude quickly and had flown at treetop level for a distance of approximately two miles when the left wing struck a tree, shearing off part of the left wing and throwing the aircraft out of control into a clearing. The aircraft was completely demolished. The pilot was unable to fully feather or unfeather the propeller on the left engine and the windmilling caused drag and a considerable loss of flying speed."

9-21-44C. Nashville, Tennessee. At 0829 CWT, a Martin B-26G crashed while attempting a take-off from Nashville Municipal Airport, Nashville, Tennessee, killing two unidentified civilians on the ground. Pilot Capt. Michael J. Morse and co-pilot 1Lt. Mark D. Chapman received minor injuries. Investigators stated, "Pilot and crew were on Operations Orders from Memphis [Tennessee] to ferry aircraft from Nashville on foreign delivery. When the crews arrived at this station [Nashville] it was found that only one pilot, Capt. Morse, was current in this type of aircraft and he had not flown this model. It was decided to have him fly one and then tell the other pilots of the handling characteristics. After a thorough cockpit check, both pilot and co-pilot felt confident they could handle the airplane. Accordingly B-26G No. 43-34529 was made ready for local flight by draining the bomb bay fuel tanks. On the morning of the accident the pilot and co-pilot again went over the aircraft very thoroughly. It was then taxied into position and checked again for take-off. Everything functioned to the complete satisfaction of the crew. However, when the carburetor air heat was checked, the pilot failed to move the handle into the 'COLD' position before returning the handle to the 'NEUTRAL,' thereby keeping it in the 'HOT' position where it remained during take-off. The ship was then cleared by the tower and then it proceeded on its take-off run. The pilot reports [lifting] the nose wheel at about 60 miles per hour. The crew noted that it did not seem to be accelerating properly but as RPM and manifold pressure were okay for take-off they thought it would finally get enough speed. The airspeed, however, never did get over 90 miles per hour and when he saw that it wasn't going to break the ground, the pilot pulled back on the stick sharply and pushed the throttles full forward in an attempt to pull [the airplane off of the ground]. The aircraft continued off of the end of the runway, crashed into a shack being built to house localizer equipment, killing two civilian workers and injuring the crew. The aircraft then caught on fire and despite efforts of the crash crew, was completely consumed."

9-21-44D. Blackfoot, Idaho. At 0813, a Republic P-47C smashed to the ground near Highway 91, eleven miles west of Blackfoot, Idaho, killing pilot 2Lt. Harry L. Goff. The airplane had taken off from Pocatello Army Air Field, Pocatello, Idaho, on an elementary formation flight. The subject airplane was flying in the number-three position of a four-ship flight. About 15 minutes after take-off, at an altitude of about 4,500 feet msl, the subject airplane's number-three piston failed and disintegrated, causing the engine to begin smoking. The pilot attempted an emergency forced landing in a field about one-half mile west and three-quarters of a mile south of Highway 91. The airplane touched down and slid for a short distance before striking an irrigation ditch, breaking up and bursting into flames. The flight leader reported that the subject airplane appeared to be on fire just before it crash-landed.

9-21-44E. Strother Field, Kansas. At 1143 CWT, a Republic P-47D crashed while attempting a landing at Strother Field, Winfield, Kansas, killing 2Lt. George D. Allison, Jr. The airplane had successfully completed a routine high-altitude training flight. The pilot radioed the control tower and reported that he had generator problems. The airplane approached the field from the southwest and made a 360-degree turn to the right and then entered the standard pattern approach. After peeling off, the pilot entered a 360-degree overhead pattern. The airplane was about one mile south of the field and at an altitude of about 300 feet agl when the tower radioed the pilot that his landing gear was still retracted. The airplane entered a shallow diving turn to the left and flew into the ground. The airplane struck on the port wing and cartwheeled across the ground, bursting into flames upon impact and killing the pilot instantly.

9-21-44F. Ephrata, Washington. At 1033, a Lockheed P-38L crashed five miles south of the Army Air Base at Ephrata, Washington, killing F/O Curtis P.

Lund. The pilot was on his first solo P-38 flight and had been airborne about 25 minutes when he began to suffer diminished performance in the port engine. He radioed the control tower that he was not getting enough manifold pressure in the port engine and was coming in for an emergency landing. The field was cleared and the pilot began a straight in approach. Rescue crewmembers aboard the crash truck, which was parked at the end of the runway, observed the wings to wobble and then the airplane was seen to nose down slightly. The pilot pulled the nose up and the airplane apparently lost airspeed, stalled and fell off in a spin to the left. The airplane burst into flames upon impact and the pilot killed instantly.

9-21-44G. Lovettsville, Virginia. At 1445, a Vultee BT-13B flying in a severe thunderstorm crashed five miles northwest of Lovettsville, Virginia, killing pilot 1Lt. William H. English and passenger Capt. Frank J. Fischer. The airplane had taken off from Camp Springs Army Air Field, Washington, D.C., on an administrative flight to Wright Field, Dayton, Ohio, and had encountered a line of severe thunderstorms. The storms contained heavy rain, high winds, abundant lightning and strong turbulence. Investigators speculated that the severe turbulence in the thunderstorm caused the pilot to lose control of the airplane. Investigation of the wreckage indicated that the airplane had struck the ground out of control and at a high rate of vertical speed.

9-21-44H. Melrose, New Mexico. At 1705 MWT, a Curtiss P-40N crashed on the AAF Bombing and Gunnery Range at Melrose, New Mexico, killing 2Lt. Forrest B. Williams. The airplane was flying in the number-two position of a four-ship flight that had taken off from Fort Sumner Army Air Field, New Mexico, on a routine dive-bombing mission at the Melrose range. The formation was at an altitude of 10,000 feet when the flight leader peeled off and began his run. The formation followed. The subject airplane failed to pull out of its dive and crashed on the range at a steep angle. The airplane exploded violently into flames upon impact. There were no witnesses to the actual crash.

9-22-44A. South Webster, Ohio. At 1225 EWT, a North American AT-6 crashed ten miles south of South Webster, Ohio, killing pilot 2Lt. James F. Dudley and engineer-rated passenger Cpl. Henry E. Rubin. The airplane had taken off from Lockbourne Army Air Field, Columbus, Ohio, on a routine pilot proficiency flight. The airplane flew to the southernmost area of the Lockbourne flying area. The airplane was being operated at very low altitude when it collided with a high-tension wire, severing the starboard wingtip. The airplane went out of control and crashed into the outhouse on the farm of William A. Ward. The airplane burst into flames upon impact with the outhouse, which was totally destroyed. After great difficulty Mr. Ward was able to extricate Cpl. Rubin, who was still alive, from the burning wreckage. Cpl.

Rubin died a very short time later while being transported to a hospital in a civilian automobile. Mr. Ward was unaware that the pilot was still trapped in the wreckage.

9-22-44B. Choteau, Montana. At an unknown time after 1030 MWT, a North American AT-6C flying in instrument conditions collided with rising terrain 35 miles west of Choteau, Montana, killing pilot 2Lt. Charles Rowland. The airplane had taken off from Gore Field, Great Falls, Montana, on a routine flight and failed to return, being declared overdue at 1200 MWT. A massive search effort was launched and the airplane was found 9-25-44 by a search plane. The subject airplane became lost and apparently encountered fog and an overcast that hung just above the mountaintops. In an attempt to escape the foggy valley where he found himself, the pilot apparently tried to out climb a mountain ridge and the airplane mushed into the ridge about 20 feet from the top. The crash occurred near the Wright Mountain Observatory. A Cessna UC-78B crashed while searching for the missing airplane and pilot, resulting in one fatality. See 9-23-44D.

9-22-44C. Independence, Kansas. At 1830 CWT, a North American AT-6C suffered a catastrophic structural failure and crashed four miles east of Independence Army Air Field, Independence, Kansas, killing pilot 2Lt. C.M. Cannoy and 1Lt. Guido Chiesa. Investigators stated, "According to statements of eyewitnesses, Lt. Cannoy had been engaged in aerobatic maneuvers, including loops, snap rolls and vertical snap rolls, for a period of approximately 30 minutes before the accident. He had just completed a loop and had started a climb when he put the nose down in a power dive. At approximately 3,000 feet, witnesses stated that when he appeared to be going rather fast [in the dive], he attempted an abrupt pull out, and started a roll to the right. A portion of the left wing (approximately six feet of the wing tip) peeled off, swept across the canopy and floated away to the right of the airplane. The airplane completed approximately two fast rolls to the right, straightened out momentarily and then dived into the ground. An inspection of the portion of the port wing which peeled off and landed approximately one and one-half miles from the scene of the crash, revealed blood, which indicted that it struck one of the occupants as it swept across the canopy."

9-22-44D. Gardner Field, California. At 2120, a Vultee BT-13A crashed three miles west of Gardner Field, Taft, California, killing A/C John L. McLaughlin. During the course of his training, the student had completed three hours and fifty minutes of his required four hours of night flying. The student took off and flew to his zone and waited for the control tower to call him in to come in for a landing, completing his night flying requirements for the course. The airplane, flying at an altitude of 400 feet agl, was seen to enter a shallow dive and then to enter a shallow spiral. The airplane nosed up and then stalled, entering a spin. The student failed to recover and the airplane spun to the ground and exploded into flames. It was speculated that the student lost the horizon and fell into a spiral. The student attempted to recover but applied too much backpressure on the stick and stalled the airplane.

9-23-44A. Selfield, Alabama. At 0015, a North American AT-6C crashed near Selfield, Alabama, killing French Air Force student Cpl. Robert P. Bouriel. The airplane had taken off from Craig Field to practice night take-offs and landings at Selfield Auxiliary Airfield. The student had attempted two landings but he had over controlled and had to go around. The airplane climbed away during the second go-around, flying on the runway heading. The student failed to turn off his landing lights and the control tower called him to warn about his lights. The airplane was seen to nose down and then to rise up. The airplane then entered a shallow diving turn to the left. The airplane continued in this attitude until it smashed into the ground and exploded.

9-23-44B. Atlantic Ocean, New Jersey. At 1130, a Republic P-47C collided with a towed target and crashed into the Atlantic Ocean seven miles off shore Townsend Inlet, New Jersey, killing 2Lt. Andrew J. Easterwood. The airplane was flying in the number-three position of a five-ship flight that had taken off from the Army Air Field at Millville, New Jersey, on a gunnery mission to be conducted at the AAF Atlantic City Aerial Gunnery Range. The flight had made about five or six passes when the subject airplane struck the target and cable. The subject airplane rolled to the right and fell into a 45-degree dive. The airplane continued in this attitude until it struck the water and exploded. The other pilots in the flight were not aware that the number-three airplane crashed. The pilot's body was never recovered.

9-23-44C. Brownwood, Texas. At 1045, a Curtiss RA-25A collided with high-tension wires and crashed into Lake Brownwood at Brownwood, Texas, killing engineer Cpl. Edward J. Petersen and seriously injuring 1Lt. Ernest F. Boruski, Jr. Investigators stated, "At approximately 1040, [the pilot] approached Brownwood Lake at about 5,000 feet indicated. He let down to 1,000 feet and circled the lake but continued to lose altitude until he approached what he considered to be approximately 200 feet. At this point he had almost completed his circle around the lake and was headed into the sun at which time he saw some high-tension wires come up dead ahead. The pilot stated that he instinctively pushed the stick forward in an effort to go under the wires rather than mush into them. The airplane struck the wires, causing the left wing to drop instantly and as the left wing struck the water the airplane cartwheeled, breaking off both wings, the engine and the tail section. The pilot was thrown clear but suffered major injuries. The crew chief that was riding in the gunner's compartment was killed instantly."

9-23-44D. Judith Basin County, Montana. At an unknown time after 0945 MWT, a Cessna UC-78B crashed into rising terrain near Big Baldy Mountain in Judith Basin County, Montana, killing co-pilot 2Lt. Virgil S. Schoeppel. Pilot 2Lt. Arch E. Perkins survived the accident and walked out of the wilderness on his own, finding civilization on 9-27-44. The UC-78B and the co-pilot remained missing until 10-18-44. The airplane had taken off at 0900 from Gore Field, Great Falls, Montana, on a mission to search for a missing North American AT-6C (See 9-22-44B). Lt. Schoeppel, who was flying the airplane from the right seat, thought that he had saw the tail of the missing aircraft and maneuvered into a valley to get a better look. Lt. Schoeppel then attempted to turn out of the canyon. The turn was too tight and the airplane began to stall. Lt. Perkins, who was looking at a map, asked Lt. Perkins if everything was okay. Lt. Perkins stated that they would make it out okay. The airplane then stalled and smashed into the ground. The airplane did not burn but was completely destroyed. Lt. Perkins had been knocked out in the crash. When he came to, he saw that Lt. Schoeppel was slumped over the control wheel in an "unnatural attitude." Lt. Perkins examined Lt. Schoeppel and he appeared dead. After a while, Lt. Perkins sought shelter near the shattered airplane and went to sleep. When he woke up he examined Lt. Schoeppel again and discovered that he had not moved during that time. Lt. Perkins then attempted to walk out of the wilderness. After four days of climbing up and down over heavily timbered ridges, he stumbled upon a civilian who was able to take him to safety and shelter.

9-24-44A. Ardmore, Oklahoma. At 1133, a Boeing B-17G suffering an in-flight fire was abandoned and crashed five miles southwest of the Army Air Field at Ardmore, Oklahoma. Three fliers were killed in unsuccessful parachute jumps; two crewmembers were injured and four others were uninjured parachuting to safety. Navigator 2Lt. Stanley Parsell, engineer Cpl. Donald L. Cooper and assistant engineer Melbourne R. Rieke were killed attempting to parachute to safety. Pilot James E. Wilsey received minor injuries and co-pilot Joseph E. Aaron sustained serious injuries parachuting to safety. Crewmembers uninjured parachuting to safety were: Cpl. Kermit W. Dunne, radio operator; Cpl. David D. Fowler, gunner; Pfc. Robert D. Armstrong, gunner; Pfc. George B. Christopher, gunner. The airplane took off from Ardmore Army Air Field on a routine training flight. After successfully completing the mission the airplane returned to the field and the pilot radioed the control tower for landing instructions. The B-17 then entered the traffic pattern on the downwind leg at an indicated altitude of 2,000 feet. The engineer informed the pilot that the number-one engine was on fire and it exploded into

flames moments later. The co-pilot attempted to feather the number-one propeller and extinguish the fire. The pilots could not put out the fire and it soon engulfed the number-two engine and portions of the wing. The pilot rang the bail out alarm bell to alert the crew to abandon the airplane. He then left his seat and went to the nose hatch. The co-pilot remained in his seat and alerted the crew to bail out over the intercom and rang the alarm bell again. The radio operator and the three gunners exited the aircraft at the main entrance door and parachuted to safety. Lt. Parsell was first to bail out of the nose hatch. He fell to his death when his parachute failed to fully deploy. Investigation revealed that the chest pack parachute was fastened to only one side of his harness. The pilot exited the nose hatch second and made a safe jump. The co-pilot was attempting to exit the B-17 via the nose hatch when he encountered Cpl. Cooper making his way to the nose. Cpl. Cooper stated that he had accidentally dropped his parachute overboard and requested that the co-pilot allow him to ride piggyback during his parachute jump. The co-pilot agreed and the two bailed out. The shock of the deploying parachute caused Cpl. Cooper to lose his grip on the co-pilot and he fell to his death. His unopened parachute was found near the body of Lt. Parsell. The co-pilot's parachute opened just above the ground, causing him to land hard and break both his legs. The co-pilot stated that he had noticed Pvt. Rieke coming toward the nose section with his parachute in his hands. The co-pilot ordered Pvt. Rieke to attach his parachute and bail out immediately. Pvt. Rieke fell to his death when he left the airplane at too low an altitude to allow his parachute to deploy. The B-17 was seen to pitch up to a near vertical climb, stall and then wing over twice, entering a nose-down attitude and smashing into the ground where it exploded into flames. Investigation revealed that the number-one cylinder of the number-one engine had failed and was blown off of the main crankcase and through the cowling. The propeller feathering mechanism was rendered inoperable because of the failure of the number-one cylinder.

9-24-44B. Casper, Wyoming. At 1305 MWT, a Consolidated B-24J crashed eight miles southeast of the Army Air Field at Casper, Wyoming, killing nine fliers. Killed in the crash were: 2Lt. Andrew W. Taraba, pilot; F/O Jerome D. Goodman, co-pilot; 2Lt. Orville A. Higgins, bombardier; Cpl. Carlton Ferguson, engineer; Cpl. Harold D. Huffman, radio operator; Pfc. James M. Laird, gunner; Pfc. Rudolph W. Sandor, gunner; Cpl. Danny W. Haynes, gunner; Pfc. Vernon E. Kandt, gunner. The airplane had taken off from Casper Army Air Field at 0630 MWT on a routine training mission. The B-24 last checked in at 1015 MWT while practicing instrument flying. The airplane, flying south at about 1,500 feet agl, was seen to enter a right turn at low airspeed in a nose-high attitude. The airplane stalled, flipped over to an inverted attitude and then entered a spin. The airplane described two turns of spin before the pilot stopped the spin. The airplane then entered a dive of 70 degrees until it struck the ground and exploded into flames.

9-24-44C. Alpine, Texas. At 1440 CWT, a North American AT-6D (42-84673) flying in instrument conditions collided with rising terrain six miles northwest of Alpine, Texas, killing pilot 2Lt. James A. Noble, Jr. and passenger Sgt. Chester Holstrom. The airplane was part of a three-ship flight that had taken off from the Army Air Field at Big Spring, Texas, on a routine navigation flight back to the home station at Williams Field, Chandler, Arizona, via Marfa Army Air Field, Marfa, Texas. The formation was not cleared to fly in instrument conditions. The formation took off from Big Spring Army Air Field on a visual clearance and turned a course direct to Marfa, flying at an altitude of 4,500 feet indicated. The ceiling was estimated to be about 6,000 feet. After flying about 40 minutes, the ceiling began to lower and the terrain began to rise in elevation. The flight came within 200 feet of the ground while maintaining visual contact. The number-two pilot, 2Lt. Charles S. Rhodes (see 9-24-44D), radioed that higher ground was approaching and that the flight should climb. 2Lt. Alma R. Flake, flying in the number-three position, radioed that the flight should turn back. The flight leader stated that the flight would continue for a little while and "see how it looks." The flight began to approach instrument conditions and the number-three ship turned back and landed safely at Pyote Army Air Field, Texas. The lead ship and the number-two ship continued on, both colliding with a mountain at an elevation of about 6,800 feet. When the two aircraft failed to return to base they were declared missing. The aircraft were not found until the next day because of poor weather conditions.

9-24-44D. Alpine, Texas. At 1440 CWT, a North American AT-6D (42-84505) flying in instrument conditions collided with rising terrain six miles northwest of Alpine, Texas, killing pilot 2Lt. Charles S. Rhodes and passenger Pfc. Robert W. Reeder. The airplane was part of a three-ship flight that had taken off from Big Spring Army Air Field, Big Spring, Texas, on a flight back to the home station at Williams Field, Chandler, Arizona, via Marfa Army Air Field, Marfa, Texas. See 9-24-44C for details.

9-24-44E. Cordova, Nebraska. At 1445 CWT, a Vultee-Stinson L-5 collided with trees and crashed five miles north of Cordova, Nebraska, killing pilot Capt. Mark J. Mathers, Jr. and WAC 1Lt. Virginia L. Weitz. The airplane had taken off from the Army Air Field at Geneva, Nebraska, to perform a simulated bombing mission on the Second Air Force Bivouac area. On the third pass the airplane stuck tall trees with the landing gear and the propeller, shattering the propeller and throwing the airplane out of control temporarily. The pilot regained control and then pulled up in a nose-high attitude and stalled the airplane. The airplane fell off on a wing and smashed into the trees, severing a wing. The airplane plunged to the ground and burst into flames. Ten AAF personnel on the ground received minor injuries attempting to rescue the two fliers from the flaming wreckage.

9-24-44F. Camp MacKall, North Carolina. At 1940 EWT, a Douglas C-47A participating in a night airborne operation collided with parachute supply bundles that were dropped from another C-47 and crashed five miles west of Camp MacKall, North Carolina, killing four crewmembers and eight paratroopers. The airplane was part of a six-ship flight that had taken off from the Army Air Field at Lumberton, North Carolina, on a night parachute drop. The airplanes were flying in a V of V's type formation at approximately 1,100 feet agl. The subject airplane collided with the parachute bundles with the port wing, the bundle's parachutes fouling the control surfaces and causing the airplane to half-roll out of control. The subject airplane spun to the ground and exploded violently into flames upon impact. Investigators noted that a V of V's formation should not be used during night parachute jumps. Published accounts of the accident reported that over 300 high-ranking army officers and several newsmen witnessed the accident. Camp MacKall public relations officers reported that approximately 30 airborne troops were injured in Waco CG-4A glider crashes or in parachute jumps during the operation. C-47 crewmembers killed in the crash were: 1Lt. Wayland W. Rose, pilot; 2Lt. George J. Helfend, co-pilot; TSgt. Victor L. Graves, engineer; SSgt. Myron K. Shnaper, radio operator. Paratroopers killed in the crash were: Sgt. Arnold J. Pappas, Cpl. Robert N. Green, Pvt. Kenneth Kirkendoll, Pvt. Elmer G. Peavely, Pvt. Emerson D. Taylor, Pvt. Luther S. Hairell, Jr., Pvt. Charlie F. Evans, Pvt. Lewis H. Lake, Jr.

9-25-44A. Mitchel Field, New York. At 0915, WAC Cpl. Anne E. Passmore was killed when she walked into the rotating starboard propeller of a Beech AT-11 operating at Mitchel Field, Hempstead, New York. The airplane had returned to the field after completing an administrative flight from Westover Field, Chicopee, Massachusetts. Investigation revealed that Cpl. Passmore was walking across the taxiway apparently to "intercept" the aircraft. She was carrying several radio facility charts and may have been blinded by the morning sun. There were several aircraft with engines running in the immediate vicinity at the time of the accident. AT-11 pilot 1Lt. Robert H. Ebenreiter and acting co-pilot Lt. Col. Herbert R. Volin were uninjured.

9-25-44B. Soda Lake, California. At 1540, a Lockheed P-38L collided with a mountain near the AAF Ground Gunnery Range at Soda Lake, California, killing pilot 2Lt. Wesley O. Johnson. The airplane was flying in the number-three position of a four-ship flight that had taken off from Santa Maria Army Air

Field on a low-altitude aerobatic training mission. After the aerobatic flight was completed the flight leader called the flight in to form up. The subject airplane performed a slow roll as he flew over to join the flight leader. The pilot attempted a second slow roll and apparently stalled the airplane in an inverted attitude. The pilot attempted a split-S type maneuver to recover. The airplane was at too low an altitude to allow recovery and the airplane dove into the ground, exploding into flames upon impact.

9-25-44C. Tampa, Florida. At 2130 EWT, a North American P-51C crashed 15 miles NNE of Hillsborough Army Air Field, Tampa, Florida, killing pilot 2Lt. Richard J. Joyce. The airplane had taken off from Hillsborough Army Air Field on a routine night training flight. The airplane smashed into the ground on the port wing and exploded into flames. Investigators could not determine the cause of the accident. It was later speculated that the pilot had lost the horizon and inadvertently flew the airplane into the ground in a diving turn to the left.

9-26-44A. Warrenton, Georgia. At 1230 EWT, a Beech AT-11 attempting an emergency forced landing crashed six miles south of Warrenton, Georgia, killing pilot 2Lt. Donald E. Dice. The airplane had taken off from Shaw Field, Sumter, South Carolina, on a ferry flight to Tuskegee, Alabama. Near Warrenton the subject airplane dropped out of the formation in a gliding turn to the left. The port engine apparently failed and the propeller was turning slowly. The pilot attempted an emergency forced landing in a cotton field. The pilot failed to make the field and the airplane smashed into a wooded area and burst into flames.

9-26-44B. Oliver, Georgia. At 1600 EWT, a Martin B-26G crashed near Oliver, Georgia, killing gunner Cpl. William E. Dudash. Pilot 2Lt. Willis R. Dakan and bombardier 2Lt. Ronald E. Nordeen sustained serious injuries; co-pilot 2Lt. Ralph M. Decker and gunner Sgt. Allen H. McKenna received minor injuries. The airplane had taken off from Hunter Field, Georgia, on an instrument calibration flight. Investigation revealed that the airplane was flying at an altitude of 200 feet agl when the starboard engine failed. The B-26 lost altitude and collided with trees and then smashed to the ground in a field. Investigators noted that the pilot had been authorized to fly at low altitude during the mission. The airplane was completely destroyed in the crash. Investigators could not determine the cause of the engine failure.

9-27-44A. Kinder, Louisiana. At 0840 CWT, a Martin TB-26 crashed 14 miles west of Kinder, Louisiana, killing the crew of four. Killed in the crash were: 2Lt. James E. Farrell, pilot; 2Lt. William J. Jackson, co-pilot; Cpl. William L. Merrill, engineer; Cpl. Delton J. Haren, radio operator. The airplane took off from Lake Charles Army Air Field, Louisiana, at 0802 on a routine formation flight. The subject airplane never made an attempt to join the formation and at approximately 0830 another pilot flying in the area contacted Lake Charles Army Air Field and notified control tower personnel that the subject airplane was apparently suffering propeller trouble. A couple minutes later the subject airplane radioed the control tower that he had a runaway propeller. The tower attempted to make further contact with the B-26 but was unsuccessful. A short time later the crash call came in to the field and rescue personnel were dispatched to the scene. Investigation revealed that the both the port engine and propeller had failed. The airplane apparently rolled to the left before smashing into the ground and exploding into flames.

9-27-44B. Tchula, Mississippi. At 0845 CWT, a Vultee BT-13A collided with a grass-cutting tractor while attempting a take-off at the AAF Auxiliary Airfield at Tchula, Mississippi, killing civilian tractor operator Fernindo C. Lishman. The airplane was flying in the number-three position of a flight of three BT-13s that had taken off from the Army Air Field at Greenwood, Mississippi, on a formation-training flight to Tchula Auxiliary Airfield. The flight leader led the flight to a successful landing at Tchula. A tractor was observed operating on the field and the airplanes had diverted from their path on the ground to avoid the tractor. The flight taxied to the south end of the field and waited for the tractor to cross the field. The flight then delayed their take-off while they waited for another airplane that was making a landing to clear the field. The flight leader then began a formation take-off to the east. The tractor began heading north, across the take-off path. The pilot of the number-three airplane observed the tractor in his path on the take-off run and attempted to pull up and over the tractor. The number-three airplane then collided with the tractor with the propeller and the port landing gear, killing the operator instantly. The BT-13A smashed back to earth and was seriously damaged. The airplane was piloted by instructor 2Lt. Gordon D. Hawk and student A/C Ira D. Coleman. Both were uninjured. Investigators noted that the flight leader was taking off "cross tee."

9-27-44C. Brownsville, Tennessee. At 0930 CWT, two Boeing B-17F airplanes collided in mid-air and crashed eight miles west of Brownsville, Tennessee, killing nine fliers. Three fliers were injured and six others uninjured parachuting to safety. Killed in the crash of B-17F #42-30787 were: Capt. Frank W. Wolfe, B-17 instructor; 2Lt. George W. Steinmetz, pilot; 2Lt. Wayne Lindquist, co-pilot; 2Lt. William E. Kittilstad, navigator; 2Lt. Jack R. Davis, bombardier; Cpl. James V. Palmer, engineer; Cpl. Stephen J. Jaksetic, radio operator; Pfc. Mack A. Watts, tail gunner. Assistant engineer/starboard waist gunner Pfc. Alfred W. Willier was able to escape from B-17F #42-30787 and parachute to safety. He received serious injuries. Crewmembers parachuting to safety from B-17F

#42-3370 were: 2Lt. Hendrik J. Mouw, pilot; F/O Leonard R. Marshall, co-pilot; Cpl. David M. Lesher, assistant engineer; Cpl. John M. Heckman, radio operator; Cpl. John W. Higgins, gunner; Pfc. James L. Moore, gunner. Assistant engineer/waist gunner Pfc. Bernard E. Lintula was seriously injured parachuting to safety and tail gunner Harold E. Layne sustained minor injuries. Bombardier 1Lt. Ray Christensen was killed in the initial collision. The airplanes had taken off from Dyersburg Army Air Field, Dyersburg, Tennessee, on a routine high-altitude formation flight. Investigators stated, "Lt. Hendrik J. Mouw, pilot of B-17F #42-3370, was flying a high altitude formation at 18,000 feet, being the [number-three airplane] of the second element in the lead squadron of a ten-ship formation. After a flight of approximately one hour and fifty minutes his airplane collided with the [number-four airplane] in the same element which was flying the low diamond position and serving as monitor ship of the formation. Lt. Mouw was called by twice by instructors and told to settle down and fly his position smoothly. Immediately prior to the accident, he was seen to slightly overrun his lead ship but momentarily drop back, then the right wing [of B-17F #42-3370] dipped and the airplane dropped farther back and down, contacting B-17F #42-30787 on top of the pilot's cabin. [Lt. Mouw] states he experienced [propeller turbulence] from [the] lead ship. All crewmembers parachuted to safety with the exception of the bombardier who received fatal injuries in the collision. There was an explosion seen at the wing of #42-30787, whereupon it dived out of control and crashed, one crewmember parachuting to safety." Investigation of the wreckage and survivor testimony revealed that the number-one and the number-three propellers of B-17 #42-3370 contacted the port and starboard wings of B-17F #42-33787; the number-two propeller of B-17F #42-3370 cut into the cockpit of B-17F #42-33787, killing the pilots instantly. The number-three propeller of B-17F #42-33787 cut into the nose section of B-17F #42-3370, killing the bombardier instantly.

9-27-44D. Moses Lake, Washington. At 1021, a Lockheed P-38L crashed eight miles north of Moses Lake, Washington, killing pilot 2Lt. Robert D. Galvin. Investigators stated, "The pilot was flying number-four position [of a four-ship flight on a routine training mission] and had been at 20,000 feet for forty minutes. The flight leader called for a gas check and received an answer from this pilot. The flight leader then dipped his right wing for the flight to go into a right echelon formation. Number-four did not go into right echelon, but continued in straight flight. [The subject pilot] began a gentle diving turn, gradually increasing to a vertical dive. The flight leader called the number-four man, but was unable to receive an answer. The airplane disintegrated in the dive. Pilot was found about 200 yards from the crash, parachute showed no signs of being used.... The statement of the flight surgeon indicates that the pilot was unconscious prior to the crash — the cause of which is unknown."

9-27-44E. Pocatello, Idaho. At 1150, a Republic P-47D crashed 24 miles north of the Army Air Field at Pocatello, Idaho, killing pilot 2Lt. Paul T. Flaharty. Investigation revealed that the pilot was engaging in simulated combat with another fighter aircraft at about 17,000 feet msl. The pilot attempted to get on the tail of the other P-47, which dove to about 13,000 feet in an attempt to escape. The targeted airplane then performed a steep chandelle to the left. The subject airplane followed and attempted to pull up and inside the maneuver. The subject airplane stalled at the top of the maneuver and snapped rolled, falling out of control. The pilot was unable to recover the airplane and it dove into the ground at high speed where it exploded into flames upon impact. The pilot had made no attempt to exit the airplane and was killed upon impact.

9-27-44F. Weatherford, Texas. At 2025, a Beech AT-7C flying in instrument conditions collided with a chimney and crashed three miles east of Weatherford, Texas, killing three fliers. Pilot Col. Orin J. Bushey, engineer SSgt. John J. Nemetz and passenger Maj. Edward E. Bailie were killed. The airplane had taken off from Hobbs Army Air Field, Hobbs, New Mexico, on an administrative flight to Fort Worth Army Air Field, Fort Worth, Texas. The airplane encountered heavy rain showers and low ceilings near Weatherford. The pilot attempted a 180-degree turn to the right to get out of the heavy showers and to stay under the overcast, striking a tall brick chimney and a metal chimney at the Knights of Pythies Orphanage. The airplane then turned to the right and flew a short distance before it crashed into a field and burst into flames about one mile north of the orphanage. Investigation revealed that the pilot had checked the weather conditions for his flight and was informed that visual flight "could but barely be maintained." The weather officer suggested that pilot get an instrument clearance and expected the pilot to return to the field weather office with the clearance but next heard the subject pilot on the radio announcing that he was over the field and outbound. The pilot asked if the weather conditions had changed and was informed that they had not. The airplane then turned on course and flew away from the field. Investigation revealed that the pilot "possessed a 1-1 rating for clearing purposes and had a Command Pilot rating [and] was able to clear himself as he wished."

9-27-44G. Tucumcari, New Mexico. At 2206, a Consolidated RB-24E flying in instrument conditions crashed 23 miles east of Tucumcari, New Mexico, killing eleven fliers. The airplane was returning to Kirtland Army Air Field, New Mexico, after completing an instrument navigation training flight to Chicago Municipal Airport, Chicago, Illinois. Investigators stated, "The pilot filed clearance at Chicago Municipal Airport for a flight under instrument conditions from Chicago to Tucumcari, New Mexico, at 10,000 feet, and from Tucumcari to Kirtland Field, Albuquerque, New Mexico, contact flight rules. Crewmembers gave position reports over each radio fix en route, making their last report over Amarillo, Texas, which is 108 miles east of Tucumcari. No further radio contact was made, and the airplane crashed." Wreckage was scattered over an area of approximately 800 feet. Investigation revealed that icing conditions existed in much of eastern New Mexico and an icing report was to be issued to the aircraft by CAA Air Traffic Control personnel when it checked in at Tucumcari. The aircraft never checked in at Tucumcari and did not receive the warning. A certified aircraft mechanic at Chicago stated that the airplane had been suffering some "inverter" problems that he had helped to correct before the take-off for Kirtland Field. It was later speculated that the airplane encountered severe icing conditions and the pilots lost control. Killed in the crash were: Capt. Willis A. Teller, B-24 instructor; 2Lt. Robert G. Anderson, pilot; 2Lt. Jewel G. West, Jr., co-pilot; Capt. Ollie F. Hastings, B-24 student; 1Lt. Donald A. Edwards, B-24 student; 2Lt. Richard D. Denison, B-24 student; 2Lt. Franklin H. Cobb, B-24 student; 2Lt. Ernest H. Alberty, Jr., B-24 student; Sgt. Frank J. McKenna, radio operator; Pvt. Robert E. Hodges, radio operator; Pfc. Frank L. Kozubik, engineer.

9-28-44A. Fort Myers, Florida. At 0815 EWT, a Curtiss P-40N crashed at Page Field, Fort Myers, Florida, killing pilot 2Lt. Laurence C. Meador. Investigators stated, "Lt. Meador's aircraft engine cut out just after becoming airborne. Lt. Meador called for an emergency landing and attempted to complete a circuit of the field to the nearest runway. He didn't have enough power to maintain altitude and tried to stretch his flight by holding the nose just above the horizontal. Lt. Meador's engine cut out again and he tightened his turn to try to reach the runway. His speed was too low and he snapped rolled into the ground from about 100 feet."

9-28-44B. Wilmington, North Carolina. At 1137, two Republic P-47D airplanes collided in mid-air and crashed 25 miles southeast of Wilmington, North Carolina, killing both pilots. Capt. Leo G. Berinati was killed in the crash of P-47D #42-22433; 2Lt. Theodore F. Petti was killed in the crash of P-47D #42-8182. The airplanes were part of a flight of four that had taken off from Bluethenthal Field, North Carolina, on an aerial gunnery mission. Capt. Berinati was the flight leader and Lt. Petti was flying in the number-two position. The weather deteriorated during the gunnery mission and the remainder of the mission cancelled. The flight was ordered to return to the area of the field and fly locally. The flight returned to the field and checked the weather. The leader climbed the flight to about 10,000 feet, passing through some broken stratus and other layers of varying thickness. Visibility was

generally "bad" during the climb, but cleared up considerably when they got on top. The number-three pilot radioed the flight leader and informed him that he was suffering engine trouble and was returning to the field. The flight leader agreed and began to lead the formation down in a "spiral turn" to the right through what turned out to be a considerable undercast. The number-three and number-four airplanes soon became separated from the leader and the wing aircraft. The leader and his wingman broke out underneath in a rain shower at about 1,800 feet agl. The airplanes dropped down to about 800 feet agl, leveled out and collided soon after. The airplanes fell to earth within one-half mile of each other near the Inland Waterway Bridge at Carolina Beach, bursting into flames upon impact.

9-28-44C. Pocatello, Idaho. At 1200, a Republic P-47D crashed 20 miles north of the Army Air Field at Pocatello, Idaho, killing pilot 2Lt. Herbert R. Etz, who fell to his death in an unsuccessful parachute jump. The airplane was part of a four-ship flight on a supervised aerobatic training mission. The formation was flying at about 20,000 feet when the leader performed an Immelmann turn. The flight followed the leader and Lt. Etz stalled out of the maneuver and fell into a spin. He recovered at about 12,000 feet and soon rejoined the formation. The leader then led the flight through another Immelmann, and Lt. Etz again stalled out at the top. The airplane entered what appeared to be a flat spin. The flight leader repeatedly radioed the pilot to bail out. The pilot attempted to parachute but he fell to earth without deploying the parachute. It was speculated that the pilot was either bailed out at too low an altitude or was struck by the tail section of the airplane upon leaving the airplane. His body was found about 200 feet from the main wreckage. The parachute, still attached to the pilot, was found unopened and the ripcord still in its holder.

9-28-44D. Philip, South Dakota. At 1439 CWT, two Boeing B-29 bombers collided in mid-air near Philip, South Dakota, killing eight fliers and injuring twelve. Killed in the crash of B-29 #42-65215 were: Capt. Billy J. Wheeler, pilot; 2Lt. Roland J. Hand, co-pilot; 2Lt. Gerald T. McGrand, navigator-bombardier; 2Lt. Robert E. Weimer, navigator-bombardier; 2Lt. Mathew C. Lutostanski, flight engineer; Cpl. Warren J. McCarthy, central fire controller; Sgt. Marshall W. Ballard, gunner; SSgt. Dallas G. Chevrie, radio operator. Gunners Sgt. William S. Clary and Sgt. Arthur J. Liberi were injured parachuting to safety. Crewmembers injured parachuting to safety from B-29 #42-63433 were: Capt. Francis J. Murray, pilot; 2Lt. John A. Gugsel, navigator-bombardier; 2Lt. Franklin L. Tannyson, flight engineer; Cpl. Franklin Moore, Jr., central fire controller; Sgt. Clifton L. Sewell, gunner; Pvt. Edward Yeager, radio operator; Sgt. Charles C. Dean, Jr., gunner; Cpl. Orville J. Grannaman, radio operator; Cpl. Dale R. Coates, gunner; SSgt. Henry W. Mingle, Jr., radio operator. Co-pilot 2Lt. Horace

S. Rich and navigator-bombardier 2Lt. Edward H. Montoney received serious injuries parachuting to safety. Investigators stated, "A formation of fourteen B-29 airplanes, [from] Kearney Army Air Field [Nebraska], took off from Smokey Hill Army Air Field, Salina, Kansas, 28 September, 1100, B-29 #42-63433 was number three and 42-65215 was number four in an element of four airplanes in a diamond formation. At 18,000 feet, No. 215 for reasons undetermined, pulled up to the left and up under No. 433. The number-four engine of No. 433 cut the right side of the fuselage of No. 215 just aft of the rear bulkhead of the pressurized cabin. The right wing of No. 433 dragged back over the vertical fin and the rudder of No. 215, smashing it down, also damaging the underside and trailing edge of the right wing of No. 433. The top of number-four engine was also damaged on No. 433. Two gunners were thrown out of No. 215 when the tail broke off, and landed safely near Philip, South Dakota. The tail section landed about one-half mile from the main part of the airplane. No. 433 attempted to return to home base, but was unable to continue due to excessive vibration of the number-four engine and being unable to keep the right wing up and hold altitude. After continuing flight for approximately 100 miles, they turned on the [automatic flight control] and decided to abandon the airplane. All crewmembers bailed out and all landed without major injury, except for the navigator who broke his ankle. The airplane crashed about six miles west of Clearfield, South Dakota."

9-28-44E. Jasper, Florida. At 1610 EWT, a North American AT-6A collided with trees and crashed at Jasper, Florida, killing 2Lt. Charles L. Konecke. Investigators stated, "Lt. Konecke, an instructor, was flying an AT-6A type airplane in the vicinity of Jasper, Florida, after having acted as a control ship for a student navigation mission from Spence Field [Moultrie, Georgia]. Two witnesses who saw the crash stated that the airplane was seen to pull away from the field, fly what amounted to a traffic pattern, come back into the field and land on the wheels, then take off again after a short roll out, retracting the landing gear as altitude was gained. The airplane then circled the field to the left in a climb and when on the east side of the field started a shallow dive across the field from east to west. After crossing the field at approximately 200 to 300 feet, the airplane appeared to start a turn to the left then complete one roll to the right and part of another roll at which time the airplane was lost from view and a crash was heard. Investigation disclosed that the airplane hit a tree with the left wing leaving parts of the left landing strut and leading edge of the left wing stuck into the tree at the point of impact. The tree that was hit was broken at the point of impact, approximately 20 feet above the ground. The airplane then slid approximately fifty yards in a northerly direction. The rear half of the fuselage was found 30 yards north

and 10 yards to the west of the path of the wreckage and the left wing from the aileron out was found approximately 10 yards to the rear and 20 yards to the east of the tree where the airplane first hit."

9-28-44F. Kingman, Arizona. At 1312 PWT, a Martin TB-26C crashed while attempting a take-off at the Army Air Field at Kingman, Arizona, killing engineer Cpl. Paul S. Morris and passenger Pvt. Wendell F. Marshall. Pilot 2Lt. Howard A. Sandberg and co-pilot 2Lt. Robert Christian, Jr. were injured. The B-26 took off to the southwest on a target-towing mission, climbing to about 300 feet agl when the port engine failed. Witnesses on the ground observed heavy smoke emitting from the area of the port engine. The pilot feathered the port propeller and trimmed the airplane for single-engine flight. The pilot then turned to the right, attempting to circle the field and land on the south runway. The airplane lost altitude in the pattern and the pilot was unable to maintain altitude during the turn from the base leg to the final approach. The pilots crash-landed the airplane on the desert about one mile from the field. The B-26 broke up and erupted into flames as it skidded along the ground. The port impeller had caused the failure.

9-28-44G. Blackfoot, Idaho. At 1555 MWT, a Republic P-47G crashed while attempting an emergency forced landing ten miles northeast of Blackfoot, Idaho, killing 2Lt. Robert M. Faber. The airplane had taken off from Pocatello Army Air Field, Idaho, on a bombing mission. After the mission was completed the subject airplane engaged in simulated combat with another P-47 at about 12,000 feet msl. After performing several evasive maneuvers, Lt. Faber's aircraft began to lose altitude because of diminished engine performance. The pilot radioed that he was having some troubles. Tower personnel stated that the pilot verbally checked off instruments and tried changing fuel tanks and priming the engine without results. The airplane was at about 9,000 feet and the pilot was able to pick out a field for a forced landing. Investigation revealed that the airplane stalled and spun to the ground while turning to land in the open field.

9-29-44A. Uvalde, Texas. At 1150, a Lockheed AT-18A crash-landed 18 miles north of Uvalde, Texas. Navigator instructor 1Lt. Robert H. Graham fell to his death in an unsuccessful parachute jump. Student navigators Cpl. Charles F. Voros, A/C Francis F. Thomasco, A/C Willard D. Weidner and A/C Willmer L. Walstad were uninjured parachuting to safety. Pilot 2Lt. Robert G. Lister was able to perform a successful crash landing and was uninjured. The airplane had taken off from Ellington Field, Houston, Texas, on a routine navigation-training mission. The airplane climbed to an altitude of 1,500 feet agl when the port engine suddenly failed. Moments later, the starboard engine failed. The pilot ordered the four student navigators and instructor to bail out. The four students bailed out immediately, but the instructor stayed with the airplane. The pilot then set up an approach for an emergency forced landing in an open field. The instructor bailed out of the aircraft when it was on the final approach for the crash landing, jumping from an altitude estimated to be less than 200 feet agl, too low to allow his parachute to fully deploy. His body, with the partially deployed parachute strung out behind, was found about three-quarters of a mile from the spot where the airplane came to rest and about four miles from the area where the students had landed. The pilot had overshot the field during the forced landing attempt and crashed into a thick stand of mesquite trees, completely wrecking the airplane. Investigation of the wreckage revealed that the port engine magneto was dirty and oily, the spark plugs were coated with excessive carbon and the port mixture control arm had come loose prior to the crash. No gasoline was found in the starboard carburetor. Investigators could not determine the cause of the starboard engine failure. Investigators speculated that the pilot had inadvertently pulled the starboard engine mixture control to idle when he intended to manipulate the port mixture control.

9-29-44B. Perry, Florida. At 120 EWT, two Curtiss P-40N airplanes collided in mid-air six miles northwest of the Army Air Field at Perry, Florida, killing 2Lt. Evral W. Kellogg in the crash of P-40N #43-24340. 2Lt. Clarence B. McComas, Jr. was able to parachute to safety from P-40N #43-24254, suffering only minor injuries. The airplanes were flying at an altitude of 6,000 feet msl. After flying straight and level at this altitude for some time, the flight encountered a cloud formation. Lt. McComas, who was leading the flight, climbed to the top of the cloud and the wing airplane followed. The wing airplane inadvertently entered the cloud and apparently collided with the lead airplane while climbing out of the cloud.

9-29-44C. Fort Sumner, New Mexico. At 1330 MWT, a Curtiss P-40N crashed 16 miles southwest of the Army Air Field at Fort Sumner, New Mexico, killing 2Lt. Robert L. Secrist. The airplane was in the number-three position of a five-ship flight that had taken off from Fort Sumner on a camera gunnery mission. The formation was flying at an altitude of 14,000 feet when the flight leader peeled off sharply and then entered a dive, losing about 2,000 feet before entering a climbing turn to the right. The flight followed and the subject airplane appeared that it was going to overrun the number-two airplane and had to pull out. The airplane was seen to flip over to an inverted position before diving away. The airplane smashed into the ground and exploded violently into flames, killing the pilot instantly. There were no witnesses to the actual crash.

9-29-44D. Avon Park, Florida. At 1657 EWT, a Boeing B-17G went out of control and crashed 19 miles east of the Army Air Field at Avon Park, Florida, killing the crew of nine. The airplane had taken off from Avon Park on a routine instrument training flight. The airplane had been airborne about 45 minutes when it went out of control and smashed to the ground. Examination of the wreckage indicated that portions of the port elevator had separated during flight. The airplane went out of control and apparently entered a steep descent. The co-pilot evidently exceeded the design limitations of the port wing while trying to pull the airplane out of the dive, causing its failure and separation. Wreckage was scattered over a wide area. Investigation revealed that the port life raft had deployed in flight and had became caught on the port horizontal stabilizer, fouling the port elevator and causing the airplane to go out of control. The airplane smashed to the ground and exploded into flames, killing the crew instantly. The pilot's body was found in the remains of the radio compartment, indicating that the pilot had left his position on the flight deck to ascertain the condition of the tail surfaces. There were no witnesses to the crash. Killed in the crash were: 2Lt. Walter J. Padlo, pilot; 2Lt. Roland C. Williams, co-pilot; F/O Hugh C. Fleming, navigator; 2Lt. James P. Huey, bombardier; Cpl. John G. Bolkovac, engineer; Cpl. Floyd Fong, radio operator; Cpl. Jerome J. Esser, assistant engineer; Pfc. Roland F. Kvarne, assistant radio operator; Pfc. Gerald Azzinnaro, gunner.

9-29-44E. Alamogordo, New Mexico. At 1509, a Consolidated B-24J crashed at the Army Air Base at Alamogordo, New Mexico, killing eight fliers and seriously injuring two others. Killed in the crash were: Capt. David R. Baker, pilot; 2Lt. George C. Jacoby, co-pilot; 2Lt. Elmore Trafford, navigator; Sgt. John A. Mandla, engineer; Cpl. Charles H. Roderick, radio operator; SSgt. Bernard P. Burford, passenger; Capt. John S. Maher, pilot-rated passenger; E.J. Hallenman, civilian passenger; TSgt. K.A. Pound, passenger; TSgt. Marvin R. Cox, passenger. The airplane had taken off from Clovis Army Air Field, New Mexico, on an administrative flight back to its base at Alamogordo Army Air Base. The airplane approached the field from the east at about 1,000 feet and requested landing instructions. The pilot advised the tower that he was having trouble with the number-one engine. The pilot did not declare an emergency so the tower instructed the pilot to go around the pattern because a B-29 was in take-off position and cleared to take-off. The pilot acknowledged and entered the pattern. The airplane continued over the south end of the field with landing gear extended and the number-one engine emitting a trail of thick black smoke. The airplane was seen to turn to the right, losing altitude. The pilot radioed the tower that the number-one propeller was windmilling and that he was going to land immediately. The tower acknowledged and authorized a downwind landing on Runway 3. The airplane continued in a right turn onto the base leg and suddenly leveled out. The port wing was seen to drop and hit the ground, causing the airplane to cartwheel into the ground and burst into flames. Investigators speculated that the pilot, while making a right hand approach with number-one engine out, had either allowed the airspeed to deteriorate to such an extent that the airplane stalled into the ground or had inadvertently pulled the number-two fuel mixture control lever to the idle position while attempting to manipulate the number-one fuel mixture control lever.

9-29-44F. Macon, Georgia. At 1728 EWT, a North American AT-6D and an AT-6C collided in mid-air and crashed ten miles southeast of Cochran Field, Macon, Georgia, killing instructor pilot 1Lt. Peter C. Langereis aboard the AT-6D (42-85433). Student A/C George T. Frankenfield was able to parachute to safety. Instructor 2Lt. Arthur G. Preacher and student A/C Lee H. Mac Alpine parachuted to safety from the AT-6C (42-48860) and were uninjured. Investigators stated, "Both aircraft were on a dual instrument ride with the students in the rear cockpit under the hood on instruments. At an altitude of between six and seven thousand feet, airplane 42-48860 was in a turn to the left (15 degree bank). Airplane 42-85433 was flying straight ahead and losing airspeed preparatory to a power-on descent. Airplane 42-85433 (Lt. Langereis and A/C Frankenfield) collided with airplane 42-48860, resulting in damage to the right wing tip, aileron, right horizontal stabilizer and rudder of airplane 42-48860 (Lt. Preacher and A/C MacAlpine). There is no evidence to determine the extent of damage to airplane 42-85433 other than the fact that smoke is known to have come form the cowl of this aircraft (neither airplane burned upon crashing).

9-30-44A. Wenatchee, Washington. At 0005, a Consolidated B-24J collided with rising terrain eight miles southwest of Wenatchee, Washington, killing the crew of six. Killed in the crash were: 2Lt. James D. Hunt, pilot; 2Lt. Ted R. Lewis, co-pilot; F/O Robert J. Hennekes, navigator; 2Lt. Francis W. Lequier, bombardier; Cpl. Calvin D. Fleming, engineer; Cpl. James R. Manthei, radio operator. The airplane took off from Walla Walla Army Air Field, Washington, at 2227 (September 29, 1944) on a local navigation training flight to Spokane, Washington, to Ellensburg, Washington, to The Dalles, Oregon, to Pendleton, Oregon, and return to Walla Walla. The airplane, flying at an indicated altitude of 9,000 feet, reported in at Ephrata, Washington, at 2355. A short time later, forest rangers observed a fire on the mountain south of Wanatchee. The wreckage was found the next day. Investigators speculated that the airplane was flying in instrument conditions at the time of the accident. Investigation revealed that the airplane collided with the mountain while in cruise flight at an elevation of 6,300 feet, exploding into flames upon impact and killing the crew instantly. The airplane was on course but 500 feet too low to safely pass over the mountain.

9-30-44B. Moses Lake, Washington. At 1331, a Lockheed P-38L crashed 27 miles south of Moses Lake, Washington, killing pilot F/O Gene L. Dyer. The airplane was part of a four-ship flight that had taken off from Moses Lake Army Air Field on a supervised aerobatic training mission. The formation climbed to 12,000 feet and entered a trail formation to do some aerobatics. There were scattered clouds in the area, topping out at 11,000 feet and with the base at about 7,000 or 8,000 feet msl. The flight leader began a "rat race" around these clouds for several minutes. The flight reformed and the subject airplane, flying in the number-two position, was discovered missing. The flight leader flew around and searched for the missing airplane, finding the crash after a short time. The airplane had struck the ground at a steep angle and exploded violently into flames. Investigators could not determine the cause of the accident.

October

10-1-44A. Randolph, Iowa. At 0410, a Cessna UC-78B crashed two miles north of Randolph, Iowa, killing pilot SSgt. Donald V. Kurtzer and engineer SSgt. James S. Heigler. The airplane, stationed at Tinker Field, Oklahoma, took off from Sherman Field, Kansas, on a navigation flight to Omaha, Nebraska. Investigation of the wreckage indicated that the airplane dove into the ground at an angle of approximately 75 degrees, completely telescoping the fuselage and crushing the occupants. The airplane burst into flames upon impact. Weather at the time of the accident was described as intermittent light drizzle, visibility eight miles and a ceiling of 800 to 1,000 feet. Investigators could not determine the cause of the accident.

10-1-44B. Courtland, Alabama. At 0827, a Consolidated RB-24E crashed while attempting a landing at the Army Air Field at Courtland, Alabama, killing five fliers and seriously injuring two others. The B-24 was one of a flight of three that had taken off from Courtland Army Air Field on a routine student formation flight. The formation flew for about one hour before returning to the field for a landing, entering the traffic pattern at an altitude of 2,000 feet for a formation landing on Runway 18. The leader ordered the subject airplane, which was flying in the number-three position of a V-formation, to peel off for the 360-degree turn and a landing. The subject airplane peeled off and the pilots extended the landing gear during the turn. As the pilot approached the final portion of the turn, in about a 45-degree bank, he noticed he was not lined up with the runway and in an attempt to shorten his turn he increased his bank, entering a high-speed stall. Investigators noted that the airspeed was 160 mph as the airplane entered the turn. The port wing dropped sharply in the turn, putting the ship in almost a vertical bank. The instructor and student were able to roll the airplane level but the B-24 entered a secondary stall. The airplane fell off and smashed to the ground nose first at an angle of approximately 30 degrees. The airplane smashed itself to pieces as it skidded along the ground for about 100 yards, all four engines separating from the wings. The rear fuselage and tail section broke off and came to rest standing on its edge with the fuselage perpendicular to the ground. The B-24 was carrying about 2,000 gallons of fuel but did not burn. Killed in the crash were: 2Lt. Joseph Kupa, pilot/B-24 student; 2Lt. Robert V. Kelsey, pilot-rated passenger; 2Lt. John G. Ronning, pilot-rated passenger; 2Lt. Robert H. Rutley, pilot-rated passenger; Pvt. Foster H. Scoboria. B-24 instructor pilot 1Lt. William R. Hill and pilot-rated passenger 2Lt. Robert C. Shad received serious injuries.

10-1-44C. El Reno, Oklahoma. At 2315, a North American TB-25D flying in instrument conditions crashed ten miles west of El Reno, Oklahoma, killing eight fliers. Passenger Sgt. John L. Carpenter received serious injuries. Killed in the crash were: Capt. Bron F. Barrett, pilot; Sub-Lt. Ghoon-Choan Chen, Chinese Air Force, co-pilot; Sub-Lt. Chia-Ho Lee, Chinese Air Force, pilot-rated navigator; Sub-Lt. Li-Geng Yang, Chinese Air Force, pilot-rated passenger; Sgt. Howard P. Loeffler, engineer; Pfc. Barnes H. Walker, passenger; Cpl. James G. Patton, passenger; Pvt. Lorin D. Miller, passenger. The airplane took off at 2046 CWT from Atlanta Army Air Field, Atlanta, Georgia, on a navigation training flight to Will Rogers Field, Oklahoma City, Oklahoma. No radio contact was made with the airplane at any time during the flight. Investigators speculated that the pilots inadvertently flew the airplane into the ground after entering instrument conditions. The ceiling was estimated to be about 200 feet agl. The airplane smashed itself to pieces as skidded along the ground, leaving a trail of fire and wreckage over 100 yards long. Civilians in the area were able to rescue Sgt. Carpenter from the flaming wreckage. He did not regain consciousness until two days after the accident.

10-2-44A. March Field, California. At 1242, a Lockheed A-29 crashed while attempting a landing at March Field, Riverside, California, killing two crewmembers and injuring eight. The airplane took off from Glendale, California, on an administrative flight to March Field. Investigators stated, "Upon entering the traffic pattern at March Field [the pilot] called the tower giving his position as downwind for Runway 30 and was advised by the tower to land on Runway 25. He extended his pattern and made his approach for Runway 25 and was cleared by the tower to land. When he made contact with the runway his aircraft bounced several times and finally went into a fast ground loop. During this ground loop the left strut, wing and fuselage were sprung, breaking open the left fuel tank and the aircraft immediately burst into flames. Several of the passengers abandoned the aircraft through the rear

turret and the hatch over the cockpit, but two of the crew went out the main door on the left side through the flames that were blowing back from the engine that was still running and were fatally burned." Engineer SSgt. Denzil A. Radabaugh and Pfc. John A. Stiles were fatally burned. Pilot 1Lt. John McGarry, radio operator SSgt. Clifford F. Burns and engineer Cpl. John J. Thursfield sustained minor injuries. Passengers injured escaping from the airplane were: 2Lt. Robert H. Hemmingsen, Cpl. Ross K. Miller, Pfc. Francisco A. Pisapia, Pfc. Merl A. Sharmo, Pfc. Raymond F. Pettek.

10-2-44B. Moultrie, Georgia. At 1615 EWT, two Curtiss P-40N airplanes collided in mid-air and crashed 23 miles northeast of Spence Field, Moultrie, Georgia, killing 2Lt. Keith R. Shull aboard P-40N #44-7053. 2Lt. James A. Kincheloe was able to safely land at Spence Field in P-40N #42-24263. Lt. Shull and Lt. Kincheloe took off from Thomasville, Georgia, on a team instrument flight to Spence Field. Lt. Shull led the two-ship flight into the Spence Field area. The two P-40s encountered a lone AT-6 and engaged it in simulated combat for several minutes. The maneuvers became violent and Lt. Shull and Lt. Kincheloe became separated. After a few minutes, Lt. Kincheloe was able to find Lt. Shull and attempted to reform on his wing. In trying to get back on the leader's wing he overshot. While he was turning back toward the leader in an attempt to reform the flight, his P-40's starboard wing collided with the port wing of Lt. Shull's airplane. Lt. Shull's airplane immediately went out of control in a diving spiral. The airplane remained in the spiral until it struck the ground at a steep angle and exploded violently into flames.

10-2-44C. Victorville, California. At 1340, a North American B-25D crashed 25 miles west of the Army Air Field at Victorville, California, killing three fliers. The airplane took off from Victorville on a routine training mission. A P-39 pilot flying in the area noticed the B-25 in a spin at an altitude of about 5,000 feet msl. The airplane continued in a spin until it struck the ground and exploded. Investigators could not determine the cause of the accident. Pilot 1Lt. George D. Rosado, WASP pilot Marie M. Michell and passenger SSgt. Gordon L. Walker were killed in the accident.

10-2-44D. Midland, Texas. At 2326, a Beech AT-11 crashed at the Army Air Field at Midland, Texas, killing the crew of four. Investigators stated, "[The pilot] was returning from a navigation training flight and when passing over [Midland Army Air Field] collided with the ground. At the time of the accident the weather was ceiling 3,000 feet, visibility unrestricted, with a thunderstorm and light rain. Witnesses in the control tower stated that this airplane approached the field from the southwest at an altitude of about 500 feet and just before reaching the center of the field it peeled off to the left and dove into the ground, exploding upon impact." It was speculated that the pilot was attempting to transition from instrument flight to visual when he was apparently blinded by an extremely bright flash of lightning, causing him to inadvertently fly the airplane into the ground. Killed in the crash were: 2Lt. Maurice C. Mace, pilot; 1Lt. Lester W. Long, bombardier; 2Lt. John M. Tompkins, bombardier; 1Lt. Joseph H. Quencas, bombardier.

10-3-44A. Chesapeake Bay, Virginia. At 0745, a Consolidated B-24J crashed into Chesapeake Bay one and a half mile off shore of New Point Comfort, Virginia, killing the crew of seven. The airplane had taken off at 0724 EWT from Langley Field, Hampton, Virginia, on a high-altitude radar navigation mission. No radio contact was made with the airplane after it took off. At 0750, Langley Field was advised that an airplane had crashed in the bay. Several witnesses observed the airplane flying normally at an altitude of about 1,000 feet. Moments later, thick smoke was seen trailing from the airplane. The airplane then went into a quarter turn to the right before diving into the bay and exploding. Killed in the crash were: 1Lt. Francis T. Bonsteel, Jr., pilot; 1Lt. David Kitko, co-pilot; 2Lt. Frank J. Carver, navigator; 2Lt. Edward R. Clauss, bombardier; 2Lt. Aaron A. Evans, navigator; Sgt. John W. Fletcher, engineer; Sgt. Charles Biederman, radio operator.

10-3-44B. Harding Field, Louisiana. At 0750, a Republic P-47D crashed 300 yards northwest of Harding Field, Baton Rouge, Louisiana, killing 1Lt. Albert B. Yearwood. The airplane was part of a two-ship flight that had completed a routine training mission and was returning to land on Runway 13 at Harding Field. The pilot of the subject airplane made a very tight pattern and he overshot when turning from the base leg to final approach. The airplane was flying at a very low airspeed with landing gear and flaps extended. The pilot tightened his turn in an attempt to line up with the runway and the airplane began mushing into the ground short of the runway. The pilot applied full throttle in an attempt to recover the airplane but he was unsuccessful. The airplane smashed into the ground violently, collapsing the landing gear and causing the engine and propeller to separate. The airplane smashed itself to pieces as it slid for 225 feet before coming to rest and bursting into flames. The pilot was killed upon impact.

10-3-44C. Morganza, Louisiana. At 0813, two Republic P-47D airplanes collided in mid-air and crashed three miles west of Morganza, Louisiana, killing 2Lt. Edward M. Tiffreau aboard P-47D #42-28191. 2Lt. Ernest R. Alexander was uninjured parachuting to safety from P-47D #42-8362. The airplane was part of a six-ship flight on a tactical formation mission. The flight leader climbed the formation to 8,500 feet and leveled off. The flight leader signaled the airplanes to assume a loose tactical formation. The formation assembled and the flight leader led it through a series of 90-degree turns. The leader turned to the left and the number-six airplane, flown by Lt. Tiffreau, fell out of position. Lt. Tiffreau attempted to regain his position

and collided with Lt. Alexander, who was flying in the number-two position in the first element. Lt. Alexander's airplane entered an extreme nose down attitude and when he was unable to regain control he bailed out. Lt. Tiffreau's airplane went out of control immediately and slammed into the ground where it exploded into flames. Lt. Alexander stated that he did not see the other airplane at any time.

10-3-44D. Rodeo, New Mexico. At 1150, a Taylorcraft L-2 crashed just after take-off from the CAA Municipal Airfield at Rodeo, New Mexico, killing F/O Carl A. Larson. The airplane had taken off on a ferry flight to Deming, New Mexico. The L-2 executed a steep climbing turn to the left just after lift-off. The airplane stalled at an altitude of 150 feet, plunging to the ground nose first. The L-2 struck the terrain on the port wing and nose section, bursting into flames upon impact and killing the pilot instantly.

10-3-44E. Santa Rosa, California. At 1200, a Lockheed P-38J crashed three miles south of the Army Air Field at Santa Rosa, California, killing F/O Donald W. Murdoch. The airplane was flying the number-three position in a four-ship training flight that had taken off from Santa Rosa Army Air Field. The flight climbed to 20,000 feet and flew formation for about an hour. The leader descended to about 10,000 feet to practice single-engine flight. The flight leader shut down his starboard engine and the rest of the formation did the same. Moments later it was discovered that the subject airplane had fallen out of formation. Several minutes later the flight leader observed thick black smoke rising from the ground and after flying to the scene discovered that it was the subject airplane. Investigation revealed that the airplane had struck the ground at a steep angle, heading about 180 degrees from the original course. It was speculated that the pilot had inadvertently shut down the port engine while attempting to manipulate the controls for the starboard engine.

10-3-44F. Shafter, California. At 1401, two Cessna UC-78B airplanes collided in mid-air and crashed two and one-half miles northeast of Shafter, California, killing four fliers. Instructor 2Lt. Lawrence K. Zeliadt and student A/C Leonard N. Mondahl were killed in the crash of UC-78B #42-71676; A/C Jeff T. Morehead and A/S Orland C. Motley were killed in the crash of UC-78B #42-71956. Investigators stated that both airplanes were on separate dual instrument training flights and were flying at an altitude of about 7,000 feet msl when the collision occurred. There were no witnesses to the actual collision. Witnesses on the ground observed a large cloud of black smoke in the sky and falling bits of wreckage. Investigators noted that both airplanes were equipped with green plexiglass in the cockpit for instrument training.

10-3-44G. Marana, Arizona. At 1520 MWT, a Vultee BT-13 crashed while attempting an emergency forced landing 11 miles SSE of the Army Air Field at

Marana, Arizona, killing WASP pilot Peggy W. Martin and seriously injuring WASP pilot Marion A. Hagen. Investigators stated, "Miss Martin's airplane was first noticed by witnesses at an altitude of approximately 6,000 feet above the ground and ten miles to the south of Avra Auxiliary Airfield, trailing heavy white smoke. Supervisory personnel at the auxiliary field tried to establish radio contact but were unsuccessful. Miss Martin headed for the field at a more or less high rate of speed and established a semi-base leg, then turned on to final approach which would give her a straight away approach to the field for about one mile. Witnesses stated that she appeared to be using approximately 30 to 40 degrees of flaps, but was gliding at an excessive speed. The airplane passed the center of the field approximately ten feet in height and still trailing dense white smoke from the exhaust stack. At this point the pilot elected to try to go around, but because the engine was not running properly she could not gain or maintain altitude. Approximately 100 yards after passing the edge of the field the pilot attempted to make a 180-degree turn back to the field, and in so doing, stalled out. The airplane struck on the left wing first, then on the engine, tearing it loose from the airplane and cartwheeling the airplane over on its back."

10-4-44A. Simsbury, Connecticut. At 0840, two Republic P-47D airplanes collided in mid-air and crashed at Simsbury, Connecticut, killing one flier. 1Lt. Junior L. Birdsong fell to his death in an unsuccessful parachute attempt from P-47D #42-22595. 2Lt. Warren F. Pierce was uninjured parachuting to safety from P-47D #42-8305. The airplanes were part of a four-ship flight that had taken off from Bradley Field, Windsor Locks, Connecticut, on a high-altitude formation-training mission. The take-off was uneventful and the flight leader climbed the formation to an altitude of about 20,000 feet. Soon after reaching altitude, the flight leader began experiencing trouble with his airplane. He led the flight back down to about 5,000 feet. As the formation approached the field, the flight leader signaled the flight to change to a right echelon formation for landing. The number-two airplane was on the leader's starboard wing and the remaining two airplanes were flying a two-ship element on the leader's port wing. Lt. Birdsong, the leader of the second element, apparently did not see the signal to change to right echelon. The flight leader signaled again and Lt. Birdsong still did not see the signal. The number-four airplane, piloted by Lt. Pierce, saw the signal and began to slide over for the right echelon. After the number-four airplane started moving over, Lt. Birdsong then began moving over to assume the echelon position. Investigation revealed that Lt. Birdsong moved to about the correct position in the formation and Lt. Pierce, flying about 50 feet above, simultaneously began to descend into the formation. Lt. Pierce then descended into Lt. Birdsong, severing the tail of Lt. Birdsong's airplane and sending it out of control. Lt. Birdsong

bailed out but apparently deployed his parachute before he was clear of the aircraft, causing him to collide with some portion of the P-47 and killing him instantly. Lt. Pierce could not hold altitude or control his airplane and he elected to parachute to safety. Investigators could not determine why Lt. Birdsong delayed his crossover to echelon.

10-4-44B. Harding Field, Louisiana. At 1555, two Republic P-47D airplanes collided in mid-air and crashed 20 miles northwest of Harding Field, Baton Rouge, Louisiana, killing 2Lt. John P. Knighton aboard P-47D #42-25347, who fell to his death in an unsuccessful parachute jump. 2Lt. B.H. Howard was able to parachute to safety from P-47D #42-8352, but he suffered serious injuries. The airplanes were part of a flight of eight P-47 airplanes on a high-altitude tactical formation-training flight. Investigators stated, "The eight aircraft joined up in two, four-ship flights after take off, the first flight of four being led by Capt. George L. Able and the second flight of four being led by Lt. John P. Knighton. The flight leader, Capt. Abel, stated that his eight-ship formation flew to Hammond and climbed to 20,000 feet. Upon reaching 20,000 feet, the Harding homing station called Capt. Abel and stated that there were a number of bombers over Harding Field and required that Capt. Abel's flight intercept the bombers and make passes. After this request from the homing station was carried out, the eight-ship formation climbed back up to approximately 20,000 feet, leveled off and the flight was put into a tactical formation by a prearranged signal from Capt. Abel. At this time, the flight was approximately 20 miles NNW of Harding Field. Capt. Abel gave radio instructions to Charley Red Flight, which was Capt. Abel's flight, to break 90 degrees. At this time, Charley Blue Flight, led by Lt. Knighton, was flying 200 to 300 feet below and approximately 400 to 500 yards behind Charley Red Flight, when the break was called. One of the wingmen in Lt. Knighton's stated that Lt. Knighton flew straight for a few seconds after the break was given over the radio, then executed a break himself to the right without giving his flight any instructions by radio or any visible signal. It was immediately after Lt. Knighton had made his break that the crash occurred, involving Lt. Howard, flying in the number-three position in Charley Red Flight and Lt. Knighton, leader of Charley Blue Flight."

10-4-44C. Casper, Wyoming. At 1847, a Consolidated B-24J crashed while attempting a take-off at Casper Army Air Field, Casper, Wyoming, killing five crewmembers and injuring three, one seriously. The airplane was scheduled to participate in a bombing mission and was cleared to take-off on Runway 21. The B-24 commenced the take-off run and began veering to the left, running off the left side of the runway about 2,500 feet from the departure end. The airplane bounced into the air twice as the pilot tried to force it into the air. Investigation revealed that the pilot had extended full flaps in an attempt to get airborne. No reduction of power was heard and no obvious attempt was made by the pilot to terminate the take-off. No violent maneuvers were observed and the engines sounded and looked to be running satisfactorily. Investigators could not determine what caused the pilot to lose control of the airplane on the take-off roll. Killed in the crash were: 2Lt. Joseph Nagel, pilot; 2Lt. Harold H. Fields, co-pilot; 2Lt. Donald M. Macsporran, bombardier; TSgt. Jack C. Davidson, engineer; Cpl. Herbert A. Gans, radio operator. Navigator F/O Paul J. Spann and gunner Pfc. John T. Simpson had sustained minor injuries and radio operator Laurin A. Harson received serious injuries.

10-4-44D. Dahlgren, Virginia. At 2120, a Cessna UC-78B flying in instrument conditions was abandoned and crashed at Dahlgren, Virginia, killing passenger Maj. Grady C. Fuller, who fell to his death in an unsuccessful parachute attempt. Pilot Capt. Lloyd B. Loggins parachuted to safety and was uninjured. The airplane had taken off from Roanoke, Virginia, on an administrative flight to Washington National Airport, Washington, D.C. Weather was reported as instrument conditions for most of the flight and at the destination airport. The aircraft arrived in the Washington, D.C. area and was advised by the control tower that there would be a delay of 60 minutes due to heavy traffic at the airport. The airplane was cleared to fly the southwest leg of the Washington, D.C. beam at 4,000 feet. A short time later the tower cleared the airplane to execute a standard instrument let-down and gave landing instructions. The pilot was unable to land on his first attempt so he aborted the landing and climbed up to 1,500 feet. The tower advised him to go around. There was not enough fuel to attempt another landing so the pilot pointed the airplane south and climbed to 6,000 feet to abandon ship. The occupants abandoned ship and the pilot was able to parachute to safety. Maj. Fuller's body was not found until 11 October 1944 when it washed ashore at Maryland Point.

10-4-44E. Gowen Field, Idaho. At 2052 MWT, a Consolidated B-24J collided with rising terrain and crashed 15 miles east of Gowen Field, Boise, Idaho, killing nine crewmembers and seriously injuring gunner Cpl. George F. Howarth, who miraculously survived. Killed in the crash were: 2Lt. Frederic S. Hulsman, pilot; F/O Francis H. Horner, co-pilot; F/O Grady L. Speir, navigator; 2Lt. James A. Lund, bombardier; Cpl. Lawrence E. Andrews, engineer; Cpl. Bernard Kandely, radio operator; Cpl. Albert K. Schehying, gunner; Cpl. William D. Clark, ball turret gunner; Cpl. Peter R. Killingsworth, tail gunner. The airplane took off from Gowen Field at 2045 on a routine night-landing practice mission. Investigators stated, "[The] pilot failed to establish a normal climb after take-off and flew out too far before attempting to turn or may have been in the process of turning when the right wing tip grazed the side of the ridge. Marks on the hillside show evidence that the right wing tip struck three separate times along the side of the ridge before the final crash and explosion. Location of the crash was at an elevation of 1,064 feet above the field elevation and at a point that was almost in a direct line with the take-off runway. The Control Tower reported that there was no evidence of trouble and the take-off appeared to normal in every respect."

10-5-44A. Valparaiso, Florida. At 0647 CWT, a North American AT-6A crashed approximately 20 miles northwest of Eglin Field, Valparaiso, Florida, killing pilot 2Lt. Mark Phillips and tow-reel operator Sgt. Joseph A. Siron. The airplane took off from Auxiliary Airfield #6 on a target-towing mission. The airplane climbed to about 1,500 feet and apparently stalled. The airplane snapped over to an inverted attitude and then smashed into the ground in an inverted dive, exploding into flames upon impact. The crash occurred about three miles north of Auxiliary Airfield #6.

10-5-44B. Sarasota, Florida. At 0900 EWT, two Curtiss P-40N airplanes collided in mid-air and crashed five miles NNW of the Army Air Field at Sarasota, Florida, killing 2Lt. Eugene D. McGuire aboard P-40N #42-105724. 1Lt. Joseph F. Wehner was uninjured parachuting to safety from P-40N #44-7835. The airplanes were part of a four-ship flight that took off at 0815 from Sarasota Army Air Field on a scheduled aerobatic training flight. Lt. McGuire was flying in the number-two position of the lead element, which was led by Lt. Wehner. The airplanes had performed various aerobatic maneuvers and were flying in a right echelon formation at about 10,000 feet msl. Lt. Wehner then peeled off and made a diving turn to the right and the flight followed in trail. At about 8,500 feet, Lt. McGuire overran the lead airplane, passing directly over him and then descending into the lead airplane from above and the right. Lt. McGuire's tail section contacted the propeller of the lead airplane, causing his port elevator, port horizontal stabilizer, vertical stabilizer, and rudder to be severed. Lt. McGuire's airplane then went out of control and entered a violent spin. Lt. Wehner's propeller separated from the airplane during the collision. Lt. Wehner determined that he would not be able to make a safe landing and he parachuted to safety moments later.

10-5-44C. Tuskegee, Alabama. At 1020, a North American AT-6C crashed six miles north of the Army Air Field at Tuskegee, Alabama, killing student A/C Raymond E. McEwen. Instructor pilot 1Lt. Joseph L. Curry received minor injuries parachuting to safety. The instructor stated that the student was on an aerobatic check ride. The instructor ordered the student to perform several barrel rolls, loops, and Immelmann turns. The instructor then ordered the student to do a half roll. As the student rolled the airplane, the instructor fell out of the rear cockpit. The instructor collided with the tail section as he fell out of the airplane, but he managed to pull the ripcord and deploy his parachute. The airplane was seen to enter a shallow spiral to the right until it struck the ground and exploded into flames. The instructor speculated that he might have struck the control stick on the way out, causing the student to believe that the instructor was taking over control of the airplane. It was later speculated that the collision of the instructor with the tail section might have damaged the flight surfaces, rendering them inoperable.

10-5-44D. Muroc, California. At 0930, a Bell P-63A smashed into the ground and exploded at Rosamond Dry Lake about 15 miles southwest of the Army Air Field at Muroc, California, killing pilot 1Lt. Robert E. Kob. Investigators stated, "Lt. Kob was number-three man in a string formation of four P-63s executing a camera-gunnery attacks on an assigned target B-24 and was in the act of completing the final pursuit curve in the rear of and at the same altitude as the target aircraft, which was at 8,500 feet. His attitude was that of a left climbing turn coordinated to approximately 40-degree bank with an estimated airspeed of approximately 250 mph at which time the number-four pursuit aircraft, which witnessed the entire procedure, noticed Lt. Kob's aircraft suddenly do a half snap roll to the left, momentarily hesitate, then almost immediately go into another violent snap roll to the left, which resulted in a power spin. 2Lt. Robert A. Windoffer, pilot of the number-four pursuit aircraft, then attempted to follow the spinning aircraft to the ground. When a speed of 400 mph indicated was noted, Lt. Windoffer stated that the spinning aircraft, which by now was describing violent gyrations, was still ahead of his and slightly gaining. Shortly after this, Lt. Kob's aircraft struck the ground on the southwest edge of Rosamond Dry Lake while in the bottom quadrant of the spin. The fact that the wreckage penetrated the relatively hard surface of the dry lake to a distance of slightly more than 15 feet indicates that contact was made at extremely high speed. Wreckage was strewn about the vicinity to an extent of 75 yards indicates an explosion immediately after contact. The aircraft struck the ground nose first at extremely high speed (estimated to be approximately 450 mph) at an angle approximately 15 degrees from the vertical and slightly inverted. The position and attitude of Lt. Kob's aircraft relative to the target B-24 aircraft indicated that [propeller turbulence] from the target aircraft was encountered, which may have initiated snap rolls and a power spin."

10-5-44E. Rockingham, North Carolina. At 1500 EWT, a Douglas A-20J crashed ten miles northwest of Rockingham, North Carolina, killing pilot 2Lt. Max D. Mecham. The airplane took off from Florence Army Air Field, Florence, South Carolina, on a routine training flight. The airplane was seen to enter a very steep left turn while flying at an altitude of 1,500 feet agl near Blewett Falls. The airplane stalled and snapped over to the right and spun to the ground, exploding into flames upon impact.

10-5-44F. Denver, Colorado. At 1430, a Curtiss P-40N crashed 25 miles ESE of Buckley Field, Denver, Colorado, killing 2Lt. Jack C. Mayer. The airplane was part of a two-ship flight that had taken off from Peterson Field, Colorado Springs, Colorado, on a camera gunnery mission. The two pilots successfully performed the mission and at an altitude of 16,000 feet msl the other P-40 made his last pass at Lt. Mayer, performing a split-S maneuver and breaking away. The other pilot stated that he observed the subject airplane roll over and attempt a split-S. The subject airplane apparently stalled out of the maneuver, falling out of control and diving into the ground in a nose-down spiral. The P-40 exploded into flames upon impact, killing the pilot instantly.

10-5-44G. Bourbon, Mississippi. At 1800, a Vultee BT-13A flying in poor weather crashed at Bourbon, Mississippi, killing pilot 2Lt. Leo McCann and passenger 2Lt. Charles E. Peery, Jr. The airplane had taken off from Greenville Army Air Field, Greenville, South Carolina, on an administrative flight to the Army Air Field at Jackson, Mississippi. The airplane encountered a powerful thunderstorm that contained violent turbulence. The thunderstorm, with a ceiling of about 3,000 feet, was confined to the local area. Visibility was only one-half mile in heavy scattered showers. The airplane was seen to spiral out of the rain shower and smash into the ground at high speed where it exploded into flames.

10-5-44H. Florence, South Carolina. At 2020, a Douglas A-20G crashed two and one-half miles south of the Army Air Field at Florence, South Carolina, killing 2Lt. Joe E. Wright. The airplane had taken off at 2017 from Florence Army Air Field on a routine night navigation flight. The airplane took off and gained about 200 feet when the starboard engine failed. The airplane veered slightly to the right as the pilot feathered the starboard propeller. Moments later the port engine failed and the airplane assumed a nose down attitude, contacting the terrain in a shallow dive and cutting through trees for about 200 feet before smashing into the ground and exploding into flames. It was speculated that the pilot inadvertently pulled the port mixture control to idle cut-off when he intended to manipulate the starboard control.

10-5-44I. Mendham, New Jersey. At 2203 EWT, a Vultee BT-13A crashed in Chester Township about four miles west of Mendham, New Jersey, killing pilot 2Lt. Peter Carr and passenger Maj. Brown H. Carpenter. The airplane took off from Pittsburgh, Pennsylvania, on an administrative flight to Newark, New Jersey. Examination of the wreckage indicated that the airplane dove into a field at a steep angle in a left-wing low attitude. The airplane burst into flames upon impact and the occupants were killed instantly. Investigators could not determine the cause of the accident.

10-5-44J. Clovis, New Mexico. At 2103, a Boeing B-29 attempting a take-off crashed one mile from the departure end of Runway 21 at Clovis Army Air Field, Clovis, New Mexico, killing three fliers and injuring nine others. The airplane made a normal take-off and the pilot retracted the landing gear. The instructor pilot "milked the flaps up too soon" and the B-29 student did not correct the flight attitude. The airplane flew into the ground in a near level attitude and slid across the ground for about 1,200 feet before smashing into a large haystack. The airplane began to break up and burst into flames before coming to rest. Bombardier/navigator 2Lt. Harry Ratmon, gunner Pfc. David R. Josephs and tail gunner Pfc. Stanley H. Kadunc were killed in the crash. B-29 instructor pilot Capt. Milton R. White and engineer SSgt. Duncan E. Stewart received minor injuries. Crewmembers receiving serious injuries were: 2Lt. Sidney Bogin, pilot; 2Lt. Dale E. Smith, co-pilot; 1Lt. Robert G. Keller, flight engineer; SSgt. Jesse K. Landkrohn, engineer; Cpl. Thomas B. Townsend, radio operator; Cpl. Arthur R. Blansett, central fire controller; Cpl. John S. Hoger, gunner.

10-5-44K. McFarland, California. At 2057, a Northrop P-61A crashed about four miles southeast of McFarland, California, killing pilot 1Lt. Albert M. Burbank and radar operator 2Lt. Fred M. Truay. The airplane took off from the Army Air Field at Visalia, California, on an air-ground gunnery mission at Soda Lake Gunnery Range, California. The airplane arrived at the range and began using the "night strafing range" and reported no trouble. The airplane left the range at 2040 and flew into the ground a short time later. Marks on the ground indicated that the airplane flew into the ground under control and with both engines producing power. Investigation revealed that visibility was good with light haze. Other pilots indicated that the haze made the horizon hard to see.

10-5-44L. Waco, Texas. At approximately 2310, a Beech AT-10BH crashed 12 miles north of Blackland Army Air Field, Waco, Texas, killing A/C Ellis H. Adair and A/C Quinn L. McKelvey. The airplane took off from Blackland Field on a routine navigation mission to Harpersville, Texas, to Wichita Falls, Texas, and return. The airplane failed to return to the field and was soon declared missing. The airplane was located on 10-7-44 at 1530 by two hunters near Tokio, Texas. Investigators could not determine what caused the cadets to fly the airplane into the ground. Weather was not considered a factor.

10-6-44A. Seymour-Johnson Field, North Carolina. At 0605, a Consolidated B-24J was abandoned and crashed six miles northeast of Seymour-Johnson Field, Goldsboro, North Carolina. Engineer Cpl. Jesse Harris was declared missing after he failed to turn up after attempting to parachute to safety. Crewmembers parachuting to safety were: 2Lt. Lee R. Patterson, pilot; 2Lt. Robert Glennon, co-pilot; 2Lt. Herbert W.

Riordan, Jr., bombardier; F/O Robert Adams, navigator; Cpl. Kenneth R. Knowles, crewmember; Cpl. Daniel W. Taylor, crewmember; Cpl. Robert D. Wood, Jr., crewmember; Cpl. Jesse J. Vaughn, crewmember; Cpl. Richard D. Hodgson, crewmember. The airplane took off at 2150 EWT (10-5-44) from Chatham Field, Savannah, Georgia, on a high-altitude night bombing mission. The pilot reported that the command radio set was inoperative but he could transmit and receive on VHF. The airplane was ordered to continue on its mission. The airplane was scheduled to fly to Jacksonville, Florida, at 20,000 feet and then fly back to the area near Chatham Field for a bombing mission. The airplane flew to what they thought was Jacksonville and returned to the bombing range only to find it obscured by clouds. The pilot radioed Chatham Field and requested that the airplane be allowed to continue on with an alternate cross-country navigation mission to Tampa, Florida, to Orlando, Florida, and return to Chatham Field. The airplane was given permission to proceed on the alternate mission. The airplane successfully navigated to Florida but the pilot became lost while attempting to find Chatham Field. The airplane was able to find the Army Air Field at Lumberton, North Carolina, by flying on radio instruments, but despite five attempts was unable to land because of heavy ground fog and a low ceiling. The pilot was advised to fly to Seymour-Johnson Field where weather conditions might be better. The pilot was able to navigate to Seymour-Johnson Field, but because of heavy ground fog and low ceilings was also unable to land, again attempting five times. The pilot stated that he had about four hours of fuel left but that he did not trust his gauges. The control tower operators told the pilot a pattern to fly so he could salvo his bombs and increase his fuel endurance. After flying around for some time, the pilot was advised to fly to Raleigh, North Carolina, which had much better visibility. The pilot cited fuel concerns and it was decided that it would be best if the crew bailed out. The pilots were then directed to an area south of the field and instructed to circle to allow the crew to bail out. The number-one engine quit just before the bail out command was given. The pilots bailed out last and the airplane circled to the left for a short period before smashing to the ground north of the bail out point. The missing airman was thought to have landed in the swollen Neuse River after parachuting, probably drowning before he could be rescued.

10-6-44B. Victorville, California. At 0735, Pvt. Richard A. Sims was killed when he walked into the rotating propeller of a Vultee BT-13A that was undergoing an engine run up on the ramp at Victorville Army Air Field, California. Pvt. Sims had just finished adjusting the idling speed on the engine, which was being run up by Pvt. Ira J. Graham.

10-6-44C. Ephrata, Washington. At 1045, a Lockheed P-38L crashed two miles northeast of the Army Air Base at Ephrata, Washington, killing pilot

2Lt. Robert W. O'Donnell. The airplane was in the number-two position of a four-ship flight that had taken off from Ephrata Army Air Base on a high altitude formation mission. The airplanes climbed to 20,000 feet msl where one airplane left the formation because of a fuel pressure problem. The remaining three airplanes climbed to 30,000 feet where they practiced routine formation flying. At 1040, Lt. O'Donnell radioed the flight leader and alerted him that he had to leave the formation because of oxygen trouble. The flight leader acknowledged and then began to take the flight down. While descending, the airspeed reached about 290 mph and buffeting was felt. The flight leader radioed the flight and alerted them to the buffeting concern. Lt. O'Donnell asked the flight leader to repeat the warning, acknowledging the second message. The flight leader leveled off momentarily and advised the flight to drop the dive flaps to facilitate the letdown. The leader then entered a "normal spiral to the left." The subject airplane was then seen to leave the formation in a slight dive. The other pilots soon lost sight of him. A column of black smoke was observed a short time later. Investigators speculated that the pilot lost consciousness due to oxygen starvation. The airplane went into a steep diving turn and smashed into the ground at high speed where it exploded into flames upon impact.

10-6-44D. Cedarville, Ohio. At 1346 EWT, a Lockheed F-5E (the photo-reconnaissance version of the P-38) crashed two and one-half miles north of Cedarville, Ohio, killing pilot F/O Wayne P. Frost. The airplane had taken off from Scott Field, Belleville, Illinois, on a ferry mission to Patterson Field, Ohio. Investigators stated, "The pilot was en route from Scott Field to Patterson Field [and he] contacted the Patterson Tower ten miles west for landing instructions. No more contact was had by the tower; although another P-38 on VHF heard the pilot, just after his first contact, say that his right engine was on fire, to clear the field, and he was coming in for [an emergency] landing. Due to heavy tower loads that day, he was blocked out, as the tower did not receive his message. About ten minutes later, a farmer phoned in that an airplane had crashed 15 miles east of Patterson Field. Crash equipment was dispatched immediately as well as a spotter plane." The subject airplane was observed flying at about 500 feet agl in a northwesterly direction with the starboard engine out, the propeller feathered. The airplane was losing altitude. The F-5E was then seen to enter a spin at an altitude of approximately 250 feet, describing about two turns of spin before smashing into the ground. The airplane exploded into flames upon impact, killing the pilot instantly.

10-6-44E. Rockport, Texas. At 1700, two North American AT-6D airplanes collided in mid-air and crashed into Copano Bay eight miles north of Rockport, Texas, killing three fliers and injuring another. Pilot 2Lt. Robert L. Landmeyer and pilot 2Lt.

Thomas M. Moore were killed in the crash of AT-6D #42-85908. Pilot 1Lt. William M. Ference was killed in the crash of AT-6D #42-85929 and pilot F/O Robert H. Hemphill received minor injuries parachuting to safety. The airplanes were part of a flight of five that had taken off from the Army Air Field at Matagorda Island, Texas, on a flight to Moore Field, Mission, Texas. Lt. Landmeyer took off fourth and Lt. Ference took off fifth. The formation was flying in a five-ship V-flight. The subject airplanes were flying on the left side of the V-formation. The flight leader began a turn to the right. The number-five ship lost altitude in the turn and in attempt to regain his position, pulled up into the number-four airplane. The two airplanes stuck together and spun into Copano Bay.

10-6-44F. Spence Field, Georgia. At 2043, a North American AT-6A crashed three-quarters of a mile northeast of Spence Field, Moultrie, Georgia, killing student A/C Howard W. Specht. The airplane took off from Spence Field on a routine training flight and crashed soon after. Investigation of the wreckage indicated that the airplane struck the ground in a diving turn to the right. The starboard wing struck a tree, forcing the airplane to veer to the right and strike another tree. The impact with the second tree, which had a diameter of about 18 inches, caused the engine to separate and the fuselage to turn 180 degrees as it careened forward for another 75 yards. The engine came to rest about 100 yards from the second tree. Investigation of the wreckage revealed that a two-inch plug in the propeller shaft had worked itself out of position prior to the crash, causing the propeller to go into high pitch due to loss of oil pressure in the propeller.

10-6-44G. Orlando, Florida. At 2233 EWT, a North American TB-25C suffering an engine failure on take-off crashed one mile north of the Army Air Base at Orlando, Florida, killing co-pilot/B-25 student Lt. Col. Steele R. Patterson and seriously injuring four others. Fliers seriously injured in the crash were: Lt. Col. Horace C. Craig, pilot; Col. Robert L. Anderson, observer; Cpl. Ivan K. Wyatt, engineer; Pvt. Clifford C. Tice, passenger. This was the pilot's first flight as a B-25 first pilot and the co-pilot had not been checked out on B-25 equipment. The pilot had been checked off as a B-25 first pilot earlier in the day. Investigation revealed that the airplane was trimmed excessively "tail heavy" by the co-pilot and during the take-off the pilot suddenly and unexpectedly found the B-25 airborne at about 80 mph. The airplane struggled into the air in a "stall take-off," climbing to about 15 feet agl. The port engine quit as the B-25 staggered to an altitude of about 125 feet agl. The pilots were unable to feather the port propeller and trim the aircraft for single-engine flight in time to avoid losing control. The airplane entered a stalling attitude and smashed into the ground where it burst into flames. Orlando Army Air Base rescue personnel were on the scene immediately. Investigators noted that low-time B-25 pilots should have about 20 hours of dual B-25 night training with a qualified B-25 instructor before being allowed to perform a solo night take-off. The engineer, who was very familiar with B-25 operations, later stated that he had cautioned the co-pilot about the excessive tail trim. Investigation revealed that the front master rod on the port engine had failed during the take-off.

10-6-44H. Ocala, Florida. At 2309 EWT, a Bell P-63A crashed five miles southeast of the AAF Bombing and Gunnery Range near Ocala, Florida, killing 1Lt. Harold R. Holland. The airplane was part of a flight of four P-63 type aircraft that took off from the Army Air Field at Kissimmee, Florida, and were participating in a secret and experimental night ground gunnery exercise. The airplanes arrived in the area of the bombing range and a flight of bombers dropped flares to illuminate a column of vehicles that were to be strafed at an altitude of 200 feet agl. The airplanes were flying at about 3,000 feet agl when they peeled off in ten-second intervals and descended to fire. The flight leader inadvertently dove too low and leveled off at about 75 feet agl. The subject airplane, flying in the number-three position, then collided with trees. The airplane cut a path through the treetops for about a quarter mile before it flew into the terrain and exploded into flames, fatally injuring the pilot. The airplane had tumbled through tall, stout trees as it went in at very high speed, smashing the airplane to bits and scattering wreckage over a large area. The pilot was found alive but unconscious several hours after the accident. He died soon after being admitted to the hospital.

10-7-44. Coushatta, Louisiana. At 1730, a Curtiss P-40N flying in poor weather crashed at Coushatta, Louisiana, killing Capt. John W. Meredith. The airplane took off from Houston Municipal Airport, Houston, Texas, on a cross-country flight to Pollock Army Air Field, Alexandria, Louisiana. The pilot flew to the north to avoid a stationary front that was parked in between Houston and the destination. The pilot could not avoid all of the weather and was next seen flying at low altitude near Coushatta. Weather was reported as instrument conditions with low ceilings, poor visibility, high winds and heavy rain. The pilot circled the area of Coushatta three times at very low altitude, apparently trying to pick up the railroad tracks between Shreveport and Alexandria. The airplane collided with a tree and then smashed into the ground where exploded into flames. Wreckage was scattered over large area.

10-8-44A. Love Field, Texas. At 1248 CWT, a Boeing B-29 went out of control and crashed five miles SSW of Love Field, Dallas, Texas, killing 15 fliers. Investigators stated, "This aircraft was on a high altitude Formation Interception and Camera Bombing, First Phase Training Mission, with B-29 #42-4432 (leader) and B-29 #42-6393 (number-two position), took off at 1025 CWT on 8 October 1944, from Pratt Army Air Field, Pratt, Kansas. The [three-ship] formation formed over Pratt, Kansas, at an altitude of 21,000 feet. The formation was loose, with approximately 200 to 300 feet between aircraft. When the formation was over Dallas, Texas, at 21,000 feet altitude, B-29 #42-6393 left the number-two position due to being low on oxygen and unable to pressurize. After B-29 #42-6393 left the formation, B-29 #42-6395 [the subject airplane] requested permission from the flight leader to change from the number-three position to the number-two position in the formation. Permission was granted, and B-29 #42-6395 dropped down and to the rear of B-29 #42-4432 and started to cross over. The subject aircraft came up within 100 feet and directly to the rear of B-29 #42-4432. At this time the [subject aircraft] started into a steep dive with a slight turn to the right. This continued for approximately 3,000 to 4,000 feet, at which time the aircraft flipped onto its back with the nose low, and started to spin. This inverted spin continued until the aircraft was approximately 3,000 feet above the ground when the nose came up, flattening the spin still more and the rate of spin decreased. The aircraft crashed, inverted, on a warehouse and machine shop of the West Dallas Field of the Texas Oil Company." Killed in the crash were: 2Lt. Earl F. Cannon, Jr., pilot; 2Lt. Lynn E. Pavitt, co-pilot; 2Lt. Walter E. Rock, navigator; 2Lt. William W. Jones, flight engineer instructor; F/O Walter A. Trymbulak, bombardier; Pfc. Daniel J. Sughrue, flight engineer student; Cpl. William W. Roberts, radio operator; Pvt. Carroll W. Smith, gunner; Pfc. Joseph M. Paniel, gunner; Pfc. Joseph M. Paetz, gunner; Pfc. Jerome M. Peters, gunner; Pfc. George S. Orr, gunner; TSgt. Steven J. Kovalich, gunnery instructor; TSgt. Herbert C. Lingafelt, gunnery instructor; Sgt. Benjamin P. Calhoun, gunnery instructor.

10-8-44B. Mountain Home, Idaho. At 1345 MWT, a Consolidated B-24J suffering an in-flight fire was abandoned and crashed 27 miles southwest of the Army Air Field at Mountain Home, Idaho, killing tail gunner Cpl. Orval B. Determan, who fell to his death in an unsuccessful parachute jump. Six fliers suffered minor injuries and three others were uninjured parachuting to safety. The airplane took off at 0806 MWT from Mountain Home Army Air Field and was part of six-ship flight on a formation-training mission. The flight flew at about 4,500 feet indicated altitude for about 45 minutes before climbing to 20,000 feet for some air-to-air gunnery training. After climbing to altitude, the number-two engine began to run rough and the airplane left the formation. The pilots shut down the engine, but could not feather the propeller. The engine soon caught on fire. The pilots were unable to extinguish the fire with the engine fire extinguishers. The pilot ordered the crew to bail out. Co-pilot 2Lt. Sherwin B. Swartz, bombardier 2Lt. William J. Hill, Jr. and engineer Cpl. Robert L. Dond were uninjured parachuting to safety. Crewmembers suffering minor injuries parachuting to safety were: 2Lt. Johnson S. Miller, pilot; 2Lt. John M. McCormick, navigator; Cpl. Robert L. Julyan, radio operator; Cpl. Worth H. Barber, gunner; Cpl. Robert H. Wahl, gunner; Cpl. Ray P. Horton, ball turret gunner.

10-8-44C. Barksdale Field, Louisiana. At 1615, a Martin B-26B crashed while taking off at Barksdale Field, Shreveport, Louisiana, killing two fliers and seriously injuring pilot 1Lt. Donald R. Luce. Co-pilot 2Lt. Allen J. Capko and engineer Archie E. Crabtree were killed. The airplane took off from Barksdale Field on a scheduled test flight. The take-off appeared to be normal. The airplane became airborne and the landing gear was retracted. Moments later, the starboard engine was seen and heard to be cutting out. The pilot was able to maintain an altitude of 50 feet for about three-quarters of a mile before the B-26 crashed into the ground and burst into flames. Investigation revealed that the starboard engine failed because of broken spark plugs in cylinder #16, in the rear bank of cylinders. The rear bank spark plugs were found to have been heat damaged by the engine running on a lean fuel mixture caused by some problem in the carburetor.

10-9-44A. Canalou, Missouri. At 1055, a Fairchild PT-19A crashed five miles SSW of Canalou, Missouri, killing A/C William E. Kennedy. The airplane took off from Sikeston, Missouri, on a routine training flight. The student evidently stalled the airplane at an altitude too low to allow recovery. The airplane dove into the ground at a steep angle and with considerable force and was completely demolished.

10-9-44B. Harlingen, Texas. At 1720, a heavily damaged Consolidated B-24H attempting an emergency crash landing at Laguna Madre Army Air Field, Harlingen, Texas, collided with an AAF Dodge WC-54 ambulance and an AAF crash truck, killing ambulance driver Cpl. Walter A. Busch and seriously injuring two unnamed enlisted men. Two unnamed civilian firemen aboard the crash truck sustained minor injuries. The B-24, which was on a gunnery training mission, and a Martin TB-26B, which was on a target-towing mission, had collided in mid-air 25 miles southeast of Harlingen Army Air Field, causing the B-24 crew to parachute to safety and the B-24 pilots to make an emergency forced landing at Laguna Madre Army Air Field. The B-24 and the TB-26B crews had finished the gunnery exercise over the Gulf of Mexico and the two airplanes flew a course of 240 degrees together toward the shoreline. The B-26 was flying slightly behind the starboard wing of the B-24. After the airplanes reached the shoreline, the B-24 began a descending turn to the left. The B-26 had fallen behind and entered a slight dive in order to overtake and zoom under the B-24 and climb ahead in a "farewell pass." The B-26 pilot misjudged the clearance between the two aircraft as he attempted to climb ahead and collided with the B-24, which was flying at an indicated altitude of 3,500 feet. The leading edge of the

B-26's starboard wing struck the B-24's port trailing edge and aileron from underneath. The two wings were momentarily stuck together and then the B-26 rolled over the top of the B-24, damaging the B-24's port wing and aileron. The aileron was jammed and the B-24 fell out of control. The B-24 lost about 1,500 feet before the pilots were able to regain control. The pilot signaled the crew to bail out. The B-24 pilots were able to maintain control by using the rudder and throttles. By the time the crew had bailed out, the altitude was too low to allow the two pilots and the gunnery instructor to bail out. The pilots radioed Laguna Madre Field (a Sub-Base of Harlingen Army Air Field) that they were coming in for a wheels-up landing. They were able to make the field and just before touching down the co-pilot noticed the crash truck and ambulance driving into their path about 300 feet from the end of the runway. The B-24 touched down on Runway 35 and soon overtook the two vehicles, smashing through both of them and sending them out of control. The B-24 slid to stop and the crew was able to escape without injury. The crash truck driver did not see the B-24 because of its low approach. The crash truck and ambulance had been too late in getting off and had failed to reach their alert stations. The tower then ordered the ambulance and crash truck to proceed to the area between Runway 35 and Runway 5 and to stand by for the emergency landing. The ambulance was following the crash truck across the runway when the accident occurred. After the mid-air collision the B-26 fell out of control. The B-26 pilot was able to regain control of his airplane at an altitude of about 2,000 feet and was able to make a normal landing on Runway 5 at Laguna Madre Army Air Field. Crewmembers uninjured aboard the B-26 were: 2Lt. Ernest J. Kinkoph, pilot; 2Lt. Leonard M. Willan, co-pilot; SSgt. Joseph J. Mauro, engineer; Pfc. Louis M. Lund, tow-reel operator. Pilot F/O Richard N. Wadman, co-pilot 2Lt. L.J. Willis Voris and gunnery instructor SSgt. Paul L. Meier were uninjured in the crash landing of the B-24H. Fliers uninjured parachuting to safety from the B-24H were: Pfc. William H. Collins, engineer; Sgt. Dwight R. Harlow, gunnery instructor; Sgt. John J. Hardy, student gunner; Pvt. Charles D. Harod, student gunner; Pvt. John A. Waynick, student gunner; Pvt. William M. Kemp, student gunner; Pvt. Leo Rezzonico, student gunner; Pvt. Charles E. Richardson, student gunner; Pvt. Coy E. Ringwald, student gunner; Pvt. Walter Kennedy, student gunner.

10-10-44A. San Pablo Bay, California. At 0906, a Beech UC-45F crashed into San Pablo Bay one-half mile northeast of McNears Point, San Rafael, California, killing two crewmembers and five passengers. Killed in the crash were: Capt. Kendall W. Shepard, pilot; Sgt. James R. Birchett, engineer; Col. Stuart G. Hall, passenger; Maj. William J. McCurry, passenger; Capt. Nicholas J. Gillisdorf, passenger; Capt. Gordon R. Tanner, passenger; Capt. Donald R. McDonald, passenger. The airplane had taken off from Hamilton Field, San Rafael, on a routine administrative cross-country flight to Bicycle Lake, California, via Fresno, California. The ceiling at Hamilton Field was about 600 feet at the time of take-off. The airplane was cleared on instruments to Fresno and contact for the flight to Bicycle Lake. The airplane took off and was cleared to climb through the overcast and instructed to contact Hamilton Field after leveling off 500 feet above the cloud tops. Moments later, U.S. Navy personnel operating a vessel in San Pablo Bay observed the subject airplane descending out of the clouds with apparent engine trouble. The engines were heard to be cutting out intermittently. The airplane glided to low altitude and the engines suddenly came back to life. The airplane climbed slightly and then entered a vertically banked turn to the left. The airplane leveled out, stalled and then went into a vertical dive, continuing in this attitude until it struck the water. Investigators could not determine the cause of the apparent engine trouble.

10-10-44B. DeFuniak Springs, Florida. At 1145 CWT, a General Motors Corporation P-75A crashed seven miles NNW of DeFuniak Springs, Florida, killing Maj. Harry R. Bolster. The airplane had taken off from Eglin Field, Valparaiso, Florida, on a "Speed and Comparative Acceleration Test" flight. The airplane was flying at high altitude when the propeller oil tank failed and the propeller lost all of its oil, causing its failure. The pilot jettisoned the cockpit canopy because it had become coated with propeller oil and he could not see. The pilot began a descending turn and he attempted an emergency forced landing, making an approach on a clear area. As the pilot made the approach, he encountered a slight rise in the terrain. The pilot raised the nose and the port wing fell. The port wing contacted the ground and the airplane cartwheeled into an area of thick tree stumps concealed by high grass. The P-75 exploded into flames, scattering wreckage over a large area.

10-10-44C. Yturria, Texas. At 1420, a Bell P-39N suffering an engine failure crashed while attempting an emergency forced landing three miles north of Yturria, Texas, killing pilot 2Lt. Harry C. Geyer. The airplane took off at 1346 from Brownsville Army Air Field, Brownsville, Texas, on a ferry flight to Kelly Field, San Antonio, Texas. The airplane was next seen circling to the left at an altitude of about 300 feet agl near Yturria, losing altitude. The pilot was apparently trying to make an approach for an emergency forced landing on a clear area and leveled off too high. The airplane stalled, fell off to the left and spun to the ground, bursting into flames upon impact. The airplane struck the ground on the port wing in a partially inverted attitude.

10-10-44D. Matagorda Peninsula, Texas. At 1701, a Lockheed-Vega RB-34A crashed nine miles east of the Army Air Field at Matagorda Peninsula, Texas,

killing pilot 2Lt. Kenneth E. Franks and engineer Pvt. Garland C. Harvey. The airplane, which was on target-towing mission, was climbing to reach 6,000 feet for a gunnery exercise. While flying at an altitude of 3,500 feet, the airplane was seen to enter a roll to the left and then dive into the peninsula about a mile northeast of Green's Bayou, exploding violently into flames upon impact. The occupants made no apparent effort to leave the airplane.

10-10-44E. Malden, Missouri. At 2109, two Waco CG-4A gliders collided in mid-air 100 yards west of the Army Air Field at Malden, Missouri, killing four fliers. Glider pilot 2Lt. Donn F. Harrington and co-pilot 2Lt. Kenneth C. Kreutner were killed in the crash of CG-4A #43-40842; Glider pilot 2Lt. Wilbur A. Kramer and co-pilot 2Lt. Robert H. Culham were killed in the crash of CG-4A #43-42516. The two glider co-pilots were both rated transport pilots. The two gliders were towed aloft by a single Douglas C-47 in a double tow operation; one glider was on a short tow-rope and the other was on long tow-rope. The gliders were to be released at an altitude of 500 feet agl and were to execute a 270-degree approach for a landing at Malden Army Air Field. The glider pilot using the short tow-rope had released his towrope late and then climbed slightly after release, disrupting the timing for the left turn he was to make and failing to clear the air space for the glider releasing from the long tow-rope. The C-47 had pulled the glider on the long tow-rope to a position abreast of the previously released glider at the time of the second release. The glider on the long tow-rope released and then made his turn to the left, colliding with the other glider. Both gliders went out of control and smashed to the ground.

10-11-44A. Fort Worth, Texas. At 1130, pilot Capt. Edward M. Woddrup was killed when he fell into the rotating number-two propeller of a Boeing B-17F that he was exiting following a small on board fire. The airplane's engines were being started in preparation for a return cross-country flight to its station at Fairmont Army Air Field, Geneva, Nebraska. The pilot was standing on the flight deck between the pilot's seats while the co-pilot and the engineer started the engines. A short time later, a fuse box shield cover came loose and fell off, contacting the fuses and causing them to arc. A shower of sparks flew and ignited some surrounding insulation. A crewmember alerted the crew to the fire. The flight crew cut the switches and all personnel abandoned the airplane. The pilot exited the airplane through the forward escape hatch and apparently fell backward into the arc of the rotating propeller. He was struck on the head and died a short time later. Co-pilot 1Lt. Perry J. Hickerson and engineer SSgt. Norman A. Salo were uninjured. The fire was extinguished immediately by a fireguard.

10-11-44B. Sioux City, Iowa. At 1231, two Boeing B-17G airplanes collided on the runway while attempting a landing at Sioux City Army Air Base, Sioux

City, Iowa, killing two fliers and injuring 14 others, three of them seriously. The B-17s were part of a six-ship flight that was returning to base after successfully completing a high-altitude formation flight. Investigators stated, "The two airplanes were participating in a formation landing [on Runway 31]. [B-17G #42-107158], following [B-17G #42-97876] in for the landing, encountered a considerable amount of [propeller turbulence] on his approach. In an effort to more effectively cope with the [propeller turbulence], the pilot kept the airspeed higher than normal, which caused him to float much further down the runway than he had probably anticipated. This floating was reducing the interval between the two airplanes and it appeared that if B-17G #876 continued landing, he might overrun B-17G #158, which had already landed just ahead; therefore, the control tower advised B-17G #876 to go around when he was approximately one-third of the way down the runway still airborne. The pilot applied power and called for the flaps to be retracted from the 'full down' to the 'one-third' position. The co-pilot misunderstood and retracted the flaps completely, which together with the low airspeed made it impossible to climb sufficiently to clear the airplane ahead [B-17G #158]. The pilot then tried to skid the airplane to the right, but was unable to clear; then his left wing struck the vertical stabilizer of [B-17G #158]." Both airplanes were smashed to pieces as they slid to a halt on Runway 31 in between Runway 35 and Runway 4. The two B-17s did not burn but were destroyed utterly. Radio operator Cpl. Philip J. Giordano was killed and gunner Pfc. Amet Chelo was seriously injured aboard B-17G #42-107158. Fliers injured in the crash of B-17 #42-107158 were: 2Lt. Gordon C. Groh, pilot; F/O Jay W. Hess, co-pilot; 2Lt. Fred Korzon, bombardier; Sgt. Richard J. Haldimann, engineer; Cpl. William C. Morris, gunner; Pfc. Thomas D. Pyburn, gunner. Radio operator Cpl. Ronald G. Fields was killed and gunners Pfc. John P. Howlett and Pfc. Joseph P. Kirwan were seriously injured aboard B-17G #42-97876. Fliers injured in the crash of B-17G #42-97876 were: 2Lt. Don C. LaMoine, pilot; F/O Carl J. Eisler, co-pilot; F/O William J. Karl, bombardier; Cpl. Robert N. Waples, Jr., engineer; Cpl. James C. Jones, gunner.

10-11-44C. Barksdale Field, Louisiana. At 2008, a Martin TB-26B crashed while attempting an emergency landing at Barksdale Field, Shreveport, Louisiana, killing four fliers. Investigators stated, "TB-26B #42-95649 ... took off from Barksdale Field at approximately 2000 CWT ... on a local night transition flight. At 2002 the subject pilot called in for an emergency landing and was given permission to land on the west runway. The pilot stated that an engine was cutting out but that he would follow the regular traffic pattern. The airplane was seen on the base leg making a short pattern. He made a turn on the final approach and appeared to be extremely low as he straightened

out on the final approach. One and a half miles from the south end of the runway, one wing appeared to dip and immediately an explosion was seen from the resultant crash. The lowering of the wheels too soon, the high wind, and losing too much altitude on his turn into his final approach, were the factors that caused the subject pilot to undershoot the runway." Killed in the crash were: 1Lt. Charles E. Roe, pilot; 1Lt. William M. Chase, co-pilot; Capt. Wilber Wendling, pilot-rated passenger; Cpl. John C. Slone, engineer/gunner.

10-11-44D. Moore Field, Texas. At 2250, a North American AT-6D crashed five miles north of Moore Field, Mission, Texas, killing A/C Charles M. Lasley, Jr. The student took off from Moore Field on a routine navigation flight to Rio Grande City, Texas, to Mercedes, Texas, to Brownsville, Texas, to Harlingen, Texas, and return. The student had successfully completed the flight and was returning to the field. The student was flying at 4,000 feet when he radioed the field and requested landing instructions. He acknowledged the landing instructions and lost control of the airplane moments later. Investigation revealed that the airplane smashed into the ground at a steep angle, exploding into flames upon impact and scattering wreckage over a wide area.

10-12-44A. La Junta, Colorado. At 0130, a North American TB-25D crashed eight miles northeast of the Army Air Field at La Junta, Colorado, killing pilot Capt. William J. Sites and co-pilot 1Lt. Bernard W. Barkley. The airplane took off at 2045 (10-11-44) on a night navigation training flight. The airplane returned from the mission and radioed the La Junta control tower at 0047 to close their flight plan. The pilots advised the tower that they would fly in the local area until about 0130 when they would return to land. The airplane crashed before the pilots could enter the traffic pattern. Investigation revealed that the airplane had dove into the ground at high speed at a 30-degree angle, exploding into flames upon impact. Investigators could not determine what caused the pilots to fly into the ground.

10-12-44B. Chamberlin, Louisiana. At 0800, a Republic P-47D crashed one mile west of Chamberlin, Louisiana, killing 2Lt. Marion C. Farrar, Jr. The airplane was flying in the number-two position of a four-ship flight that had taken off from Harding Field, Baton Rouge, Louisiana, on a tactical formation flight. The flight climbed to an indicated altitude of 32,000 feet. The number-four ship dropped out because of oxygen problems. The flight executed several turns and others maneuvers before the flight leader led the formation down to about 26,000 feet. The flight leader then executed a "wing over and barrel roll." The flight followed and the number-two airplane never recovered and dove into the ground at high speed. Investigators speculated that the pilot could not regain control of the airplane because of compressibility.

10-12-44C. Buckley Field, Colorado. At 0930, a Curtiss P-40N dove into the ground at the AAF Bombing Range 20 miles southeast of Buckley Field, Denver, Colorado, killing pilot 2Lt. Oran D. McGrew. The airplane was part of a six-ship flight that had taken off from Peterson Field, Colorado Springs, Colorado, on a dive-bombing mission. The subject airplane did not rejoin the formation after the second bombing pass. Investigation revealed that the airplane had dove into the ground at a steep angle and exploded into flames upon impact. Investigators could not determine why the pilot failed to pull out of the dive after making his pass.

10-12-44D. Norfolk, Nebraska. At 1215 CWT, a Republic P-47D suffering an apparent engine failure crashed while attempting an emergency forced landing five miles southeast of the AAF Gunnery Range near Norfolk, Nebraska, killing pilot 2Lt. John L. Brace. The airplane had taken off from Bruning Army Air Field, Bruning, Nebraska, on a ground gunnery mission. The subject airplane began suffering engine trouble and the pilot attempted to get to the airfield at Norfolk for an emergency landing. The engine quit and the pilot was forced to make an emergency forced landing. The pilot chose a relatively flat field to shoot his emergency forced landing but allowed the airspeed to diminish, causing the airplane to stall and fall off on the starboard wing. The airplane smashed into the ground and burst into flames, killing the pilot instantly.

10-12-44E. Santa Rosa, California. At 1048, a Lockheed P-38H crashed on Runway 18 at the Army Air Field at Santa Rosa, California, killing 2Lt. Clark E. Zimmerman, Jr. The airplane was part of a four-ship flight that was taking off on an elementary formation flight. The subject airplane was taking off fourth. Investigators stated, "From the time he got into the air the right wing started to drop and continued down gradually until a near vertical bank was reached. The nose then dropped and the airplane hit a slight [rise in the terrain] approximately 100 yards to the right of the runway with its right wing and propeller. The airplane cartwheeled and as it came apart it caught fire and burned."

10-13-44A. Hernando, Mississippi. At 1035, a Boeing B-17G and a Bell P-63A collided in mid-air and crashed ten miles southwest of Hernando, Mississippi, killing three B-17 crewmembers and injuring another. Seven B-17 crewmembers were uninjured parachuting to safety. Flying debris damaged another B-17 flying in the formation. P-63A pilot Capt. Frank L. Gaunt received serious injuries but was able to parachute to safety. The B-17s were part of a wing formation (usually comprised of 54 aircraft) that had been intercepted by a flight of P-63 fighters. Investigators stated, "Capt. Gaunt ... was participating in a pursuit interception of fifty-two B-17 airplanes flying a wing formation at 16,000 feet. At the time of the mid-air collision the pursuit airplanes were concentrating their attacks on the lead group of the formation, which was flying above and to the left of the low group. Capt. Gaunt proceeded to make his second attack on the lead group of the formation, and upon diving below the entire formation, pulled up and collided with the nose section of B-17G #44-6043 with the right wing, causing severe damage to both airplanes. All but three crewmembers of the two airplanes were able to parachute to safety. Three crewmembers that were in the nose compartment of the B-17G airplane received fatal injuries in the collision. B-17G airplane #42-102408, flying the lead position in the second element of the low group received major damage from flying debris." Navigator Capt. Vincent Brodeur, navigator 2Lt. Robert W. Swanson and bombardier F/O Nathan Jones were killed in the collision. Crewmembers uninjured parachuting to safety were: 2Lt. Joseph A. McCoy, pilot/B-17 student; 2Lt. Gayle E. Jory, co-pilot/B-17 student; Capt. Reuben W. Neie, B-17 instructor; Cpl. Lawrence E. Goodale, engineer; Cpl. Robert F. Ennis, radio operator; Cpl. George A. DeLorie, gunner; Cpl. Walter W. Whitaker, gunner. Gunner Cpl. Seymour K. Buckner received minor injuries. B-17 instructor Capt. Arthur L. Nelson, pilot/B-17 student 2Lt. John B. Morat and co-pilot/B-17 student 2Lt. Eugene I. Agnew were uninjured and were able to safely land B-17G #42-102408 at the Army Air Base at Dyersburg, Tennessee.

10-13-44B. Naples, Florida. At 1204 EWT, a Bell P-39Q crashed while attempting a landing at the Army Air Field at Naples, Florida, killing pilot 2Lt. David V. Jacobson. Investigators noted that the airplane entered the traffic pattern at 1,200 feet instead of the required 2,000 feet pattern. The airplane stalled and smashed to the ground near the field, bursting into flames and killing the pilot instantly. The airplane was part of a two-ship flight that was returning from a routine camera-gunnery mission.

10-13-44C. Gulf of Mexico, Florida. At 1345 EWT, a Curtiss P-40N crashed into the Gulf of Mexico five miles southwest of St. Marks, Florida, killing 2Lt. Henry W. Robison. The airplane was part of a flight that had taken off from Dale Mabry Field, Tallahassee, Florida, on a ground gunnery mission. The formation was flying at about 3,000 feet when the engine of the subject airplane was observed to begin smoking. Moments later the engine erupted into flames and the pilot bailed out. The pilot was either incapacitated by smoke and flames while bailing out or was struck by some part of the airplane as he exited. The pilot parachuted into the sea, but he apparently never released his parachute upon entering the sea and never activated his dye marker. His body was never recovered.

10-13-44D. Cochran Field, Georgia. At 1600 EWT, a North American AT-6D suffered a catastrophic structural failure and crashed 20 miles northwest of Cochran Field, Macon, Georgia, killing student A/C

John W. Taylor. Flying student Cpl. Kingdon M. Stover received minor injuries parachuting to safety. The airplane had taken off from Cochran Field on an instrument team ride. Cpl. Stover was flying in the front cockpit and A/C Taylor was flying in the rear cockpit as the safety pilot. After flying on instruments at 10,000 feet indicated altitude for about 50 minutes, A/C Taylor and Cpl. Stover agreed to go "cloud hopping" on the way back to Cochran Field. Cadet Stover dove the airplane at a cloud, reaching an airspeed of 200 mph before pulling out and flying over the top of the cloud. Cpl. Stover climbed the airplane back up to 10,000 feet and then A/C Taylor took over the controls and dove at a cloud. The airplane had built up considerable speed as the cadet attempted to pull out at approximately 5,000 feet. The surviving cadet stated that the windscreen tore away during the pull up and the airplane began disintegrating. The surviving cadet stated that he could not regain control of the airplane so he bailed out. Investigation revealed that both horizontal stabilizers and elevators had failed and separated from the aircraft. The airplane went out of control and smashed to the ground. A/C Taylor did not parachute to safety and went in with the airplane.

10-13-44E. Camp Lee, Virginia. At an unknown time between 2000 and 2030 EWT, a North American AT-6D flying in instrument conditions collided with terrain two miles east of Camp Lee, Virginia, killing pilot 2Lt. Stewart M. Tweed and passenger Sgt. Max Cohen. The AT-6 had taken off from Columbia, South Carolina, on an instrument flight to Langley Field, Hampton, Virginia. Over Raleigh, Virginia, the pilot requested that the flight plan be changed to a visual clearance. Raleigh Radio approved the new flight plan. Near Camp Lee the pilot attempted to let down out of the overcast and collided with terrain at cruise speed, crashing into a heavily wooded area and bursting into flames. Both occupants were killed instantly. Investigation revealed that the overcast extended all the way down to the treetops at the accident scene at the time of the accident.

10-13-44F. McNary, Arizona. At 2030 MWT, a Vultee BT-13B collided with rising terrain 20 miles ESE of McNary, Arizona, killing pilot 2Lt. Kenneth J. Ray and injuring passenger SSgt. Edgar M. Rutter. The airplane had taken off at 1840 from Albuquerque, New Mexico, on a navigation flight back to its home station at Marana Army Air Field, Marana, Arizona. En route to Marana the pilot encountered scattered thunderstorms and attempted to go around them. The airplane collided with trees and then crashed into the side of a mountain while flying in instrument conditions. The airplane broke up and smashed to a halt. The passenger miraculously survived. The pilot was apparently attempting to let down to visual conditions when the accident occurred.

10-13-44G. Vanderbilt, Texas. At 2245, a North American AT-6C crashed six miles northwest of Vanderbilt, Texas, killing 2Lt. Luther G. Kent, Jr. The airplane took off at about 2200 from Foster Field, Victoria, Texas, on a weather checking flight. Witnesses on the ground reported that the airplane performed two complete loops before entering what appeared to be an almost vertical climb. The airplane apparently stalled at the top of the maneuver and fell off on a wing, entering a steep dive. The airplane continued in this dive until it smashed into the ground and exploded. Investigation revealed that the landing lights were illuminated when the airplane struck the ground at a steep angle with considerable force.

10-14-44A. Gulf of Mexico, Florida. At 1600 EWT, a Curtiss P-40N crashed into the Gulf of Mexico two miles southeast of the AAF Bombing Range at Alligator Point, Florida, killing pilot 2Lt. Thomas Liddle. The airplane was part of a five-ship flight that took off from Perry Army Air Field on a skip bombing mission. The formation made three passes and the leader climbed to 1,000 feet following the third pass. The subject airplane did not rejoin the formation and failed to return to base. A search effort revealed that the airplane had crashed into the gulf. Investigation of the wreckage revealed that the airplane struck the water in a steep bank to the left. A pair of pliers were found in the wreckage, but investigators stated that the pliers were not likely to have fouled the controls in such a manner to cause the airplane to enter the steep left bank from which it crashed. The pilot's body was never recovered.

10-14-44B. Pocatello, Idaho. At 1539 MWT, a Republic P-47D crashed five miles north of the Army Air Field at Pocatello, Idaho, killing pilot 2Lt. Raymond V. Frye. The airplane was flying in the number-two position of a four-ship flight that had taken off from Pocatello Army Air Field on an operational ceiling flight. After being airborne for about ten minutes, the subject pilot reported to the flight leader that he was experiencing engine trouble. Other pilots radioed the pilot that his airplane was on fire. The pilot turned back toward the field but he could not make it. The pilot fell to his death when he apparently bailed out at an altitude too low to allow his parachute to deploy. His body was found about 100 yards from the wreckage. The airplane smashed into a bog and exploded. Wreckage was embedded over 20 feet below the surface. Efforts to recover the engine and propeller were abandoned. The number-three aircraft, which was circling the crashed airplane, stalled and crashed into the ground nearby, killing pilot 2Lt. Troy W. Stewart. See 10-14-44C.

10-14-44C. Pocatello, Idaho. At 1539 MWT, a Republic P-47C crashed five miles north of the Army Air Field at Pocatello, Idaho, killing pilot 2Lt. Troy W. Stewart. The airplane was flying in the number-three position of a four-ship flight that had taken off from Pocatello Army Air Field on an operational ceiling flight. After being airborne for about ten minutes, the pilot of the number-two airplane reported to the flight leader that he was experiencing engine trouble. Other pilots observed the airplane to be on fire. Lt. Stewart began circling the airplane that was in trouble until it crashed, killing pilot 2Lt. Raymond V. Frye. Lt. Stewart's airplane lost airspeed while circling and stalled, smashing to the ground seconds later. The airplane exploded into flames upon impact. Investigators speculated that the pilot was so enthralled with the impending crash of Lt. Frye's airplane that he neglected his own airspeed until the airplane entered a stall. See 10-14-44B.

10-14-44D. Tonopah, Nevada. At 1630 PWT, a Consolidated B-24D crashed five miles northeast of the Army Air Field at Tonopah, Nevada, killing the crew of eight. The airplane took off from Tonopah on a routine training mission. The airplane was last seen flying at low altitude in normal flight. The airplane exhibited no signs of trouble. Moments later, the port wing was seen to drop until the airplane entered a near vertical bank to the left. The nose dropped and the airplane smashed into the ground, exploding into flames upon impact. Investigators could not determine the cause of the crash. Killed in the crash were: 1Lt. George R. Gilpin, pilot; 2Lt. Joseph R. Hoye, co-pilot; 2Lt. Jack McChesney, navigator; 2Lt. Edward R. Schindler, bombardier; Cpl. Fred Rains, engineer; Cpl. Junior F. Ott, gunner; Pfc. Lynn R. Steele, gunner; Cpl. Harry A. Morse, gunner.

10-14-44E. Newton, New Jersey. At 2128 EWT, a Cessna UC-78B crashed while attempting an off-field landing three miles east of Newton, New Jersey, killing passenger 2Lt. John Weissheimer. Pilot 1Lt. Daniel N. Walls was seriously injured and passenger Capt. Lambert Agin received minor injuries in the crash. The airplane took off from Langley Field, Hampton, Virginia, on an administrative flight to Mitchel Field, Long Island, New York. The pilot became lost and was attempting a landing in a field that was illuminated by automobile headlights. As the pilot circled the field at very low altitude in an attempt to land, the airplane struck trees and went out of control. The airplane crashed to the ground in an inverted attitude and erupted into flames. The survivors were able to escape the wreckage immediately after the crash. The flames became too intense for the survivors to rescue the passenger that was trapped in the wreckage.

10-15-44A. Alamogordo, New Mexico. At 0600 MWT, a Boeing B-29 crashed while attempting a take-off from the Army Air Base at Alamogordo, New Mexico, killing 12 fliers and injuring two others. Killed in the crash were: Capt. Earl L. Hammond, pilot; Capt. Herbert C. Davis, co-pilot; 2Lt. William E. Camp, co-pilot; 1Lt. John W. Wilczewski, bombardier; 2Lt. William L. Wagner, flight engineer instructor; 2Lt. Salvadore J. Martinez, student flight engineer; 1Lt. Jack W. Griffith, student flight engineer; Sgt. Theodore T. Powell, radio operator; TSgt. Franklin J. Harrison, student engineer; Pvt. Robert E. Stephens, student radio

operator; Pvt. Allen M. Zart, central fire control; Pfc. Warren G. Weeg, gunner. Gunners Pfc. Gordon R. Mountford and Cpl. R.O. Dinsmore were injured. The airplane was taking off on an instrument training flight. The B-29 made a normal take-off, the landing gear and flaps were retracted and the airplane climbed to about 50 feet agl. The airplane then flew back into the ground about one-half mile from the end of the runway, breaking up and bursting into flames as it skidded across a highway and came to rest about 300 yards from the point of impact.

10-15-44B. Lake City, South Carolina. At 1100, two Douglas A-20J airplanes collided in midair and crashed ten miles southeast of Lake City, South Carolina, killing six fliers. Killed in the crash of A-20J #43-21561 were: 1Lt. Joseph J. Brady, pilot; Cpl. William J. Collier, gunner; Sgt. William L. Mann, gunnery instructor. Killed in the crash of A-20J #43-22089 were: 2Lt. Robert M. Nuckolls, pilot; Cpl. Richard M. Lavynder, gunner; SSgt. William E. Rentz, gunnery instructor. Investigators stated, "Flight of three A-20 type aircraft took off [from Florence Army Air Field, South Carolina] at approximately 1000 EWT on a camera-gunnery mission. On this mission, a P-40 makes passes at the formation, and the camera gunner shot at the fighter airplane. The formation flew to the north for 15 or 20 minutes, then turned and flew south for approximately 30 minutes, thus completing the gunnery phase. Upon completion of the gunnery portion, Lt. Brady, in the lead ship, called Lt. Nuckolls, number-three man in the formation, and said his radio compass was out. Lt. Nuckolls called back and said his compass was working, and did he want him to take the lead position? Lt. Brady replied OK. No more radio conversation followed. At that time the formation was flying very loose, having approximately 50 to 100 feet between wing tips and flying back several ship lengths. After Lt. Brady had replied OK on the radio, Lt Nuckolls' airplane closed in rapidly. Lt. Nuckolls flew into the lead ship, his wing striking the tail of the lead ship. Just before the impact or just after, Lt. Nuckolls (the number-three man) was seen to bank sharply to the left, apparently realizing too late that a collision was imminent. Upon impact, both ships went out of control, one diving straight in and one spinning. No parachutes were used and both ships were a total wreck."

10-15-44C. Monroe, North Carolina. At 1210, a Douglas A-20J crashed 12 miles south of Monroe, North Carolina, killing four crewmembers. Killed in the crash were: 2Lt. David W. Edmondson, pilot; F/O William H. Hackney, bombardier/navigator; Cpl. Sidney Lampert, gunner; Cpl. Lawrence J. Cuff, gunner. The airplane was one of three that had taken off at 1055 EWT from Morris Field, Charleston, North Carolina, for a bombing mission at Sand Hill Bombing Range. Investigators stated, "Upon arriving at the range, the pilots found that the range was closed and [returned] to Morris Field. On the return flight to Morris Field,

Lt. Edmondson pulled up in formation alongside another A-20 flown by 2Lt. Warren H. Camp, who was also returning from the bombing range. The lead aircraft, not desiring to participate in the formation, peeled off to the left, and made three turns — one to the right, and another to the left — losing altitude to about 3,500 feet. During this last turn, [Lt. Camp] glanced back and noticed Lt. Edmondson attempting to rejoin by cutting inside the lead aircraft, using a steep left bank. At this point, the lead pilot [Lt. Camp] observed [Lt. Edmondson] in a flat spin to the right [describing two or three turns of spin]. The aircraft apparently partially recovered from the spin when a few hundred feet from the ground, but [the pilot] failed to gain complete control and the aircraft crashed, killing the entire crew and demolishing the airplane."

10-15-44D. Tubbs Island, California. At 1705, a Lockheed P-38J crashed at the AAF Gunnery Range at Tubbs Island, California, killing pilot 2Lt. LuVerne C. Minnich. The airplane was the number-four airplane of a flight of four that took off from the Army Air Field at Santa Rosa, California, on a dive-bombing training mission. The flight arrived at Tubbs Island and began a series of diving attacks from 6,500 feet, pulling out at 2,500 feet. On the third pass, the number-four ship was seen to explode at an altitude of about 3,000 feet. The port wing was blown off near the engine nacelle and the airplane began shedding numerous pieces as it erupted in flames. The airplane entered a flat spin before smashing into the ground and exploding. Pieces of the airplane were scattered over an area of about one mile. Investigators were not able to determine with any certainty the cause of the in-flight explosion.

10-16-44A. Camel's Hump, Vermont. At 0158, a Consolidated B-24J collided with rising terrain at Camel's Hump, Vermont, killing nine fliers and seriously injuring gunner Pfc. James W. Wilson, who miraculously survived. Killed in the crash were: 1Lt. David E. Potter, pilot; F/O John Ramasocky, co-pilot; 2Lt. Robert W. Geoffrey, navigator; 1Lt. David C. McNary, bombardier; Cpl. Luther N. Hagler, engineer; Cpl. James Perry, radio operator; Cpl. Robert E. Denton, gunner; Pfc. Richard C. Wynne, gunner; Pfc. Casper Zacher, gunner. The airplane took off at 2214 EWT (10-15-44) from Westover Field, Chicopee, Massachusetts, on a routine instrument flight. The airplane had made several contacts with ground stations while in flight. The last contact was at 2342. When the airplane failed to return to base, a search effort was commenced. The airplane was found at 1430 and the surviving crewmember transported to the station hospital at Westover Field. The surviving airman gave conflicting statements so he was allowed to recuperate and give a statement at a later date. Investigators stated that the pilot had let down to about 4,000 feet msl in an effort to relieve the crew of the intense cold at altitude. The pilots were supposed be flying at an altitude

of 8,000 feet and were apparently unaware that they were flying over mountainous terrain. The airplane collided with a 4,083-foot mountain about 150 feet from the top while in normal cruise flight. The airplane broke up on the slope and only a small portion of the wing had burned. Investigation revealed that nine crewmembers were concentrated at the front of the aircraft and the survivor had gone aft just before the initial contact.

10-16-44B. Coffeyville, Kansas. At 0840 CWT, a Lockheed F-5E crashed at the Army Air Field at Coffeyville, Kansas, killing 2Lt. James M. Mulkey. Investigators stated, "[The airplane] took off at 0820 CWT and engine trouble developed causing the pilot to feather the left engine. Aircraft was equipped with drop tanks. These tanks were dropped east of the airfield in a field. The airplane then turned on the base leg and put the wheels down. As the airplane turned on final approach, it appeared to be high. As it approached the end of the runway it was quite evident he was going to overshoot. Tower told him to pull up and go around. The pilot was slow in putting the wheels up in preparing to go around. Wheels came up at about the middle of the runway at an altitude of 300 to 400 feet. Airplane leveled off and flew level for about 1,000 feet further. At this point the [port] wing dipped to the left and leveled out again, then banked sharply to the left and went into a spin. Aircraft hit in an open area north of the barracks behind the hangar line." The F-5E burst into flames upon impact and the pilot was killed instantly.

10-16-44C. Winnfield, Louisiana. At 1020, a Martin B-26C crashed ten miles WSW of Winnfield, Louisiana, killing eight fliers. The B-26 was part of a three-ship flight that took off from the Army Air Field at Lake Charles, Louisiana, at 0815 on a high-altitude bombing and formation mission. About 20 minutes after take-off, the pilot of the subject airplane reported that he was having engine trouble. The pilot reported that he was having an oil pressure and engine temperature problem but did not specify what engine. The airplane continued with the flight. The formation reached 4,500 feet msl when the pilot of the subject airplane reported that he was breaking the formation and returning to the field. The formation broke up and the remaining two airplanes continued on the mission to the Kisatchie Bombing Range. They made several runs and completed the mission. Instead of returning to the field, the subject airplane continued on the mission individually and at 0945 was heard over the radio asking for clearance for a bomb run at 10,000 feet msl. Personnel at Kisatchie cleared the airplane to bomb. The airplane was last heard from at 1012 when the pilot reported a course of 180 degrees for another bomb run. Several minutes later the airplane was seen in a steep spin to the right, momentarily going inverted while describing one and a half turns of spin. The airplane leveled off for an instant before smashing into the ground nose-first and exploding. The B-26 slammed into a small hillside about 75 yards from a group of civilian workmen and their truck. Investigators speculated that the port engine had failed and the airplane entered a spin to the left. The pilot recovered after losing a few thousand feet but entered a secondary spin in the opposite direction. The pilot failed to recover and the airplane slammed into the ground and exploded, scattering wreckage over a wide area. Investigation revealed that the B-26, which had been previously stationed at Kellogg Field, Battle Creek, Michigan, was flown only one hour between 3-2-44 and 7-6-44 before being transferred to Lake Charles Army Air Field. The airplane had been flown very little up until the time just prior to the accident and no special storage procedures had been undertaken. Killed in the crash were: 2Lt. Phillip L. McGill, pilot; F/O James M. Morran, co-pilot; 1Lt. Thomas J. McLaughlin, bombardier; F/O William L. Doyle, student bombardier; 2Lt. Albert A. Levesque, student bombardier; SSgt. Frank G. Mesaros, engineer; Cpl. Billie J. Williamson, radio operator; Cpl. William O'Kelley, gunner.

10-16-44D. Maxton, North Carolina. At 1450 EWT, a Waco CG-4A crashed at Laurinburg-Maxton Army Air Base, Maxton, North Carolina, killing pilot-rated passenger F/O Harold L. Hamilton and injuring two glider pilots. The glider stalled while on final approach and spun in from an altitude of 75 to 100 feet. The glider struck the ground on the starboard wing and then on the starboard fuselage and was completely demolished. Glider instructor pilot F/O Dale W. Taylor and student F/O Benjamin E. Hagmann were injured in the crash.

10-16-44E. Lake Charles, Louisiana. At 1415, a Martin TB-26C crashed while attempting an emergency forced landing ten miles east of the Army Air Field at Lake Charles, Louisiana, killing the crew of six. The airplane was part of a six-ship flight that took off from Lake Charles at 1400 on a "Striker Mission." Several minutes later the pilot radioed the control tower that he was returning to the field. The airplane was seen flying at about 150 feet agl with the starboard propeller feathered and the engines smoking. The B-26 was losing altitude and barely clearing trees and power lines. It became clear that the airplane would not make the field and the pilot attempted an emergency landing. The B-26 initially made a good wheels-up landing on the property of B.T. Wait. The airplane hit a ditch and bounded back into the air for a short distance before smashing back into the ground where it broke up and burst into flames, scattering wreckage over a wide area. Four crewmembers had been hurled from the wreckage. Two of these crewmembers were found alive but both died soon after the accident. Investigation of the failed engine revealed that the spark plugs were in poor condition. Investigation revealed that the B-26, which had been previously stationed at Kellogg Field, Battle Creek, Michigan, was flown only six hours between

1-11-44 and 7-28-44 before being transferred to Lake Charles Army Air Field. The B-26 had been flown very little up until the time just prior to the accident and no special storage procedures had been undertaken. Killed in the crash were: 2Lt. Kenneth B. Thompson, pilot; 2Lt. Norman E. Skinner, co-pilot; F/O Michael Sideria, bombardier; Cpl. Harlon C. Smith, engineer; Cpl. Harvey E. Perkins, radio operator; Cpl. Carl H. Carlson, gunner.

10-16-44F. Sumter, South Carolina. At 1615 EWT, a Vultee BT-13A crashed six miles north of Sumter, South Carolina, killing WASP pilot Jeanne L. Norbeck. Investigation revealed that the airplane had been grounded at Shaw Field, South Carolina, on 10-13-44 because of "dangerous stall characteristics." The airplane was thoroughly checked by AAF mechanics and no serious damage or obvious anomalies could be found. The airplane had lost a wheel during a take-off the previous July, but the landing gear and wing tip were only lightly damaged during the incident. The airplane was cleared "in error" for student flight on 10-16-44 and was taken up by two students on an instrument team ride. They reported only a worn tire, which was replaced that day. No unusual flying characteristics were noted. The airplane was released for a test flight and the WASP pilot was assigned to carry out the mission. She flew in the local area for some time and did not report any trouble. The pilot flew away from the local area and flew at 4,000 feet with another WASP pilot who was testing a North American AT-6 airplane. After flying with the AT-6 for a while, the subject airplane broke off and performed several diving turns, leveling out about 1,000 feet agl. A short time later, the airplane was observed in a shallow diving spiral. The airplane continued in this attitude until it smashed into the ground and burst into flames, killing the pilot instantly. Investigators could not determine the cause of the accident.

10-16-44G. Walnut, Mississippi. At 1845, a North American AT-6C crashed one mile south of Walnut, Mississippi, killing WASP student Marjorie Davis. The airplane and student, based at Avenger Field, Sweetwater, Texas, took off from Stuttgart, Arkansas, on a routine training flight to Courtland, Alabama. The student apparently got lost and was circling the area of Walnut when the airplane struck power lines. The airplane circled two more times before striking additional power lines. The airplane circled the area one more time in an attempt to land. The airplane then crashed in a gully and the pilot was killed when she struck the left side of her head violently against the right side of the cockpit windscreen frame. Investigators stated that the pilot was leaning out of the right side of the cockpit and not wearing a shoulder harness at the time of the accident.

10-17-44A. Ukish, Oregon. At 0130 PWT, a Consolidated B-24J collided with rising terrain 15 miles northwest of Ukish, Oregon, killing ten fliers. The B-24

initially took off at 2200 PWT (10-16-44) from Walla Walla Army Air Field, Walla Walla, Washington, on a bombing mission. The airplane made a normal take-off, climbed to about 20,000 feet, and then flew to the bombing range but was unable to bomb. The airplane returned to the field and landed in order to deplane the bombardier instructor. The airplane then took off at 0045 on an instrument practice mission, using the Walla Walla radio range. When the airplane failed to return to base and the time of the estimated fuel endurance had passed, a search effort was commenced. At 1830, three hunters alerted AAF authorities that they had found a crashed airplane and ten bodies on a wooded hillside. Investigation revealed that the airplane collided with the wooded hillside while in level flight with all four engines operating at cruise rpm. It was speculated that the pilot had been flying on the radio range at an altitude too low to allow safe flight over known obstacles along the flight path. Weather was not considered a factor in the crash and there was no evidence of in-flight fire or mechanical or structural failure. Investigators noted that the airplane had an "outstanding maintenance record." The B-24 exploded violently into flames upon impact and the crew was killed instantly. Killed in the crash were: 2Lt. Gerald A. Walker, pilot; F/O Otis R. Smith, co-pilot; F/O Sheldon A. Silver, bombardiers; F/O Herbert R. Keene, navigator; Cpl. Robert B. Baker, engineer; Cpl. Maurice J. Huset, radio operator; Cpl. Edwin D. Feese, gunner; Pfc. Michael H. Szucs, gunner; Sgt. Roy H. Johnson, gunner; Pfc. R.D. Balter, gunner.

10-17-44B. Fort Myers, Florida. At 1636 EWT, a Bell P-39Q exploded in mid-air and crashed ten miles SSW of Page Field, Fort Myers, Florida, killing pilot 2Lt. Clifford I. Jayson. The airplane took off at 1622 EWT from Naples Army Air Field, Naples, Florida, on a hurricane evacuation flight to Turner Field, Albany, Georgia. The airplane was observed flying at low altitude NNW of Page Field when the starboard wing was seen to explode and separate from the airplane. The P-39, estimated to be at about 100 feet agl, went out of control and slammed to the ground where it exploded into flames, scattering wreckage over a wide area. The pilot was killed instantly. Investigators could not determine the cause of the accident.

10-17-44C. Dale Mabry Field, Florida. At 1749 EWT, a Curtiss RP-40N crashed two miles southwest of Dale Mabry Field, Tallahassee, Florida, killing pilot 1Lt. Frederick H. Beaver and passenger SSgt. Samuel A. Waldren. The two-place P-40 took off from the Army Air Field at Sarasota, Florida, on a hurricane evacuation flight to Dale Mabry Field. Investigators stated, "It appears that his engine had been running excessively hot during the entire flight from his home station. After circling Dale Mabry Field for approximately 20 minutes due to very heavy traffic, Lt. Beaver called in for an emergency landing, the field was cleared and he began his approach. Witnesses stated

that during his approach his engine began to smoke and then burst into flames, immediately afterwards his airplane rolled over on its back and crashed approximately two miles southwest of the field."

10-17-44D. Casper, Wyoming. At 1610 MWT, a Consolidated B-24J crashed two miles north of the Army Air Field at Casper, Wyoming, killing three fliers and injuring seven. The airplane was taking off from Casper on a high-altitude bombing mission when it suffered the failure of the number-one engine at an altitude of 50 feet agl over the departure end of Runway 34. The pilots attempted to feather the number-one propeller and the first pilot advanced the remaining mixture controls to "emergency rich" in an effort to get more power. Investigators stated that this move actually decreased the available horsepower on the remaining three engines. The pilot was unable to maintain altitude and he performed a wheels-up emergency forced landing straight ahead. The ship broke up and burst into flames. Rescue personnel were on the scene immediately and extinguished the fire. Investigation revealed that a rough running number-one engine had delayed the airplane's departure. The airplane was taxied back to the line and a mechanic ran up and examined the engine, but no apparent problems could be found. The pilot then taxied back out to the runway and commenced the unsuccessful take-off. Bombardier 2Lt. Robert E. Stokely, gunner Cpl. Edward C. Boczar and gunner Cpl. Harry W. McManus were killed in the crash landing. Crewmembers injured in the crash landing were: 2Lt. Richard J. Mardis, pilot; F/O Harry R. Palmer, co-pilot; 2Lt. Frank J. Olive, navigator; Cpl. Frank C. Rygiel, engineer; Cpl. Samuel L. Diamond, radio operator; Cpl. Elbert L. Kellin, gunner; Cpl. John J. Lewisky, gunner.

10-18-44A. Gardner Field, California. At 1035, a Vultee BT-13A crashed 18 miles north of Gardner Field, Taft, California, killing student A/S Leonard L. Johnson. The airplane was seen to enter a spin to the right at an indicated altitude of about 5,500 feet. The student stopped the spin but entered a secondary stall and a spin to the left, describing several turns. The student stopped the spin, but the airplane stalled and then entered another spin to the right, continuing in this attitude until it struck the ground and exploded. The student was killed instantly.

10-18-44B. Samsula, Florida. At 1445 EWT, a Vultee RA-35B flying in instrument conditions crashed five miles southwest of Samsula, Florida, killing F/O Charles H. Sloan. The airplane took off at 1352 from Morrison Field, West Palm Beach, Florida, on a hurricane evacuation flight to Jacksonville, Florida. The airplane was observed flying at very low altitude, just under the low hanging overcast and in heavy rain. The pilot observed trees in his path and pulled up abruptly, stalling the airplane and entering a spin. The pilot was unable to recover and the airplane struck the ground at a steep angle, exploding into flames upon impact. The pilot was killed instantly.

10-18-44C. Matagorda Bay, Texas. At 1625, a Bell P-39L crashed into Matagorda Bay one-quarter mile off shore north of Runway 18 at Matagorda Peninsula Army Air Field, Texas, killing 2Lt. Lyman W. Tarbet. The airplane was returning to the field after completing a "Training-Camera Research" mission. The airplane was flying south at 1,500 feet after turning on the final approach about one and one-half miles out. The airplane was not lined up with the runway and the pilot attempted a steep bank to the right at slow speed with the landing gear extended. The airplane stalled, lost altitude and then rolled to the left, entering an inverted attitude at about 150 feet above the surface. The P-39 crashed into the bay in an inverted dive, sinking immediately and killing the pilot instantly.

10-18-44D. Granite, Oklahoma. At 1630, a Cessna UC-78C and a Cessna AT-17 collided in mid-air and crashed one and a half miles south of Granite, Oklahoma, killing A/C Kelton L. Smith aboard the AT-17. A/C Norman E. Spencer parachuted to safety and was uninjured. A/C Roger B. Vipond and A/C Francis Venneman were uninjured parachuting to safety from the UC-78C. The airplanes were part of a three-ship flight on a student formation mission. The AT-17 was flying in the number-three position and the UC-78C was flying in the number-two position. The collision occurred while the flight was executing a cross over to an echelon formation to the left. The UC-78 overran the AT-17 while moving over and passed underneath it. The pilot of the UC-78 inadvertently ascended into the AT-17, the UC-78's port tail striking and damaging the port engine and port fuselage of the AT-17. The UC-78 went out of control and the students bailed out immediately. As he was on the wing in an effort to bail out, Cadet Spencer noticed that Cadet Smith opened his parachute prematurely and was fouled on the tail surfaces. Cadet Spencer got back in the airplane and attempted to fly the airplane. The airplane would not respond to any control input so the cadet left the ship. As he parachuted to safety, he noticed that Cadet Smith's parachute was no longer attached to the tail section and he observed three parachutes descending below him.

10-18-44E. Pennsville, New Jersey. At 1859, a Fairchild AT-21 crashed at Pennsville, New Jersey, killing five fliers. Killed in the crash were: 2Lt. Richard F. Storburg, pilot; 1Lt. James A. Finfera, co-pilot; 1Lt. Robert C. Ryan, bombardier; Pfc. Charles R. White, passenger; Lt. Col. Leslie B. Cooper, passenger. The airplane took off from New Castle Army Air Base, Wilmington, Delaware, on an administrative flight to Wright Field, Dayton, Ohio. The airplane made a normal take-off and a short time later the pilot contacted the control tower requesting an emergency landing. The airplane was observed with fire trailing the starboard engine and wing. The AT-21 entered a steep bank to the right, lost altitude and then crashed into the ground, exploding into flames upon impact.

10-18-44F. DeRidder, Louisiana. At 2231, a Douglas F-3A (photo-reconnaissance version of the A-20 type) attempting a take-off crashed one and a half miles north of the Army Air Field at DeRidder, Louisiana, killing pilot 2Lt. Phillip J. Barton and navigator 2Lt. Daniel Montgomery. Photographer/gunner Cpl. Elmer C. Mifler received serious injuries. The airplane was scheduled to fly three photographic/bombing missions for the night. The airplane flew two successful missions and landed to bomb up. The airplane was loaded with ordinance and then taxied out to take off for the third mission. The airplane took off and climbed to about 100 feet and then began to steadily lose altitude until it smashed into the terrain. The airplane bounded back into the air for a short distance before it smashed back into the ground on the starboard wing and nose section. The airplane smashed itself to pieces as it skidded to the right and burst into flames.

10-19-44A. DeRidder, Louisiana. At 1455, two Curtiss P-40N airplanes collided in mid-air and crashed eight miles WSW of the Army Air Field at DeRidder, Louisiana, killing 2Lt. Paul Bradshaw aboard P-40N #44-7945. 2Lt. Joseph F. Kot received minor injuries parachuting to safety from P-40N #44-7997. The airplanes were part of a six-ship flight that took off from DeRidder Army Air Field on a formation training flight. Lt. Kot was flying in the number-two position and Lt. Bradshaw was flying in the number-three position of a large V-flight. Three P-40s were flying on the starboard wing of the leader and two P-40s were flying on the leader's port wing. The flight leader signaled the flight to cross over to an echelon to the right. Lt. Bradshaw crossed over too rapidly and overshot. Lt. Bradshaw kept his eye on the leader but did not keep Lt. Kot in view and passed under him. Lt. Bradshaw pulled up into Lt. Kot's airplane, causing both airplanes to go out of control and crash. It is likely that Lt. Bradshaw was killed in the collision.

10-19-44B. Deming, New Mexico. At 2145 MWT, a Boeing B-17G suffering an in-flight fire was abandoned and crashed 30 miles southeast of Deming, New Mexico, killing six fliers. Four crewmembers were able to parachute to safety. The B-17 took off from Biggs Field, El Paso, Texas, on a cross-country navigation flight to Blythe Army Air Field, Blythe, California, and return. About 20 minutes after leaving Biggs Field, the number-three engine exploded and caught fire. The explosion evidently destroyed the propeller feathering oil line, preventing the pilots from feathering the propeller. The pilots could not extinguish the fire with the number-three engine fire extinguisher and it raged out of control, threatening to consume the starboard wing. The pilot ordered the crew to bail out and then he and the co-pilot immediately abandoned the airplane. Killed in the crash were: 2Lt. James E. Briggs, co-pilot/B-17 student; SSgt. Earl F. Hegerle, engineer; Cpl. Charles S. Crockett, radio operator; Cpl. Henry G. Reul, gunner; Cpl. John G. Olson, gunner; Cpl. Stanley J. Kling, gunner. Crewmembers parachuting to safety were: 1Lt. Shelby F. Vaughn, B-17 instructor pilot; 2Lt. Robert G. Bakula, pilot/B-17 student; 2Lt. Gerald W. Wilburn, navigator; 2Lt. Reny A. DeLotto, bombardier.

10-20-44A. Harlingen, Texas. At 1100, a Consolidated B-24H suffering an in-flight fire was abandoned and crashed three miles north of the Army Air Field at Laguna Madre Sub-Base, Harlingen, Texas, killing two fliers. Three crewmembers were injured and nine were uninjured parachuting to safety. The airplane took off from Harlingen Army Air Field on a student gunnery mission. The airplane successfully completed the air-to-air gunnery mission and let down from 7,500 feet to about 700 feet above splash targets at Laguna Madre for air-to-surface firing. Shortly after, the engineer alerted the pilot that a fire had erupted in the forward bomb bay. The pilot added power and climbed, turning the airplane toward the mainland. The pilot alerted the crew to stand by to bail out. The engineer was successful in extinguishing the fire and he ordered the pilot to open the bomb bay doors. The pilot complied and then set up an approach for the field at Laguna Madre Sub-Base. The pilot extended the landing gear and a second fire erupted at the same location in the forward bomb bay. Hydraulic fluid began spraying near the fire and then an explosion occurred in the forward bomb bay. The engineer, Pfc. Laverne B. Behrens, was covered with hydraulic fluid. A fire then erupted in the after bomb bay and the engineer continued to fight the fire, but was unsuccessful in putting it out. The engineer stepped back from the fire and his flight clothes were ablaze. The engineer, without his parachute, then fell out of the open bomb bay in flames and plunged to his death. The remaining crewmembers bailed out and the airplane entered a steep, diving spin to the left, exploding into flames about 100 feet before striking the ground, scattering flaming wreckage over a wide area. The pilot, 1Lt. Lyda A. Burnett, parachuted to safety but landed in the flaming wreckage and was seriously injured. Gunnery Instructor TSgt. John C. Webb and student gunner A/C Henry W. Fulks were seriously injured parachuting to safety. Crewmembers uninjured parachuting to safety were: Sgt. Robert G. Keeshan, gunnery instructor; Sgt. Preston V. McDonald, gunnery instructor; A/C William C. Gardner, student gunner; Pvt. Archie D. Lynch, student gunner; Pvt. Joseph J. Macaluso, student gunner; Pvt. Robert C. Young, student gunner; Pvt. Edmund A. Woronics, student gunner; Pvt. Adam J. Sypitkowski, student gunner; Pvt. John J. Takach, student gunner. Co-pilot 2Lt. Chester G. Mangrum fell to his death because his parachute harness was incorrectly fastened to his body and his parachute was damaged when he struck the airplane while bailing out. The fire was caused by faulty fittings in the hydraulic system in the forward bomb bay, allowing atomized fluid to be ignited by electric current in the main power switch in the auxiliary hydraulic pump while it was being operated. Investigation revealed that the operation of the bomb bay doors caused enough current to restart the fire.

10-20-44B. Fort Sumner, New Mexico. At 1115, a Curtiss P-40N crashed at Fort Sumner, New Mexico, killing pilot 2Lt. James T. Hicks. Investigators stated, "Airplane took off normally and climbed normally to approximately 25 feet where it started a gradual turn to the left. The turn and bank to the left increased until a 90-degree turn had been accomplished and [the airplane] was in a vertical bank. The nose came up abruptly and the airplane stalled out and crashed on the parking ramp, bursting into flames when it hit. Inspection of the airplane proved all controls to be connected and workable to a small extent after the wreck. The engine and propeller were operating correctly and no material failure could be found."

10-20-44C. Manzanar, California. At 1015, a Consolidated B-24J suffering an apparent instrument failure and double engine failure made an emergency forced landing in the Saline Valley 35 miles east of Manzanar, California. Engineer Cpl. Robert C. Dunlop was fatally injured when he fell through the rotating number-two propeller while exiting the airplane after the forced landing. Eight other crewmembers were uninjured. The airplane had taken off from the Army Air Field at Muroc, California, on a bombing and aerial gunnery mission. While on the air-to-air gunnery portion of the mission, the airplane began to experience a loss of power on the right side. The number-four engine instruments indicated that the oil pressure was dropping and the fuel pressure was at maximum. The pilot assumed an "Autosyn Inverter failure" with a simultaneous engine failure. The pilot feathered the propeller, trimmed the airplane for three-engine flight and increased the rpm on the remaining three engines. Engine oil temperature began to climb considerably on the remaining three engines. The number-three engine then began to lose oil pressure until it went to zero. The pilots feathered the number-three propeller and trimmed the aircraft for two-engine flight. The B-24 was hemmed in by mountains and could not climb over them because the number-one and number-two engines began to lose oil pressure and power. The pilot attempted to restart the number-four engine but was unsuccessful in getting it to produce power. The number-four propeller was re-feathered and the pilot decided to make a landing in the valley. The airplane, with landing gear extended, made a successful emergency forced landing on what appeared to be a dry lakebed. The airplane rolled for about 1,000 feet before it broke through the saline crust and into the underlying mud. Mud was sprayed all over the airplane. The airplane came to a halt as it became mired in the mud, but was not seriously damaged in the forced landing. After the B-24 came to a halt, the engineer exited the airplane from the upper hatch and announced "all clear." The engineer slipped on the muddy airplane surface and fell backwards into the still rotating number-two propeller, severing his left leg at mid-thigh. First aid was administered immediately. Search planes located the downed B-24 at about 1630, but help did not come for hours. Civilian law officers and a civilian doctor and nurse arrived on the scene at about 2300. Cpl. Dunlop died of "sustained shock" while being transported to a hospital about 17 hours after the accident. Investigators speculated that the pilots might have misread the instruments that had been changed to "direct reading" from "autosyn." Investigation revealed that the autosyn inverters were in working order and the cause for the instruments indicating the loss of oil pressure in all four engines could not be determined. The B-24 suffered damage to the landing gear and could not be taken off from the valley surface even if repaired. Because of the remote location, B-24J #42-51444 was abandoned in place and was eventually salvaged by a civilian operation in the late 1950s. Crewmembers uninjured in the crash landing were: 2Lt. John T. Hansen, pilot; 2Lt. Byron C. Avgerings, co-pilot; 2Lt. Philip P. Murphy, bombardier; Cpl. Loren W. Pargen, radio operator; Cpl. Victor W. Bleau, gunner; Cpl. Edmund Cullen, gunner; Cpl. Dean J. Blackwood, gunner; SSgt. Edward W. Shaulinski, gunnery instructor.

10-20-44D. Abilene, Texas. At 1434, a Republic P-47D crashed one-quarter mile south of the Army Air Base at Abilene, Texas, killing pilot F/O Amando Gomez. Investigators stated, "F/O Gomez was in traffic pattern after an orientation flight from this field. He had overshot the runway twice by not anticipating his turn onto the final approach. Both times Runway Control had sent him around. On his third pattern, F/O Gomez stalled out in a nose-high turn and rolled on his back. He went practically straight down from an altitude of approximately 300 feet."

10-20-44E. Ozark, Alabama. At 1600, a Curtiss P-40N suffered a catastrophic structural failure and crashed seven miles west of Ozark, Alabama, killing F/O William T. Morgan. The P-40 was observed to enter a steep dive at high altitude, performing a roll while in the dive. The airplane recovered from the dive at about 2,000 feet agl and then climbed back to the original altitude. The pilot pitched the airplane into another steep dive, but as he attempted to pull out the starboard wing failed and separated from the aircraft. The airplane began to break up and exploded violently into flames when it slammed into the ground. Wreckage was scattered over a distance of over three-quarters of a mile. The pilot had made no attempt to parachute to safety. Investigation revealed that the pilot had stated a few days earlier that he had performed a vertical dive maneuver in a P-40 airplane and was "disappointed" that he had achieved a speed of only 450 mph.

10-20-44F. Walesboro, Indiana. At 1635 CWT, a Beech AT-10GL and an AT-10 collided in mid-air five miles southwest of Walesboro Auxiliary Airfield, Walesboro, Indiana, killing instructor 1Lt. Charles L. Knuth and injuring student pilot 2Lt. Richard I. Fricke, who parachuted to safety from the AT-10. Pilot Capt. Gordon H. Fleisch and co-pilot 2Lt. Kendall R. Moore were able to successfully land the AT-10GL at Freeman Field, Seymour, Indiana. The AT-10 was flying at 3,000 feet with the student under the hood and flying on instruments on the Seymour Range. The AT-10 was then struck by the AT-10GL, severing the AT-10's tail section. The AT-10 went out of control and the instructor ordered the student to bail out. The instructor, unable to escape the spinning airplane, was flung from the AT-10 at low altitude and he was unable to use his parachute. The pilots in the AT-10GL were flying at 4,000 feet with one student under the hood. The airplane made a descending turn and collided with the AT-10. The AT-10GL went out of control but the pilots were able to recover the airplane and make a safe landing at Freeman Field.

10-20-44G. Burbank, California. At 1709, a Lockheed YP-80A jet fighter crashed just after take-off from the Lockheed Air Terminal at Burbank, California, killing Lockheed Chief Test Pilot Milo G. Burcham. The airplane, which was on its first flight, took off to the west on the East-West Runway. The pilot retracted the landing gear and flaps and the P-80 climbed to about 300 feet agl when the engine failed. The pilot turned to the north and entered a steep glide in an attempt to make it to large open area for a belly landing. The large open area contained a gravel pit that was 400 feet wide and about 125 feet deep. The wall of the gravel pit was sloped at about 60 degrees. The airplane did not have enough speed for the pilot to extend his glide over the gravel pit and land on the other side. The airplane crossed over the gravel pit and then pancaked into the north wall of the gravel pit about 50 feet below ground level. The airplane burst into flames upon impact and the pilot was killed instantly. Investigation revealed that the General Electric I-40 jet engine failed due to a fuel stoppage of unknown origin. Investigators noted that this was the second P-80 type airplane flown from the Lockheed Air Terminal.

10-20-44H. Coffeyville, Kansas. At 2020 CWT, a Lockheed F-5 suffering an in-flight fire crashed seven miles south of the Army Air Field at Coffeyville, Kansas, killing 2Lt. John S. Stock, who fell to his death in an unsuccessful parachute jump. The airplane took off from Coffeyville Army Air Field at 1900 CWT on a routine night flight. At 2023, the pilot radioed the Coffeyville control tower and alerted them that he was flying south of the field at 3,000 feet and that he had excessively high engine temperature in one of his engines. Tower personnel observed the airplane on fire and diving straight down toward the ground. The airplane smashed into the ground at a steep angle and exploded. Investigation revealed that the starboard engine had thrown a rod, causing it to catch fire. The pilot either was struck by the airplane upon exit or had bailed out at so low an altitude that he did not have time to pull the ripcord.

10-20-44I. Lemmon, South Dakota. At 2220 MWT, a Consolidated RB-24E suffering a multiple engine failure made a forced landing 25 miles southeast of Lemmon, South Dakota, killing four fliers. Three fliers were killed in unsuccessful parachute jumps. Three fliers were uninjured parachuting to safety. Four crewmembers stayed with the airplane, one being killed in the crash landing. The B-24 had taken off from Casper Army Air Field, Casper, Wyoming, on a routine night cross-country flight to Fort Peck, Montana, to Bismarck, North Dakota, and return. The B-24 took off at about 1720 MWT and successfully navigated to Fort Peck and then to Bismarck. Investigators stated, "About one hour out of Bismarck, North Dakota, the trouble started.... [The pilot stated] that one of the engines on the right side cut out, so he gave the emergency three-engine procedure, mixture auto-rich, propellers high rpm. The mixture remained at auto-lean. He thought number-one engine went out, so he feathered it. When realizing that the trouble was in the number-three engine, because they were only pulling 20 inches of manifold pressure and 2,000 rpm on it, and that the cylinder head temperature 280 degrees, he feathered it. After feathering number-three, the pilot and co-pilot stated that the number-four engine went out. All of the engine instruments were reading normal on this engine and the reason they knew it went out was from the added control pressure. Pilot gave order to bail out and the co-pilot repeated the order. They were able to hold the ship in an attitude approximating level flight for about 30 seconds. Then the ship went into a right spin. Shortly after the entry into the spin, number-two engine was seen to explode and shoot blue flames. The throttles were pulled back and the ship was brought under control. The co-pilot then cut number-two magneto. Some of the crewmembers still had not jumped and the aircraft was now at 4,300 feet, after going into the spin at 8,600 feet. The pilot again gave the order to bail out. The co-pilot tried to get out [but] he could not make it so he got back into his seat. The pilot then switched on the lights, and the co-pilot lowered the flaps [and apparently the landing gear] and the ship made a crash landing with wheels down, although neither pilot nor co-pilot indicated that they were lowered. The aircraft was landed on top of a slight ridge and skidded approximately 75 yards with full brakes on. The ship then skidded off a slight knoll and dropped approximately 10 feet, blowing the left landing gear tire. This caused the ship to skid somewhat sideways, breaking the nose wheel landing gear, letting the nose fall to the ground. The ship then skidded approximately 135 yards before coming to a full stop. The pilot and co-pilot crawled through the right

front window and with the help of a farmer rescued the lower turret gunner, who was pinned under the top turret. The navigator was killed in the bomb bay. His parachute was open and was found 40 feet behind the wreckage of the aircraft." Investigators stated that they found no evidence of engine failure and speculated that the pilots became confused. Bombardier 2Lt. Milan J. Mikulec, engineer Cpl. William S. Goczewski and gunner Cpl. Ralph L. Kizer fell to their deaths in unsuccessful parachute jumps. Navigator F/O Alex J. Kalmavek was killed in the crash landing. Pilot 2Lt. Robert A. Meyer, co-pilot William A. McManus were uninjured and gunner Cpl. Carson A. Bass received light injuries in the crash landing. Radio operator Cpl. Dernard W. Day, gunner Cpl. Sheldon B. Kaiser and gunner Cpl. Sherwood W. Long parachuted to safety and were uninjured.

10-21-44. Hillsborough Auxiliary Airfield, Florida. At 1257, a North American P-51D suffered a catastrophic structural failure and crashed on the AAF Bombing Range one mile northeast of Hillsborough Auxiliary Airfield, Tampa, Florida, killing Capt. Russell F. Mimmack. The airplane was leading a four-ship flight that had taken off from Hillsborough Auxiliary Airfield on a dive-bombing mission. The airplane was observed to enter its fourth bombing run at 9,000 feet msl, pulling out at about 2,500 feet at an estimated speed exceeding 400 mph. The airplane was seen to enter a slow roll to the left. As the airplane became inverted, the pilot rolled the airplane back to a level position at about 1,500 feet. Moments later the port wingtip was seen to separate from the airplane, and as the airplane rolled to the left the starboard wing separated from the airplane, followed by the port wing and pieces of the tail section. The P-51D went out of control and crashed on the range, exploding into flames upon impact.

10-22-44A. Lubbock, Texas. At 0155, glider pilot F/O Richard J. Engles was killed when he was run over by a Waco CG-4A that was being towed to the take-off position at South Plains Army Air Field, Lubbock, Texas. CG-4A glider pilot and AAF rated airplane pilot 2Lt. Gerald L. Ovel was uninjured. The two glider pilots had just landed at the field and had been towed back to the marked position of the officer in charge of grading the landings and pilot assignments. F/O Engles' flying period had ended and he was slow in exiting the glider. The officer ordered F/O Engles to remain in the glider until it was towed to the take-off position where he could safely exit. A short time after the tug began towing the glider back to the take-off position F/O Engles shouted for the tug driver to stop and let him out. Lt. Ovel ordered him to remain on board until the glider was towed into take-off position, but F/O Engles exited the glider, which apparently was still moving. F/O Engles lost his balance and fell into the path of the glider's starboard main wheel and was run over, causing fatal injuries.

10-22-44B. Denver, Colorado. At 1320, a Consolidated TB-24D suffered a catastrophic structural failure and crashed five and one-half miles southeast of Lowry Field, Denver, Colorado, killing pilot Capt. Howard P. Wentling, Jr. and engineer SSgt. Donald L. Schuermann. The airplane had taken off from Lowry Field on a test flight to check on the wings, which had received repair. The pilot performed two "small preliminary dives" before climbing to about 20,000 feet. The pilot radioed the tower and asked them to observe his dive to 10,000 feet. The control tower personnel were unable to observe the maneuver due to the brightness of the sun. Moments later, personnel observed a cloud of smoke southeast of the field where the airplane slammed into the ground. Investigation revealed that the rudders had failed and the bomb bay doors had separated from the airplane in the dive or during the attempted pullout. Wreckage was scattered over a great distance. Fabric from the rudders was found about three miles from the main wreckage. Investigators noted that this airplane had returned from a combat tour and was war weary. The new wing rivets and repairs were satisfactory and held during the extreme maneuver and the subsequent accident. The repairs were not considered a factor in the accident.

10-22-44C. Davis-Monthan Field, Arizona. At 1504 MWT, a Consolidated B-24J attempting a take-off crashed 300 yards from the departure end of Runway 12 at Davis-Monthan Army Air Field, Tucson, Arizona, killing nine fliers and seriously injuring two others. Killed in the accident were: 2Lt. Ernest S. Conley, pilot/B-24 student; 2Lt. John P. Hackman, co-pilot/B-24 student; 2Lt. Stanley Danziger, navigator; 2Lt. Albert C. Jolley, bombardier; Cpl. Robert L. Tweet, radio operator; Pfc. Carl F. Skyton, gunner; Pfc. Alexander W. Stewart, gunner; Cpl. Alfred E. Leannais, gunner; Pfc. Howard A. Waeltz, gunner. B-24 instructor pilot 2Lt. Lloyd O. Windham and engineer Cpl. William C. Hunt were seriously injured. Investigators stated, "During a formation take-off on a routine training mission, it is believed that one of the pilots pressed the brakes. This could have been voluntary to stop the rotation of the wheels prior to retraction of the landing gear or involuntary on the part of either pilot. However, it is believed that the action of breaking the wheels was voluntary and that the Landing Gear Selector Handle was placed in the 'Up' position, the pilots believing that the aircraft was airborne. The aircraft not being airborne and the brakes being applied caused the puffs of smoke as seen by the witnesses and as the gear started to retract, the left wing dropped, probably due to the [propeller turbulence] of the airplane which had previously taken off; the number-one propeller striking the runway caused the loss of the number-one engine, and part of the number-one propeller being thrown into the number-two engine, probably hitting the nose section of that engine, allowing the number-two propeller to become uncoupled from the engine, causing the loss of that engine also. Although the ship was airborne by this time, the pilots were unable to maintain control with two engines out on one side and the airplane struck the ground, crashed and burned." A motorist observed that an accident was imminent and he accelerated his automobile to high speed to escape the path of the crashing airplane, which clipped a salvage yard with the port wing before cartwheeling to the ground and exploding. The motorist ran to the scene of the accident and was able to rescue the two crewmembers that had been thrown clear.

10-22-44D. Sioux City, Iowa. At 1628 CWT, a Boeing B-17G crashed into a fire station while attempting a three-engine landing at the Army Air Base at Sioux City, Iowa, killing five army fliers and two civilian firefighters on the ground. Six B-17 crewmembers and two civilians on the ground were injured. Investigators stated, "The crew took off for a high-altitude formation flight. The pilot was having difficulty in keeping up with the formation even by using excessively high power settings, so he finally left the formation and went to a bombing range for high altitude bombing. He had arrived at the bombing range and was still using the high power settings when the number-one engine started throwing oil. The pilot tried to feather the propeller, but it would not feather. The pilot returned to the field and came in for a three-engine landing on Runway 17. The airplane bounced hard on the landing, so the pilot applied power with all three engines to ease the airplane down. The uneven power — with the number-one engine out — turned him to the left toward the ramp and hangar line. [The pilot] did not hold the airplane straight with the rudder. He then decided that he would not have enough room to land without colliding with some parked aircraft and hangars so he applied full power on his three engines and pulled up to go around. He called for the wheels to be retracted and the flaps to be raised to one-third down. The co-pilot did not respond, so the engineer started the wheels and flaps up. The pilot was trying to clear a hangar, the roof of which he scraped with his left wing. An instant later the airplane stalled and crashed. The left wing struck [the roof of a parked automobile and then the B-17 cartwheeled into Fire Station #2] adjacent to the hangars, which was destroyed by the crash and subsequent fire. In addition, several privately owned automobiles parked near the fire station were [destroyed]." A rescue vehicle in the fire station was also destroyed. Crewmembers killed in the crash were: F/O Gordon L. Shepherd, bombardier; Cpl. Russell W. Hill, radio operator; Pfc. Robert H. James, gunner; Sgt. Wilmer R. Flowers, gunner; Cpl. Curgus J. Garrison, Jr., gunner. Crewmembers injured in the crash were: 2Lt. Kenneth A. Wait, pilot; 1Lt. Frank E. Covington, Jr., navigator; Cpl. Joe D. Silva, engineer; Pfc. Joseph P. Costa, gunner. Co-pilot Edwin C. Pyle received serious injuries.

Civilian firefighters George W. Brown and Garret J. Davelaar were killed on the ground; firefighters Irvin A. Schindel and Oscar L. Oder received minor injuries.

10-23-44A. Fort Worth, Texas. At 0735, a Consolidated B-24D attempting a landing crashed one mile south of the Army Air Field at Fort Worth, Texas, killing the crew of four. Killed in the crash were: 2Lt. Tom A. Garwood, B-17 instructor; 2Lt. Rubert E. Rogers, pilot/B-17 student; F/O Orville E. Roush, co-pilot/B-17 student; Pfc. Victor Degutis, engineer. Investigators stated, "The aircraft was being flown on a student transition flight. Approximately one hour after take-off, the tower was called and the aircraft entered the traffic pattern. Upon turning on the base leg, the tower was called and notified of a gasoline leak aboard the aircraft. On making a steep turn to the final approach, considerable altitude was lost and the aircraft rolled out of the turn and banked steeply to the right. It then rolled back to the left and seemed to enter a spin, striking the ground and exploding."

10-23-44B. Canfield, Arkansas. At 1145 CWT, a Martin B-26B crashed two miles west of Canfield, Arkansas, killing six fliers. The B-26 took off at 0800 from Barksdale Field, Shreveport, Louisiana, on a compass swing/airspeed calibration flight. At about 1130, the B-26 was observed flying over the calibration course at low altitude. The course was a set of railroad tracks between the towns of Benton, Louisiana, and Canfield, Arkansas. The airplane was authorized to fly at 500 feet agl on the calibration mission. The B-26 was seen to enter a steep bank while apparently attempting to turn around for another run on the course. The airplane stalled and entered a spin at an altitude too low to allow recovery. The B-26 smashed into the ground and exploded into flames, killing all on board instantly. Killed in the crash were: 2Lt. Arthur E. Anderson, pilot; 2Lt. Grady L. Carter, co-pilot; 2Lt. Joseph A. Jenschke, bombardier-navigator; Cpl. William F. Berardi, engineer; Sgt. Vernon E. Renfroe, radio operator; Cpl. Gale J. Cianci, gunner.

10-23-44C. Gulf of Mexico, Mississippi. At 1220, a Boeing B-17F ditched in the Gulf of Mexico four miles west of Gulfport, Mississippi, fatally injuring navigator 2Lt. Nicholas C. Koshivos. Seven crewmembers were injured, one seriously; the two pilots were uninjured. The airplane was flying in formation at 1,500 feet when it suffered the failure of the number-three and number-four engines. The pilot dropped out of formation and was flying at about 700 feet agl when he began to attempt a turn back toward Gulfport Army Air Field. The airplane had lost about 400 feet in the turn toward the field. The pilot was unable to maintain altitude and he attempted a ditching. The pilot turned the airplane toward the gulf to avoid trees and houses on the shore. The starboard wing struck the water first, causing the airplane to turn right sharply about 180 degrees and slam into the water about 200 yards off shore. The airplane broke in two at the radio room as it smashed into the shallow water. The crewmembers that were in the radio room were thrown clear when the airplane broke up. The airplane caught fire a short time after the ditching. Pilot 2Lt. George W. Mueller and co-pilot F/O Clifton W. Nippert were uninjured and gunner Pfc. Clarence E. Enghadl sustained serious injuries. Crewmembers injured in the ditching were: F/O Donald E. Proue, bombardier; Cpl. Emery D. Carlisle, engineer; Cpl. James A. Brickley, assistant engineer; Cpl. Edmond G. Lombardo, radio operator; Pfc. Hardy E. Stohler, assistant radio operator; Pfc. William J. Kuespal, gunner. The navigator died a short time after being rescued.

10-23-44D. Wilson, North Carolina. At 1440, a Republic P-47D and a P-47G collided in mid-air and crashed ten miles southeast of Wilson, North Carolina, killing P-47D pilot 1Lt. L.W. Juliano. F/O Alfred J. Imiola was able to parachute to safety from the P-47G. The airplanes were part of two separate flights that had taken off from Seymour-Johnson Army Air Field, North Carolina, on a high-altitude formation flight. The formations climbed to 25,000 feet and individually completed the mission. The airplanes began to descend to return to the field. Lt. Juliano's was leading a three-ship flight that was descending in a wide, shallow spiral. Lt. Juliano was unable to avoid striking the P-47G, which was flying in the number-two position of a flight that was descending in a very slight dive. The P-47D struck the P-47G from the right rear and underneath, sending both airplanes out of control at about 7,000 feet. It was likely that Lt. Juliano was killed in the collision when the P-47G propeller cut into the P-47D cockpit as the P-47D slid across the bottom of the P-47G.

10-23-44E. Chico, California. At 1425, a Lockheed P-38L suffering an engine failure entered a spin and crashed one mile west of Chico, California, fatally injuring pilot F/O Walter T. Bryan in an unsuccessful parachute jump. The airplane took off from Long Beach, California, on a ferry flight to Chico Army Air Field. Before take-off, the pilot had trouble retracting the dive brakes and was delayed about 50 minutes. The flaps were eventually retracted and the airplane checked out so the pilot took off on his mission. The pilot radioed that he felt a "lurch" in the ailerons so he turned the aileron boosters off. Hydraulic pressure then dropped to zero and the pilot could not restore it with the hand pump. A short time later, the pilot radioed that he could not draw fuel from either of his outer wing tanks, which were both full. The P-38 arrived at Chico and the pilot radioed tower personnel that he had a hydraulic problem and wanted to perform a fly-by for a landing gear check. The pilot was given permission and he buzzed the tower at about 500 feet agl. Tower personnel alerted the pilot that the landing gear was not fully down. The pilot then attempted to recycle the landing gear and get it to extend fully and lock. The tower requested another fly-by and the pilot stated that he would fly by at 2,000 feet agl because he was very low on fuel. The pilot reported that he had 25 gallons total remaining. The landing gear still had not fully extended and the pilot continued to maneuver in an attempt to "shake" the landing gear down. The pilot entered the traffic pattern at 2,000 feet on the downwind leg. The starboard engine quit moments later. The pilot switched fuel tanks and the port engine failed moments later. The airplane was seen to roll over and enter a spin, describing two full turns. The pilot bailed out and the P-38 smashed to the ground and exploded, scattering flaming wreckage. The pilot parachuted into an area of large flames on the ground and was seriously burned. He died shortly after being pulled from the fire.

10-23-44F. Corona, California. At 1837, a Consolidated B-24J crashed while attempting an emergency landing six miles northwest of Corona, California, killing four crewmembers in an unsuccessful parachute jump and injuring the two pilots who went in with the airplane. The B-24 took off at 1751 from March Field, Riverside, California, on a routine training mission to Fresno, California, and return. The airplane was flying near Santa Ana at 9,000 feet when the number-three propeller ran away. The pilot feathered the propeller but soon noticed flames in the number-three engine. The propeller was unfeathered and the flames were blown out. The propeller then was feathered again and the pilot turned the airplane back toward March Field. The pilot ordered the engineer to cut off the fuel supply to the number-three engine. The engineer manipulated what he thought was the fuel valve for the number-three engine, but the number-four engine failed seconds later. The pilots were unable to maintain altitude and ordered the crew to bail out. The crew bailed out and the airplane was losing altitude rapidly. The pilot put the landing gear handle in the down position and the airplane flew into the ground moments later with the landing gear partially extended. The airplane careened over the terrain for about 400 feet until it smashed into a large ditch. The airplane crashed to a halt with the wings resting on the banks of the ditch and fuselage hanging over the ditch. Pilot 1Lt. Benjamin W. Hill and co-pilot 2Lt. Phillip H. Ball crawled out of the shattered front fuselage and fell about 30 feet to the bottom of the ditch. Both were seriously injured. Investigation revealed that the fuel valves for the number-three and number-four valves were mislabeled. When the engineer turned the number-three fuel valve, he unintentionally shut off the fuel supply to the number-four engine. The four crewmembers fell to their deaths when they bailed out at an altitude too low to allow their parachutes to deploy. Killed in the accident were: 1Lt. William A. Wirz, navigator instructor; 2Lt. Gerhard K. Kaden, navigator; Cpl. Harlan L. Byran, engineer; Cpl. Antonio F. Mancina, radio operator.

10-23-44G. Madill, Oklahoma. At 2135, a Vultee BT-13A crashed six miles south of Madill, Oklahoma, killing instructor 2Lt. Clinton M. Jacob and student A/C Carlton L. Mullis. The airplane took off from Perrin Field, Sherman, Texas, on a routine student night flight. The instructor was scheduled to show the student prominent local landmarks for night flying in the local solo area. The airplane flew away from the local practice area and the instructor flew low over Madill in an apparent effort to identify the town. The airplane was observed to fly over a lighted oil derrick at low altitude. The airplane was seen to enter a turn, continuing in the turn until it flew into the ground and burst into flames. It was speculated that the instructor, flying from the rear cockpit, had lost the horizon in the turn and inadvertently flew the airplane into the ground.

10-24-44A. Barksdale Field, Louisiana. At 0406, a Boeing B-29 crashed after it collided with trees just after take-off from Barksdale Field, Shreveport, Louisiana, killing 11 fliers. The airplane was taking off on a navigation mission back to its home station at Harvard Army Air Field, Harvard, Nebraska. The Accident Classification Committee stated, "The subject pilot took off from Harvard, Nebraska, 23 October 1944, at about 0830, en route to Whalerock, to Galveston, Texas, and back to Harvard, Nebraska, on a routine training flight. Due to a faulty engine instrument [the B-29 landed] at Lake Charles, Louisiana, and the subject pilot was instructed over the telephone by some authority at Harvard, Nebraska, to complete his mission and return to Harvard. The pilot took off from Lake Charles at about 1700, 23 October 1944, and apparently finished his mission and was on his way back to Harvard, Nebraska, when he called into Barksdale Field at 0115, 24 October 1944, and asked for landing instructions in order to land and refuel. The purpose for landing at this station, as stated by the pilot, was because he could not transfer fuel from the bomb bay fuel tanks to the main fuel tanks, and therefore needed more gasoline to complete the mission to Harvard. After the subject pilot landed, his aircraft was serviced with 3,700 gallons of 100-octane gasoline, and he immediately proceeded to take off from Barksdale Field at 0405 CWT, 24 October 1944. It was approximately 50 seconds after the actual take-off that the subject aircraft crashed and burned two miles south of Barksdale Field. It is believed ... that the subject pilot took off from this field, using his usual take-off procedure, and that he did not pay close attention to his altitude. As a result he hit a tree with his left wing and this threw the subject airplane out of control and it crashed." Killed in the crash were: Capt. Paul T. Dowling, pilot; 2Lt. Robert E. Powers, co-pilot; 2Lt. Melvin R. Plunkett, passenger; 2Lt. Louis Fueyo, passenger; TSgt. C.F. Olsen, engineer; Sgt. Cyril T. Martin, radio operator; Cpl. Werner F. Schmidt, gunner; Pfc. John I. Nowak, gunner; Pfc. Harold F. Newfield, gunner;

Pfc. Kent A. Lentz, gunner; Pfc. Joaquin J. Ramos, gunner.

10-24-44B. Russell, Kansas. At 1050 CWT, a Boeing B-17F suffering apparent engine trouble crashed while attempting an emergency forced landing one-quarter mile west of Russell, Kansas, killing 12 crew-members. The airplane took off from Walker Army Air Field, Walker, Kansas, on a routine training mission. The airplane was observed flying west, north of Highway 40, toward Walker Army Air Field at an altitude of about 1,000 feet with the engines laboring. The airplane was observed to go into a spin to the left. The pilot stopped the spin and leveled off momentarily, but the altitude was too low to allow a complete recovery. The airplane smashed into the ground at a 45-degree angle, burying the four engines several feet into the terrain and telescoping the fuselage and throwing it ahead of the impact point. The airplane exploded into flames upon impact and the crew was killed instantly. Killed in the crash were: Capt. Charles E. Gibson, Jr., B-17 instructor pilot; Capt. Herbert J. Keading, pilot; 2Lt. Fred B. Swank, co-pilot; 2Lt. John T. Olakayia, bombardier; 2Lt. James N. Rector, navigator; TSgt. Arnold P. Simmons, Jr., engineer; Sgt. Marce Bartino, engineer; Cpl. Lloyd Sheets, radio operator; Cpl. Jack W. Smith, gunner; Cpl. Peter J. Smith, gunner; Cpl. Andrew L. Snyder, gunner; Cpl. Donald A. Palmer, gunner.

10-24-44C. Florence, South Carolina. At 1605, a Douglas A-20J crashed five miles southeast of the Army Air Field at Florence, South Carolina, killing Capt. James H. Ullett, who fell to his death in an unsuccessful parachute jump. Investigators stated, "Pilot was on a test hop and at an altitude of approximately 14,000 feet and made a right bank of about 50 or 60 degrees. The left wing stalled out and the airplane flipped over to the left and on its back. At this time the escape hatch flew off and airplane went into a tail up [attitude of about] 30 to 35 degrees [and then entered] a flat spin to the left. It spun approximately five turns and then the pilot bailed out. Apparently he was caught by the propeller and his parachute ripped open as the ripcord was still intact upon examination."

10-24-44D. March Field, California. At 2222, a Consolidated B-24J crashed three miles northwest of March Field, Riverside, California, killing six fliers. The airplane took off from March Field at 2215 and about three minutes later the pilot radioed the control tower that he was suffering a fire in the number-two engine and was requesting clearance for an emergency landing. The tower cleared the airplane to land but it crashed into the ground and exploded moments later. A motorist observed the airplane to enter a steep bank at low altitude. The airplane then dropped into the ground at a steep angle and exploded about 500 yards from the South Radio Range. The airplane was on a scheduled cross-country navigation mission. Killed in the crash were: 2Lt. Roy J. Cole, pilot; 2Lt. William

A. Deverell, co-pilot; 2Lt. Anthony V. Galioto, navigator; 1Lt. Joseph E. Bowman, navigator instructor; TSgt. Harold Bentley, engineer; Cpl. James L. Bertrand, radio operator.

10-25-44A. El Reno, Oklahoma. At 0845, two Lockheed F-5E airplanes collided in mid-air 13 miles north of El Reno, Oklahoma, killing 2Lt. Clarence T. Moore, Jr. who was killed in the crash of F-5E #44-24556. 2Lt. Archie N. Nix was able to parachute to safety from F-5E #44-24703. He was uninjured. Lt. Nix later stated to investigators, "About 0830 I headed northwest and came in from the southeast to the northwest, between Okarche and Kingfisher. This other airplane came from the north, headed a little bit southwest. I turned and followed about 500 feet above him and approximately one mile behind him. The aircraft in front turned left and then turned back to the original heading. I followed him in this maneuver at the same distance. I then made a turn to the right at which time he too turned [and then] I lost sight of the airplane while watching the inside of my turn. While in the turn I looked down between the nacelle and the gondola, and the leading edge of the wing on the left side, and noticed a wing of an aircraft right under me. I did not know what happened after that until finding myself in a dive, upside down. Immediately reaching for the escape hatch with my left hand the safety belt with my right hand, I pulled them simultaneously. I then pulled the ripcord and the parachute opened immediately. While floating down in my parachute, I watched an aircraft diving and burning, which was losing altitude at a steep angle, leaving a black trail of smoke. This was the aircraft that Lt. Moore was in."

10-25-44B. Morris Field, North Carolina. At 1145, a Douglas A-20G crashed five miles northeast of Morris Field, Charlotte, North Carolina, killing pilot 2Lt. Merle E. Rudy, Jr. The airplane was part of a flight that had taken off from Morris Field on a routine formation flight. The flight climbed to about 4,000 feet msl and flew formation for a short time. Lt. Rudy radioed the flight leader and alerted him that his starboard cylinder head temperature was getting high and he had to return to base. The flight leader advised Lt. Rudy to open the lower cowl flaps. This helped lower the cylinder head temperature and the flight soon arrived over Morris Field. The subject pilot closed the cowl flaps halfway and the cylinder head temperature returned to normal. The flight leader elected to continue with the formation mission. Reliable observers on the ground reported that the starboard engine was heard to cut out. The airplane began skidding to the right and then out of the formation. Lt. Rudy then radioed that his "left" engine had cut out and he had feathered the propeller. The airplane entered a spiral and crashed to the ground where it exploded violently into flames. Investigators speculated that the pilot had either inadvertently feathered the port engine, or feathered it on purpose to facilitate a safe bail out, and then changed his mind and attempted to restart the port engine. The pilot failed to restart the engine and lost considerable airspeed, causing the airplane to stall and smash into the ground.

10-25-44C. Madras, Oregon. At 0850, a Lockheed P-38L crashed ten miles SSE of Madras, Oregon, killing pilot 2Lt. Jack R. Kirby. The airplane was part of a four-ship flight that had taken off from the Army Air Field at Redmond, Oregon, on a routine gun-camera and formation flight. The flight had successfully completed the gun-camera portion of the mission and the flight leader ordered the flight to go on single-engine flight. After flying on single engine for a short time the leader ordered the pilots to restart their feathered engine. The flight began to join up at 12,000 feet and practice formation maneuvers when the number-two ship of the flight lost its cockpit canopy. The number-two ship left the formation and began to lose altitude. The flight leader signaled the flight to form up in a V-flight. The flight leader then put the formation in a slight dive in an attempt to catch up with the number-two ship. Members of the flight never saw Lt. Kirby's airplane again and only learned that it was missing from the flight when they returned to land at Redmond. Investigation revealed that the airplane entered a dive at an altitude of about 10,000 feet and failed to recover in time to avoid smashing into the ground at high speed. The P-38 exploded violently into flames upon impact.

10-25-44D. Seymour-Johnson Field, North Carolina. At 1200, a Republic P-47G crashed one mile southeast of Seymour-Johnson Field, Goldsboro, North Carolina, killing pilot 2Lt. James J. Paunoff. The P-47 was part of a three-ship flight that was returning to the field after completing a navigation training flight. The subject airplane was seen to overshoot the runway on final approach and then the pilot advanced the throttle for an attempted go-around. It was noted that the pilot was slow in retracting the landing gear and flaps. The airplane was seen in a nose-high position as the pilot attempted a steep left turn at low airspeed. The airplane stalled and snapped over on its back before entering a spin at an altitude too low to allow recovery. The airplane did not burn but was destroyed utterly. The pilot was killed upon impact.

10-25-44E. Pocatello, Idaho. At 1225 MWT, a Republic P-47D crashed three miles east of the Army Air Field at Pocatello, Idaho, killing 2Lt. George S. Manos, who fell to his death in an unsuccessful parachute attempt. Investigators stated, "Lt. Manos took off ... at 1115 on a camera gunnery mission. After completing the mission he returned to the field and called in for landing instructions. When at approximately 7,000 feet, his wingman informed him that he was trailing white smoke. He was then instructed by the tower to jump. His airplane was seen to have gone into a dive and struck the ground at a 45-degree angle. The airplane was completely destroyed, preventing a thorough

investigation. The pilot was seen to leave the ship at a very low altitude and made an unsuccessful jump."

10-25-44F. Webster, South Dakota. At 1550 CWT, a Douglas A-26B crashed while attempting an emergency forced landing five miles west of Webster, South Dakota, killing pilot 1Lt. Ray D. Schott and seriously injuring engineer SSgt. Donald R. Asper. The airplane took off at 1352 CWT from Watertown Army Air Base, Watertown, South Dakota, on a routine training flight. The crew discovered that the forward cabin heaters were unsatisfactory due to excessive smoking. The airplane landed at 1502 and switched engineers and then took off for a second time at 1525. The airplane climbed to 4,000 feet msl and leveled off in cruise flight. The ducting heater was inoperative but the recirculating heater was apparently operating satisfactorily. This heater was operated without apparent problem for about 15 minutes and then it was turned off. The heater continued to get hot even after being turned off. Despite numerous efforts to shut down the heater by the pilot and engineer, the heater continued to get hotter until it finally erupted into flames. The engineer was unable to extinguish the flames with hand-held fire extinguishers so the pilot set up an approach for an emergency landing on what appeared to be flat terrain. The terrain was actually rough and rolling with an uneven upslope of about 35 degrees. The pilot was able to set the airplane down gently despite the cockpit filling up with smoke. The airplane touched down and careened across the terrain, digging in and flipping over on its back. The airplane broke up and burst into flames. A civilian was able to rescue the engineer from the shattered wreckage.

10-25-44G. Victoria, Texas. At 1820, a North American AT-6C suffered a catastrophic structural failure and crashed nine miles north of Aloe Army Air Field, Victoria, Texas, killing A/C Richard O. Kolley. The airplane took off from Aloe Field on a scheduled aerobatic flight. The student climbed to 5,000 feet msl and flew to the assigned aerobatic area and began performing a series of rolls and loops. A short time later, civilian witnesses on the ground observed the airplane falling to the ground minus the starboard wing. The airplane smashed to the ground and burst into flames.

10-25-44H. Tonopah, Nevada. At 2315, a Consolidated B-24J crashed while attempting an emergency landing at the Army Air Field at Tonopah, Nevada, killing five fliers and seriously injuring four others. Killed in the crash were: F/O John O. Mauldin, navigator; 2Lt. Elmer M. Heckman, bombardier; Cpl. Louis Sabatasac, engineer; Cpl. Evert C. Sewall, gunner; Pfc. Herbert Fitzgerald, gunner. Crewmembers seriously injured in the crash were: 2Lt. Henry T. Rogers, pilot; 2Lt. Wilburn J. Norman, co-pilot; Cpl. William M. Mogan, radio operator; Cpl. Emery I. Bryant, gunner. Investigators stated, "Lt. Rogers and crew were on a night bombing mission at 20,000 feet. As they approached their assigned altitude, the number-one engine was noticed to be running a [high] cylinder head temperature.... The pilot feathered the number-one engine to practice feathering and to cool it off. After feathering the number-one engine the pilot stated [later] that the airplane was hard to hold on course therefore he feathered the number-four engine. With both outboard engines feathered at 20,000 feet, the ship began to lose altitude very fast and the pilot called the tower for an emergency landing and [then began] to descend over the field. The pilot stated that he feathered and unfeathered both number-one and number-four engines a couple of times on the way down and also stated that he wasn't getting much power from any engines. The airplane was losing altitude so fast that after turning from base leg he was unable to turn onto his final approach and was forced to straighten out and make a crash landing [straight ahead] resulting in five fatalities and complete destruction of the airplane." Investigation revealed that the pilot never leveled out during the rapid glide and the B-24 smashed into the terrain in a nose-down attitude about one mile north of the field, causing the airplane to break up. The airplane did not burn but was destroyed utterly.

10-26-44A. Reno, Nevada. At 0603 PWT, two Curtiss C-46A transports collided in mid-air and crashed at Reno Army Air Base, Reno, Nevada, killing 12 fliers. Killed in the crash of C-46A #42-96716 were: 2Lt. Olin Thomas Jackson, C-46 instructor; 1Lt. Irvin Frank Hoeper, pilot; F/O Robert F. Johnson, co-pilot; Pfc. Donald H. Kraft, engineer; Cpl. Henry J. Long, engineer; TSgt. James F. Smith, radio operator; SSgt. Frederick E. Urweider, Jr. radio operator. Killed in the crash of C-46A #41-12357 were: 1Lt. Paul V. Frykman, C-46 instructor; F/O Frederick W. Dulfer, Jr., pilot; 1Lt. Robert H. Heintz, co-pilot; Cpl. Leland F. Heckman, engineer; Cpl. Robert Y. Goebel, engineer. Investigators stated, "According to eyewitnesses and the location of the debris it is apparent that the collision occurred near the center of the field [at 2,000 feet agl] from east to west but near the north edge of the field which indicates that Lt. Jackson was following the prescribed instrument let down.... The location of the two airplanes on the ground indicated that Lt. Jackson was on the prescribed heading and that Lt. Frykman was heading approximately 225 degrees. This is apparent by the fact that both aircraft fell approximately one-half mile west from the point of collision.... [Evidence] indicates that Lt. Frykman ran into Lt. Jackson. All members of both aircraft were killed instantly. It is assumed that [a propeller from] Lt. Frykman's airplane went through the right side of the cockpit of Lt. Jackson's airplane since the crewmembers in the cockpit of Lt. Jackson's airplane were badly cut up and found approximately one-half mile from the main wreckage. The pilot seats, radio equipment, and other material from the cockpit of Lt. Jackson's airplane were strewn along the path of flight, which further indicates that the propeller of Lt. Frykman's airplane went through cockpit of Lt. Jackson's airplane. The nose section of

Lt. Jackson's airplane fell off and was shedding parts as it fell earthward."

10-26-44B. Albuquerque, New Mexico. At 1415, a Vultee BT-13A crashed ten miles west of Kirtland Army Air Field, Albuquerque, New Mexico, killing pilot 2Lt. William C. Shelton, Jr. and passenger Sgt. George S. Race. The pilot took off from Kirtland Field on a proficiency flight. Investigation revealed that the airplane entered a spin to the right at a safe altitude. The pilot was able to stop the spin to the right, but the airplane entered a spin to the left. The pilot was able to stop the spin to the left, but the recovery was made at an altitude too low to allow recovery.

10-26-44C. Guadalupe, California. At 1430, a Lockheed P-38J crashed four miles west of Guadalupe, California, killing pilot 2Lt. Edward B. Hall. The airplane was part of a four-ship flight that had taken off from Santa Maria Army Air Field, Santa Maria, California, on an aerial camera-gunnery mission. The subject airplane was delayed on the ground for a short time because of radio problems and the pilot was not able to take off with his flight. The subject airplane eventually took off and joined the flight in the gunnery pattern as the number-four ship. After completing the exercise, the flight leader ordered the flight to assemble in a normal tactical formation. The flight leader ordered Lt. Hall to attack one of the elements from the rear. Lt. Hall acknowledged and broke away from the formation to conduct the attacks. The formation flew between Orcutt, California, and Nipomo, California, waiting for Lt. Hall's attack. After waiting ten minutes, the flight leader ordered the flight to form up for the flight back to base. The flight leader called Lt. Hall on the radio but he received no response. The flight leader was alerted to the crash and the pilots searched the Guadalupe area but they were unable to locate Lt. Hall's airplane. The airplane had struck the ground at a steep angle and was completely destroyed. The cause of the crash was unknown.

10-26-44D. Zorn, Texas. At 1720, a North American TB-25D suffered a catastrophic structural failure and crashed two miles east of Zorn, Texas, killing four crewmembers. The airplane had taken off from Randolph Field, San Antonio, Texas, on a routine training flight. Investigators stated, "Inspection of the wreckage disclosed that both wings had broken off at a point about where the landing lights are located, and that both rudders and vertical stabilizers were missing. These parts were found approximately one-half mile from the location where the fuselage and engines were located. Inspection disclosed that the right wing, which was torn from the airplane, was in two sections, front and rear. The left wing, which was torn from the rest of the airplane, disclosed the wingtip to be torn off, and it was found approximately 500 yards from the wing. The right vertical stabilizer appeared to have been struck by part of the wing. The left rudder appears to have been struck by the left wing, probably when the tip was torn off. The right flap was found about 300 yards away from the right wing. Both propellers were in the full-feathered position.... It is believed that this airplane was quite high when the wings, vertical stabilizers and rudders tore loose because of the length of time it took the wings to float down to earth. It is also believed that the aircraft was traveling at high speed and had an excessive load in units of gravity placed on it, for all parts examined that floated to the ground, and which were subject to very little impact with the ground, showed evidence of rivets torn loose, stringers twisted and a general all-around tensile torsion strain.... No witnesses could be found who saw the failure of the wings in flight." Killed in the crash were: 1Lt. Alvin C. Lawrence, pilot; F/O Henry L. Thompson, co-pilot; 2Lt. Robert P. Rhyne, passenger; Sgt. John P. McNulty, engineer.

10-26-44E. Location Unknown. At an unknown time after 1545 PWT, a North American P-51D disappeared after taking off from Mines Field, Los Angeles, California. WASP pilot Gertrude V. Tompkins was missing and presumed killed. The P-51D was one of 50 that had taken off from Mines Field individually and were being ferried to an undisclosed location east. Investigators stated, "The aircraft was not reported as missing for several days as the Tower had not received the flight plan correctly, nor had a flight plan been filed with Airway Traffic Control. The error occurred in that the ship number was never received by the Tower, either by an error in calling in the list of flight plans, or in copying them at the tower. No departure time was recorded by the Tower operator; however, the aircraft was cleared into position and cleared for take-off by the tower, as shown by transcripts on file at the Mines Field Tower." Investigators noted that a heavy haze was prevalent in the area at take-off time and that other pilots had to climb to 2,500 feet agl to break out of it.

10-27-44A. Daggett, California. At 0804, a Lockheed P-38J crashed four miles west of the Army Air Field at Daggett, California, killing 2Lt. Paul J. Schwegler. The airplane was part of a five-ship flight on an aerial gunnery mission. After completing the mission, the formation flew toward Daggett Field for a landing. A short time later, Lt. Schwegler dropped back and feathered his port propeller. He radioed Daggett Field control tower for landing instructions and then headed for the field. The subject airplane was seen to snap roll and then enter a spin, describing about five turns before smashing to the ground and exploding. The pilot made no apparent effort to parachute to safety.

10-27-44B. Gunter Field, Alabama. At 1515, two North American AT-6D airplanes collided in mid-air and crashed three miles north of Gunter Field, Montgomery, Alabama, killing A/C Theodore J. Burge. Pilot Lt. Col. Carver T. Bussey and pilot-rated passenger Maj. Robert S. Moon were able to parachute to safety from AT-6D #42-86010. They were uninjured. Investigators stated, "Lt. Col. Bussey and Maj. Moon ... had been flying in the local flying area for an hour and forty minutes observing student flying and checking on conditions of the auxiliary airfields and were approaching Gunter Field from the northwest at an altitude of approximately 2,200 feet. A/C Burge had just taken off into the north from Gunter Field and was climbing in the general direction of northwest, heading for Deatsville Auxiliary Airfield [Alabama]. The first Lt. Col. Bussey was aware of Cadet Burge's airplane it was slightly below and to the left of his flight path but climbing directly toward him. At this time Lt. Col. Bussey immediately pulled up and away from the oncoming airplane but Cadet Burge went into a bank to the right and continued his climb with the result that Lt. Col. Bussey's evasive action was nullified and the two airplanes collided, striking wings. Lt. Col. Bussey's airplane continued straight ahead [losing altitude and then going out of control] and both occupants used their parachutes successfully. A/C Burge's airplane went into a spin from which it did not recover. A/C Burge either made no attempt to use his parachute or delayed action too long before deciding to use it, his body being thrown clear of the wreckage but evidence indicated that he was still in the airplane when it hit the ground and exploded."

10-28-44A. Kansas City, Kansas. At 1350 CWT, a Douglas A-26B suffering engine trouble struck a building and crashed near Fairfax Field, Kansas City, Kansas, killing 2Lt. John V. Postemsky. The airplane had taken off from Fairfax Field on a ferry flight to Chatham Field, Savannah, Georgia. Investigation revealed that the pilot started the port engine after three unsuccessful attempts. The pilot then taxied out to Runway 22 and performed the engine run-up and the take-off checklist. The engines checked out so the pilot taxied out, lined up and began his take-off on Runway 22. The airplane became airborne about two-thirds of the way down the runway and climbed slightly before the port engine lost significant power. The pilot was barely able to maintain altitude and the starboard propeller tip struck a two-story building near the field, severing about four inches of the propeller tip. The A-26 circled the field to the right at very low altitude; the pilot had feathered the starboard propeller and the port engine performance was seriously degraded. The control tower operator stated later that the pilot was calm in his transmission to the tower and he had patiently flew the traffic pattern in preparation to landing. The tower cleared the pilot to land on Runway 33 or 35 and gave him pertinent wind information. The pilot lost altitude gradually and extended the landing gear. The airplane continued in a gradual right turn to final approach. The pilot overshot slightly and attempted to skid the airplane to the right in an effort to line up with the runway. The airspeed was very low and the airplane stalled and entered a spin at an altitude too low to allow recovery. The airplane smashed into the ground nose-first and exploded into flames upon impact. The pilot was killed instantly.

10-28-44B. Chatham Field, Georgia. At 2117 EWT, a Consolidated B-24J crashed while attempting an emergency landing at Chatham Field, Savannah, Georgia, killing four crewmembers and injuring six others. The airplane had taken off at 2053 on a 1,000-mile over-water navigation flight. After reaching an altitude of about 5,000 feet, the number-three engine oil pressure dropped to zero. The pilot attempted to feather the propeller but the feathering mechanism malfunctioned, causing the propeller blades to continue to rotate until they passed through the feathering position. The propeller did not feather and began to windmill. The pilot attempted to feather the number-three propeller a second time but was unsuccessful. The pilot returned to the field and requested landing instructions. As the airplane turned from the downwind leg, the number-four engine oil pressure began to drop rapidly. The engineer quickly hit the feathering button and the number-four propeller feathered properly. The pilots were able to trim the airplane and maintain control. The co-pilot extended the landing gear as the airplane turned from base to final. The pilot noticed that he was going to undershoot the runway and added power to the two remaining engines. The pilot was unable to maintain altitude and the airplane descended into the ground, striking a railroad roadbed about 1,000 feet from the end of the runway. The airplane broke up as it slid toward the runway, the main center section erupting into flames as it came to rest. Killed in the crash landing were: Pfc. Stephen G. Penna, assistant engineer; Pfc. Harlan H. Wyman, assistant radio operator; Pfc. Richard D. Hodgson, gunner; Cpl. Alexander P. Tochman, gunner. Crewmembers seriously injured were: F/O Edwin A. Pearson, bombardier; F/O Francis Harding, navigator; Cpl. Clayton E. Kukuk, engineer; Cpl. John J. Kashino, radio operator. Pilot 2Lt. William E. Davis, Jr. and co-pilot F/O Gordon H. Smith, Jr. received minor injuries.

10-29-44A. Grandview, Missouri. At 0830, a Douglas C-47 crashed three and one-half miles southeast of Grandview Municipal Airport, Grandview, Missouri, killing three fliers. The airplane had taken off from Sedalia Army Air Field, Warrensburg, Missouri, on a routine training flight. The subject airplane was in a right-hand traffic pattern for a landing on Runway 24 at Grandview. The airplane turned from downwind to the base leg and encountered another C-47 on the base leg, but flying in a left hand pattern. The subject pilot turned off of the pattern to allow the other C-47 to make its turn from base to final and make their landing. The subject pilot attempted to maneuver back into the traffic pattern and was seen in a steep bank to the right. The airplane stalled and fell into a spin at an altitude too low to allow recovery, smashing into the

ground and exploding into flames. C-47 instructor pilot 1Lt. George P. Tierman, pilot F/O Raymond C. Ernst and co-pilot F/O Edward L. Geim were killed in the crash.

10-29-44B. Chatham Field, Georgia. At 1423 EWT, a Consolidated B-24J attempting an emergency landing crashed two miles east of Chatham Field, Savannah, Georgia, killing the crew of ten. The airplane had taken off from Chatham Field on a high altitude formation flight. Soon after take-off the number-one engine was lost and the propeller feathered. Moments later, the number-three engine was seen to be trailing thick smoke. The pilot radioed the tower for landing instructions and then returned to the field for an emergency landing. The pilot entered the pattern and made an approach but overshot his landing. The pilots retracted the landing gear, entered a gradual left turn and attempted to go around. The airplane continued in the turn, losing altitude until it mushed into the ground and burst into flames. Killed in the accident were: 2Lt. Robert G. Clegg, pilot; F/O Edwin W. Smith, co-pilot; F/O James T. Wilson, bombardier; 2Lt. James H. Vogel, navigator; Cpl. Aldro J. Stedman, engineer; Pfc. Michael S. Minerivini, assistant engineer; Cpl. Eugene Kargher, radio operator; Pfc. Robert P. Uhlhorn, assistant radio operator; Cpl. Walter E. Tannehill, gunner.

10-29-44C. Lake Arbuckle, Florida. At 1613 EWT, a Boeing B-17G crashed into Lake Arbuckle three miles northwest of the Army Air Field at Avon Park, Florida, killing nine crewmembers. Tail gunner Sgt. Harold H. Barar received minor injuries parachuting to safety. Killed in the crash were: 2Lt. James C. Murphy, pilot; F/O George P. Gaudet, co-pilot; F/O John G. Evernden, navigator; 2Lt. Mortimer A. Alnwick, bombardier; Sgt. Harvey R. Ford, engineer; Cpl. Salvatore C. Bongiovanni, radio operator; Pfc. Frank W. Bannerman, assistant radio operator; Cpl. George L. Heffington, gunner. Investigators stated, "Lt. Murphy took off [from Avon Park] at 1604 EWT on a routine training mission. Following take-off a normal climb was assumed. After leaving the traffic pattern and upon reaching an estimated altitude of 3,000 feet, the tail gunner noticed that the number-four engine was smoking. At this time the airplane was on an approximate heading of west. The tail gunner stated that smoke died out several times momentarily and then reoccurred. Each time it reappeared it had grown in intensity, until finally the inspection cover on the upper part of the number-four nacelle blew open and flames appeared. The co-pilot ordered the crew to abandon the airplane immediately. While attaching the parachute pack, the tail gunner stated that he felt terrific vibrations and heard a series of explosions. Shortly afterwards number-four engine and nacelle fell off; whereupon the airplane began a steep spiral to the right. The tail gunner managed to leave the airplane only a few seconds before the airplane hit the water and exploded.... The definite cause of the accident cannot be fully determined because of the lack of sufficient evidence."

10-29-44D. Petrolia, Texas. At 1610, a Fairchild PT-19B crashed near the AAF Auxiliary Airfield at Petrolia, Texas, killing instructor 2Lt. Mitchell Gralnick and flying student Pvt. Claude E. Howard. Investigation revealed that the airplane, flying at about 300 feet agl, stalled and entered a spin while turning from the base leg to final approach. The instructor stopped the spin, but there was not enough altitude to allow a full recovery and the airplane smashed into the ground about one-quarter of a mile east of the field.

10-30-44A. Shelby, Mississippi. At 0830, a Vultee BT-13A crashed six miles WNW of Shelby, Mississippi, killing instructor 2Lt. Henry B. Stimson, Jr. and student A/C Joseph A. LaRosa. The airplane had taken off at 0800 from the Army Air Field at Greenville, Mississippi, on a routine student instrument training flight. Lt. Stimson was in the front cockpit as the safety pilot and the student was in the rear cockpit as the instrument pilot. The airplane was seen circling the Brooks Plantation at about 500 feet. The airplane circled twice before entering a shallow dive and then pulling up in a steep climb. The airplane stalled and entered a spin to the left at an altitude too low to allow recovery. The airplane smashed into the ground and burst into flames. Investigation revealed that the instructor pilot and friends had spent the previous weekend at the plantation and it is likely that the pilot was attempting to attract the attention of people he knew on the ground at the plantation. It was also noted that the airplane was 14 miles outside of the authorized instrument training area.

10-30-44B. Xenia, Ohio. At 1229 CWT, a Lockheed P-38H broke up in mid-air and crashed six and one-half miles east of Xenia, Ohio, killing pilot 1Lt. Leigh S. Hall, Jr. The airplane had taken off at 1215 from Wright Field, Dayton, Ohio, on a test flight of a new type of oil cooler pump installed on the starboard engine. Investigation revealed that the airplane had crashed about 14 minutes after take-off. Examination of the wreckage indicated that the airplane crashed into the ground in a relatively flat attitude with the starboard wing separated. The tail section and port engine had separated in flight and were found some distance from the main wreckage. The pilot had made no attempt to parachute to safety. Investigators speculated that the pilot might have suffered from oxygen starvation and had inadvertently allowed the airplane to go out of control.

10-31-44. Abilene, Texas. At 1405, a Republic P-47D crashed one-half mile southeast of the Army Air Base at Abilene, Texas, killing 2Lt. Robert L. Kuhl. The airplane was flying in the number-two position of a two-ship flight that had taken off from Abilene AAB on a camera gunnery mission. The subject pilot radioed the flight leader that he was returning to the field because of propeller trouble. Investigators stated, "Upon

reaching the vicinity of the field, he called the tower to clear the pattern so he could make a straight-in approach. Upon interrogation he stated that it was not an emergency. His approach was very poor and instead of a correct straight-in approach, he was high enough so that he 'peeled off.' He put his wheels down too late so that when he was in a position to land, the wheels were not yet fully down. Runway control advised him to go around. He started his wheels up and started climbing. Then it appeared he started to lose altitude and at the same time he made a left turn. After completing the turn, the airplane stalled in a nose-high attitude, rolled on its back and hit the ground at a steep angle. The airplane burned and the pilot was killed instantly."

November

11-1-44A. Laredo, Texas. At 1023 CWT, a Consolidated B-24H exploded in mid-air and crashed nine miles northeast of Laredo Army Air Field, Laredo, Texas, killing seven fliers. The B-24 took off from Laredo Army Air Field at 0938 on a routine engineering test flight after a number-four engine change. After successfully performing the test flight, the pilot radioed the tower at 1013 for permission to shoot landings. Permission was granted and the airplane flew toward the field to enter the traffic pattern. At 1023 an explosion occurred in the forward bomb bay, blowing off the forward bomb bay doors and causing them to collide with the tail section. Parts of the vertical fin, rudders and the forward bomb bay doors were found about one-half mile from where the airplane smashed into the ground at an angle approaching 90 degrees. The airplane exploded violently into flames upon impact with the ground and all seven on board were killed instantly. Investigators speculated that gasoline fumes had gathered in the forward bomb bay and were possibly ignited by the operation of the auxiliary hydraulic pump. It was later speculated that the airplane might have suffered from faulty hydraulic fittings that allowed atomized hydraulic fluid to be sprayed out and ignited by current produced by the operation of the auxiliary hydraulic pump main switch. There were no witnesses to the accident and its cause remains undetermined. See 10-20-44A for a similar B-24H accident. Killed in the crash were: 2Lt. Richard F. Bancroft, pilot; 2Lt. Earl W. Cantebury, co-pilot; SSgt. George T. Malone, engineer; 2Lt. Horace P. Crudo, pilot-rated passenger; Sgt. Edwin E. Helton, passenger; Sgt. Ralph Owens, passenger; Pfc. Leon H. Yauk, passenger.

11-1-44B. Amarillo, Texas. At 1330, a Vultee BT-13B crashed two miles north of the Army Air Field at Amarillo, Texas, killing pilot F/O Grosvenor Gilbert and pilot-rated passenger 2Lt. Richard O. Waibel. The airplane was seen to go into a spin at an altitude of approximately 2,000 feet agl. The pilot stopped the spin, but the airplane entered a second spin in the opposite direction. Investigators stated that the airplane continued in the spin in a "loose, mushy, and flat manner, and crashed into the ground." The airplane burst into flames upon impact with the ground.

11-1-44C. Victoria, Texas. At 1725, a North American AT-6D crashed seven miles northwest of Foster Field, Victoria, Texas, killing A/S Francis J. Ford. Instructor 2Lt. Rivon E. Jones parachuted to safety and was uninjured. The airplane was part of a three-ship flight on an authorized aerobatic mission. The instructor climbed the airplane to 6,000 feet and waited for the other two airplanes to arrive. The other two airplanes failed to arrive at the designated point and the instructor proceeded on the mission and began to perform aerobatic maneuvers. The instructor had trimmed the rudder to the left to allow an ease of control at speeds above cruise. The instructor and student performed various aerobatic maneuvers in the authorized area and at a safe altitude. After the session, the instructor took the controls and flew toward the field. The instructor decided that he would turn the controls back over to the student and then perform an unannounced simulated engine failure by turning the fuel selector valve to the off position. The instructor turned the airplane over to the student and then cut the engine moments later. The instructor was looking down in the cockpit at the fuel pressure light when he felt pressure on his lap from the safety belt. The pressure increased and the instructor suddenly noticed that the airplane was pointing nearly straight down. The instructor attempted to get control of the aircraft but was unsuccessful. He then ordered the student to bail out. He looked back and noticed that the student was looking right at him with a "vacant stare." The instructor again ordered the student to bail out. The instructor bailed out and parachuted to safety. The instructor did not recognize that the airplane was seriously out of trim and it was learned that the instructor had not re-trimmed the rudder before turning over control of the airplane to the student. Investigation revealed that the student mistakenly moved the elevator trim instead of the rudder trim during the simulated engine failure. When the airplane entered an extreme nose-down attitude, the student apparently froze on the controls and prevented the instructor from regaining control. Investigation revealed that the student disliked aerobatic flight and had become ill on nearly every aerobatic session.

11-2-44A. Menlo, Iowa. At 0430 CWT, a Boeing B-17G suffering a fire in flight crashed six miles southwest of Menlo, Iowa, killing six fliers. Two crewmembers were injured and another uninjured parachuting to safety. The airplane took off at 0335 CWT from Lincoln Army Air Field, Lincoln, Nebraska, on the first leg of a flight overseas via Grenier Field, New Hampshire. The heavily loaded airplane climbed to 9,000 feet and soon afterward smoke and a light spray of oil was seen trailing from the number-one engine. The pilots turned back for Lincoln but did not shut down the engine and feather the propeller. The oil that was trailing was set on fire by the supercharger flame. The fuel lines behind the engine ruptured and caught fire inside the engine nacelle. The fire became so intense that the number-one engine and part of the nacelle fell off of the airplane. The airplane rolled out of control and spun to earth, exploding into flames upon impact. Killed in the crash were: 1Lt. Robert C. Jordan, III, pilot; 2Lt. Melvin C. Ehrman, co-pilot; 1Lt. Alan I. Rohrbach, navigator; 1Lt. Harold L. Ray, bombardier; Cpl. William A. Hudgens, gunner; Leroy D. Crick, gunner. Gunner Cpl. John M. Kerrigan was seriously injured parachuting to safety. Gunner Sgt. William B. Hensley sustained light injuries and gunner Cpl. Everett A. Covington was uninjured parachuting to safety.

11-2-44B. Pampa, Texas. At 1005, a Lockheed F-5E crashed 18 miles southeast of the Army Air Field at Pampa, Texas, killing pilot F/O Robert H. Schaner. The airplane took off from Oklahoma City, Oklahoma, on a photo mission. The airplane struck the ground at high speed at an angle of 50 degrees. The pilot's body was found one mile southeast of the main wreckage. The pilot's leather jacket and parachute harness were found some distance from the pilot's body. Numerous pieces of the parachute were found scattered over the terrain. Investigators could not determine the cause of the accident.

11-2-44C. Chatham Field, Georgia. At 1237 EWT, a Consolidated B-24J crashed while attempting an emergency landing at Chatham Field, Savannah, Georgia, killing three fliers and injuring six, two of them seriously. One crewmember escaped injury. The B-24 took off on a scheduled high-altitude formation/camera-gunnery mission at 0741 EWT. The airplane returned to the field because of an improperly seated fuel tank cap. Investigators stated, "At 0920 [the airplane] departed on the originally scheduled mission. During the mission the pilot stated that he could not get but 27 inches Hg from the number-one engine but was able to complete the mission. While letting down from altitude the pilot reduced the power settings on all four engines.... Between 6,000 and 8,000 feet the number-three and number-four propellers ran away. The pilot and co-pilot both decided it was one of these two engines so without attempting any check to ascertain which engine propeller was giving the trouble, the pilot feathered the number-four propeller. The [propeller] definitely feathered and as the noise continued the pilot then knew that the runaway propeller was on the number-three engine and continued to unfeather the number-four propeller. The pilot was attempting to feather number-three engine, but was unsuccessful.... The armorer gunner in the tail section called the pilot via the interphone and stated that the number-four engine was on fire. Immediately the pilot feathered the number-four engine and ordered the engineer to cut the gasoline off to the number-three and number-four engines.... The armorer gunner stated that he definitely saw flames coming from number-four engine prior to his calling the pilot and that the flames were approximately two feet long coming from around the supercharger and around the cowl flaps.... The airplane continued to lose altitude and at approximately 3,000 feet over the field, the co-pilot called Chatham tower and reported being over the field with two engines out and was going to land on Runway 5 to the northeast. The tower cleared the field for the landing. The airplane made a base leg, which was estimated to be at least 1,000 feet by witnesses. After starting the turn from base leg to final approach the pilot put the landing gear lever in the down position and the left gear and nose wheel came down; the right gear stayed in the up position. Chatham Tower called the airplane three times warning it of the condition but received no reply. After the turn was completed to final approach the airplane appeared to be aligned with the runway but the airplane was in a diving attitude losing altitude rapidly. Just before striking the ground, the airplane appeared to level out then crash into the ground 3,000 feet short of the runway of intended landing.... The airplane skidded on the ground for 225 feet from point of initial contact. The number-one, number-two, and number-four propellers were torn from these engines; the nose section was demolished, the underside of the fuselage was demolished, and the fuselage buckled about the waist section." Gunner Cpl. Donald F. Behler was uninjured. Bombardier 2Lt. Billie L. Pike, navigator 2Lt. Paul E. Bowen, Jr. and engineer Cpl. Robert D. McCracken were killed in the crash. Engineer Cpl. John W. Bruce, Jr. and assistant radio operator Cpl. Marvin N. Krumholz received serious injuries. Crewmembers injured in the crash were: 2Lt. Edward A. Kauffman, pilot; 2Lt. Charles D. Walker, co-pilot; Cpl. Albert S. Adley, radio operator; Cpl. Walter I. Modzewski, gunner.

11-3-44A. Glennie, Michigan. At 1025 EWT, a Martin TB-26C crashed ten miles north of Glennie, Michigan, killing four fliers. The airplane took off at 1015 EWT from the Army Air Field at Oscoda, Michigan, on a flight to check out aerial engineers. The airplane was seen flying at 1,000 feet agl when it went into a dive to an altitude of about 200 feet, pulling up in a gradual climb. As the airplane climbed to 400 feet, it was seen to roll over to an inverted attitude and then smash into the ground vertically, exploding into flames in the Huron National Forest. Investigators could not determine the cause of the accident. Killed in the crash were: 1Lt. Roy E. Yturria, pilot; SSgt. Donald W. Jaton, engineer; SSgt. Willis A. Dunn, engineer; SSgt. Emanuel Mokol, engineer.

11-3-44B. Kingman, Arizona. At 0752 PWT, a Boeing B-17G and a Bell P-39Q collided in mid-air 20 miles northwest of the Army Air Field at Kingman, Arizona, killing P-39Q pilot 2Lt. Martin H. Campbell and fourteen fliers aboard the B-17G. The airplanes took off from Kingman Army Air Field at 0732 on a routine gunnery-camera training mission. The B-17 was flying in the number-two position of a three-ship flight that was to be intercepted and attacked by the P-39Q. The flight climbed to altitude at the authorized air-to-air gunnery area and the P-39 made several successful passes from varying positions around the clock. The P-39 then attacked the lead B-17 from the 4:30 O'clock high position, which would be about the 3 O'clock high position of the subject B-17, which had been "lagging behind." Investigators noted that the accepted method of attack was for the P-39 to fire at the nearest outside B-17 one side at a time as opposed to attacking the lead or the furthest outside aircraft. The B-17s in the flight would change positions after their respected firing runs in order that all gunners will have fired from every position in the formation; co-pilots are also given flying practice from all positions in the formation. The gun cameras of the B-17 gunners rolled as the P-39 bore in on the lead B-17 from the 4:30 high position and attempted to dive through behind it. The P-39 pilot evidently did not see the number-two B-17 as he made this attack, apparently seeing the subject B-17 at the last moment and attempting to turn away. The P-39 passed over the center of the number-two B-17 and while in a steep bank to the right its starboard wing collided with the port B-17 wing, severing the starboard P-39 wing and the port B-17 wing 15 feet from the tip. Both airplanes immediately went out of control. The B-17 flipped over to an inverted position and spun to the ground where it exploded violently into flames. The P-39 apparently broke up in mid-air and slammed to the ground and exploded near where the B-17 hit. One parachute was seen. Investigators stated, "The pilot of the P-39 was thrown out [of the airplane] and his parachute opened by some undetermined means, the rip cord not being pulled. The P-39 pilot was evidently killed in the collision since a portion of his head was found in his aircraft, his right foot and lower leg were found 500 feet from the P-39 and the remainder descended with the parachute and landed approximately one-half mile from the P-39, equidistant from and between the B-17 and P-39 wreckage." It was noted that the P-39 pilot was a high-time four-engine pilot who had recently completed a combat tour overseas. Killed in the crash were: 1Lt. Oliver E. Wright, pilot; 2Lt. Richard B. Coyte co-pilot; Pvt. Roy C. Benbow, engineer; Cpl. Robert J. Baribeau, gunnery instructor; Pfc. Charles R. Surface, gunnery instructor. Student gunners killed in the crash were: Pvt. James W. Burner, Pvt. Claude B. Busta, Pvt. Jackson B. Campbell, Pvt. Herman H. Castile, Pfc. Augustine J. Cioffi, Pvt. Elmo E. Churchill, Pvt. John F. Clinch, Pfc. Calvin E. Connolly, Pvt. Theodore L. Cook.

11-3-44C. Lancaster, California. At 0855, a Lockheed P-38J crashed five miles east of Lancaster, California, killing pilot F/O Edward G. Wild, Jr. The airplane was part of a four-ship flight that had taken off from the Army Air Field at Van Nuys, California, on a dive-bombing and aerobatic mission. Investigators stated, "After completing the dive-bombing mission, the flight climbed to 12,000 feet and prepared to perform their aerobatics. The flight leader took the number-two man out away from the flight to check him out in aerobatics. The number-three and number-four were instructed to circle in that area while the flight leader checked the number-two man. After the flight leader left, the three and four man in trail started a so-called rat race that consisted of several loops, Immelmanns and slow rolls, ending up in a tight Luftbery circle with F/O Wild as number-four man. The circle continued for ten or twelve turns, with each circle getting tighter and tighter, the airplanes getting into more of a power stall all the time and at the same time steadily losing altitude. During this circle the two pilots were so concentrated on getting on the other's tail that they neglected to read their altimeter or note how close they were getting to the ground. Finally the number-three man noted how close he was to the ground and leveled out but then number-four man, F/O Wild, either did not see the ground at all, or saw it too late as the ship struck the ground in a mushing attitude, exploding upon contact with the ground."

11-3-44D. Anniston, Alabama. At 1107, a Douglas RA-20B crashed 13 miles west of the Army Air Base at Anniston, Alabama, killing pilot 2Lt. Julius F. Iole and passenger 1Lt. Arthur M. Hartman. The airplane took off from the base on a local flight. Just after take-off, the pilot radioed the tower and asked for a fly-by to check that the landing gear had fully retracted. The tower operator authorized the fly-by and advised the pilot not to fly below 1,000 feet agl during the fly-by. The airplane passed over the field at about 200 feet agl and the tower acknowledged that the landing gear was up. The airplane flew away from the field. A couple minutes later the pilot radioed the tower and requested landing instructions. The tower acknowledged and the airplane entered the traffic pattern on the downwind leg. The airplane, with landing gear retracted, was then seen to perform a snap roll while flying in the pattern. The landing gear was then seen to come down as the pilot performed a second roll. The airplane stalled and fell off on a wing. The pilot could not recover the airplane and it pancaked to the ground and burst into flames, scattering wreckage over wide area. Rescue personnel immediately extinguished the small fires that had erupted after the crash. Before the flight, the pilot was overheard telling the passenger that the flight would be a "thrill."

11-3-44E. Del Rio, Texas. At 1530, a Martin B-26C crashed one mile northwest of Laughlin Field, Del Rio, Texas, killing three fliers. The airplane was returning to the field after completing a routine local training flight. The airplane stalled while turning from the base leg onto final approach, entering a spin to the right at an altitude too low to allow recovery. The airplane smashed into the ground and exploded into flames, killing pilot 2Lt. Gordon H. Wicks, co-pilot 1Lt. Ronald A. Bowerstock and engineer Sgt. Wayne E. Chinn.

11-3-44F. Valparaiso, Florida. At 1715 CWT, a North American AT-6A performing a landing collided with a Jeep that was operating at Auxiliary Airfield #2, six miles NNE of Eglin Field, Valparaiso, Florida, killing jeep driver SSgt. James J. Moon, Jr. AT-6A pilot 2Lt. Delmar S. Stone and pilot-rated passenger 2Lt. Clayton C. Solberg were uninjured. The airplane was returning to the field after completing a routine gunnery-training mission. After making a low approach, the airplane made a "good" three-point landing on Runway 6. The driver of the jeep was detailed to pick up targets that were dropped on the field by target-towing airplanes and apparently did not wait for a green light from the tower when he crossed Runway 6 in front of the rolling airplane. The jeep driver spotted the airplane at the last moment and turned violently to the right to avoid the collision. The airplane, which had just touched down, had rolled at landing speed for about seven yards before it struck the jeep on the left rear side with the port landing gear and wing, causing the jeep to roll over. The driver was killed instantly in the collision. The port landing gear fell off and airplane went out of control, skidding to a halt a short distance later. The jeep windshield frame had been impaled on the port wing; it was sticking out of the leading edge of the port wing about halfway at a 45-degree angle toward the engine.

11-3-44G. Victoria, Texas. At 1715, a Curtiss TP-40N crashed six miles west of Aloe Field, Victoria, Texas, killing pilot 2Lt. Gerald E. Holliday. The airplane took off from Aloe Field at 1636 CWT on a routine transition flight. The airplane was seen diving straight down with the engine roaring. No fire or smoke was observed trailing the airplane. The airplane was not seen spinning or turning. The airplane smashed into the ground and exploded violently. The pilot made no apparent effort to use his parachute.

11-4-44. Oakland, Maryland. At 2050 EWT, a North American AT-6D flying in instrument conditions collided with trees and crashed 12 miles southeast of Oakland, Maryland, killing pilot 2Lt. Harry S. Farrow and pilot-rated passenger 2Lt. Kenneth M. Haight. The airplane took off from Roanoke, Virginia, on an instrument flight to Pittsburgh, Pennsylvania. The airplane was about 25 miles off course when it struck a tree with the port wing. The airplane went out of control and smashed into a wooded area, busting into flames upon impact. It was speculated that the pilot had attempted to let down through instrument conditions and collided with terrain. The airplane and pilots were stationed at Shaw Field, Sumter, South Carolina.

11-5-44A. Muroc, California. At 0640, bombardier 2Lt. Robert J. Wolfe was killed when he walked into the rotating number-three propeller of a Consolidated B-24J operating at the Army Air Field at Muroc, California. The airplane had been idling on the ramp in preparation for a take-off on a bombing and instrument training flight. Investigation revealed that after the briefing for the bombing mission, the crew, with the exception of the bombardier, was transported out to the aircraft. Pre-flight checks were completed and the engines were started and run up. Investigators stated, "This particular B-24 was the only operational aircraft in this section of the parking ramp. Engines had been running five or ten minutes and were idling when the bombardier arrived. Bombardier had left the briefing room with the rest of the crew but difficulty in drawing his equipment and insufficient transportation from the flight line to the aircraft probably caused an excessive delay. He arrived at aircraft on a tow tug and after placing bombing computers in the nose section of the airplane through the nose wheel door, tried to walk between fuselage and the number-three propeller while carrying bombing camera and notebooks. The number-three propeller struck the bombardier on the left side of the head and fatality resulted within five minutes." Pilot 2Lt. Eugene H. Paules, co-pilot 2Lt. Joseph R. Enloe, Jr., and six other crewmembers were uninjured.

11-5-44B. Camden, Arkansas. At 1559, a Vultee BT-15 crashed at Harrell Field, Camden, Arkansas, killing pilot 2Lt. Harold E. Seiser. The airplane, which was on a test flight prior to being ferried to an unknown location, took off to the north on the North-South Runway. Investigators stated, " According to witnesses standing on the ramp, and the firemen who were standing at the southern end of the North-South runway at a point 250 yards west of where the aircraft crashed, the airplane came in over the west end of the field, flew over the northern edge of the ramp in an easterly direction at approximately 15 feet above the ramp between two rows of parked airplanes. Upon reaching a point approximately 100 feet south of the end of the North-South runway the aircraft was pulled up in a steep climb to approximately 500 feet at which point it stalled and fell off into a left spin. The spin was stopped at a westerly heading about 200 feet high and almost immediately the airplane started a right spin, continuing on down and striking the ground with the right wing and nose, resulting in complete destruction of the aircraft and the instant death of the pilot."

11-5-44C. Pacific Ocean, California. At 1738, two Lockheed P-38L airplanes collided in mid-air and crashed into the Pacific Ocean one-quarter mile offshore of the Auxiliary Airfield at Mendocino, California. 2Lt. Dennis M. Lake (P-38L #44-24534) and 2Lt. Jules E. Lambert (P-38L #44-24751) were both missing and presumed killed. The P-38s were part of a five-ship flight that had taken off from the Army Air Field at Santa Rosa, California, on a low-level navigation flight. While flying near Mendocino, the flight leader called for a gentle 180-degree turn to the left. Investigators stated, "The number-three (Lake) and the number-four (Lambert) airplanes started to swing wide and to the outside of the turn. The number-four man continued across to take the outside position on the turn and while so doing hit the number-three airplane and severed the tail section just behind the coolant radiators. The number-three airplane went down [and crashed into the sea] immediately and then number-four airplane, its right engine and propeller dead, changed from a left to a right turn, nosed down and also went into the water. Neither airplane exploded upon impact and both sunk immediately, leaving no debris or telltale marks on the water. Search aircraft and rescue boats were unable to locate personnel or wreckage before darkness. Continuation of the search the following day brought no results and emergency rescue boats abandoned search shortly thereafter."

11-5-44D. Garden City, Kansas. At 2109 MWT, a Vultee BT-13A crashed six miles north of the Army Air Field at Garden City, Kansas, killing student A/C Melvin M. Courtney. The cadet took off from Garden City Army Air Field at 1900 MWT on a local night transition flight and flew directly to Auxiliary Airfield #3. The student shot six successful landings at the auxiliary airfield and was then assigned to an overhead pattern at 5,000 feet to complete his night flying in the upper traffic pattern. The mission was scheduled to terminate at 2145. At 2110 a local civilian telephoned the field and reported an airplane crash near his farmhouse. Investigators stated, "Apparently the cadet, while flying on the downwind leg, lost control of his airplane and probably spiraled or spun into the ground. The wreckage was strewn to the north from the point the airplane hit the ground. At the point of impact there was a clear front profile of the airplane imprinted in the ground. The motor was completely imbedded and the broken end of the pitot tube was driven into the ground at about a 70-degree angle. It was evident that the cadet was free of the airplane when it crashed, as his body first struck the ground 50 feet south of where the airplane hit, and was thrown through the air for 120 feet in the same direction that the wreckage was strewn. Apparently the cadet endeavored to bail out. The parachute was free of its pack and the ripcord was fully pulled."

11-6-44A. Benton, Louisiana. At 0835, a Martin TB-26B exploded in mid-air and crashed six miles northwest of Benton, Louisiana, killing seven fliers. The airplane took off at 0747 from Barksdale Field, Shreveport, Louisiana, on an instrument check ride. The B-26 instructor was flying in the co-pilot seat as safety pilot and the B-26 student was flying in the pilot seat under the hood as the instrument pilot. The airplane was flying at about 3,000 feet when witnesses on the ground observed it to explode in mid-air and begin shedding pieces. The B-26 entered a spin and dove into the ground at high speed, exploding violently into flames upon impact. Killed in the crash were: 1Lt. Ralph W. Walker, B-26 instructor; 2Lt. James S. Drew, pilot/B-26 student; 2Lt. Salvatore J. Diana, B-26 student; 2Lt. Billie M. Stoneking, bombardier/navigator; SSgt. Roland L. Weller, engineer; Cpl. Paul R. Dronski, radio operator; Cpl. Theodore L. Graham, gunner.

11-6-44B. Phoenix, Arizona. At 1015 MWT, a North American AT-6D and an AT-6C collided in mid-air and crashed 25 miles northwest of Luke Field, Phoenix, Arizona, killing A/C Robert Howard Boyles aboard the AT-6D. A/C Wallace Neal Brown parachuted to safety from the AT-6C and was uninjured. A/C Boyles was flying in the number-three position and A/C Brown was flying in the number-five position of a six-ship flight that had taken off from Luke Field on a routine formation flight. The flight had climbed to 13,000 feet and performed various maneuvers while in formation. The flight entered a right echelon formation and a short time later, the flight leader, instructor 2Lt. Benjamin E. Bell, signaled the formation to peel off to the left. The leader peeled off and the flight followed. As the flight leader leveled out of the turn, the number-four airplane threatened to overrun the third airplane so the number-four ship dove down and out to avoid a collision. The number-five ship, piloted by A/C Brown, almost overran the number-three ship and pulled up left of and nearly abreast of number-three. The number-five ship attempted to slip down and to the right to break off and regain his proper position. He could not go forward or down to the left because the number-four ship was there and apparently climbing back up. A/C Boyles, in the number-three ship, apparently did not see the other two airplanes and entered a steep bank to the left to close the gap between the number-two ship. A/C Boyles' port wing struck the fuselage of A/C Brown's airplane as A/C Brown passed under and A/C Boyles banked left. Both airplanes flipped over to an inverted spin. The tail section of the number-five ship failed and separated. Both airplanes spiraled to the ground and burst into flames upon impact. A/C Brown parachuted to safety. Investigators speculated that A/C Boyles was either killed or incapacitated in the collision and had made no attempt to parachute to safety. Investigators faulted both students for failing to see and avoid each other.

11-6-44C. Perrin Field, Texas. At 1735, a Vultee BT-13A crashed eight miles northwest of Perrin Field, Sherman, Texas, killing student Pfc. Frank O. Eisenhart, Los Angeles, California, who fell to his death in an unsuccessful parachute jump. Investigation revealed that the student had entered a spin at a safe altitude and was unable to recover. The student bailed out at an altitude of about 300 feet agl; he cleared the airplane and pulled the ripcord but the parachute failed to open before he struck the ground. The airplane slammed to the ground and burst into flames nearby. Investigators noted that the student had been authorized to perform solo spins.

11-6-44D. Tallahassee, Florida. At 1857 EWT, a Curtiss P-40N suffering an in-flight fire crashed while attempting an emergency forced landing five miles north of Dale Mabry Field, Tallahassee, Florida, killing pilot 2Lt. Guy R. Sheppard, Jr. The airplane was part of a flight that had taken off from the field on a night formation training flight. Soon after take-off the subject airplane peeled off of the formation and the pilot radioed the tower for an emergency landing. The tower cleared the pilot to land on Runway 36 immediately, but the pilot refused because he was northeast of the field and out of position. Moments later the pilot was heard "screaming" on the radio that his airplane was on fire. The airplane entered a glide and was seen trailing white smoke. Witnesses observed small flashes of fire from the engine. The flight leader yelled over the radio for him to bail out but the airplane, trailing thick smoke, entered a steep spiral and smashed into the ground and exploded. Investigation of the wreckage revealed that the engine fire was caused by the failure of the number-three and number-four crankshaft rod bearings.

11-6-44E. Bend, Oregon. At 1705 PWT, a Lockheed P-38L crashed ten miles east of Bend, Oregon, killing pilot 2Lt. Byron R. Greenway. The airplane was part of a flight that had taken off from Redmond, Oregon, on a formation flight. The flight climbed to 11,000 feet and began to perform formation maneuvers. The flight leader performed a slow roll and Lt. Greenway attempted to follow the maneuver but failed to complete the roll. The subject airplane fell into an inverted nose-down position and the pilot attempted to recover by performing a split-S maneuver. The P-38 had excess speed and insufficient altitude to safely complete the split-S and the airplane slammed into the ground at a shallow angle with wings level, bursting into flames upon impact. Investigators speculated that the subject airplane encountered propeller turbulence and was unable to roll through after becoming inverted.

11-6-44F. Syracuse, Kansas. At 1825 MWT, a Cessna UC-78B flying in poor weather crashed ten miles NNW of Syracuse, Kansas, killing four fliers. The airplane had taken off from Fairfax Field, Kansas City, Kansas, on an administrative flight to Walker Army Air Base, Walker, Kansas. The airplane became lost and landed at Garden City Municipal Airport, Garden City, Kansas, with only three gallons of fuel on board. The airplane was refueled and the pilot resumed his flight. Soon after take-off the pilot radioed Garden City that he had flew into a front that contained strong thunderstorms and severe turbulence and he was turning back. The airplane never returned and was soon overdue. A search effort failed to locate the airplane; a civilian found the airplane at about 1400 MWT on 11-8-44.

Investigators speculated that the pilot had lost control of the airplane due to extreme weather conditions. Killed in the crash were: 2Lt. Benjamin M. Harris, pilot; 1Lt. Chester S. Bowles, passenger; Pvt. Phillip G. Starrett, passenger; Pvt. August B. Venturia, passenger.

11-7-44A. Washington, D.C. At 0935, a Douglas RA-24A crashed one-eighth of a mile WSW of Camp Springs Army Air Field, Washington, D.C., killing pilot 2Lt. John C. Miller and pilot-rated passenger 2Lt. Clarence W. Slutter. The airplane was turning from the downwind leg to the base leg when it entered a stall and spin at an altitude too low to allow recovery. The airplane was returning to the field after a routine proficiency flight.

11-7-44B. Waco, Texas. At 1500, two Beech AT-10 airplanes collided in mid-air and crashed two miles north of Blackland Army Air Field, Waco, Texas, killing three fliers. A/C Harold T. Ore and A/S James M. Willis were killed in the crash of AT-10 #41-9417. Instructor 2Lt. Daniel C. Utton was killed aboard AT-10 #41-2372 and A/C Donald L. Cass parachuted to safety, escaping injury. Investigators stated, "[At] approximately 1450, AT-10 41-9372 flew over Blackland Army Air Field in a southerly direction on the Waco Radio Range. After hitting the cone, Lt. Utton instructed A/C Cass to make a turn to the right. A/C Cass held this heading for a short time and then made another turn to the right. After holding this heading for a short time they reached a point which was apparently just north of the base leg of the traffic pattern. Lt. Utton told the student to take down the hood and he [Lt. Utton] took over the airplane. A/C Cass stated that he had taken down the piece of polarized glass while his instructor was making an approach to the traffic pattern at a 45-degree angle. He stated that the instructor made a turn to the right in traffic and called in upper traffic. At this time A/C Cass had just finished removing the instrument glass and was storing this glass behind his seat. At this time A/C Cass states that he felt the aircraft jar, then violently skid to the right and fall into a spin to the right. Both he and the instructor attempted to regain control but the controls had no effect. A/C Cass pulled back the canopy, unfastened his safety belt and parachuted. The aircraft crashed with Lt. Utton. AT-10 #41-9372 was hit by AT-10 #41-9417 which had been flying formation but had peeled off and was making a normal entry into traffic." The collision occurred at an indicated altitude of 3,000 feet. The two airplanes had apparently entered the traffic pattern on the downwind leg at the same point and at the same time, one airplane turning into the other. AT-10 #41-9372 fell out of control and dove straight into the ground where it exploded into flames.

11-7-44C. Troupsburg, New York. At 1605, a Boeing B-17F crashed one and one-half miles northeast of Troupsburg, New York, killing three fliers and injuring three others. The airplane took off from Rome Army Air Field, Rome, New York, on a routine test flight and a search mission for a missing aircraft. The airplane was detailed to search an area between Rome and Elmira, New York, to Jamestown, New York, back to Elmira and then back to Rome. The airplane was flying at 2,800 feet on the return leg to Elmira when the pilot noticed that the number-two engine was on fire. Engine instruments did not indicate any trouble. The pilot began to feather the number-two propeller. The propeller feathered and then unfeathered itself and then began to windmill. The airplane was not equipped with an on-board engine fire extinguisher and the fire grew in intensity. The fire began to spread to the after engine nacelle and wing so the pilot ordered the crew to abandon ship. The co-pilot, Capt. Vernon K. Cammack, bailed out of the forward hatch and the remaining occupants had gathered in the after fuselage when the bail out alarm bell and verbal command was given. Survivors stated that the two civilians that failed to parachute to safety had been reluctant to bail out. Civilian instructor H. Weed exited first and A. Fiduccia bailed out second, and both sustained minor injuries parachuting to safety. Both stated that their parachutes opened just before they hit the ground. Pilot Capt. Arthur R. Friesz and civilian instructor A. Wright were killed when they bailed out at an altitude too low to allow their parachutes to deploy. Civilian H.L. Drew did not bail out and went in with the airplane. The airplane smashed into the ground and exploded, scattering wreckage for over 100 yards on a hillside meadow.

11-7-44D. Van Nuys, California. At 1530, a Lockheed P-38J crashed ten miles northeast of Van Nuys, California, killing pilot F/O John R. Field. The P-38 was in the number-three position of a four-ship flight that took off from Van Nuys Municipal Airport on a dive-bombing mission. About ten minutes after take-off, the number-four man noticed white smoke coming out of the subject airplane's port engine. Moments later, flames were seen trailing the same engine. The subject airplane dropped out of the formation and the flight leader ordered the pilot to pull the fuel mixture to idle cut-off on the bad engine and feather the propeller. The fire, which flowed from the leading edge at the engine nacelle, continued and grew in intensity until it was streaming past the supercharger and along the tail boom. The airplane lost altitude, dropped through some scattered clouds and then was seen to roll twice to the left before smashing into the side of a mountain and exploding. The pilot made no apparent effort to parachute to safety. Investigators could not determine the cause of the engine fire.

11-7-44E. Marianna, Florida. At 1930 CWT, a Douglas A-26B crashed five miles northeast of the Army Air Field at Marianna, Florida, killing 1Lt. Edward C. Anthon. The A-26 was taking off on a routine training flight when the port engine failed just after the ship left the ground. The pilot retarded both throttles in an attempt to control the lateral stability. The airplane then smashed into the ground at a steep angle, exploding into flames upon impact. It was speculated that the pilot might have lost the horizon or had his head in the cockpit, causing him to inadvertently fly the airplane into the ground.

11-8-44A. Midland, Texas. At 1212, a Lockheed P-38L crashed five and one-half miles northeast of Midland, Texas, killing pilot Lt. Col. Russell C. Larsen. The airplane took off at 1210 from Midland Municipal Airport on Runway 34. The airplane climbed to 300 feet agl and made a right turn to leave the traffic pattern. As the airplane climbed and turned, thick black smoke was seen trailing the port engine. The airplane stopped climbing at about 500 feet agl and then began to lose altitude as it turned to the southeast. At about 250 feet agl, the airplane stalled and dropped off on the starboard wing. The pilot was able to recover at about 100 feet and then fell off on the port wing, describing one-quarter of a turn before smashing into the ground and exploding. Wreckage was scattered over an area of 150 feet. The pilot was thrown clear of the airplane but killed instantly upon hitting the ground.

11-8-44B. Easton, Maryland. At 1503, a Martin B-26G crashed five miles northwest of Easton, Maryland, killing civilian engineer R.C. Knox. Civilian pilot H.R. Popalawski and civilian engineer R.L. Woody parachuted to safety. Mr. Woody sustained minor injuries. All were employees of the Glenn L. Martin Aircraft Company. The airplane took off at 1425 from the Martin Factory Airfield on a production test flight. The B-26 climbed to an indicated altitude of 9,000 feet and proceeded to the test area. A short time later, two Republic P-47 airplanes arrived and began maneuvering near the subject airplane. The P-47s were behind the subject airplane and it entered a 45-degree bank to the right in an attempt to maneuver away. The pilot then noticed he was losing altitude in his turn with an airspeed of 230 mph. The pilot then felt the airplane jerk through the ailerons and then enter a spin. The pilot pulled back on the throttles and attempted a recovery using the rudder. The pilot leveled the airplane out but the ailerons were vibrating badly. Any movement of the control wheel caused the port wing to drop. The pilot retarded the starboard throttle and increased the port throttle to level the airplane. The airplane shook violently and the pilot was able keep the airplane level only with great difficulty. The airplane continued to lose altitude and it continued to drop to the left. The pilot opened the bomb bay doors and then ordered the crew to bail out. Mr. Woody bailed out but Mr. Knox remained in the co-pilot seat. The pilot attempted to trim the airplane so he could get out. The pilot again ordered Mr. Knox to bail out and he finally left his seat. The pilot lowered the landing gear and then left his seat; he observed Mr. Knox kneeling near the forward bomb bay and he hollered at him to bail out. The pilot then exited

the airplane through the nose wheel door at an altitude of about 2,000 feet. Mr. Knox did not leave the airplane and was killed in the crash. Investigators could not determine the cause of the control anomaly.

11-8-44C. Harlingen, Texas. At 1610, two Bell TP-39Q airplanes collided in mid-air and crashed five miles northeast of Laguna Madre Sub-Base, Harlingen, Texas, killing 2Lt. Richard M. Goldberg aboard TP-39Q #42-20711. 2Lt. James D. Moore was able to parachute to safety from TP-39Q #42-19631. The two airplanes had taken off from Laguna Madre on a mission to intercept and simulate an attack two Consolidated B-24 airplanes. The P-39 airplanes had made several successful attacks from both sides of the bomber flight. Lt. Goldberg had made an attack from the left and was changing over to an attack from the right. He began his attack before Lt. Moore had cleared from his attack from the left. The pilots lost track of each other and collided nearly head on. The airplanes went out of control and Lt. Moore bailed out. Investigation revealed that Lt. Goldberg had fallen free of the P-39 but had evidently struck his head on part of the airplane and was unable to pull his ripcord, causing him to fall to his death.

11-8-44D. Canton, Oklahoma. At 1700, a North American TB-25J collided with a windmill and crashed 11 miles west of Canton, Oklahoma, killing A/S Samuel O. Allen and A/C Grover B. Bjarnason. The airplane took off from the Army Air Field at Pampa, Texas, on a routine training flight. A/S Allen was to fly the first 45-minute period of flight as the pilot with A/C Bjarnason as co-pilot. The airplane was to return to Pampa for a landing and the students would then switch seats and repeat the exercise. The airplane never returned to land at Pampa. Witnesses on the ground observed the airplane flying at very low altitude just before the crash. The airplane was seen to climb to avoid striking objects in its path. The airplane collided with a windmill on the farm of M.V. Anson and crashed into the ground a short distance later, exploding into flames upon impact. It was unknown if the airplane was suffering engine trouble.

11-8-44E. Tarriffville, Connecticut. At 1817, a Consolidated B-24J crashed one mile north of Tarriffville, Connecticut, killing five fliers and injuring seven others. The airplane was on a combat crew training mission when the number-one propeller ran away. The gunnery instructor on board noticed smoke trailing the number-one engine and the pilot was alerted. The pilot attempted to feather the propeller but it would not feather. The number-one engine then burst into flames and the propeller continued to windmill. The pilot increased power on the remaining engines but was unable to maintain altitude. The pilot attempted to make an emergency forced landing in what he thought was a clear area. The airplane collided with a telegraph pole and went out of control, smashing violently into the upslope of a small hill. The airplane did not burn but was destroyed utterly. Crewmembers seriously injured were: 2Lt. Roland C. Curtiss, pilot; F/O Reese A. McClennahan, Jr., co-pilot; F/O Vincent M. Vallaro, bombardier; Cpl. Francis A. Crawford, gunner; Cpl. Cono A. Galliani, gunner; SSgt. Charles J. Nigro, gunnery instructor. Killed in the crash were: Cpl. Henry C. Fay, gunner; Pfc. Lester L. Shoemaker, gunner; Cpl. Charles W. Powell, gunner; Cpl. Gastano L. Fastiggi, gunner; Cpl. Furman Watson, gunner. Navigator 2Lt. Douglas D. Dettle sustained minor injuries parachuting to safety.

11-8-44F. Ontario, California. At 1612, a Vultee BT-13B crashed while attempting an emergency forced landing three and one-half miles northeast of the Army Air Field at Ontario, California, killing pilots 2Lt. Aubrey D. York and 2Lt. Kent H. Soper. The BT-13 took off from Ontario on an instrument training flight. The airplane suffered a fire in flight and the pilot attempted an emergency forced landing. The airplane struck a large tree, went out of control and crashed into the ground where it burst into flames.

11-8-44G. La Junta, Colorado. At 2035, a North American TB-25C crashed three miles northwest of the Army Air Field at La Junta, Colorado, killing A/C Pietro R. Fragale and A/C Ralph C. Fougler. The airplane had taken off at 2018 MWT from La Junta on a routine student night navigation flight. Investigation revealed that the airplane was headed back toward the field because of some kind of problem. The student had radioed the tower but his transmission could not be understood. The student attempted an emergency forced landing when it became evident that he could not make the field. The wheels-up landing was initially good, the tailskid hitting the ground first at a very low speed. The airplane skidded along for about 60 yards until it struck a 15-foot deep ditch and embankment. The airplane smashed to a halt and was completely destroyed. Investigators could not determine the cause of the trouble that prompted the students to return to the field.

11-9-44A. Marathon, Florida. At 0250 EWT, a Douglas C-54A crashed into a small cay near Marathon Bay three and one-half miles ENE of Marathon, Florida, killing the crew of four. Killed in the crash were: 1Lt. James B. Waltz, C-54 instructor; F/O David K. Edwards, pilot/C-54 student; F/O Roy W. Huddleston, co-pilot/C-54 student; MSgt. Robert A. Bumpass, engineer. The C-54A, which was from Homestead Army Air Field, Homestead, Florida, had taken off on Runway 7 from Marathon Cay Airfield and crashed soon after becoming airborne. A Consolidated B-24D that took off from Runway 7 crashed about one mile from the end of the runway at 0115; three fliers were seriously injured. The C-54 passed over the B-24 wreck, flashed its landing lights and then crashed into the water a short distance later. Investigators could not determine why the C-54 flew into the water.

11-9-44B. Indian Rocks, Florida. At 1415 EWT, a North American P-51C crashed into the Gulf of Mexico surf at the AAF Ground Gunnery Range two miles south of Indian Rocks, Florida, killing F/O Groover W. Partin. The P-51 was part of a four-ship flight that had taken off from Bartow Army Air Field, Bartow, Florida. Investigators stated, "The pilot, as number-four man in a ground gunnery flight, had made about nine passes on his target. Some of his dives were too steep and finished with a sharp pullout. The range officer warned the pilot and was acknowledged once. Subsequent warnings were not acknowledged and it was discovered immediately after the crash that, due to a malfunction of the radio generator, the range officer was not transmitting. He had attempted to warn the pilot on two more passes. On [the pilot's] last pass his dive was again too steep and his pullout so sharp that the aircraft made a high speed stall and crashed into the water about 50 feet beyond the target, killing the pilot and totally demolishing the aircraft."

11-10-44A. Weslaco, Texas. At 1220, a Bell P-39Q crashed one and a half miles north of Weslaco, Texas, killing 2Lt. Lambert A. Wood. The airplane took off from Brownsville Municipal Airport, Brownsville, Texas, on a routine training flight. The pilot was practicing climbing 180-degree turns when he allowed the airplane to enter a stall. The airplane fell off on a wing and into a spin. The pilot made a partial recovery, but the airplane stalled again and entered a flat spin from which he was unable to recover. The airplane smashed into the ground vertically in a flat spin to the left. The airplane fell into an orange grove but did not damage any of the trees. The starboard escape door was seen to leave the airplane at an altitude of about 1,200 feet agl. The pilot was unable to escape the airplane and his body was found in the wreckage. There was no fire.

11-10-44B. Porum, Oklahoma. At 1507, a Curtiss P-40R crashed five miles NNE of Porum, Oklahoma, killing 2Lt. Douglas Thomas, Jr. The airplane was part of a three-ship flight that had taken off from DeRidder Army Air Base, DeRidder, Louisiana, on a navigation mission to Love Field, Dallas, Texas, to Muskogee, Oklahoma, and return to DeRidder. The flight landed at Dallas to refuel and then took off for Muskogee. The flight arrived at Muskogee on schedule and refueled for the trip back to DeRidder. The flight leader formed the flight at about 2,000 feet. Moments later Lt. Thomas was seen turning away from the formation. The flight leader checked the map and then heard a transmission by Lt. Thomas, stating that he was having some trouble. The flight leader began to look around for Lt. Thomas. A short time later, the flight leader noticed a fire on the ground. He descended to low altitude and ascertained that the fire was the remains of Lt. Thomas's airplane. Investigation revealed that the airplane moved out of the formation to the right in a nose-high position. The airplane stalled and flipped over on its back. The pilot attempted to recover by performing a split-S type maneuver but the altitude was insufficient. The pilot was unable to recover and the airplane smashed into the ground at a 45-degree angle, exploding into flames upon impact.

11-10-44C. Clovis, New Mexico. At 2130, a Boeing B-29 crashed 25 miles southeast of the Army Air Field at Clovis, New Mexico, killing 15 fliers. The airplane took off from Clovis Army Air Field at 2005 MWT on a local night transition flight. The airplane returned and made one successful landing and took off again at 2113. The airplane was heading east to west at the time of the crash. The airplane smashed into the ground on the starboard wing at a steep angle and then cartwheeled into the ground and exploded. Investigation revealed that the airplane was on fire prior to it striking the ground. Pilot-rated passenger 2Lt. Anselmo Quian bailed out at extremely low altitude and fell to his death. His body was found about one mile from the main wreckage. Killed in the crash were: Capt. Thomas R. Opie, B-29 instructor; 1Lt. Noble J. Klink, pilot/B-29 student; 2Lt. Robert J. Creecy, co-pilot/B-29 student; 2Lt. Charles H. Ziegel, flight engineer instructor; TSgt. Walter Lewkowicz, flight engineer; Sgt. James B. Perry, radio operator instructor; Pvt. Franklyn B. Jackson, radio operator; Pfc. Max Freund, radio operator; TSgt. John K. Crowther, central fire control instructor; Cpl. Wilber E. Miller, gunner; Pfc. John A. Nelson, gunner; Pfc. William E. Harris, gunner; Pfc. Duane W. Duman, tail gunner; Capt. John S. Baldwin, pilot-rated passenger.

11-11-44A. Sioux City, Iowa. At 0720, engineer Cpl. Eli Trbovich was killed when he walked into the rotating number-four propeller of a Boeing B-17F that was operating at Sioux City Army Air Base, Sioux City, Iowa. The airplane was preparing to taxi out and take off on a routine high altitude formation and bombing mission. The engineer was seen to engage in conversation with the crew chief at the front of the airplane while the engines were running. The engineer then walked under the starboard wing in an effort to get back to the main entrance door. The engineer misjudged the location of the number-four propeller and walked into it, killing him instantly. The B-17 was piloted by 2Lt. Lewis B. Berggren and F/O John P. Byrnes. The two pilots and six other crewmembers were uninjured.

11-11-44B. Glendale, California. At 2017, a Douglas C-47B flying in instrument conditions crashed into Strawberry Peak 14 miles north of Glendale, California, killing 13 crew and passengers. The airplane had taken off from Hamilton Field, San Rafael, California, on a transport flight to Mines Field, Los Angeles, California, via Bakersfield, California. The airplane took off from Bakersfield and soon afterward encountered bad weather. The airplane inadvertently entered an area where weather conditions consisted of low ceilings, hard rain, turbulence, and severe icing at altitudes exceeding 8,500 feet msl. The pilot lost radio contact soon after leaving Bakersfield. The pilot used dead reckoning for the remainder of the trip and then, believing that he had cleared the mountainous area, attempted to let down for his landing in Los Angeles. The airplane was under control when it collided with trees in a straight-and-level attitude. The C-47 smashed its way through trees for about 30 yards and then careened over the top of the ridge and bounded across a deep canyon about 200 yards wide before smashing into the side of the mountain just below the peak, exploding into flames upon impact. All personnel on board were killed instantly. Investigation revealed that the pilot was given inadequate weather and wind information at Bakersfield prior to take-off. The erroneous wind information caused the airplane to be off course and over mountainous terrain when the pilots began to let down through instrument conditions. Pilot Capt. Rae C. Kelley and co-pilot 2Lt. Carl Hankle were killed in the crash. Passengers killed in the crash were: Maj. John G. Anderson III, Maj. Thomas W. Mackey, 1Lt. Hugh G. Cramer, 1Lt. Dan B. Field, 2Lt. Gertrude Maschoff, 2Lt. Alva Eickmeyer, MSgt. Harold Peterson, Jr., SSgt. Francis E. Ward, Cpl. Manuel C. Garcia, Cpl. Kenneth C. Bedford, Pharmacist Mate 3rd Class Buford B. Chism (U.S. Navy).

11-12-44A. Atlantic Ocean, South Carolina. At 2038, a Lockheed-Vega R-37 (similar to the B-34 type) crashed into the Atlantic Ocean one mile south of the Army Air Field at Myrtle Beach, South Carolina, killing six fliers. Killed in the crash were: 2Lt. Cloyd R. Reynolds, pilot; 1Lt. Walter H. Dahl, pilot-rated passenger; 2Lt. James W. Wright, pilot-rated passenger; SSgt. Walter J. Prell, engineer; Sgt. Karl A. Artman, engineer-rated passenger; Pfc. Wayne G. Fisher, radio operator. The airplane was cleared for a flight to its home station at Walterboro Army Air Field, Walterboro, South Carolina. The airplane took off on Runway 17 and disappeared from the view of control tower personnel. Departing aircraft were using Runway 26, but the subject airplane used Runway 17 because the pilot was in a hurry to take off. Traffic was not using Runway 17 because this necessitated taking off toward the ocean, which provided no horizon or ground reference. The airplane failed to arrive at its home station and became overdue. The ship declared missing and a massive search conducted. Sgt. Artman's body washed ashore on 11-13-44 in front of the summer Officers Club, which is located off of the end of Runway 17. Two oil slicks and small bits of wreckage were located just to the right of the runway heading about one mile from the end of the runway; Lt. Wright's body was found in this area on 11-13-44. A pilot seat, clothing, and life vests belonging to the crew were also found in this area. U.S. Navy divers found the aft fuselage under water and positively identified it as belonging to the missing aircraft. The main wreckage was found and brought ashore. Lt. Dahl's body washed ashore on 11-21-44. Investigators could not determine the cause of the crash. It was speculated that the pilot had inadvertently flew the airplane into the sea just after take-off.

11-12-44B. Dyersburg, Tennessee. At 2236 CWT, a Boeing B-17F crashed just after take-off from Dyersburg Army Air Base, Dyersburg, Tennessee, killing eight fliers and seriously injuring two others. The airplane took off at 2235 on a high-altitude bombing training mission. The airplane made a normal take-off on Runway 35 and then flew into the ground 2,000 feet from the end of the runway. The airplane broke up upon impact, sliding for 700 feet before coming to rest. Wreckage was scattered over a large area of a local farm. Some wreckage erupted in flames. Investigators noted that despite a row of "horizon reference lights" located near the take-off runway, the area is very dark at night and provides no discernable light horizon or other visual reference. Two pregnant cows, a pig and a mule were killed on the ground; a cornfield and a cotton field were destroyed in the crash and by rescue vehicles. Pilot 2Lt. Max O. Anderson and co-pilot F/O Frederick N. Samuels sustained very serious injuries in the accident and were unable to make a statement about the accident. Killed in the crash were: 2Lt. Michael F. Egan, navigator; 2Lt. Walter S. Bardwell, bombardier; Cpl. William P. Hess, gunner; Cpl. Samuel B. Hall, engineer; Cpl. Frederick L. Needles, radio operator; Cpl. Donald Bestelmeyer, gunner; Cpl. James T. Ellis III, gunner; Cpl. Howard B. Rayford, gunner.

11-13-44A. March Field, California. At 1340 PWT, a North American B-25J flying in instrument conditions broke up and crashed at March Field, Riverside, California, killing six fliers. The airplane took off from Coolidge Field, Coolidge, Arizona, at 1121 MWT on a ferry flight to McClellan Field, Sacramento, California. The B-25 encountered severe icing conditions at about 12,000 feet and called March Field controllers, requesting a descent to 9,000 feet. The March Field controller advised the subject airplane to stand by while he checked the traffic and local weather information. The pilot radioed controllers that he was commencing his descent without clearance because of severe ice. A U.S. Navy pilot flying in the area stated over the radio that there was less icing at 13,000 feet. The controller radioed the subject airplane but received no reply. Moments later, personnel at March Field heard the sound of an airplane flying at what sounded like high speed. The airplane broke up and began falling from the overcast. The main wreckage fell in the barracks area and various pieces were scattered over the entire field. Investigators stated that the pilot had lost control of the airplane while attempting to let down and stressed the aircraft beyond design limitations in an attempt to recover control. The pilot did not have a current instrument card. Killed in the crash were: 2Lt. George F. Tobols, pilot; 2Lt. Duane J. Hardy, co-pilot; 2Lt. James

S. Mercier, bombardier/navigator; Cpl. Joseph Koss, engineer; Cpl. Harry R. Kough, radio operator; Cpl. Frederick A. Severt, gunner.

11-13-44B. Wetumka, Oklahoma. At 1610, a Douglas F-3A (the photo-reconnaissance version of the A-20) crashed seven miles WNW of Wetumka, Oklahoma, killing bombardier/navigator 2Lt. Charles F. Anderson and seriously injuring pilot F/O Paul L. Hooper. The airplane took off from Will Rogers Field, Oklahoma City, Oklahoma, at 1335 CWT on a low-altitude round robin navigation flight to Waynoka, Oklahoma, to Wyandotte, Oklahoma, to Lake Murray, Oklahoma, and return to Will Rogers Field. Investigators stated, "Pilot Hooper was on course between Wyandotte and Lake Murray when his left fuel pressure suddenly dropped and the engine cut out completely. He immediately switched from bomb bay [fuel tank to the number-one tank]. Finding his engine would not start he then switched to the number-three tank and then attempted to start the engine but failed in this so he feathered [number-one propeller]. Going on single engine he applied all rudder trim and also stood on the right rudder in an attempt to keep the aircraft flying straight. Pilot Hooper then attempted to gain altitude and managed to gain 200 feet before his airspeed dropped. His right engine began heating up and realizing that he was liable to lose his good engine he began looking for a place to make a forced landing. He spotted a field and made his approach to it at an indicated airspeed of 130 to 140 mph, wheels up and no flaps. The airplane first contacted the ground when the tail, which was low, struck a small rise. This impact threw the nose down destroying the nose section of the aircraft back to the pilot's cockpit. Lt. Anderson was thrown clear but fatally injured. F/O Hooper was still in his seat but received serious injuries, including a broken jaw, broken nose, broken arm and multiple lacerations of the face. The pilot was removed by eyewitnesses who reached the scene of the crash."

11-13-44C. Hammer Field, California. At 1622, a Douglas P-70B crashed while taxiing at Hammer Field, Fresno, California, killing pilot 1Lt. Gerald L. Cushing. The airplane landed after completing a successful "slow time" flight after a double engine change. The airplane made a normal landing on the runway and slowed to taxi speed by the end of the 7,200-foot runway. The airplane was then seen to veer sharply to the right, making a turn of about 70 degrees. Ground witnesses stated that the pilot had apparently advanced the throttle on the port engine to make the turn. The airplane then taxied at high speed for 800 feet across rough terrain at the end of the runway. The throttles were retarded and the airplane careened over a ten-foot embankment, ripped through a wire fence and then smashed into the far bank of a 15-foot wide irrigation canal. The pilot was killed in the crash. Investigation revealed that the airplane suffered a loss of braking action because of a hydraulic failure.

11-13-44D. Casper, Wyoming. At 2328 MWT, a Douglas C-47 crashed three miles northwest of the Army Air Base at Casper, Wyoming, killing pilot 1Lt. Roy L. Barnhill and co-pilot 1Lt. Eugene L. Kramer and nine passengers. The airplane had taken off from Cheyenne, Wyoming, to Billings, Montana, with Casper, Wyoming, as an alternate. The airplane landed at Casper Army Air Base at 1221 MWT on 11-12-44 and was trapped there by poor weather. The pilot was refused a contact clearance for Great Falls, Montana. At 1800 on 11-13-44, the pilot was cleared on a contact flight to Wardell Municipal Airport at Casper. The starboard engine would not start and the flight was delayed until 2300 when the engine was finally started. The airplane was given another clearance for Wardell Airport. The pilot taxied to Runway 7 and performed his engine run up. The tower advised the pilot that he was on the wrong end of the runway for take-off. The pilot acknowledged and taxied down the runway to the correct end and turned around. The airplane was cleared for take-off on Runway 25 and the pilot began his roll. The airplane made a normal take-off and then entered a turn to the right to leave the traffic pattern. Moments later tower personnel noticed a flash in the distance where the airplane crashed. Investigation revealed that the airplane struck the ground in a vertical bank to the left, the port wingtip contacting the ground first. The airplane then cartwheeled into the ground and exploded into flames. Passengers killed in the crash were: 1Lt. Eugene L. Kramer, passenger; 2Lt. Homer I. Martin, passenger; Sgt. John E. Carnahan, passenger; Pvt. Robert A. Wightman, passenger; Pvt. Leonard J. Hutchins, passenger; Cpl. Herman E. Bennett, USMC passenger; Pfc. Willoughby C. Nason, USMC passenger; AMM 3C Harry W. Siden, USN passenger; Yeoman 3C Anna Majoros, WAVE passenger.

11-14-44A. Moody Field, Georgia. At 0215 EWT, a North American TB-25J crashed one mile northeast of Moody Field, Valdosta, Georgia, killing A/C John V. Petterson and A/C Josh Warrington. The students had successfully completed a routine training mission and had entered the traffic pattern to land at Moody Field. The students extended the landing gear but the nose gear did not lower. The students left the traffic pattern and advised the control officer in the tower of the problem. The control officer advised the students to lower the nose landing gear manually. The students lowered the nose gear manually but still could not get the "down lock" in place. The students were ordered to fly past the control tower at 200 feet agl so the control officer could have a look at the landing gear. Investigators stated, "One of the tower officers on duty stood on the outside of the control tower and watched the student pilots during the procedure of letting down. The airplane was observed to make a sharp 180-degree spiraling descent from 2,000 feet to approximately 200 feet to a heading in the direction of the field. The student pilots reported that they were turning on the approach during the sharp spiraling descent. After the student pilots had completed the turn, the tower officer warned them of their critically low altitude and instructed them to pull up. The students did not acknowledge the last instruction but leveled off momentarily at about 200 feet and started a slow descent until they struck the trees.... From the available evidence it was determined that the airplane struck the trees in a shallow descent in a wings-level attitude. Thirty-seven pine trees of about 10 inches in diameter breast high were broken off at varying heights from the tops of the trees to the base of the trees during the descent."

11-14-44B. Lake Louisa, Florida. At 1710 EWT, a North American P-51B crashed into Lake Louisa three miles SSE of Clermont, Florida, killing pilot 1Lt. Dean R. Gilmore. The airplane was part of a flight of five that took off from Bartow Army Air Field, Bartow, Florida, on a low-altitude navigation mission. The flight made one complete circuit around the prescribed course and then began a second circuit. As the flight approached the vicinity of Clermont at an altitude of about 200 feet, the subject airplane entered a shallow dive and crashed into Lake Louisa, exploding upon impact.

11-14-44C. Location Unknown. At an unknown time after 1908 EWT, a Beech AT-11 disappeared and is presumed to have crashed with its crew of three. Pilot 2Lt. Harold J. Leveilles, engineer SSgt. Huland T. Harris and passenger Cpl. Donald J. Benson were missing and presumed killed. The airplane took off at 1905 EWT from Dover Army Air Field, Dover, Delaware, on a cargo mission to Olmsted Field, Middletown, Pennsylvania. The airplane made a normal take-off and was last heard from on the tower frequency at 1908. The airplane was scheduled to arrive at Middletown at about 1945. The airplane became overdue at 2000 and was declared missing soon after. No trace of the airplane or crew has been located.

11-14-44D. Kokomo, Mississippi. At 2240, a Beech AT-7 crashed two miles north of Kokomo, Mississippi, killing five fliers. The airplane, which was stationed at Selman Army Air Field, Monroe, Louisiana, was part of an 11-ship flight that took off from Tallahassee, Florida, on a navigation training flight back to Selman Field via Baton Rouge, Louisiana. The subject airplane became separated from the formation and crashed a short time later. Investigation revealed that the airplane flew into the ground in a shallow descent at cruise speed. Investigators could not determine the cause of the crash. Killed in the crash were: 2Lt. Marion E. Reed, pilot; Pvt. Herman D. Armstrong, engineer; A/C Robert W. Schott, student navigator; A/C William J. Clark, student navigator; A/S George W. Shematek, student navigator.

11-15-44A. Ontario, California. At 1002, a Lockheed P-38L crashed nine and a half miles northeast of Ontario Army Air Field, Ontario, California, killing pilot 2Lt. David R. Phillip. The airplane took off on Runway 25, circled the field and then climbed to about 3,000 feet. The airplane was seen to snap to an inverted position while attempting a turn to the right. The airplane described two or three turns of spin to the right. The airplane stopped spinning, straightened out and then plunged into a vineyard in a vertical dive, exploding violently into flames upon impact and killing the pilot instantly. Investigators noted that this was the pilot's second solo flight in a P-38 type airplane.

11-15-44B. Chico, California. At 1113, a Lockheed P-38G crashed ten miles northwest of the Army Air Field at Chico, California, killing pilot 2Lt. Charles R. Towry, Jr. The airplane was part of a two-ship flight that was taking off from Chico Army Air Field on a ferry flight to Ephrata Army Air Field, Ephrata, Washington. The subject airplane was attempting to catch up with the lead aircraft when it began suffering engine problems. The airplane apparently stalled and plunged to the ground in a vertical dive, exploding violently into flames upon impact and killing the pilot instantly.

11-15-44C. Roswell, New Mexico. At 2257 MWT, a Boeing B-17G flying in poor weather crashed into a hill 45 miles southwest of Roswell Army Air Field, Roswell, New Mexico, killing three fliers. The airplane took off from Roswell Army Air Field at 2015 MWT on a routine night navigation flight to Clovis, New Mexico, to Lubbock, Texas, to Hobbs, New Mexico, and return to Roswell. At 2250 the airplane reported in at 6,000 feet over Hobbs. The airplane soon encountered rainstorms and the pilots elected to turn back for Hobbs. The pilot was cleared to fly the Roswell Radio Range back to Roswell and the pilot acknowledged, reporting that he was about 15 minutes out. The airplane did not answer repeated radio calls and failed to arrive at Roswell. The airplane was declared missing soon after and a search was commenced at daybreak. The aircraft was located at about 1700. Investigation revealed that the aircraft struck the hill in full flight at cruise speed about 400 feet below the crest of the hill. The airplane exploded upon impact at an elevation of about 5550 feet msl, scattering wreckage to the top of the hill. B-17 instructor pilot 1Lt. John F. Moriarty, B-17 student 2Lt. William H. Troop and B-17 student 1Lt. Gerald B. Trotman were killed in the crash.

11-16-44A. Geyserville, California. At 0912, a low-flying Lockheed P-38J collided with structures and trees and crashed into a hillside one-quarter mile southwest of Geyserville, California, killing pilot 2Lt. James B. McMinimes, 20. The P-38 was part of a six-ship flight that had taken off from Santa Rosa Army Air Field, Santa Rosa, California, on an instrument training flight. The flight broke up into three pairs with one pilot flying on instruments and the other pilot flying as the safety ship. After flying on instruments for about 30 minutes the subject airplane went off instruments

and descended to low altitude in the vicinity of Geyserville. The safety airplane did not descend with the subject airplane but circled the local area at high altitude. The subject airplane made three passes over the town at extremely low altitude. The first pass was from north to south, parallel to the main highway. The airplane then circled around to the left and made a low level pass from east to west, perpendicular to the main highway. The airplane made a 360-degree left turn and buzzed the town at very low altitude from approximately northeast to southwest at about a 45-degree angle to the main highway. The airplane mushed down as the pilot attempted to pull up from a slight dive, striking a house chimney with the starboard propeller, engine nacelle and tail boom. The airplane then collided with a telephone pole and telephone lines before smashing through some treetops. The airplane climbed away with a small flame trailing the starboard engine. The starboard tail boom failed, causing the entire tail section to tear away at the tail booms, sending the airplane out of control. The airplane passed over a small hill, barely missing colliding with it before smashing into a small hill and exploding about one-quarter mile away. Investigation revealed that on 11-5-44 the pilot had visited a cousin living in Geyserville and that the pilot reportedly told the cousin that he would find an occasion to fly over. The airplane collided with the chimney and trees about 100 yards from the house the pilot had visited.

11-16-44B. Myrtle Beach, South Carolina. At 1220, a Douglas A-20J crashed ten miles north of Myrtle Beach Army Air Field, Myrtle Beach, South Carolina, killing three fliers. The airplane had taken off from Myrtle Beach Army Air Field on a routine gunnery mission and then flew to the AAF Gunnery Range eight miles NNW of Myrtle Beach. The airplane had completed a run at the target and had crashed while making the first turn on the gunnery range traffic pattern. Investigation of the wreckage indicated that the airplane struck the ground at an 80-degree angle with both engines operating at high rpm. It was speculated that the airplane stalled in the turn at an altitude of about 800 feet agl, too low to allow the pilot to make a recovery. Pilot 2Lt. Harold L. Patterson, gunnery instructor TSgt. James R. Thompson and gunner Cpl. Jack McClymont were killed in the crash.

11-16-44C. Placitas, New Mexico. At 1330 MWT, a Douglas C-47B flying in poor weather collided with rising terrain and crashed five miles south of Placitas, New Mexico, killing seven fliers and injuring five others. The airplane took off on an instrument clearance from Wichita, Kansas, on a cargo flight to Los Angeles, California, via Long Beach. Investigation revealed that the airplane flew on top of the overcast for some time and also flew at low altitude for a portion of the flight. The airplane encountered poor weather so the pilot changed course and stated that he intended to land at Albuquerque. The airplane entered snow showers and the pilot apparently attempted to let down in an effort to maintain visual contact. Survivors stated that the airplane had encountered strong vertical winds just prior to the crash. The airplane collided with the crest of a mountain and smashed itself to pieces as it came to rest on the peak. The survivors were stranded on the mountain for three days before rescuers could reach them. Three passengers who had survived the initial crash died on the first night while waiting to be rescued. On the third night surviving passengers noticed lights in their vicinity. Two men left the crash scene and made their way to a road nearby. The two walked a short distance and were able to reach an inhabited area. Pilot Lt. Col. Robert I. Dittrich was killed. Passengers killed in the crash were: 2Lt. Joseph R. Zazvorka, Jr., Sgt. Gilbert L. Reinhardt, Cpl. Murray Rosenberg, Coxswain Mate Frank J. Klinger (USN), Coxswain Mate Marion J. Houser (USN), Richard L. Aitken (civilian technician for Beech Aircraft Company). Passengers injured in the crash were: 2Lt. Lloyd A. Bennett, Sgt. Rubin S. Goldberg, Ensign Fred R. Jenks (U.S. Merchant Marines), Cpl. Frederick G. Hall, Cpl. Clifford F. Wagner.

11-16-44D. Ontario, California. At 1605, a Lockheed P-38L crashed one-quarter mile northeast of the Army Air Field at Ontario, California, killing F/O James L. McGurk. The airplane took off at 1444 on a routine training flight and returned to the field at 1603, flying over the field before peeling off to enter the traffic pattern. The airplane flew the traffic pattern and made a good approach, letting down to about 10 feet agl over the first third of the runway. The pilot then advanced the throttles and retracted the landing gear, apparently attempting to go around. The airplane was climbing away straight ahead when the port engine began backfiring and suffering diminished performance. The airplane climbed to about 800 feet agl before attempting a turn to the left. Both engines were backfiring and flames were observed trailing the port turbo-supercharger. The airplane stalled and entered a spin to the left at an altitude too low to allow recovery, plunging into a vineyard and exploding into flames upon impact. Rescue personnel were on the scene immediately. Investigation revealed that the pilot had mistakenly feathered the starboard propeller instead of the port propeller, causing the airplane to lose all power and enter a stalling condition. Investigators noted that the initial approach was "very good" and it was not known why the pilot failed to complete the landing.

11-16-44E. Hope, New Mexico. At 1944 MWT, a Vultee BT-13B flying in poor weather collided with rising terrain and crashed ten miles southwest of Hope, New Mexico, killing pilot 2Lt. Edward C. Bailiff and pilot-rated passenger 2Lt. Raymond V. Johnson. The airplane had taken off from Liberal, Kansas, on a flight to Carlsbad, New Mexico. The airplane failed to arrive and was declared missing. The initial search was hampered by poor weather. The airplane was found on 11-22-44.

Investigation revealed that the pilots were mistakenly given weather information for Scott Field, Belleville, Illinois, instead of Carlsbad as they had requested. The airplane entered instrument conditions and while apparently letting down collided with the mountaintop in a flying attitude with power. The landing gear collapsed and the airplane skidded along the terrain for approximately 250 feet and then bounded back into the air for about 50 feet before crashing back to earth. The airplane skidded along for another 100 feet before flying off the edge of the hill, falling about 300 feet vertically as it crashed into a bank of earth surrounding an arroyo about 400 feet further on.

11-16-44F. Peyton, Colorado. At 2200 MWT, a Consolidated B-24J crashed 15 miles northeast of Peterson Field, Colorado Springs, Colorado, killing two fliers and injuring two others. Five crewmembers were uninjured. Investigators stated, "B-24J #42-73514 bellied into a cornfield one mile southeast of Peyton, Colorado. The aircraft was on a routine night transition flight from Pueblo Army Air Field, Pueblo, Colorado. After encountering low clouds the instructor pilot decided to let down and get under it and also started a left turn to get out and away from the clouds. The airplane was flying in a westerly direction in a nearly level course when it first struck the ground. The left wing was slightly lower than the right. Number-one and number-two propellers struck the ground soon after the fuselage of the airplane. It then skidded for approximately 500 yards shedding equipment along the course. It came to a rest in a low place headed in the opposite direction with the fuselage broken in half. The propellers and propeller marks show that all engines were operating normally." Co-pilot 2Lt. James E. Leitch and engineer Cpl. Irving D. Cornell were killed. Co-pilot 2Lt. Charles V. Wortman and gunner Pfc. Romalo Capoville sustained minor injuries. Crewmembers who were uninjured were: 1Lt. Cyrus K. Rickel, pilot; 2Lt. George P. Katibah, navigator; 2Lt. Charles W. Donahoo, bombardier; SSgt. Louis F. Strohe, engineer instructor; Cpl. Raymond L. Bartlett.

11-16-44G. Tonopah, Nevada. At 2356 PWT, a Consolidated B-24J collided with rising terrain 30 miles NNE of the Army Air Field at Tonopah, Nevada, killing the crew of eight. Killed in the crash were: 1Lt. Earl J. Jacobes, pilot; F/O Dale R. LeBlanc, co-pilot; F/O Harry J. Holsten, navigator; F/O Hohn H. Williams, bombardier; Cpl. Allan W. Gransee, engineer; Cpl. Charles G. Miller, radio operator; Cpl. Harold A. Ellis, gunner; Cpl. Edward E. Zimmerman, gunner. The airplane took off at 2154 PWT from Tonopah Army Air Field, Tonopah, Nevada, on a routine training flight. At 2226 PWT the airplane returned to the field and the pilot reported that gasoline was siphoning out of the starboard fuel caps. Ground crewmembers and the crew chief installed new fuel caps. The installation checked out and the airplane took off at 2323 PWT. There was no other contact with the aircraft after the second take-off. Several witnesses in the air and on the ground observed an explosion on a mountainous plateau at 2356. Investigation revealed that the airplane smashed into the terrain in a flying attitude with the engines producing power. Investigators could not determine the cause of the crash and it was noted that the pilot had been a B-24 instructor pilot and had been recently assigned a combat crew in preparation to being shipped overseas to an active war theater.

11-17-44A. Wynnwood, Oklahoma. At 1100, a Lockheed F-5E crashed three miles south of Wynnwood, Oklahoma, killing 2Lt. George V. Rollow. The airplane took off from Will Rogers Field, Oklahoma City, Oklahoma, on a routine transition flight. The airplane appeared over Wynnwood and buzzed the area at an altitude of about 500 feet agl. The airplane was seen to enter a cloud and then quickly emerge. The F-5E then entered a second cloud and inadvertently entered a steep diving turn to the right. The airplane was seen to dive straight down out of the 1,400-foot overcast. The pilot was unable to recover and the airplane struck the ground in a vertical dive, exploding into flames upon impact and killing the pilot instantly. Investigator's noted that Wynnwood was the pilot's hometown. There was no evidence of extremely low flying.

11-17-44B. Fort Myers, Florida. At 1325 EWT, a Curtiss P-40N crashed 25 miles NNE of Page Field, Fort Myers, Florida, killing 2Lt. Henry J. Virkler. The airplane took off from Page Field on a ground gunnery mission. Investigators stated, "Lt. Virkler was firing ground gunnery and evidently misjudged his distance or became too absorbed in his target, and was not observant enough of his altitude. The propeller of Lt. Virkler's airplane struck a log on the ground behind the targets. He pulled up and prepared for a crash-landing straight ahead. His airplane struck a tree and the airplane exploded, striking the ground aflame."

11-17-44C. Balmorhea, Texas. At 1653, a Cessna UC-78B flying in instrument conditions collided with rising terrain 16 miles southwest of Balmorhea, Texas, killing instructor 2Lt. Clifford B. Mason and student A/C James A. Kennedy. The airplane took off at 1613 from Pecos Army Air Field, Texas, on a student instrument training flight. The ceiling was at about 700 feet agl at the time of take-off. Investigation revealed that the airplane had collided with the top of a small ridge at cruise speed while in a gentle bank to the right. Both engines were producing power. The airplane struck about 25 feet below the top of the ridge at an elevation of approximately 5,000 feet msl, bursting into flames upon impact. Wreckage was scattered to the top of the ridge. Apparently the airplane flew through a break between two high peaks before colliding with the ridge. The terrain was totally obscured by clouds at the time of the accident.

11-17-44D. Pampa, Texas. At 2053, a North American TB-25J crashed 24 miles southwest of the Army Air Field at Pampa, Texas, killing three fliers. B-25 instructor pilot Capt. Ray M. Flake and B-25 students A/C Ira L. Dixon and A/C Charles B. Collins were killed in the crash. The airplane took off at 2015 from Pampa Army Air Field on a routine night transition mission. The instructor had radioed the control tower at Pampa and reported the weather, stating that the weather was above minimums and that student flights could continue. At 2045, the control tower officer requested that the airplane fly to the area of Clarendon for a weather check. The instructor acknowledged. The airplane never checked in at Clarendon and failed to return to Pampa. The control tower attempted to contact the airplane at about 2300, but without success. The airplane was found the next day. Investigation of the wreckage indicated that the airplane struck the ground while in a shallow dive while flying to the southeast. The port wingtip had struck the side of a small hill, causing the airplane to cartwheel into the terrain. The airplane burst into flames upon impact, scattering wreckage over an area of 200 yards. It was speculated that the instructor encountered a lowering ceiling and had apparently attempted to remain below the overcast, inadvertently flying the airplane into the terrain.

11-17-44E. George Field, Illinois. At 2310, a Douglas C-47F crashed just after take-off from George Field, Lawrenceville, Illinois, killing pilot 2Lt. Irving A. Johnson and co-pilot 2Lt. Kenneth H. Klinger. Investigation revealed that the airplane had completed a student mission earlier in the evening and that the subject pilots, two other C-47 students, and an instructor had switched seats several times during the course of the night flying, which had commenced at about 1930. The airplane had landed at George Field and the instructor and two C-47 students deplaned. The subject pilots were ordered to fly the traffic pattern and make a couple landings before picking the instructor up for more training. The subject airplane took off directly behind a nine-ship flight of C-47s. The subject airplane made a normal take-off and then entered a diving turn into the ground, exploding into flames upon impact. Investigators speculated that a landing gear problem might have distracted the co-pilot, allowing the throttles to creep away from take-off rpm. The power fall off was uneven, causing the airplane to turn into the ground.

11-18-44A. Location Unknown. At a time after 0015 PWT, a Douglas P-70B disappeared and is presumed to have crashed with its pilot. 2Lt. Burton W. Stuart was declared missing and presumed killed. Investigators stated, "Pilot took off [from Salinas Army Air Base, Salinas, California] at approximately 2200 PWT 17 November 1944 on a local night transition flight. Tower contacted the airplane at about 2245 and told pilot to return to the field. Pilot told tower he was lost and would change to Salinas Homer to get a steer. Reports of Salinas Homer show pilot was given several headings to fly and remained in contact with the Homer for approximately 30 minutes, then they lost contact with the airplane. Pilot had been instructed to climb to 9,500 feet and was without contact for about 20 minutes when he was picked up by Mt. Tamalpais Homer. Mt. Tamalpais Homer worked the airplane for about 25 to 30 minutes, giving him several headings to bring him to San Francisco [California]. The last homing the pilot was given was 010 degrees at 0015 PWT 18 November 1944. This was the last contact with the airplane." Investigators noted that a ceiling of 10,000 feet moved into the area but that conditions underneath were generally good.

11-18-44B. Fort Green Spring, Florida. At 0725 EWT, a Boeing B-17F suffered a catastrophic structural failure and crashed 12 miles WSW of Fort Green Spring, Florida, killing nine crewmembers. The airplane took off at 0635 EWT from MacDill Field, Tampa, Florida, on a routine formation flight. The crew was instructed to climb on instruments to 15,000 feet and then rendezvous at 0745 EWT over MacDill Field with other B-17s for high-altitude formation training. The airplane apparently entered a stalling attitude that progressed into a steep dive with excessive speed. The pilots stressed the airplane beyond its design limits in an attempt to re-establish level flight. Both wings, horizontal stabilizers and elevators separated from the airplane during the pull-out. The airplane burst into flames as it broke up and fell, smashing to the ground 28 miles southeast of MacDill Field. Wreckage was scattered over a wide area. Killed in the crash were: 2Lt. Lawrence S. Scofield, pilot; 2Lt. Farlan L. Spencer, co-pilot; 2Lt. Alonza L. Winder, navigator; 2Lt. Lowell E. Smith, bombardier; Cpl. Lynn H. McClure, Jr., engineer; Cpl. John F. Brady, radio operator; Pfc. Edward J. Strainer, assistant radio operator; Cpl. Adrian W. Armour, gunner; Pfc. Gene Sine, gunner.

11-18-44C. Perry, Florida. At 0900 EWT, a Vultee BT-13A crashed 11 miles south of the Army Air Field at Perry, Florida, killing pilots 2Lt. William G. McAdoo and 2Lt. Roger B. Vail. The airplane took off at 0845 from Perry Army Air Field on a routine instrument training flight. Lt. McAdoo was flying in the front cockpit as the safety pilot and Lt. Vail was flying in the rear cockpit as the instrument pilot. The airplane collided with three pine trees while apparently attempting an actual or simulated forced landing in a 650-foot field. The airplane went out of control and smashed to the ground in the field, bursting into flames upon impact. Investigation of the wreckage indicated that the propeller was turning at the time of impact. Examination of the engine failed to reveal if the engine was malfunctioning.

11-18-44D. Stockton, California. At 1425, two fliers were killed and a passenger was lightly injured when they mistakenly bailed out of a Consolidated B-24J attempting an emergency landing at Stockton Army Air Field, Stockton, California. The B-24 had taken off from Hamilton Field, San Rafael, California, on the return leg of a navigation training flight back to its home station at March Field, Riverside, California. The B-24 suffered a fire in the number-two engine while flying near Stockton Army Air Field. The pilot ordered the passengers and crew to secure their parachutes and stand by for an emergency landing at Stockton. Paratrooper-rated passenger T/5 Ernest L. Jackson was ordered to bail out by gunner Cpl. William E. Donnelly and gunner Cpl. Raymond F. Lewis. T/5 Jackson protested because of the low altitude. The gunners again ordered the passenger to bail out. T/5 Jackson bailed out of the bottom fuselage hatch and parachuted to safety, suffering only minor injuries. Moments later Cpl. Lewis and Cpl. Donnelly followed T/5 Jackson out of the bottom hatch. They bailed out at an altitude too low to allow their parachutes to deploy and they fell to their deaths. Pilot 2Lt. John F. Daly and copilot 2Lt. Peter G. McCabe stated to investigators that they had not ordered any crewmembers or passengers to bail out. The pilots, six crewmembers and one passenger escaped injury in the successful emergency landing.

11-18-44E. Wilmington, California. At 1520, a Douglas A-20H crashed east of Wilmington, California, killing pilot Capt. Grover Huppel, Jr. The airplane took off from Long Beach, California, on a local flight. Investigators stated, "[Ensign R.J. Chitwood, U.S. Navy] stated that he was flying a [Grumman F6F] at the time of the accident, and was above the A-20 watching his actions, as the A-20 seemed to be flying at reduced power at an altitude of about 3,500 feet. He further stated that the aircraft seemed to be pulled up at a steep angle and on reaching approximately 5,000 feet, the aircraft stalled and whipped over on its back (to the left). The aircraft entered a spin (not inverted or flat) to the left and continued to spin for about five to seven turns. The aircraft seemed to recover for a few seconds, then enter a secondary spin to the left and continued to spin until the ship crashed." The aircraft exploded violently into flames upon impact. Investigators speculated that the pilot had been practicing stalls and slow speed flight just before the accident. Investigators noted that the pilot, a twin-engine instructor, had been incorrectly cleared on the flight. Investigation revealed that the pilot had not received a cockpit check off on A-20 type airplanes. Investigators noted that the pilot had successfully completed an A-20 ground school course and passed a written examination on the A-20.

11-18-44F. Colorado Springs, Colorado. At 1825, a North American TB-25J flying in instrument conditions crashed three miles northeast of Peterson Field, Colorado Springs, Colorado, killing pilot 1Lt. Leslie G. Wolbrink. On the morning of 11-18-44, forty-eight B-25s landed at Peterson Field due to poor weather conditions at the home station, La Junta Army Air Field, La Junta, Colorado. In the afternoon the weather cleared at La Junta and 46 B-25s returned to La Junta. Two remained on the ground at Peterson Field due to engine trouble. At 1801 MWT, two B-25s were dispatched from La Junta to Peterson Field to pick up the crews of the two grounded B-25s and return them to La Junta Army Air Field. The subject airplane, TB-25J #43-4617, and TB-25J #43-5103 joined into a two-ship formation with 43-5103 leading. As the airplanes approached the area of Peterson Field the pilots observed a formation of clouds in their flight path. The clouds blocked out the view and lights of Peterson Field. The lead pilot decided to turn back. Before the lead pilot began the 180-degree left turn, he looked back to check the wing ship and noticed that it was not there. The lead ship began the turn to the left and soon entered instrument conditions, which included heavy snowfall. The lead pilot leveled out on the reciprocal heading and soon broke out of the clouds. The lead pilot was unable to locate the subject airplane after breaking out in the clear. The lead ship returned to La Junta Army Air Field and landed without incident. The lead ship and the Peterson Field control tower were unable to contact the subject airplane by radio. Investigation of the wreckage revealed that the airplane smashed into the west side of a small hill in an inverted dive of approximately 75 degrees. Investigation of the propellers indicated that they were both turning at the time of impact. Investigators speculated that the pilot lost control of the airplane while flying in instrument conditions and was unable to recover before smashing to the ground.

11-18-44G. March Field, California. At 1900, a Consolidated B-24J crashed five miles south of March Field, Riverside, California, killing 11 fliers. The airplane had taken off from March Field on a bombing and gunnery mission. After successfully completing the mission, the airplane returned to the area of March Field and the pilot was instructed to fly around the local area at 12,000 feet. The pilot had reported to tower personnel that he had had some overheating in one of his engines earlier in the flight but the trouble had cleared up and the engine was performing satisfactorily. Approximately ten minutes later the B-24 was seen in a steep inverted dive, passing close to an airplane flying at 8,000 feet before smashing into the ground and exploding into flames. Investigators stated, "The aircraft went out of control at approximately 12,000 feet for an undetermined reason.... Examination of the wreckage indicated that the aircraft had been in a spin when it struck the ground, in that the position of the wings indicated a rotating impact. The tail turret was thrown backwards from the rest of the wreckage and indicated that it had struck the ground with great force, probably caused by the aircraft striking in a nose-down, inverted position. It was also determined that the left elevator had disintegrated long enough before the crash for parts

of fabric and ribs to have come off and land 1,800 to 2,000 feet away." Killed in the crash were: 1Lt. Donald G. Hopkins, pilot; F/O Harold M. Gibble, Jr., co-pilot; 2Lt. James E. McLaughlin, navigator; 1Lt. Lucien E. Lyons, navigator; Sgt. Vernon W. Anderson, engineer; Sgt. Thomas J. Sheils, engineer; Cpl. John W. Riley, radio operator; Cpl. Russell H. Allison, assistant radio operator; Cpl. Coy L. Keys, gunner; Pfc. Carl Carlson, Jr., gunner; SSgt. Troy E. Dahmer, gunnery instructor.

11-19-44. Fort Stockton, Texas. At 1525, a Cessna UC-78B crashed 20 miles west of Fort Stockton, Texas, killing instructor 2Lt. Arthur R. Malone and student A/C Reinhard W. Roman. The airplane took off at 1500 from Pecos Army Air Field, Pecos, Texas, to patrol a student low-level navigation mission to McCamey, Texas, to Fort Stockton, Texas, to Balmorhea, Texas, and return to Pecos. The airplane was observed to enter a steep chandelle type maneuver, stalling out at the top of the maneuver and entering a spin at an altitude too low to allow recovery. The airplane struck the ground at a 90-degree angle, exploding into flames upon impact.

11-20-44A. Oracle, Arizona. At approximately 0945 MWT, a North American AT-6D collided with a tall cactus and crashed near Oracle, Arizona, killing instructor 2Lt. Alfred S. Farnsworth and seriously injuring student A/C John P. Purser, Jr. The airplane took off from Williams Field, Chandler, Arizona, on a day interception problem. The airplane was early for its checkpoint at Tucson. The student was going to circumnavigate Mt. Lemmon in attempt to use up four minutes so that the airplane would arrive on time at Tucson. The instructor took control of the airplane and flew the airplane between two peaks. The instructor then began buzzing a ranch at the bottom of the mountains. The instructor flew the airplane at very low altitude and collided with a tall cactus. The student stated that he attempted to take over control and pull up over the cactus but was unsuccessful. The airplane collided with the cactus and went out of control, crashing on the Sigel Ranch about 30 miles southeast of Phoenix. It was noted that the instructor was flying the airplane from the rear seat.

11-20-44B. Williamsport, Ohio. At 1955 CWT, a Boeing B-17F suffering engine trouble crashed five miles southwest of Williamsport, Ohio, killing five fliers. The airplane took off at 1940 CWT from Lockborne Army Air Base, Columbus, Ohio, on a cross-country navigation mission. The airplane crashed fifteen minutes after take-off. Investigators stated, "Upon investigation the following facts were disclosed: The aircraft had developed engine trouble and was turning back toward Lockbourne AAB. It made a wide circle over Atlanta, Ohio, then came east toward Williamsport, turned north and then west, struck trees on the top of a ridge and came to rest in the declivity at the bottom of the ridge approximately 400 feet from where it struck the trees. It was determined from the path of the aircraft and the position of the wreckage that the left wing struck the trees, causing it to bank steeply to the left. The airplane struck the ground at the bottom of the decline in this attitude and cartwheeled, swinging the fuselage around 180 degrees from the direction of the approach." Killed in the crash were: 1Lt. Robert G. Fontaine, B-17 instructor; 1Lt. Bernard J. McGuire, B-17 student; 1Lt. Jack B. Reimland, B-17 student; 1Lt. Wilbur R. Priess, navigator; Pvt. James E. Sanders, engineer. Investigators noted that both B-17 students had served combat tours overseas in another type of aircraft.

11-20-44C. Gowen Field, Idaho. At 2245 MWT, a Consolidated B-24J crashed two and one-half miles southeast of Gowen Field, Boise, Idaho, killing four fliers. The airplane, which was on a routine training flight, made a normal take-off and climbed away. Control tower personnel diverted their attention to other aircraft and approximately 90 seconds later observed a fire on the ground about 15 degrees to the right of the take-off heading. Investigation revealed that the airplane descended into the ground in a flying attitude in a slight bank to the right. Landing gear and flaps were in the retracted position. The airplane had passed over a bank of horizon lights just prior to flying into the ground. Investigators speculated that the pilots had established a slow rate of climb after passing over the horizon lighting and that the pilots had been distracted by something in the cockpit, causing them to inadvertently fly the airplane into the ground. Killed in the crash were: 2Lt. Albert A. Amodeo, pilot; 2Lt. Ernest J. Seymour, co-pilot; Cpl. Matthew Hirshman, engineer; Cpl. Harry N. Sampsell, radio operator.

11-20-44D. Chandler, Arizona. At 2323 MWT, a North American TB-25C crashed 14 miles southwest of Williams Field, Chandler, Arizona, killing A/C William A. Mahoney and A/C Samuel L. Mahoney. The airplane took off from Douglas Army Air Field, Douglas, Arizona, on a scheduled night formation and navigation training flight to Tucson, Arizona, to Phoenix, Arizona, to Palm Springs, California, to Tucson, and return to Douglas. The subject airplane, flying in the number-three position, assembled with two other airplanes and began the flight. The flight encountered a massive and "towering cumulus cloud" between Tucson and Phoenix. Investigators stated, "Just prior to entering this cloud, [the instructor], flying in the number-two position, had directed his flight to descend in hopes of passing under any clouds. On entering the cloud, [the instructor] radioed the aircraft in his flight to make a climbing turn to the right of 180 degrees to come out of the cloud area. The instructor's aircraft broke out of the clouds at 17,000 feet about 15 miles south of Phoenix, Arizona. Shortly after this, the lead aircraft also broke out at a slightly higher altitude.... [The subject aircraft] was never seen to break out, but crashed in the desert about 12 miles south of Chandler, Arizona.... [When the airplane failed to return to base it was declared missing.] A ground search located the aircraft, which appeared to have crashed from a sweeping circle, and apparently at high speed. Following impact, the aircraft began to burn and exploded when the fire reached the fuel cells. It is believed that the sudden flight into clouds at night may have resulted in confusion and vertigo, and that aircraft may have fallen into a diving spiral from which the students could not recover." See 11-20-44E.

11-20-44E. Chandler, Arizona. At 2341 MWT, a North American TB-25D crashed 17 miles WSW of Chandler, Arizona, killing A/C Thomas P. Henry and A/C Ole A. Hendrickson. The airplane took off from Douglas Army Air Field, Douglas, Arizona, on a scheduled night formation and navigation training flight to Tucson, to Arizona, Phoenix, Arizona, to Palm Springs, California, to Tucson, and return to Douglas. The flight encountered a massive "towering cumulus cloud" between Tucson and Phoenix. The students inadvertently flew into the cloud and apparently lost control of the aircraft. Investigation of the wreckage indicated that the airplane had struck the ground in a 45-degree dive, exploding into flames upon impact and scattering wreckage over a wide area. The students were killed instantly. All student airplanes were called in following the crash. See 11-20-44D.

11-21-44. Esparto, California. At 1813, a Curtiss C-46A broke up in flight and crashed at Esparto, California, killing eight fliers. The airplane took off from Reno Army Air Field, Reno, Nevada, on a routine training flight to McClellan Field, Sacramento, California, and return to Reno. The airplane took off from McClelland Field at 1612 PWT on the return trip to Reno Army Air Field. Investigation of the wreckage revealed that both elevators and horizontal stabilizers and the outer portions of both wings had separated from the airplane prior to impact. The airplane struck the ground in an inverted spin to the left, exploding into flames upon impact. Investigators speculated that the pilots had lost control of the airplane, causing the airplane to enter a steep dive and to build up excessive airspeed. The pilots stressed the airframe beyond its design limits in an attempt to re-establish level flight. Wreckage was scattered over an area of two and a half miles. Killed in the crash were: F/O Henry J. Drescher, Jr., C-46 instructor; F/O James W. Kelly, pilot/C-46 student; 2Lt. Irving W. Jones, Jr., copilot/C-46 student; Cpl. Thomas C. Anderson, radio operator; Sgt. John S. Janowicz, radio operator; Pvt. John F. Boland, radio operator; Pfc. Casimir Olszenwski, engineer; Pvt. Ross F. Judd, engineer.

11-22-44A. Albany, Texas. At 0925, a Republic P-47D and a U.S. Navy Grumman TBM-3 (a Grumman Avenger manufactured by Eastern Aircraft) collided in mid-air and crashed ten miles WSW of Albany, Texas, killing three fliers. Killed in the crash of TBM-3 Bureau No. 68403 were Ensign James M.

Bayley, U.S. Navy, and passenger 1Lt. Frederick R. Woodward, U.S. Marine Corps. 2Lt. James A. Segawa was killed in the crash of the P-47. The P-47 was flying in the wing position of a two-ship flight that took off at 0800 CWT from Abilene Army Air Field, Abilene, Texas, on an instrument-training mission. The P-47s climbed to 8,000 feet and then the lead P-47 pilot went on instruments with Lt. Segawa acting as the observer pilot. The subject P-47 was flying behind and to the right of the instrument P-47. After about 30 minutes the two P-47s changed positions. Lt. Segawa attempted to fly a beam problem but despite several attempts was unsuccessful. The P-47s changed positions again and Lt. Segawa dropped back to the observer position. The lead P-47 flew instruments for several minutes before the pilot decided to go back to visual flight. As the lead pilot folded his maps and began retrimming his airplane, an Avenger type airplane was seen to pass close to the two P-47s in a zooming dive. The lead pilot did not notice the Avenger dive past. The lead pilot looked back and observed that Lt. Segawa was engaged in simulated combat with a "grayish blue" airplane. Lt. Segawa and the TBM separated and the P-47 gained altitude. The subject P-47 winged over in an attempt to make a pass at the TBM. The TBM began a climbing turn in an attempt to engage the subject P-47. The airplanes closed in on each other and at about 7,000 feet the P-47's port wing collided with the starboard horizontal stabilizer of the TBM. The P-47 went in a diving turn to the south and crashed, exploding violently into flames upon impact. The TBM immediately nosed over into a steep dive, smashing into the ground and exploding upon impact. The TBM was part of a two-ship flight that had taken off from Fort Worth, Texas, on a routine training flight to El Paso, Texas, and return.

11-22-44B. Orlando, Florida. At 1550 EWT, a North American P-51D crashed 11 miles southeast of Orlando, Florida, killing pilot 1Lt. Francis J. Nutter. The airplane was part of an eight-ship flight that had taken off from Kissimmee Army Air Field, Kissimmee, Florida, on a dive-bombing demonstration at the AAF Bombing Range at Pinecastle, Florida. The subject airplane was flying in the number-four position in the lead four-ship flight. The formation appeared over the range and the first three airplanes of the lead flight peeled off at 6,000 feet and commenced their bombing runs. The first three ships made 60-degree dives, released their bombs and began their pull-outs at 2,500 feet. The subject airplane made an identical dive to the preceding three airplanes but did not release his bomb until about 2,000 feet, starting his pull-out at 1,500 feet. As the airplane began to pull up at 300 feet agl the starboard wing failed. The airplane pitched up abruptly and the wing separated and collided with the tail section. The tail section failed and separated. The airplane winged over to the left and gyrated out of control, crashing onto the range and bursting into flames. The pilot was killed upon impact.

11-22-44C. Matagorda Peninsula, Texas. At 1458, a North American AT-6C crashed nine miles east of Matagorda Peninsula Army Air Field, Texas, killing pilot 2Lt. Walter K. Sims and passenger Pvt. George Mitchell. The AT-6C had taken off from Matagorda Peninsula Army Air Field on an aerial gunnery mission. A pilot and passenger flying in the area observed the subject airplane in a spinning attitude. The airplane stopped spinning and entered a steep dive at about 50 feet agl. The airplane then smashed into the ground nose first and exploded into flames upon impact, killing both occupants instantly. Investigation revealed that the passenger was not authorized to be on board the airplane. Pvt. Mitchell had asked another pilot if he could ride along as a passenger. That pilot asked the Squadron Operations Officer if it was okay to take Pvt. Mitchell along as a passenger and was told that passengers were not allowed on training flights. The other pilot advised Pvt. Mitchell that he could not ride along as a passenger. Apparently Lt. Sims agreed to take Pvt. Mitchell up and had signed for the enlisted man's parachute.

11-22-44D. Biggs Field, Texas. At 1430 MWT, a Culver PQ-14A crashed 32 miles ENE of Biggs Field, El Paso, Texas, killing F/O Loren A. Stevens. The airplane was part of a two-ship flight that had taken off from Biggs Field on a routine training flight. The subject airplane was flying in the lead position and the wing airplane was flying in the number-two position. The two airplanes were climbing at 110 mph in order to pass over 5,000-foot mountains in the flight path. The subject airplane then pitched up abruptly and then winged over to the left past vertical. The wing airplane entered a 30-degree turn to the left and the pilot observed the subject airplane smash into the ground. Investigators noted that the pilots were advised of the limited performance capabilities of PQ-14 type aircraft.

11-22-44E. Biggs Field, Texas. At 1728 MWT, a Boeing B-17G crashed ten miles southeast of Biggs Field, El Paso, Texas, killing three fliers. B-17 instructor 1Lt. Mont V. Beckstrand and engineer Cpl. Charles D. Owens were uninjured parachuting to safety. Pilot/B-17 student 1Lt. Laurence P. Otto, Jr., co-pilot/B-17 student 2Lt. James William Langill and radio operator Cpl. James J. Monetta were killed in the crash. The airplane took off from Biggs Field on an instrument training flight. The airplane made a normal take-off. After about ten minutes of flight, the number-one cylinder on the number-three engine exploded and blew off the engine cowling. The engine caught fire immediately so the pilot cut the throttle and attempted to feather the propeller but was unsuccessful. The fire became uncontrollable so the instructor ordered the crew to bail out. The pilot, co-pilot and engineer went to the back of the ship. The engineer pulled the emergency release on the main exit door but the door remained in place. The engineer attempted to jettison the door several times without success. The pilot pulled the emergency release and then kicked the door free. The pilot ordered the engineer to bail out, which he did immediately. The instructor, thinking that the airplane had been abandoned, left the co-pilot seat and attempted to bail out of the front hatch. The instructor had trouble releasing the nose hatch so he returned to the flight deck to fly the airplane down. The instructor noticed that the pilot and co-pilot and the radio operator were still on board. The pilot successfully released the nose hatch. The instructor began to fly the airplane but the flames had increased to such an extent that he feared that the B-17 would explode, so he again ordered the crew to bail out. The B-17 students went to the back of the airplane and the instructor bailed out of the nose hatch, parachuting to safety. He was uninjured. The airplane entered a slow spiral to the right before smashing to the ground and exploding into flames. The bodies of the remaining crewmembers were found in the wreckage. Investigation revealed that the exploding number-one cylinder head had severed the oil line for the feathering mechanism, preventing the pilots from feathering the number-three propeller.

11-22-44F. Homestead, Florida. At 2052 EWT, a Consolidated RB-24E attempting a take-off crashed three-quarters of a mile north of Runway 32 at Homestead Army Air Field, Homestead, Florida, killing four fliers. The airplane made a normal take-off and two minutes later the lights at the field went out suddenly and an explosion was seen north of the runway. Investigators stated, "The flight path of the airplane through the trees indicated that the aircraft had [made] a steep right turn from the end of Runway 32 and the condition of the clipped pine trees show that the right wing was down approximately 45 degrees when the first trees were topped. Then the aircraft slid through the trees on its right wing and across a road running due north, hitting a high tension wire on the other side of the road with the left wing. The aircraft then exploded and scattered wreckage for approximately 250 yards. The distance from the initial point of contact with the trees to the farthest bit of wreckage measures approximately 700 yards." Investigation of the wreckage reveals that the landing gear was in the down and locked position. Investigators speculated that the airplane suffered diminished performance or the failure of one or both starboard engines. This differential power condition coupled with slow airspeed and extended landing gear rendered the aircraft impossible to fly. The pilots lost control of the aircraft and it crashed into the trees. Investigators noted that the subject airplane was the "best flying" B-24 on the base. Killed in the crash were: Capt. John L. Cousins, B-24 instructor; 2Lt. Albert H. Winkeler, pilot/B-24 instructor trainee; co-pilot/B-24 instructor trainee; Pvt. Marshall I. Morgan, engineer.

11-23-44A. Bradenton, Florida. At 1115 EWT, a North American P-51D crashed 15 miles east of Bradenton, Florida, killing 2Lt. Donald F. Heydinger.

The airplane took off from Sarasota Army Air Field, Sarasota, Florida, on a transition training flight. The airplane apparently suffered unknown engine problems, forcing the pilot to bail out. Investigators stated, "Pilot, upon bailing out, either struck the tail group or opened parachute prematurely, allowing it to become caught on elevator and tearing the parachute canopy, resulting in the death of the pilot. Civilian witness who saw crash from afar stated that no smoke or sign of fire was apparent previous to the crash. Airplane dived into the ground and was completely demolished, making it impossible to determine cause of the accident."

11-23-44B. Miles City, Montana. At 1245 MWT, a Bell P-63A crashed one mile south of the Miles City Airport, Miles City, Montana, killing F/O Elbert R. Partridge, Jr. The airplane took off from Fargo, North Dakota, on a ferry flight to Billings, Montana. Investigators stated, "[The airplane] cleared Fargo at 1120 and at 1233 MWT [the pilot] called Miles City Radio stating that he was having engine trouble and must land at Miles City. He was given landing instructions, direction of wind, etc. He was seen to pass over the field going west, and to make a left turn. The engine was sputtering badly [and] gray smoke was seen coming from the airplane. He started another left turn to get back to the field, but seemed to keep falling off to the left. He crashed and burned one mile south of the airport at 1245 MWT. Due to the fact that the airplane was completely demolished it was impossible to determine the cause of the engine failure."

11-23-44C. Great Falls, Montana. At 1408 MWT, two Bell P-63A airplanes collided in mid-air and crashed five miles east of Great Falls, Montana, killing WASP pilot Hazel Young Lee aboard P-63A #42-70412. 1Lt. Charles H. Russell was seriously injured in the crash of P-63A #42-70393. The airplanes took off from Bismark, North Dakota, on a flight to Great Falls, Montana. Investigators stated, "A number of P-63s were arriving at East Base, Great Falls, Montana, at 1400 MWT ... simultaneously on individual flight plans. Approximately ten P-63s had been, or were, in the traffic pattern at the time of the accident. Aircraft involved in the accident were first observed on final approach two or three miles from the boundary of the field. Upper aircraft lowered landing gear on final; as they approached the field the vertical distance between the two airplanes gradually diminished until the two ships collided just short of the runway [with P-63A #70393], the upper ship, straddling [P-63A #42-70412], the right auxiliary wing tank of [P-63A #70393] exploded, enveloping both ships in flames. After colliding, [the airplanes] struck the ground about 500 feet from the north end of the Runway 20 and slid to a stop 300 feet further down the runway. Investigation revealed that [P-63A #42-70393] had not established radio contact at any time with the tower. [P-63A #42-70412] called in on the downwind leg for landing instructions; pilot received instructions and was so acknowledged. Ships were told verbally to 'go around' while still well out on final approach, and since no acknowledge was received, a red light was given to both ships, accompanied by further verbal warning that they should proceed around the field. Three or four minutes prior to the accident, the pilot of [P-63A #42-70393] had signaled to the WASP pilot that he was unable to contact the tower after which she called in for landing instructions for him. Tower was at no time able to establish the position of the aircraft without radio on final or determine whether or not they had any radio communication. An officer who witnessed the accident extricated the WASP pilot from the burning wreckage. Immediately thereafter the fire crews arrived and extinguished the flames while the other pilot was being removed from his airplane."

11-24-44A. Pueblo, Colorado. At 0725, Chinese Air Force bombardier Sub-Lt. Yu-Kwei Tsao was killed when he walked into the rotating number-two propeller of a Consolidated B-24J that was operating at Pueblo Army Air Base, Pueblo, Colorado. The airplane was piloted by B-24 student Capt. Hsi-Jui Liu and co-pilot/B-24 student Capt. Shu-Chiao Chang, both of the Chinese Air Force, and were supervised by AAF B-24 instructor 1Lt. Kenneth C. Morse. The three pilots, two AAF gunnery instructors and seven other Chinese Air Force airmen were uninjured. The airplane was preparing to taxi to the take-off position from the parking area in preparation to taking off on a high-altitude gunnery and formation mission. The airplane had been completely pre-flighted and all four engines were started. The nose gunner was installing the nose turret gun barrels and was having trouble with the installation of one of the barrels. He asked the bombardier in the nose of the aircraft to summon an instructor from the waist section for help with the gun installation. The bombardier exited the aircraft through the nose wheel hatch and then walked directly into the rotating number-two propeller, killing him instantly. There were no witnesses to the incident.

11-24-44B. Kinston, North Carolina. At 1115, a Republic P-47D crashed five miles northwest of Kinston, North Carolina, killing pilot F/O Gordon R. Sellar, 19. The P-47 was part of a three-ship flight that took off at 1020 EWT from Seymour-Johnson Field, North Carolina, on a high-altitude aerobatic mission. The flight climbed to 25,000 feet and commenced maneuvering. The airplane had completed a series of barrel rolls and slow rolls and had lost about 4,000 feet during the maneuvers. The airplane was then seen to perform a loop. The airplane began yawing to the right as it leveled out at the bottom of the loop. The airplane fell away and was not seen again. The P-47 failed to return to base and a short time later it was learned that it had crashed near the Marine Corps Air Station near Kinston. Investigation revealed that the airplane struck the ground in a steep, spiraling dive. The airplane exploded violently into flames upon impact. Pieces of the airplane were found embedded several feet below the surface. Investigators could not determine the cause of the accident with any certainty.

11-24-44C. Trinity, Texas. At 1805, a Beech AT-11 suffering engine trouble while flying in poor weather crashed eight miles west of Trinity, Texas, killing six fliers. Ground school instructor Pvt. Edward W. Sappo parachuted to safety, receiving only minor injuries. The airplane was part of a four-ship flight that had arrived at Lake Charles Army Air Field, Lake Charles, Louisiana, while on a cross-country training flight. At 1708 the airplanes took off separately for the flight back to base at Byron, Texas. The subject airplane was flying in and out of instrument conditions at 6,000 feet when the starboard engine suddenly quit. The pilot trimmed the airplane for single-engine flight and then began a rapid descent. The airplane flew into worsening weather conditions as it descended, encountering heavy rain at about 1,500 feet. The pilot leveled off and advanced both throttles abruptly and the remaining engine quit. The pilot was unable to restart the engines so he ordered the crew to bail out. One crewmember attempted to open the door in the conventional manner but the slipstream prevented operation of the door. Pvt. Sappo told the flier to move out of the way and then he pulled the emergency release and kicked the door out. The airplane went out of control and apparently entered an inverted spin that forced Pvt. Sappo to the floor, holding him there for a few moments. The force let up and he was then able to crawl over to the open door and bail out. The airplane crashed moments later. Killed in the crash were: Capt. Ray S. Thompson, pilot; Sgt. Irving G. Masin, ground instructor; SSgt. Paul J. Munsch, ground instructor; Pvt. Lloyd E. Realph, ground instructor; Cpl. Orio D. Anderson, ground instructor; Sgt. Merle B. Porter, ground instructor.

11-24-44D. Benson, Arizona. At an unknown time after 1736 MWT, a Cessna UC-78B collided with rising terrain 20 miles northwest of Benson, Arizona, killing pilot 1Lt. Donald R. Jackson and passengers Col. Thomas P. Hawthorne and TSgt. Charles E. Darrow. The airplane, which was based at Pecos, Texas, took off at 1405 MWT from Yuma Army Air Field, Yuma, Arizona, on a flight to Deming, New Mexico. The pilot radioed Tucson, advising that he was picking up ice at 9,000 feet about ten miles north of Tucson and was requesting a lower altitude. Tucson radio asked the pilot to stand by while they got clearance from Tucson air traffic control, which the airplane was unable to contact. The airplane transmitted a signal at 1736 but was never heard from again. The airplane failed to return to base and was declared missing. The airplane was found on 11-28-44. Investigation revealed that the airplane smashed into the northeast side of Spud Peak at an elevation of 6,000 feet while in a flying attitude with both engines producing power, exploding into flames upon impact and killing the occupants instantly.

11-24-44E. Roseville, California. At 2345, a Douglas P-70B crashed five miles north of Roseville, California, killing pilot F/O Robert A. Metzger, 20. The airplane took off from Salinas Army Air Field, Salinas, California, on a training flight to Sacramento, California, and return. Witnesses stated that the airplane spun to the ground and exploded. Investigation revealed that the airplane struck the ground in a flat spin with no forward speed. The pilot had unbuckled his safety belt and had released the cockpit canopy but was unable to parachute to safety. He was thrown clear of the wreckage upon impact. Investigators could not determine the cause of the accident. Two civilians had observed the accident but they were unfamiliar with aviation matters and investigators considered their testimony unreliable.

11-25-44A. Ontario, California. At 1520, a Lockheed P-38J crashed one mile east of Ontario Army Air Field, Ontario, California, killing pilot 2Lt. John W. Stewart, 20. Civilian John Allen was killed on the ground. The airplane was flying in the number-three position of a four-ship flight that took off from Ontario Army Air Field on a high-altitude formation mission and camera-gunnery mission. The flight successfully completed its mission and had returned to the field for a landing. The aircraft passed over the field and then peeled off for landing. The subject airplane extended the landing gear on the downwind leg and made a normal turn onto the base leg. The airplane began losing altitude and the landing gear was seen to retract at an altitude of about 250 feet agl. The airplane stalled and fell off into a spin to the left. The airplane smashed into a small wooden building on a grape vineyard, bursting into flames upon impact. The pilot and the civilian were killed upon impact. Investigation of the wreckage indicated that a baffle plate in the carburetor ducting for the port engine was blown into the carburetor screen by a violent backfire. The baffle plate and the carburetor screen smashed into the carburetor throat, disrupting the fuel/air mixture in the carburetor and causing the engine to lose power at a critical time in the traffic pattern. Investigators could not determine what caused a backfire violent enough to blow out the baffle plate. Investigators noted that the subject airplane had lagged about a quarter mile behind the formation when the formation was approaching the area of the field.

11-25-44B. Richmond, Virginia. At 2030, a Republic P-47C collided with a jeep while performing a landing at Richmond Army Air Base, Richmond, Virginia, killing jeep driver Pvt. Robert Henry Lewis. Pilot 2Lt. Ross R. Matheson was uninjured. The airplane was part of a two-ship flight that was returning to the field following a routine night flight. The subject airplane was flying in the number-two position. The lead airplane landed and taxied to the end of the runway; a jeep pulled out on the runway to provide lighting for the lead airplane as it taxied. The pilot of the lead airplane had a hard time unlocking the tail wheel and was delayed but he had exited the runway. While the jeep driver was waiting for the lead P-47 to begin taxiing, he stopped the jeep directly on the runway and was standing at a right angle to the runway heading. The subject pilot radioed the control tower that he was having trouble lowing the flaps; the port flap remained retracted and the starboard flap extended fully. The pilot retracted the flaps and requested permission for a "flap-up" landing, which was granted. The pilot made his approach and flew over the first third of the runway at about 140 mph. The airplane touched down fast; the pilot realized that he was not going to make the landing so he advanced the throttle and began to take off for a go around. The pilot did not see the jeep parked at a right angle in the middle of the runway. The airplane struck the jeep and driver with the tail wheel and the lower fuselage, killing the driver instantly. The jeep remained upright and was relatively undamaged. The pilot felt a slight jar upon impact but was unaware that he had collided with the jeep. The pilot was able to make a safe landing despite the loss of the tail wheel in the collision.

11-25-44C. Blue Mounds, Wisconsin. At 2350 CWT, a Douglas C-47B flying in poor weather crashed into rising terrain two miles WNW of Blue Mounds, Wisconsin, killing four fliers. Killed in the crash were: Capt. Allen L. Swinton, pilot; 2Lt. John A. Oppreicht, co-pilot; Pfc. John E. Nolan, passenger; Pvt. Stanley A. Schlesinger. The airplane took off at 2210 CWT from Chicago Municipal Airport, Chicago, Illinois, on a cargo flight to Minneapolis, Minnesota. The C-47 encountered poor weather conditions as it flew toward its destination. The pilot reported that he was picking up ice at 4,000 feet indicated altitude and had requested a descent to 2,000 feet. The request was denied because of conflicting traffic. The pilot radioed that he was three minutes from Lone Rock at 3,500 feet and that he could not hold altitude due to severe ice. The pilot stated that he was going to try to turn out of the icing conditions. The pilot was advised that the minimum safe altitude for the area was 3,000 feet, but he was given authorization to fly at 2,000 feet to Minneapolis, mostly because the airplane was unable to fly any higher. The pilot elected to turn around and head back to Chicago because of the severe icing problem. The pilot was advised to land at Lone Rock but he declined. At 2348 CWT the airplane was cleared to fly back to Chicago and was assigned an altitude of 2,000 feet at the request of the pilot. The pilot was advised to check in when over Rockford. Two minutes later the C-47 collided with a hill while in a flying attitude with power at an elevation of 1,716 feet, about 150 feet from the top, bursting into flames upon impact.

11-26-44A. Martinsburg, Pennsylvania. At 1856, a Curtiss RA-25A suffering an in-flight fire crashed one mile east of Martinsburg, Pennsylvania, killing pilot 2Lt. Samuel A. Hart and passengers 2Lt. Robert W. Phillips and Pvt. Claude A. Carr. The airplane, which was based at Olmstead Field, Pennsylvania, took off from the U.S. Naval Air Station at Columbus, Ohio, on a routine flight to Middletown, Pennsylvania. The A-25 began suffering electrical problems during the flight. The airplane was cruising at about 7,000 feet when an explosion occurred in the fuselage. Flames filled the front cockpit and small pieces of the airplane were seen falling away. The airplane pitched down and dove into the ground where it exploded into flames upon impact. The airplane produced a ten-foot deep crater that was about 30 feet wide. Investigation revealed that pilots who had flown the aircraft on previous flights had complained about a strong odor of gasoline fumes in the cockpit following the servicing of the bomb bay fuel tank. It was discovered that the voltage regulator had caused an arc that ignited the gasoline fumes, causing the cockpit to burst into flames and incapacitate the pilot.

11-26-44B. Georgiana, Alabama. At 1815, a Boeing B-17F flying in instrument conditions suffered a catastrophic structural failure and crashed 17 miles west of Georgiana, Alabama, killing pilot Maj. Harwell P. Tilly III, co-pilot 2Lt. William Thatch and engineer SSgt. Frank J. Geraci. The airplane took off on an instrument clearance at 1744 CWT from Brookley Field, Mobile, Alabama, on a ferry flight to Fairfield Air Depot, Fairfield, Ohio, via Patterson Field, Ohio. Large pieces of the airplane were seen hurtling out of a 400-foot overcast, the various pieces smashing to the ground on the Blackman Farm. Wreckage was scattered over an area of 2,000 feet. Investigators speculated that the pilot had lost control of the airplane while flying in turbulent weather conditions and had stressed the airplane beyond its design limitations in an effort to re-establish level flight. The port wing failed first, followed by both horizontal stabilizers and elevators and then the starboard wing. Weather in the area at the time of the crash was reported as ceiling 400 feet with visibility 200 yards in heavy rain.

11-26-44C. New Carlisle, Ohio. At 2045 CWT, a North American AT-6 flying in poor weather crashed two and a half miles southwest of New Carlisle, Ohio, killing WASP pilot Katherine Dussaq. The airplane took off on an instrument clearance from National Airport, Washington, D.C., on a flight to Lunken Field, Cincinnati, Ohio. The pilot became lost in the area of Cincinnati. Despite the efforts of ground stations to guide the pilot to Patterson Field, she was unable to orient herself. At 2035 CWT, the pilot advised Lunken Field control tower that she was lost and low on fuel. This was the last contact with the airplane. Investigation of the wreckage indicated that the engine quit due to fuel exhaustion. The AT-6 descended into the ground after breaking out of the low overcast. Pilots arriving in the area at the time of the crash reported the weather as ceiling 500 feet with visibility of one mile in light rain.

11-27-44A. Tipp City, Ohio. At 1455 CWT, a Vultee BT-13 crashed four miles ESE of Tipp City, Ohio, killing pilot 2Lt. Thomas A. Hitrick and passenger MSgt. Everette F. Robinson. The airplane took off from Wright Field, Dayton, Ohio, on a local flight. The airplane was observed flying in a northeasterly direction near Tipp City at about 1,000 feet agl. The airplane entered a steep turn to the right and as it rolled out on a southerly heading stalled and flipped over to an inverted attitude. The pilot recovered momentarily and re-established level flight but entered a secondary stall and spin at an altitude of about 300 feet. The airplane plunged to the ground, bursting into flames upon impact.

11-27-44B. Gulf of Mexico, Mississippi. At 1714 CWT, a Boeing B-29 suffering an in-flight fire in the number-one engine was abandoned by eleven crewmembers over the Gulf of Mexico 38 miles south of Keesler Field, Biloxi, Mississippi. Eight crewmembers drowned before they could be rescued from the sea. Bombardier 1Lt. Fred H. Bigelow, gunner Cpl. William K. Jones and gunner Cpl. Ned C. Johnson parachuted to safety and were rescued from the sea, sustaining only minor injuries. The pilot, 1Lt. Eugene W. Hammond, was able to make an emergency landing at Keesler Field, Mississippi, and was uninjured. Crewmembers who drowned before they could be rescued were: 2Lt. James C. Clark, navigator; 2Lt. John Kelley, flight engineer; Capt. Stacy Kelland, observer; Pfc. Robert D. Rife, radio operator; Cpl. John B. Schaeverly, crewmember; Pfc. Richard H. Little, gunner; Cpl. George D. Kennedy, gunner. Co-pilot 2Lt. Edwin H. Gresham was missing and presumed lost at sea. The airplane had taken off from the Army Air Field at Great Bend, Kansas, on a bombing mission at the AAF Bombing Range at Freemason Island, Mississippi, and a cross-country celestial navigation mission to Havana, Cuba, and return to Great Bend. The airplane had been airborne for about five and a half hours and had just finished its third bombing run, flying in a southerly direction at about 12,000 feet. About three minutes after the completion of the third bombing run, Cpl. Jones, who was flying in the port "scanner" position, observed heavy smoke pouring from the number-one engine. His intercom system was inoperative so he moved over the starboard side and informed the pilot. The pilot looked out of his window and observed the number-one engine in flames. The pilot and engineer began the usual emergency procedure but were unable to feather the propeller. The pilot ordered the bombardier to jettison the bombs and the bomb bay fuel tank; the pilot also pulled his emergency release. The pilot and engineer were unable to feather the propeller or extinguish the fire. The fire continued out of control as the airplane descended through 10,000 feet. The pilot extended the landing gear, steered the airplane toward shore and then ordered the crew to bail out. Seven crewmembers bailed out immediately from the after part of the ship. The pilot put the airplane on autopilot and then moved over to the co-pilot seat. The airplane began to go out of control as the five remaining crewmembers prepared to bail out. The airplane fell into a steep spiral to the left. The pilot turned off the autopilot, regained control of the airplane and then flew the B-29 by hand until the remaining crewmembers bailed out. The pilot noticed that the number-one engine had separated from the wing during the steep spiral and that the fire had extinguished itself. An airman noticed the engine fall past him as he parachuted to safety. The pilot determined that he could fly the airplane, so he retracted the landing gear and then radioed Keesler Field, requesting an emergency landing. The pilot made a hard landing with landing gear extended, full flaps, and open bomb doors. The airplane was slightly damaged in the landing, which was described as being similar to having stalled the airplane at an altitude of six feet. Investigators noted that search planes, AAF crash boats and other vessels were immediately dispatched to the bailout area 15 miles west of Freemason Island. It was also noted that darkness fell on the area 15 minutes after the crew bailed out. Crewmembers on board search planes spotted seven dye markers in the water before darkness fell. Investigation revealed that the fuel line in the number-one engine broke loose and allowed gasoline to flood the engine nacelle. The gasoline fumes were ignited by hot exhaust ducting located in the nacelle. The number-one fuel tank was found to be completely empty.

11-27-44C. Wright Field, Ohio. At 1748 CWT, a Beech C-45 flying in instrument conditions crashed at Wright Field, Dayton, Ohio, killing pilot Col. Carl A. Cover and civilian passenger Max Stuper. The airplane took off at 1500 CWT on an instrument clearance from Marietta, Georgia, on an administrative flight to Wright Field. The airplane arrived in the vicinity of Wright Field and was flying the southwest leg of the Patterson Radio Range. Air traffic control authorized the airplane to begin an instrument let-down in preparation for landing at Wright Field, where the ceiling was estimated to be about 1,000 feet agl. The pilot apparently stalled the airplane while flying in instrument conditions at an altitude too low to allow recovery. The airplane crashed onto Wright Field in a steep spiral to the right. It was noted that the pilot had over 6,700 hours of flying time. The civilian passenger was an employee of Bell Aircraft Company.

11-28-44A. Bruni, Texas. At 1700, a Bell TP-39K flying in poor weather crashed ten and a half miles northeast of Bruni, Texas, killing pilot 2Lt. David S. Daniels. The airplane was part of a three-ship flight that took off from Harlingen Army Air Field, Harlingen, Texas, on a ferry flight to Spokane, Washington, via Laughlin Field, Del Rio, Texas. The airplanes had been cleared from Harlingen Army Air Field at 1545 CWT on a contact flight plan. Soon after taking off from Harlingen the flight encountered a lowering ceiling. The airplanes attempted to stay underneath the lowering ceiling, which was estimated to be at about 1,000 feet and included rain showers. The flight leader observed that the weather had extended to ground level in the flight path so he decided to return to Harlingen. The flight leader was unable to contact the subject airplane and it flew into the weather. The subject airplane was observed to come out of the overcast in a diving turn to the right, striking the ground at a steep angle and exploding into flames.

11-28-44B. Assiria, Kansas. At 1840 CWT, a Boeing B-29 suffering an in-flight fire crashed at Assiria, Kansas, killing nine crewmembers. Bombardier 1Lt. William J. Bennett received serious injuries parachuting to safety. Crewmembers escaping injury parachuting to safety were: Capt. William F. Barthel, navigator; TSgt. Perry F. Shope, engineer; SSgt. Victor M. Vanderpoole, radio operator instructor; Cpl. Henry G. Garner, gunner. Bombardier instructor 1Lt. Edward T. Charles, Jr. fell to his death in an unsuccessful parachute jump. Killed in the crash were: Capt. Alan M. Miller, pilot/B-29 instructor; 2Lt. Lewis S. Phillips, co-pilot/B-29 student; Pvt. John J. McCauley, radio operator; Cpl. John E. Nandino, gunner; Cpl. Leslie P. Fries, gunner; Cpl. John H. Gavin, gunner; Cpl. Earl J. Gilbert, gunnery instructor; Sgt. John W. Anderson, central fire control gunner; Cpl. Ralph C. DeMoss, gunnery instructor. The airplane took off at 1819 CWT from Smoky Hill Army Air Field, Salina, Kansas, and was scheduled to fly an instrument-training mission on the Salina Radio Range and then the instructor was to allow the co-pilot to perform several night landings. The airplane made a normal take-off and began to climb, and upon reaching an altitude of 800 feet the number-three engine began running rough. The pilots reduced rpm on this engine and advanced the remaining three and the airplane climbed into the overcast at about 870 feet. When the airplane broke out at 5,000 feet, the pilot and engineer decided to shut down the number-three engine and feather the propeller. The pilot radioed the field and requested emergency landing instructions. The control tower gave the pilot permission to land and cleared the field. The airplane descended back into the overcast and moments later the left scanner advised that the number-two engine was on fire. The pilot and engineer began emergency procedures; the pilot then ordered the crew to stand by to abandon ship. Officers in the control tower overheard this message. Control tower personnel asked the pilot for his position and altitude. The pilot replied that he was flying at 2,600 feet but did not know his position. Personnel in the control tower observed an explosion on the ground about 20 seconds later. The B-29 had struck the ground in a steep inverted dive, exploding violently into flames upon impact. Investigation revealed that the number-three engine had suffered "pounded" and deformed valves in five of its

18 cylinder heads. The number-three engine also suffered from "pounded" and deformed valves in four of its cylinder heads. One piece of valve was swallowed by the number-three engine. Investigators speculated that the fire could have been caused by a faulty fitting or connection of the fuel line from the fuel pump to the carburetor, causing gasoline to escape into the engine nacelle where it was ignited by hot exhaust ducting. Testimony of surviving crewmembers indicated that some confusion existed during the bail out attempt. Investigators suggested that all crews be trained in proper bail out procedure and that parachutes should be worn in flight whenever possible.

11-29-44A. Nottingham, New Hampshire. At 0947, a Consolidated B-24L flying in instrument conditions suffered a catastrophic structural failure and crashed into a heavily wooded area on the Pawtuckaway State Reservation near Nottingham, New Hampshire, killing the crew of nine. The airplane took off at 0929 EWT from Grenier Field, Manchester, New Hampshire, on a ferry flight to the European Theater of Operations via Goose Bay, Labrador. The airplane climbed into instrument conditions as it flew away from the field. The pilots apparently lost control of the B-24 while flying in instrument conditions and had stressed the airplane beyond its design limitations attempting to re-establish level flight. The starboard vertical fin and rudder failed and separated at the point where they meet the starboard horizontal stabilizer, which failed at the point where it is attached to the fuselage. The airplane went out of control and crashed near Round Pond, exploding into flames upon impact. Investigators noted that light rime ice was discovered on the tail surfaces that were found on the line of flight. Investigators dismissed the ice as the cause of the structural failure. Investigators noted that the pilot had had instrument training but had not flown in actual instrument conditions. Killed in the crash were: 2Lt. Paul L. Hackstock, pilot; 2Lt. Wilbur C. Stephenson, co-pilot; F/O Russell E. Jones, navigator; Cpl. Calvin R. Rickenbach, engineer; Cpl. Thomas L. McDougall, radio operator; Cpl. Robert H. Wells, gunner; Cpl. William E. Swarner, gunner; Cpl. Preston K. Smith, gunner; Cpl. Kenneth J. Young, gunner.

11-29-44B. Quinebaug River, Connecticut. At 1000, a Republic P-47D crashed into the Quinebaug River two miles east of Putnam, Connecticut, killing 1Lt. Robert W. Anderson, 27. The airplane was part of a four-ship flight that took off from Hillsgrove Army Air Field, Hillsgrove, Rhode Island, on a gunnery mission. The airplanes arrived at the local gunnery range and it was discovered that poor visibility had closed the range. The flight leader proceeded to lead the flight in a basic combat formation flight, which was the alternate mission. The flight had encountered a thin layer of scud and the leader attempted to climb over it. The formation entered instrument conditions for a few seconds and when it emerged, the subject airplane was missing. Witnesses on the ground observed the subject airplane dive straight into the river. The pilot had apparently lost control of the airplane in instrument conditions and bailed out. The horizontal stabilizer struck the pilot on the head, killing him instantly. His body was found over two miles from where the airplane crashed. Investigators noted that Lt. Anderson had bailed out of a P-47 airplane after similarly losing control in instrument conditions while on a combat tour in New Guinea. It was also noted that the pilot had made comments expressing his dislike for instrument flying.

11-29-44C. Mayo, Florida. At 1048 EWT, a Vultee BT-13A flying in instrument conditions crashed six miles southwest of Mayo, Florida, killing pilot 2Lt. Robert G. McGuire, 22. The airplane took off at 0912 EWT from Buckingham Army Air Field, Fort Myers, Florida, on an instrument clearance to Turner Field, Albany, Georgia. The airplane deviated from the instrument flight plan when it arrived in the vicinity of Cross City, Florida, setting a direct course for Turner Field. The airplane never arrived at Turner Field and was declared missing. Search planes found the airplane wreckage the next day. Investigators did not reach the crash site until 12-1-44. Investigators speculated that the pilot apparently lost control of the airplane while flying in instrument conditions. The pilot bailed out but was struck in the head and decapitated by part of the empennage. His headless body was found about 100 yards from the spot where the airplane smashed to earth in a swampy wooded area and exploded into flames. The parachute was found partially deployed and strung out behind the pilot. Wreckage was scattered over an area of 300 yards.

11-29-44D. El Reno, Oklahoma. At 1008, a Lockheed F-5E crashed at El Reno, Oklahoma, killing Chinese Air Force pilot 2Lt. Chao-Yu Shih, 23. The airplane took off from Will Rogers Field, Oklahoma City, Oklahoma, on a photographic training flight. The airplane was seen flying at an altitude of approximately 2,000 feet when it entered a spin from which the pilot was unable to recover. The airplane entered a steep nose-down attitude at about 800 feet agl and smashed to the ground where it exploded in flames.

11-30-44A. Tucson, Arizona. At 0847 MWT, two Consolidated B-24J airplanes collided in mid-air and crashed eight miles east of Davis-Monthan Army Air Field, Tucson, Arizona, killing 18 fliers. Killed in the crash of B-24J #42-73344 were: 2Lt. Harold D. Ballard, pilot; F/O Alexander F. Tuttle, Jr., co-pilot; 2Lt. Joseph N. Shober, navigator; 2Lt. Earl J. Viney, bombardier; Cpl. Henry A. Smith, engineer; Cpl. James G. Spagnoli, radio operator; Cpl. Eunice W. Young, gunner; Pfc. Raymond J. Link, gunner; Pfc. Leonard S. Schultz, gunner. Killed in the crash of B-24J #42-73357 were: 2Lt. Theodore V. Glock, pilot; 2Lt. William J. Turner, co-pilot; 2Lt. Charles F. Gavilsky, navigator; 2Lt. Joseph D. Stoudemire, bombardier; Cpl. David H.

Lorenz, engineer; Cpl. Harry C. Meade, Jr., radio operator; Pfc. Arnold L. Vanderbuilt, gunner; Cpl. Julian J. Magiera, gunner; Pfc. William T. Dixon, gunner. Investigators stated, "The aircraft involved in this crash were part of a 19-ship formation which started its initial take-off at [0830 MWT] with 30 second intervals on Runway 12. The aircraft involved were members of the second six-ship flight referred to as B-Baker flight. [Aircraft #42-73344, (Field #41) was briefed to fly as Baker-2 and aircraft #42-73357 (Field #29) was briefed to fly as Baker-5.] The flights were flying separately at the time of the accident. Take-off was as briefed into the sun.... Take-off was normal and a 180-degrgee turn to the left was made after the ships went out on the take-off heading for four minutes. In this turn all pilots were blinded by the sun, and in most cases were only able to see the aircraft ahead of them if any at all. Baker-3 experienced difficulty in getting his under-carriage up and turned inside of the formation causing some confusion to the pattern of following aircraft, and when the pilots rolled out of their turns to proceed on the downwind leg at approximately 1,500 to 1,700 feet above ground level, confusion existed as to the identity of their element leaders and Baker-2 mistakenly formed on Baker-4's right wing. Baker-1 at this time was to the left, low and practically abreast of Baker-4 with some distance separating them. Baker-5 was low and behind the formation and as the aircraft proceeded on the downwind leg, attempted to climb and proceed to the right to assume his proper position, which at the time was occupied by Baker-2. In so doing, Baker-5 collided with Baker-2. Baker-5 struck Baker-2's left rudder and vertical stabilizer with his wing in the area between the number-three engine and fuselage. The sections of empennage were seen to leave Baker-5's tail and fly back through the formation. The two aircraft proceeded in this locked position momentarily and then broke apart. Baker-5 dove straight into the ground and Baker-2 did a maneuver, described as a chandelle to the right, and went into the ground at an angle. The aircraft crashed approximately 200 yards from each other almost simultaneously with an immediate explosion and fire, which destroyed both aircraft entirely and instantly killed all aboard. No parachutes or bodies were seen to leave the aircraft while in flight."

11-30-44B. Winchester, California. At 1045, a Stearman PT-13A and a PT-13H collided in mid-air and crashed four miles southwest of Winchester, California, killing both students. A/S Harold L. Hyink was killed in the crash of the PT-13A; A/C Samuel P. Jones was killed in the crash of the PT-13H. The airplanes had taken off from the AAF Primary Training School at Hemet, California, on separate student training flights. The PT-13A was flying west and the PT-13H was flying east, in a direct collision course with each other. Neither student saw the other and the airplanes collided head-on, each airplane's propeller striking the

port wing of the other airplane. The airplanes broke up in the collision and went out of control, smashing to the ground and bursting into flames.

11-30-44C. Mountain Home, Idaho. At 2030 MWT, a Consolidated B-24J crashed ten miles northwest of Mountain Home Army Air Base, Mountain Home, Idaho, killing five fliers and injuring three others. Investigators stated, "Aircraft departed Mountain Home Army Air Base at 2010 MWT on a local night mission to fly the Burley Radio Range. While climbing over the field at approximately 10,000 feet a flash and explosive report was thought to come from the right side of the airplane. A check of the instruments revealed that the manifold pressure for number-three engine dropped from 35 to 28 inches. The pilot at once assumed he had lost the engine and feathered the propeller. Shortly thereafter, a flash and explosive report appeared to come from the left side of the aircraft. While attempting to analyze the situation, the pilot lost vertical control of the airplane, which at this time was making a steep spiral to the right. On recovering from this position, number-three propeller was unfeathered. In the meantime the airplane lost considerable altitude and drifted away from the vicinity of the field. Finally, when the pilot was aware that a crash landing was imminent, he called for wheels and lights. A few seconds later, before the landing gear had time to come down, the airplane struck the ground with landing lights on and in a steep glide attitude." Pilot 1Lt. Gerald E. Prentice, bombardier F/O Donald Troiano and gunner Pfc. Robert Richeson were seriously injured. Killed in the crash were: F/O Anthony A. Bonchi, co-pilot; Cpl. Robert S. Lovell, radio operator; Cpl. Joe Maglin, engineer; Cpl. Francis J. Pronesti, gunner; Pfc. Raymond A. Bonello, gunner; Pfc. John W. Grider, gunner.

December

12-1-44A. Merced, California. At 0015, a Northrop P-61A collided with trees and crashed 15 miles south of Merced, California, killing F/O Horace B. Stanfield and radar operator F/O Robert O. Knight. The P-61A took off from Hammer Field, Fresno, California, at 2250 PWT and was scheduled to fly a high-altitude (21,000 feet) cross-country navigation mission to Palmdale, California, to Desert Center, California, to Muroc, California, and return to Hammer Field. Investigation revealed that the airplane deviated from its intended flight plan and assigned altitude and it was noted that investigators discovered the flight crew's high altitude gear and oxygen masks in their lockers after the crash. The airplane was flying over the terrain at low altitude when it encountered low-hanging fog and collided with trees in a peach orchard. The pilot pulled up abruptly and the tail section collided with the trees, causing the vertical fins, horizontal stabilizer and elevator to separate. The airplane went out of control and smashed to the ground near the edge of the orchard, smashing trees down in its path. The airplane did not burn but was destroyed utterly. Investigation revealed that the fog was thick laterally but not vertically. A civilian witness who was a quarter-mile away from the crash stated that the moon was partially visible overhead through the fog but forward visibility was less than 150 feet. There was no evidence of engine failure. Investigation revealed that the airplane was flying at high speed in a level attitude when it initially collided with the trees.

12-1-44B. Sumter, South Carolina. At 1050, two North American AT-6D airplanes collided in midair five miles southeast of Shaw Field, Sumter, South Carolina, killing instructor 2Lt. Fred J. Sutton aboard AT-6D #42-86442. Student A/C Loren F. Driver was uninjured parachuting to safety. Instructor 2Lt. George F. Herrity and student A/C David Lambert were uninjured and able to safely land AT-6D #42-44622. Investigators stated, "Both students were receiving instruction in pre-solo transition, and investigation revealed that both instructors had informed their students to level off at 4,000 feet. The standard procedure for leveling off is to climb 150 feet above the desired altitude, then use the extra 150 feet to gain cruising speed and trim the airplane. [AT-6D #42-44622] had reached its desired altitude and was just completing this leveling off procedure when [AT-6D #42-86442] appeared off to the right and in front of [AT-6D #42-44622]. [AT-6D #42-86442] was still in a climbing attitude (approximate airspeed 110 mph) and was flying in the same direction as [AT-6D #42-44622]. The leading edge of the right wing of [AT-6D #42-44622] struck the left aileron of [AT-6D #42-86442], damaging the aileron and causing it to stick in the up position.... Neither airplane was rendered uncontrollable immediately after the accident. The student riding in [AT-6D #42-86442] opened his canopy as soon as the collision occurred, but was told to close it by his instructor. The student stated that the airplane was in a shallow bank to the left and after approximately one and a half minutes, the instructor told the student to bail out.... Witnesses stated that [AT-6D #42-86442] continued to lose altitude in a spiral to the left. The spiral gradually became steeper, until at an estimated altitude of 1,500 feet the airplane almost became inverted, but was seen to recover from the inverted attitude just before crashing." The other airplane returned to the field and landed safely.

12-1-44C. Moro, Arkansas. At 1015, a Beech AT-10BH suffered the catastrophic structural failure of the starboard wing and crashed two miles southeast of Moro, Arkansas, killing A/C Alex Witkovich, who fell to his death in an unsuccessful parachute jump. A/C Stanley Woodward, Jr. parachuted to safety and escaped injury. Investigation revealed that the wing failed when A/C Woodward put the airplane in a diving attitude and then executed an abrupt pull-up maneuver. This maneuver stressed the wing structure beyond its design limitations, causing it to fail and separate. The airplane suddenly rolled twice to the right and the students bailed out. Investigation revealed that A/C Witkovich pulled his ripcord prematurely and the parachute was fouled on the radio antennas. The parachute never successfully deployed and the student fell to his death.

12-1-44D. Robins Field, Georgia. At 1215 CWT, a Douglas RB-18B crashed while attempting to take-off from Robins Field, Warner-Robins, Georgia, killing pilot 1Lt. Wilmer H. Boozer and engineer SSgt. Ova V. Vaughn. Investigators stated, "[The airplane] swung onto the runway and began the take-off run. After the airplane cleared the ground and during the time the landing gear was retracting, the right wing was seen to drop. There was no apparent corrective action taken on the part of the pilot, instead the right wing was seen to drop to the point of dragging the ground and the airplane began a turn to the right off of the runway. After the wing had dragged for approximately 150 feet the airplane climbed approximately 30 feet and then fell off on the right again, crashed head-on into the ground, exploded and burned." Investigation revealed that the aileron controls were operating in reverse of normal, causing a loss of control.

12-1-44E. Newberry, Florida. At 1606 EWT, a Sikorsky XR-5A helicopter (serial number 43-28237) suffered a catastrophic material failure and crashed seven miles NNW of Newberry, Florida, killing pilot Capt. William J. McGuire and civilian technician Gustav W. Heiden, Sikorsky Aircraft Company. The helicopter took off from Orlando Army Air Field, Orlando, Florida, on a cross-country flight to Alachua Army Air Field, Gainesville, Florida, to Cross City Army Air Field, Cross City, Florida, and return to Orlando. The helicopter took off from Alachua Army Air Field just before the crash. Investigation revealed that the main rotor blade developed a fracture at the weld holding the rib fittings to the rotor blade spar. This fracture caused a complete failure of the rotor blade spar. The rotor blade failed and the helicopter went out of control and smashed to the ground in an abandoned quarry, bursting into flames upon impact and killing the occupants instantly. Wreckage was scattered over an area of 300 yards. The helicopter had been flying in a straight and level attitude in good weather conditions with normal turbulence. This crash is thought to be the first fatal helicopter accident suffered by the U.S. Army during World War II.

12-1-44F. Mico, Texas. At 2000, a Cessna UC-78B collided with rising terrain and crashed five and a half miles northeast of Mico, Texas, killing instructor 2Lt. Emerson L. Morris and student A/C James C. Lamb, Jr. The airplane took off at 1707 CWT from Marfa Army Air Field, Marfa, Texas, on an instrument cross-country flight to Kelly Field, San Antonio, Texas. The pilot checked in at Hondo, Texas, stating his position as five miles north of Hondo Army Air Field at

7,000 feet. The airplane was next seen about 20 miles northwest of Kelly Field at about 2000. The airplane approached from the east at about 500 feet agl with landing lights illuminated. The airplane was then seen to circle to the right until it was heading south. The airplane cleared a small range of hills before smashing into a hill several yards from the top. The airplane exploded into flames upon impact, killing the occupants instantly.

12-1-44G. Keene, California. At 1945, a Lockheed C-60A flying in instrument conditions collided with rising terrain four miles west of Keene, California, killing nine fliers. Killed in the crash were: 1Lt. Robert D. Simpson, Jr., pilot; 2Lt. Jimmie E. Hoch, co-pilot; 2Lt. Clarence W. DeGroff, navigator; F/O Michel D. Pavich, navigator; 2Lt. James F. Applewhite, navigator; 2Lt. Louis J. Hoekstra, navigator; F/O Otto F. Illias, pilot-rated passenger; Sgt. Ira F. Ratliff, engineer; Pfc. Walter H. Denk, passenger. The airplane took off from Hondo Army Air Field, Hondo, Texas, on a navigation flight to Fresno, California, and return. The airplane, flying from east to west, collided with the east side of the north slope of Bear Mountain at an elevation of 4,300 feet msl. The airplane failed to arrive at Fresno and was declared missing. A Civil Air Patrol aircraft found the airplane wreckage at 0930 on 12-4-44. Investigators speculated that the pilot was attempting to let-down through instrument conditions.

12-2-44A. Tonopah, Nevada. At 1018 PWT, a Consolidated B-24D crashed 16 miles northeast of Tonopah, Nevada, killing six fliers. Three fliers escaped injury parachuting to safety. The airplane took off from Tonopah Army Air Field at 1005 PWT on a training flight. Investigators stated, "The engineer stated that the take-off was normal and all engines were functioning normally. The aircraft was turning on the downwind leg and had reached an altitude of approximately 1,600 feet above the terrain when the number-two engine caught fire. The engineer stated that the engine caught fire immediately after the second power reduction. The pilot attempted to extinguish the fire by putting the mixture control in idle cut off position and pumping the throttles, but the pilot did not give the engineer instructions to cut off the fuel to the number-two engine. Shortly after the fire broke out, the pilot ordered the crew to bail out. The engineer stated that there was a great deal of confusion among the crewmembers after the pilot gave the bail out order. The gunner had unbuckled his leg straps and was unable to buckle them again. The co-pilot left his seat and was very excited. The pilot remained at the controls at all times. The engineer then felt the ship lurch as if an explosion had occurred and immediately thereafter the aircraft seemed to go into a flat spin. In the meantime the navigator had gone into the nose with the apparent intention of bailing out through the nose wheel doors. Evidently he was unable to open the nose wheel doors, and while trying to return to the bomb bay he

became lodged in the passage way. The engineer answered the navigator's call for help and freed him. Both the engineer and the navigator made their way around the gunner and reached the bomb bay together. The engineer then jumped and pulled the ripcord. Immediately after the parachute opened fully, he hit the ground. The aircraft appeared in a flat spin at the time and crashed approximately three seconds before the engineer reached the ground. The navigator was lying in the wreckage on the cat walk and the engineer was making his way to the navigator when a fuel cell ignited and the entire airplane started to burn." Killed in the crash were: 2Lt. Wayne C. Hurst, pilot; 2Lt. Edward M. Piwowarski, co-pilot; 2Lt. Lloyd E. Warner, bombardier; 2Lt. Myron E. Smith, navigator; Pfc. Charles W. Lantz, radio operator; Pfc. John F. Devlin, assistant engineer. Engineer Cpl. James N. Gross, gunner Cpl. Joseph M. Roach and gunner Pfc. Daniel C. Brzezinski parachuted to safety and escaped injury.

12-2-44B. Wright Field, Ohio. At 1418 CWT, a Republic P-47D crashed eight and a half miles south of Wright Field, Dayton, Ohio, killing pilot Capt. Zed D. Fountain, Jr., 27. The airplane took off from Wright Field on a test flight. The objective of the test was to dive the airplane from 30,000 feet at a shallow angle, gradually increasing throttle to full as the dive progressed and pulling out at about 15,000 feet. For the second test dive, the pilot was to enter an abrupt dive at 30,000 feet with a rapid application of power, pulling out at 15,000 feet after attaining a maximum speed of 430 mph. The pilot successfully completed the first dive test and climbed back to 30,000 feet for the second test. The airplane entered a dive at high altitude and the pilot failed to recover. The airplane smashed into the ground at high speed three miles north of Bellbrook, Ohio, exploding violently into flames upon impact.

12-2-44C. Fresno, California. At 1352, a Northrop P-61B crashed four miles east of Fresno, California, killing pilot 2Lt. Richard G. Lester and injuring radar operator 2Lt. Stanley J. Komarowski. The airplane was taking off on Runway 29 at Hammer Field, Fresno, California, on a routine training mission. The take-off appeared normal, but soon after becoming airborne and before the end of the runway had been reached the airplane began a shallow turn to the left. Traffic at the field was to the right. The P-61B appeared to be circling back to the left for an approach for Runway 7, but the ship was unable to gain altitude as it continued in the left turn. The pilot attempted to stretch his glide and stalled the airplane at an altitude too low to allow recovery. The airplane smashed into the ground in a flat attitude and burst into flames. The radar operator was able to escape the burning airplane with minor injuries but the pilot's foot was trapped in the wreckage and he was burned to death before he could free himself or be rescued. Investigators speculated that the airplane had suffered an engine failure or diminished performance just after becoming airborne.

12-2-44D. Haswell, Colorado. At 1727 MWT, a North American TB-25C suffered a catastrophic structural failure and crashed 22 miles northwest of Haswell, Colorado, killing student A/C James F. Costello, Jr. and injuring A/C Daniel G. Cowen, who parachuted to safety. The airplane took off at 1355 MWT from La Junta Army Air Field, La Junta, Colorado, on a rendezvous problem at Eads, Colorado. The airplane made a successful rendezvous at 1655. The airplanes flew together for a short time before breaking up at 1715. The subject airplane was instructed to fly around until 1830, which was the end of the flying period. Investigation revealed that the airplane was observed making 60 and 70-degree banked turns at a speed of 225 mph when the starboard wing failed and separated at a point near the landing light. The wing separated and struck the tail section, causing its progressive failure and separation. The airplane went out of control and smashed to the ground, exploding into flames upon impact. Investigators did not think that the maneuvers were violent enough to cause a structural failure. All B-25 aircraft on the station were checked for fatigue in the wing leading edges.

12-2-44E. Sioux City, Iowa. At 2103 CWT, a Boeing B-17G crashed into a vacant lot at 220 S. Judd Street, Sioux City, Iowa, killing the crew of nine. The airplane took off at 1609 CWT from Sioux City Army Air Base on a formation and bombing mission. The airplane successfully completed the mission, returned to the field and radioed for landing instructions at 2057. The pilot was told to land on Runway 17. The airplane was observed flying at extremely low altitude. The B-17 collided with power-lines and then descended into the ground. The airplane broke up as it skidded along the ground, destroying a wooden structure before coming to rest in the vacant lot and bursting into flames. Wreckage was scattered over an area of 100 yards. Investigators did not know why the airplane descended into the ground. Killed in the crash were: 2Lt. Frank M. Barne, pilot; F/O Donald W. Anderson, copilot; F/O James M. Bell, navigator; 2Lt. David M. Brinn, bombardier; Cpl. John W. Lancaster, engineer; Cpl. Elton C. Modine, radio operator; Cpl. Jack R. Reed, gunner; Pfc. Allan W. Chadbourne, gunner; Pfc. Robert E. Klassen, gunner.

12-2-44F. Courtland, Alabama. At 2106 CWT, a Consolidated RB-24E crashed two miles southwest of the Army Air Field at Courtland, Alabama, killing three fliers and seriously injuring engineer Pvt. Fred T. Duke, Jr. B-24 instructor 1Lt. Michael R. Searby and B-24 students 2Lt. Thomas Y. Mattox, Jr. and 2Lt. Theron B. Rust were killed in the crash. The airplane had taken off at 2047 from Courtland Army Air Field on a navigation flight to Mobile, Alabama, and return. The airplane returned to the field a short time later and requested an emergency landing. The pilot alerted the

control tower that he thought he had a massive fuel leak behind the number-two engine. The B-24 was cleared to land on Runway 5 and the pilot acknowledged. The airplane descended into a wooded area and broke up as it smashed a path through the trees for 175 yards. The airplane burst into flames as it smashed itself to pieces. Wreckage was scattered over a wide area. Investigators speculated that the pilot had undershot the field and inadvertently flew the airplane into the ground.

12-4-44A. Brunswick, Georgia. At 1732 EWT, a Beech AT-10BH crashed three miles NNW of Brunswick, Georgia, killing students A/C Kenneth E. Bird and A/C Arthur L. McCavran. The airplane took off at 1415 EWT from Moody Field, Valdosta, Georgia, on a student instrument-training flight. A/C McCavran was assigned to fly as the instrument pilot and A/C Bird was assigned to be the safety pilot. The airplane was landed on schedule at 1630 so the students could exchange roles. The airplane took off at 1645 with A/C Bird flying as the instrument pilot and A/C McCavran as safety pilot. The students were ordered to practice instrument approaches to the field. The instrument work was to be done on the Moody Field Radio Range and not near the coast. The airplane was observed flying in a southerly direction at about 2,000 feet in the area north of Brunswick. The port engine apparently failed and the students increased the power on the starboard engine. The airplane rolled to the left until it was in an inverted attitude and then it entered a spin. The students were unable to regain control of the airplane and it spun to the ground, smashing into the muddy bank of a tidal pond. The students were killed instantly. The airplane did not burn but was destroyed utterly. Investigators noted that the students were over a hundred miles from the assigned practice area.

12-4-44B. Courtland, Alabama. At 1952, a Consolidated RB-24E crashed three miles northwest of the Army Air Field at Courtland, Alabama, killing four fliers. The airplane took off at 1945 from Runway 9 at Courtland Army Air Field on an instrument training flight. The instructor pilot radioed the control tower at 1948, reporting that his number-three engine had failed and requesting an emergency landing on Runway 13, which was the nearest runway. The tower cleared the field and approved the landing on Runway 13. The pilot radioed the tower at 1950 and reported that the number-four engine had failed. The airplane began losing altitude while turning back to the field and the pilot was unable to raise the starboard wing. The B-24 slammed into the ground and exploded, killing all on board instantly. Investigators speculated that the pilots had inadvertently cut off the fuel supply to the number-four engine instead of the number-three engine as they had intended. The pilots allowed the starboard wing to drop and lost control of the airplane. Killed in the crash were: 1Lt. Harvey C. Steadman, B-24 instructor pilot; 2Lt. Joseph L. Huelster, pilot/B-24 student; 2Lt. Stanley E. Jones, co-pilot/B-24 student; Sgt. Cleardon Oxendine, engineer. Investigators noted that the instructor pilot had more than 1,300 hours on B-24 type airplanes.

12-5-44A. Walterboro, South Carolina. At 0825 EWT, a Republic P-47D and a Consolidated B-24J collided in mid-air and crashed six miles west of Walterboro Army Air Field, Walterboro, South Carolina, killing thirteen fliers. The subject B-24 was the lead airplane of a three-ship flight that took off at 0800 EWT from Charleston Army Air Base, Charleston, South Carolina, on a high-altitude camera-gunnery mission. The subject B-24 flight was assigned to fly at 20,000 feet. The P-47D was flying in the number-four position of a four-ship flight that took off at 0816 EWT from Walterboro Army Air Field and was to rendezvous with another four-ship B-24 flight, which was a separate flight flying at 8,000 feet. The subject B-24 flight took off, assembled and climbed out on a course of 270 degrees. The P-47s took off, circled the field once and began climbing out on a course of 250 degrees. The two formations were flying on converging courses; the B-24s were to the left and slightly ahead of the P-47 flight. The P-47s were climbing at 500 feet per minute at 170 mph and the B-24s were climbing at 400 feet per minute at about 160 mph. Investigation of the wreckage indicated that the subject P-47, while in a nose-high, left wing high position, collided with the underside of the number-four engine nacelle of the lead B-24. The number-four propeller of the subject B-24 passed through the top of the fuselage of the P-47, passing from right to left across the rear of the cockpit and across the leading edge of the port wing. Investigation revealed that the lead B-24 was climbing on autopilot. Both formations apparently climbed on their original courses until it was too late to avoid a collision. The lead ships of both formations were attempting to turn away from each other when the collision occurred at about 5,500 feet indicated altitude. As the number-four (subject) P-47D passed beneath the B-24 formation, the fighter was turning to the right in an attempt to follow the lead P-47, colliding with the lead B-24's number-four engine. The B-24's starboard leading edge, number-four engine, nacelle and propeller separated and fell away. The B-24 went into a flat spin and smashed to earth, exploding into flames upon impact. The tail section and port wing of the P-47 separated and the airplane went out of control, slamming to earth and exploding a short distance from the B-24. Investigators speculated that the P-47 pilot, because he was watching his wingman while flying formation, did not see the B-24 formation until the instant that the collision occurred. It was noted that the P-47 pilot had been lagging behind early in the flight. Killed in the crash of the B-24 were: Capt. R.W. Miller, B-24 instructor; 2Lt. E.C. Knowlton, pilot/B-24 student; 2Lt. R.R. Coates, co-pilot/B-24 student; 2Lt.

D.C. Estes, bombardier; 2Lt. W.R. McDermott, navigator; Cpl. J.A. Dacek, engineer; Cpl. L.R. DuBreevil, radio operator; Pfc. H.D. Sidman, assistant radio operator; Cpl. L.T. DuGovic, gunner; Pfc. G.L. Simms, gunner; Pfc. O.R. Rhodes, gunner; Sgt. T.C. Conway, gunnery instructor. Pilot F/O Robert M. Johnson, 19, was killed in the crash of the P-47D.

12-5-44B. Albuquerque, New Mexico. At 1003 MWT, a Lockheed RP-38E crashed ten miles southeast of Albuquerque, New Mexico, killing pilot 1Lt. Cletus J. Gombold, 25. The airplane made a normal take-off from the airfield at AAF Convalescent Hospital. The airplane climbed to about 300 feet agl and then entered a gradual turn to the right, climbing out to about 1,000 feet agl with a very thin trail of black smoke issuing from the port engine. The airplane then snapped over to the left and entered a spin to the left. The pilot was unable to recover and the airplane smashed to the ground in a diving spin to the left. The airplane burst into flames upon impact and the pilot was killed instantly. Investigators could not determine the cause of the apparent engine trouble.

12-5-44C. McNeal Auxiliary Airfield, Arizona. At 1205 MWT, two North American TB-25D airplanes collided on the runway and crashed at McNeal Auxiliary Airfield, 20 miles north of Douglas Army Air Field, Douglas, Arizona, killing two fliers. Four fliers escaped injury. B-25 instructor pilot 1Lt. Richard H. Dievendorff and B-25 student A/C William F. Schulze were killed in the crash of TB-25D #41-29874; B-25 student A/C Henry J. Sattler was uninjured. B-25 instructor pilot 1Lt. Owen B. Dubell, B-25 student 1Lt. Frank H. Scherer and B-25 student 1Lt. Elmer E. Tackage escaped injury aboard TB-25D #41-30169. Investigators stated, "McNeal Auxiliary Airfield has three runways which form a triangle. At the time of the accident, aircraft were landing on Runway 21, to the southwest, and taking off on Runway 35, to the north. Runways 21 and 35 intersect at the southwest corner of the field with a stub of approximately 300 feet extending on each runway beyond the intersection. Lt. Dubell, in TB-25D #41-30169, was in take-off position on Runway 35. His student, Lt. Scherer, in the pilot seat, was to practice simulated instrument take-off. Lt. Dubell called the Control Officer, who was located at the downwind end of Runway 21, for permission to take-off, and was cleared for take-off. The Control Officer stated that he checked traffic and saw only one aircraft, which was on the base leg. As Lt. Dubell's aircraft started to move toward the intersection, Lt. Dievendorff, in TB-25D #41-29874, landed on Runway 21 at a fairly high speed. Both aircraft continued on their course, Lt. Dubell taking off on Runway 35, and Lt. Dievendorff completing his landing roll on Runway 21, the two aircraft colliding at the intersection of the two runways. Lt. Dubell's aircraft passed in front of Lt. Dievendorff's in such a manner that the right wing of Lt. Dubell's aircraft struck the plexiglass nose of Lt. Dievendorff's aircraft, completely demolishing the upper nose section and the pilot's compartment, and causing the death of Lt. Dievendorff and [fatally injuring] A/C Schulze.... Following the impact, Lt. Dubell's aircraft, which had the right wing sheared off, veered off the runway to the right and was stopped after a run of about 75 feet. Lt. Dievendorff's aircraft also curved slightly to the right, crashed through a fence, and came to a stop about 100 yards from the point of impact."

12-5-44D. Edgerton, Wyoming. At 1600 MWT, a Consolidated B-24E suffered a catastrophic structural failure following an in-flight fire and crashed east of Edgerton, Wyoming, killing B-24 instructor 1Lt. Donald E. Trail and B-24 student 2Lt. Norwood Sisson. Four crewmembers were injured parachuting to safety. Co-pilot/B-24 student 2Lt. Moses A. Lane suffered serious injuries parachuting to safety. Aerial engineer instructor TSgt. Homer A. Kerr, student engineer Cpl. Richard W. Makarado and radio operator Cpl. James Nielson sustained minor injuries parachuting to safety. The B-24 took off at 1525 MWT from Casper Army Air Base, Casper, Wyoming, on an aerial engineer training flight. The object of the training flight was to execute emergency procedures in the local area. The airplane was flying at 10,000 feet indicated altitude and the number-three engine was shut down and its propeller feathered so that the student engineer could lower the landing gear under emergency conditions. The student engineer successfully completed the operation and the engineer instructor turned on the fuel valve for the number-three engine and told the pilot to restart the number-three engine. The airplane had been running on three engines for about 20 minutes. A moment later the number-four engine began to "whine" and was observed to be on fire. The instructor took control of the airplane, shut down the number-four engine and attempted to feather the propeller. The instructor was unable to feather the propeller or restore normal power to the engine so he ordered the crew to bail out. The pilot and co-pilot remained at the controls until the crew bailed out. The starboard wing failed at the number-four engine nacelle and the airplane went out of control and smashed to the ground, exploding violently into flames and killing the two pilots instantly. The number-four oil cooler, which was heavily damaged by fire, was found about five miles from the scene of the crash. Investigators speculated that the number-four engine suffered a fuel fire in the accessory section behind the number-four engine. This fire destroyed electrical and oil connections and prevented the pilot from feathering the propeller. Investigators noted that the "whine" that was heard was the sound of the engine over-speeding.

12-6-44A. Pacific Ocean, California. At 1014, a Douglas P-70A crashed into the Pacific Ocean eight miles off shore of Carmel, California, killing pilot 2Lt. John V. Folsom, Jr. The airplane was part of a three-ship flight that had taken off from Salinas Army Air Field, Salinas, California, on an air-to-air gunnery mission. The flight rendezvoused with an AAF target tow-ship and commenced the gunnery mission at 8,000 feet indicated altitude. The subject airplane was the first airplane to make a firing pass at the target tow-ship. The subject airplane maneuvered back into firing position for a second pass. The subject airplane banked to the left during its second attack, the bank increasing to a vertical position. The airplane stalled and snapped over to the right to an inverted position, rolling through to an upright position and appearing to stand still momentarily before entering a spin to the right. The airplane entered a flat spin and continued in the spin to the right until it struck the water, remaining afloat for approximately 15 seconds before sinking below the waves. The pilot's body was recovered by crash boats a short time after the accident.

12-6-44B. Hayward, California. At 1519, ground crewman SSgt. Charles Petro was killed when he walked into the rotating port propeller of a Lockheed P-38L operating at Hayward Army Air Field, Hayward, California. Pilot 2Lt. Edgar A. Hendricks was uninjured. Sgt. Petro walked forward into the port propeller after unchocking the port main wheel and was killed instantly.

12-6-44C. Albany, Georgia. At 2012 CWT, a North American TB-25G crashed one and a half miles northeast of Turner Field, Albany, Georgia, killing three French Air Force students. Student pilots Sgt. Roger R.A. Ducrocq, Cpl. Roger M. Caze and Cpl. Paul J. Lagarde were killed in the crash. The airplane took off from Turner Field on a student night training flight. The take-off appeared normal and the airplane drifted slightly to left as it climbed to about 500 feet agl before entering a shallow turn to the right. The airplane lost altitude as it continued to turn to the right. Moments later the airplane flew into the ground in a descending right turn, exploding into flames upon impact. The students were killed instantly. Investigators speculated that the students had lost the horizon while turning to the right over dark terrain. Investigators noted that this was the first night training flight without an instructor for these students.

12-6-44D. Marianna, Florida. At 2040 CWT, a Douglas A-26B crashed while attempting to take off from the Army Air Field at Marianna, Florida, killing pilot 2Lt. James B. Nicholson and gunner Cpl. Earl G. Drown. Investigators stated, "Pilot took off from Runway 8 in an apparently normal manner. At a point approximately 3,000 feet off the end of the take-off runway, and directly in line with the take-off runway, the airplane struck the tops of two large oak trees with the left wing, at an altitude of approximately 40 feet. It is evident from the swath cut through the oak tree foliage that the aircraft was in about a ten degree dive and continued straight ahead in this constant descent for approximately 200 yards at which point the aircraft struck the trunks of two other large trees, shearing off some 15 feet of the left wing. From this point the airplane rotated to the left around its longitudinal axis through approximately 270 degrees, striking the ground on the right wing tip in a 40-degree dive. Aircraft exploded on impact and burned. The distance from the point where the airplane sheared the left wing to the point of impact with the ground is approximately 200 yards. Landing gear was found in the up position and flaps were in the quarter down position."

12-6-44E. Old Tampa Bay, Florida. At 2210 EWT, a Boeing B-17G crashed into Old Tampa Bay four and a half miles WSW of Drew Field, Tampa, Florida, killing the crew of nine. The airplane took off from Drew Field on a routine gunnery-training mission. When the gunnery mission was completed, the airplane landed at Drew Field and the gunnery instructor was dropped off. The airplane took off to complete the second portion of its mission. At 2150 EWT the airplane returned to the area of Drew Field and requested landing instructions. The airplane entered the traffic pattern; a pilot of a B-17 that was in the pattern just behind the subject B-17 reported that the subject airplane flew a wide traffic pattern. The B-17 turned from base to final and the pilot realized that he was going to undershoot the runway. The pilot pulled up the nose and added power. The airplane lost considerable flying speed and began to mush. The pilot was unable to recover and the airplane stalled. The airplane smashed into the surface on the port wing, causing it to cartwheel into the sea. The airplane broke up and burst into flames upon impact. Killed in the crash were: 2Lt. Frank V. Letterman, pilot; 2Lt. George T. Hadley, Jr., co-pilot; 2Lt. Raymond W. Delo, bombardier; F/O John D. Kumcrek, navigator; Sgt. Hubert J. Iffrig, engineer; Cpl. Leroy T. Hanford, radio operator; Cpl. Wilford G. Hurst, gunner; Pfc. Robert Demeulenaere, gunner; Sgt. Victor J. Barchenger, gunner.

12-6-44F. Randsburg, California. At 2105, a North American TB-25J and a Lockheed XF-14 (an early photo-reconnaissance version of the P-80 Shooting Star) collided in mid-air and crashed seven miles SSW of Randsburg, California, killing three AAF crewmembers and two Lockheed employees. The B-25 and the XF-14 took off from Muroc Test Base, Muroc, California, at about 2030 on a test flight "to determine the degree of visibility of the P-80's jet flame from another aircraft at night." The XF-14 made four passes at the B-25, flying past the bomber on a parallel course from the six o'clock position. The four passes from behind were made at about 10,000 feet indicated altitude and were uneventful. The XF-14 was scheduled to make two head-on passes at the bomber at a slightly higher altitude. The jet was apparently maneuvering in order to break away and take a position to make a head-on pass when the collision occurred. Both airplanes went out control immediately and slammed to ground, exploding into flames upon impact. Wreckage

was scattered over a wide area. Investigation of the wreckage indicated that the port wing of the XF-14 had struck the tail assembly of the B-25. The port wing of the XF-14 and the B-25 tail section were severed in the collision and were found some distance from the main wreckage. It was noted that this was the first night flight attempted in an XF-14/P-80 type airplane. Killed in the crash of the B-25 were: 1Lt. Henry M. Phillips, pilot; Capt. Benjamin Van Doren, Jr., observer; TSgt. William L. Eckert, engineer; Robert Eickstadt, civilian technician/engineer, Lockheed Aircraft Company. Lockheed Test Pilot Perry E. Claypool was killed in the crash of the XF-14.

12-7-44A. Odessa, Texas. At 0920, a Republic P-47D crashed one-half mile northeast of Odessa, Texas, killing F/O Billy R. Dahl, 20. The subject airplane was the lead ship in a two-ship flight that had taken off at 0820 from Abilene Army Air Field, Abilene, Texas, on a high-altitude camera-gunnery mission. The mission was to be conducted at 25,000 feet. The flight took off and flew to the area of Sweetwater, Texas. The subject pilot, who was leading the flight, indicated to the wingman that he was going to fly to Odessa, which was the lead pilot's hometown. The two airplanes flew to Odessa at 8,000 feet indicated altitude. When the airplanes arrived over Odessa, the lead airplane broke off and lined up to make a low-level pass over the residential area. Witnesses confirmed that the wing ship remained at about 2,500 feet while the subject airplane dropped down to treetop level. After the first pass, the subject airplane pulled up steeply and circled around to the right, making a circle north of town. The wing ship remained at 2,500 feet and circled to the right in order to meet up with the lead ship. The subject airplane again dropped to very low altitude and made another pass over the residential district. The subject airplane, flying at about 200 feet agl, performed a slow roll as it passed over, losing altitude as it recovered and immediately entering another slow roll at an altitude of about 100 feet agl. The airplane fell out of the roll while in an inverted position and smashed into the ground before the pilot could recover. The airplane exploded violently into flames when it slammed to earth in a pasture northeast of town, killing the pilot instantly.

12-7-44B. Hoboken, Georgia. At 1300 EWT, a Curtiss P-40N suffering an engine failure crashed two miles south of Hoboken, Georgia, killing 1Lt. John G. Clore, 25, in an unsuccessful parachute jump. The airplane was the lead ship of a three-ship flight that had taken off at 1230 EWT from Waycross Army Air Field, Waycross, Georgia, on a camera-gunnery mission, which was to be performed at the AAF Aerial Gunnery Range between Waycross and Folkston, Georgia. The pilot led the flight to an area southeast of Waycross and rendezvoused with target-towing ship. The flight began its gunnery passes and a short time later the subject airplane dropped out and the pilot radioed the tow ship that he was suffering engine trouble and was heading back to Waycross. The tow-ship dropped down and flew with the subject airplane. No smoke was observed at first but a short time later white puffs of smoke were observed coming from the engine section. The smoke became thicker as the airplane flew toward the field and moments later the engine quit altogether. The subject pilot radioed that he was bailing out. The pilot bailed out at about 1,500 feet but was struck in the head by the horizontal stabilizer, fracturing his skull. The pilot was unable to pull his ripcord and he fell to earth. His body was found about 175 yards from the spot where the airplane crashed to the ground and exploded. Investigation revealed that the engine had overheated because of a failure of the coolant system and the loss of the coolant, causing the engine to fail.

12-7-44C. College Park, Georgia. At 1325 EWT, a Beech AT-11 crashed while attempting an emergency forced landing one mile west of College Park, Georgia, killing pilot Capt. Frank L. Brady. Five passengers were injured. The airplane took off at 0939 EWT from Hendricks Field, Sebring, Florida, on an instrument flight plan to Atlanta Army Air Field, Atlanta, Georgia. The airplane was assigned to fly at an altitude of 6,000 feet. The airplane contacted Atlanta air traffic control at 1205 EWT. Atlanta control tower cleared the airplane to drop from 4,000 feet to 3,000 feet and cleared it to land on an instrument approach. The airplane descended out of the overcast directly over Atlanta and passengers were able to easily pick out familiar landmarks. The pilot failed in his instrument approach and was attempting to go around when he noticed that the airplane was running low on fuel. The pilot realized that he could not make the field so he attempted an emergency forced landing on a golf course. The pilot overshot the fairway that he was attempting to land on and the airplane collided with trees. Passengers injured in the crash landing were: Maj. Joseph L. Roberts, Capt. Robert S. Sneed, 1Lt. Charles T. Lombardo, Cpl. Thomas J. Boyle, Sgt. Frank J. Yabrosky, engineer.

12-7-44D. Victorville, California. At 1305, a Lockheed P-38L crashed 12 miles south of the Army Air Field at Victorville, California, killing F/O Richard S. Secor, 20. The airplane was flying in the number-two position of a four-ship flight that took off at 1232 from Ontario Army Air Field, Ontario, California, on an aerial gunnery mission. The four airplanes were scheduled to rendezvous with a Martin B-26 type airplane that would tow a target over the aerial gunnery range. The B-26 was unable to make the flight because of mechanical difficulty. The flight's mission was changed to an individual combat flight. The flight climbed to 7,500 feet and broke up into pairs. The airplanes engaged in simulated combat for several minutes and performed various aerobatic maneuvers. The airplanes had lost considerable altitude while maneuvering and the flight leader called for the airplanes to

begin to climb back up to 7,500 feet. Moments later the subject airplane entered a tight turn and stalled, snapping over to an inverted position before entering a steep dive. The pilot was unable to recover and the airplane smashed to the ground and exploded.

12-8-44A. Hayward, California. At 1113, a Lockheed P-38L crashed one-half mile south of the Army Air Field at Hayward, California, killing pilot 2Lt. Allan Baurain, 25. The subject airplane was flying in the number-two position of a three-ship flight that had taken off from Hayward Army Air Field at 1055 on a scheduled ground-gunnery mission. The number-three airplane had just made the first turn away from the field when the number-three pilot radioed the flight leader and alerted him that the he had a serious oil leak in the port engine and was returning to the field. The flight leader acknowledged. The lead pilot and the subject airplane circled the field in a climbing turn, and as they passed through 2,000 feet indicated altitude they were "jumped" by two U.S. Navy Chance-Vought F4U type fighters, which attempted to engage them in simulated combat. The subject flight continued to circle the field to the left in an attempt to keep the number-three ship in sight and observe the pilot's single-engine landing attempt. The number-three ship made two attempts to land but was unsuccessful and then made a crash landing on the third attempt. The flight leader looked back and observed that the subject airplane was in a skidding turn to the left. The wings approached a vertical position and the nose pitched down about 30 degrees. The subject airplane apparently stalled and snap rolled through inverted before it entered a spiral. The subject airplane momentarily leveled out after describing about 270 degrees of spiral. Immediately after leveling out, the subject airplane violently mushed into a stand of eucalyptus trees, exploding into flames upon impact and killing the pilot instantly. The P-38 flight leader and witnesses on the ground stated that the lead P-38 and his wingman did not engage the Corsairs prior to the accident. Investigation revealed that the same pair of Corsairs had attempted to engage a three-ship P-38 flight that had taken off from Hayward Army Air Field at 1045 without success. The F4U airplanes had taken off from Livermore Naval Air Station, Livermore, California, at 0954 and were piloted by Ensign Homer A. Morrow and Ensign Frank L. Nelson.

12-8-44B. Sioux Falls, South Dakota. At 1502 CWT, two Lockheed AT-18 airplanes collided in midair and crashed at Sioux Falls Army Air Field, Sioux Falls, South Dakota, killing seven fliers and three army personnel on the ground. Two other army personnel on the ground were seriously injured. Seven crewmembers were injured parachuting to safety. AT-18 #42-55636 crashed into Barracks #631 at Sioux City Army Air Field and AT-18 #42-55691 crashed five miles northeast of the field after it was abandoned in flight at about 1515 CWT. Cpl. Frank M. Bogdan, Pvt. Richard J. Barber and Pvt. Martin Brickner were killed on the ground and Sgt. Marven J. Curteman and Cpl. Edward J. Zoski received serious injuries. Killed in the crash of AT-18 #42-55636 were: 1Lt. Paul S. Curry, pilot; SSgt. James D. Simmons, engineer; Pfc. Orville J. Spellman, radio operator instructor; Cpl. Thomas F. Hart, Jr., student radio operator; Pvt. Floyd R. Terral, student radio operator; Pvt. Edward M. Gregory, student radio operator; Pvt. Kent H. Rhodes, student radio operator. Crewmembers injured parachuting to safety from AT-18 #42-55691 were: Capt. Willard B. Woody, pilot; Pfc. Alex C. Galvan, engineer; Sgt. Johannes E. Vanderkerk, radio operator instructor; Pvt. George Caron, student radio operator; Pvt. Robert A. Faring, student radio operator; Pvt. Frederick M. Graves, student radio operator; Pvt. Joseph P. Eichholzer, student radio operator. Capt. Woody took off on Runway 33 at 1457 and Lt. Curry took off at 1458 on separate radio training flights. Capt. Woody made a left turn just after take-off and then made a right turn out of the traffic pattern. After climbing to about 1,000 feet agl, Capt. Woody then turned to the left and circled back and around the field. Lt. Curry took off a minute after Capt. Woody and turned left just after becoming airborne, making another left turn moments later, flying a course parallel to the take-off runway. He also was flying at an altitude of about 1,000 feet agl. The airplanes continued on these respective courses until the collision occurred. Investigation revealed that Capt. Woody's starboard wing, engine and propeller struck the starboard vertical fin and rudder of Lt. Curry's airplane. Lt. Curry's airplane flew for about a quarter mile before going out of control and entering a spiral to the right, smashing into Barracks #631 and bursting into flames. Capt. Woody was unable to control his airplane when speed dipped below 165 mph. The starboard engine began smoking and the airplane was vibrating excessively, making the airplane almost uncontrollable. A portion of the starboard wing leading edge of Capt. Woody's airplane had been torn away and pieces of the other airplane's tail section were lodged in the starboard wing. Capt. Woody climbed the airplane, held the speed steady and ordered the crew to bail out.

12-8-44C. Fort Sumner, New Mexico. At 1845 MWT, a Curtiss P-40N crashed ten miles ENE of Fort Sumner Army Air Field, Fort Sumner, New Mexico, killing pilot 2Lt. Clinton R. Dickinson, 20. The airplane took off from Fort Sumner Army Air Field at 1752 MWT on a local night transition flight. The airplane was assigned to fly in "Zone 1" at 13,000 feet. At 1843, the control tower radioed the subject pilot and called him in for his landing practice. The pilot acknowledged and stated that he was going to begin to let down. The subject airplane failed to arrive at the field and the control tower called for it over the radio several times and requested that a pilot flying in the area attempt to contact the pilot. All radio calls went unanswered. At 1900, an airplane that was flying in the

area reported that the subject airplane had crashed. Investigation revealed that the P-40 had descended into the ground under control in a flying attitude at high speed. Investigators noted that the pilot had just been assigned duty at Fort Sumner and had been previously stationed at Galveston Army Air Field, Galveston, Texas. Investigators noted that the elevation at Galveston Army Air Field is six feet and the elevation at Fort Sumner Army Air Field is 4,130 feet msl. It was speculated that the pilot forgot about the field elevation at Fort Sumner and had descended into the ground while letting down for his first night landing at the station.

12-8-44D. Marianna, Florida. At 2022 CWT, a Douglas A-26B crashed one and a half miles north of the Army Air Field at Marianna, Florida, killing gunner SSgt. Horace W. Moore, Jr. and seriously injuring pilot 1Lt. Ellard J. Connors, 28. The airplane made what appeared to be a normal take-off and just after becoming airborne the port engine began losing power. The pilot was going to feather the propeller but the engine came back up to power momentarily. The engine then began surging and ultimately failed. The pilot was able to maintain directional control of the airplane but could not prevent it from descending into a stand of tall trees about one nautical mile from the end of the runway. The airplane smashed itself to pieces and burst into flames as it careened through the trees, finally coming to rest in a cleared field adjacent to the woods.

12-8-44E. Barksdale Field, Louisiana. At 2213 CWT, a Martin B-26B crashed while attempting to take-off from Barksdale Field, Shreveport, Louisiana, killing three French Air Force fliers and seriously injuring three others. The airplane made what appeared to be a normal take-off to the south and descended into the ground about a mile and a half from the end of the runway. Marks on the ground indicate that the port main wheel made contact first, leaving a rut about 60 feet long before both propellers began striking the ground. The nose wheel collapsed and the airplane nosed into the ground and went completely out of control. The airplane smashed itself to pieces and broke into flames as it careened across the terrain at high speed. Investigators noted that the landing gear and flaps were in the down position at the time of the crash. It was speculated that the pilot had allowed the airplane to descend into the ground after take-off. French Air Force personnel killed in the crash were: SSgt. Pierre L. Cartereau, pilot; Aspt. Fernand G. Veith, bombardier/navigator; Sgt. Maximo Juncuennet, radio operator. Co-pilot Aspt. Max F. Cluzel, engineer SSgt. Marcel M. Mouly and gunner SSgt. Andre P. Capin were seriously injured.

12-8-44F. Princeton, California. At 2108, a Lockheed P-38L crashed two miles west of Princeton, California, killing pilot F/O Byron K. Duke, Jr., 21. The airplane was flying in the number-two position of a four-ship flight that had taken off at 2025 from Chico Army Air Field, Chico, California, on a scheduled night formation mission. The flight climbed to 6,000 feet and flew straight and level in a loose echelon formation to the right. The flight leader radioed the second element and ordered them to cross over to the left. This maneuver was completed without incident. A few minutes later, the subject airplane (number-two position) fell out of formation and began falling back. The number-four pilot observed flames on the ground a few minutes later. Investigation revealed that the airplane smashed into the ground at an angle of 80 degrees in a left wing down attitude. The pilot's body was found about 60 yards from the point of impact along the path of the wreckage. The airplane exploded into flames upon impact. Witnesses on the ground stated that both engines were "roaring" as the plane dove into the ground. Investigators could not determine the cause of the crash.

12-9-44A. Wilson Creek, Washington. At 1019, a Lockheed P-38L crashed four miles north of Wilson Creek, Washington, killing pilot 2Lt. Albert J. Hogan, 21. Investigators stated, "2Lt. Hogan took off from Ephrata Army Air Base [Ephrata, Washington] at 0945 in the number-two position of a four-ship flight on a scheduled camera gunnery and aerobatic flight. This flight had been properly briefed on this mission before take-off, and had also visually checked their gas supply. After take-off, good radio communication was established among the members of the flight. After climbing to 10,000 feet, the flight leader put the flight in string formation and executed three slow rolls and one loop, with all members of the formation following in order. About five minutes after making the loop, the flight leader executed an Immelmann, rolling out between 10,000 feet and 11,000 feet. Lt. Hogan was seen ... to reach the top of the Immelmann and from which point he apparently lost control and went into a spin. The airplane spun to the ground, crashed and burned."

12-9-44B. Claremore, Oklahoma. At 1659, a Cessna UC-78B crashed one and a half miles NNE of Claremore, Oklahoma, killing three fliers. Pilot 2Lt. George W. Crowe, co-pilot Mary L. Webster (WASP) and passenger Sgt. Melvin A. Clark were killed in the crash. The airplane took off at 1432 from Frederick, Oklahoma, on a flight to Tulsa, Oklahoma. The airplane was cleared on an instrument flight plan and assigned to fly at an altitude of 5,000 feet. The pilots were cleared by Fort Worth, Texas, control to maintain at least 500 feet of clearance above the cloud tops. The pilot checked in at Oklahoma City at 1602, reporting that he was "on top" at 9,000 feet. The pilot was then instructed to fly on the northeast leg of the Tulsa radio range and await further instructions. An icing warning was also given. At 1642, the airplane was finally cleared to descend for a landing at Tulsa. The airplane descended to about 5,500 feet and at 1648 the pilot requested permission to climb back to the on-top position. This request was granted at 1656 and acknowl-

edged by the pilot immediately. There was no further contact with the airplane. Witnesses on the ground observed the airplane plunge out of the overcast in an inverted vertical dive. The airplane smashed into the ground in this position and exploded into flames upon impact. Pilots who had landed just prior to the subject airplane arriving in the Tulsa area reported that icing conditions were severe.

12-10-44. Quitman, Georgia. At 1625 EWT, a North American AT-6A collided with trees and crashed eight miles south of Quitman, Georgia, killing A/S Richard W. Twichel, 23. The airplane took off at 1540 EWT from Spence Field, Moultrie, Georgia, on a "200-foot" navigation mission to Climax, Georgia, to Rocky Ford, Georgia, and return. The student had successfully navigated to Climax and at Rocky Ford, checking in with the control ship at 1620. The airplane collided with trees a short time later, shearing off the port wing and sending the airplane out of control. The airplane smashed into the ground at high speed in a vertical bank to the left. Investigation revealed that the engine was turning at cruise speed at the time of impact. Investigators speculated that the student had misjudged his height above the trees while flying at very low altitude.

12-11-44A. Rome, New York. At 1001, a Cessna UC-78B crashed five miles east of Rome, New York, killing pilot F/O Stanley F. Sakser, 26. Co-pilot 1Lt. Lamar Peeples and engineer Pfc. Joseph A. Papa were seriously injured. The airplane took off at 1000 EWT from Runway 10 at Rome Army Air Field on a local instrument training flight. Investigators stated, "After reaching an altitude of approximately 100 feet agl, the right engine cut out. It did not backfire or throw any smoke at all. Lt. Peeples switched gas tanks, pumped the wobble pump, and applied carburetor air heat to the right engine, but the engine would not start again. Lt. Peeples radioed the Control Tower requesting an emergency landing and permission was granted to make a 180-degree turn and landing on Runway 28. After the right engine cut out, F/O Sakser continued to gain altitude to approximately 300 feet with the intention of returning to the field for a landing. He tried to turn to the left, but the aircraft would not respond to the controls and the left engine seemed to slowly lose power. He informed Lt. Peeples that he could not make the field, but would have to make a crash landing. He chose an open field directly in front of him, but due to the loss of power in the left engine he did not make it to the field. In trying to stretch his glide over the tops of the trees on the edge of the field, the airplane stalled and started a spin to the right. The right wing hit a tree, then a telephone pole. The airplane hit the ground on its right wing and nose, slid approximately forty feet and came to rest heading east." Investigators speculated that severe carburetor icing had caused the loss of the starboard engine and the diminished performance in the port engine.

12-11-44B. Muroc, California. At 1834, a Consolidated B-24J crashed while attempting an emergency forced landing at Muroc Army Air Field, Muroc, California, killing tail gunner Cpl. Billie Burnett. Pilot 2Lt. Roy L. Reed received minor injuries and bombardier F/O Robert F. Cudney was seriously injured. Crewmembers uninjured in the crash landing were: 2Lt. Donald F. Esser, co-pilot; Cpl. John F. Clarke, engineer; Cpl. Robert B. Kirkman, assistant engineer; Cpl. William W. Rowlette, radio operator; Cpl. Courtney N. Rea, Jr., asst radio operator; Cpl. Charles A. Phillips, gunner. Investigators stated, "The subject aircraft had completed an air-to-air gunnery mission and landed to discharge three instructor personnel. Immediately thereafter Lt. Reed taxied out for the second take-off. At this time the control tower directed immediate take-off due to congested traffic. The pilot applied power without running up the engines for the reason that the aircraft had just been in flight and all equipment was functioning normally. During the take-off roll, [the number-two engine] backfired severely; however, aircraft executed a normal take-off. Immediately after take-off the pilot observed number-two manifold pressure in excess of 60 inches and rotating counter-clockwise to stop at approximately 50 inches. At this time fire was observed issuing from lower cowl flap and more rearward portions of the engine nacelle. The number-two engine was promptly feathered. Altitude at this point was approximately 2,600 feet indicated or slightly over 100 feet above the terrain in this area. Take-off power was retained on engines numbers one, three and four as were one-half flap and extended landing gear due to the possibility of impending crash landing. Upon attaining approximately 2,900 feet indicated, pilot called for retraction of the landing gear and co-pilot advised Control Tower of emergency conditions. On downwind leg of traffic pattern, number-one engine roared and tachometer indicated 3,000 rpm and a runaway propeller.... [Before] remedial action could be taken number-one engine completely lost power. The aircraft, whose airspeed was below 130 mph indicated with one-half flap still down (the pilot still endeavoring to gain altitude), veered sharply to the left shortly past mid-field. Pilot was unable to effect any corrective maneuver, due to low altitude and night visibility conditions, and landed aircraft in a clearing and revetment area approximately 50 yards from the edge of the building line. Aircraft burst into flames on impact (the landing gear had not been extended since retraction) and bombardier and tail gunner who were riding in the waist were thrown from the aircraft when it broke in two just aft of the center section. The tail gunner was killed instantly; the bombardier suffered extensive lacerations and additional superficial injuries; all others escaped injury. Due to the fact that the subject aircraft was one of a certain few recently being subjected to engine dilution, it is entirely possible that excessive carbon and foreign material is being dislodged and introduced into the propeller governing and pitch actuating mechanisms as a result of the dilution activity. The propeller, which ran away and thereby caused total engine failure, had not been inspected for propeller dome sludge and excessive carbon for approximately 90 days previous to failure, and subsequent inspection of all diluted engines in other aircraft at this station [Muroc AAB] evidenced extremely heavy carbon deposits of large, granular nature approximately the size of rock salt in and about the oil screens and sumps."

12-12-44A. Richmond, Virginia. At 1120, a Republic P-47G crashed at Richmond Army Air Base, Richmond, Virginia, killing 1Lt. Charles T. Hancock, 23. The airplane took off at 1000 from Richmond AAB on an engine "slow time" flight. The airplane returned to the base and entered the traffic pattern at 1,000 feet for a landing on Runway 24. The airplane passed over the runway at 1,000 feet and at the halfway point broke away into a 360-degree pattern. The airplane was observed with flaps and landing gear extended on the downwind leg. An officer in the tower radioed the pilot and advised him to bring up the flaps up incrementally before beginning a turn. The pilot retracted the landing gear and immediately radioed the tower that he was having "trouble" with his airplane. The airplane, with landing gear retracted and flaps down, rolled to a vertical bank and began losing airspeed and altitude. The airplane fell off on the port wing and crashed to the ground, smashing into some bleachers near the field. The airplane bounced back into the air momentarily before smashing back to the ground on the port wing and cartwheeling into a wooded area. The airplane burst into flames as it smashed through the trees. Investigators stated that the airplane apparently stalled at an altitude too low to allow recovery.

12-12-44B. Minot, North Dakota. At 1905 MWT, a Boeing B-17G crashed four miles northwest of Minot, North Dakota, killing five fliers and seriously injuring four others. Killed in the crash were: 2Lt. Clifford L. Higgins, pilot; 2Lt. Koester E. Johnston, co-pilot; Cpl. Jack H. Campbell, engineer; Cpl. Iver C. Everts, radio operator; Cpl. Phillip H. Simonetti, gunner. Crewmembers injured in the crash were: F/O Charles E. Atkinson, navigator; 2Lt. Chester J. Teich, flight engineer; Cpl. Jesse D. Stubblefield, gunner; Sgt. Rudolph E. Ekstrom, gunner. The airplane took off at 1506 CWT from Sioux City Army Air Base, Sioux City, Iowa, on a 1,000-mile navigation flight to Ainsworth, Nebraska, to Minot, North Dakota, and return to Sioux City AAB. At 1942 CWT, the pilot radioed Minot that the number-three engine was out and that he would fly to Bismarck, North Dakota. At 2000, the pilot radioed Minot that the number-two engine was also out and requested flares and an emergency landing at Minot Municipal airport. The pilot was unable to feather the number-three propeller, and about ten miles south of Minot the number-two engine began throwing sparks. The pilot had great difficulty locating the field at Minot, which did not have adequate lighting. The pilot tried to locate the field by blinking his landing lights and requesting that Minot radio advise him of his position. The pilot was apparently maneuvering in an effort to find the airport. The airplane was in a steep bank to the right at very low altitude when the starboard wing collided with the terrain. The airplane cartwheeled into the ground, breaking up and bursting into flames. Investigators speculated that the dark conditions prevented the pilot from realizing his proximity to the ground. Investigation revealed that the number-three propeller would not feather on the ground prior to the final flight because of a faulty propeller-feathering switch. The propeller-feathering switch was replaced and the propeller operated satisfactorily on the pre-flight inspection. Investigators could not determine the cause of the double engine failure.

12-13-44A. Zelma, Missouri. At 0215 CWT, a Douglas C-47A crashed two miles northeast of Zelma, Missouri, killing four crewmembers. The airplane took off at 0030 CWT from Malden Army Air Field, Malden, Missouri, on a night transition flight. The airplane and crew had just completed three hours of glider tow training and the mission that they had taken off on was an "optional" two-hour night training flight. Investigators stated, "The plane, flying from north to south at a speed estimated to be between 180 and 200 mph hit the treetops in a left wing low attitude of 45 degrees or more, sliced through 36 yards of trees, glanced from the top of the first ridge, and exploded upon impact with the next ridge. While the apparent attitude of the airplane was left wing low, the direction of travel was straight forward, more or less eliminating the possibility of a spiral." There were no witnesses to the crash and investigators could not determine why the airplane smashed into the wooded terrain. Killed in the crash were: 2Lt. Robert M. Moser, pilot; F/O Emil P. Ratay, co-pilot; Cpl. William H. Schauss, engineer; Cpl. Kenneth H. Newell, radio operator.

12-13-44B. Lake Park, Georgia. At 1137 EWT, a Curtiss P-40N crashed one mile east of Lake Park, Georgia, killing student A/C Harold T. Prichard, who fell to his death in an unsuccessful parachute jump. The airplane took off at 1029 EWT from Perry Army Air Field, Perry, Florida, on an escort mission. The subject airplane was the lead ship of the second element of a four-ship flight that was to rendezvous with a formation of bombers in the vicinity of Valdosta, Georgia. As the subject airplane flew at about 9,000 feet, the student radioed the flight leader and advised that his engine had quit completely. The subject airplane dropped out of formation and his wingman followed him down. The student radioed that he was going to attempt an emergency landing at Lake Park Auxiliary Airport, which was a few miles away. The subject airplane glided toward the field and when at a point about

two and a half miles northeast of the auxiliary field at an altitude of 1,000 feet the student announced that he was bailing out. The student bailed out but his parachute failed to fully deploy and he slammed to the ground, coming to rest about 100 feet to the rear of the wreckage and about 30 feet to the left of the flight path. His parachute had strung out behind him but never unfolded. The airplane crashed into a heavily wooded area and exploded into flames.

12-13-44C. Wasco, California. At 1550, a Cessna UC-78B crashed 11 miles west of Wasco, California, killing A/C Everett L. Morningstar and A/S Anthony W. Milano. The airplane took off at 1450 from Minter Field, Bakersfield, California, on a routine training flight. The airplane was flying at about 6,000 feet when it was observed to go into a 45-degree dive. The airplane continued in the dive until it struck the ground and burst into flames. Investigators found pieces of plexi-glass about a mile from the area of the main wreckage. Investigation revealed that the plexi-glass was from the subject airplane's "sky window." The sky window was broken out when A/S Milano's head smashed into the sky window during an apparent stall recovery. Investigation of the wreckage revealed that A/S Milano's safety belt was not secured and fastened, allowing him to be forced into the sky widow during an unknown maneuver. A/C Morningstar's safety belt was found in a fastened position.

12-13-44D. El Paso, Texas. At 1735 MWT, a Boeing B-17F suffering an in-flight fire was abandoned and crashed four miles east of Biggs Field, El Paso, Texas, killing pilot 2Lt. Charles J. Kartz, 22, who fell to his death in an unsuccessful parachute jump. B-17 instructor pilot 1Lt. George L. Furney and navigator 2Lt. Edward E. Schultz were seriously injured parachuting to safety. Crewmembers uninjured parachuting to safety were: 2Lt. Earl Peterson, co-pilot; TSgt. Vernon R. Black, engineer; Cpl. Leroy G. Krieger, radio operator; Cpl. Edward J. McDermott, gunner; Sgt. Erric V. Lemons, gunner; Cpl. David W. Goddin, gunner. The airplane took off at 1630 MWT from Biggs Field on a high-altitude formation mission. The airplane was flying at 10,000 feet and had been airborne about 45 minutes when the engineer smelled something like "burning metal." The airplane had just been intercepted by fighter airplanes and the engineer thought that maybe the turret was hot. The engineer stepped back for a few minutes and then entered the turret to cut the switches. At that time the engineer noticed a fire in the base of the turret. The turret fire increased in intensity and was described as being like a "welding torch." The engineer alerted the pilot of the fire and grabbed a fire extinguisher. But before he could use it, the pilot, noticing the intensity of the fire, immediately ordered the crew to bail out. There were several minor explosions and the fire spread with each blast. The blazing airplane entered a spiral to the left and smashed to earth, exploding into flames upon impact.

Investigation revealed that the top turret fire was caused by a failure of the oxygen lines that feed the top turret and the chafing of wiring that pass through the same area of the top turret electrical junction box located in the base of the turret. The arcing wires ignited the leaking oxygen, causing the intense fire. Investigators noted that the failed parachute attempt and the two serious injuries were caused by crewmembers sitting on chest-pack type parachutes. The compression of the chest type parachutes prevents the parachute from readily leaving the pack and fully deploying. It was noted that all crewmembers had been trained to hand feed chest-pack type parachutes out of the pack that were slow in deploying. The body of the pilot was found about two miles from the area of the main wreckage.

12-13-44E. Raleigh, North Carolina. At 2221, a North American B-25C crashed at Raleigh Army Air Field, Raleigh, North Carolina, killing five fliers and seriously injuring two others. Killed in the crash were: 1Lt. Bernard R. Langlois, B-25 instructor; 2Lt. Jay Steele, pilot/B-25 student; F/O Stanley Stevenson, co-pilot/B-25 student; 2Lt. David Bell, bombardier/navigator; Cpl. Wilson Klein, engineer. Gunner Cpl. Rudolph Valenzuela and radio operator Cpl. Norman Meliorato sustained serious injuries. The B-25 took off at 1947 EWT from Greenville Army Air Field, Greenville, South Carolina, on a night navigation and "strange field" landing mission to Aiken, South Carolina, to Raleigh-Durham, North Carolina, and return to Greenville. Investigators stated, "At approximately 2159 EWT, [the pilot] requested landing instructions from Raleigh-Durham AAB Tower. At approximately 2209, [the aircraft] landed on Runway 23 and taxied back for take-off. While taxiing back, this aircraft requested radio check and advised tower he would make another touch-and-go landing. Pilot was advised to stay in the pattern and notify the tower on the base leg. Aircraft took off on Runway 23 at approximately 2221 EWT, and was last seen in flight over the boundary of the field." The airplane had apparently descended into a heavily wooded area about 1,800 feet from the field. The airplane did not burn but was completely destroyed. Fire fighters on the field noticed that the airplane sound "disappeared" sooner than usual. The fire fighters contacted the tower and asked about the airplane, expressing concern that a crash had occurred. Because of a communication mix up, the control tower erroneously advised the firefighters that the subject airplane had flown away from the field. No immediate search was made for the aircraft. Investigation revealed that tower personnel communicated with the wrong airplane, which had advised them that it was flying away from the area. One of the surviving crewmembers from the subject airplane crawled out of the wreckage at day break and at approximately 0750 was able to get the attention of a passing civilian motorist, who gave him a ride to the operations office at Raleigh-Durham AAB. Rescuers arrived at the wreck at about

0820, saving one crewmember. Investigators speculated that the pilot had lost the horizon over the dark terrain just after take-off and had inadvertently descended into the terrain.

12-14-44A. Nashville, Georgia. At 1237 EWT, a Curtiss P-40N crashed six miles south of Nashville, Georgia, killing pilot 2Lt. Walter H. Spence, 21. The pilot and aircraft, which were stationed at Spence Field, Moultrie, Georgia, were part of a three-ship flight that took off from the Army Air Field at Tifton, Georgia, on a routine training mission. The three ships were seen maneuvering in the area of Nashville prior to the crash. The subject airplane became separated from the flight and was observed in a spin to the right at high altitude. The airplane entered a high-speed dive and the pilot was unable to recover. The P-40 dove into the ground and exploded violently into flames, killing the pilot instantly. The starboard aileron had separated from the aircraft during the attempted recovery and was found about one-half mile from the main wreckage. Witnesses stated that the engine was running at high rpm during the spin and attempted recovery.

12-14-44B. College Park, Georgia. At 1250 EWT, a Lockheed P-38L crashed into farm structures four miles south of College Park, Georgia, killing a civilian on the ground and seriously injuring pilot F/O Frederick L. Baughman. Investigators stated, "Pilot took off from Jackson, Mississippi, at 1102 EWT on a ferry flight to Greensboro, North Carolina. He crashed approximately four miles southeast of Atlanta Army Air Base, Georgia, at 1250 EWT, hitting a silo before making contact with the ground. Subject airplane was seen three miles west of Atlanta AAB at an altitude of 750 feet with the landing gear in the down position. Apparently the pilot had the intention of landing at Atlanta due to mechanical difficulties with the aircraft. The wheels on the plane were then observed to retract and the aircraft made a 180-degree gliding turn to the right and crashed.... The aircraft apparently stalled into the silo with no power being applied; the tail being the only part of the aircraft to make contact with the silo." The airplane smashed to the ground and wreckage struck other buildings in the immediate area. Civilian Almon Mealer, College Park, Georgia, was killed when he was struck by the careening aircraft and flying wreckage. Investigation revealed that the airplane suffered a double engine failure because of the mismanagement of the fuel system.

12-14-44C. Cuero, Texas. At 1420, a Curtiss P-40N crashed ten miles southeast of Cuero, Texas, killing student A/C Robert D. Barton, 19, who fell to his death in an unsuccessful parachute jump. The airplane was flying in the number-three position of a three-ship flight that took off from Foster Field, Victoria, Texas, on a routine student training mission. The flight climbed to 12,000 feet and began practicing various formations. The flight leader (instructor) signaled for an echelon right formation and then signaled for a peel-off and trail formation aerobatics. The formation fell apart and the instructor ordered the students to reform the echelon right for more work. The flight leader signaled for a peel-off and trail formation aerobatics. The flight leader then performed a lazy-eight and then a barrel roll. After completing the barrel roll, the flight leader noticed that the number-three ship was missing. After a short search, the flight leader spotted a fire on the ground and upon further investigation discovered that the fire on the ground was the subject airplane. Investigation revealed that the student had stalled the airplane while in an inverted position during the roll. The student recovered after losing considerable altitude but pulled up too early, causing a high-speed stall. The student recovered again but discovered that the altitude was inadequate to allow a pull-out from the dive. The student bailed out at an altitude too low to allow him to pull his ripcord. His body was found about 100 yards from the wreckage of the airplane, which struck the ground at a steep angle and exploded.

12-14-44D. Tifton, Georgia. At 1545 EWT, a North American AT-6A crashed seven miles south of Tifton, Georgia, killing student A/C Evan H. Orme. The airplane took off from Spence Field, Moultrie, Georgia, on a contour-flying mission. On the leg from Quitman, Georgia, to Fender, Georgia, the subject airplane was seen to begin a gradual climb to an altitude of about 1,500 feet. A/C Augie T. Ong stated that the subject airplane then entered a steep, climbing turn to the right. The subject airplane stalled as it climbed and turned to a heading about 90 degrees to the right of the original heading. The subject airplane then snapped over to an inverted position and fell off. The student was able to roll out of the maneuver but was unable to regain control of the airplane as it entered a dive at about 700 feet agl. The airplane continued in the dive until it smashed to the ground at a 60-degree angle about 20 feet from a barn, bursting into flames upon impact and damaging several farm structures.

12-14-44E. Lubbock, Texas. At 1447, a Beech AT-10BH crashed 15 miles northwest of the Army Air Field at Lubbock, Texas, killing A/C George W. LaMontaine, Jr. and A/C Vernon J. Landon. The airplane took off from Lubbock Army Air Field on a routine student training flight. The students were scheduled to practice at North Auxiliary Airfield. When the student airplanes arrived at the field they discovered that the control ship was not on station. The three student airplanes circled to the left near the field while they waited for the control ship to arrive. The subject airplane was seen to snap over and enter a spin to the right. The students were unable to recover and the airplane smashed to the ground and exploded into flames, killing both instantly.

12-14-44F. Mission, Texas. At 1740, a North American AT-6C and an AT-6D collided in mid-air and crashed nine miles northeast of Moore Field, Mission,

Texas, killing four fliers. Instructor 2Lt. Byard B. Hubbard and student A/C Arthur D. Mills were killed in the crash of the AT-6C (lead ship); A/C Robert A. Nicholson and A/C James J. Flaherty were killed in the crash of the AT-6D (number-two position). The airplanes were part of a flight of four that had taken off from Moore Field on a flight to the Army Air Field at Matagorda Island Bombing Range, Matagorda Island, Texas, to pick up four students who had been undergoing gunnery training. On the return flight back to Moore Field, the airplanes were in a loose V-formation with Lt. Hubbard leading the flight. Two other ships were in echelon on the leader's right wing and the remaining ship was on the leader's port wing. The formation made a gradual descent from 4,000 feet to 3,000 feet about 12 miles from Moore Field. Investigators stated, "Shortly after that, the lead plane peeled off, up and over the left wing man, with the planes still in a very loose V-formation. The left wingman attempted to hold his wing position, but soon realized the safe thing to do was to go down and to the right. The two airplanes on the right side followed the leader [the subject AT-6D followed first]. The leader turned approximately 90 degrees in the peel off, then leveled off and started a dive straight ahead. Between 2,000 feet and 3,000 feet he leveled off and started a gentle turn to the right. In this turn to the right, the [AT-6D in the number-two position] crashed into the lead airplane from the inside of the turn. [The following airplane] then turned abruptly to the right to avoid pieces of the two airplanes [that had collided]. At this time the left wingman was approximately one-half mile to the right of the formation and flying abreast of the lead airplane.... From the examination of the wreckage, apparently the first point of contact was the right wing of [the AT-6D] on the topside of the right wing of [the AT-6C]. The propeller of [the AT-6D] entered the front cockpit and accessory section of [the AT-6C] almost simultaneously, and the center section of [the AT-6D] crushed the canopy and cockpit section of [the AT-6C]. The fuselage of [the AT-6D] was offset slightly to the left of the fuselage of [the AT-6C], so that more damage was done to the left side of [the AT-6C]. However, the canopy and turnover bracing of [the AT-6C] was completely crushed. A fire followed after the impact, with flames extending back past the tail surfaces, and probably an explosion followed soon afterward, when the airplanes came apart in their vertical fall." Both airplanes smashed to the ground in close proximity to each other and burst into flames.

12-14-44G. Pueblo, Colorado. At 2014, civilian firefighter Edwin Powers was killed when he walked into the rotating number-three propeller of a Consolidated B-24J operating at the Army Air Base at Pueblo, Colorado. The B-24 had taxied to the run-up area at the west end of the East-West Runway and was preparing to take-off on a routine training mission when the pilots noticed smoke in the cockpit. The smoke seemed to be originating from the nose section and smelled like an "electric wire burning." The pilot alerted the control tower that he was postponing his take-off until he could investigate the origin of the smoke. The control tower officer sent fire trucks out to the airplane in case they were needed. The lead fire truck stopped at a point about 60 feet in front of the B-24 and firefighter Powers grabbed a hose and dragged it to the front of the starboard wing area of the airplane. Powers was dragging the hose in a stooped position and remained in this position without looking up, walking right into the rotating propeller. He was killed instantly. A fellow firefighter yelled a warning to firefighter Powers but he evidently did not hear it. The B-24 was piloted by 2Lt. Gerald D. Timberlake and co-pilot 2Lt. Morris Mersky. The pilots and eight crewmembers were uninjured.

12-14-44H. Las Vegas, Nevada. At 1920 PWT, a Boeing B-17G collided with power-line poles and crashed one and one-quarter miles northeast of Las Vegas Army Air Field, Las Vegas, Nevada, killing four fliers and seriously injuring two others. Killed in the crash were: 1Lt. David K. Nauman, pilot; Sgt. Erlgin G. Lorentz, engineer; T Sgt. Charles D. Wright, radio operator; Pfc. Donald L. Schultz, radio operator. B-17 student co-pilots A/C Donald J. Mayhew and A/C Thomas P. Gergurich sustained serious injuries. The airplane took off from Las Vegas Army Air Field on a co-pilot training flight. Because of poor weather conditions in the region, all airplanes were restricted to the local area. The subject airplane made a touch-and-go landing at 1919 and failed to report after that time. Southern Nevada Power Company service personnel discovered the wreck while investigating the downed power-lines and reported the crash to the field at about 2245. Investigators stated, "This aircraft had struck the poles of a 60,000-volt high-tension line, approximately 25 feet above the ground. Two poles were completely sheared off by the right wing, a third pole was split, a fourth was pulled out of line and the [conductors were sheared off]. The aircraft had apparently, thereafter, struck the ground, broke in two at the radio room with the cockpit ending up in a position 180 degrees from the line of flight. The radio compartment and the rear section were approximately 100 feet from the front section. It was evident that the pilot had gained very little altitude, as the terrain at the scene of the accident was approximately 100 feet above field elevation. The wreckage was one and a half miles from the end of, and in line with, the runway from which the aircraft departed." Wreckage was scattered over an area of a quarter mile.

12-15-44A. Lake Pepin, Minnesota. At 1350 CWT, a Consolidated B-24L crashed into frozen Lake Pepin approximately 60 miles southeast of St. Paul, Minnesota, killing three fliers. Pilot Capt. Dan D. Mitchell, co-pilot F/O Buddy B. Beasley and engineer Sgt. Edward A. Demski were killed in the crash. The airplane took off on an instrument clearance from St. Paul on a flight to Billy Mitchell Field, Milwaukee, Wisconsin. There was no further radio contact with the airplane after it took off. Weather was described as "marginal contact" conditions. Investigators speculated that the pilots had lost control of the airplane while flying in instrument conditions. The airplane crashed through the ice and settled to the bottom of the lake.

12-15-44B. Tampa, Florida. At 1550 EWT, a North American P-51K crashed two miles north of MacDill Field, Tampa, Florida, killing pilot F/O Anthony Maula. The airplane was flying in the number-four position of a four-ship flight that took off from Sarasota Army Air Field, Florida, on a camera-gunnery mission. The airplane was maneuvering in the area east of MacDill after successfully completing its gunnery mission. Investigators could not determine the cause of the crash.

12-15-44C. Deming, New Mexico. At 1855 MWT, two Beech AT-11 airplanes collided while taxiing at Deming Army Air Field, Deming, New Mexico, killing student bombardier A/C Lendell L. Woody aboard AT-11 #41-27532. Investigation revealed that AT-11 #41-27532 had taxied to the run-up area for Runway 8R and had halted. The pilot performed his engine run-up while the students secured themselves for take-off. AT-11 #42-37241 then taxied into the rear of AT-11 #41-27532, killing the student bombardier, who had been sitting near the hatch in the after part of the ship. Uninjured aboard AT-11 #41-27532 were: F/O Henry S. Hudspeth, pilot; F/O Clarence R. Wilhoit, bombardier instructor; A/C Frank E. Wineinger, student bombardier; A/C Sidney C. Woolever, student bombardier. Uninjured aboard AT-11 #42-37241 were: 2Lt. Robert W. Huffman, pilot; 2Lt. William C. Keasler, bombardier instructor; A/C Virgil J. Gebhart, student bombardier; A/C Herbert Gelfand, student bombardier; A/C Paul A. Geschwind, student bombardier.

12-15-44D. Hardeeville, South Carolina. At 2120, a Consolidated B-24J crashed three miles east of Hardeeville, South Carolina, killing seven crewmembers. Radio operator Cpl. William W. Davis received minor injuries parachuting to safety. The B-24 took off from Chatham Field, Savannah, Georgia, on a night navigation mission to Fort Myers, Florida, to Tampa, Florida, and return to Chatham Field. The surviving radio operator stated that the airplane climbed to an indicated altitude of 5,000 feet when a gunner alerted the pilot that the number-one engine was on fire. The pilot ordered a crewmember to cut off the fuel to the number-one engine and then he feathered the propeller. The pilot rang the bail out bell and alerted the crew to stand by. The radio operator went to the bomb bay area and stood on the catwalk. The blaze in the number-one engine spread to the entire port wing. The B-24 then climbed abruptly until the survivor felt the airplane stall. The bombardier ordered the radio operator to jump. The radio operator bailed out of the forward starboard bomb bay at very low altitude and pulled the ripcord immediately. The airplane then banked sharply to the left and dove into a wooded area, exploding into flames upon impact. The surviving crewmember hit the ground moments later. Killed in the crash were: 2Lt. Francis W. Johnson, pilot; 2Lt. David F. Bheam, co-pilot; 2Lt. Charles E. Downing, Jr., navigator; F/O Alfred Algeria, bombardier; Cpl. Barney B. Purdon, engineer; Cpl. Phillip S. Lange, Jr., assistant engineer; Pfc. William D. Brooky, gunner.

12-15-44E. Del Rio, Texas. At 2028 CWT, a Martin TB-26C crashed two miles south of Laughlin Field, Del Rio, Texas, killing three fliers. Pilot 2Lt. Francis T. Hagemann, co-pilot 2Lt. Robert C. Besselman and engineer SSgt. Lester L. Lippold were killed in the crash. The airplane made what appeared to be a normal take-off on Runway 13L. Almost immediately after becoming airborne the airplane entered a steep bank to the right and climbed to about 200 feet agl. The airplane held this altitude for about ten seconds, still holding the steep bank to the right. The airplane then began descending in a steep bank to the right, remaining in this attitude until it smashed to the ground and exploded into flames. Investigators speculated that the pilots had lost the horizon while taking off over dark terrain. The take-off direction had been changed; the previous take-off direction provided adequate horizon lighting. Investigation revealed that on the day of the accident Lt. Hagemann had performed six night take-offs and Lt. Besselman had performed ten night take-offs and had also passed an instrument check flight earlier in the day, making one instrument take-off. Both officers were regarded as capable pilots.

12-15-44F. Napier Field, Alabama. At 2049 CWT, a Curtiss RA-25A crashed one-half mile north of Napier Field, Dothan, Alabama, killing pilot Lt. Col. Eugene T. Yarbrough. The airplane took off at 1627 CWT (1727 EWT) from Bolling Field, Washington, D.C., on a flight to Eglin Field, Valparaiso, Florida, via Atlanta, Georgia. The first leg of the flight was an instrument clearance and the final leg was a contact clearance. At 2047, control tower personnel at Napier Field observed the airplane enter the downwind leg for the Southeast Runway. The airplane had not radioed the control tower and did not answer calls from the tower. The airplane turned in the pattern and appeared to be attempting a landing on the sod portion of the field. The control officer gave the airplane the red light because the sod portion of the field was not for use by transient aircraft. The airplane continued on its heading and the engine quit moments later. The airplane then crashed from low altitude. The pilot bailed out at an altitude too low to allow his parachute to deploy and he was killed when he struck the ground. Investigators noted that the pilot should have been able to safely complete the flight with the fuel that was

known to be on board at Bolling Field. It was later speculated that the pilot had either deviated from his original flight path or had gotten lost, causing him to expend his fuel supply. It was also noted that the airplane's rate of fuel consumption had increased as the airplane aged over time.

12-16-44A. Location Unknown. At 0038 EWT, a Consolidated B-24J disappeared and is presumed to have crashed into the Atlantic Ocean about 70 miles northeast of Daytona Beach, Florida, with its crew of nine. Missing and presumed lost at sea were: Capt. Charles K. McClure, pilot; 2Lt. Robert F. Glennon, co-pilot; F/O Robert Adams, navigator; 1Lt. Stuart W. Cowan, bombardier; Cpl. Rubio Edwards, engineer; Cpl. Paul D. Demarest, radio operator; Cpl. Peter S. La Rocca, assistant engineer; Cpl. Jessie J. Vaughn, gunner; Cpl. Gordon C. Lawie, assistant radio operator. The airplane took off at 2238 EWT (12-15-44) from Chatham Field, Savannah, Georgia, on a 1,000-mile night navigation mission. The airplane transmitted a routine position report at 2351 EWT. At 0005 EWT, the airplane transmitted in the clear that it had lost the number-four engine and was turning for shore. The airplane transmitted a position report at 0007, heading 270 degrees at 6,000 feet indicated altitude. At 0012, Chatham Field radio operators requested that the airplane send out a continuous series of dashes and call sign so that ground radio stations could get a fix on the aircraft. Several ground stations began receiving the airplane's transmissions and began to calculate fixes. At 0038 EWT, radio transmissions from the aircraft suddenly ceased. The last position of the aircraft was estimated to be "30:05N — 80:35W." A massive search effort failed to find any trace of the airplane or crew. Investigators noted that the radio operator had failed to activate the IFF equipment.

12-16-44B. George Field, Illinois. At 0135 CWT, a Douglas C-47 on final approach collided with trees and crashed 300 yards from George Field, Lawrenceville, Illinois, killing three fliers. The airplane had taken off with several others from George Field at about 1900 CWT (12-15-44) on a night formation-training mission. At 0115 the George Field control tower called in the airplanes because poor weather was approaching the flying area. Several C-47s had landed and one was about to land when tower personnel noticed the subject airplane in the pattern. The subject airplane had not contacted the control tower and was in the number-two position on final, flying extremely low. The pilot was S-turning in an effort to squeeze his airplane in between two others on long final. The subject airplane's starboard propeller collided with a tree as the pilot attempted a second S-turn to the right on the final. The pilot added power but the landing gear and flaps were in the extended position, diminishing the airplane's performance. The airplane was unable to clear a tall tree about 60 yards further along the flight path, colliding with the tree and shearing off the starboard wing outboard of the engine nacelle. The airplane went out of control and rolled to the right until the port wing tip struck the terrain. The airplane burst into flames upon impact with the ground about one mile east of the east end of Runway 27. Pilot F/O Howard L. Warren, co-pilot 2Lt. Richard Q. Otterbacher and engineer Pfc. Stephen E. Heefner were killed in the crash. It was not known why the pilot entered the traffic pattern without permission and out of turn.

12-16-44C. Orangeburg, South Carolina. At 0900, a Stearman PT-17 crashed seven miles west of Orangeburg, South Carolina, killing French Air Force student pilot Pvt. Henri Soreda. The airplane was seen to enter a spin to the left at an altitude of about 4,000 feet. The airplane spun three times and then began spinning to the right, describing three turns of spin before smashing to the ground. The airplane did not burn but was destroyed utterly. Investigation revealed that the wooden seats installed in later production PT-17 airplanes could not withstand the shock of an otherwise survivable crash. The student was killed when his head struck the airframe and instrument panel.

12-16-44D. Arcadia, Florida. At 1455 EWT, a North American P-51D crashed seven miles east of Arcadia, Florida, killing pilot 2Lt. John A. MacMullen, 20. The airplane was part of a two-ship flight that had taken off from Lakeland Army Air Field, Lakeland, Florida, on a high-altitude aerobatic mission. The subject airplane was flying in the wing position at 34,000 feet indicated altitude. The subject pilot reported that his starboard fuel tank was empty and that he only had 45 gallons of fuel remaining in the port fuel tank. The flight leader began to descend and at about 30,000 feet he noticed that the subject airplane was no longer with him. The subject airplane broke up on the way down, scattering wreckage over a wide area. The pilot had made no apparent effort to parachute to safety. Investigators were unable to determine the cause of the accident.

12-16-44E. Starkville, Mississippi. At 1420, a North American B-25D crashed 14 miles west of Starkville, Mississippi, killing three fliers. The airplane had taken off from Berry Field, Nashville, Tennessee, on a cross-country navigation flight back to its home station at Lafayette Army Air Field, Lafayette, Louisiana. The starboard engine apparently failed during the return flight so the pilot feathered the starboard propeller. The airplane stalled and entered a spin to the right, smashing to the ground and exploding into flames upon impact. Investigation of the wreckage revealed that excessive amounts of oil had been deposited on the starboard tail surfaces prior to the crash. Pilot 1Lt. Melvin L. Vancie, co-pilot 2Lt. Edward P. Herring and engineer SSgt. Elwood E. Tompkins were killed in the crash.

12-16-44F. Glendale, Arizona. At 1425 MWT, a Stearman PT-13D crashed five miles ENE of Glendale, Arizona, killing Chinese Air Force student A/C Si-Che Lu, 23, who fell to his death in an unsuccessful parachute jump. The airplane was flying at 3,000 feet agl when it was seen to go into a half roll and then a split-S type maneuver. The student was unable to recover and the airplane smashed into the ground at high speed. The pilot bailed out at about 300 feet agl, but he was unable to deploy his parachute in time to avoid striking the ground.

12-16-44G. Dallas, Texas. At 1534, U.S. Army paratrooper Sgt. John T. Parrow, Jr. was killed attempting to parachute from a Douglas C-47A operating at Hemsley Field, Dallas, Texas. The airplane was crewed by pilot Capt. William M. Campbell, co-pilot 2Lt. Robert A. Elliot and Sgt. Albert Mazalewski. The flight crew and eight paratroopers were uninjured. No other details are available.

12-16-44H. Abilene, Texas. At 1715, a Republic P-47D crashed at Abilene Army Air Field, Abilene, Texas, killing Capt. Calvin C. Moody, 28. The airplane stalled while attempting a steep turn from base to final. The airplane snapped over to the left but the pilot was able to make a partial recovery. Moments later, the airplane snapped over again to the left and entered a spin at an altitude too low to allow recovery (about 50 feet agl). The airplane smashed to the ground and exploded into flames.

12-17-44A. Madera, California. At 0155, a Douglas P-70B crashed three miles west of Madera, California, killing pilot F/O Paul M. Harper and radar operator 2Lt. Joe B. Currie. The airplane took off at 0052 from Hammer Field, Fresno, California, on a routine night training mission. The airplane rendezvoused with another P-70 over Madera for an interception problem. After intercepting the airplane, the subject airplane turned away and snapped over into a spin to the left. The pilot was unable to make a full recovery. Investigation of the wreckage indicated that the airplane struck the ground in a flat spin to the left. Investigators speculated that the pilot had entered a secondary spin after a partial recovery.

12-17-44B. Fort Worth, Texas. At 0715, two Consolidated B-24E airplanes collided in mid-air and crashed five miles southwest of the Army Air Field at Fort Worth, Texas, killing eight fliers. Killed aboard B-24E #42-7240 were: 2Lt. William H. Magee, B-24 instructor; 2Lt. Douglas M. McAllister, pilot/B-24 student; 2Lt. Walter D. McPherson, co-pilot/B-24 student; Cpl. Fritz M. Waldal, engineer. Killed aboard B-24E #42-6988 were: 2Lt. Vauclain W. Barnes, Sr., B-24 instructor; 2Lt. Millard J. Walline, pilot/B-24 student; 2Lt. Frederick C. Webster, co-pilot/B-24 student; SSgt. Everett A. Gaines, engineer. B-24E #42-6988 took off from Fort Worth Army Air Field at 0700 on an instrument training flight; B-24E #42-7240 took off at 0707, also on an instrument training flight. Investigators stated, "From the location of various parts of both aircraft, which were scattered over an area of approximately two and a half miles, and because of the right wing and elevators of B-24E #42-7240 showing definite propeller cuts, it was determined that B-24E #42-7240 was flying a southerly course at approximately 4,500 feet when struck by B-24E #42-6988, which was flying a southeasterly course.... Exact bearings of each aircraft at the time of collision are unknown. The propellers of [B-24E #42-6988] badly sliced the outer right wing and completely cut off the empennage of [B-24E 42-7240].... B-24E #42-7240 exploded immediately, losing the greater part of its wing and its fuselage from the rear bomb bay section back, including the empennage, and fell to the ground in flames. B-24E #42-6988 stayed partially intact but dropped two of its engines, which caused the aircraft to spiral to the left — possibly under partial control — until it struck the ground. It exploded approximately three-quarters of a mile south of the burning wreckage of B-24E #42-7240. Because of the explosion in the air of B-24E #42-7240 and the explosion of B-24E #42-6988 following its impact with the ground and the resulting intense fires, it was impossible to determine the position of any crewmembers aboard either aircraft.... From the position of the wreckage and the propeller marks on each aircraft, it is assumed that B-24E #42-6988 collided with B-24E #42-7240 in an approximate 80 degree nose to nose attitude approaching toward B-24E #42-7240's right wing."

12-17-44C. Memphis, Tennessee. At 1009 CWT, a Douglas C-49J crashed while attempting to take-off from Memphis Municipal Airport, Memphis, Tennessee, killing three fliers and injuring three others. Investigators stated, "Pilot was taking off on ... Runway 17, and statements show that he was airborne at approximately 2,500 feet down the runway.... [Immediately] after the aircraft became airborne, it started drifting to the right in a stalled attitude with the nose high and tail low. Aircraft struck four parked gasoline trailers, tearing out the tail wheel assembly and causing the aircraft to lose enough speed to make it stall completely. A small brick storehouse containing kerosene was struck by the aircraft, and the impact demolished both the storehouse and the aircraft.... The primary cause of the accident was [starboard] engine failure. Examination of the [starboard] engine showed that several connecting rods were broken and some piston skirts were flared." Investigation revealed that the pilot had never been officially checked off on C-49 type transport aircraft and had only nine total C-49 hours, all logged on the subject airplane. The pilot took possession of the airplane at Romulus, Michigan. Passenger Capt. John K. Harrison and civilian technicians Philip M. Perry and John G. Martin, both of the Ethyl Corporation, were killed in the crashed. Pilot Capt. Buford H. Morris was seriously injured; engineer Cpl. Granville W. Kleinjohn and passenger Sgt. Isidore E. Weiss received minor injuries.

12-17-44D. Waco, Texas. At 1500, two Vultee BT-13A airplanes collided in mid-air and crashed at Waco Army Air Field, Waco, Texas, killing three fliers.

Student A/C Harold W. Moore, Jr. was killed in the crash of BT-13A #41-9928; Instructor 1Lt. Eugene C. Craig and student A/C Andrew J. Keefe were killed in the crash of BT-13A #42-88842. The airplanes were part of a three-ship student flight on a routine formation training flight. A/C Moore was flying in the number-two position and the instructor was flying in the lead position. Investigators stated, "A gentle turn to the left was started as is the practice at this station. A few minutes later, Lt. Craig started a medium climbing turn to the right and Cadet Moore's airplane either over banked or under banked and moved forward out of position. In attempting to correct, Cadet Moore probably over controlled, placing his airplane in a severe skid and skidded over the top of Lt. Craig's airplane. A view of the wreckage indicated that the propeller of Lt. Craig's airplane cut the fuselage of A/C Moore's airplane at approximately the second bulkhead, from slightly below and on the left hand side.... Immediately upon impact, the tail of Cadet Moore's airplane was severed and Cadet Moore's airplane went into a spin, crashed to the ground and burned. Sometime during the descent of Cadet Moore's airplane, he was either thrown out or attempted to jump as his body was found some 45 feet from the wreckage of his airplane. Lt. Craig's airplane went into a spin and crashed some 400 feet from Cadet Moore's airplane, exploding upon contact with the ground and then burned. Both bodies were thrown clear of the wreckage."

12-17-44E. Sebring, Florida. At 2330 EWT, a Boeing B-17F crashed three miles SSW of Hendricks Field, Sebring, Florida, killing three fliers. The two B-17 students had performed several landings and take-offs with two instructors aboard. The instructors deplaned at 2305 and the students took off to complete their session. The airplane made one touch-and-go landing and entered the traffic pattern for another landing. The pilot radioed the tower after turning on the base leg. There was no further radio contact with the subject airplane and it crashed about a minute later. The B-17 had passed close to an airplane flying near the pattern area. The two aircraft blinked landing lights. The landing lights remained illuminated on the subject airplane. The subject airplane entered a descending right turn, remaining in this attitude until it struck a tall pine with the starboard wing. Ten feet of the starboard wing was sheared off. The airplane continued flying for a mile and a half before smashing to the ground and exploding into flames. Pilot 2Lt. John W. Gadjo, co-pilot 2Lt. DeRay Gladman and engineer Pfc. Benjamin L. Hale were killed in the crash.

12-18-44A. Rapid City, South Dakota. At 0935 MWT, a Boeing B-17G made an emergency landing one mile southeast of the Army Air Field at Rapid City, South Dakota. Gunner Cpl. Robert W. Corley was killed in an unsuccessful parachute jump. Three crewmembers parachuted to safety and were uninjured. Six crewmembers were uninjured in the crash landing.

Investigators stated, "Aircraft was climbing to altitude in formation when pilot and crewmembers noticed oil spurting from the number-four engine. Pilot immediately aborted and attempted to feather the [propeller]. [Propeller] could not be feathered and [engine] began to smoke. Pilot and co-pilot went through 'Fire in flight' procedure and also dove the ship in an attempt to put out the flame, but were unsuccessful. He cut ignition switch and generator, and told co-pilot to call in for an emergency landing. Immediately after call, [number-four] engine burst into flames. Pilot then rang the alarm bell, three short rings, one long ring, and left it ringing. He told the co-pilot and engineer to bail out, left his seat and went down in the nose. He opened the [forward] escape hatch. Cpl. Corley, togglier, jumped first, pilot second, navigator third, and engineer last. Engineer, after receiving orders from the pilot to bail out, stayed on interphone a few seconds longer to tell crew to bail out. He thought that everyone was on interphone at the time. When he jumped he thought that he and the co-pilot were the only ones left in the ship. When the four crewmembers bailed out, the engine fell off. Co-pilot took over the controls, called the bombardier to come up and assist him, had assistant engineer to make 'before landing check,' and brought the ship in for a successful landing. Cpl. Corley's jump was unsuccessful. His body was found approximately two miles east of the field.... A complete investigation of the engine revealed that the master rod had collapsed in flight.... When the master rod collapsed, the articulating rods also collapsed. All rods were torn from the piston skirts, and most of the pistons were battered into small chunks. When the number-nine rod broke, it came through the cylinder wall and into the number-two cylinder, crossing from number-one to number-two cylinder. The number-nine rod struck the feathering pump line, causing it to snap at the high-pressure oil inlet.... When the number-nine rod hit the feathering line it resulted in a [diminished] flow of oil. This was the reason for the pilot's inability to feather the propeller." Pilot 2Lt. Lionel G. Haugen, navigator 2Lt. Bernard M. Singer and engineer Cpl. Robert G. Anderson were uninjured parachuting to safety. Crewmembers uninjured in the emergency landing were: F/O Robert L. Gill, co-pilot; 2Lt. Clifford H. Markuson, bombardier; Cpl. James M. Kunz, radio operator; Cpl. Ivison R. McKee, gunner; Cpl. Roscoe L. Pearce, gunner; Cpl. Thomas J. Brown, gunner.

12-18-44B. Santa Maria, California. At 0850, two Vultee BT-13B airplanes collided in mid-air seven and a half miles northwest of Santa Maria Army Air Field, Santa Maria, California, killing 2Lt. Max A. Sommerlot aboard BT-13B 42-90394. 2Lt. Robert L. Moon was uninjured. 2Lt. Roy B. Williams and 2Lt. Wilkie G. Swenson were uninjured aboard BT-13B #42-90631. Investigators stated, "Lt. Moon took off on a scheduled local transition flight at 0830. Pilot made a normal take-off and climbed for altitude. He then saw several other BT-13s in the air and decided to join and fly formation with them. [Lt. Moon] dove his ship down to fly formation with BT-13B #42-90361, piloted by Lt. Williams. Lt. Williams was unaware that another BT-13 was diving down on his right to fly formation with him and [Lt. Williams] made a turn to the right. Lt. Moon's airplane then tried to avoid hitting Lt. Williams' airplane by diving underneath but was unable to avoid the collision. Lt. Moon saw an object coming towards him and he bent his head and crouched down in the cockpit and stated that he felt a sudden jar and shock of the other airplane striking his. Lt. Moon stated that his ship was uncontrollable and the only possible thing to do was make a crash-landing on the beach. Lt. Williams' airplane, which was then vibrating and pounding, was flown back to the field and crash landed due to the fact that one wheel had been partially knocked off in the mid-air collision."

12-18-44C. Marana, Arizona. At 1105 MWT, a North American AT-6C and an AT-6D collided in mid-air and crashed at Rillito Auxiliary Airfield ten miles SSE of the Army Air Field at Marana, Arizona, killing both students. A/C William A. Zvolanek was killed in the crash of the AT-6C; student 1Lt. Thomas N. Gerhart was killed in the crash of the AT-6D. Investigators stated, "1Lt. T.N. Gerhart, Student Officer, was shooting a stage at Rillito Auxiliary Airfield. A/C Zvolanek had been left at the main airdrome to transfer an airplane to Rillito Auxiliary Airfield. Witnesses stated that he entered the traffic pattern at Rillito approximately 150 feet lower than the other airplanes in traffic pattern and just ahead of Student Officer 1Lt. Gerhart. A/C Zvolanek placed his base leg further away from the field than did 1Lt. Gerhart and was on his final approach when Lt. Gerhart made his final turn into the field. Cadet Zvolanek was below [Lt. Gerhart] and when [Lt. Gerhart] made his turn onto final approach he turned down and into the tail of Cadet Zvolanek's airplane. The tail assembly [of A/C Zvolanek's airplane] was completely severed just behind the rear cockpit and both airplanes immediately crashed."

12-18-44D. Charleston, Kansas. At 1430 MWT, a Martin B-26F crashed four and a half miles north of Charleston, Kansas, killing the crew of three. The airplane took off from Dodge City Army Air Field, Dodge City, Kansas, on a routine training mission. Several civilian witnesses on the ground observed the airplane in the minutes before the crash, but their statements were contradictory. There were no witnesses to the actual crash. Investigation of the wreckage indicated that the airplane struck the ground in an inverted dive. The airplane exploded into flames upon impact. Investigation revealed that both propellers were turning under power at the time of impact. Pilot 1Lt. George D. Grubb, Jr., co-pilot 1Lt. Robert R. Hribel and engineer Pfc. George W. Birdsong were killed in the crash.

12-18-44E. Elkhart, Texas. At 1605, two Beech AT-7 airplanes collided in mid-air and crashed two miles north of Elkhart, Texas, killing eight crewmembers and seriously injuring student navigator/rated bombardier 2Lt. James E. Nall, who parachuted from AT-7 #41-1174. AT-7 #41-1174 was part of a 15-ship flight that took off at 1415 CWT from Hondo Army Air Field, Hondo, Texas, on a day celestial navigation mission to Jacksonville, Texas. The flight was cleared to fly at 7,000 feet to Jacksonville and at 6,000 feet for the return flight to Hondo. AT-7 #43-50028 was part of a 15-ship flight that took off at 1345 CWT from San Marcos Army Air Field, San Marcos, Texas, on a familiarization flight for new student navigators at the station. The flight was cleared to fly at 7,000 feet to Lufkin, Texas, and then cleared to fly at 8,000 feet to Cleburne, Texas, and return to San Marcos. The two formations were to cross paths about four miles south of Palestine, Texas — on the first leg of the Hondo flight and the second leg of the San Marcos flight — if the flights went as briefed. The flights actually crossed paths about eight miles south of Palestine, resulting in the collision at 6,700 feet. AT-7 #41-1174 caught fire at the time of the collision and went out of control. The airplane "oscillated" out of control and smashed to the ground, bursting into flames upon impact. AT-7 #43-50028 went into a spin and then entered a vertical dive toward the ground. The airplane remained in this attitude until it struck the ground and exploded. Killed in the crash of AT-7 #43-50028 were: Maj. Richard L. Strickland, pilot; 2Lt. William H. Bither, bombardier instructor; A/C Russell E. Ricker, student bombardier; A/C Edmund L. Roth, student bombardier; A/C Morris E. Rivers, student bombardier. Killed in the crash of AT-7 #41-1174 were: F/O Rollins C. Wellman, pilot; 2Lt. John Newmyer, Jr., student navigator and rated bombardier; 2Lt. Miles B. Taber, student navigator and rated bombardier. Lt. Taber had attempted to parachute to safety but was unsuccessful. Lt. Nall's clothing was in flames as he descended in his parachute into a farmer's back yard. The farmer came on the scene immediately and extinguished his flaming clothing.

12-19-44A. Amarillo, Texas. At 0912, a Douglas A-26B crashed onto railroad tracks near Amarillo Army Air Field, Amarillo, Texas, killing pilot 1Lt. Carl E. Samuelson, 24. Pilot-rated passenger 2Lt. Edward E. Dixson and engineer Cpl. Patrick H. Moriarty received minor injuries. Harry Jasper, a brakeman for the Santa Fe Railroad, received minor injuries when his freight locomotive and several railroad cars were derailed. The airplane took off at 0906 on Runway 21. Immediately after take-off the starboard engine began backfiring and losing power. The pilot feathered the propeller and trimmed the airplane for single-engine flight. The pilot began a shallow turn to the left in an attempt to get back to the field for a landing on Runway 31. As the airplane circled the field, the pilot lowered the landing gear and at that instant an explosion

occurred and a fire erupted in the starboard engine nacelle. Pieces of the starboard engine nacelle began to fall away. The pilot retained control and attempted a landing on Runway 31 with excessive speed, estimated to be about 150 mph. The airplane touched down on the runway and bounced several times because of the excessive speed. The brakes failed and the airplane rolled off the end of the runway, through the airfield fence, across a highway and then smashed down a telegraph pole before plowing through railroad tracks and flipping over onto its back. Two crewmembers scrambled to safety but were unable to free the pilot from the wreckage. A Santa Fe Railroad freight train was underway about 500 yards away from the airplane's path at the time of the crash. The engineer slowed the locomotive down as much as he could before it passed over the damaged track and derailed. The locomotive went off of the track and came to a sudden halt in an upright position. Eleven freight cars derailed and piled up behind the locomotive and tender. The A-26 wreckage came to rest about 150 feet away from the derailed train. Investigation revealed that a fuel line leak behind the starboard engine allowed gasoline and fumes to collect in the nacelle.

12-19-44B. Van Nuys, California. At 1407, a Lockheed P-38J crashed south of Van Nuys, California, killing pilot 2Lt. Douglas V. Jolley, 21. The airplane was flying in the number-three position of a four-ship flight that took off at 1307 from Metropolitan Airport, Van Nuys, California, on a low-altitude navigation mission. The subject airplane gained about 200 feet of altitude before entering a descending turn to the left. Puffs of black smoke were seen emitting from the port engine area. The airplane remained in this attitude until it smashed into the ground and exploded into flames. Investigators noted that the subject pilot had to switch airplanes at the last minute because of a problem with the initial airplane. The pilot was the last to take-off. Investigation revealed that the port engine "left magneto plugs were not firing."

12-20-44A. Waco, Texas. At 1530, two Beech AT-10 airplanes collided in mid-air and crashed three miles south of Blackland Army Air Field, Waco, Texas, killing three fliers and seriously injuring A/C Dan M. Paradies aboard AT-10 #41-27159. A/C Vernon D. Pearson was killed. Student A/C Arthur H. Richter and A/S John A. Scholl were killed in the crash of AT-10 #41-27180. Both airplanes took off from Blackland Army Air Field on separate instrument training flights. Investigation revealed that AT-10 #41-27180 approached AT-10 #41-27159 from the front right quarter. Both pilots put their airplanes in a diving maneuver to avoid collision. A/C Paradies stated that he attempted to reverse the maneuver but before he could act the collision occurred. He was thrown clear and was able to use his parachute. The other three pilots were killed in the collision. Both airplanes broke up into numerous pieces as they fell to earth, the main sections bursting into flames upon impact.

12-20-44B. Madison, Wisconsin. At 1831 CWT, aerial engineer Pfc. Joseph C. La Mantia was killed when he walked into the rotating number-one propeller of a Consolidated B-24M operating at Truax Field, Madison, Wisconsin. The airplane was piloted by Capt. James S. Soloman and co-pilot 1Lt. Ray C. Newson, who were uninjured. The airplane had landed at the field following a routine flight and had taxied back to the parking area. The engineer exited the airplane in an effort to watch for wingtip clearance. The engineer exited through the bomb bay and walked behind the port landing gear. He then stooped over and began to walk forward from behind the number-one engine, walking into the propeller from behind.

12-20-44C. Granby, Massachusetts. At 2012, a Consolidated B-24J crashed two miles south of Granby, Massachusetts, killing three fliers and injuring seven, three of them seriously. Navigator 2Lt. George E. Bennett, bombardier 2Lt. Julian Berger and gunner Cpl. Stanley Saffer were killed in unsuccessful parachute jumps. Crewmembers injured parachuting to safety were: Cpl. Kenneth A. McPhail, engineer; Cpl. Robert H. Risdon, radio operator; Pfc. Fred T. Lopez, gunner; Pfc. Walter E. Oparowski, gunner. Pilot 2Lt. James E. Webster, co-pilot F/O George H. Slack and gunner were seriously injured in a crash landing. The airplane took off from Mitchel Field, Long Island, New York, on a cross-country flight to Westover Field, Chicopee, Massachusetts. Upon arriving at Westover Field, the pilot overshot the runway because of poor visibility. The pilot aborted the landing, applied power, retracted the landing gear and climbed to go around. The airplane was climbing to 2,000 feet when the co-pilot alerted the pilot that the number-three and four engines were "running away." The pilot was unable to reduce the rpm and a fire was reported in the number-three engine moments later. The pilot feathered the number-three engine. The number-one engine began losing power, so the pilot feathered the number-one propeller. The pilot began to turn for the field when he lost power on the remaining two engines. The pilot ordered the crew to bail out and several men left the ship. The pilot realized that he could not make the field so he aimed the airplane at a clear area. Investigators stated, "The airplane skidded along the snow for about 100 yards and then hit a tree with the right wing between the number-three engine and the fuselage. This sheared the wing off completely. The airplane continued farther on and smashed into the side of a farmhouse. Number-two engine continued to run minus its propeller. The pilot could not reach the ignition switch because of the way the nose of the airplane was smashed in, but he finally was able to shut the engine off by bending the mixture controls so that they would pass under the smashed instrument panel. It is remarkable that the airplane did not catch fire for the number-one engine had been torn off from its mount and there was a large quantity of dripping gasoline in the vicinity."

Investigators determined that a bad inverter caused the overspeeding engines and apparent instrument failure. Investigators speculated that the engines were torching excessively at the exhaust ring, causing the flight crew to think that a fire was present.

12-20-44D. Ironton, Missouri. At 2220 CWT, a Douglas C-47A collided with rising terrain and crashed six miles south of Ironton, Missouri, killing five fliers. Killed in the crash were: 2Lt. James E. Gibson, pilot; 2Lt. Edgar Lougee, Jr., co-pilot; 2Lt. Vincent R. Olgos, pilot-rated passenger; 2Lt. Julian Kaufman, pilot-rated passenger; 2Lt. Harold Silverman, pilot-rated passenger. The airplane took off from Malden Army Air Field, Malden, Missouri, on a low-altitude (500 feet agl) cross-country flight to Marion, Illinois, to Farmington, Missouri, and return to Malden. The airplane was observed circling the area of Ironton for several minutes before the wing collided with a ridge at an elevation of 1,400 feet msl. The airplane went out of control and smashed into the ground, bursting into flames upon impact. Investigators speculated that the pilots were lost and were circling the area in an effort to identify a landmark on the ground and determine their position when the accident occurred.

12-21-44A. Foster Field, Texas. At 1425 CWT, a North American AT-6C being taxied at Foster Field, Victoria, Texas, struck and killed civilian electrician Louis T. Metcalf. Airplane mechanic Pvt. Roger L. Buchanan was taxiing the AT-6C near fuel pit #10, which was inoperative, when it struck the electrician, who was the field electric foreman. The starboard machine gun struck the electrician in the neck and severed two main arteries. He died moments after the accident.

12-21-44B. Cuervo, New Mexico. At 1345 MWT, two Curtiss P-40N airplanes collided in midair three miles southeast of Cuervo, New Mexico, killing 2Lt. Michael T. Murray, 21, aboard P-40N #44-7284. Pilot 2Lt. William W. Wyper received minor injuries crash landing P-40N #43-24456. Investigators stated, ""Capt. Arthur W. Hill led a flight of six ships on a low-altitude formation mission. About two or three miles southeast of Cuervo, Capt. Hill led his flight in a gentle climbing turn to the left. Lt. Wyper, leading the second element, flying on Capt Hill's left, started a cross under. Lt. Murray, flying Lt. Wyper's left wing also started a cross under. Lt. Murray overshot on this maneuver and on the recovery crashed into Lt. Wyper's airplane. Lt. Murray's airplane was damaged so badly that it spun in immediately and exploded and burned on impact with the ground. Lt. Wyper's airplane also sustained damage to the propeller and right wing but he was able to crash-land it approximately one mile north of Lt. Murray's crash."

12-21-44C. Palmdale, California. At 1347, a Beech AT-11 and a Vultee BT-13A collided in mid-air and crashed ten miles west of the Army Air Field at

Palmdale, California, killing four fliers. Pilot Capt. Delbert L. Phifer and co-pilot 2Lt. Joseph A. Harris, Jr. were killed in the crash of the AT-11; Pilot 1Lt. Arthur R.L. Heise and 1Lt. William G. Housman were killed in the crash of the BT-13A. The AT-11 took off from Victorville Army Air Field, Victorville, California, on a routine instrument training flight. The BT-13A took off at 1300 from Minter Field, Bakersfield, California, on a routine cross-country flight back to its home station at Hemet, California. Investigators stated, "At 1347, these two airplanes collided approximately ten miles from the cone of the Palmdale Range on the southeast leg. Immediately afterwards, the AT-11 exploded and both aircraft fell to the ground, scattering wreckage over an area a half mile square. The BT-13 remained fairly intact, but evidently upon impact a gas tank ruptured on the AT-11, causing aforementioned explosion with the result that aircraft separated into about five major pieces — tail section, center fuselage, nose section and the two wings. Examination of the parts disclosed that the right wing and engine of the AT-11 sustained initial impact with the BT-13A. A deep gouge in the right wing, three feet outboard of the engine, contained fabric from the rudder of the BT-13A. From the condition of the cockpit of the BT-13A and its occupants, it is believed that the right propeller of the AT-11 caused all the damage. The probable method of impact was practically head-on, but as no witnesses were available, this is only a conclusion drawn from the examination of the wreckage."

12-22-44A. Union Lake, New Jersey. At 1439 EWT, a Republic P-47D crashed into Union Lake two miles north of the Army Air Field at Millville, New Jersey, killing F/O Ceylon R. Morrison, 23. The airplane was in the number-two position of a three-ship flight that was taking off on a routine training flight. The flight's take-off was delayed because of several airplanes that were landing. The flight leader took off and the two wing ships followed. The flight leader retracted the landing gear and began a shallow turn to the left to leave the traffic area. The subject airplane became airborne and the pilot retracted the landing gear. At about 300 feet agl, a small trail of black smoke was seen trailing the subject airplane, which then made a shallow turn to the right. A stream of thick white smoke trailed the engine as the airplane began descending under control. The starboard wing contacted the surface of Union Lake, causing the airplane to cartwheel into the lake. The airplane came to an immediate halt in about four feet of water. A pilot who had flown the airplane on 12-16-44 reported that the engine was "cutting out." The spark plugs were changed and some push rod seals were replaced on that day and that appeared to clear up the problem for the next twelve flights. A temperature gauge and two spark plugs were changed on 12-17-44. The airplane passed an inspection on 12-20-44. Investigators stated that the accident was caused by an "Engine failure from an unknown cause."

12-22-44B. McAlester, Oklahoma. At 1400 CWT, a Vultee BT-13B on an unauthorized flight crashed two miles southeast of the municipal airport at McAlester, Oklahoma, killing pilot 2Lt. Ivie Gus Prickett, 27, McAlester, and seriously injuring civilian passenger James W. Davis. Investigation revealed that pilot 2Lt. Max L. Telhaird took off at 0930 CWT from Laughlin Field, Del Rio, Texas, on a navigation mission to Tarrant Field, Fort Worth, Texas, and return. Pvt. Jack F. Brown was listed as passenger. Lt. Telhaird stated that he could not raise Fort Worth on the radio so he diverted to Perrin Field, Sherman, Texas, where he landed at 1136 CWT. Lt. Telhaird filed an arrival form and filed a direct clearance to McAlester, Oklahoma, stating on the form that the flight was "Official Business." The airplane was refueled and Lt. Telhaird and Pvt. Brown took off. At 1310 the airplane landed at McAlester, Pvt. Brown's hometown. Pvt. Brown telephoned his mother and she picked up the two fliers at the field and drove them into McAlester. At McAlester, Pvt. Brown's mother exited the vehicle and turned it over to Pvt. Brown. Pvt. Brown and Lt. Telhaird then drove to the local prison, where they had lunch with Pvt. Brown's former co-workers. A short time after Lt. Telhaird and Pvt. Brown left for McAlester, Lt. Prickett and Mr. Davis arrived at the airfield. Lt. Prickett, who was a B-17 instructor officially on leave from Hobbs Army Air Field, New Mexico, had initially arranged to fly a civilian-owned Taylorcraft airplane located at the airfield. Lt. Prickett wanted the Taylorcraft owner to fly with him after his flight with Mr. Davis. The Taylorcraft owner told Lt. Prickett that he would be delayed because he had a chore to do and gave Lt. Prickett permission to fly the Taylorcraft at that time. The Taylorcraft owner stated that when he returned to the hangar area ten minutes later he observed Lt. Prickett and Mr. Davis take off in the BT-13B that was flown in by Lt. Telhaird. The airplane flew around the field at very low altitude and made four touch-and-go landings, entering a steep climb after each. After the fourth touch-and-go the airplane entered a steep climb; ground observers stated that they feared that the pilot would stall the airplane every time he climbed away. The airplane climbed to about 300 feet agl and then nosed down slightly. The airplane then pulled up into a chandelle type maneuver. The airplane stalled and fell off to the left and smashed to the ground, killing the pilot instantly. Mr. Davis was rescued immediately. Investigation revealed that Pvt. Brown's mother telephoned Pvt. Brown at the prison and informed him of the crash.

12-22-44C. La Crescenta, California. At an unknown time between 1611 and 1617, a Consolidated B-24J broke up in mid-air and crashed onto Mt. Gleason eight and one-half miles NNE of La Crescenta, California, killing the crew of ten. The airplane took off at 0802 PWT from Muroc Army Air Field, Muroc, California, on an over-water instrument flight to Alijos

Rocks and return. The B-24 was assigned to fly at 12,000 feet indicated altitude. The airplane made eight position reports during the flight. At 1603 PWT, the airplane radioed Muroc ground station and reported that it was experiencing atmospheric interference; at 1611, the crew radioed Muroc ground station and reported that they were lost and were requesting a vector. Muroc ground station told the crew to stand by; at 1618 PWT Muroc ground station radioed the subject airplane but received no reply. The airplane did not answer repeated radio calls and failed to return to Muroc Army Air Field. The airplane was found the next day. Investigators stated, "An examination of the wreckage at the scene of the crash indicated that the aircraft had disintegrated in the air. The wreckage was strewn over an area of 3,000 feet by 2,500 feet on both sides of a ridge. The ridge on which the wreckage was distributed was lower that the surrounding mountains, but the wreckage showed no indication of contact with the ground before disintegration." Killed in the crash were: 2Lt. Barrett D. Corneille, pilot; 2Lt. Robert H. Asplund, co-pilot; F/O Donald M. Pipkin, navigator; F/O Alan P. Dondero, bombardier; Cpl. Eugene L. Ferrin, engineer; Cpl. Elliot R. French, radio operator; Cpl. Peter P. Bobovich, assistant engineer; Cpl. Rueben F. Welk, gunner; Cpl. Jerry K. Stillinger, gunner; Cpl. Robert E. Jacoby, gunner.

12-23-44A. Bailey, Mississippi. At 0129 CWT, a Martin B-26B flying in instrument conditions crashed two miles northwest of Bailey, Mississippi, killing six crewmembers. The B-26 took off at 2010 (12-22-44) from Barksdale Field, Shreveport, Louisiana, on a navigation flight to Memphis, Tennessee, to Little Rock, Arkansas, and return. The airplane had made several routine position reports to ground stations and then was not heard from again until 0058 CWT, when the radio operator was heard by CAA ground stations attempting to contact Key Field, Meridian, Mississippi, on VHF. Meridian tower did not answer so the CAA ground station established contact with the subject airplane. The pilot was lost and reported that he had about 30 minutes of fuel left. The CAA radio operator was able to contact Key Field, which had been closed down, and alerted them to the B-26 coming in. Key Field personnel were able to get a fix on the airplane and guide it to the Key Field radio range. The airplane was given landing instructions and permission for an instrument let-down. The pilot was able to let down to an area north of Key Field. Key Field personnel were able to hear the subject airplane's motors and directed the pilot to fly a heading of 220 degrees for the field. The message was acknowledged by the pilot, who reported that he had less than ten minutes of fuel left. There was no further communication with the airplane and it crashed ten miles north of Key Field a short time later. Investigation revealed that the airplane descended into trees while under control with no power. Weather conditions at the scene were reported

as ceiling 400 feet with one mile visibility in light fog. There was no fire but the airplane was destroyed utterly. Killed in the crash were: F/O Arthur L. Haugland, pilot; F/O Howard B. Beatty, co-pilot; F/O Harold H. Brush, bombardier/navigator; Cpl. Stewart E. Hazelgrove, engineer; Cpl. John R. Riesen, radio operator; Cpl. Lynn T. Nelson, gunner.

12-23-44B. Napier Field, Alabama. At 1051 CWT, a North American AT-6D crashed three-quarters of a mile southwest of Napier Field, Dothan, Alabama, killing pilots F/O James W. Mitchell and 2Lt. Marvin H. Rowland. The pilots were scheduled to fly an instrument team ride with F/O Mitchell as the instrument pilot and Lt. Rowland as the safety pilot. Investigators stated, "[Immediately] after take-off, F/O Mitchell ... held the airplane at an altitude of 50 to 75 feet above the ground until an abnormal speed had been reached. The aircraft was next observed as it made a nearly vertical turn to the left at an altitude of approximately 500 feet. At this point, the aircraft made a violent bank to the right, was then leveled, and then started a very steep and violent turn to the right. While still in a climbing turn to the right, the aircraft made a violent turn to the left and then to the right to an inverted position. From this position, the ship dove into the ground. It is the conclusion of the board that the operator of the aircraft attempted a very steep turn with insufficient airspeed, which resulted in the aircraft becoming uncontrollable, thereby resulting in an inverted position with insufficient altitude to recover from the ensuing dive." The airplane exploded violently into flames upon impact with the ground.

12-23-44C. Clyde, Ohio. At 2230 EWT, a Vultee BT-13A flying in poor weather crashed six miles SSE of Clyde, Ohio, killing pilots 2Lt. Francis J. Foley and 2Lt. Ronald J. Jansonis. Investigators stated, "Aircraft was cleared IFR direct from Chanute Field, Rantoul, Illinois, to Cleveland, Ohio. Instrument conditions existed over the route and icing conditions were reported.... [Ground witnesses] saw the aircraft come out of the clouds and make two circles over an area of approximately four miles square. On the second circle, which was much lower than the first, the aircraft suddenly dived straight into the ground. No lights, other than one green and one red light on the wings, were seen burning. The engine seemed to be running at full throttle when it crashed."

12-23-44D. Ridgely, Tennessee. At 2215 CWT, a Beech AT-11 flying in instrument conditions crashed five miles east of Ridgely, Tennessee, killing five fliers. The airplane took off from the Army Air Field at Greenwood, Mississippi, on a navigation training flight to Blytheville, Arkansas. The airplane had taken off from Boca Raton, Florida, earlier in the day on the return flight to its home station at Freeman Army Air Field, Seymour, Indiana. The airplane was under control and operating with power when it descended into a heavily wooded area about 37 miles northeast of

Blytheville. Weather at the scene of the crash was described as ceiling 100 feet with one-quarter mile visibility in light drizzle and heavy fog. Killed in the crash were: 2Lt. Robert L. Rogers, pilot; 2Lt. Elton V. Thompson, co-pilot; Cpl. John E. Daffis, engineer; Pvt. James E. Delaney, student engineer; Pvt. Raymond G. Wirick, passenger.

12-24-44. Sardinia, Ohio. At 1804 CWT, a Douglas A-26B flying in poor weather crashed five miles south of Sardinia, Ohio, killing three fliers. Passenger TSgt. Ralph W. Heckler was seriously injured parachuting to safety. The airplane took off from Marianna, Florida, on a flight to Cleveland, Ohio, via Lunken Field, Cincinnati, Ohio. The airplane encountered instrument conditions when it arrived in the area and the pilot flew into the ground while attempting to get underneath the clouds. Heavy ground fog was prevalent at the time of the crash. The airplane was declared overdue after 1729 CWT and the crash site was not found until 2230. Pilot Capt. Benjamin F. Schoenfeld, bombardier-navigator 1Lt. James M. Pollock and passenger Pfc. Joseph P. Jankovic were killed in the crash. TSgt. Heckler bailed out at about 4,500 feet but he was seriously injured when he was struck in the legs by the horizontal stabilizer.

12-25-44A. Harrisburg, Pennsylvania. At 0427, a Douglas C-47A crashed three miles southwest of the municipal airport at Harrisburg, Pennsylvania, killing four crewmembers and eight passengers. Fifteen passengers were seriously injured. The airplane took off at 1440 EWT (12-24-44) from Jacksonville, Florida, on a flight to Stout Field, Indianapolis, Indiana. When the airplane arrived at Stout Field, the pilot discovered that the field was closed. The airplane diverted to Patterson Field, Ohio, and landed there at 2059 EWT. The airplane was serviced and took off for Minneapolis, Minnesota, at 2221 EWT (12-24-44). The airplane was cleared to fly at 12,000 feet from Dayton, Ohio, to McCool, Indiana, and then over Chicago, Illinois, to the Madison, Wisconsin, range and then to Minneapolis. The airplane was east of Dayton when the pilot radioed Cincinnati Flight Control and reported that he was picking up 54 mph headwinds and was requesting a different altitude. Control personnel advised the pilot to return to Patterson. The pilot flew on the original course for a short time and then began to pick up ice. The pilot elected to turn back for Patterson Field at 2318, reporting that he was suffering some radio problems. The airplane returned to Patterson Field but the weather had deteriorated somewhat since departure. The pilot overshot the field on his first landing attempt. The ceiling began lowering and the pilot was given an alternate of Harrisburg, Pennsylvania, which was reporting a ceiling of 2,500 feet and good visibility. The airplane arrived in the area of Harrisburg amid deteriorating weather conditions and was cleared for a straight-in approach on Runway 12. The airplane made five unsuccessful landing attempts and

was given an alternate. The pilot replied in the negative, stating that he was low on fuel and had a good amount of ice on the airframe. The airplane made four more landing attempts and was unsuccessful. Control tower personnel had the airplane in sight and attempted to guide the airplane in, but the pilot disregarded instructions for a base leg turn that would have lined him up perfectly with Runway 12. The ceiling gradually lowered until the point where tower personnel would have the airplane in sight only intermittently. Ceiling was 700 feet agl and freezing rain had moved into the area. The airplane was observed to make its last approach, coming in at 600 feet agl with excessive speed. The airplane was seen to make a shallow turn to the right before disappearing into the weather. The tower advised the airplane to turn left and head to Middleton, Pennsylvania. The airplane did not reply although tower personnel could still hear the airplane. Moments later the airplane was heard to crash. The airplane had collided with 900-foot Reeser's Summit at an elevation of approximately 700 feet msl. The airplane broke up and burst into flames upon impact, smashing to pieces as it careened across the terrain. Killed in the crash were: 2Lt. Eugene Hoskins, pilot; 2Lt. James J. Ryan, co-pilot; Sgt. Martin J. Oswald, engineer; Cpl. George Rosenblatt, radio operator. Passengers killed in the crash were: Pvt. Moneer Assam, Pvt. Philbert O. Kjelden, Pvt. Donald A. Krogh, Pvt. John P. Morris, Pvt. Ernest M. Monger, Pvt. Helmuth Mudder, Pvt. Howard H. Underwood, Pvt. Kenneth A. Wallenstein. Passengers seriously injured in the crash were: Pvt. John W. Audius, Pvt. Lou I. Binder, Pvt. Harold J. Kleinschmit, Pvt. Eli C. Larson, Pvt. Alva J. Meyers, Pvt. Henry Miller, Pvt. Pryle W. Stahl, Pvt. Vernus T. Thompson, Pvt. Marion M. Tracy, Pvt. Allan C. Tufteland, Pvt. Clarence J. Wang, Pvt. Arthur J. Wiedmann, Pvt. Charles R. Zard.

12-25-44B. Salome, Arizona. At 1155, a Lockheed P-38L flying in instrument conditions collided with rising terrain and crashed in a remote area 20 miles SSE of Salome, Arizona, killing 1Lt. John J. Donlin. The airplane had disappeared after taking off at 0912 PWT from Long Beach, California, on a flight to Coolidge, Arizona. The pilot checked in at Palm Springs, California, and was never heard from again. The airplane was found on 9-8-45 by aircraft from Luke Field, Arizona, searching for another lost airplane. The airplane crashed into a mountaintop at a steep angle while flying in a westerly direction 53 miles SSW of Luke Field. The airplane burst into flames upon impact. Investigators noted that thick clouds obscured the mountains at the time of the accident and the airplane was off course.

12-25-44C. Quartzite, Arizona. At 1215 MWT, a Douglas C-47B flying in instrument conditions crashed into Black Mesa Mountain 13 miles southeast of Quartzite, Arizona, killing seventeen passengers and crew. The airplane took off from Tucson Municipal Airport, Tucson, Arizona, at 1048 MWT on a flight to Palm Springs, California. The flight originated at El Paso, Texas, and the pilot was scheduled to transmit position reports at Phoenix, Arizona, and Blythe, California, en route. The airplane apparently entered instrument conditions while flying over mountainous terrain. The airplane collided with the mountain about 15 feet from the peak. The airplane broke up and burst into flames upon impact. The airplane was not found until 12-28-44. Killed in the crash were: Capt. Ben H. Gibson, pilot; 2Lt. George S. Winzler, co-pilot; Sgt. Jack M. Maier, engineer; Pvt. Frank T. Byrne, radio operator. Passengers killed in the crash were: A/S Donald L. Brakebill, Pvt. Bruce J. Carden, Pfc. George R. Francis, Pfc. Dudley A. Hutton, Pfc. George A. Jones, F/O A.C. Middleton, Chief Pharmacist Mate E.C. Milliron (USN), Seaman 2C Edgell C. Powell (USN), Pfc. Haig S. Senain, Pvt. Leroy E. Thomas, Sgt. Clyde C. Wikadel, Pvt. James L. Woodley, Cpl. William V. Yates.

12-26-44. Cincinnati, Ohio. At 1207 CWT, a North American F-10 (the photo-reconnaissance version of the B-25D) crashed while attempting a take-off at Lunken Field, Cincinnati, Ohio, killing three fliers and injuring six others. The airplane began its take-off run on Runway 6. About 1,000 feet from the point of take-off the propellers began striking the runway. The airplane was suddenly pulled into the air about 100 feet from the end of the runway. The starboard landing gear appeared to be half way up and the port and nose landing gear already retracted. The airplane gained minimum altitude and began a shallow turn to the right. The airplane remained in this attitude and after traveling about a quarter mile it crashed into the ground in a right-wing low position. The airplane careened over the terrain and smashed into a dike about 200 feet from the point of impact, breaking up and bursting into flames. Three airmen were discovered alive in the tail section, which flames had not yet reached. The surviving crewmembers attempted to rescue the three airmen trapped in the after part of the ship but they were unable to free them before flames consumed the wreckage. Investigators noted that snow and icy terrain prevented fire fighting equipment from reaching the wreck sooner. Crewmembers injured in the accident were: Capt. Thomas F. Lambie, pilot; F/O Samuel H. Macy, co-pilot; F/O Albert H. Hollenbeck, navigator; Sgt. Paul M. Cline, engineer; Sgt. Lloyd I. Rush, engineer; 2Lt. Eugene C. Stano, pilot-rated passenger. Passengers 2Lt. Theodore G. Martin, TSgt. Edward P. Singer and Sgt. Louis E. Heywood were killed in the crash.

12-27-44A. Langley Field, Virginia. At 0639 EWT, a Consolidated B-24H crashed three miles southwest of Langley Field, Hampton, Virginia, killing six fliers. The B-24 took off from Langley Field at 0630 EWT on a high-altitude H2X radar-training mission. At 0631 the pilot contacted the tower and alerted personnel that he was returning to the field because he had

an engine "out." Control tower personnel gave the pilot landing instructions and attempted to visually observe the airplane without success. At 0636 the pilot radioed the tower that he was on the downwind leg for Runway 7, flying a right-hand traffic pattern, away from the bad engine. The pilot then alerted the tower that he had another engine "going out." The field was cleared and the pilot was requested to transmit a check call when on the base leg and on final approach. Tower personnel were unable to see the airplane nearly the whole time that it maneuvered in the traffic pattern. Then the pilot radioed the tower and said that he could not see the field. Tower personnel blinked the field lights and then observed the aircraft lights approaching the field. Moments later an explosion was observed. Investigation revealed that the pilot was unable to hold altitude and the airplane smashed into the ground. It was speculated that the pilot realized that he was not going to make the field and in an apparent effort to orient himself and make an emergency landing had illuminated the landing lights just prior to crashing. Investigators speculated that the pilot inadvertently flew the airplane into the ground. Wreckage was scattered over an area of 300 yards. Weather was not considered a factor in the accident. Investigators noted that the first pilot had been shot down over occupied Europe and had "walked out." Killed in the crash were: 1Lt. French M. Russell, pilot; F/O Frederick T. Libertino, co-pilot; 1Lt. John J. Pfieffer, navigator instructor; 2Lt. James E. Fletcher, student navigator; TSgt. John P. Martin, radio operator; TSgt. Donald W. Card, engineer.

12-27-44B. Wilmington, North Carolina. At 1420, a Republic P-47G crashed 13 miles northeast of Bluethenthal Field, Wilmington, North Carolina, killing pilot 2Lt. Harry L. Colwell, Jr., 20. The subject airplane was part of a 12-ship flight that had taken off from Bluethenthal Field on a routine formation mission. The flight assembled at 2,000 feet, and after about 15 minutes of flying the pilot alerted the flight leader that his engine temperature was reading high. The flight leader ordered the subject pilot to return to the field. The subject airplane peeled off of the formation and his wingman followed. The subject pilot advised the wingman not to fly on his wing. The wingman dropped back and throttled back, calling the tower and alerting them to the emergency landing. The wingman was about to transmit a message to the control tower when the subject airplane winged over to the left and entered a descending turn. The airplane remained in this attitude until it struck the ground and burst into flames. The pilot was killed upon impact.

12-27-44C. Kissimmee, Florida. At 1610 EWT, a North American P-51D crashed one and a half miles southeast of the Army Air Field at Kissimmee, Florida, killing pilot Capt. Neill A. Bollus, 23. The airplane had taken off from Kissimmee Army Air Field on a routine training flight. The airplane was apparently returning to the field and was diving to pick up speed for a fly over and peel-off for a landing on Runway 33. The airplane was observed in a dive at about 1,000 feet agl with the engine running normally. Then the canopy was observed to fall away from the airplane as it dove toward the ground. Investigators stated, "Somewhere in the dive the canopy came off, apparently striking the pilot hard enough to daze him momentarily. Witness' statements show that at the last moment before the crash an attempt was made to pull the airplane up to normal flight and a sudden application of throttle was heard. As the airplane was in a roll to the left, an inherent tendency of this aircraft when left to fly itself, the pilot evidently became conscious enough to realize his position and tried to correct it. However, the right wing caught on the ground, indicating that nearly a complete roll was made before the airplane struck the ground. The airplane was apparently nearly out of the diving position as it made contact with the ground. It scooped out a fairly large hole in the ground, bounced back into the air and exploded. Parts of the pilot's body and pieces of windshield glass were found in the hole indicating that the aircraft hit on the right wing and right side of the fuselage in more or less an inverted position."

12-27-44D. Mattituck, New York. At 1815, a Consolidated B-24J flying in poor weather crashed two miles west of Mattituck, New York, killing eleven crewmembers. The airplane took off from Westover Field, Chicopee, Massachusetts, on a cross-country flight to Montauk Point and return. The airplane was observed circling the area of Suffolk County Army Air Field, Suffolk County, New York, at about 1735 EWT. Ceiling was about 1,200 feet agl with light snow showers. The control tower attempted to contact the airplane by radio but was unsuccessful. Control tower personnel gave the airplane the green light for a landing but still did not receive a reply from the crew. The weather was deteriorating as the airplane circled and at 1750 it was seen to fly away from the area of the field on a heading of 50 degrees, apparently toward Westover Field. Some minutes later a large explosion was observed. Investigators stated, "The position of the wreckage indicated that the aircraft hit the ground with its left wing low at approximately an 80-degree angle with the ground, then flipped over on its back and exploded. The fuselage broke away from the engines." Investigators speculated that the pilot had lost control of the airplane when he suddenly encountered instrument conditions at low altitude and was not able to make the transition to instrument flight from visual flight. Killed in the crash were: F/O Victor F. Belotti, pilot; F/O William D. Sanders, co-pilot; F/O Clifford J. McElwee, navigator; F/O Louis P. Pernala, bombardier; Cpl. Lawrence L. Tench, engineer; Cpl. George H. Reis, Jr., radio operator; Cpl. John H. Benner, gunner; Cpl. Vito D. Ferrano, gunner; Cpl. Roger D. Westervelt, gunner; Cpl. Joseph Mastorana, gunner; SSgt. Nicholas M. Carusone, gunnery instructor.

12-27-44E. Stanton, Texas. At 1805, a Beech AT-11 crashed on the C.M. Houston Ranch 18 miles northwest of Stanton, Texas, killing five fliers. The airplane took off from Midland Army Air Field, Midland, Texas, on a routine cross-country navigation flight to Amarillo, Texas, via, Lubbock, Texas, and return. Investigation revealed that the ceiling in the flying area was at 4,000 feet and had progressively lowered to about 1,200 feet. Student flying was cancelled and all students were ordered to return to the field. The ceiling had lowered to 900 feet agl and Midland was below minimums. The pilots had been briefed to land at Lubbock Army Air Field or Roswell Army Air Field, Roswell, New Mexico, if poor weather was encountered. The subject airplane was observed to fall out of the overcast out of control with both engines running and landing gear extended. The airplane smashed into the ground at high speed and at a steep angle. Ceiling at the scene of the accident was 200 feet with light rain. Killed in the crash were: F/O John D. Jones, pilot; 1Lt. Alvin R. Wong, bombardier instructor; A/C Elbert E. Wood, student bombardier; A/C Walter J. Woycik, student bombardier; A/C Joseph W. Yorkey, student bombardier.

12-27-44F. Ontario, California. At 1714, a Lockheed P-38J and TP-38L collided in mid-air and crashed three miles southeast of the Army Air Field at Ontario, California, killing both pilots. 2Lt. Donald R. Wright, 22, was killed in the crash of the P-38J; 2Lt. Clarence J. Jones, 22, was killed in the crash of the TP-38L. The airplanes were part of a four-ship flight that had taken off from Ontario Army Air Field on a camera-gunnery mission. Lt. Jones was leading the formation and Lt. Wright was flying in the number-three position. The formation was flying at 7,500 feet about four miles south of Ontario Army Air Field when the flight leader radioed air traffic control for permission to land. The flight leader signaled his flight to switch to a spread formation. The flight leader made a slight turn to the right before entering a steep "roll over to the left" in "a sort of 45-degree split-S to lose altitude before making his initial approach to the field." Lt. Jones continued in this roll and started losing altitude as he entered inverted flight. Lt. Jones' airplane was directly over Lt. Wright's airplane and began descending on it vertically. The port wing of Lt. Jones' airplane collided with Lt. Wright's starboard wing as he rolled through inverted, sending both airplanes out of control. The port wing of Lt. Jones' airplane was severed and the port engine and center section erupted into flames. Lt. Jones' airplane spiraled out of control and smashed to the ground where it exploded into flames in close proximity to the spot where Lt. Wright's airplane spun to the ground.

12-28-44A. Tiana Bay, New York. At 1105, two Republic P-47D airplanes collided in mid-air and crashed into Tiana Bay six miles southeast of the Army Air Field at Suffolk County, New York, killing pilot

2Lt. Raymond D. Clarke aboard P-47D #42-27828. 2Lt. Harold L. Cadwell, Jr. parachuted to safety from P-47D #42-23118. The airplanes were part of two, four-ship flights that took off from Suffolk County Army Air Field on an aerial gunnery mission. The tow-ship was delayed and the two flights became engaged in simulated combat at about 12,000 feet indicated altitude. The two airplanes, one from each flight, collided when they both attempted to rejoin what each pilot thought was his formation. Investigation revealed that Lt. Cadwell was joining the wrong flight and had collided with Lt. Clarke's airplane. The airplanes stuck together for a short time before breaking up and spinning to the surface. Lt. Cadwell parachuted into Shinnecock Bay and was rescued a short time after the collision by a U.S. Coast Guard vessel operating in the area. Lt. Clarke's body was not recovered until 1-1-45.

12-28-44B. Corcoran, California. At 1015, a Vultee BT-13B spun to earth 20 miles south of Corcoran, California, killing pilot F/O John J. Shockey. Navigator-rated passenger F/O Chester F. Low parachuted to safety and was uninjured. The airplane had taken off at 0930 from Lemoore Army Air Field, Lemoore, California, on a routine instrument training flight. Investigation revealed that the pilot had performed a series of aerobatic maneuvers and had to stop because the passenger was feeling airsick. The pilot allowed the passenger to fly the aircraft and allowed him to perform a chandelle. The pilot took over and demonstrated the maneuver properly and then allowed the passenger to perform another chandelle. The pilot then took over and performed several snap rolls and loops. The passenger again complained of being airsick. The pilot again allowed the passenger to fly the aircraft. The passenger climbed the airplane back up to 6,000 feet and the pilot took over. It apparently had been agreed that the passenger would perform a loop when he felt "able." After a short time of straight and level flying, the passenger stated that he was ready to take over and attempt a loop. The pilot allowed the passenger to take control. The passenger dove the airplane slightly until he reached an airspeed of 170 mph and then he advanced the throttle and pulled back on the stick simultaneously. The passenger apparently stalled the airplane while in an inverted position. The pilot took over the controls but the airplane "slipped" out of a recovery. The airplane began spinning to the left, describing at least three turns of spin. The spin tightened up for two more turns and then the pilot recovered momentarily at 4,000 feet, but the airplane entered a spin to the right. The pilot ordered the passenger to bail out after failing to recover after three turns. The pilot was unable to make a full recovery and the airplane "pancaked" into the ground with no evidence of forward motion. The airplane burst into flames upon impact.

12-28-44C. Dover, Delaware. At 1332, a Lockheed A-29 crashed while attempting to take-off from Dover Army Air Field, Dover, Delaware, killing four fliers. Killed in the crash were: 2Lt. Richard A. Miller, pilot; 2Lt. Richard P. Peoples, co-pilot; SSgt. Elbert A. Maywald, engineer; Cpl. Manuel Martinez, tow-reel operator. The airplane took off at 1330 on a scheduled target-towing mission. Investigators stated, "Lt. Miller took off on Runway 31, reaching an altitude of approximately 700 feet and was approximately one mile northwest of the field when he called the Tower for an emergency forced landing saying he was having trouble with his left engine. This was at 1331 EWT. Pilot was advised by the Tower that he was cleared to land on Runway 31, if possible, and that the field was being held open for his use. After the initial call the Tower did not hear any acknowledgment or [voicing] of the pilot's intentions. Lt. Miller made a tight pattern to the left. It appeared to the Tower he was making a downwind leg for Runway 31. Lt. Miller called the Tower asking, "How about using [Runway] 36 instead of [Runway] 31?" This call was not heard by the Tower but by another tow-target aircraft that had taken off right after Lt. Miller. As Lt. Miller was approaching the south end of the field near the end of the Runway 36 at an altitude of approximately 600 feet, the aircraft turned over on its left wing and spun into the ground at 1332 EWT, exploding immediately upon contact with the ground. Witness stated that the propeller on the left engine was feathered and the wheels were fully retracted." Investigators noted that Lt. Peoples was a fighter pilot who was flying as co-pilot to log some twin-engine time.

12-29-44A. Greenbush, Michigan. At 1301 EWT, a Republic P-47D crashed three miles west of Greenbush, Michigan, killing French Air Force pilot Sgt. Jacques J. Martin. The airplane was part of a two-ship flight that took off from Oscoda, Michigan, on a formation training flight. An AAF instructor pilot was leading the flight and the subject airplane was flying in the wing position. The control tower asked the flight leader to check the weather in the local flying area. The flight leader performed the weather check and radioed back to the tower. The flight leader then performed a medium turn to the left and then performed a medium turn to the right. After leveling out of the right turn at about 2,500 feet indicated altitude, the flight leader looked back and noticed that the wing airplane had fallen out of formation. The flight leader then noticed a fire on the ground and smoke rising. The flight leader descended to about 800 feet and was able to determine that the smoke was coming from the remains of the subject airplane. Investigation revealed that the pilot had stalled the airplane and was unable to effect a recovery in time to avoid striking the ground.

12-29-44B. Santa Ana, California. At 1230, a Lockheed TP-38J crashed two miles east of Orange County Army Air Field, Santa Ana, California, killing 2Lt. Robert F. Paryzek, 21. The airplane was flying in the number-four position of a four-ship flight that had taken off from Orange County Army Air Field on an authorized simulated combat flight. The four-ship flight broke up into two, two-ship flights and began engaging in simulated combat at 10,000 feet. Investigators stated, "After several turns, Lt. Paryzek stalled out of a turn to the right and started to spin. The nose of the aircraft came up a few degrees and the airplane continued spinning alternately to the right and left. At about 3,000 and 4,000 feet the aircraft began a flat spin with the nose nearly level, and [then spun] to the ground. The aircraft hit the ground in a perfectly flat or level attitude and began to burn, killing the pilot instantly."

12-29-44C. Walterboro, South Carolina. At 1600 EWT, a Republic P-47D crashed at Walterboro, South Carolina, killing pilot 2Lt. Cornelius D. Dowling, 26. The airplane took off from Walterboro Army Air Field on a routine training flight. The pilot flew around for an hour and then returned to the field to land. The pilot received landing instructions and entered the traffic pattern. He overshot his turn from the base leg to final and began to S-turn in an effort to get lined up with the runway. The pilot attempted a steep turn at low altitude at too low an airspeed, causing the airplane to stall. The airplane entered a spin at an altitude of about 800 feet agl and smashed to the ground at a 90-degree angle after describing one-half turn of spin. The aircraft exploded into flames upon impact, killing the pilot instantly. It was noted that this was the pilot's first solo flight in a P-47 aircraft.

12-29-44D. Meridian, Idaho. At 1907 MWT, a Consolidated B-24J flying in poor weather suffered a catastrophic structural failure and crashed two and one-half miles northeast of Meridian, Idaho, killing the crew of ten. The airplane took off at 1850 MWT from Gowen Field, Boise, Idaho, on a local bombing mission. At 1855, the pilot contacted the bombing range control officer for a range assignment. The control officer informed the pilot that the range was closed due to poor weather. The pilot was advised to fly contact in the local area and the pilot acknowledged. At 1904, the pilot was instructed to return to the field and land. A few minutes later a pilot flying in the area observed the explosion on the ground and reported it to the tower. Investigators stated, "[The] aircraft was apparently in the overcast coming down in a power dive. Airplane was out of control in a spiraling dive when it was observed coming out of the overcast and just before crashing was seen to 'flip over' in the nature of a snap roll. There was no fire issuing from the aircraft prior to the crash. Aircraft struck the ground in a partially inverted position. Violent use of the controls at excessive speed was evidenced by fabric torn off and ribs of elevators sheared off in addition to the entire tail assembly.... Pieces of fabric were first found in a field about 2 miles from the scene of the crash. Then continuing in southeasterly direction towards the crash, additional pieces of fabric parts of rudder and parts of

elevator, ribs, etc, were found at intervals. These were followed by other parts of the airplane such as: right front bomb bay door, left rear bomb bay door, section of the right wing, portion of the top section of the right vertical stabilizer, and finally, the entire tail section, (horizontal stabilizer with vertical stabilizers attached), which was found 300 yards from the main wreckage." Weather at the scene of the crash was reported as ceiling of 700 feet agl with a rainsquall in the area and was described as "bad." Killed in the crash were: 2Lt. James L. Fisher, pilot; F/O John H. Sondermann, co-pilot; F/O Edwin D. Groce, navigator; F/O Morton S. Reiter, bombardier; Cpl. Burr T. Cahill, engineer; Cpl. Melvin R. Strong, radio operator; Pfc. Earl C. Wickman, gunner; Pfc. Robert D. Brown, passenger; Cpl. Jonathon H. Bayress, Jr., gunner; Pfc. Roger E. Hall, tail gunner.

12-30-44A. Walker, Kansas. At 1025 CWT, a Boeing B-29 suffering an engine fire crashed 13 miles northwest of Walker Army Air Field, Walker, Kansas, killing ten fliers. Three crewmembers parachuted to safety and received minor injuries. The airplane took off from Walker Army Air Field at 0915 on a high-altitude bombing mission. The airplane climbed to 25,000 feet. During the climb, the number-three engine began to run hot. Investigators stated, "Apparently the number-three engine swallowed a valve causing a backfire and forcing the manifold pressure to 65 inches and immediately after a fire broke out on the underside of the engine. In the meantime the throttle had been retarded as soon as the fire was reported and the [propeller] was feathered, the fuel, mixture, booster pump, and ignition switch were shut off. One fire extinguisher was pulled and the flames were reported to have subsided. However a few seconds later they burst forth again and the second extinguisher was pulled. The warning signal was given, the emergency bomb release and emergency cabin air pressure release were pulled, and the wheels were put in the down position. Only the right main wheel extended according to the indicator lights. Examination of the wreckage proved the right gear to be extended and the left gear and flaps to be retracted in that it normally extends first and here the right gear had extended first. Next the bail out execution signal was given and almost at the same instant a severe explosion occurred blowing the number-three engine from its mounting, tearing the right gear and oleo strut off, and weakening the wing structure to such an extent that it immediately broke off. Close examination revealed the main spars to be broken and torn but not burned. The left wing and the fuselage, with the exception of the tail, remained generally intact and crashed together with the number-one and number-two engines, which, it is believed, had been pulling normal power up until the crash. The tail probably tore loose at a high altitude soon after the spin started in that it was found a great distance from the wreckage. A second explosion somewhere between 3,000 and

6,000 feet caused by a fuel cell is believed to have blown several of the bodies out of the airplane and into the propellers. The airplane fire was caused by a fire in the induction system of the number-three engine." Killed in the crash were: 1Lt. Rufus C. Anderson, pilot; 2Lt. Glenn V. Welander, co-pilot; F/O Thomas H. Joyce, navigator; 2Lt. Stanley M. Franklin, bombardier; Sgt. Dale M. Thompson, flight engineer; Pfc. William R. Fiorini, radio operator; Cpl. Robert F. Rich, radio operator; Pfc. Kenneth L. Bryant, engineer; Cpl. Harry Bochichio, tail gunner; Capt. Loy G. Coffee, bombardier instructor. Flight engineer 2Lt. Verne E. Roycraft, gunner Pfc. Nicholas G. Brando and gunner Cpl. Richard R. Berg were flung from the airplane as it broke up and were able to parachute to safety, suffering minor injuries.

12-30-44B. Gulf of Mexico, Florida. At 1420 EWT, a North American P-51A crashed into the Gulf of Mexico one mile west of Estero Bay, Florida, killing pilot F/O David R. Breakwell, 21. The airplane was part of a two-ship flight that had taken off from Lakeland Army Air Field, Lakeland, Florida, on a low-altitude navigation flight. The flight leader testified that he was flying at 200 feet above the surface and when he looked back over his shoulder and observed the subject airplane in a steep climb. The subject airplane's propeller was observed flying through the air behind the subject airplane. The subject airplane plunged into the sea and sank in 27 feet of water. A pilot flying in the area observed a "white streak" on the surface of the water, like a boat wake, just before the airplane began the steep climb. Investigators speculated that the subject airplane's propeller struck the water and the pilot pulled up abruptly. The propeller flew off and struck the canopy, seriously injuring the pilot. The pilot was unable to escape the airplane and was recovered from the wreckage by divers.

12-31-44A. Fairbanks, Texas. At 1018 CWT, a North American F-10 (the photo-reconnaissance version of the B-25D) flying in poor weather conditions broke up and crashed three miles northwest of Fairbanks, Texas, killing five fliers. The airplane took off from Houston Municipal Airport, Houston, Texas, on a cross-country flight to Love Field, Dallas, Texas. The airplane made a normal take-off and climbed to about 5,500 feet msl. The pilot reported that he was on top of the overcast and on course. Some time later, witnesses on the ground heard a loud report and observed the airplane spinning out of the overcast shedding pieces. Investigators speculated that the airplane entered a huge front and encountered powerful storms and turbulence. The pilot lost control of the airplane and apparently stressed the airframe beyond its design limitations in an effort to re-establish level flight. The airplane spun crazily to earth and exploded upon impact. Pieces of the airplane were scattered over a wide

area. Killed in the crash were: 1Lt. Everett M. Taylor, pilot; F/O Benjamin N. Rosen, co-pilot; Cpl. Allen W. Carpenter, engineer; SSgt. Henry B. Waller, passenger; SSgt. Harry R. Shepardson, passenger.

12-31-44B. Galveston, Texas. At 1101, a Martin TB-26B crashed into a bayou one-half mile northeast of the Army Air Field at Galveston, Texas, killing the crew of three. The airplane took off at 1049 from Galveston Army Air Field on a routine tow-reel gunnery mission. The pilot radioed the field at 1055 and alerted tower personnel that his port engine had failed and that he could not feather the propeller. The airplane, flying at 1,000 feet agl, turned on the runway heading of 170 degrees about five miles out. As the airplane neared the field the pilot extended the landing gear. The airplane lost altitude rapidly so the pilot retracted the landing gear and tried to hold altitude. The pilot realized that he would not be able to make the field so he elected to attempt a crash landing in a bayou about one mile from the field. The airplane approached the bayou at about 20 feet agl. The pilot observed that he could possibly collide with civilian huts on the shore of the bayou if he continued his flight path so he attempted a slight turn to the left. The airplane stalled as it turned and the port wing dropped into the bayou. The airplane cartwheeled into the bayou and broke up. The nose section dove into the bayou and separated. The center section slammed into the water and sank. The starboard wing broke off. The tail section broke off and traveled about 50 feet past the main wreckage. The tail section stayed afloat on the bayou for about 15 minutes before sinking. The main wreckage sank in 30 feet of water in the bayou. Pilot 1Lt. Elery L. Bush, engineer Cpl. Baynard L. Chesire and radio operator Cpl. Kenneth A.E. Gooch were killed in the crash. Investigators noted that it took several days to recover the crew's bodies.

12-31-44C. Fort Sumner, New Mexico. At 1910 MWT, a Curtiss P-40N crashed three miles east of the Army Air Field at Fort Sumner, New Mexico, killing F/O Adolphus Doring. The airplane was part of a three-ship flight that took off from Fort Sumner on a night formation-training mission. After completing its mission, the flight returned to the field and the first two airplanes flew over the field and then peeled off for landing. The control tower instructed the subject airplane, which was trailing the other two, to complete a 360-degree turn and then fly over the field for an individual peel-off. The subject pilot acknowledged and flew away. When the airplane failed to return to land, the control tower attempted to contact the pilot but was unsuccessful. A couple minutes later, tower personnel noticed a fire on the ground east of the field. Investigators speculated that the airplane flew into the ground at high speed.

1945

January

1-1-45A. Smoky Hill Army Air Field, Kansas. At 1845 CWT, flight engineer TSgt. Jess J. Ausburn was fatally injured when he walked into the rotating number-two propeller of a Boeing B-17F operating at Smoky Hill Army Air Field, Salina, Kansas. The airplane, which was piloted by 1Lt. Benjamin L. Powell and co-pilot 2Lt. William F. Blackington, was being prepared for a high-altitude training mission. Witnesses stated that the engineer walked directly into the number-two propeller after signaling the pilot to shut down the number-three engine, which had caught on fire during the starting process. The pilots and four other crewmembers were uninjured. The engineer died about 35 minutes after the incident.

1-1-45B. Casper, Wyoming. At 2205 MWT, a Consolidated B-24J crashed 25 miles south of the Army Air Field at Casper, Wyoming, killing six fliers. The airplane took off at 1634 MWT from Casper Army Air Field on a night navigation flight to Grand Island, Nebraska, to Ainsworth, Nebraska, and return to Casper. The airplane was heard at 2030 to transmit a message indicating that it was going to climb over the overcast. The airplane failed to return to Casper and was declared missing. The airplane was found on 1-3-45. Investigation revealed that the airplane descended into the ground at cruise speed with all four engines producing power. The airplane struck the ground initially on the bomb bay, the propellers digging in once the fuselage collapsed. The airplane smashed itself to pieces as it careened across the terrain. The airplane passed over a wide, 40-foot deep gully and smashed into the ridge on the other side, the main section bursting into flames. The four engines, a wing panel and the tail section were found in the gully. Killed in the crash were: 2Lt. Robert E. Murchison, pilot; 2Lt. Harold B. Paulk, co-pilot; 2Lt. Reede L. Bludworth, navigator; 2Lt. Reuben J. Clark, bombardier; Cpl. Eugene J. Opala, engineer; Cpl. Robert S. Hillard, radio operator.

1-2-45A. Pond, California. At 0445, a Northrop P-61B crashed nine miles WSW of Pond, California, killing pilot F/O Charles R. Northcutt and radar operator 2Lt. Grady F. Shytles, Jr. The airplane took off from the Army Air Field at Palmdale, California, on a high-altitude interception mission via Trona, California. The subject airplane failed to intercept its target airplane over Trona. Radio contact could not be established with the subject airplane. Investigation revealed that the airplane was off course at the time of the crash and the ceiling in the area was 600 feet agl. The subject airplane, flying in instrument conditions and attempting to let down, apparently circled the area of Pond for several minutes before the crash. The airplane collided with the terrain with the starboard wingtip. The airplane cartwheeled into the ground, exploding into flames upon impact and scattering wreckage over an area of 200 yards. Investigators noted that the crew had flown three training missions in the seven hours prior to the accident. It was also noted that the pilot was unhappy about having to fly three training missions on New Years Eve. The crew was three flight hours from completing night fighter training.

1-2-45B. Selma, Alabama. At 1100, a Curtiss P-40N crashed 20 miles southwest of Craig Field, Selma, Alabama, killing pilot F/O Darrel D. Copple, 20. The airplane had taken off from Craig Field on a routine training flight. The airplane was observed in a 60-degree dive before it smashed into the ground at a steep angle. The airplane exploded violently upon impact, producing a ten-foot deep crater that was 20 feet wide. The pilot was killed instantly.

1-2-45C. Charleston Army Air Field, South Carolina. At 2040, radio operator Cpl. Donald P. Dalton was killed when he walked into the rotating number-three propeller of a Consolidated B-24J that was operating at Charleston Army Air Field, Charleston, South Carolina. The airplane was scheduled to fly a routine training mission and the radio operator had pre-flighted the radios on the flight deck. The pilots entered the cockpit and began to commence pre-flight checks and engine starting. The radio operator then went to the nose compartment and pre-flighted the radio compass and radio connections. The radio operator completed his work and exited the aircraft via the nose wheel door, walking directly into the rotating number-three propeller and killing him instantly. The airplane was piloted by instructor 1Lt. William E. Rotert, pilot 2Lt. Julian L. McDonald and co-pilot 2Lt. Robert M. Looney. No other crewmembers were injured.

1-3-45A. Pyote, Texas. At 0850, 2Lt. John S. Jamison was killed when he walked into the rotating number-two propeller of a Boeing B-17G operating at Pyote Army Air Field, Pyote, Texas. Lt. Jamison, a rated pilot, apparently approached the aircraft without the knowledge of the crew and either fell into or walked into the rotating propeller. The airplane, which was preparing to take off on a routine training mission, was piloted by Capt. John H. Holmes and co-pilot 2Lt. Allan M. Michelson. The pilots and crew were uninjured. Lt. Jamison was not scheduled to fly on the mission and it was unknown why he was approaching the aircraft.

1-3-45B. St. Augustine, Florida. At 1240 EWT, a Curtiss P-40N crashed at the U.S. Navy Auxiliary Airfield at St. Augustine, Florida, killing pilot F/O Ray M. Skelton, 20. The airplane, which had taken off from Dale Mabry Field, Tallahassee, Florida, on a routine training flight, was observed flying north of the auxiliary airfield at about 4,000 feet. Investigators stated, "Pilot circled the field and the tower gave him a flashing green light. He came around to land, wheels and flaps extended. His approach was very high but he continued down to about 15 feet off the runway. He was landing on Runway 2, which is 3,000 feet in length. He would have touched down near the halfway mark. At about 15 feet [above the surface of the runway] he applied throttle and started to go around. According to naval personnel engine was normal, however, the airplane seemed to be mushing as he was just clearing the trees north of the field. Civilian witnesses north of the field stated that the airplane came over at treetop level in an increasing left bank, apparently trying to get to Runway 13 or Runway 16. Plane struck a tall pine tree at edge of the wooded area with left wing and cartwheeled into the ground. The pilot was killed instantly."

1-3-45C. Las Vegas, Nevada. At 1115 PWT, a Bell P-39Q suffering an engine failure crashed 20 miles west of Las Vegas Army Air Field, Las Vegas, Nevada, killing pilot 2Lt. Ernest A. Morris, 20. The airplane took off from Las Vegas Army Air Field on a camera gunnery mission. The subject airplane rendezvoused with a Boeing B-17 airplane for the gunnery mission about ten miles south of Indiana Springs, Nevada. The crew of the B-17 reported that the P-39 was trailing a thick stream of black smoke. The subject airplane broke away and was next seen flying at an altitude of about 3,000 feet. The airplane, still trailing black smoke, stalled and then snapped over into a spin to the right. The pilot was able to recover momentarily before the airplane entered a spin to the left. The pilot failed to recover and the airplane smashed to the ground, exploding into flames

upon impact. Investigation revealed that the engine suffered the failure of one or more connecting rods, causing the engine to fail and catch fire.

1-4-45A. Blytheville, Arkansas. At 1034, a North American TB-25G crashed near Blytheville Army Air Field, Blytheville, Arkansas, killing three fliers. The airplane took off from Blytheville Army Air Field at 0940 on a routine training flight. The airplane returned to the area of the field and the pilot advised that he was having trouble with the port propeller, which apparently would not feather. The pilot was cleared for an emergency landing on the field, but he refused, asking the tower to get somebody of "authority" in the tower. A major and lieutenant colonel were summoned to the tower. At this time the subject airplane was observed flying at about 800 feet agl five miles southwest of the field with the landing gear extended and the port propeller windmilling. The pilot stated to the major that he was not getting power from his starboard engine. The major advised the pilot to make an emergency single-engine landing on the field. The airplane, flying at about 500 feet agl with landing gear extended, was on final approach about one mile from the runway when it was seen to go into a gentle bank to the left, the bank increasing and the nose dropping until the airplane went partially inverted. The airplane entered a spiral to the left and smashed into the ground, exploding into flames upon impact. The port propeller feathering mechanism was totally destroyed in the crash and could not be examined. Investigation revealed the pilot had inadvertently placed the starboard propeller in low rpm in an effort to manipulate the propeller pitch of the port propeller. Investigators speculated that the airplane's slow speed and extended landing gear compounded the mistake. The airplane stalled and entered a spiral at an altitude too low to allow recovery. B-25 instructor pilot Capt. Miles Blunt Jr., B-25 student 2Lt. James A. Rose and pilot-rated passenger 2Lt. Walter B. Beisinger were killed in the crash.

1-4-45B. Luke Field, Arizona. At 1030 MWT, a Curtiss P-40R crashed 35 miles west of Luke Field, Phoenix, Arizona, killing instructor pilot 1Lt. William R. Doughty, 28. The pilot was leading three Chinese Air Force sub-lieutenants on a combat training flight. The four-ship flight took off from Luke Field at about 0900 MWT and flew to the practice area. The instructor engaged in simulated combat with each Chinese pilot individually at least twice. The instructor was engaged in simulated combat with a student at about 10,000 feet, and was able to stay on the tail of the student for about five minutes. Now flying at about 5,000 feet, the instructor allowed the student to get in the six o'clock position. The instructor then entered a slight dive to pick up speed, pulling up into a steep climbing turn to the left. The airplane stalled, snapped over to the right and entered a spin at an altitude of about 6,500 feet. The airplane remained in the spin to the right until it struck the ground and exploded into flames. There was no evidence of structural failure. Investigators could not determine why the pilot, an experienced P-40 instructor pilot, could not recover the airplane.

1-4-45C. Laurinburg-Maxton Army Air Base, North Carolina. At 2115 EWT, a Waco CG-4A glider crashed at Laurinburg-Maxton Army Air Base, Maxton, North Carolina, killing glider pilot F/O Karl F. Kuehn and seriously injuring glider pilot F/O Keith M. Browning. The glider was on final approach when it stalled at an altitude too low to allow recovery.

1-4-45D. Columbus, Mississippi. At 2345, a Beech AT-10GF crashed one-half mile west of Columbus Army Air Field, Columbus, Mississippi, killing A/C Jess B. Oldham, Jr. and A/S Edward P. Sanders, Jr. The students had performed several take-offs and landings with an instructor on board and were cleared to fly solo. The instructor deplaned and the airplane taxied for take-off. The airplane took off but then collided with trees about one-half mile from the end of the runway. The airplane went out of control and smashed to the ground, exploding into flames and killing the students instantly.

1-5-45A. Midland, Texas. At 0028, a Beech AT-11 crashed one and a half miles northeast of Midland Army Air Field, Midland, Texas, killing Capt. Robert B. Rickerstaff and 1Lt. James W. Foster. The airplane had made several take-offs and landings prior to the accident. The airplane made its eighth take-off at 0023 and entered the traffic pattern. A pilot flying in the area noticed the subject airplane's starboard wing drop momentarily. The wing came up but the airplane entered a 45-degree dive, remaining in this attitude until it struck the ground and exploded. Investigators could not determine what caused the pilot to lose control of the airplane.

1-5-45B. Smyrna Army Air Field, Tennessee. At 0117 CWT, student flight engineer A/C Joseph A. Urban was killed when he walked into the rotating number-two propeller of a Consolidated RB-24E that was operating at Smyrna Army Air Field, Smyrna, Tennessee. The airplane, which had returned to the parking area following a routine training mission, was piloted by 2Lt. Roger A. Ohman and co-pilot 2Lt. Elmer W. Perry. The pilots and crew were uninjured. Investigation revealed that the student engineer, who was assisting the aerial engineer in placing the wheel chocks, was struck when he attempted to walk between the number-two propeller and the port side fuselage.

1-6-45A. Carson City, Nevada. At 0715 PWT, a Curtiss C-46A suffering a fire in the port engine and wing crashed 30 miles east of Carson City, Nevada, killing five fliers. Killed in the crash were: F/O Donovan J. Hogan, pilot; 1Lt. Walter E. Schmidt, co-pilot; Pvt. Fred R. Davis, engineer; Pvt. William R. Carl, engineer; Capt. William A. Bevon, pilot-rated passenger.

The airplane took off at 0650 PWT from Reno Army Air Field, Reno, Nevada, on a routine training flight. Investigation revealed that the port wing and port engine were on fire prior to port wing separating near the port engine nacelle. The port wing panel separated at a point outside of the port engine nacelle and struck the tail section, causing its progressive failure. The starboard wing failed and separated moments later. The airplane smashed to the ground out of control and exploded. Wreckage was scattered over a wide area. Investigators speculated that the "riveted" fuel tank design contributed to the fire in the port wing and engine accessory section.

1-6-45B. Bend, Oregon. At 1045, a Vultee BT-13B crashed ten miles southeast of Bend, Oregon, killing pilots Capt. Ralph A. Thiessen and 2Lt. Max E. Smith. The airplane took off at 1030 from Redmond, Oregon, on a hooded instrument training flight. There was no other contact with the airplane with ground stations. Investigation of the wreckage indicated that the airplane smashed to the ground while in a spin. The pilots made no apparent effort to parachute to safety.

1-7-45A. Taft, California. At 1032, a Vultee BT-13A crashed 19 miles southwest of Gardner Field, Taft, California, killing instructor 2Lt. Kenneth A. Hull and student A/C Earl W. Stephenson. The airplane, which was based at Gardner Field, took off from Cuyama Auxiliary Airfield on an instrument training flight. The wind had shifted during the flight and the instructor in the control ship transmitted the new wind direction and new landing instructions on the field frequency. The subject airplane arrived at the auxiliary airfield and the pilots, apparently unaware of the change in wind direction and landing direction, entered the pattern against traffic. The pilots observed an airplane taking off and climbing toward them while they were on their approach. The subject airplane began a climbing left turn. The airplane turned through 90 degrees and then climbed straight ahead toward the hills. The tower operator advised the pilot to get the nose down and turn left or right. The airplane continued straight ahead in a climb and stalled moments later. The airplane snapped over to the left and entered a spin, remaining in this attitude until it struck the ground and exploded.

1-7-45B. Horn Lake, Mississippi. At 1830, a Beech C-45F flying in instrument conditions crashed at Horn Lake, Mississippi, killing six fliers. Killed in the crash were: Lt. Col. Edward P. Dimmick, pilot; Capt. Vincent J. Rose, co-pilot; Sgt. Phillip H. Lauch, engineer; 2Lt. Adron S. Hoover, glider pilot-rated passenger; Maj. Wilber T. Jones, passenger; Maj. William L. Gladfelter, passenger. The airplane took off from Tinker Field, Oklahoma City, Oklahoma, on a flight to Memphis, Tennessee, with Jackson, Mississippi, as an alternate. Investigation revealed that the airplane struck the ground in a near vertical dive, smashing to the ground in between two tall trees. The airplane exploded violently into flames upon impact, killing the occupants instantly.

1-8-45A. San Antonio, Texas. At 1214, a Douglas A-20G crashed 17 miles southeast of Kelly Field, San Antonio, Texas, killing pilot Capt. John A. Urick, 26. The pilot took off from Kelly Field at 1152 on his first transition flight in an A-20 type aircraft. Witnesses on the ground observed that an engine was "cutting out" as the A-20 flew over at about 600 feet agl. It was speculated that the starboard engine was not producing full power. The port engine was seen smoking and investigators attributed this to the pilot using an excessively rich mixture on this engine. The pilot evidently stalled the airplane at an altitude too low to allow recovery. The airplane entered a spin to the right and smashed to the ground, exploding into flames upon impact. It was speculated that the pilot was attempting to clear the engines and maintain level flight when he stalled the airplane. Examination of the wreckage revealed that the propellers were not feathered and the main switch was found in the "on" position. It was noted that the pilot was an experienced multi-engine pilot.

1-8-45B. Gulf of Mexico, Texas. At 1318, a Republic P-47D crashed into the Gulf of Mexico 20 miles northeast of Matagorda City, Texas, killing pilot 2Lt. George L. Murray, 22. The airplane was flying in the number-two position of a four-ship flight that had taken off from Palacios Army Air Base, Palacios, Texas, on an aerial gunnery mission. The airplanes were flying in a string formation and returning to base when the subject airplane pulled out of the formation. Other pilots noticed that the subject airplane's propeller was not turning and the cockpit canopy was in the open position. The pilot rolled the airplane over on its back in an apparent effort to bail out. The airplane fell into a spin. The pilot righted the airplane and it then entered a dive. The airplane remained in this attitude until it smashed into the Gulf of Mexico and exploded. The pilot was not seen to bail out. The pilot's body was not recovered and it was speculated that he was killed upon impact.

1-9-45A. Lubbock, Texas. At 0016, a Beech AT-10BH crashed 13 miles southwest of Lubbock Army Air Field, Lubbock, Texas, killing A/S John E. Mayer and A/S Richard P. McDaniel. The airplane took off from Lubbock Army Air Field on a night transition flight. The students had participated in a student night formation flight earlier in the evening and were flying in the local area shooting landings at the field prior to the accident. Investigation of the wreckage indicated that the airplane smashed into the ground in a near straight and level flying attitude with both engines producing power. It was speculated that the students had lost the horizon and inadvertently flew the airplane into the ground.

1-9-45B. Location Unknown. At an unknown time after 1445 CWT, a Boeing B-29 disappeared and is presumed to have crashed into the sea. Sixteen passengers and crew were missing and presumed lost at sea. The airplane had taken off from Pratt Army Air Field, Pratt, Kansas, on a navigation training flight to San Juan, Puerto Rico, and return. There was no further contact with the airplane after 1445 CWT. Missing and presumed lost at sea were: Capt. Irving H. Ward, pilot; 2Lt. Roy L. Winzer, co-pilot; 2Lt. John W. Osner, navigator; 2Lt. William R. Sage, bombardier; 2Lt. Shelva F. Harbour, radar operator; TSgt. Steven R. Pesuth, engineer; Sgt. Rex O.F. Ford, radio operator; Cpl. Albert J. Mueller, central fire controller; Sgt. William J. Geary, starboard gunner; Sgt. Dominick Destefano, port gunner; Cpl. William M. Duncanson, tail gunner; MSgt. John N. Matkins, passenger; MSgt. Warren G. Allen, passenger; SSgt. Henry G. Schenker, passenger; 2Lt. Borden F. March, passenger; Maj. William T. Boren, passenger.

1-9-45C. Savannah, Georgia. At 1711 EWT, a Consolidated B-24J suffered a catastrophic structural failure and crashed seven miles south of Chatham Field, Savannah, Georgia, killing 11 crewmembers. Gunner Cpl. Joseph S. Lanzi parachuted to safety and was uninjured. Investigators stated, "B-24J AAF #42-50524 departed Chatham Field, Georgia, at 1507 EWT on a high-altitude formation gunnery mission. After taking off three routine radio position reports were received by the ground radio station. None of these reports gave any report of malfunctioning of the airplane. Cpl. Joseph S. Lanzi, the ball turret gunner, stated that his airplane joined the formation but the position of his plane in the formation and how long plane #524 stayed in the formation he did not know. Plane #524 was in the number-three position of the lead flight. After starting to load the turret guns, the left waist gunner informed Cpl. Lanzi that the pilot was going down to 500 feet for splash gunnery. Cpl. Lanzi did not know why the mission was being changed. While letting down, Cpl. Lanzi stated that he did not know whether a cloud coverage was beneath them and did not know whether the plane let down through cloud coverage. After letting down for some time, Cpl. Lanzi stated the airplane made a violent steep left turn or a violent maneuver of a nature [that] resulted in the three men in the tail to be thrown about in the plane. This violent maneuvering continued, resulting in equipment and the men in the waist being thrown about in this section of the plane. Cpl. Lanzi stated that he attempted to get his parachute but it was tossed away by the violent maneuvering and finally after reaching a parachute bailed out through the right waist window in which the glass had broken during the maneuvering. After the parachute opened, Cpl. Lanzi stated that he was below the cloud level and could see the ground. Witnesses near the crash stated that a very unusually loud roar was first noticed but none could see the airplane at the time due to the airplane being above the clouds. After breaking through the overcast, which was estimated to have been 2,000 feet by pilots' reports, the airplane was diving almost vertically with part of the tail section off and falling free of the airplane. During the dive, witnesses also stated that a section of one of the wings ripped or broke off and the airplane immediately started into a spin to the left then crashing. The airplane appeared to have crashed nose first then to have fallen over on the back of the fuselage. All witnesses stated that parts and pieces of the airplane were falling while the airplane was spinning and continued to fall after the plane had crashed. The right wing broke off approximately 12 feet from the wing tip and was found approximately one and one-half miles from the crash proper. The right aileron and part of the tail section were found three-quarters of a mile from the crash proper. Pieces of metal were found over an area on one and one-quarter miles from the crash proper. Upon crashing the airplane exploded and burned completely." Killed in the crash were: 2Lt. Jack W. Milkey, pilot; 2Lt. Ernest R. Records, co-pilot; 2Lt. Jack A. Gadhue, bombardier; 2Lt. Donald E. Wickert, navigator; Cpl. Robert S. Norris, engineer; Sgt. Harvey C. Burchfield, engineer; Cpl. Peter J. Glynn, radio operator; Cpl. Peter V. Martelli, radio operator; Cpl. Richard V. Williams, gunner; Cpl. William H. Garmen, radio operator; SSgt. Angelo M. Sergi, gunnery instructor.

1-10-45A. Greenville, Mississippi. At 0825, a Vultee BT-13A crashed near Greenville Army Air Field, Greenville, Mississippi, killing A/C Edwin S. Bundy, 19. The airplane took off from Greenville Army Air Field, Greenville, Mississippi, on a routine student training flight. The airplane took off to the southwest and climbed to about 300 feet agl when the engine quit. The student, in an apparent effort to get back to the field, entered a steep turn to the left. The airplane slowed down and lost considerable altitude while in the turn. After turning through 100 degrees from the take-off heading, the airplane smashed into the ground while still in a steep bank to the left. Investigation of the engine revealed that the engine stopped because of ice in the carburetor.

1-10-45B. Fordoche, Louisiana. At 1215, a Republic P-47D crashed at Fordoche, Louisiana, killing pilot 2Lt. Joseph A. Doherty, 22. The airplane was part of a two-ship flight that took off from Harding Field, Baton Rouge, Louisiana, on a local individual combat training flight with G-suit test. The subject airplane and the instructor's ship climbed to 16,000 feet and the subject pilot began his first dive. The subject pilot recovered from the dive and radioed the instructor that he had pulled eight Gs and the G-suit was working fine. The subject airplane climbed back up and the two P-47s engaged in simulated combat for several minutes. The subject airplane pulled up in a steep climb and then rolled inverted, entering a steep dive. The subject pilot attempted to pull out at about 6,000 feet but was unable to recover the airplane and it dove into the ground and exploded.

1-10-45C. Madison, Georgia. At 1600 EWT, a North American B-25C crashed seven miles southwest of Madison, Georgia, killing five fliers. The airplane

took off from Greenville Army Air Base, Greenville, South Carolina, on a low-altitude rendezvous mission at 500 feet. Investigators stated, "The formation broke and the subject aircraft flew in the vicinity of Monticello and Madison, Georgia. Witnesses stated that the airplane flew altitudes ranging from 50 feet to 500 feet. The airplane was seen to descend from 300 feet to an altitude of approximately 100 feet, at which time the wings of the airplane were perpendicular to the ground. At that time the right wing, which was the lowest, struck a tree and the airplane hit in an inverted position." The airplane burst into flames upon impact. Killed in the crash were: 2Lt. Calvin E. Stocking, pilot; 2Lt. S.J. Strong, co-pilot; 2Lt. F.H. Bartels, navigator/bombardier; Cpl. W.E. Trash, engineer; Cpl. R.D. Rodgers, radio operator.

1-11-45A. Princeton, North Carolina. At 1020, a Republic P-47D crashed three miles west of Princeton, North Carolina, killing F/O Mervyn L. Lasater, 22. The airplane was flying in the number-three position of a three-ship flight that had taken off from Seymour Johnson Field, North Carolina, on a formation training flight. The airplanes were maneuvering at about 1,500 feet agl when the subject airplane fell out of formation and then flew into the ground. Investigators were unable to determine why the airplane flew into the ground at flying speed with the engine producing power.

1-11-45B. Earlsboro, Oklahoma. At 1027 CWT, a Boeing B-29 suffering an in-flight fire was abandoned and crashed two miles south of Earlsboro, Oklahoma, killing four fliers. Eight crewmembers were injured parachuting to safety. Co-pilot 2Lt. Harold M. Price and gunner Cpl. Anthony T. Cerino did not bail out and were killed in the crash. Bombardier 1Lt. Paul E. Hecker and radio operator SSgt. Albert W. Schmid fell to their deaths in unsuccessful parachute jumps. Flight engineer 2Lt. Robert H. Adams, flight engineer 1Lt. Herbert J. Gere and central fire control Pvt. Emmanuel J. Ballos suffered serious injuries parachuting to safety. Crewmembers suffering minor injuries parachuting to safety were: Capt. John H. Cannon, pilot; 1Lt. Albert F. Kempf, navigator; 2Lt. Anthony A. Ruggeri, radar operator; Cpl. Edward A. Bulcavage, gunner; Cpl. Garvin S. Butler, Jr., tail gunner. The airplane took off at 0857 from Walker Army Air Field, Walker, Kansas, on a long-range navigation and bombing mission to Batista Field and return to Walker Army Air Field. At 1023, the crew noticed a fire in the number-one engine nacelle. After failing to extinguish the fire, the pilot ordered the crew to bail out. Investigation revealed that a fuel line became detached from its fitting behind the number-one engine, causing fuel to spill out and collect in the number-one engine nacelle. The gasoline fumes were ignited by very hot engine ducting, causing the number-one engine nacelle accessory section to erupt in flames. The fire weakened the structure of the port wing and engine nacelle. The number-one engine separated from the wing and fell away. Pieces of the airplane were strewn across several miles. The port wing eventually failed near the number-one engine and separated, sending the airplane out of control. The B-29 smashed to earth and exploded into flames.

1-11-45C. Salinas, California. At 2210, a Douglas TP-70B collided with rising terrain 30 miles east of Salinas, California, killing pilot 2Lt. Robert I. Fletcher, 23, and radar operator F/O Willkie L. Cunningham. The airplane took off from Salinas Army Air Base, Salinas, California, on a night navigation flight to Chico, California, to Stockton, California, to Bakersfield, California, and return to Salinas. The airplane failed to return to base and it was discovered that the airplane smashed into a mountainside at an elevation of 2,000 feet msl. The airplane exploded into flames upon impact, killing the fliers instantly. Investigators speculated that the pilot had mistaken Hollister, California, for Salinas and apparently was attempting to let down through overcast conditions while flying over mountainous terrain.

1-12-45A. Moody Field, Georgia. At 0802 EWT, two North American TB-25J airplanes collided on Runway 9L at Moody Field, Valdosta, Georgia, killing three fliers and injuring three others. B-25 instructor 2Lt. August Geib, Jr. and B-25 students A/C Donald C. Walker and A/C Kenneth E. Warren were killed aboard TB-25J #44-29475. B-25 instructor 2Lt. Charles C. Mann and B-25 students A/C James R. Sampson and A/C Peter G. White were injured aboard TB-25J #44-29076. Both airplanes were scheduled to fly on separate student training flights. Two other B-25 airplanes were standing by, preparing to take-off on Runway 9R. The take-off position for Runway 9R is about 1,500 feet forward of the take-off position for Runway 9L. TB-25J #44-29076 was cleared to take off on Runway 9L and TB-25J #44-29475, proceeding to Runway 9R on taxi strip 8, was ordered to hold at the intersection with Runway 9L, which is about 1,200 feet from the take-off position of Runway 9L. TB-25J #44-29076 began its take-off and after rolling about a thousand feet the pilot observed TB-25J #44-29475, taxiing on taxi strip 8 at the intersection with Runway 9L, crossing directly in front of him from left to right. The pilot pulled up and applied right rudder in an attempt to avoid a collision but was unsuccessful. TB-25J #44-29076's port engine, wing and landing gear collided with the starboard engine and starboard fuselage — forward of the wing leading edge — of TB-25J #44-29475. Both airplanes burst into flames upon impact. TB-25J #44-29076 smashed through TB-25J #44-29475 and crashed in flames onto the runway. The crew was able to escape with their lives. TB-25J #44-29475 was smashed to pieces and pilots Lt. Geib and A/C Walker were killed instantly when they were blown out of the cockpit by the force of the collision. A/C Warren was also killed instantly; his body

was found in the shattered fuselage. Control tower personnel anticipated the runway incursion and attempted to contact both airplanes on the field frequency but were unsuccessful. Tower personnel gave TB-25J #44-29475 a red light to the cockpit but the airplane continued to taxi across the active runway. Control tower personnel ordered TB-25J ## 44-29076 to cut throttles and abort the take-off but the airplane continued on the take-off run. Investigation revealed that the pilots of TB-25J #44-29076 had switched to interphone after being cleared for take-off and did not hear the tower's warning of the runway incursion. Investigation of the wreckage of TB-25J #44-29475 revealed that the radio was also set in the intercom position and not on the tower frequency.

1-12-45B. Morgantown, West Virginia. At 1215, a North American B-25J attempting a landing crashed three miles east of Morgantown Municipal Airport, Morgantown, West Virginia, killing four fliers and injuring four others, one of them seriously. The airplane took off from Greenville Army Air Base, Greenville, South Carolina, on a flight to Morgantown. The airplane arrived in the area of Morgantown after an uneventful flight. Investigators stated, "The aircraft circled Morgantown Municipal Airport twice before making the first approach. The first approach was made for a landing on Runway 36 but the approach was too high and the aircraft went around for another try. The second approach was for Runway 36 and the aircraft touched the ground approximately 2,000 feet from the south end of the North/South Runway. The aircraft plunged off of the end of the North runway, went through some electric and telephone wires, took off the tops of some small trees and struck the ground at the bottom of the ravine, immediately bursting into flames. It is the opinion of the board that the Morgantown Municipal Airport at the time of this accident was not suitable for the landing of B-25 type aircraft. All runways were covered with approximately one-half solid ice and one-half slush ice." Killed in the crash were: Capt. Leonidas V. Keck, B-25 instructor; 2Lt. G.L. Bilby, co-pilot; Cpl. R.Q. Ortiz, engineer; 1Lt. L.B. Durgin, pilot-rated passenger. Pilot-rated passenger 2Lt. Manley A. Mason and radio operators TSgt. S. Mata and Cpl. L. Rice suffered minor injuries and navigator 2Lt. H.D. Altman received serious injuries.

1-12-45C. Dodge City, Kansas. At 1115 CWT, a Martin TB-26B crashed at Dodge City Army Air Field, Dodge City, Kansas, killing three fliers. Instructor pilot 2Lt. Ernest Peyke, French Air Force student pilot Sgt. Thierry E. DeSevin and engineer Cpl. Elmore S. Gregorik were killed in the crash. The airplane was part of a three-ship flight that took off from Dodge City Army Air Field at 0904 on a routine formation training flight. The airplanes flew formation for two hours and then returned to the local flying area. The airplanes began to maneuver in an effort to enter the traffic pattern for Dodge City Army Air Field. The subject airplane was observed in a skid to the left. The starboard wing then was seen to drop. The airplane was righted, but in a nose-high position. The starboard wing dropped again and the airplane rolled through until nearly inverted. The airplane then entered a spiraling descent, describing one and a half turns of spiral before striking the ground and exploding into flames.

1-12-45D. Meridian, Mississippi. At 1635 CWT, a North American F-6K crashed seven miles south of Key Field, Meridian, Mississippi, killing pilot 2Lt. Thomas H. Berry, 22. Investigators stated, "Lt. Berry was leading a flight of two ships on a return flight to Key Field from a photo-reconnaissance mission. Approaching Key Field ten miles to the south, Lt. Berry was observed by his wingman to make an extremely tight 90-degree turn to the right and during this turn Lt. Berry's aircraft became inverted. At this point, [the wingman] lost sight of Lt. Berry. [A witness on the ground] observed the aircraft turning from side to side until it crashed into the ground. As Lt. Berry was flying at 2,000 feet altitude at the time of his turn, it is apparent that he put his aircraft into a series of high speed stalls in his effort to recover the airplane before crashing into the ground."

1-12-45E. Guadalupe, California. At 1500, a North American P-51K and a Lockheed P-38L collided in mid-air four miles northeast of Guadalupe, California, killing P-38L pilot 2Lt. Rex H. McMillen, 20. P-51K pilot F/O Jack E. Spencer, 26, was able to make a successful emergency landing with his damaged airplane. The P-38 was flying in the number-three position of a three-ship flight that took off from Santa Maria Army Air Field, Santa Maria, California, on an authorized aerobatic flight and a camera-gunnery mission. The P-38 flight had been performing rolls and loops at about 9,500 feet while waiting to rendezvous with other airplanes for the camera-gunnery mission. The P-38s were flying in string formation when the leader began a series of wingovers. The flight leader stated that after a wingover to the left he looked back and observed a P-38 in a descending spiral, trailing smoke from the port engine. The P-38 straightened out into a steep dive, remaining in this attitude until it struck the ground and exploded. The P-38 pilot made no obvious effort to leave the airplane. The P-51K had taken off from Palm Springs, California, on a flight to Oakland, California. The P-51K was flying at 9,000 feet in the middle of the airway on the north leg of the Santa Maria range when the pilot felt a sudden jolt and then discovered that three feet of his port wing was missing. The pilot, after checking stalling characteristics and determining the controllability of the aircraft, made a successful emergency landing at Santa Maria Army Air Field.

1-12-45F. Douglas, Arizona. At 1640 MWT, a North American TB-25D crashed near Douglas Army Air Field, Douglas, Arizona, killing pilot Capt. James R. Stout, 23. Engineer Pvt. James K. McLaughlin

received minor injuries parachuting to safety. The airplane took off from Douglas Army Air Field on a routine local flight. Soon after becoming airborne, it became evident to the pilot that the control column would not move forward past the neutral position. The pilot and the engineer tried to the force the column forward and free it of the apparent obstruction without success. The pilot was able to descend by reducing power. The pilot approached the area of Douglas Army Air Field and ordered the engineer to bail out. The engineer pulled the emergency release for the forward bottom hatch and jettisoned the hatch. He also secured the floor cover plate in the upright position so that it would not fall back down and impede the egress of Capt. Stout. The engineer bailed out at about 600 feet agl. The pilot apparently left the controls and attempted to bail out. Before the pilot could safely exit, the airplane went into a climbing turn to the right and fell off on a wing. The airplane entered a steep descending spiral, exploding into flames upon impact with the ground. Investigators were unable to determine what was interfering with the free movement of the controls.

1-12-45G. Albany, Georgia. At 2340 CWT, a North American TB-25J crashed three miles northeast of Turner Field, Albany, Georgia, killing three French Air Force fliers and seriously injuring another. The airplane was in the traffic pattern for Turner Field, flying a wide pattern. Other pilots in the pattern stated that the subject airplane appeared to be flying normally. The subject airplane, flying at 1,000 feet agl, made the turn from base leg to final and then entered a descent until it flew into the ground and burst into flames. Pilot Lt. Olivier R. Mantoux, 26, and B-25 students Sgt. Jacques D. Scheer and Cpl. Georges M. Vassal were killed in the crash. B-25 student Cpl. Armand E. Roux received very serious injuries.

1-13-45. McComb, Ohio. At 2219 CWT, a Beech UC-43 crashed while attempting to make an emergency landing four and a half miles southwest of McComb, Ohio, killing pilot 1Lt. Robert H. Weisberger, 24. The airplane took off at 1946 CWT from Romulus Army Air Base, Romulus, Michigan, on a flight to Wright Field, Dayton, Ohio. The airplane had flown from Wright Field to Romulus earlier in the day. The airplane arrived at Findlay, Ohio, at 2138 and began circling. The pilot, believing he was over Columbus, Ohio, attempted to contact the control tower. The pilot was unable to orient himself and the airplane began running low on fuel. The pilot attempted to make a wheels-down landing on an 80-acre field. The airplane rolled for a few yards before nosing over and smashing itself to pieces. Wreckage was scattered over an area of 200 yards.

1-14-45A. Port Huron, Michigan. At 1655 EWT, a Republic P-47D dove into the ground eight miles north of Port Huron, Michigan, killing 2Lt. Harold E. Grathwohl, 23. The airplane was flying in the number-three position of a three-ship flight that took off at 1630 EWT from Selfridge Field, Mt. Clemens, Michigan, on an aerial gunnery mission. The airplanes were flying at 9,000 feet when the flight leader ordered the airplanes into a trail formation. The flight leader spotted a flight of fighter airplanes and led his flight on a dive past the other formation, passing on the right and below. After pulling up from the fly-by, the flight leader noticed that the number-three airplane was missing. The flight leader searched for the subject airplane for ten minutes without success. Witnesses on the ground observed the subject airplane in a steep dive. The airplane remained in this attitude until it struck the ground and exploded into flames. The airplane dug a ten-foot deep hole in the frozen earth. Investigation revealed that the pilot had blacked out while performing a dive on a previous flight and had not reported it to the flight surgeon.

1-14-45B. Towncreek, Alabama. At 2120 CWT, a Consolidated RB-24E crashed one and a half miles ENE of Towncreek, Alabama, killing four fliers. The airplane took off from Courtland Army Air Field, Courtland, Alabama, at 2100 on a routine night training mission. The airplane entered the traffic pattern and on the downwind leg the pilot radioed the tower for permission to make a touch-and-go landing. The tower advised the pilot to stay in the pattern because of numerous airplanes on the end of the runway waiting for take-off. The pilot acknowledged and went around. There was no further communication with the airplane and the tower. A short time later, witnesses on the ground observed the airplane north of the traffic area in a steep bank to the left. The airplane stalled in the turn and then fell off on a wing. The pilot rolled level but the airplane entered a steep dive. The pilot attempted to level the airplane but it mushed into the ground, bounding back into the air before crashing back to the ground 100 feet further on. The airplane burst into flames as it smashed itself to pieces, coming to rest about 100 yards from the point of the second impact. Killed in the crash were: Capt. Saul Fineman, pilot; 2Lt. William W. Miller, Jr., co-pilot; 2Lt. Theopil C. Polakiewicz, co-pilot; Cpl. Irving E. Barrington, engineer.

1-14-45C. Drummonds, Tennessee. At 2130 CWT, a Beech AT-10BH flying in a severe rainstorm crashed three miles west of Drummonds, Tennessee, killing students A/C Hugh J. Sisson IV and A/C Donald R. Smith. The airplane took off from Blytheville Army Air Field, Blytheville, Arkansas, on a flight to Little Rock, Arkansas. Investigation revealed that the airplane dove into the ground and exploded into flames. Investigators were unable to determine the cause of the accident. Weather at the time of the accident was reported as heavy rain with abundant lightning.

1-15-45A. Santa Elena, Texas. At 1400, two Curtiss P-40N airplanes collided in mid-air six miles west of Santa Elena, Texas, killing 2Lt. Robert L. Garnett, 19, aboard P-40N #42-105882. 1Lt. George H.

Fittell, 27, was able to make a wheels-up landing at Moore Field, Mission, Texas, in P-40N #44-7735, which was seriously damaged. Lt. Fittell was uninjured. Lt. Fittell was flying the lead ship of a three-ship flight that took off at 1335 from Moore Field on a routine formation training flight. The airplanes climbed to 10,000 feet and leveled off. The flight leader ordered the two wingmen to connect their oxygen masks and test their oxygen systems. The two airplanes dropped back and after testing oxygen systems pulled back into a V-formation on the lead ship. The formation climbed to 15,000 feet and the flight began a series of 90-degree crossover turns. The subject airplane was flying in the number-two position when the leader began a 90-degree turn to the right. The flight leader felt a sudden jolt and then his airplane fell into a spin to the right. The flight leader was able to regain control of the airplane despite a seriously damaged starboard wing. The flight leader determined that the airplane was flyable and was able to make a successful belly landing at Moore Field.

1-15-45B. Corona, California. At 1455, a Noorduyn C-64 crashed into Modjeska Peak ten miles south of Corona, California, killing pilot 2Lt. Robert J. Letscher and engineer Pfc. Lee R. Mayes. The airplane took off from Santa Ana Army Air Field, Santa Ana, California, on a flight to San Bernardino Army Air Field, San Bernardino, California. Investigators speculated that the pilot had inadvertently flew up Santiago Wash instead of Santa Ana Wash while trying to navigate visually in poor visibility. Santiago Wash ends in a blind canyon. The airplane collided with trees before smashing into the north slope of Modjeska Peak at an elevation of 3,700 feet.

1-15-45C. Chattanooga, Tennessee. At 1935 CWT, a Boeing B-17G flying in instrument conditions collided with rising terrain and crashed 12 miles northwest of Chattanooga, Tennessee, killing three fliers. The airplane took off at 1715 CWT from Warner-Robins Field, Macon, Georgia, on a flight to Chattanooga. The airplane, flying on instruments at 9,500 feet, contacted Chattanooga at 1837 CWT. The airplane was cleared to descend to 8,000 feet and was told to hold on the northeast leg of the radio beam. At 1856, the pilot reported that he was picking up ice and was requesting a change in altitude because he had no de-icing equipment. The airplane was cleared to descend to 6,000 feet or climb to 300 feet on top of the overcast. The pilot elected to climb to get on top. At 1906, the pilot reported on top at about 8,000 feet. A short time later the airplane was cleared to descend on the north leg of the beam. The airplane began descending over mountainous terrain and collided with Signal Mountain. The airplane struck several trees, severing the port wing and the port horizontal stabilizer. The airplane bounded over a ravine and then smashed into the terrain, exploding into flames upon impact. Wreckage was scattered over a wide area. Pilot 2Lt.

Hugh B. Cannon, 20, co-pilot F/O Roy E. Lavan and engineer Sgt. John I. Waring were killed in the crash.

1-15-45D. Dyersburg, Tennessee. At 2100 CWT, a Boeing B-17F crashed eight miles WNW of Dyersburg Army Air Field, Dyersburg, Tennessee, killing ten fliers. The airplane took off from Runway 31 and a short time later smashed into a swamp at a steep angle, exploding into flames upon impact. The crash site was covered by five feet of water, hampering recovery and investigation. Investigators could not determine the cause of the crash. Killed in the crash were: 2Lt. Charles C. O'Bryan, pilot; 2Lt. George H. Miller, co-pilot; 2Lt. Kirk A. Powers, navigator; 2Lt. Ralph E. Daniels, bombardier; Cpl. James V. Coltrin, engineer; Cpl. Edward J. Lavin, radio operator; Cpl. Robert E. Therrien, gunner; Pfc. Kenneth C. Kalert, gunner; Pfc. Earl E. Taylor, gunner; Pfc. John T. Mullins, tail gunner.

1-15-45E. Sabine Lake, Texas. At 2108 CWT, five fliers were drowned when they bailed out of a Boeing B-17F suffering an engine fire while flying over Sabine Lake near Port Arthur, Texas. Gunner Pfc. John F. Thacker safely parachuted to safety and was rescued from the lake by a civilian craft in the area. Four fliers made a safe landing in the B-17 at Ellington Field, Houston, Texas. The airplane took off from Gulfport, Mississippi, on a practice-bombing mission near Houston, Texas. The crew made a navigational error and the airplane was off course. The airplane headed to the north to pick up its proposed course, but the pilots and the navigator were unable to determine the airplane's exact position. The airplane approached a large town and the navigator speculated, correctly, that it was Port Arthur. Minutes later the pilots were alerted that the number-one engine was on fire. The pilots were unable to extinguish the fire or feather the propeller. The pilot ordered the crew to stand by to bail out. Moments later the engineer notified the pilot that men were bailing out of the rear of the airplane. It was then discovered that the navigator and bombardier had bailed out as well. The engineer advised the pilot that the men had bailed out without their life vests. The number-one engine continued to burn and fell off over the lake. The pilots determined that the airplane was controllable and requested an emergency landing at Ellington Field. Crewmembers killed attempting to parachute to safety were: 2Lt. Robert L. Richardson, navigator; 2Lt. George R. Vaughn, bombardier; Pfc. Louis G. Heavrin, gunner; Cpl. Francis M. Burch, radio operator; Cpl. Gabriale Villano, gunner. Crewmembers uninjured making a successful emergency landing were: Pilot 2Lt. Robert S. MacDonald, co-pilot 2Lt. William F. Keenan and engineer Cpl. Alfred E. Tucker were uninjured and were able to make a safe wheels-down landing at Ellington Field.

1-16-45A. Homestead, Florida. At 0702 EWT, a Consolidated B-24D flew into the ground 12 miles southeast of Homestead Army Air Field, Homestead, Florida, killing engineer Sgt. Fred Mrosko. Co-pilot/B-24 student 2Lt. Arthur H. Scott received serious injuries. B-24 instructor 1Lt. Curtis R. Hovde and co-pilot/B-24 student 1Lt. Joseph S. Higgins were uninjured. Investigators stated, "Take-off was at 0657 EWT and as the aircraft moved down the runway it was slow getting airspeed. After take-off first power reduction was made at 300 feet agl and it was noticed that the number-two propeller governor was out. Climb was continued and at approximately 900 feet flaps were raised. At approximately 1,400 feet, number-two propeller was controlled momentarily by the feather button. It was then decided ... to return to the field ... since nothing could be done for controlling the number-two propeller. A 180-degree turn was made and the control tower was contacted and told of the propeller trouble and [the pilot] asked for landing instructions. The tower acknowledged and gave pilot landing instructions. At this time altitude was approximately 1,000 feet according to reading of altimeter, speed approximately 155 mph. Speed was slow enough so landing gear was lowered. All this time the instructor pilot was controlling number-two propeller intermittently with the feathering button. Student pilot was completing the landing checklist when the aircraft came in contact with the ground. The aircraft bounced once and started to turn to the left when it stopped. The terrain where the crash occurred is a swampy area with scrub pine trees that grow approximately 20 feet high." The airplane did not burn but was totally destroyed.

1-16-45B. Brownfield, Illinois. At 1100 CWT, a Douglas C-47A crashed two miles west of Brownfield, Illinois, killing four fliers. The airplane took off at 0925 CWT from Malden Army Air Base, Malden, Missouri, on a low-altitude (500 feet) navigation flight to Harrisburg, Illinois, to Fulton, Kentucky, and return to Malden Army Air Base. The airplane collided with houses and crashed while apparently attempting an emergency landing on the second leg of the flight. The airplane burst into flames upon impact with the terrain. Investigation revealed that the airplane was on fire before it crashed into two houses. Killed in the crash were: 2Lt. Floyd E. Williams, pilot; 2Lt. Emerson L. Weller, co-pilot; Pvt. Saul Z. Silver, radio operator; Pvt. Frank C. Lee, radio operator.

1-16-45C. Raymondville, Texas. At 1130 CWT, a Consolidated B-24J crash landed 19 miles ENE of Raymondville, Texas, killing five fliers and injuring ten others, five of them seriously. Pilot 1Lt. Stanley M. Brain and B-24 student F/O Joseph Burey were killed in the crash landing. Engineer SSgt. John J. Shea, gunnery instructor SSgt. James B. Fineran and student gunner Pvt. Kenneth R. Herbert were seriously injured in the crash landing. Co-pilot/B-24 student F/O Raymond D. Bell, Jr., student gunners Pvt. Robert G. Heath and Pvt. Frank F. Hill II suffered minor injuries in the crash landing. Gunnery instructor TSgt. Earl L. Love and student gunner Pvt. Howard Easterling were injured parachuting to safety. Student gunner Pvt. William J. Dobes and Pvt. Kenneth R. Gordon received serious injuries parachuting to safety. Gunnery instructor 1Lt. Wayne M. Ralston and student gunners Pvt. Allen W. Fritzsche and Pvt. Kenneth G. Hicks were killed when they bailed out at an altitude too low to allow their parachutes to deploy. The airplane was on a high-altitude camera-gunnery mission. The crew had completed the high-altitude portion of the mission and the airplane had let down to 7,000 feet. The pilot alerted the crew that he was going to practice propeller-feathering procedure for the student co-pilots on board. The pilot feathered the number-three propeller and then feathered the number-four propeller. The pilot trimmed the airplane for differential powered flight. The pilot maneuvered the airplane and went over the procedure with the students. The pilot, in an effort to unfeather the number-three and number-four propellers, inadvertently began feathering the number-one and number-two propellers. The student gunner in the top turret stated that when he looked outside of the aircraft all four propellers appeared to be feathered. The airplane began losing altitude as the instructor began to restart the engines. The number-three and number-four engines were successfully restarted but the pilot was unable to get adequate power from the number-one and number-two engines. The airplane, in a left-wing low attitude, continued to lose altitude and at 2,000 feet the pilot ordered the crew to bail out. After a few crewmembers had bailed out the airplane had reached an unsafe altitude and the order was rescinded. The pilot was unable to arrest the descent and alerted the crew to prepare for a crash landing. The airplane continued to drop in a left-wing low attitude as the pilots struggled to depress the right rudder pedal. They were unsuccessful because the airplane was trimmed to fly on the two port engines. Full right aileron was unable to bring up the port wing and the airplane continued toward the ground in a descending left bank. The pilot extended the landing gear just before impact. Investigators stated, "The airplane struck the ground in a crabbing attitude to the left and rolled approximately 200 yards where it went over a sand dune and dropped about 10 or 12 feet where it came to rest heading almost 90 degrees to the left of its path on the ground. The left main landing gear and the nose wheel sheared off when the airplane went over the sand dune. [The] nose section was demolished and the fuselage split at about the Martin Turret upon contact with the ground. The pilot, who evidently did not have his safety belt fastened, was thrown from his seat and killed. One of the student co-pilots who was sitting in the seat on the left side of the flight deck was [thrown from his seat and] killed." The airplane burst into flames as it smashed itself to pieces and was completely destroyed.

1-16-45D. Nelson, New York. At 1315 EWT, a Boeing TB-17F, lost and nearly out of fuel, crash-landed one mile northwest of Nelson, New York. Pilot

1Lt. Alfred Kramer and co-pilot 2Lt. William M. Boothby were uninjured in the crash landing. Seven crewmembers parachuted to safety, suffering only minor injuries. Bombardier 1Lt. Charles F. Blanke fell to his death in an unsuccessful parachute jump. Crewmembers parachuting to safety were: 1Lt. Coleman Sanders, navigator; SSgt. Henry F. Penna, engineer; Pvt. Michael M. Wall, student engineer; Pfc. Thomas M. Cox, radio operator; 2Lt. Ralph E. Anderson, pilot-rated passenger; 2Lt. Howard F. Amrhein, pilot-rated passenger; Capt. Norman W. Riddulph, pilot-rated passenger. The airplane took off at 0143 EWT from Hendricks Field, Sebring, Florida, on a navigation-training flight to Jacksonville, Florida, to Nashville, Tennessee, to Cincinnati, Ohio, to Pittsburgh, Pennsylvania, to Stewart Field, Newburgh, New York, and return to Hendricks Field. The airplane climbed to 11,000 feet and was "on top" of the overcast for the majority of the flight. When the airplane arrived at Pittsburgh the crew was advised that weather east of Syracuse, New York, was very bad and that Stewart Field was closed due to bad weather. The pilot requested a landing at and was cleared to land at Syracuse, which was his alternate. The airplane arrived at Syracuse at 1024 EWT and found that the field was closed due to poor visibility. The pilot was advised to continue to Burlington, Vermont, which was still open. The pilot was unable to find Burlington and returned to the area of Syracuse. The pilot was still unable to find Syracuse and, with approximately an hour and a half of fuel left on board, began to circle the area of the field at an altitude of 5,000 feet. The pilot was then cleared to fly to Rome Army Air Field, Rome, New York, on a special clearance. The airplane arrived in the area of Rome but the crew was unable to locate the field. The pilot was advised to continue to look for the field until his fuel was nearly exhausted. The pilot was advised that in the event that he was unable to find the field he was to point the airplane to the north before ordering the crew to bail out. The pilot was unable to find the field so he climbed the airplane to 2,800 feet and ordered the crew to abandon the airplane. The crew abandoned the airplane in good order. The pilots determined that all of the crewmembers had bailed out but because of diminishing altitude they elected to make a wheels-up crash landing. The crash landing was made successfully 22 miles southeast of Syracuse Army Air Base and the pilots were uninjured.

1-16-45E. Thomasville, Georgia. At 1845 EWT, a Curtiss P-40N crashed 15 miles northwest of the Army Air Field at Thomasville, Georgia, killing pilot F/O George B. Vidrine, 28. The airplane had been on an instrument training flight. The pilot had been practicing "homings" for some time when he requested a vector back to Thomasville. The airplane was seen flying at high altitude before it entered a 45-degree dive. The airplane remained in the dive until it struck the ground and exploded. Investigators speculated that the pilot's life raft had inadvertently inflated in the cockpit during flight, incapacitating the pilot. The cause of the accident was undetermined.

1-17-45A. Shaw Field, South Carolina. At 0005 EWT, a North American AT-6C crashed one mile west of Shaw Field, Sumter, South Carolina, killing student A/C Robert J. Kohanek, 19. The student apparently inadvertently flew the airplane into the terrain soon after taking off from Shaw Field on a scheduled night-training flight. The airplane was found at about 1300 EWT.

1-17-45B. Elmore, Alabama. At 0210 CWT, a North American AT-6D crashed three miles southwest of Elmore, Alabama, killing French Air Force student pilot Sgt. Louis J. Saly, 24. The airplane took off from Gunter Field, Montgomery, Alabama, on a routine training flight. The student had performed several night take-offs and landings at Elmore Auxiliary Airfield with his French instructor on board. The student demonstrated proficiency in night flying and was allowed to fly a night solo. The student flew to his assigned zone and circled to the left at 2,500 feet. The airplane circled for a short time and then was seen to fall vertically to the ground. The engine was heard operating at high rpm just before the airplane smashed to the ground.

1-17-45C. Victorville, California. At 1040, a Lockheed P-38J crashed 15 miles west of the Army Air Field at Victorville, California, killing 2Lt. Robert J. Sleske, 26. The airplane was part of a four-ship flight that took off from Van Nuys Airport, Van Nuys, California, on a dive-bombing mission and individual combat mission. The flight completed the dive-bombing mission and assembled south of the bombing range at 10,500 feet for the combat portion of the mission. After about ten minutes of maneuvering, the flight leader observed the subject airplane explode as it smashed into the ground. Investigation revealed that the airplane struck the ground at high speed in a nose down attitude, exploding violently into flames upon impact.

1-17-45D. Page Field, Florida. At 1445 EWT, a Curtiss P-40N attempting a landing crashed at Page Field, Fort Myers, Florida, killing F/O Hugh C. Walker, 22. The pilot made a normal approach but leveled off too high. The tower officer advised the pilot to add throttle and drop the nose. The pilot applied some power and "dropped" the airplane in. The airplane collided with the runway with the port wing tip and landing gear, sending the airplane off of the pavement approximately 80 degrees to the take-off heading, assuming a three-point position. The tower officer advised the pilot to cut throttle but the pilot added power and attempted to get airborne. The pilot pulled up sharply to avoid trees near the edge of the field. The airplane fell off on a wing and the port wing "dug in," causing the airplane cartwheeled into the ground. The airplane burst into flames as it smashed to pieces.

1-17-45E. Lancaster, California. At 1250, a Lockheed P-38J crashed 12 miles northwest of Lancaster, California, killing 2Lt. Austin G. Strickland, 24. The airplane was flying in the number-four position of a four-ship flight that took off from Van Nuys Municipal Airport, Van Nuys, California, on a skip bombing mission and individual combat mission. The flight completed the skip bombing mission and assembled at 13,000 feet for the individual aerobatics. The flight leader noticed the subject airplane in a flat spin at an altitude of 8,000 feet. The airplane remained in the flat spin until it stuck the ground and exploded into flames. The pilot made no apparent effort to bail out.

1-17-45F. Quantico, Virginia. At 2010, a Cessna UC-78B was abandoned and crashed five miles northwest of Quantico, Virginia. Passenger 1Lt. Henry J. Warke disappeared after parachuting from the airplane and is presumed lost at sea. Pilot Capt. Gerard S. Meacham and co-pilot Capt. Stanley K. Hanson were uninjured. The airplane became lost and low on fuel while flying from Raleigh-Durham, North Carolina, to Camp Springs Army Air Field, Washington, D.C. and return. The pilots became lost and descended to low altitude to find a landmark. The pilots spotted the hangars and flight line at Quantico Marine Corps Air Station, buzzing the field six times in an apparent attempt to find the runway. The pilot was unable to locate runway for an approach. The Quantico tower advised the pilot to climb, point the aircraft east and abandon the aircraft. The pilots complied and were rescued a short time later. Despite a great search effort, Lt. Warke remained missing.

1-17-45G. Bay Minette, Alabama. At 2145 CWT, a Boeing B-17F suffering an in-flight fire crashed out of control six miles northwest of Bay Minette, Alabama, killing eleven crewmembers. The airplane took off from Gulfport Army Air Field, Gulfport, Mississippi, on a 1,000-mile cross-country flight to Savannah, Georgia, and return. The airplane transmitted a routine position report at 2133, reporting 40 miles east of Mobile, Alabama. At 2136 the crew reported that the aircraft was on fire. The airplane crashed into a swamp a short time later, exploding violently into flames upon impact. Pieces of the airplane's engine ring cowl and accessory section were found strewn over a mile from the point of impact. Investigation revealed that the number-one engine had caught fire due to a failed cylinder. Killed in the crash were: 2Lt. Richard P. Poe, pilot; 2Lt. Douglas W. Mallory, co-pilot; 1Lt. Sydney Abramson, navigator instructor; 2Lt. Ivan P. Richey, navigator; 2Lt. Randall A. Weaver, bombardier; Pfc. Clayton E. Plopper, engineer; Cpl. James B. Fleming, radio operator; Cpl. Thomas M. Rutledge, gunner; Pfc. James C. White, gunner; Pfc. George H. Smith, gunner; Pfc. John R. Maffei, gunner.

1-17-45H. Marana, Arizona. At 2220 MWT, a Vultee BT-13A crashed into structures at Avra Auxiliary Airfield, 11 miles SSE of Marana Army Air Field,

Marana, Arizona, killing Pvt. Thomas N. Burch on the ground. Chinese Air Force student A/C Fei-Peng Wu suffered minor injuries. The Accident Classification Committee stated, "A/C Wu landed at Avra Auxiliary Field and was taxiing back to the take-off position preparatory to taking off for a second landing. The student was in the take-off position lined up with the take-off runway when given the green light by the control tower. After receiving the green light the student turned the airplane approximately 45 degrees to the left and gave it the throttle for take-off. At this point the airplane was lined up in a direct path with a stage house in the center of the field. Control tower operators were unable to take action soon enough to prevent the airplane from crashing directly into the stage house." Pvt. Burch, an army truck driver, was inside the stage house when the airplane smashed completely through the structure, killing him instantly.

1-18-45A. Valparaiso, Florida. At 1030 CWT, a North American AT-6B crashed eight and one-half miles northeast of Eglin Field, Valparaiso, Florida, killing pilot F/O Harry G. Modes, 21. The airplane was flying in the number-five position of a five-ship flight that took off from Auxiliary Airfield #2, Eglin Field Military Reservation, on a ground gunnery mission. The flight made one pass at the ground gunnery range without incident. On the second pass, the subject airplane made a very steep approach on the target. The airplane cleared the target by 20 feet during the pull-out. The airplane disappeared behind the target as it mushed downward. The airplane was seen to emerge and then to climb steeply to 75 feet. The airplane rolled over to an inverted attitude and then dove into the ground, exploding into flames upon impact.

1-18-45B. Mountain Home, Idaho. At 1130 MWT, a Consolidated B-24J crash-landed seven and one-half northwest of the Army Air Field at Mountain Home, Idaho, killing three crewmembers and injuring two others. Four other crewmembers escaped injury. Investigators stated, "A normal take-off was made by [the subject aircraft] on Runway 30. The pilot states that he thought he was in [propeller turbulence] and at around 300 feet, he skidded the plane to the right. The plane was climbing at 160 mph and 700 feet per minute. Due to the [propeller turbulence], the pilot was using 40 inches [Hg of manifold pressure] and 2,500 rpm. The plane continued to vibrate after becoming free from all influence of [propeller turbulence]. Ten degrees of flaps were raised. When the plane reached an altitude of 700 feet, it began to lose altitude. A turn of 90 degrees was made to the left in an effort to get back to the field. The pilot realized that he could not make the field and crashed landed the plane. In trying to come out of the turn to the left, the pilot had to use full rudder control because the ailerons were inoperative. Elevators were used in an effort to raise the nose of the plane but no reaction was received from them. The nose was raised prior to crashing by using

full power on all engines." The airplane hit the ground violently at a flat angle and was completely destroyed. Investigators speculated that the airplane had excessive ice on the airframe and control surfaces that was not removed prior to the take-off. Navigator F/O Ronald K. Peterson, radio operator Cpl. Robert L. Anderson and gunner Pfc. Walter H. Walker were killed in the crash. Pilot 1Lt. Saul Milcham and gunner Cpl. Richard P. Reed received minor injuries. Uninjured in the crash landing were: F/O Fred J. Masek, co-pilot; Cpl. Samuel R. Bressler, engineer; Pfc. Charles R. Herrell, gunner; Cpl. Emanuel P. Audino, gunner.

1-18-45C. Atlantic Ocean, New Jersey. At 1735 EWT, a Republic P-47G crashed into the Atlantic Ocean ten miles east of Cape May, New Jersey, killing 2Lt. Howard S. Frankel, 21. The airplane was part of a five-ship flight that took off from Dover Army Air Field, Dover, Delaware, on a gunnery mission. The subject airplane made four successful passes on the towed-target. On the fifth pass, the subject airplane collided with the target. Witnesses stated that the airplane pulled up with the guns still firing. The airplane entered a steep vertical spin, remaining in this attitude until it struck the sea. The body of the pilot was not recovered.

1-18-45D. American Corners, Maryland. At 1740, a Republic P-47D crashed one-half mile west of American Corners, Maryland, killing 2Lt. Stuart D. Kestenbaum, 22. The airplane was flying in the number-three position of a four-ship flight that took off from Camp Springs Army Air Field, Washington, D.C., on a high-altitude formation mission. The flight climbed to 30,000 feet and maneuvered for a short time. The flight leader maneuvered the formation to 25,000 feet where they flew for a few minutes. The flight leader then began a series of diving turns until the formation reached 16,000 feet. The flight leader entered a steep diving turn and leveled out at 10,000 feet at 350 mph. The subject airplane, which had assumed the number-two position, had dropped out of the formation and was last seen in an inverted dive. The airplane failed to return to base and it was later found to have crashed. Investigation revealed that the airplane struck the ground inverted at a 30-degree angle, exploding violently into flames upon impact.

1-18-45E. Alamogordo, New Mexico. At 1602 MWT, a Boeing B-29 suffering an in-flight fire crashed into large sand dunes while attempting an emergency forced landing ten miles southwest of Alamogordo Army Air Field, Alamogordo, New Mexico, killing six fliers and injuring four others. Investigators stated, "The airplane was taking off in a normal manner and shortly after the wheels were retracted, number-one engine caught fire and the crew was unable to extinguish the flames with the fire extinguisher after the propeller was feathered. The pilot continued straight ahead and landed in extremely rough terrain, apparently being more interested in landing immediately due to the fire, than in making any determined or prolonged effort to remain airborne or to get back to the field or selecting a more suitable landing spot." Killed in the crash-landing were: F/O William T. Raynes, co-pilot; 2Lt. Louis R.K. Smith, navigator; 2Lt. Frank Marek, navigator instructor; TSgt. Clarence V. Peterick, flight engineer; Pvt. Elbert R. Stratton, radio operator; SSgt. Orville H. Robertson, radio operator instructor. Crewmembers seriously injured were: 1Lt. Jack S. Hart, pilot; 2Lt. John C. Myers, Jr., bombardier; Pvt. Carter H. Lippincott, central fire controller; Cpl. Eugene M. Wheeler, central fire controller. Gunners Sgt. Robert W. Jones, Pfc. Hale D. Hampton and Pfc. Robert E. Jones, Jr. were uninjured.

1-18-45F. Tylertown, Mississippi. At 1910 CWT, a Vultee BT-13B flying in poor weather crashed 12 miles northwest of Tylertown, Mississippi, killing pilot Capt. Bernard J. Shapiro and passenger Capt. Harry D. Mason. The airplane took off from Ellington Field, Houston, Texas, on a flight to Key Field, Meridian, Mississippi. The airplane was seen flying north in a heavy rainstorm at about 700 feet agl. The engine was heard to stop suddenly. The airplane was then seen to make a "sharp gliding turn" to the right, crashing into the ground at a 45-degree angle heading southeast.

1-18-45G. Seymour, Indiana. At 2346 CWT, a Beech AT-10BH crashed four miles west of Freeman Field, Seymour, Indiana, killing students A/C Kenneth J. Pocta and A/C John S. Rush. The airplane was returning to Freeman Field after completing a routine student-training mission. The students overshot the runway on the approach and went around. The airplane flew into the ground a short time later.

1-18-45H. Casper, Wyoming. At 2254 MWT, a Consolidated B-24J crashed while attempting a take-off at Casper Army Air Field, Casper, Wyoming, killing engineer Pvt. Milton B. Long. Bombardier 2Lt. James F. Benjamin and co-pilot F/O Mitchell W. Mateik received serious injuries. Crewmembers injured in the crash were: 2Lt. Donald R. Bachelder, pilot; 2Lt. William H. White, Jr., navigator; Cpl. Donald W. Brogen, radio operator; Cpl. Irving L. Francisco, gunner; Pfc. Martin R. McCoy, Jr., gunner; Pfc. Waldemar V.W. Radesk, gunner; Pfc. Everett J. Stroud, gunner. The airplane made a normal take-off from Runway 34 on a local transition mission. Investigators stated, "Just after [the landing gear] had been brought up and instruments were read and found normal, the pilot noticed a flash to his left and found the number-one engine on fire. He immediately called the tower and informed them he was going to make an emergency landing. He called for number-one fuel valve to be turned off. The tower was unable to turn on all runway lights at once, [the tower personnel] followed the path of the aircraft and when [the pilot] headed towards the field, [tower personnel] turned on lights for Runway 7. Pilot did not put landing gear down as he

felt it would be safer to belly land, as he was landing quite far down the runway. He did not attempt to feather [the number-one] propeller as he felt the drag of the windmilling propeller would aid him in getting in. The engineer cut the fuel just before the aircraft struck the ground and the fire had gone out. The pilot turned past the runway on his approach and then, rather than make a low turn back to the runway, chose to set it down on the left side of it. Aircraft struck on a small rise and slid for about 200 feet." The airplane burst into flames as it smashed to pieces. Investigation revealed that the number-one engine suffered a failed number-twelve cylinder head. The cylinder head failed and split in halve because of small stress cracks located around the spark plug holes.

1-19-45A. Goldsboro, North Carolina. At 0620, a Republic P-47D crashed five miles southeast of Goldsboro, North Carolina, killing pilot F/O Rolf H. Maakestad, 22. The airplane was the last of an 18-ship flight that took off from Seymour-Johnson Field, North Carolina, on a night formation mission. The subject airplane made a normal take-off and seemed to be flying normally as it passed over the end of the runway. The subject airplane entered a gentle left turn in an effort to join up with the formation. The airplane lost altitude as it turned and collided with the ground with the port wing. The airplane cartwheeled into the ground and exploded into flames. The pilot evidently lost the horizon and inadvertently flew into the ground while in the turn.

1-19-45B. Hunter Field, Georgia. At 0721 EWT, a Boeing B-17G crashed just after take-off from Hunter Field, Savannah, Georgia, killing six fliers and seriously injuring four others. The subject airplane was in the second position of an 18-ship flight that was taking off from Runway 10 on an "Aerial Port of Embarkation" flight. Weather at the time was reported as 3,200 feet ceiling, visibility four miles with light smoke and light rain. Wind was calm and "night conditions prevailed." Investigators stated, "The pilot was observed to make a normal take-off, continued straight ahead and gained an altitude of 50 feet to 100 feet. He continued straight ahead and flew his airplane at such an attitude that altitude was lost and he crashed into three trees in rapid succession between 2,400 feet and 2,500 feet from the end of Runway 10. Prior to contacting the trees he passed over a power-line, which was 2,200 feet from the end of runway and 21 feet higher than the end of the runway.... [His] flight path continued until his left wing hit the first of three trees 300 feet beyond the power-line at a point 13.77 feet above the elevation of the runway, which indicates that his flight path was in a descending attitude at that point. Upon contacting the trees the airplane started to disintegrate and proceeded 600 feet beyond the point of contact with the trees at which point it came to a stop and exploded and burned at a point approximately 3,000 feet from the end of the runway." Killed in the crash were: 2Lt.

Henry M. Hale, Jr., pilot; Cpl. Wallace E. Matthews, engineer; Cpl. John E. Sheerin, gunner; Cpl. Edward J. Dul, assistant radio operator; Sgt. Lee G. Ross, gunner; 2Lt. Richard E. Maynard, pilot-rated passenger. Crewmembers seriously injured in the crash were: F/O Ray E. Eby, co-pilot; 2Lt. Richard E. Cribbs, navigator; Cpl. Theodore Michalak, radio operator; Cpl. Frank F. Fleischer, assistant engineer. The remaining sixteen airplanes took off without incident.

1-19-45C. Gulf of Mexico, Florida. At 1155 EWT, two Curtiss P-40N airplanes collided in mid-air over the Gulf of Mexico 20 miles west of Page Field, Fort Myers, Florida, killing 2Lt. Alexander S. Robinette, 25, aboard P-40N #44-7023. 2Lt. Paul H. Skogstad, 21, was uninjured and able to make a safe crash-landing in P-40N #44-7045. The airplanes were part of a nine-ship flight that took off from Page Field on an aerial gunnery mission, which was to be conducted at 5,000 feet. The first four-ship flight completed its firing passes at the towed target. Investigators stated, "Lt. Skogstad, leading the second flight, was firing on the tow target with his number-two man. The number-two man left the pattern because he had finished firing and the number-three pilot, Lt. Robinette, was making his first pass. Lt. Robinette's plane was behind and below Lt. Skogstad, who was therefore unable to see him. Evidently, Lt. Robinette was concentrating on the target so intently that he never saw the first airplane until he had flown into him. The collision occurred as Lt. Skogstad finished firing. Lt. Robinette's plane struck Lt. Skogstad's aircraft from beneath. Lt. Robinette's plane then rolled over and dove into the Gulf of Mexico at the north end of Sanibel Island about two miles from Red Fish Pass. Lt. Skogstad was able to bring his plane back and landed wheels-up on [Page Field]."

1-19-45D. Santa Rosa, California. At 1210, a Lockheed TP-38J crashed seven miles southwest of Santa Rosa Army Air Field, Santa Rosa, California, killing 2Lt. Edward C. Carpenter, 21. The subject airplane was the number-four ship in a four-ship flight that took off from Santa Rosa Army Air Field on a high-altitude formation flight. The airplanes climbed to 32,000 feet and leveled off. The flight leader called for an oxygen check. All pilots replied and all acknowledged that everything was all right. The flight made a 360-degree turn to the left, leveling out in a slight climb. The subject pilot then thought he had "fumes" in the cockpit, reporting it to the flight leader. The flight leader advised the pilot to turn oxygen to "pure oxygen." The pilot replied that he had. The subject airplane was then seen to climb abruptly out of the formation, entering a gentle turn to the right that progressed into a steep spiral to the right. The airplane remained in this attitude until it struck the ground at high speed and at a steep angle, exploding into flames upon impact. The pilot made no apparent effort to bail out. Investigation speculated that the pilot might have

affixed his oxygen hose on his parachute leg strap, causing the connection to be placed in "tension" and the connection might have failed while the mask was in use. Investigators stated that the sudden loss of oxygen caused the pilot to think that he had fumes in the cockpit. The pilot was quickly incapacitated and lost control of the airplane.

1-19-45E. Maybrook, New York. At 1647, a North American AT-6D crashed two miles northeast of Maybrook, New York, killing pilot 2Lt. Maurice D. Campbell and pilot-rated passenger 2Lt. George F. McElroy. The pilots were part of a group of ten that were waiting for overseas assignment at Camp Kilmer, New Jersey. The pilots were ordered to Stewart Field, Newburgh, New York, for proficiency flying. The two pilots flew earlier in the day with Lt. McElroy as pilot and Lt. Campbell as passenger. After a rest period of an hour and a half, Lt. Campbell and Lt. McElroy switched positions. A short time later the airplane smashed into a wooded area at 30-degree angle. The only witnesses to the crash were two children under the age of eight. Their testimony was considered "limited" in scope and largely "unreliable."

1-19-45F. Santa Maria, California. At 1513, a Lockheed P-38L suffering an in-flight fire crashed seven miles northwest of Santa Maria Army Air Field, Santa Maria, California, killing pilot 2Lt. Richard V. Durham, 21. The airplane was flying in the number-four position of an eight-ship flight that had taken off from Santa Maria Army Air Field on a ground gunnery mission and unit combat exercise. After completing the ground gunnery mission, the flight climbed to 11,000 feet for the combat exercise. The formation split up into two, four-ship elements. Moments later, the subject pilot radioed that his starboard engine was on fire. The pilot began to shut down the engine and feather the propeller when he radioed that his port engine was "running away." The port engine then burst into flames. The subject airplane passed over Santa Maria Army Air Field at 7,000 feet. Witnesses on the ground observed the port engine on fire and trailing thick smoke. The airplane entered a shallow descent until it struck the ground and exploded. Investigation revealed that the port engine suffered a rod failure because of the failure of connecting bolts, causing uneven stress on the "wrist pin."

1-19-45G. El Paso, Texas. At 2238 MWT, a Beech AT-7 and an AT-7C collided near El Paso Municipal Airport, El Paso, Texas, killing four fliers aboard the AT-7C. Student navigator A/C Edward J. Struckus was seriously injured in the crash of the AT-7C. The AT-7 was able to make a safe landing at El Paso Municipal Airport and the four fliers on board were uninjured. Killed in the crash of the AT-7C were: 2Lt. Richard H. Lass, pilot; SSgt. Samuel K. Hollis, engineer; A/C Gordon T. Sugden, student navigator; A/C William F. Stump, student navigator. Uninjured aboard the AT-7 were: F/O Robert E. Raper, Jr., pilot; A/C

Crosby J. Vail, student navigator; A/C Armand A. Tougas, student navigator; A/C John E. Vanden-Bosch, student navigator. The airplanes were two of a number of student AT-7s that took off at 2009 CWT from San Marcos Army Air Field, San Marcos, Texas, on separate navigation flights to El Paso Municipal Airport. The subject airplanes arrived in the area of El Paso at the same time and neither contacted the control tower prior to entering the traffic pattern. Both ships were observed on long final approach to Runway 4. The control tower operator observed that a potential conflict was developing and attempted to contact the airplanes without success. One aircraft was on a high approach and the other was on a lower approach. It became apparent that the aircraft on the higher approach was overtaking the lower aircraft and that a dangerous situation was developing. The control tower operator gave the upper aircraft the red light to prevent a landing or a mid-air collision but was unsuccessful. The upper airplane came down on the lower airplane, which crashed onto the field and exploded.

1-20-45A. Muroc, California. At 0115, a Consolidated B-24J crashed two and a half miles southeast of Muroc Army Air Field, Muroc, California, killing ten fliers. The airplane took off from Muroc Army Air Field on a bombing training mission. The crew completed the bombing mission on the AAF range east of Muroc and returned to the area of the field for a landing. The airplane entered the traffic pattern and encountered instrument conditions while flying on the downwind leg for Runway 24. The airplane failed to land at Muroc and the pilot of another B-24 in the pattern observed a fire on the ground southeast of the field. Investigation revealed that the airplane flew into the ground on the base leg for Runway 24. The airplane struck the ground at a flat angle in a left-wing low position and burst into flames as it smashed itself to pieces. Wreckage was scattered over an area of 1,000 feet. Investigation revealed that the weather in the area of the field had deteriorated at about 0100. Lowering ceilings and light snow flurries had moved into the area, hampering visibility. Other pilots attempting landings at the field reported that they had encountered instrument conditions and snow while flying in the traffic pattern. Killed in the crash were: 2Lt. Frederick J. Vieson, pilot; 2Lt. Joseph C. Vanderdecker, co-pilot; F/O Joseph C. Syper, navigator; 2Lt. William R. Livermore, bombardier; Cpl. Jack T. Seymour, engineer; Sgt. John J. Hardy, engineer; Cpl. Leslie J. Raber, radio operator; Pfc. Jack C. Newcomb, radio operator; Cpl. John J. Vera, gunner; Pfc. George B. Well, gunner.

1-20-45B. San Bernardino, California. At 0955, a Lockheed TP-38L crashed 12 miles NNW of San Bernardino Army Air Field, San Bernardino, California, killing pilot 1Lt. Walter J. Carroll, 27. The subject airplane was the lead ship of a four-ship flight that took off from Ontario Army Air Field, Ontario, California, on a low-level navigation flight and skip-bombing mission. The flight leader, leading the flight at very low altitude, collided with a high-tension wire. The airplane went out of control, rolling to the left three times before smashing to the ground and exploding into flames. The pilot was killed instantly.

1-20-45C. Benson, Arizona. At 1200 MWT, a North American B-25D flying in instrument conditions collided with rising terrain 15 miles northwest of Benson, Arizona, killing five fliers. The airplane had taken off at 0917 CWT from Kelly Field, San Antonio, Texas, on a routine transition flight to Tucson Municipal Airport No. 2, Tucson, Arizona. Investigators stated, "At 1144 MWT the pilot called Tucson Radio, said he was 13 miles east and requested permission to make a CFR [visual] approach to Tucson Municipal Airport No. 2. At about 1200 MWT, three ranchers were driving along Happy Valley Road at a point approximately 4,300 feet in elevation about five miles slightly south of east of Rincon Peak. A B-25 passed over their heads at what they estimated to be 300 feet. The plane was in a climbing turn to the left. At that time there were clouds obscuring the mountains and snow flurries obscuring the lower hills and valleys. The cloud bases were estimated at that time to be at about 6,000 feet. Visibility was restricted by snow flurries to one-quarter to one-half mile. The ranchers immediately left the automobile and estimate that approximately 30 to 40 seconds after they saw the plane pass overhead they heard the crash. There was no diminution of the roar of the engines until the sound of the crash. The crash occurred at about 6,900 feet against the north face of the south side of a large canyon, approximately 250 feet from the top of the canyon. The airplane struck against the bare rock face of the canyon wall, disintegrated to a large extent, and fell approximately 50 feet to a ledge below. The largest section remaining was the tail surface and approximately ten feet of the fuselage in front of the tailskid. Wreckage was scattered below the point of impact over an area extending 300 feet wide and 600 feet down the canyon." Investigators speculated that the pilots were in error in their estimation of their distance from the field. The pilots evidently discovered their error and attempted to climb out. Killed in the crash were: Maj. Lorin D. Geil, pilot; 2Lt. Floyd W. Halonen, co-pilot; Pvt. Frank J. Tomczak, engineer; Maj. Francis R. Lawther, passenger; 1Lt. D. Loveman, passenger.

1-20-45D. Bend, Oregon. At 1245, a Lockheed P-38L crashed 33 miles southeast of Bend, Oregon, killing 2Lt. Richard J. Trienen, 21. The airplane took off from Redmond Army Air Field, Redmond, Oregon, on an individual combat mission. The subject airplane engaged in simulated combat with another aircraft at about 11,000 feet, making several passes. The subject airplane made a pass at the target airplane and then entered a roll as it broke away. The airplane went out of control and fell away. The airplane was reported to be in a continuous "barrel roll" as it dove to earth at a

60-degree angle. The airplane remained in this attitude until it struck the ground and exploded into flames.

1-20-45E. Arroyo Grande, California. At 1712, two Lockheed P-38L airplanes collided in midair and crashed two miles northeast of Arroyo Grande, California, killing F/O Benjamin B. Dutton aboard P-38L #44-23903. Flight leader1Lt. Frank C. Shearin, Jr. parachuted to safety from P-38L #44-25984 and was uninjured. The airplanes were part of a four-ship flight that took off from Santa Maria Army Air Field, Santa Maria, California, on a ground gunnery and unit combat mission. The flight completed their gunnery mission at the Point Sal Gunnery Range, reassembled at 6,000 feet, and then climbed to 9,500 feet. The airplanes broke up into two elements and then engaged in simulated combat. The two elements maneuvered until they were approaching each other head-on. The surviving pilot stated that he observed the other element approaching from the one o'clock position. The pilot maneuvered in an attempt to avoid the collision but was unsuccessful. The two element leaders collided at 10,500 feet. The tail of Lt. Shearin's airplane was sheared off so he bailed out. The other airplane went out of control and spun to earth. The pilot made no effort to leave the airplane.

1-21-45A. Myrtle Beach, South Carolina. At 1200, a Vultee-Stinson L-5 crashed into a swampy area 18 miles southwest of Myrtle Beach Army Air Field, Myrtle Beach, South Carolina, killing pilot 1Lt. Clyde Doyle, Jr. Passenger TSgt. Robert E. Flynn was very seriously injured and was unable to give a statement to investigators. There were no witnesses to the accident.

1-21-45B. Hendricks Field, Florida. At 2150 EWT, a Vultee BT-13A crashed seven and one-half miles northwest of Hendricks Field, Sebring, Florida, killing 1Lt. Connor C. Myers, Jr., 27. The airplane took off from Jacksonville Army Air Field, Jacksonville, Florida, on an instrument training flight to Hendricks Field. The airplane was observed flying at about 2,000 feet agl about ten miles from Hendricks Field. The engine was heard to be "cutting in and out" as the airplane flew away. The airplane was observed to enter a gentle turn to the left and was last observed in a glide with the engine cutting out. The pilot apparently stretched his glide while attempting an emergency forced landing, stalling the airplane at an altitude too low to allow recovery. The airplane smashed to the ground and burst into flames upon impact, killing the pilot instantly.

1-22-45A. Waco, Texas. At 0010 CWT, a Beech AT-7 crashed while attempting an emergency forced landing six miles south of Waco, Texas, killing navigator 2Lt. Lyman L. Watrous, II. Pilot 2Lt. James R. Milton and navigator 1Lt. David W. Baris received serious injuries. The airplane had taken off from Smyrna Army Air Base, Smyrna, Tennessee, on an instrument navigation flight to Waco Army Air Field and return. The airplane became lost and low on fuel so the pilot attempted an emergency forced landing.

1-22-45B. Langley Field, Virginia. At approximately 0700 EWT, navigator 2Lt. Mark E. Daniel disappeared in-flight while on board a Consolidated B-24L that was operating over the Atlantic Ocean near Langley Field, Hampton, Virginia. The navigator's parachute was found on board and all aircraft hatches were secure and found to be in the closed position. The airplane, which had taken off from Langley Field at 0640 on a radar bombing training mission, was piloted by 1Lt. Sidney W. Paul and co-pilot F/O Perry B. Chance. Lt. Daniel was last observed going aft to lower the radar dome for the mission. He was never seen again.

1-22-45C. Carolina Beach, North Carolina. At 1050 EWT, a Douglas A-20G crashed while attempting landing on a civilian airfield four miles south of Carolina Beach, North Carolina, killing pilot F/O Clarence H. Olsen, 19. Gunners Cpl. William J. Russell and Cpl. Arthur E. Elvin received serious injuries. Investigators stated, "F/O Olsen and crew took off from [Myrtle Beach Army Air Field] Myrtle Beach, South Carolina, on a bombing mission at 0900 EWT. Arriving over the target at approximately 0915, he commenced his bomb run but could not release his bombs. He could not locate the master bomb switch, and spent several minutes locating it. After jettisoning his bombs, F/O Olsen turned back to base. This was at approximately 0930 EWT. After cruising along the beach for several (undetermined) minutes, contact was made with Myrtle Beach — conversation unknown — but this contact faded after a while, and the pilot did not locate the base. After circling and cruising along the beach for an undetermined number of minutes, he became hopelessly lost. He sighted a small light plane landing strip near Carolina Beach, and after looking it over for several minutes, landed wheels and flaps down. After making a normal landing, and near the completion of his landing run the aircraft, upon contacting a small high spot, left the ground and upon settling to earth again, the nose wheel collapsed, throwing the aircraft on its back causing major injures to the crew and fatal injuries to the pilot." Investigators noted that the subject airplane's radios were not functioning correctly. The airplane could transmit but could not receive.

1-22-45D. Cleburne, Kansas. At 1005 CWT, a Vultee-Stinson L-5 crashed one mile south of Cleburne, Kansas, killing liaison pilot Pfc. Herman L. Eason, 22. The airplane had taken off from Marshall Field, Kansas, on a local orientation flight. Investigators speculated that the airplane had suffered carburetor icing and had lost power, causing the pilot to attempt and emergency landing. The airplane collided with power lines and smashed into a small embankment. There were no witnesses to the accident.

1-23-45A. Peterson Field, Colorado. At 1330, a Curtiss P-40N crashed ten miles ENE of Peterson Field, Colorado Springs, Colorado, killing 2Lt. Robert E. Evans, Jr., 22. The airplane was flying in the number-two position of a two-ship flight that took off from Peterson Field on a camera-gunnery mission. The airplanes climbed to 11,500 feet, and after leveling off the flight leader radioed the subject airplane and informed the pilot that he was returning to the field because of generator trouble. The wingman acknowledged the leader and then began the let-down with the lead airplane. When about five miles from Peterson Field, the subject airplane broke off and turned back toward the practice area. The lead airplane made a successful emergency landing at Peterson Field. The subject airplane failed to return to the field and it was discovered that it had crashed. Investigation revealed that the subject airplane had entered a spin at an altitude too low to allow recovery. There were no witnesses to the crash.

1-23-45B. Lemoore, California. At 1500, a Northrop P-61A crashed ten miles WNW of Lemoore Army Air Field, Lemoore, California, killing pilot Capt. Peter B. Keene, 25, and WAC passenger Pfc. Viola J. Smith. The airplane was part of a three-ship flight that had taken off from Hammer Field, Fresno, California, on a local flight. The formation was flying at 7,500 feet when it was noticed that a stream of white vapor was trailing the subject airplane. The vapor stream seemed to be coming from the area near the radar operator's compartment. Moments later the fuselage burst into flames. The airplane dropped to about 4,000 feet, leveled off for a few seconds, and then dropped down to about 1,000 feet, again leveling off for a few seconds. The airplane then entered a descending right turn, remaining in this attitude until it struck the ground and exploded. Investigation revealed that the improper use of the gasoline cross-feed valve caused overfilling that resulted in the overflow of fuel. This fuel collected in the fuselage and was ignited by some part of the electrical system. Investigators noted that several P-61 pilots had noticed gasoline fumes and the collection of fuel when the cross feed valve is not shut down in time to prevent overflow.

1-23-45C. Greenville, Texas. At 1733, a Republic P-47D crashed while attempting to take-off from Majors Field, Greenville, Texas, killing Mexican Air Force pilot 2Lt. Crisoforo Salido, 25. The pilot was scheduled to lead a two-ship camera gunnery mission. Investigators stated, "The flight left the parking area and taxied down the E-W (8-26) Runway. At the east end of the runway the flight parked and completed the pre-take off run up. Lt. Salido taxied out on the taxi strip between Runway 26 and Runway 30 and called for take-off instructions. The tower instructed him [to taxi] to the end of Runway 30. Lt. Salido started taxiing down the taxiway, stopped and turned around, starting back to Runway 26. The tower called him and instructed him to make another 180-degree turn and continue taxiing to Runway 30. This message was [acknowledged]. The pilot turned as instructed, but lined up on the taxi strip, and attempted to take off, running off the end of the taxi strip into the mud and flipping over on his back."

1-24-45A. Fort Sumner, New Mexico. At 1000 MWT, a Curtiss P-40N crashed three miles northeast of Fort Sumner, New Mexico, killing 2Lt. David Miller, 22. Investigators stated, "The pilot overshot on initial approach and had to go around. On the second approach the pilot overshot the runway on the final turn into the approach and was attempting to turn back to the runway when the airplane stalled and started to spin. The airplane struck the ground upside down and exploded before a complete turn of a spin had been completed."

1-24-45B. Rock Hill, South Carolina. At 1230, a Douglas A-20J crashed eight miles southeast of Rock Hill, South Carolina, killing pilot F/O Arthur J. Ulfers, 22. The airplane took off from Morris Field, Charlotte, North Carolina, on a low-altitude (200 feet) navigation mission. The airplane was on course and on time when it apparently began suffering engine trouble. The airplane was observed to make a turn to the right in an attempt to make an emergency forced landing in a large field. The pilot was unable to make the field. The airplane descended into trees at the edge of the field at a 45-degree angle. The airplane broke up and burst into flames, killing the pilot instantly.

1-24-45C. Dyersburg, Tennessee. At 1610 CWT, two Boeing B-17F airplanes collided in midair and crashed northeast of Dyersburg, Tennessee, killing sixteen fliers. Four fliers were able to parachute to safety. The two airplanes were part of a five-ship flight that took off at 1500 from Dyersburg Army Air Field on a high altitude formation flight with pursuit ship interception. B-17F #42-6177, which crashed six miles northeast of Dyersburg, was flying in the number-three position of the lead element; B-17F #42-5325, which crashed four miles northeast of Dyersburg, was flying in the lead position of the second element. The formation was flying north at 10,000 feet when the number-three airplane dropped back out of position and then veered to the right. The number-three airplane continued to move to the right and stopped directly over the lead ship of the second element. The number-three airplane "hovered" over the lead ship for several seconds before dropping down on top of the lead ship. The number-three ship's nose collapsed on the port wing and its starboard propellers cut into the fuselage of the lead ship near the radio room. The lead ship broke up as it fell away from the number-three ship, the port wing was severed and the fuselage broke in two near the radio room. The starboard wing failed and separated as the lead ship fell to earth. None of the ten crewmembers aboard the lead ship were able to parachute to safety. The number-three ship went into a slight climb and then entered a spiral to the right, remaining in this attitude until it crashed to ground and exploded into flames. Four crewmembers able to parachute to safety from B-17F #42-6177 were: 2Lt. William L. Blair, pilot; F/O Nicholas J. Digiro, co-pilot; Cpl. Ellis B. Sizemore, radio operator; Pfc. Jack R. Fyne, gunner. Engineer Cpl. Corlyss J. Paulus fell to his death in an unsuccessful parachute jump. Killed in the crash of B-17F #42-6177 were: F/O John J. Walther, navigator; 2Lt. Richard I. Moffitt, bombardier; Cpl. Robert L. Martin, gunner; Pvt. Victor W. Primo, gunner; Cpl. Raymond H. McCoy, gunner. Killed in the crash of B-17F #42-5325 were: 2Lt. John R. Sedivic, pilot; F/O James R. Seloover, co-pilot; 1Lt. Salvatore L. Catanese, B-17 instructor; F/O William F. Simpson, navigator; David E. Shelley, bombardier; Sgt. Claude L. Ogle, engineer; Cpl. William C. Spalding, Jr., radio operator; Pfc. George W. Eldridge, gunner; Cpl. Lawrence J. Manoleff, gunner; Cpl. John E. Westfall, gunner.

1-25-45. Van Nuys, California. At 1645, a Lockheed P-38L crashed into homes four miles northeast of Van Nuys Metropolitan Airport, Van Nuys, California, killing pilot 2Lt. Emil B. Feaster, 29, and a child on the ground. The airplane was in the number-three position of a four-ship flight that took off from Van Nuys Metropolitan Airport on a gunnery mission. The subject airplane's take-off was described as "erratic" and the landing gear came up slowly as the airplane climbed to about 600 feet agl. The airplane turned slightly to the right and then rolled to the right three or four times before rolling into the ground. The airplane smashed into two civilian homes at 9363 Burnett Street, exploding into flames. The two houses were destroyed. Dennis Walker, age 3, was killed on the ground. Investigation revealed that the aileron boost control unit had failed.

1-26-45A. Westover Field, Massachusetts. At 1115, pilot-rated passenger F/O James E. Sonnenmeir was killed when he walked into the rotating number-two propeller of a Consolidated B-24J operating at Westover Field, Chicopee, Massachusetts. Investigation revealed that F/O Sonnenmeir had stood fireguard as the instructor and student co-pilot started the engines. After the engines were started F/O Sonnenmeir returned to the flight deck and stowed the fire extinguisher. F/O Sonnenmeir then exited the aircraft and was struck by the number-two propeller. Investigators speculated that the B-24 student had either fell or slipped on the icy ramp, causing him to fall into the arc of the propeller. It was not known why he had exited the airplane. It was noted that the wheel chocks had been removed but the aircraft had not yet started to taxi. Investigators noted that F/O Sonnenmeir had just returned from emergency leave to visit his father, who was very ill.

1-26-45B. Alpha, Ohio. At 1040 CWT, a Vultee-Stinson L-1 crashed three miles south of Alpha, Ohio, killing pilot 1Lt. Paul A. Hobe, 27. The airplane had taken off from Wright Field, Dayton, Ohio, on a test flight. The airplane was flying at 5,000 feet when it fell into an inverted spin. The pilot was unable to right the airplane. The starboard wing failed at about 400 feet agl during the spin and the airplane crashed to the ground. The pilot had either bailed out or was hurled from the airplane at an altitude of 100 feet agl. His ripcord had not been pulled. The airplane had been scheduled to undergo stall testing because of the type's "unpredictable" stall characteristics.

1-26-45C. Birmingham, Alabama. At 1506 CWT, a Douglas C-47, towing two Waco CG-4A gliders, crashed while attempting to take off from Birmingham Army Air Field, Birmingham, Alabama, killing four fliers. Investigators stated, "The pilot came over the field, requested to land his two gliders downwind on the take-off runway in order to expedite take-off. He was given N, NW wind and advised to use the NE-SW runway, 5,760 feet long with minimal obstructions. However, he decided on the N-S runway. The gliders and airplane were landed without incident. In taking off to the north the pilot made a three-point take-off and held this nose high attitude, indicated airspeed 80-90 miles per hour, to an altitude of 100-150 feet. About 300 feet north of the field, immediately over a cut in the hill (made to reduce the flying hazard), the airplane stalled. The tow-ropes were released from the gliders and from the airplane at this instant. The left wing dropped and the airplane did a half turn into the side of the cut, immediately bursting into flames. The gliders made 180-degree turns and landed safely on the field. The airplane had 2,100 pounds of cargo. The gliders were empty except for the pilots. The runway is 4,855 feet long [and made] of concrete. The elevation of the crash is 55 feet above the point of take-off." Investigators stated that the pilot had used the wrong runway. The airplane/gliders tandem took off from Laurinburg-Maxton Army Air Field, Maxton, North Carolina, on a ferry flight to South Plains Army Air Field, Lubbock, Texas. Killed in the crash were: 1Lt. Fred T. Heise, pilot; 2Lt. John N. Ricker, co-pilot; TSgt. E. M. Bechtold, engineer; A/C Lawrence W. Schloss, passenger.

1-26-45D. Ogden, Kansas. At 1825 CWT, a Boeing B-29 returning from a long-range navigation mission crashed one mile south of Ogden, Kansas, killing five fliers and injuring nine others, two of them seriously. Gunners Cpl. Denton O. Calyer and Cpl. James H. Young escaped injury. Bombardier-rated passenger Capt. Don W. Beroset and navigator 2Lt. Edwin E. Courter received serious injuries. Injured in the crash were: 2Lt. Charles D. Zimmer, co-pilot; 2Lt. Ray O. Jones, bombardier; Cpl. George W. Bise, radio operator; Pfc. Harry V. Hoetesler, gunner-rated passenger; Cpl. Paul Wilder, gunner-rated passenger; 2Lt. Harland E. Fraley, navigator-rated passenger; TSgt. Clifford C. Glenn, passenger. Killed in the crash were: Capt. George P. Erwin, pilot; 2Lt. Donald C. Tarr, bombardier-rated passenger; SSgt. Anthony P. Tomaini, flight engineer; Cpl. Joe F. Horn, radio operator; SSgt. Francis J. Merdan, engineer. The airplane took off from Grand Island Army Air Field, Grand Island, Nebraska, on a navigation and camera-bombing flight to Borinquen Field, Puerto Rico. The airplane successfully navigated to Borinquen Field and then took off for the trip back to its home field. The airplane, flying at an altitude of 20,000 feet, returned to Grand Island via Jamaica; Batista Field, Cuba; Homestead, Florida; Miami, Florida; West Palm Beach, Florida; Orlando, Florida; Jacksonville, Florida; Moody Field, Georgia; Birmingham, Alabama; Sedalia, Missouri. A large front was encountered near Jacksonville and winds aloft between Orlando, Florida, and Valdosta, Georgia, averaged over 100 mph and lasted well over an hour. After the B-29 climbed to 26,000 feet and passed over the front, the pilot made a gradual let-down, and by the time he was over Birmingham, Alabama, the airplane was flying at 10,000 feet. The airplane then descended to 8,000 feet and flew at this altitude from Alabama to Kansas. Winds averaged about 30 mph. The airplane encountered a weak cold front that stretched from Missouri to Kansas. Ceilings were 500 to 1,000 feet, the clouds topping off at about 3,000 feet. Visibility was three to four miles in light fog. At 1745 CWT, the number-one and number-four engines stopped because of fuel exhaustion. The pilot ordered the crew to put on parachutes and to assume the crash positions. At about 1820, the number-two engine stopped because fuel exhaustion. The pilot had Marshall Field in view when the number-two engine stopped. The airplane, flying on only the number-three engine and unable to maintain altitude, began mushing toward the ground. The pilot activated the crash alarm. The port wing struck trees and the airplane smashed to the ground. The airplane broke up as it smashed itself to pieces.

1-26-45E. Bell Center, Ohio. At 2015 EWT, a Curtiss C-46 suffering an in-flight fire crashed two and a half miles south of Bell Center, Ohio, killing three crewmembers. Pilot Capt. James W. Hartzog, co-pilot 2Lt. F.F. Muncharath and MSgt. Jacob Oswalt, engineer. The airplane, which was based out of Romulus Army Air Field, Michigan, took off from Romulus Army Air Field and flew to Buffalo, New York. At Buffalo the airplane loaded 8,000 pounds of cargo and took off on a cargo flight to Patterson Field, Ohio. A fire broke out in the fuselage near the fuel cross-feed valve in an area where the auxiliary power unit and gasoline heater are. The fire burned intensely, burning through the fuselage and causing the floor to fail. Cargo, the auxiliary power unit and the gasoline heater fell from the airplane. Witnesses on the ground observed the airplane trailing flames at an altitude of about 1,000 feet. A few minutes later the airplane began dropping toward the ground. The starboard wing struck trees and was severed, sending the airplane out of control.

The airplane smashed to the ground and burst into flames. Wreckage and cargo was found scattered over a four-mile area.

1-26-45F. Sebring, Florida. At 2130 EWT, a Boeing B-17G crashed two miles southwest of Hendricks Field, Sebring, Florida, killing ten crewmembers. The airplane had taken off from Hendricks Field on a high-altitude bombing mission at the AAF Bombing Range at Avon Park, Florida. The airplane, flying at 20,000 feet, received permission to bomb on the range. The airplane never arrived on the range and did not drop its bombs. Investigators stated, "At 2130 EWT several witnesses from Hendricks Field heard a terrific roar as of over-speeding engines. This was followed by a large flash in the air at approximately 1,000 feet, which was in turn followed by a second explosion as the airplane crashed into the ground. The two explosions were almost simultaneous and indicate that the airplane was traveling at high speed and [diving] nearly straight down. The scene of the accident also indicated a terminal velocity dive or high-speed spiral as, prior to the crash, the airplane began to shed parts. All parts torn loose were found south of the point of impact, with some small pieces of wing skin being found six miles away. The heavier parts of the control surfaces, wing flaps, ailerons, trailing edges of the wings, horizontal and vertical stabilizers and elevators were all within one mile of the point of impact.... The entire fuselage was completely demolished and no information could be gathered from the wreckage due to the fire after crashing. Parts of the fuselage and engines were buried six to eight feet in the ground. Due to no evidence of fire or explosion on the scattered parts, it is the opinion of the [investigators] that the first explosion seen in the air was a result of the airplane coming apart rather than a cause [of the crash]. All bodies were recovered in their approximate flight positions and as near as could be determined, all had oxygen masks on." The cause of the accident could not be determined. Killed in the crash were: 2Lt. Frank L. Young, pilot; F/O Neil M. McLean, co-pilot; F/O Neale L. Edelen, navigator; 2Lt. Samuel J. Wilson, bombardier; Cpl. Howard Wertz, engineer; Cpl. Lewis E. Pulliam, radio operator; Cpl. Robert LaBounty, gunner; Pfc. Raymond E. Bekins, gunner; Pfc. Wilson M. Pearce, ball turret gunner; Pfc. Tom B. Jones, tail gunner.

1-27-45. Panama City, Florida. At 1225 CWT, a Martin TB-26C suffering a fire in flight broke up and crashed four miles north of Tyndall Field, Panama City, Florida, killing six fliers. The airplane took off from Tyndall Field on a flight back to its base at Appalachicola Army Air Field (a sub-base of Tyndall Field), Appalachicola, Florida. The airplane made what appeared to be normal take-off from Runway 9. A short time after taking off, the pilot radioed the field requesting permission for an emergency landing. The airplane was cleared for the landing but it never made it back to the field. The airplane, trailing fire from the starboard wing, was observed maneuvering to make an emergency landing in shallow water on the shore of West Bay. The starboard wing, weakened by the intense fire, failed and separated. The airplane went out of control and smashed to the ground, exploding into flames upon impact. Killed in the crash were: 1Lt. Thomas J. Lewnes, pilot; 2Lt. James R. Criscoe, co-pilot; Pfc. George Storch, engineer; 2Lt. Harold J. Kinsey, pilot-rated passenger; SSgt. Harry I. VanWelsenaere, passenger; Pfc. Henry G. Burfeind, passenger.

1-28-45. Hernando, Mississippi. At 1950, a North American AT-6D crashed six miles east of Hernando, Mississippi, killing 1Lt. Wade C. Waller, 22. The airplane took off from Jackson, Mississippi, on a flight to Memphis, Tennessee. At 1946 CWT, the pilot reported that he was over Memphis at 7,000 feet. He was cleared to descend. The airplane crashed minutes later. The airplane struck the ground in a vertical dive at high speed, exploding violently into flames upon impact. Investigators could not determine the cause of the crash.

1-29-45A. Bradenton, Florida. At 1130 EWT, a North American P-51D crashed five miles southeast of Bradenton, Florida, killing pilot 2Lt. Robert W. MacChesney, 23. The airplane was flying in the number-three position of a six-ship flight that took off from Sarasota Army Air Field, Sarasota, Florida, on an aerial gunnery mission, which was to be conducted at 9,000 feet. The airplanes finished their firing passes and the subject airplane and another airplane were ordered by the flight leader to practice "homings." The airplanes flew near an approaching front as they attempted to use their radios. Interference was so great that no homing practice could be done. The subject airplane and the wing ship separated, the wing ship returning to the field and making a safe landing. The subject airplane failed to return to the field and was discovered to have crashed. Investigation revealed that the airplane smashed into the ground at high speed and at a steep angle, exploding into flames upon impact and killing the pilot instantly.

1-29-45B. Peoria, Arizona. At 1235 MWT, a Curtiss P-40R crashed two miles SSW of Peoria, Arizona, killing pilot 2Lt. William G. DeBrun, 19. The airplane was part of a three-ship flight that took off from Luke Field, Phoenix, Arizona, on a formation training flight. After the completion of the formation exercise the flight returned to Luke Field for landing. The flight leader signaled for a peel-off to enter the traffic pattern. The subject airplane did not enter the pattern and was later observed flying at about 300 feet agl with black smoke trailing. The airplane was observed in a nose-high attitude, remaining in this position until it snapped over to the left and smashed into the ground. The airplane exploded into flames upon impact, killing the pilot instantly.

1-29-45C. Charleston, South Carolina. At 1510 EWT, armorer/gunner Cpl. Steven W. Johnston was killed when a port waist fifty-caliber machine gun mount failed and allowed the gun to fire into a Consolidated B-24J operating 35 miles southeast of the Army Air Field at Charleston, South Carolina. Investigators stated, "During a routine air to water gunnery mission, the left waist gun mount locking pin slid out of its proper position, releasing the upper portion of the gun mount. The gun, which was being fired at the time, swung back inside the airplane, shooting the armorer/gunner in the abdomen and causing damage to the airplane." Investigation revealed that subject armorer/gunner knew of the defective side gun firing post lockpin handle and also of a broken safety pin retaining bracket. The armorer rigged the firing post lockpin in a way that he thought would allow the safe use of the gun. The B-24 was piloted by 2Lt. John F. Anderson and co-pilot F/O Leonard H. Loos. The pilots and ten other crewmembers were uninjured.

1-29-45D. Greenville, Texas. At 1812, a Republic P-47D crashed at Majors Field, Greenville, Texas, killing pilot F/O Stanley J. Gay, 26. The airplane was part of a two ship flight that took off from Majors Field on a high-altitude training mission. The two airplanes completed their mission and returned to the field for landing. The airplanes were cleared to land on Runway 17L. The airplanes performed a steep peel-off to enter the traffic pattern. The airplanes came in too fast and too low to make a safe landing, so they were sent around for another landing. The two airplanes reformed and then performed another steep peel-off to enter the pattern. The airplanes flew a tight pattern and the flight leader was able to make a safe landing. The subject airplane again came in too low and too fast. The subject pilot overshot the approach on the turn from base to final. He tightened his turn and stalled the airplane. The airplane snapped to an inverted position and smashed to the ground, bursting into flames upon impact.

1-30-45A. Sioux City, Iowa. At 0758 CWT, engineer Sgt. Paul C. Harris was killed when he walked into the rotating number-two propeller of a Boeing B-17G operating at Sioux City Army Air Field, Sioux City, Iowa. The airplane was piloted by 2Lt. Jeremiah Callahan, co-pilot 2Lt. Gilbert F. Peerey and B-17 instructor Capt. Leo H. Flowers. The pilots and eight other crewmembers were uninjured. Investigation revealed that the engineer had been standing fireguard during engine start-up. The pilots started the number-one and two engines. The engineer was about to take position at the number-three engine when the number-one engine stopped. The pilots attempted to restart the number-one engine, but were having trouble with the mesh switch in the cockpit, so they ordered the engineer to operate the external mesh switch. The engineer acknowledged the order and then walked directly into the arc of the rotating number-two propeller. He was killed instantly.

1-30-45B. Whitney, Texas. At 1130, a North American AT-6C crashed six miles south of Whitney, Texas, killing student A/C Thomas D. Denson, 21, who fell to his death in an unsuccessful parachute jump. The airplane took off from Waco Army Air Field, Waco, Texas, on a routine training flight. The airplane was seen in a 70-degree dive before it struck the ground and exploded into flames. Investigators speculated that the student had put the airplane in a steep diving attitude and was unable to recover. The student had maneuvered that airplane in such a manner that negative pressure blew out parts of the cockpit canopy. The student bailed out at an altitude too low to allow his parachute to deploy. His body was found about 30 feet from the point where the airplane crashed with about ten feet of his parachute strewn out behind him.

1-30-45C. Santa Maria, California. At 1227, a Lockheed P-38J crashed four miles south of Santa Maria Army Air Field, Santa Maria, California, killing pilot F/O Elmer R. Steffey, Jr., 20. Civilians Tillie Rusconi and John Doff were killed on the ground and Arthur Beall was injured. The airplane was flying in the number-five position of a five-ship flight that took off from Santa Maria Army Air Field on a camera-gunnery mission. The flight practiced feathering technique, maneuvering in a single-engine configuration for a short time. The pilots went back to normal two-engine operation and reformed their flight at about 5,600 feet. The flight leader looked back and observed the subject airplane falling in a spin to the right. The flight leader advised the subject pilot to bail out. The airplane continued in the spin until it smashed to the ground in the center of Santa Maria and burst into flames. The airplane smashed into a lunch café, killing two civilians who were eating inside. The structure was completely destroyed. Investigation revealed that the port engine was not producing power at the time of impact.

1-30-45D. Lake Mead, Nevada. At 1410 PWT, a Bell P-63A crashed into Lake Mead 40 miles northeast of the Army Air Field at Las Vegas, Nevada, killing pilot 2Lt. Stanley Clark, Jr., 26. Investigation revealed that the airplane was part of a two-ship flight that took off from Las Vegas Army Air Field on a routine training mission. The two airplanes maneuvered for over an hour before flying over the west shore of Lake Mead on their way back to the field. The airplanes were flying in formation at about 1,000 feet agl when the subject airplane descended until it was flying over the lake at very low altitude. The wing ship remained at about 1,000 feet. An instant later the subject airplane struck the lake, which was described as "glassy smooth." The subject airplane climbed to about 600 feet, trailing flames from the starboard side. The airplane leveled out momentarily before diving into the lake. The pilot's body was not recovered. Investigators speculated that the pilot had misjudged his altitude above the lake.

1-30-45E. Santa Maria, California. At 1415, a Lockheed P-38L crashed two miles southwest of Santa Maria Army Air Field, Santa Maria, California, killing pilot F/O Robert E. Pettigrew, 20. The airplane was flying in the number-four position of a five-ship flight that took off from Santa Maria Army Air Field on a camera-gunnery mission. The airplane made what appeared to be a normal take-off and climbed to about 500 feet and then was seen to yaw violently. The airplane snapped over to the left and smashed to the ground, exploding into flames upon impact. Investigation revealed that the aileron booster until had failed, causing the aircraft to go out of control.

1-30-45F. Helendale, California. At 1420, a Consolidated B-24L crashed ten miles west of Helendale, California, killing four fliers. Bombardiers 2Lt. Carl F. Hansen and 2Lt. John R. Palin parachuted to safety and were uninjured. Killed in the crash were: 1Lt. James G. Wright, pilot; 2Lt. Norbert J. Vehr, co-pilot; 2Lt. Herbert A. Perry, bombardier; TSgt. Harvey L. Cook, engineer. The airplane took off at 1410 on Runway 34 from Victorville Army Air Field, Victorville, California, on a routine radar mission. The airplane, heading north, climbed to 1,000 feet agl when the radar instructor smelled smoke coming from the bomb bay area. The pilot sent the engineer to investigate the smoke. Moments later, the pilot began to experience trouble with the number-two engine. The pilot attempted to feather the number-two propeller without success. The airplane was still heading north but had climbed to about 2,000 feet agl. The number-two propeller began to over speed, so the pilot again attempted to feather the propeller and again was unsuccessful. The pilot ordered the crew to bail out. The number-two engine began to vibrate excessively and began to trail flames, but the airplane continued to fly straight and level. Two men bailed out while the rest of the crew gathered and fastened their parachutes. The airplane entered a gentle turn to the left and began losing altitude. The airplane continued in the descending turn to the left, smashing to the ground as it passed through 180 degrees of turn. The airplane cartwheeled into the ground and exploded into flames. Investigation revealed that the number-six cylinder of the number-two engine blew, damaging the electrical conduits to the number-two propeller's "fast feathering pump" and preventing the pilot from feathering the propeller at a critical time.

1-30-45G. Tuskegee, Alabama. At 1725, two Stearman PT-13D airplanes collided in mid-air and crashed three miles northwest of Tuskegee, Alabama, killing A/C Alfred S. Barclay aboard PT-13D #42-17772. A/C Leonard C. Hall, Jr. received minor injuries aboard PT-13D #42-17121. A/C Hall took off from the Tuskegee Institute Airfield at 1615 on a scheduled student solo flight; A/C Barclay took off from the same field at 1620 on his first unsupervised student solo flight. The airplanes returned to the field separately at about 1725, and both attempted to enter the downwind leg of the traffic pattern from the traffic entry leg. Both airplanes were flying parallel paths with A/C Barclay's airplane flying 200 yards ahead, 50 feet above and slightly to the left of A/C Hall's airplane. A/C Hall turned his ship to the left for the downwind leg as A/C Barclay began to descend into the downwind leg. A/C Hall's propeller collided with the tail section of A/C Barclay's airplane, severing the tail section and sending A/C Barclay's airplane out of control. A/C Hall was able to land his aircraft on the field even though pieces of A/C Barclay's airplane were lodged in the front of A/C Hall's airplane.

1-31-45. Marana, Arizona. At 0830 MWT, a Vultee BT-13A crashed eight miles WSW of the Army Air Field at Marana, Arizona, killing Chinese Air Force student A/C Siu-Yu Chang. The airplane took off from Marana Army Air Field on a routine student solo flight. A witness of the ground observed the airplane in a spin at about 1,000 feet agl. The rotation of the spin slowed as the airplane neared the ground. The student was unable to recover and the airplane smashed into the ground and burst into flames.

February

2-1-45. Thomasville, Georgia. At 1304 EWT, two Curtiss P-40N airplanes collided in mid-air and crashed three miles south of the Thomasville Army Air Field, Thomasville, Georgia, killing 2Lt. Paul R. Hippensteel, 21, aboard P-40N #43-23127. 2Lt. Grover C. Carico parachuted to safety from P-40N #43-24442 and was uninjured. Lt. Hippensteel was flying in the number-three position of a four-ship flight on an aerial gunnery mission; Lt. Carico was leading a four-ship flight on a separate aerial gunnery mission. Both flights originated at Thomasville Army Air Field. After the completion of the their gunnery mission, Lt. Hippensteel's flight practiced "homings" for a short time. Lt. Hippensteel and his wingman reformed after their instrument practice and headed for Thomasville Army Air Field. Lt. Carico's flight was flying below and to the left of Lt. Hippensteel's two-ship flight, which was flying at about 5,500 feet. Lt. Carico entered a climbing turn to the right until he collided with Lt. Hippensteel's airplane. Both airplanes broke up and fell to earth, bursting into flames upon impact. Lt. Hippensteel made no apparent effort to avoid the collision and was likely unaware of Lt. Carico's position.

2-2-45A. Gaston, South Carolina. At 1110, a North American B-25J crashed two miles west of Gaston, South Carolina, killing four fliers. Navigator F/O John H. Lumpkin was injured parachuting to safety. Killed in the crash were: 2Lt. Everett E. Thompson, pilot; 2Lt. John E. Powell, co-pilot; Cpl. Amund R. Utne, engineer; Cpl. William H. Strong, radio operator. The airplane was the flying in the number-two position of a three-ship flight that took off at about 1050 from Columbia Army Air Field, Columbia, South Carolina, on a formation training flight. The flight climbed to about 5,500 feet when the subject airplane's port engine "coughed" twice and then burst into flames. The pilot was unable to extinguish the fire, so he ordered the crew to bail out. The navigator bailed out and then the aircraft entered a descending turn to the left. The turn evolved into a steep spiral from which the aircraft never recovered. The airplane smashed to the ground at a steep angle and exploded into flames. Investigators could not determine the cause of the engine fire.

2-2-45B. Greenville, South Carolina. At 1152, a North American B-25C crashed one-quarter mile southwest of Greenville Army Air Base, Greenville, South Carolina, killing six fliers. The airplane was flying in the number-two position of a flight that took off from Greenville Army Air Base on a routine formation training flight. The formation returned to the base and entered the traffic pattern. The subject airplane turned from base to final and crashed soon after. The airplane fell off on the left wing and smashed to the ground at a steep angle, exploding into flames upon impact. It was not known whether the airplane stalled out of the turn from base to final or had crashed after leveling off following the turn. Investigators also speculated that the port flap might have failed to extend, causing the airplane to go out of control. Killed in the crash were: 2Lt. Harry L. Davis, pilot; 2Lt. Edward L. Heinlen, co-pilot; 2Lt. Roy T. Jaynes, Jr., bombardier-navigator; Pvt. Joseph G. Magovern, engineer; Pvt. Nicholas F. Marzano, radio operator; Cpl. Cecil M. Hirral, gunner.

2-2-45C. Shaw Field, South Carolina. At 1455, a North American AT-6D crashed while attempting a landing at Shaw Field, Sumter, South Carolina, fatally injuring A/S Glenn Boustead, 21. The student died of his injuries on 2-4-45. Control tower personnel stated that the airplane made a normal approach but leveled off too high. The starboard wing dropped and the student added throttle. The airplane, in a nose-high attitude, climbed slightly before rolling to the right and into the runway. The airplane struck the ground on the engine and the port wing, bounding back into the air for a short distance before landing on its wheels and skidding to a halt. The airplane did not burn but was completely destroyed. Investigation revealed that the elevator trim tab was in the "full back position," causing an elevator stall.

2-2-45D. Elkhart, Kansas. At 1430 CWT, a Republic P-47D crashed eight and one-half miles southeast of Elkhart, Kansas, killing 1Lt. Leander L. LaFlex, 27. The airplane was part of a two-ship flight that took off from Dalhart Army Air Field, Dalhart, Texas, on a scheduled "hooded" instrument-training mission. The two airplanes flew on the instrument mission for about an hour; the subject airplane was flying as the instrument ship and the wingman flew as the safety observer. Weather was reported as "patches" of low stratus with light fog underneath. After flying above the low stratus deck on instruments for an hour and fifteen minutes, the subject pilot began performing lazy-eights and other aerobatic maneuvers. The subject airplane then performed a half-roll and split-S maneuver at an altitude of 6,000 feet msl, disappearing into the cloud deck. The subject airplane failed to recover from the maneuver and smashed into the bank of a drainage ditch at a steep angle, exploding violently into flames upon impact. The ceiling at the scene of the accident was reported to be 200 feet above ground level. Elevation was about 4,000 feet msl and the top of the low ceiling was estimated to be about 5,000 feet msl. The wingman was unable to locate the subject airplane under the low-hanging ceiling and returned to Dalhart.

2-2-45E. Romulus, Michigan. At 1959 EWT, a Consolidated B-24M crashed while attempting to take-off from Romulus Army Air Field, Romulus, Michigan, killing B-24 student 1Lt. John J. Dillin. Engineer Cpl. Arthur P. Mueller and B-24 student Capt. Norman S. Harna received minor injuries; B-24 student 1Lt. Bruce F. Jones sustained serious injuries. B-24 instructor 1Lt. John P. Taylor and B-24 students F/O Calvin E.J. Schroeder and 1Lt. Edward J. Cloutier escaped injury. The B-24, which was taking off on a scheduled night training mission, began to veer to the left while on the take-off run, running off the edge of the runway and heading into snow and uneven terrain. The instructor pilot cut the engines, but the nose wheel collapsed and the airplane went completely out of control. The airplane smashed itself to pieces as it careened across the airfield, coming to a rest with the tail pointing up at about a 45-degree angle. The nose section was completely smashed and collapsed. The B-24 did not burn but was destroyed utterly. Investigation revealed that the B-24 took off left of the center of the runway.

2-2-45F. Ludlow, Illinois. At 2000, a Vultee BT-13A crashed four and one-half miles west of Ludlow, Illinois, killing 2Lt. Ernest R. Gribble, 26. The airplane took off from Chanute Field, Rantoul, Illinois, on a routine night training flight. The airplane made one circle of the area before smashing into the ground and bursting into flames. Investigation revealed that the flight indicator was not operating correctly on the airplane's previous flight. The incident was properly written up but no corrective action had been taken. Investigators speculated that the pilot did not fully read the Form 1A and was unaware that the flight indicator was still inoperative.

2-2-45G. Yuma, Arizona. At 2350 MWT, a Boeing B-17G crashed two miles southeast of the Yuma Army Air Field, Yuma, Arizona, killing four fliers. Killed in the crash were: 2Lt. Peter K. Guillen, B-17 instructor; 2Lt. George K. Myers, B-17 student; 2Lt. Daniel A. Rice, B-17 student; Sgt. Donovan U. Oseth, engineer. The airplane was one of eight B-17s that were flying around the area of Yuma Army Air Field on scheduled student co-pilot training flights. Four airplanes were ordered to fly in the local area and four others, including the subject airplane, were ordered to practice take-offs and "follow-through" landings. The subject airplane made ten take-offs and landings on Runway 26 before stopping and dropping off two student co-pilots and picking up two more. The landings were changed to Runway 8 (with the traffic pattern to the left), and the subject airplane made one successful follow-through landing and take-off on Runway 8. The subject airplane successfully landed on Runway 8 a second time and climbed away straight ahead after taking off the second time, this time on instruments. The subject airplane suddenly made a right-hand turn away from the traffic pattern, remaining in this right turn until it descended into the ground. The airplane cartwheeled into the ground and burst into flames. Wreckage was scattered over 400 yards. Investigation revealed that the pilots became confused by lights in the area and thought that a mid-air collision was imminent if a right-hand turn was not made.

2-4-45. Tehachapi, California. At 1730, a North American P-51D suffered a catastrophic structural failure and crashed ten miles southwest of Tehachapi, California, killing pilot F/O George J. Greaves, 22. The airplane took off from Mines Field, Los Angeles, California, on a ferry flight to Bakersfield Municipal Airport, Bakersfield, California. Investigation revealed that the pilot had engaged in aerobatic maneuvers over the Cummings Valley and Cummings Mountain. The pilot attempted an abrupt pull-up from a high-speed dive, stressing the starboard wing beyond its design limitations and causing it to fail just outboard of the wing root. The starboard wing separated and the airplane went out of control, going into a "horizontal roll" and then into a tight spin to the right, remaining in this attitude until it struck the ground vertically. The airplane did not burn. The pilot did not make an effort to parachute to safety and was killed instantly upon impact. The starboard wing, the starboard flap, pieces of cockpit canopy and assorted pieces of wreckage were found scattered over an area of a mile. Investigators noted that the airplane had just been accepted by the AAF from the North American Aircraft Company factory and had a total time of 1.5 hours.

2-5-45. Charleston, South Carolina. At 0825, a Consolidated B-24J crashed four and a half miles northeast of the Army Air Field at Charleston, South Carolina, killing eight fliers and seriously injuring four others. Killed in the crash were: 2Lt. Frederick J. Kienzle, pilot; F/O Frederick C. Bellemore, co-pilot; 2Lt. Eugene B. Nelson, navigator; 2Lt. Hal A. Calhoun, bombardier; Pvt. Armando J. Landolina, engineer; Cpl. Pasquale Zimbardi, radio operator; Pfc. Richard P. Thomas, assistant engineer; Sgt. Anthony D. Albondy, armorer/gunner. Seriously injured in the accident were: Pfc. Victor J. Levatino, gunner; Pfc. Ralph H. Hendrix, gunner; Cpl. Kenneth C. Turcott, gunner; SSgt. Arley F. Aten, gunnery instructor. The airplane took off from

Charleston Army Air Field at 0813 EWT on Runway 21 on a routine training mission. The airplane was attempting to climb after take-off when an armorer/gunner noticed a huge fuel leak on the port wing. The armorer attempted to inform the pilot over the intercom but was unsuccessful. The assistant engineer went forward to inform the pilot of the fuel leak. A short time later the airplane entered a gentle turn to the left. The tail section was felt to flutter violently for several seconds, so the gunnery instructor ordered the men to fasten their parachutes. The airplane entered a "very tail low attitude" and the pilot had "difficulty keeping the plane in the air." The gunner instructor opened both waist windows and jettisoned hatches. The fuel leaking from the port wing began to enter the waist section through the open waist window, so the instructor ordered the port window closed. The airplane continued to lose altitude so the instructor ordered the men aft to assume crash positions. The airplane remained in a descending turn to the left, barely missing colliding with the masts of U.S. Navy vessels moored at Charleston Navy Yard. The airplane banked sharply to the right to avoid colliding with a tall radio antenna near the navy yard, missing it by only several feet. The radio operator transmitted a message to Charleston Army Air Field that the airplane was returning to the field because of a huge fuel leak. The field was cleared and rescue personnel alerted to the impending emergency landing. The airplane leveled out somewhat, prompting the men who were aft to think that the airplane was gaining altitude. Several men stood up but the airplane continued to lose altitude, entering a tail-low altitude. The men in the waist section again assumed crash positions when it seemed that a crash was imminent. The airplane entered a steep bank to the right before it collided with trees and then crashed to the ground in swampy terrain, smashing itself to pieces as it came to a rest. The airplane did not burn but was destroyed utterly. Investigation revealed that the assistant engineer and engineer had failed to properly secure and safety the port inboard fuel cell cap. Investigators speculated that the airplane suffered diminished performance of either or both the number-one and number-two engines.

2-7-45A. Pueblo, Colorado. At 1115 MWT, armorer/gunner SSgt. Eugene E. Conrad was killed when he was crushed in the top turret of a Consolidated B-24J operating on an aerial gunnery mission 45 miles west of the Pueblo Army Air Field, Pueblo, Colorado. Investigation revealed that the armorer had failed to shut off the turret's main switch while attempting to make repairs to the turret's port gun "booster motor," which was not functioning properly. The armorer inadvertently pushed the upper-lower gun traverse switch and simultaneously rotated the turret to the right while reaching into the turret to make a repair to the booster motor. The guns elevated fully and the turret rotated to the right, crushing the armorer's head. Sgt. Conrad

was found unconscious with his head trapped in the turret. Only after great difficulty were members of the crew able to free Sgt. Conrad from the turret. It was then discovered that the armorer was dead. The bomber returned to Pueblo Army Air Field without further incident. The airplane was piloted by 2Lt. Edwin W. Allen and B-24 instructor Capt. Eugene J. Mark. The pilots and ten other crewmembers were uninjured.

2-7-45B. Longview, Texas. At 1430, a Martin TB-26C crashed after colliding with a flock of birds four miles northeast of Longview, Texas, killing French Air Force pilot SSgt. J.E. Pastourel and French Air Force bombardier/navigator SSgt. Paul Bonnet. French Air Force fliers seriously injured in the crash were: SSgt. Georges J. Bernard, co-pilot; SSgt. Max Bruch, engineer; SSgt. Jean Ducoureau, radio operator; Cpl. Jean Niek, gunner. The airplane had taken off from Barksdale Field, Shreveport, Louisiana, on a routine training flight. The airplane headed west after take-off and the co-pilot left his position to view maps at the navigator's table. The radio operator had misinterpreted a radio message and alerted the co-pilot that the airplane had been ordered back to base. The co-pilot went forward and alerted the pilot and then returned to the navigator's table. The airplane entered a steep turn to the right, colliding with a large flock of birds at about 400 feet agl. The birds struck the airplane on the nose and starboard wing, but most of the birds collided with the port side and port engine, entering and clogging the port engine air intake ducts. The windscreen was not shattered. Bird remains could be observed on the port engine and cowling. The port engine lost power and the pilot had trouble controlling the airplane. The co-pilot assumed his position and fastened his safety belt just an instant before the airplane crashed. The airplane crashed to the ground in a "tail-first" attitude, bounding back into the air and over a slight rise before slamming back to the ground violently and smashing itself to pieces. Both engines were torn loose and were hurled ahead of the wreckage. The airplane bust into flames as it came to rest. Surviving crewmembers attempted to rescue the pilot and bombardier, but they were driven back by intense flames. Investigation revealed that the airplane had not been ordered back to the field by any ground station and that the radio operator was in error.

2-8-45A. Randolph Field, Texas. At 0855, a North American AT-6C and an AT-6D collided in midair and crashed three and a half miles northeast of Randolph Field, San Antonio, Texas, killing two fliers and injuring two others. 1Lt. James J. Lash and Capt. Thomas C. Brownfield were killed in the crash of the AT-6C; Capt. Hugh W. McLane was injured and Capt. John M. O'Hare received serious injuries parachuting to safety from the AT-6D. The airplanes smashed to earth about two miles northeast of Schertz, Texas. The AT-6D took off from Randolph Field at 0840 on a routine training flight. The AT-6C took off from Randolph

Field on an aerobatic flight. The AT-6D, flying east, climbed to about 4,000 feet agl, the pilot leveling out about 200 feet higher than anticipated. Capt. McLane, who was flying the AT-6D, put the nose down slightly to glide back down to the desired altitude. Moments later he collided with the AT-6C. The AT-6C entered a spin from which it did not recover, remaining in the spin until it struck the ground and exploded into flames. The AT-6C pilots were unable to parachute to safety. The pilots of the AT-6D were flung free of the wreckage as the airplane came apart in the air and were able to parachute to safety.

2-8-45B. Dalhart, Texas. At 0955, a Republic P-47D crashed at Dalhart Army Air Field, Dalhart, Texas, killing pilot F/O Jack K. Anderson, 22. Investigators stated, "At 0955 CWT, F/O Anderson ... started his take-off run to the northwest on Runway 30. The aircraft was equipped with two (2) one hundred and sixty-five (165) and one (1) seventy-five (75) gallon external gasoline tanks that were full. This was the fourth take-off for F/O Anderson under these weight load conditions from this field. After an apparently normal or slightly extended take-off run due to the added weight the aircraft became sluggishly airborne. Several witnesses claim hearing the engine cough once before the plane left the ground. All witnesses agree the aircraft became airborne from five to fifty feet before the pilot retarded the throttle or the engine failed. The aircraft returned to the ground [in a] three point [position] and continued a straight ahead roll for approximately 2,000 feet before nosing over approximately 800 feet west of the field perimeter. At this point the aircraft burst into flames and all witnesses agreed this was the first sign of smoke or flames that was seen."

2-8-45C. Choctawhatchee Bay, Florida. At 1015 CWT, a North American P-51D dove into shallow water at Choctawhatchee Bay three miles northeast of Fort Walton, Florida, killing Lt. Col. Isadore W. Toubman, 29. The aircraft took off from Eglin Field, Valparaiso, Florida, at 0930 CWT on a bombing mission at Bombing Range #53, Eglin Field Military Reservation. The airplane arrived at the range at 0936 and was cleared to enter the range traffic pattern. The airplane then made a dry run on the range. The airplane came back for the second pass; the pilot made his bomb run and then buzzed the tower. Tower personnel advised the pilot that his wings were clear of ordinance, and the airplane flew off the range at 0957. At approximately 1015, a civilian witness on the ground observed the airplane in a 40-degree dive. The angle of the dive increased until the airplane slammed into a sand bar in three feet of water in a near vertical attitude. The cause of the accident could not be determined.

2-8-45D. Florence, South Carolina. At 2120 EWT, a Douglas A-26B crashed two miles SSW of the Army Air Field at Florence, South Carolina, killing pilot 2Lt. James P. Waits and engineer/gunner Cpl.

Ervin Dziekan. The airplane took off from Florence Army Air Field, climbed slightly and then entered a left turn. The airplane continued in the left turn until it descended into the ground at a 30-degree angle and exploded into flames. The cause of the accident could not be determined.

2-9-45A. Oklahoma City, Oklahoma. At 1340 CWT, a Boeing F-13A (photo-reconnaissance version of the B-29) crashed while attempting an emergency landing at Tinker Field, Oklahoma City, Oklahoma, killing pilot 1Lt. Charles Gorham, Jr. and co-pilot 2Lt. Francis M. Agnew. Four crewmembers were seriously injured and six others escaped injury. Seriously injured in the crash landing were: 2Lt. George L. Rookstool, navigator; 2Lt. Robert B. MacDonald, navigator; 2Lt. Harold L. Cates, flight engineer instructor; SSgt. John F. Schott, engineer. Crewmembers escaping injury were: Cpl. Eino R. Olander, radio operator; SSgt. Edwin L. Stroud, photographer; Sgt. Donald H. Thompson, photographer; Pfc. Clifford J. Knipple, gunner; Pfc. Harry Beaman, Jr., gunner; Pfc. Harold F. Southworth, gunner. The airplane took off at 1206 CWT from Smoky Hill Army Air Field, Salina, Kansas, on a photo-reconnaissance mission to Havana, Cuba, and return. About an hour after take-off, the right scanner noticed that the number-three engine was leaking oil badly. The heavy flow of oil appeared to coming from the oil reservoir filler cap, the oil coating the starboard horizontal stabilizer after it streamed back over the wing. A few minutes after the oil leak was spotted, the number-two engine began to lose rpm and manifold pressure. The port scanner then reported that the number-two engine was backfiring; small bursts of flames and puffs of smoke were observed coming from the exhaust stacks. The pilot attempted to arrest the backfire but was unsuccessful. The backfiring in the number-two engine became "rhythmical" and strong vibrations set in. The vibration became so severe that the pilot decided to feather the propeller. The propeller went past the full feather position and was windmilling "backwards." The pilot radioed Tinker Field for an emergency landing. Permission was granted for an emergency landing on Runway 17 and the field was cleared. The airplane approached the area of the field and entered the traffic pattern. The pilot turned onto the base leg and then extended the landing gear, but the B-29 began to lose altitude rapidly. Just before the turn to final was to be made, the landing gear was seen to be retracting. The airplane did not turn on the final approach and the pilot radioed that he was going to jettison the bomb bay fuel tanks and then come back around for the landing. The airplane continued to fly to the west on an extension of the base leg. The airplane was flying at about 1,000 feet agl approximately three and a half miles from the field when it was cleared to land on Runway 12. Moments later, the number-one engine began backfiring and losing power. The pilot was unable to get full power from the remaining engines and the airplane began to lose altitude gradually. The airplane began a gentle turn to the left and the pilot was able to line up with Runway 12. The pilot jettisoned the bomb bay tanks and ordered the co-pilot to extend the landing gear. The airplane continued to lose altitude until it collided with trees. The pilot was able to pull up and hold the airplane level for a few moments before it descended back into the ground tail first. The tail skid contacted the ground and dragged through two fences. The airplane skidded for about 150 yards before crashing into an embankment that surrounded a drainage ditch near the field. The airplane broke and burst into flames as it crashed to a halt. The men in the after part of the airplane were able to escape before flames engulfed the wreckage. The cockpit was crushed in the crash and the pilots killed. The surviving crewmembers helped pull injured men from the shattered forward fuselage.

2-9-45B. Bend, Oregon. At 1200, a Lockheed P-38L crashed 60 miles southeast of Bend Oregon, killing pilot 2Lt. Max J. Clark, 25. The airplane was part of a four-ship flight that took off from Redmond Army Air Field, Redmond, Oregon, on a gunnery mission. Investigators stated, "The pilot was flying in the number-four position on an aerial gunnery mission. He started his pass at the Tow Target with a vertical bank. The airplane rolled over and went into a barrel roll to the right. The airplane stopped rolling and went into a spin to the left. The pilot was told to bail out by the flight leader. The flight leader saw Lt. Clark jettison the canopy, but the pilot failed to leave the airplane. The airplane spun to the ground and exploded."

2-9-45C. Burley, Idaho. At 2116 MWT, a Consolidated B-24J crashed into Mt. Harrison 19 miles southeast of Burley, Idaho, killing nine fliers. The airplane took off from Mountain Home Army Air Field, Mountain Home, Idaho, on a routine training mission. The airplane completed the camera-bombing portion of its mission and then went on to fly instruments at 9,500 feet msl on the Burley Range. The pilot radioed Mountain Home Army Air Base at 2034 and advised that he was going to start flying on the range on the northwest leg. This was the last contact with the airplane before it slammed into 9,000-foot Mt. Harrison 100 feet from the top. The airplane was flying straight and level in a northwesterly direction at cruise speed when it collided with the mountain. Killed in the crash were: 2Lt. Clinton R. Madeley, pilot; 2Lt. James T. Sanders, co-pilot; F/O Frank J. Pryor, Jr., navigator; F/O Stuart R. McMaster, bombardier; Sgt. Donald J. McClure, engineer; Cpl. William G. Doyle, radio operator; Cpl. Charles R. Tucker, gunner; Cpl. William J. Little, gunner; Pfc. George M. Ellett, tail gunner.

2-9-45D. Denio, Oregon. At 2200 MWT, a Consolidated B-24D crashed five miles northwest of Denio, Oregon, killing 11 crewmembers. The airplane took off at 1744 MWT from Gowen Field, Boise,

Idaho, on a navigation training flight to Hamilton Field, San Rafael, California, and return. The airplane successfully navigated to Hamilton Field and was flying on the leg back to Gowen Field. The airplane was observed flying to the northeast, safely clearing a ridgeline. The airplane flew in a normal manner for a short distance and then began a 180-degree turn, coming back over the previous line of flight. The airplane approached the ridgeline at a lower altitude and the pilot had to pull up abruptly to avoid a collision with terrain. The airplane cleared the mountain but stalled and fell off on the other side of the ridge. The pilot was unable to regain control of the airplane and it slammed into a "rocky shelf" at an elevation of 7,500 feet msl, exploding into flames upon impact. Killed in the crash were: 2Lt. Edward A. Bucek, pilot; 2Lt. Allen D. Blake, co-pilot; F/O Anton Pecnik, navigator; F/O Roy L. Pierson, bombardier; SSgt. Dean H. Sautter, engineer; Cpl. Arthur G. Manchester, gunner; Cpl. Edmund J. Green, radio operator; Pfc. George G. Wilkins, gunner; Pfc. Carl E. McDaniel, gunner; Pfc. Sidney M. Davis, gunner; TSgt. Ernest Kampinen, instructor.

2-10-45A. Valdosta, Georgia. At 0055 EWT, a North American TB-25J crashed six miles west of Moody Field, Valdosta, Georgia, killing students A/C Kenneth Parkin and A/C Charles R. Ohsberg. The airplane had taken off from Moody Field at 0037 on a routine student night training flight. The airplane was in the traffic pattern and was turning from base to final when it flew into the ground in a 45-degree bank and a dive of about 15 degrees. Investigators speculated that the students had lost the horizon and inadvertently flew the airplane into the ground. The airplane exploded into flames upon impact and wreckage was scattered over an area of 150 yards.

2-10-45B. Altus, Oklahoma. At 1405, a Cessna AT-17B and a Cessna UC-78B collided in mid-air ten miles south of Altus Army Air Field, Altus, Oklahoma, killing instructor 1Lt. Leroy C. Bowman and student A/C John J. Curry in the crash of the AT-17B. Students A/C Jerry G. Foreman and A/C Frederick J. Flurschutz were uninjured and able to safely land the UC-78B Altus Army Air Field. Both airplanes had taken off from Altus Army Air Field on separate hooded instrument flights. The students aboard the UC-78 were flying on instruments when they felt a slight bump. The safety pilot took over the controls. The two students then observed an AT-17 flying below and to the left. The AT-17 then went into a steep dive. The students in the UC-78 maneuvered to keep the AT-17 in sight. The AT-17 had entered a vertical dive, remaining in this attitude until it smashed into the ground and exploded into flames.

2-10-45C. Waco, Texas. At 1428 CWT, a North American AT-6C crashed three miles ENE of the Army Air Field at Waco, Texas, killing instructor 2Lt. Chester A. Davis and student A/C Carl M. Paulson, Jr. The airplane took off from Waco Army Air Field on a routine training flight. Investigation revealed that the airplane stalled at an altitude too low to allow recovery. Investigation of the wreckage revealed that the propeller was not turning when the crash occurred.

2-11-45A. Scott, Mississippi. At 1045, a Vultee BT-13A crashed five miles north of Scott, Mississippi, killing student A/C Albert J. Reick, 21. The airplane had taken off from Greenville Army Air Field, Greenville, Mississippi, on a routine training flight. Investigators stated, "The airplane struck the ground in practically a vertical dive with engine throttled back, completely demolishing the airplane. The cadet attempted to parachute, but left the airplane at too low an altitude to open his parachute.... The cause of the accident is unknown, but it is believed that, since the student was on a acrobatic mission, he may have attempted some acrobatic maneuver from which recovery was made in a dive, such as a half-roll followed by a split-S recovery, at too low an altitude for completion." The airplane did not burn.

2-11-45B. Dodge City, Kansas. At 1055 CWT, a Martin TB-26C crashed five miles southwest of Dodge City, Kansas, killing three fliers. The airplane had been landed at Jetmore Auxiliary Airfield, Dodge City, Kansas, on a positioning flight to Dodge City Army Air Field. The airplane had been landed at Jetmore Auxiliary Airfield because of a malfunction of the starboard engine. The engine was repaired and checked off by a maintenance crew. The next day, a flight crew was sent to the field to retrieve the B-26. The flight crew ran up the engines and nothing unusual was found. The airplane entered the runway and commenced a take-off run. The airplane made a "dry run" and then taxied back to the take-off position. The pilots did not inform the control tower of the reason for the dry run. The airplane made a successful take-off and climbed away. Tower personnel watched the aircraft closely with field glasses because of the apparent aborted take-off attempt. The airplane climbed to about 2,000 feet and was eight miles from the field when the tower personnel stopped observing the airplane. Moments later the airplane was seen to enter a steep spin, remaining in this attitude until it struck the ground and exploded into flames. The nose wheel door was found one mile southwest of the point of impact. Investigators speculated that the port nose wheel door had separated and collided with the propeller or tail assembly. Pilot Capt. Robert Lind and engineers SSgt. Vern M. Veach and SSgt. Lionel Fincke were killed instantly in the crash.

2-11-45C. El Morro, New Mexico. At 1800 MWT, a Republic P-47N flying in poor weather crashed 20 miles south of El Morro, New Mexico, killing pilot Capt. Harold W. Morrow, 43. The airplane took off from Kirtland Army Air Field, Albuquerque, New Mexico, on a ferry flight to Long Beach, California. A rancher had heard the airplane flying in the overcast and moments later heard a "dull" crash.

Investigation of the wreckage indicated that the airplane dove into the ground at a steep angle and exploded violently into flames. The pilot had filed a visual flight plan but had encountered instrument conditions en route. The pilot changed course to avoid bad weather and changed to an instrument flight plan; he was flying on the El Morro Radio Range station at the time of the accident. The cause of the accident was undetermined. Investigators speculated that the airplane might have encountered severe icing conditions, causing the pilot to lose control.

2-12-45A. Hunter Field, Georgia. At 0713 EWT, a Boeing B-17G crashed while attempting to take off from Hunter Field, Savannah, Georgia, killing seven fliers and seriously injuring three others. Killed in the crash were: 2Lt. John F. Bradley, pilot; 2Lt. Marion R. Crowe, co-pilot; 2Lt. Leslie N. Becker, navigator; Cpl. Donald Green, engineer; Cpl. Ralph N. Leverette, radio operator; Cpl. Don W. Van Atta, gunner; 2Lt. John J. Greggo, passenger. Gunners Sgt. William D. Knight, Cpl. Milan F. Vukelich and Cpl. Raymond T. Waddell received serious injuries. The airplane was taking off from Hunter Field on a flight to Dow Field, Bangor, Maine, en route to an active war theater when it collided with trees 4,000 feet from the end of the runway. The starboard wing contacted the trees first, and the outer tip was severed after striking several. The port wing came up and the airplane rolled to the right into the trees; the starboard horizontal stabilizer and elevator were severed and the starboard engines and propellers separated, and the vertical fin sheared off as the airplane smashed through more trees. The airplane burst into flames as it broke up, the main wreckage coming to rest about 1,200 feet from the initial point of impact. Investigators could not determine why the airplane flew into the trees.

2-12-45B. Santa Rosa, California. At 1115, a Lockheed P-38J crashed one mile southwest of the Army Air Field at Santa Rosa, California, killing 2Lt. Joseph Troise, 25. The airplane was flying in the number-two position of a flight that was taking off on an aerial gunnery mission. The airplane began its take-off run and puffs of black smoke were observed emitting from the port engine. The airplane continued with the take-off and became airborne. The port engine continued "popping" and puffs of smoke continued to emit from the engine as the airplane climbed away with landing gear retracting. The starboard wing dropped and the pilot re-established level flight. The airplane continued to climb and its speed was observed to slow appreciably. The airplane then snapped over to an inverted attitude, rolled through and slammed into the ground in an upright position, exploding into flames upon impact. The pilot was killed instantly. Investigators speculated that the airplane suffered from an induction system problem in the port engine.

2-12-45C. Troy, Oregon. At 1300, a Consolidated B-24D crashed 15 miles southwest of Troy, Oregon, killing ten fliers. Radio operator Cpl. Walter R.

Burnside received minor injuries parachuting to safety; bombardier instructor 1Lt. Richard C. Fies escaped injury parachuting to safety. Killed in the crash were: 2Lt. Haynes E. Peery, pilot; 2Lt. Clair E. Roberts, co-pilot; 2Lt. Arnold Heermann, bombardier; F/O James A. Ross, navigator; Cpl. Francis A. Brouski, engineer; Cpl. Alan P. Glover, gunner; Cpl. John V. Ormond, gunner; Cpl. Maurice C. Peterson, Sr., gunner; Cpl. Lawrence L. Wren, gunner; Cpl. Martin E. Labaig, passenger. The airplane took off at 1138 PWT from Walla Walla Army Air Field, Walla Walla, Washington, on a training mission to Blythe, California, and return. The airplane climbed on instruments to 25,000 feet where it broke out on top of the clouds, leveling off just above the tops. The airplane had picked up some ice during the climb but it was minimal. The airplane was flying in and out of the cloud tops when the port wing dropped. The pilot righted the airplane, but moments later the starboard wing dropped and the airplane fell out of control into the overcast. The pilot was able to right the airplane momentarily but it stalled violently and fell out of control again. The instructor and radio operator bailed out immediately and both stated to investigators that after they bailed out they observed the airplane falling in a steep spiral to the right. The airplane continued in the steep spiral until it smashed into the ground and exploded. No other crewmembers were able to bail out.

2-12-45D. Norman, Oklahoma. At 1502, a Lockheed F-5E crashed four miles west of Norman, Oklahoma, killing Chinese Air Force pilot 2Lt. Tsou-Min Lee, 22. The airplane had taken off from Will Rogers Field, Oklahoma City, Oklahoma, on a training flight. The pilot was to familiarize himself with the P-38 type aileron boost system. After flying for an hour and a half, the pilot was instructed to practice "beam" flying. The airplane slammed to the ground at a steep angle and exploded a short time later. The cause of the crash could not be determined.

2-12-45E. Silurian Lake, California. At 1445, two Lockheed P-38L airplanes collided in mid-air and crashed ten miles south of Silurian Lake, California, killing both pilots. Capt. William D. Horton, 26, was killed in the crash of P-38L #4424790; Capt. Donald L. Webber was killed in the crash of P-38L #44-24807. The airplanes were part of a five-ship flight that took off from Daggett Municipal Airport, California, on a scheduled camera-gunnery mission. Capt. Horton was flying in the number-three position; Capt. Webber was flying in the number-two position. The flight rendezvoused with the target ship and the aircraft began making their camera-gunnery passes. The two subject airplanes entered into a pass at the target ship simultaneously, with Capt. Webber in front and slightly below. The airplanes peeled off of the target and collided. Both airplanes went out of control and smashed to the ground. Neither pilot was able to parachute to safety.

2-12-45F. Pacific Ocean, California. At 1820, a Martin TB-26C was abandoned and crashed into the Pacific Ocean 11 miles west of Jenner, California. Pilot 1Lt. Leroy E. Young parachuted to safety and was uninjured. Engineer SSgt. Curtis J. Myhre and passenger Cpl. Gene A. Wheeler parachuted from the airplane but were missing and presumed lost at sea. The airplane took off from Santa Rosa Army Air Field, Santa Rosa, California, on a target-towing mission. The subject airplane rendezvoused with a flight of P-38s, which were to fire on the towed target. The subject airplane headed out to sea on a course of 250 degrees and the P-38s made their firing runs. The subject airplane made a 180-degree turn and headed back toward the coast. As the P-38s continued with the gunnery mission, the subject pilot smelled "something like burning wire." The odor abated and the crew was unable to find its source, so the flight continued. The first flight of P-38s completed their firing runs and the subject airplane rendezvoused with a second flight of P-38s. The subject airplane again turned toward the sea and the second flight of P-38s began their firing runs. As the subject airplane headed back toward the coast the odor in the cockpit persisted, so the pilot sent the engineer to the bomb bay to investigate. The engineer returned to the flight deck and advised the pilot that all of the hydraulic fluid had leaked from the main reservoir. The subject airplane again turned back out to sea in order to give a third flight of P-38s opportunity to fire at the target. After a few minutes, smoke began to infiltrate the cockpit. The engineer traced the smoke to the ship's heating system. The smoke began to become heavy and the bomb bay filled with smoke. Moments later the engineer reported to the pilot that the starboard engine appeared to be on fire in the accessory section. The pilot turned the airplane toward the coast. The fire began burning through the top of the starboard wing. The pilot ordered the crew to fasten parachutes and stand by to bail out. The engineer controlled the airplane while the pilot attached his parachute. The pilot ordered the two crewmembers to bail out. The bomb bay doors opened and the two airmen jumped. The pilot left his station but the bomb doors had closed. The pilot regained his position and then attempted to lower the nose wheel but was not successful. The pilot then jettisoned the cockpit upper hatch and deliberately stalled the airplane. The airplane snapped over to an inverted position and the pilot dropped out and parachuted to safety. A U.S. Navy blimp dropped a life raft to the pilot and he was able to retrieve it. After several hours, a navy craft rescued the pilot from the sea.

2-13-45A. Lawrenceville, Illinois. At 2119, a Waco CG-4A glider crashed after prematurely releasing its tow cable as it was towed aloft at George Field, Lawrenceville, Illinois, killing glider pilot 2Lt. David B. Goldin and seriously injuring rated airplane pilot 1Lt. Kenneth J. Sire. The subject glider was one of two that were being towed aloft by a Douglas C-47 type airplane for a glider-training mission. Investigation revealed that the subject glider's pilot intentionally released from the tow ship because of heavy condensation on the glider windscreen. The subject pilot climbed his glider steeply in an effort to gain altitude so he could return to the field for a landing. The glider stalled at the top of the climb and fell off on the port wing, smashing to the ground in a steep left bank near the take-off runway. The second glider released from the tow ship and landed safely on the field.

2-13-45B. Ellijay, Georgia. At 2330 EWT, a Douglas A-26B collided with rising terrain ten miles east of Ellijay, Georgia, killing pilot 2Lt. Harald W. Gilbert and engineer-gunner Cpl. Anthony J. Simnowski. The airplane took off from Florence Army Air Field, Florence, South Carolina, on a routine training flight. The pilot became lost and apparently was unaware of his proximity to the terrain he was flying over. The airplane smashed into the 3,950-foot mountain at an elevation of 3,852 feet while in a straight and level attitude with both engines producing power. The airplane exploded into flames upon impact and the occupants killed instantly.

2-13-45C. Grandfield, Oklahoma. At 2338 CWT, a North American TB-25J crashed three and a half miles southwest of Grandfield, Oklahoma, killing students A/C Walter E. Brongiel and A/C William K. Briggs. The airplane took off from Frederick Army Air Field, Oklahoma, on a routine student night training mission. Investigation revealed that the students had inadvertently flown the airplane into the ground in a near-level attitude with both engines producing power. Investigators noted that a new flight indicator had been installed on the airplane the day before.

2-14-45A. Coffeyville, Kansas. At 1207 CWT, a Boeing RB-17G crashed into a large tree while attempting an emergency forced landing one-third of a mile north of Coffeyville Army Air Field, Coffeyville, Kansas, killing navigator instructor 1Lt. John Prokop, Jr. Co-pilot 2Lt. Richard K. Larson received serious injuries; flight engineer 2Lt. Jack R. Morris sustained minor injuries. Six fliers escaped injury. The airplane took off from Clovis Army Air Field, Clovis, New Mexico, on a navigation-training mission to Coffeyville via Kansas City. The pilot changed the flight plan to Wichita, Kansas, instead of Kansas City. The airplane was flying at 7,000 feet near Independence, Kansas, when the pilot saw a hole in the clouds and elected to dive through to get underneath. The airplane performed a descending 270-degree turn and was able to get underneath the overcast. The pilot leveled out and then advanced the throttles, but the engines did not respond. The pilot noticed that the fuel pressure was at zero; the booster pumps were turned on and a small increase in fuel pressure resulted. The fuel pressure dropped back to zero moments later. The pilot was unable to get appreciable power from the engines, so he turned toward Coffeyville Army Air Field and lowered the landing gear. The B-17 began losing altitude rapidly and it became evident that the ship was not going to make it to the field. The pilot landed the airplane in a three-point attitude but the airplane rolled over rough terrain. The airplane collided head on with a large tree, crushing the nose section and killing the navigator instructor. Crewmembers uninjured in the crash landing were: 2Lt. Gerald K. Kunkle, pilot; 2Lt. W.M. Barner, navigator; 2Lt. B.J. Weinberg, navigator; 2Lt. J.D. Campbell, navigator; Sgt. Alfred A. Bonzo, radio operator; Sgt. Jerard J. Kelleher, radio operator.

2-14-45B. Oblong, Illinois. At 2250, a Douglas C-47A suffering an engine failure crashed four miles southeast of Oblong, Illinois, killing three fliers and seriously injuring two others. The airplane was part of a 21-ship flight that had taken off from George Field, Lawrenceville, Illinois, on a para-pack drop exercise. The formation was made up of two, nine-ship flights and third flight of three airplanes in three V-formations. Because the flight leader maneuvered the formation unexpectedly while battling a strong tailwind that disrupted the timing of the drop, the subject airplane was among a few that fell out of formation after the parachute run over the drop zone. The subject airplane then began experiencing trouble with the port engine. The pilot mistook the trouble for a runaway propeller and apparently did not realize that the engine had failed. The pilot advised the tower of the emergency and attempted to maneuver back to the field. The pilot was unable to maintain altitude and the airplane descended until it collided with trees, going out of control and smashing to the ground. Pilot 2Lt. Monte R. Hamilton, co-pilot 2Lt. Alexander Cairney and engineer Pfc. Clarence C. Orr were killed in the accident. Radio operator Cpl. Jack L. Bowers and passenger Sgt. Eugene G. Harrison survived with serious injuries.

2-15-45A. Gulf of Mexico, Florida. At 1410 CWT, a North American AT-6A and an AT-6B collided in mid-air and crashed 18 miles southwest of Eglin Field, Valparaiso, Florida, killing AT-6B pilot 2Lt. Clarence V. Robey, 20. AT-6A pilot 2Lt. John Prodan parachuted to safety, suffering only minor injuries. The airplanes were part of a four-ship flight that took off from Auxiliary Airfield #7, Eglin Field Military Reservation, on an aerial gunnery mission on Water Range 32 at 5,000 feet. Lt. Robey was flying in the number-three position and Lt. Prodan was flying in the number-four position. On the first firing pass on the towed target, it was discovered that Lt. Prodan's gun sight was inoperative. He flew on three firing passes with the flight but did not fire. On the fourth pass, Lt. Prodan mistook the number-two ship for the number-three ship. Lt. Prodan maneuvered into position and began his fourth pass. Other pilots observed that an unsafe situation was developing and attempted to warn the subject pilots. Lt. Prodan's airplane descended onto Lt. Robey's airplane. Lt. Prodan's engine and propeller collided with the cockpit area of Lt. Robey's airplane.

The airplanes stuck together momentarily before separating and spinning to earth. Lt. Robey's airplane went out of control and smashed to the ground. The airplane did not burn but was destroyed utterly. It was speculated that Lt. Robey was "stunned," injured or killed by the collision, preventing him from bailing out. Lt. Prodan was not able to regain control of his airplane, and with some difficulty he was able to parachute to safety. He was rescued from the sea by an AAF crash boat.

2-15-45B. Flushing Bay, New York. At 1553 EWT, a Boeing B-29 crashed into Flushing Bay near LaGuardia Field, New York, killing five crewmembers and injuring five others, two of them seriously. Killed in the crash were: Maj. Billy B. Southworth, Jr., pilot; 1Lt. C.D. Magee, co-pilot; 1Lt. M.J. Licursi, navigator; SSgt. Joseph Yarbroudi, port scanner; 2Lt. Ralph L. Stickle, observer. Flight engineer 2Lt. R.N. Worcester, Maj. W.L. Anken and civilian observer W.A. Burkus received minor injuries. Starboard scanner Pvt. Howard Card and radio operator TSgt. Lewis Munford were seriously injured. The airplane took off at 1530 EWT from Mitchel Field, Long Island, on a flight to Morrison Field, West Palm Beach, Florida. Investigators stated, "Approximately one minute after take-off, the left scanner reported a thin stream of white smoke coming from the top of the number-one engine. The pilot gave instructions to keep a constant check. Maj. Anken [seated in the central fire control position] reported a heavier volume of smoke, but no fire, a few minutes later. The pilot feathered the number-one engine and the smoke stopped immediately. However, he notified the crew he would make an emergency landing at LaGuardia Field. At this time the pilot reported to LaGuardia Tower ... that he was flying at 6,000 feet northeast of the field and wanted an emergency landing. He was cleared to descend and given Runway 13. After circling the field to lose altitude, the pilot passed up turning for Runway 13 and lined up for Runway 9. He was advised by the Tower that he was landing on Runway 9, which was 4,580 feet in length but did not acknowledge, and continued his approach with wheels and full flaps down. The aircraft crossed the field boundary and when about 1,000 feet of runway was used, it was still 20 feet in the air. At this point the pilot evidently decided to go around, applied power, and began a slow climbing turn to the left with the angle of bank steadily increasing, crossed the boundary of the field headed northeast. In this position the pilot reduced power and the airplane stalled, hitting the water of Flushing Bay with its left wing, cartwheeling into the water. Fire started immediately upon impact; the crash occurred at 1553 EWT."

2-15-45C. Hollister, Oklahoma. At 1515, a Vultee BT-13A crashed six miles southeast of Hollister, Oklahoma, killing pilots 1Lt. Frederick W. Watson and F/O Darrill P. Elkins. The airplane took off from Frederick Army Air Field, Frederick, Oklahoma,

on a local flight. Investigators stated, "The aircraft was first observed flying east at a low but not unreasonable altitude. It was seen to turn toward the south, then to the west losing altitude and making S-turns. As the airplane turned to the north it was observed to be at an altitude of approximately 100 feet. It continued in the right turn and seemed to steepen the bank until a nearly vertical bank was attained. The aircraft continued in this steep turn until the time of the crash. There was no fire."

2-16-45. Atlantic Ocean, Virginia. At 1430 EWT, a Republic P-47C and a Martin TB-26B collided in mid-air and crashed into the Atlantic Ocean 45 miles northeast of Norfolk Army Air Field, Norfolk, Virginia, killing P-47 pilot F/O Robert A. Cain and four crewmembers aboard the bomber. Killed in the crash of the B-26 were: Capt. Wesley E. Barbour, pilot; TSgt. Carl H. Bundy, engineer; Pvt. John J. Termotto, radio operator; Pvt. Edward E. Altimus, tow-reel operator. The subject P-47C was flying in the number-three position of a three-ship flight that took off from Norfolk Army Air Field on an aerial gunnery mission. The P-47 flight rendezvoused with the subject B-26 target tow-ship and commenced the gunnery exercise, making about eight to ten passes at the target ship. The subject P-47 collided with the B-26 while pulling up after a firing pass. The P-47's starboard wing collided with the port wing of the B-26, severing the wings of both aircraft. The airplanes went out of control and fell into the sea. Pilots flying in the area observed three parachutes descending, but no airmen were found d when rescue craft arrived in the area. Investigation revealed that Pvt. Termotto and Sgt. Bundy were struck by turning propellers when they bailed out, killing both instantly. Their parachutes apparently deployed because of the force of the impact. The bodies of Capt. Barbour and Pvt. Altimus were not recovered.

2-17-45A. Key Field, Mississippi. At 0940 CWT, a Vultee BT-13A crashed 14 miles east of Key Field, Meridian, Mississippi, killing pilots 2Lt. Lewis T. Davis, Jr., 27, and 2Lt. William E. Sheetz. The airplane took off from Key Field on an instrument-training flight. Investigation revealed that the pilots flew over an area that had been devastated by tornados and were maneuvering at low altitude, apparently observing the damage to terrain and structures. Investigators speculated that the pilots had become distracted by something on the ground and allowed the airplane to enter a stall at an altitude too low to allow recovery. The airplane, which "fluttered like a leaf" as it fell to earth, smashed to the ground and burst into flames.

2-17-45B. Page Field, Florida. At 1557 EWT, a Curtiss P-40N crashed one-half mile southwest of Page Field, Ft. Myer, Florida, killing pilot F/O Liston L. Rochelle, 21. The pilot returned to the field and radioed the tower that he was having engine problems. The airplane entered the traffic pattern and the pilot attempted the landing. The airplane flew level over Runway 22 for a few seconds before the pilot appeared to dive the airplane toward the runway. The airplane overshot the runway, crossed the highway and smashed into a field of tree stumps. The airplane, which had landed with the landing gear in the retracted position, broke up in the rough terrain. The airplane did not burn but was destroyed utterly. Investigation revealed that the engine was cutting out because the pilot was flying on a nearly empty fuel tank.

2-19-45A. Warwick River, Virginia. At 0615, a Consolidated B-24L suffering a loss of power was abandoned and crashed into the Warwick River 12 miles WNW of Langley Field, Hampton, Virginia. Pilot 2Lt. Joseph H. Burton, 23, parachuted from the airplane but was not rescued. He was declared missing and presumed lost. Five crewmembers parachuted to safety. The airplane took off from Langley Field at 2104 with an instructor aboard so the pilot could be checked off on night landings. The pilot performed his night take-off and landing satisfactorily so the instructor deplaned. The airplane took off again at 2138 to fly a celestial navigation mission. The airplane flew its mission, returned to the area of Langley Field and began flying on the Langley radio range. The engineer "leveled" the number-one and number-two main fuel tanks before the co-pilot flew the airplane over the field, circling at 4,000 feet. The pilot took over and descended to traffic altitude. Moments later the number-one and number-two engines cut out. The co-pilot ordered the engineer to turn the bomb bay fuel valves for the fuel tanks to the "tank to engine" position. The engineer complied and the engines came back on, but according to the surviving co-pilot they didn't sound right. The pilots and engineer attempted to get the airplane back to normal performance but were unsuccessful. The pilot ordered the crew to bail out and they did in good order. The pilot and the radio operator bailed out last. The airplane was abandoned and it crashed into the Warwick River. Crewmembers parachuting to safety were: 2Lt. Edward T. Purcell, co-pilot; F/O John R. Evans, navigator; 2Lt. Carl E. Brown, navigator instructor; Cpl. Robert P. Redden, engineer; Cpl. Glenn R. Eckstrom, radio operator.

2-19-45B. Barstow, California. At 0830, two Lockheed P-38L airplanes collided in mid-air and crashed 22 miles north of Barstow, California, killing both pilots. 2Lt. Walter E. Mogensen, 24, was killed in the crash of P-38L #44-25637; 2Lt. Earl A. Morgan, 20, was killed in the crash of P-38L #44-23861. The airplanes were part of a four-ship flight that took off at 0750 from Daggett Municipal Airport, California, on an aerial gunnery mission. Lt. Mogensen was flying in the number-three position; Lt. Morgan was flying in the number-four position. The flight had rendezvoused with a target tow ship on the Daggett Gunnery Range north of Barstow and commenced the exercise, flying to the south on the west end of the range

at 10,000 feet. The tow ship was approaching the southern end of the range and the pilot radioed the P-38 leader that the tow ship was going to have to make 180-degree turn to the left in order to fly north on the range. The P-38 flight leader informed his pilots of the maneuver. The P-38 pilots acknowledged, and at the time the two subject airplanes were finishing their passes. The P-38s collided just in front of and to the left of the target ship as they broke off their pass. Both airplanes exploded and went out of control, shedding pieces as they fell to the ground. The pilots were unable to bail out and investigators speculated that they were killed in the collision. Wreckage was scattered over an area of several square miles. It was speculated that bright sunlight might have been a factor in the collision.

2-19-45C. Marana, Arizona. At 1105 MWT, a Vultee BT-13B crashed 14 miles southwest of Marana Army Air Field, Marana, Arizona, killing students A/S James J. Farrell and A/S Alvin W. Harroun. The airplane took off at 1050 MWT from Sahuaro Auxiliary Airfield, Arizona, on an instrument flight. A/S Farrell was flying in the front cockpit as the safety pilot and A/S Harroun was flying in the rear cockpit as the instrument pilot. At 1110 MWT, Marana Army Air Field control tower personnel observed a column of black smoke rising in the vicinity of Sahuaro Field. An airplane was sent over to investigate and it was confirmed that an airplane had crashed. Rescue personnel from Marana Field were immediately dispatched, but because of rough terrain arrived at the crash scene after 50 minutes. The fire was put out in minutes. The students were killed in the crash. There were no witnesses and the cause of the crash remains undetermined.

2-19-45D. Punta Gorda, Florida. At 1718 EWT, two North American P-51C airplanes collided in mid-air while maneuvering at Punta Gorda Army Air Field, Punta Gorda, Florida, killing pilot 2Lt. Vance Burson, 21, aboard P-51C #42-103664. F/O Rudolph J. Habe, 21, was uninjured aboard P-51C #42-103665 and was able to make an emergency landing near the field. The airplanes were part of a five-ship flight that took off from Venice Army Air Field, Venice, Florida, on a ground gunnery mission. Lt. Habe was flying in the number-three position; Lt. Burson was flying in the number-four position. After the completion of the gunnery mission, the flight was to practice strange field landings. The formation flew to Page Field, Ft. Myers, Florida, and performed a landing and a take-off. The airplanes then flew to Punta Gorda Army Air Field and requested landing instructions. The flight leader was told to circle at 2,000 feet and await instructions. The subject flight was at 1,500 feet, so the leader ordered the flight to echelon right and then climbed the formation to 2,000 feet. The flight circled to the left for a short time before the tower cleared the flight to land on Runway 9. Visibility was hazy. The flight leader mistakenly began a pattern and approach for Runway

3. The flight leader realized his mistake while on the approach. The tower cleared the flight to go around and land on Runway 9. The flight pulled up and began a climbing turn to the left. The flight leader peeled off and the number-two ship continued straight ahead. Lt. Burson began to overrun Lt. Habe and maneuvered to keep spacing. He was unsuccessful and collided with Lt. Habe's aircraft from underneath. Lt. Burson's airplane was cut in two at the coolant shutter and spun to earth, exploding into flames on the field. Lt. Habe was able to make a successful wheels-up landing near the field.

2-19-45E. Norway, South Carolina. At 1735 EWT, a North American B-25H collided with trees and crashed one mile northeast of Norway, South Carolina, killing four fliers and seriously injuring another. The airplane took off from Columbia Army Air Base, Columbia, South Carolina, on a chemical spraying mission. The airplane arrived at the target area at Lake Murray and completed its mission, which was conducted at 100 feet agl. The airplane flew away to the northeast, continuing to fly at about 100 feet agl. The airplane arrived over Norway and flew over the high school and the city water tower at very low altitude. The airplane was next observed flying below treetop level. The airplane was observed flying at extremely low altitude and then climbing to clear trees in the flight path, then descending back to very low altitude in open fields after clearing the trees. The pilot failed to pull up soon enough and the airplane collided with a stand of trees. The port wing and tail section were sheared off and the airplane burst into flames. The airplane went out of control, smashing to earth and bursting into flames in an open field about 1,000 feet from the point it collided with the trees. Killed in the crash were: F/O Robert B. Tuskey, pilot; F/O Donald D. Crosby, co-pilot; Cpl. Robert Hanneman, engineer; Cpl. Robert B. Bogdewiecz, gunner. Gunner Sgt. Sigmunt W. Makowski received very serious injuries.

2-20-45A. Litchfield Park, Arizona. At 1025 MWT, a Curtiss P-40R dove into the ground four miles west of Litchfield Park, Arizona, killing pilot F/O William J. Laney, 19. The airplane took off from Luke Field, Phoenix, Arizona, on a short-duration oxygen mission. The airplane climbed to 8,000 feet and rendezvoused with two other airplanes for the exercise. The airplanes joined in formation, the pilots went on oxygen, and then the formation climbed to 15,000 feet. The formation maneuvered for approximately 15 minutes at 15,000 feet, and then the exercise was terminated as planned. The subject airplane appeared normal as it was turning away from the formation for the flight back to Luke Field. A short time later, two other pilots flying in the area observed the airplane in a vertical dive at about 6,000 feet, trailing a stream of black smoke. The port wing and the propeller were seen to separate as the airplane continued to dive toward the ground in a vertical manner. The airplane, shedding

pieces, remained in this attitude until it struck the ground and exploded violently into flames. Wreckage was scattered over a wide area.

2-20-45B. Key Field, Mississippi. At 2104, a North American AT-6C flying in instrument conditions crashed two miles east of Key Field, Meridian, Mississippi, killing pilot 2Lt. Albert R. Vesper and observer Sgt. Sol R. Greenburg. The airplane took off on an instrument clearance from Alexandria Army Air Field, Alexandria, Louisiana, to Key Field on a weather checking flight that originated at Bryan Army Air Field, Bryan, Texas. The pilot arrived in the area of Meridian and reported to Key Field control tower personnel that he was flying on the southeast leg of the Key Field radio range and was requesting landing instructions. The pilot was given the current weather and landing instructions. At 2030 the pilot advised that he was starting his let-down on the southeast leg. A short time later, the pilot radioed that he was returning to Alexandria Army Air Field, which was one of his alternates. The pilot was issued an instrument clearance and he flew away toward Alexandria. At 2101 the airplane returned to the area of Key Field and the pilot requested landing instructions from Key Field. The pilot was advised of current weather and the pilot acknowledged, reporting that he was flying contact and had the field in sight. The airplane was cleared to land and the pilot acknowledged. There was no other radio contact with the subject airplane. The airplane failed to land at the field and a short time later it was discovered to have flown into the side of a small hill. The airplane exploded into flames upon impact and the occupants killed instantly.

2-21-45A. Laredo, Texas. At 1101, a Curtiss P-40R crashed two miles northeast of Laredo Army Air Field, Laredo, Texas, killing pilot 2Lt. William M. Adin, 21. The airplane was part of a two-ship flight that took off from Laredo Army Air Field on a local flight. The subject airplane made a normal take-off and climbed to about 700 feet when it began experiencing engine trouble. The airplane turned around and headed back toward the field. Moments later, puffs of smoke were observed emitting from the exhaust stacks on both sides. The pilot decided that he could not make the field so he turned into the wind and attempted to shoot an emergency landing in a clear area near a road. The pilot attempted to stretch his glide and stalled the airplane at an altitude too low to allow recovery. The airplane spun to the ground and struck with the port wing, cartwheeling into an inverted position and bursting into flames. The pilot was killed instantly. Investigation revealed that the impeller bearing failed, causing the diminished engine performance.

2-21-45B. Phillipsburg, Montana. At 1530 MWT, a Consolidated B-24L suffering multiple engine failures crashed 30 miles west of Phillipsburg, Montana, killing pilot F/O Jarrett E. Howard and co-pilot 2Lt. Kenneth H. McComb. Engineer SSgt. Charles A. Rose parachuted to safety and was uninjured. The airplane took off at 1243 PWT from Spokane Army Air Field, Spokane, Washington, on a ferry flight to Gore Field, Great Falls, Montana. The airplane climbed to 13,000 feet and leveled off for the flight. Moderate to light icing conditions were present in the clouds along the line of flight. Radio reception was very poor and there was no radio contact during the flight. A short time after leveling off, the number-four engine made a "loud roar" and then the engine quit. The pilot remained on course, hoping to re-establish radio contact and spot landmarks through the scattered clouds underneath. At approximately 1530 MWT, the number-one engine roared and then quit. A few moments later, the number-two engine roared and quit. The engineer opened the bomb bay doors and bailed out immediately. He deployed his parachute and the airplane crashed a few seconds later. The engineer parachuted to safety and was rescued from the wilderness by a civilian working in the area. Investigators noted that the airplane had been in storage since September 1944. It was also noted that Wright Field personnel were preparing a memorandum "giving more thorough instructions in the use of turbo power settings on the B-24 type aircraft in icing conditions." It was also noted that all B-24 type aircraft that were being taken out of storage at Spokane Army Air Field be given a more thorough examination and a longer "test hop" before being cleared for transfer off station.

2-21-45C. Dallas, Texas. At 1714 CWT, a Consolidated B-24J spun to the ground eight miles WNW of Love Field, Dallas, Texas, killing the crew of four. The airplane took off at 0830 CWT from Smyrna Army Air Field, Smyrna, Tennessee, on a cross-country navigation flight to Hensley Field, Dallas, Texas. The crew successfully navigated to Hensley Field, arriving at 1450 CWT. The crew secured quarters in Dallas for the night and then returned to the field to fly a three-hour local flight, taking off at 1641 CWT. At 1714 CWT, two AAF officers observed the airplane in a spin to the left in an extreme nose-down attitude. The airplane remained in the spin to the left and smashed to the ground in a flooded gravel pit. Three bodies were recovered immediately and the fourth was recovered on 2-25-45. Heavy precipitation and overflow from the Trinity River hampered the initial investigation. Most of the wreckage remained submerged in the flooded gravel pit until the floodwaters receded. Investigators could not determine how the airplane entered the fatal spin. Killed in the crash were: 2Lt. Arthur P. Bick, pilot; Capt. Albert W. Gillespie, B-24 student; 1Lt. Frederick D. Gaulke, B-24 student; Pfc. Herman N. White, engineer.

2-21-45D. Blytheville, Arkansas. At 2055, a Beech AT-10GL crashed three miles WNW of Blytheville Army Air Field, Blytheville, Arkansas, killing students A/C Robert C. Higginson and A/C Robert J. Hine. The airplane took off from Blytheville

Army Air Field on a night formation training flight. The airplane successfully completed the night formation flight and the flight returned to the field for landing. The formation broke up and the airplanes entered the traffic pattern. The subject airplane was number-two to land, behind the flight leader. The subject airplane flew a low approach and was too close to the lead airplane. The control tower officer ordered the subject airplane to go around. The student acknowledged and climbed away for the go around. The airplane never returned to the field and was found to have crashed. The airplane descended while in the traffic pattern and struck a tree, causing the airplane to go out of control and smash to the ground. The airplane did not burn but was destroyed utterly. Investigation revealed that the altimeter was not correctly set, causing the students to inadvertently fly the airplane into the terrain.

2-22-45. Leroy, Texas. At 1730, a North American AT-6D crashed two miles southeast of Leroy, Texas, killing student A/C John R. Forgey. Instructor 2Lt. Haywood P. Stockton was seriously injured. The airplane took off from Waco Army Air Field, Waco, Texas, on a routine training flight. The student and instructor were practicing spins. The instructor entered a deliberate spin to the right at an altitude of 7,000 feet. The student was unable to recover after five turns of spin so the instructor took over. The instructor stopped the spin momentarily, but the student pulled the stick back and the airplane stalled again. A partial recovery was made before the airplane struck the ground.

2-23-45A. Savannah, Georgia. At 0701 EWT, a Consolidated B-24J collided with trees and crashed just after take-off from Chatham Field, Savannah, Georgia, killing four crewmembers and seriously injuring six others. Engineer Pfc. James L. Dorr escaped injury. The airplane was scheduled to fly on an Instrument Calibration Mission. The airplane made what appeared to be a normal take-off, lifting off after a 3,000-foot take-off roll. The pilot went on instruments immediately following the take-off because no horizon was visible. When the airplane reached 100 feet agl the pilot called for the landing gear to be retracted. The co-pilot apparently complied and the airplane, which was climbing, immediately became nose-heavy. The pilot pulled back on the control column and the airplane began to "mush along," losing altitude until it collided with the trees about 3,200 feet from the end of the runway. The starboard wing was severed when it struck a 56-foot tree and the airplane went out of control, smashing to the ground and bursting into flames about 5,000 feet from the end of the take-off runway. Investigators speculated that the co-pilot might have inadvertently retracted the flaps after take-off, causing the airplane to sink back to the ground despite the effort of the pilot to keep the airplane climbing. Killed in the crash were: 1Lt. George R. Chernowski, navigator instructor; Cpl. Kenneth M. Alexander, radio operator; Cpl. David S. Harvey, engineer; 1Lt. William J. Spangler. Crewmembers seriously injured in the crash were: 2Lt. Morton W. Madsen, pilot; 2Lt. Brazel P. Busick, co-pilot; 2Lt. Robert H. Schaefer, bombardier; Cpl. Walter J. Dever, radio operator; Cpl. James A. Demonthreun, gunner; Pfc. Lewis W. Agee, gunner.

2-23-45B. Hartselle, Alabama. At 0715 CWT, a Consolidated RB-24E crashed ten miles southeast of Hartselle, Alabama, killing the crew of five. The airplane took off from Courtland Army Air Field, Courtland, Alabama, on a routine orientation flight. The airplane was flying at about 6,000 feet when it stalled, fell off on wing and then entered a spin to the right. Power was applied and a partial recovery was made at about 4,000 feet. The spin stopped momentarily, but then the airplane fell off again and entered steep vertical spin to the right. The airplane remained in this attitude until it smashed to the ground in a wooded area with the right wing low, exploding into flames upon impact. Killed in the crash were: 1Lt. Anthony C. Yenalavage, pilot; 2Lt. Robert H. Garvin, co-pilot; 2Lt. Marion J. Seaney, pilot-rated passenger; Pfc. Jackson G. Stewart, engineer; Pvt. Gaston P. Whitsey, student engineer.

2-23-45C. Freeport, Texas. At 1420, a Lockheed P-38L crashed 14 miles southwest of Freeport, Texas, killing Lt. Col. Frank E. Adkins, 30. The airplane was flying in the number-three position of a four-ship flight that took off from Kissimmee Army Air Field, Kissimmee, Florida, on a special long-range simulated combat flight to Galveston, Texas, and return. The airplanes, flying at 30,000 feet, arrived in the vicinity of Freeport and then simulated combat maneuvering by flying in a loose circle at full power settings. The pilots discussed the various power settings over the radio as they maneuvered. The airplanes began to turn toward the sea so that they could drop their auxiliary fuel tanks. At this time it was noticed that two small puffs of smoke came from the subject airplane's engines. It was thought that the subject airplane's auxiliary fuel tanks ran out of fuel at that moment. The pilot switched to the main fuel tanks. Moments later, the subject airplane was seen to enter a climbing turn to the left. The airplane passed through about 180 degrees of turn before entering a slight diving attitude. The airplane then entered an abrupt wing over to the left and then plunged to the ground in a high-speed dive. The airplane remained in this attitude until it struck the ground and exploded into flames. The pilot made no apparent effort to leave the airplane. Investigators speculated that the pilot had inadvertently disconnected his oxygen hose as he attempted to switch fuel tanks, causing him to lose consciousness.

2-23-45D. Kingman, Arizona. At 1432 PWT, a Bell P-39Q crashed 12 miles southwest of Kingman, Arizona, killing 1Lt. Dean E. Schoenfeldt, 25. The airplane took off at 1340 PWT from Kingman Army Air Field on a local formation flight. The airplane joined in formation at 10,000 feet with a Bell P-63 airplane at 1350. The two airplanes flew north up the Las Vegas Valley and then returned on a southern heading back to the area of Kingman. The two airplanes, still at 10,000 feet, then flew to the area southwest of Kingman. At this time the subject airplane, which was flying in the lead position, was observed to enter a climbing turn to the left. The other pilot lost the subject airplane during the maneuver. The wingman maneuvered to regain visual contact with the subject airplane. He spotted the subject airplane flying in a nose-high attitude at about 13,000 feet. The airplane then stalled and entered a tumbling end-over-end spin with starboard wing down. The airplane fell in this manner for three or four revolutions before entering a conventional spin, describing three or four turns of spin to the right. The pilot recovered momentarily, but the airplane entered a flat spin to the left. The pilot never recovered and the airplane smashed to the ground, bursting into flames upon impact.

2-24-45A. Selfridge Field, Michigan. At 1440 EWT, a Republic P-47D stalled while turning from base to final while attempting a landing at Selfridge Field, Mt. Clemens, Michigan, killing pilot 2Lt. Charles W. Hassner, 22. The airplane was part of a 16-ship flight that had taken off from Selfridge Field at 0933 EWT on a 1,000-mile cross-country training mission. The flight was returning to the field after being airborne for five hours and seven minutes. The subject airplane was flying in the number-two position in the fourth four-ship flight. The airplanes maintained proper spacing while in the pattern, but then the subject airplane began to gain on the lead ship while flying on the base leg. Investigators stated, "The aircraft then seemed to slow up while in the turn on approach. Suddenly it faltered in the air and spun to the ground while in a left bank. The airplane spun twice and hit the ground, bursting into flames. Two 165-gallon wing tanks were used on this mission."

2-24-45B. Lorraine, Texas. At 1415, a Stearman PT-17 crashed eight miles south of Lorraine, Texas, killing pilot F/O Thomas E. Dolphin, 22. The airplane was part of a 14-ship flight that took off at 1330 from Abilene Army Air Field, Abilene, Texas, for Pyote Army Air Field, Texas, while on a ferry flight from Moody Field, Valdosta, Georgia, to San Francisco, California. The airplane was observed to spin to the ground after flying through the propeller turbulence of a transient U.S. Navy Consolidated PBY-5 airplane that had passed in front of it.

2-25-45. Forester, Arkansas. At 1719 CWT, a Boeing B-17G collided with rising terrain and crashed seven and a half miles SSE of Forester, Arkansas, killing three fliers. The airplane, which was on an instrument clearance, took off at 1130 MWT from Cheyenne Army Air Field, Cheyenne, Wyoming, on a ferry flight to Memphis, Tennessee, via Tulsa, Oklahoma. The flight had originated at Long Beach, California. The airplane made routine position reports over Pueblo, Colorado; La Junta, Colorado; Garden City, Kansas. The last radio contact with the airplane was at 1540 CWT, reporting "500 feet on top" at 8,000 feet five miles southeast of Wichita, Kansas. The airplane failed to arrive at Memphis and was not found until 3-13-45. The airplane had collided with the side of a 1,500-foot wooded ridge in the Ouachita Mountains while in a flying attitude with all four engines producing power. Investigation revealed that the airplane was heading southeast when the propellers cut a 100-yard path uphill through the trees. The port wing collided with a stout 100-foot pine, severing the wing and sending the airplane completely out of control. The airplane broke up and burst into flames as it smashed through the trees. Wreckage was scattered over several hundred yards. Pilot 1Lt. John E. Cayea, co-pilot 1Lt. Walter J. White and engineer TSgt. Truman W. Carter were killed in the crash. The AAF had accepted delivery of the B-17 on 2-23-45.

2-26-45A. Geyserville, California. At 0835, a Lockheed P-38L crashed ten miles west of Geyserville, California, killing F/O Robert O. Hall, 21. The airplane was flying in the number-two position of a three-ship flight that took off from Santa Rosa Army Air Field, Santa Rosa, California, on a high-altitude aerobatic flight. The flight climbed to 22,000 feet and leveled off. The flight leader called for an oxygen check. All pilots checked in that everything was all right. Investigators stated, "About five minutes later the flight leader called for a right echelon and asked the pilots if they both had the P-38L model and they replied that they had. The flight then peeled off and dove to 18,000 feet. The flight leader stated that when he leveled off he had an airspeed of 400 mph. He did two slow rolls and then peeled up again. When he looked down he saw a column of smoke below so he asked the flight to check in but the number-two ship did not answer. The number-three man stated that F/O Hall peeled off with the proper interval and then he, the number-three man, followed F/O Hall down. Shortly after the flight leader pulled out of his dive the number-three man began pulling out of his dive slowly, waiting for F/O Hall to come out. When F/O Hall's airplane disappeared under the number-three man's plane it was in the same attitude as when it started the dive.... All evidence indicates that he hit the ground in a steep dive." Investigators speculated that the pilot might have failed to deploy his dive flaps before entering the maneuver or that the dive flap system failed, causing the airplane to suffer severe buffeting caused by compressibility and preventing the pilot from recovering from the dive. The airplane exploded violently into flames upon impact and was completely destroyed.

2-26-45B. Van Nuys, California. At 1842, a Lockheed P-38J crashed one-quarter mile east of Van Nuys Metropolitan Airport, Van Nuys, California, killing pilot 2Lt. Tom Throssell, Jr., 22. The airplane was flying in the number-three position of a three-ship flight that took off at 1735 from Van Nuys Metropolitan Airport on an aerial gunnery mission. The flight successfully completed the gunnery mission and returned to the field to land. The flight leader was ordered to go on search mission so he flew away from the field. The two remaining airplanes entered the traffic pattern for Runway 17. The subject airplane was observed to be getting extremely close to the number-two airplane on the approach leg and was ordered to go around by tower personnel. The subject airplane continued to float the down the runway just above the number-two ship with landing gear and flaps extended. About two-thirds down the runway the pilot pulled up the gear and began to go around. The tower called the pilot while on the downwind leg and advised him that he was number one to land. The pilot did not acknowledge and tower personnel became occupied with other tasks. Moments later the tower operator looked up and observed the subject airplane heading straight for the control tower with the port engine turning very slowly; the landing gear was retracted and the flaps were extended. The pilot pulled up into a steep climbing turn into the dead engine, gaining a few hundred feet before falling off on the port wing and entering a very flat spin. The airplane remained in this attitude until it struck the ground in a flat attitude, killing the pilot instantly. The airplane, which had crashed onto a road, burst into flames a few minutes after the crash. Investigation revealed that the pilot had failed to switch from the reserve fuel tank to the main fuel tank when ordered to do so before entering the traffic pattern. Investigators stated that the port engine had apparently lost power when the airplane was turning onto the base leg on the initial aborted landing. The pilot feathered the engine momentarily and had to shorten his pattern, causing him to get too close to the lead airplane. The pilot restarted the port engine on the approach and was able to get the engines to respond when he needed to pull up and go around. The port engine apparently failed due to fuel exhaustion as the airplane maneuvered in the traffic pattern during the second attempt to land.

2-27-45A. Fairhaven, Michigan. At 1800 EWT, a Republic P-47D crashed three miles NNE of Fairhaven, Michigan, killing pilot 1Lt. Andrew J. McConaughey, 25. The airplane was part of a two-ship flight that had taken off from Selfridge Field, Mt. Clemens, Michigan, on a high-altitude gunnery mission. The airplanes had completed their gunnery mission and assembled at 20,000 feet over Port Huron. The airplanes maneuvered for a short time when the subject airplane was observed to enter a spin to the right. The spin was observed to briefly stop on two occasions but the airplane reentered a spin to the right. The airplane remained in the spin to the right until it struck the ground and exploded into flames. The pilot had jettisoned the cockpit canopy in an attempt to bail out, but he was unable to escape the airplane and was found in the wreckage.

2-27-45B. Athens, Oregon. At 1800, a Bell P-63A crashed four miles northwest of Athens, Oregon, killing pilot F/O Floyd L. Goodrich, 23. The airplane took off from Walla Walla Army Air Field, Walla Walla, Washington, on a camera-gunnery mission. The airplane rendezvoused with the bomber formation and simulated attacks. A short time later, the subject aircraft was observed in a steep dive, trailing flames. The pilot jettisoned the door, stepped out on the wing and apparently deployed his parachute before clearing the aircraft. The pilot collided with the tail section but the parachute deployed successfully. The pilot drifted down in his parachute and was found immediately, but he died about 20 minutes after the accident. Investigation revealed that the engine failed and caught fire due to the loss of the oil supply.

2-28-45A. Thermal, California. At 0745, a Consolidated B-24J crashed 10 miles SSE of the Army Air Field at Thermal, California, killing nine crewmembers. Engineer Cpl. Paul J. McKeever was seriously injured parachuting to safety. Investigators stated, "The aircraft was on a high altitude camera-bombing mission. The statement by the engineer indicates that the aircraft was flying at 20,000 feet on instruments when the air suddenly became turbulent and the aircraft started into a dive. The pilot and co-pilot were fighting the controls to no avail. At 19,000 feet the order was given to bail out. The aerial engineer snapped on his parachute and jumped into the bomb bay. While attempting to open the bomb bay doors, the right bomb bay door blew off and the aerial engineer was thrown out of the aircraft. When the engineer's parachute opened he saw the aircraft come out of the clouds without the tail assembly. The aircraft was tumbling and small pieces of tail were falling. The aircraft continued to tumble and crashed into the ground and exploded. In the opinion of the Aircraft Accident Board the aircraft was power-stalled, and went into a power spin on instruments. As a result of the power spin, the tail assembly was torn off. The tail assembly appears to have hit the left wing, and approximately 15 feet of the outer wing panel was torn off." Killed in the crash were: 2Lt. Norris C. Hall, pilot; 2Lt. Joseph C. Rischman, co-pilot; 2Lt. William M. Shipp, navigator; 2Lt. Carl J. Currat, bombardier; Cpl. William J. Yoe, assistant engineer; Cpl. Robert R. Noles, radio operator; Sgt. Robert L. Myers, radio operator; Cpl. Frank B. Kerrigan, gunner; Cpl. Harry L. O'Bryan, gunner.

2-28-45B. Mena, Arkansas. At 1625, a Beech AT-7C flying in instrument conditions collided with rising terrain and crashed nine miles northwest of Mena, Arkansas, killing three crewmembers. Two fliers were able to parachute to safety. The airplane took off at 1401 CWT from Fairfax Army Air Field, Fairfax, Kansas, on a navigation proficiency flight to Tulsa, Oklahoma. The airplane climbed to 4,000 feet. A short time later the crew requested and was granted a change of destination to Monroe, Louisiana. Cloud cover

closed up underneath the airplane as it passed over Stigler, Oklahoma, and the airplane entered instrument conditions near Heavener, Oklahoma. The airplane began picking up heavy ice, so the pilot descended and was able to break out into contact conditions underneath. The crew decided to follow the Kansas City and Southern Railroad right-of-way, which runs through a valley between Page, Oklahoma, and Mena. The ceiling dropped and the airplane followed the valley at an altitude of 150 feet agl. The ceiling dropped to zero as the airplane passed over Eagleton, Arkansas, so the pilot added power began a climbing turn on instruments. The airplane collided with trees and then crashed into 2,681-foot Rich Mountain and burst into flames. Pilot 2Lt. Jack W. Perrin, co-pilot 2Lt. James L. Peil and navigator 2Lt. Wayne Roberts were killed in the crash. Navigator 2Lt. Alden V. Ellison and engineer Sgt. Carl D. Capra survived the crash with serious injuries.

March

3-1-45. Victorville, California. At 0918, two Bell P-59A jet fighters collided in mid-air and crashed 20 miles west of Victorville Army Air Field, Victorville, California, killing both pilots. 2Lt. Howard L. Wilson, 23, was killed in the crash of P-59A #44-22626; 2Lt. Robert W. Murcock, 25, was killed in the crash of P-59A #44-22620. The airplanes were part of a two-ship flight that took off at 0846 from Palmdale Army Air Field, Palmdale, California, on a scheduled low-level anti-aircraft "tracking" mission over Victorville Auxiliary Airfield #4. The mission was conducted to give anti-aircraft gunners the opportunity to track low-flying aircraft operating at high speed. The P-59s made several separate passes each, passing over the airfield at about 500 feet agl at over 300 mph. The airplanes then made passes from opposite directions. Lt. Murdock was flying near the airfield from north to south at about 500 feet agl and Lt. Wilson was flying near the airfield from southeast to northwest at about 500 feet agl. The airplanes, estimated to be traveling at about 350 mph, collided at a point over a set of power lines two miles south of Victorville Auxiliary Airfield #4. Investigation revealed that both pilots attempted evasive action by banking to the right just before the collision. Lt. Murdock's wing collided with the tail of Lt. Wilson's airplane, sending both airplanes out of control. Lt. Murdock's airplane climbed steeply, trailing flames and shedding pieces before diving into the ground and exploding into flames. Lt. Wilson's airplane tumbled end-over-end, breaking in half as it spun crazily to earth and exploded.

3-2-45A. Ogden, Utah. At 1325 MWT, a Consolidated C-87 and a Republic P-47D collided in mid-air at Ogden, Utah. The P-47 crashed and exploded in the southwest section of Ogden, killing P-47D pilot Capt. Robert G. Hodson, 26. C-87 crewmembers uninjured in the collision were: Capt. Jacob E. Manch, pilot; 1Lt. Allen E. Baldwin, co-pilot; Capt. Richard N. Cosman, passenger; Sgt. R.L. Frankovitch, passenger; Robert Greenman, civilian passenger; Allen Wells, civilian passenger. The P-47D took off from Hill Field, Ogden, Utah, on a routine service test flight. The C-87 also took off from Hill Field, but was on a flight to check the A-3 autopilot. The pilot of the C-87 and the pilot of the P-47 agreed to meet in the air and fly formation over the Great Salt Lake. The airplanes flew together in formation for a while and then separated. The C-87 then began suffering trouble with the number-one propeller. The pilot feathered the propeller but was unable to unfeather it. He headed back for Hill Field and requested a landing. The P-47 then appeared on the right wing of the C-87, which was flying at 6,800 feet msl, and flew in this position for a few minutes. The P-47 then crossed under the C-87 and pulled up before clearing the ship, the P-47's starboard elevator and horizontal stabilizer colliding with the leading edge of the C-87's port wing. The P-47 immediately went out of control and fell to earth, gyrating three times in a "falling leaf" fashion. The P-47 then entered an inverted 70-degree dive and remained in this attitude until it struck the ground and exploded. The C-87 suffered only minor damage and the pilot was able to make a safe landing at Hill Field.

3-2-45B. Hapeville, Georgia. At 1734 EWT, a North American P-51K crashed one mile southwest of Hapeville, Georgia, killing pilot F/O William A. Quinn, 21. The airplane took off from Greenwood, Mississippi, on a flight to Atlanta, Georgia. The airplane arrived in the vicinity of Atlanta and radioed the control tower at Atlanta Municipal Airport for landing instructions. The pilot was cleared to land on Runway 15 and was advised of a slight crosswind. The airplane entered the pattern and flew a normal approach. The airplane bounced on the landing and became airborne again, stalling just above the runway. The port wing fell and began scraping on the runway. The pilot added power and apparently attempted to get airborne. The airplane continued to the left, aided by the torque of the full throttle, and the port wing began digging in. The airplane then nosed over and the propeller separated. The engine separated as the airplane smashed to a halt, and the wreckage burst into flames moments later.

3-2-45C. Charleston, South Carolina. At 2113 EWT, a Consolidated B-24J attempting a take-off crashed one-half mile southwest of Charleston Army Air Field, Charleston, South Carolina, killing nine fliers. The airplane was scheduled to fly a combat training mission from Charleston Army Air Field to Morrison Field, West Palm Beach, Florida, and return. The B-24 took off on Runway 21 and climbed to about 400 feet when it was seen to roll to the left and then smash to the ground in a wooded area. The airplane burst into flames upon impact and all on board were killed instantly. Investigation revealed that the number-one

engine was in the feathered position when the crash occurred, indicating that the number-one engine had failed or had been shut down. The fuel valve lever for the number-one engine was found in the "Off" position. Investigators could not determine why the number-one propeller was feathered. Killed in the crash were: F/O Paul V. Medolo, Jr., pilot; F/O Peter T. Flaherty, co-pilot; F/O William F. Latham, Jr., navigator; Cpl. Ervin A. Bloch, engineer; Cpl. Anthony T. Loscalso, gunner; Cpl. Connie F. Green, gunner; Pfc. William A. Hieder, radio operator; Pfc. John W. Scott, gunner; Pfc. Richard C. Gibbs, gunner.

3-3-45A. Coahuila, Mexico. At 0900 CWT, a Martin TB-26C crashed 12 miles northwest of Villa Acuna, Coahuila, Mexico, killing three fliers. The B-26 took off from Laughlin Field, Del Rio, Texas, on a routine training flight. The pilot had completed a combat tour in a unit that flew Douglas A-20 type airplanes and was transitioning to B-26 airplanes. The airplane was observed flying at about 1,000 feet agl when it suddenly went into a spin to the right. The spin became very steep but the airplane rolled to an upright position before striking the ground on the port wing and exploding violently into flames. Both engines were producing power at the time of impact. The cause of the accident could not be determined. Investigators speculated that the pilots had inadvertently entered Mexican airspace. B-26 student Capt. Richard B. Mitchell, B-26 instructor Capt. James M. Peters and engineer SSgt. Charles A. Ojajarvi were killed in the crash.

3-3-45B. Gate, Oklahoma. At 1020 CWT, a Martin TB-26B collided with a house and crashed three miles east of Gate, Oklahoma, killing four fliers. The airplane took off from Dodge City Army Air Field, Dodge City, Kansas, on a routine training flight. Investigators stated, "The first take-off was made at 0815 CWT, and subsequent landing at 0917. Second take-off was made at 0926. At approximately 1020 CWT, [the subject airplane] struck the roof gable of a farm house three miles east of Gate, Oklahoma, some 65 miles south of Dodge City Army Air Field. The aircraft headed south when it struck the house, started to disintegrate immediately after [the collision], parts of the airplane being strewn in a narrow path extending for one-half mile south of the house. The main body of the wreckage was one-quarter mile south of the house, with fire destroying the major portion of the same. All personnel aboard were killed instantly." Killed in the crash were: 2Lt. Arthur E. Takkunen, pilot/B-26 student; Aspirant Roland H. Veillard (French Air Force), co-pilot/B-26 student; Pfc. Charles B. Roberts, engineer; Pfc. Arnold C. Knutsen, engineer.

3-3-45C. Farmingdale, New York. At 1604 EWT, a Republic P-47N crashed one mile east of Farmingdale, New York, killing pilot F/O Russell T. Culbreth, 22. The airplane took off at 1602 from the southwest runway at Farmingdale Army Air Field on a flight to Wilmington, Delaware. The airplane climbed to about 300 feet before the engine began "cutting out." The pilot radioed the tower and advised that he was returning to the field. The tower cleared the aircraft, which was flying at very slow speed, to land on any runway. The pilot extended the landing gear as he began a turn back toward the field. The airplane stalled and spun to the ground, bursting into flames upon impact.

3-3-45D. Savannah, Georgia. At 1825 EWT, a Consolidated B-24J suffering an engine fire and attempting an emergency landing crashed into a small creek 18 miles southeast of Chatham Field, Savannah, Georgia, killing engineer Cpl. Dolan J. Rogers and gunner Cpl. Joseph E. Kuduk. Eight crewmembers were injured, two of them seriously. The airplane took off from Chatham Field at 1806 EWT on a Splash Gunnery Mission. Investigators stated, "During take-off and immediately thereafter, the engines were functioning properly as stated by the pilot and co-pilot. After arriving at the gunnery range, at an altitude of approximately 700 feet the pilot ordered the crew to begin firing. At the same time the number-three propeller ran away and the throttle to this engine was retarded and the co-pilot tried to reduce the RPM by the toggle switch. The mixture control was not put in idle cut off. The pilot then lowered ten degrees of flaps and turned back to return to Chatham Field and at the same time the co-pilot was attempting to feather number-three engine propeller without success. The pilot attempted to gain altitude and while doing so the engineer and tail gunner reported that the number-three was on fire. The engineer was instructed to cut the gasoline switch to this engine off. The time between the moment that the propeller started running away and until the engine was first observed on fire was approximately three minutes. The number-two engine began backfiring, exploding and vibrating resulting in the airplane losing airspeed and altitude. The warning bell was rung and the pilot ordered the crew to bail out at an altitude of approximately 300 feet. The crew, except for Cpl. Kuduk, remained in the airplane. The airplane crash-landed in a creek, resulting in bursting open the nose of the airplane and the fuselage was bent at station #7. The wings and engines remained intact. The crew swam out of the creek except for the engineer, Cpl. Rogers, who was missing until 7 March 1945. Cpl. Kuduk was found approximately 50 yards from the plane with the parachute partially opened." Pilot 2Lt. John W. Gemmell and navigator 2Lt. David J. Manheim received serious injuries. Crewmembers injured in the crash were: 2Lt. Edward B. Robertson, co-pilot; F/O Robert W. Jacks, bombardier; Sgt. Joseph G. Stevenson, engineer; Cpl. Norman A. MacDonald, radio operator; Pfc. Robert S. Durbin, assistant radio operator; Pfc. Grant C. Fox, gunner.

3-3-45E. Winnfield, Louisiana. At 1956 CWT, a Boeing B-17F flying in poor weather collided with trees and crashed nine miles WSW of Winnfield, Louisiana, killing ten fliers. The airplane took off at 1615 CWT from Alexandria Army Air Field, Alexandria, Louisiana, on an instrument training flight, which was to be conducted at 8,000 feet on the Alexandria Radio Range. Because of deteriorating weather conditions, flying was called off at 1900. The airplane checked in at 1945 and was ordered to return to base. The pilot acknowledged the order and was given weather information. At 1956, the airplane was observed flying at very low altitude in a heavy rainstorm. The airplane was headed straight for a house, but before a collision could occur the airplane turned to the right and collided with a 65-foot tree, severing the port wing. The airplane rolled out of control and crashed to the ground, exploding into flames upon impact. Killed in the crash were: 2Lt. Russell L. Cobbs, pilot; 2Lt. Robert J. Keavy, co-pilot; 2Lt. Omar H. Lauridsen, navigator; 2Lt. Bruno J. Wieclaw, bombardier; Cpl. Jay E. Crite, engineer; Sgt. Harold D. Bush, radio operator; Cpl. N. Anderson, gunner; Pfc. Bruce C. Rue, gunner; Cpl. Lester J. Smejkal, gunner; Pfc. Millard B. Beard, gunner.

3-4-45A. Fort Worth, Texas. At 0429, a Beech AT-7C was abandoned and crashed five miles northwest of Fort Worth, Texas, killing one crewmember. Engineer Sgt. Clement M. Blomer was missing and presumed killed. Student navigator A/S Roland E. Schuenemann was seriously injured parachuting to safety. Pilot F/O Crawford M. Scott, Jr. and student navigators A/C Raymond W. Sebens and A/S John R. Taylor parachuted to safety and were uninjured. The airplane took off from Coffeyville Army Air Field, Coffeyville, Kansas, on a navigation-training flight back to the home station at Hondo Army Air Field, Hondo, Texas. The airplane had successfully navigated from Hondo Army Air Field to Carlsbad, New Mexico, and to Denver, Colorado, before landing at Coffeyville Army Air Field. The airplane took off from Coffeyville Army Air Field at 2240 CWT, climbed to 5,000 feet and then proceeded on course to Hondo, checking in by radio to ground stations at various times along the way. About 48 miles west of Waco, the airplane began experiencing radio problems. Soon afterward, the airplane began to encounter a thickening undercast. The pilot made a 180-degree turn to return to Waco. The airplane passed over Lake Buchanan at 0302. When the airplane arrived in the area of Waco, it was learned that the ceiling had dropped to 100 feet so the pilot set a course for Hensley Field, Dallas, Texas. By the time the airplane arrived in the area south of Dallas, Hensley Field had closed down. Fort Worth advised that it was open so the pilot flew northwest toward Fort Worth. The pilot was unable to find the field because of some scattered low-hanging clouds. The pilot then tried Meacham Field, but the ceiling was about 200 feet. The pilot was unable to get in to a clear airport and the fuel situation was becoming critical, so he ordered the crew to bail out, which they did in good order. The pilot set the autopilot and bailed out last at about 3,000 feet. The airplane landed in a flat attitude, coming to rest nearly intact. The airplane did not burn and was only slightly broken up.

3-4-45B. Laredo, Texas. At 1718, a Consolidated B-24E crashed while attempting an emergency landing at Laredo Army Air Field, Laredo, Texas, killing three crewmembers and seriously injuring nine others. The airplane had taken off from Laredo Army Air Field on a gunnery mission. The airplane radioed the field at 1715 and reported that the number-four propeller was feathered and the engine shut down. The airplane was cleared to land and the pilot was advised to report in on the base leg. The airplane entered the traffic pattern and the pilot radioed in on the base leg. Tower personnel observed the airplane flying at slow speed and at low altitude on the base leg. The airplane, with landing gear extended, turned from base to final. It became evident that the airplane was not going to be able to make the field. The pilot retracted the landing gear, but the airplane continued to lose altitude and crashed just short of the field. Investigation revealed that the number-one, rather than number-four, magneto switch had been inadvertently turned to the "off" position. The number-four magneto switch was found in the "on" position. Pilot 2Lt. O.E. Ronning, gunnery instructor TSgt. K.B. Lundin and gunnery student Pvt. R.M. Brown were killed in the crash. Crewmembers seriously injured in the crash were: F/O R.A. Clausen, co-pilot; Pfc. B.L. Crawford, engineer; Pvt. R.B. Avery, student gunner; SSgt. J.P. Hansel, gunnery instructor; SSgt. Hugh Brooks, gunner; TSgt. L.I. Schooneover, gunner; TSgt. M.M. Gardner, gunner; SSgt. S.L. Schwartz, gunner; Pvt. H.J. Blackburn, gunner.

3-4-45C. Lawrenceville, Illinois. At 2008, a Douglas C-47A and a C-47B collided in mid-air three miles west of George Field, Lawrenceville, Illinois, killing two fliers and injuring two others, one of them seriously. Three fliers escaped injury. Pilot 2Lt. Douglas C. Woodbury and co-pilot 2Lt. Robert W. Stroh were killed in the crash of the C-47A. Co-pilot 2Lt. Walter G. Ziegler was lightly injured and radio operator Sgt. George J. Koermer received serious injuries aboard the C-47B; pilot 2Lt. Hugh F. Walti, co-pilot 2Lt. Donald B. McCoy and engineer Sgt. William L. Mitchell were uninjured. Both airplanes took off from George Field on separate local training flights. The C-47A was on a routine night training flight; the C-47B was on a double glider tow mission. After the completion of their respective exercises, the airplanes were to land and the pilots were to exchange missions. The airplanes had approached the area of the field and were endeavoring to enter the traffic pattern for a landing when the collision occurred at about 2,400 feet msl. The C-47A was letting down in a descending right turn when its starboard wing collided with the port engine of the C-47B. The C-47A's starboard wing was severed and the C-47A went out of control. The C-47A rolled toward the C-47B, and as it did its tail section collided with the cockpit compartment of the C-47B near the escape door. The C-47A fell away in a spin to the right and slammed to the ground, exploding into

flames upon impact. The pilots of the C-47B were able to maintain control of the aircraft and make a successful emergency landing at George Field. The port landing gear remained in the retracted position during the landing. Pieces of the C-47A's rudder and tail structure were found wedged in the cockpit area of the C-47B.

3-5-45A. Pyote, Texas. At 0810 CWT, a Boeing B-29 crashed 16 miles south of Pyote, Texas, killing two crewmembers and injuring 12 others, three of them seriously. Radio operator/instructor TSgt. A.F. Russell and gunner Pfc. Harold D. Melincore were killed. Navigator F/O Leonard W. Penke, bombardier F/O Manny Greenburg and flight engineer SSgt. Earl A. Peek were seriously injured. Crewmembers injured in the crash were: 1Lt. Allen M. Michelson, B-29 instructor; 1Lt. Nicholas Sienkienwicz, pilot; 2Lt. Philip Tobias, co-pilot; Pvt. Henry J. Kelly, radio operator; Cpl. Earl S. Hebbeler, gunner; Sgt. James F. Wallace, gunner; Cpl. Keith E. Nelson, gunner; SSgt. Joyce R. Williams, gunnery instructor; 2Lt. Gerald Van Fleet, flight engineer. The airplane took off at 0750 CWT from Pyote Army Air Field on a routine training mission. The airplane made a normal take-off and began climbing out. When the airplane reached an altitude of 5,000 feet the number-three engine began backfiring and throwing sparks and trailing flames, which seemed to originate under the upper cowl flaps. The flight engineer shut down the engine and feathered the propeller. Moments later, the number-four engine began losing power. The airplane began losing airspeed and altitude. The instructor, who was flying from the left seat, called for more power. The two port engines responded satisfactorily but there was no noticeable power increase in the number-four engine. The airplane continued to lose altitude so the pilot ordered the crew to stand by for a crash landing. The pilot leveled the airplane off and was able to make the crash landing. The airplane broke up as it slid across the terrain. The port wing erupted into flames moments later. The tail gunner was killed upon impact and the radio operator instructor was trapped under the upper turret, which had twisted out of position as the fuselage broke up, and could not be rescued. Investigation revealed that the number-one exhaust valve had failed and burned, causing "induction fire" in the number-three engine.

3-5-45B. Kissimmee, Florida. At 1340 EWT, a Lockheed P-38H and a Martin B-26F collided in mid-air and crashed at the intersection of the Runway 15 and Runway 24 at Kissimmee Army Air Field, Kissimmee, Florida, killing eight fliers. Capt. Warren L. Breinig was killed in the crash of the P-38H. Killed in the crash of the B-26F were: Capt. Thomas J. James, pilot; Capt. Louis Saul, co-pilot; TSgt. John L. Rannells, engineer; 1Lt. William R. Haenzi, navigator; 1Lt. Adam J. Pyzyna, bombardier; David Korn, civilian technician; Pryor Watts, civilian technician. The B-26, which was returning to the field following a local flight, was landing on Runway 15 when it was struck by the P-38H, which was taking off on Runway 24. Both airplanes were several feet off the ground when the collision occurred at the intersection of the two runways. The airplanes exploded into flames in the collision, smashing to the ground and scattering wreckage all over the field. The fliers aboard both airplanes were killed instantly. Investigators stated, "The B-26 called for landing instructions and was cleared to land on Runway 15. While the B-26 was in the process of entering traffic and setting his pattern, [the] P-38 called the tower for taxi instructions. These were given. As the P-38 approached Runway 24 he stopped and requested permission to enter runway to attach the tow target. The tower then gave this permission. From this point on there is an uncertainty on the part of the witnesses and statements as to further instructions given the P-38 or the B-26. It is known that the P-38 finished making the tow target attachment and the mechanic stepped back and signaled the pilot that all was OK. The P-38 then proceeded to take off as the B-26 was on the last stages of his approach for landing. It is evident that neither pilot saw the other plane until just before the collision.... [It] is evident that both planes tried to hurdle each other and struck a few feet off the ground, exploding and burning."

3-5-45C. Wright Field, Ohio. At 1337 CWT, a Lockheed C-60A crashed into a hangar while attempting to take-off from Wright Field, Dayton, Ohio, killing five fliers and three ground crewmembers working in the hangar. Ten ground crewmembers were seriously injured in the hangar. The airplane was cleared to take off from Runway 16 on a cargo parachute flight, being scheduled to drop four 300-pound cargo canisters in the Wright Field drop zone. The Accident Classification Committee stated, "The aircraft proceeded down the runway with both engines apparently operating in a normal manner. The aircraft left the ground approximately 1,800 feet from the starting point in an attitude ... considered about normal. Immediately or shortly after leaving the ground the aircraft began a steep climb estimated as approximately 40 degrees to 45 degrees from the horizontal. This attitude was maintained to approximately 50 feet altitude at which time it apparently increased an additional 15 degrees to 20 degrees and [the airplane] continued climbing to between 200 and 300 feet. At this time the airplane had apparently reached a power-on stall condition, and fell into the initial phase of a left spin, crashing nose down at the base of the steel door of Hangar #4, with the path of the airplane approximately 70 degrees from the horizontal. The airplane struck on its left wingtip and nose and completed a cartwheel motion, which broke the fuselage and threw the tail section into the hangar." The C-60 crashed into Hangar 4 and exploded violently into flames, totally gutting the hangar and destroying eight aircraft that were parked there. The cause of the power-on stall

above the runway could not be determined with any certainty. Investigators speculated that the pilot might have incorrectly set the elevator trim tabs in the full nose-up position, or that he had neglected to unlock the elevator controls, or that the load, which was not secured, had shifted aft during the take-off. Lockheed P-38G #42-12687, Republic P-47D #42-8009, North American P-51F #43-43333, Bell P-59A #44-22609, Bell XP-63A #42-78015, Boeing B-29B #42-63598, Consolidated B-32 #42-108473 and Douglas C-47A #42-23357 were destroyed in the fire. Killed in the crash of C-60A #42-55962 were: 1Lt. Elwin I. Brawner, Jr., 25, Blacksburg, Virginia, pilot; 2Lt. Robert E. Jackson, 24, Thomasville, Alabama, co-pilot; Sgt. Albert B. Weathers, 27, Ensley, Alabama, passenger; Pvt. Browder A. Richmond, 25, Kansas City, Kansas, passenger; Pfc. Eugene L. Fisher, 34, Tyringham, Massachusetts, passenger. Killed on the ground were: TSgt. Russell P. Schryer, hometown unknown; Sgt. William W. Wilson, 30, Owensville, Indiana; Pfc. John Ravegum, 24, Lititz, Pennsylvania. Ground crewmembers seriously injured in Hangar 4 were: Pvt. John D. Aufiery, 22, Philadelphia, Pennsylvania; Pfc. Charles J. Gabrielli, 34, Mill Valley, California; TSgt. George A. Hudak, 31, Canton, Ohio; TSgt. Harold D. Klein, 29, Dayton, Ohio; Pfc. Harold E. Long, hometown unknown; Sgt. Harry Statman, 43, Bronx, New York; TSgt. Kenneth D. Waggoner, hometown unknown; Pvt. Thomas L. Williams, 22, Malad City, Idaho. Civilian technicians Henry P. Gillen and George W. Bennett received serious injuries when the C-60 smashed through the hangar.

3-5-45D. Ajo, Arizona. At 1635 MWT, a North American AT-6A crashed at the AAF Auxiliary Gunnery Range 14 miles northwest of Ajo Army Air Field, Ajo, Arizona, killing pilots 2Lt. Alden A. Olston and 2Lt. Darth R. Hanson. Investigators stated, "Lt. Olston ... [flying in the front seat] was giving a demonstration ground gunnery ride to Lt. Henson, who was in the rear seat and classified as a qualified dual. Lt. Olston was flying a left hand pattern and firing on [Target #1] on the right range. His base leg was at 900 feet above the terrain. Upon making a level 110 mph turn onto the target from the base leg, the airplane stalled, flipped to the left, slightly past inverted.... The airplane went into a spin to the left and flattened out momentarily after three-quarters of a turn at an altitude of approximately 200 feet. The gun at this point started firing and continued to firing until contact with the ground. The airplane appeared to go into a secondary spin to the left, spinning about 90 degrees. The left wing made contact first, followed by the fuselage at a 40-degree angle. The right wing was sheared off at the center section, coming to rest 75 feet from the main body of the airplane. The airplane ignited ... 15 seconds after contact with the ground and was extinguished in four minutes by the fire truck and crash crew."

3-5-45E. Venice, Florida. At 2138 EWT, a North American P-51C crashed at Venice Army Air Field, Venice, Florida, killing 2Lt. Robert L. Chodrick, 22. The airplane took off at 1930 EWT from Venice Army Air Field on a night training flight. The pilot returned to the field at about 2130 and requested landing instructions. The airplane was cleared to land on Runway 23. The airplane approached the field and peeled off to the left for a landing. The airplane remained in a diving turn until it smashed onto the field at the southeast end of Runway 31, killing the pilot instantly. The airplane did not burn but was destroyed utterly. Wreckage was scattered over an area of 150 yards. Investigators speculated that the pilot had lost the horizon while peeling off to the left. Investigators noted that a "Z-shape" riveting bucking bar was found in the wreckage of the tail section, but was not thought to have contributed to the accident.

3-5-45F. Battle Mountain, Nevada. At 2230 MWT, a Consolidated B-24J crashed at Battle Mountain, Nevada, killing navigator 2Lt. Gene L. Hamilton. Engineer Cpl. Hamod Sallie and radio operator Cpl. Lawrence J. Clark were injured in the crash. Crewmembers uninjured in the crash were: 2Lt. Walter R. Graig, pilot; 2Lt. Robert R. Eaton, co-pilot; Cpl. John R. Frye, gunner; Sgt. John G. Dean, gunner; Cpl. Richard J. Berryman, gunner; Cpl. Haynes M. Brantley, gunner. The airplane took off at 1706 MWT from Mountain Home Army Air Field, Mountain Home, Idaho, on a round robin navigation mission to Pendleton Field, Pendleton, Oregon, to Humboldt, Nevada, and return to Mountain Home. The airplane, which was on a contact clearance, entered a large frontal system on the Pendleton Field to Humboldt leg. The pilot radioed ground stations that he was picking up light ice. The airplane began becoming heavy with ice and the pilots attempted to land at the small airstrip at Battle Mountain. The airplane, with flaps down and landing gear extended, was heading southeast when it flew into the ground two miles from the CAA Radio Station at Battle Mountain. The airplane broke up upon impact, but did not burn.

3-6-45A. Cope, South Carolina. At 1030, two Stearman PT-13B airplanes collided on the runway at Kennedy Auxiliary Airfield, Cope, South Carolina, killing French Air Force student Lt. Paul R. Lebargy aboard PT-13B #40-1643. French Air Force student Cpl. Elysse Crisias was uninjured aboard PT-13B #40-1732. Investigation revealed that Lt. Lebargy had landed and then came to a complete stop on the runway. Lt. Lebargy was still stationary on the runway when Cpl. Crisias' airplane landed on top of Lt. Lebargy's airplane from behind. Cpl. Crisias' propeller struck the cockpit area, cutting Lt. Lebargy's airplane in half; Lt. Lebargy was killed instantly. Investigation revealed that Cpl. Crisias had landed against a red light from the control tower. It was not known why Lt. Lebargy had failed to exit the runway in timely fashion.

3-6-45B. Harlingen, Texas. At 1013 CWT, a Consolidated B-24H crashed while attempting to take-off from Harlingen Army Air Field, Harlingen, Texas, killing two fliers and injuring three others, two of them seriously. The airplane was taxied up to Runway 35E, and instead of going all the way to the end of the taxiway to the beginning of the runway — an additional 500 feet — the airplane entered the runway from the taxi-way intersection at Runway 35 and commenced its take-off from there. The airplane, with flaps still retracted, began its take-off roll. The pilot was not getting the take-off performance that he desired so he advanced the manual turbo boost full forward and pushed the nose down. The airplane began to pick up speed, advancing to 115 mph. The pilot pulled the airplane into the air and, thinking he was airborne, moved the landing gear lever to the "up" position. The pilot pulled the column back and the airplane reached an altitude of ten feet before settling back to the runway. The port wheel touched down 350 feet from the end of the runway, skidding slightly before the airplane bounded back into the air momentarily and then slammed back to the ground. The pilot hit the brakes but the airplane went off the end of the runway. The nose wheel collapsed and the starboard main landing gear began to fold up. The starboard wing went down and began dragging, and the number-three and four propellers began striking the ground. The airplane continued across the ground out of control until it crashed into a large drainage ditch surrounding the field. The airplane burst into flames soon after coming to rest. Investigators stated that the failure of the co-pilot/instructor to extend 20 degrees of flaps and the failure to start the take-off at the very end of the runway contributed to the failed take-off. B-24 student Capt. Fritz R. Sandburg and engineer Sgt. Herman M. Boeker were killed in the crash. Pilot and B-24 student Capt. Clifford D. Marburger and engineer SSgt. Edward Paris were seriously injured. B-24 instructor/co-pilot Capt. Kenneth E. Booth sustained minor injuries.

3-6-45C. Mountain Home, Idaho. At 1111 MWT, a Consolidated B-24D attempting a landing crashed one-quarter mile west of Mountain Home Army Air Field, Mountain Home, Idaho, killing ten crewmembers. The airplane was part of a five-ship flight that was returning to the field after successfully completing a routine training flight. The airplane peeled off and entered the traffic pattern. The airplane flew a normal pattern and radioed in on the base leg. The airplane, flying at about 800 feet agl, was observed in a vertical bank to the left while in the turn from base to final. The airplane was righted but moments later it again entered a vertical bank. The airplane rolled to an inverted attitude and then dove into the ground, exploding into flames upon impact. Investigators speculated that the B-24 experienced an aileron stall from which the pilot was unable to recover. Killed in the crash were: 2Lt. Roy B. Harris, pilot; F/O Curtis R.

Hayes, co-pilot; 2Lt. Thomas C. Jones, navigator; F/O Henry R. Bettis, bombardier; Cpl. William E. Gates, engineer; Cpl. Maynard Evans, radio operator; Cpl. Wilbur V. Giles, gunner; Pfc. Joseph R. DeVitto, gunner; Pfc. David Geleer, gunner; Pfc. John P. Bezirium, gunner.

3-6-45D. Pecos, Texas. At 2108, a North American AT-6A crashed five miles east of Pecos Army Air Field, Pecos, Texas, killing pilot 1Lt. Kenneth G. Ferguson and seriously injuring pilot-rated passenger 2Lt. Ralph R. Malanga. The airplane took off at 2005 from Pecos Army Air Field on an authorized local night flight. The airplane smashed to the ground in a spin to the right. The airplane did not burn but was destroyed utterly. Lt. Malanga woke up in the wreckage about six hours after the accident. He extricated himself from the wreckage, wrapped himself in a parachute, and then went back to sleep. He woke up a little after sunrise and noticed airplanes taking off from Pecos Army Air Field. Lt. Malanga began walking toward the field and was rescued about two miles from the wreckage.

3-6-45E. Charleston, South Carolina. At 2322, a Consolidated B-24J crashed while attempting to take-off from Charleston Army Air Field, Charleston, South Carolina, killing four fliers. Co-pilot 2Lt. Victor E. Smith and navigator 2Lt. William J. Finan sustained very serious injuries. The airplane was taking off on Runway 21 on a routine night transition flight. The airplane rolled straight down the runway for about 1,000 feet before it began to veer to the right. The airplane ran off of the runway to the right but continued the take-off run. The airplane continued to the right for about 1,100 feet and then crossed Runway 10-28, continuing for about 850 feet before becoming airborne. The airplane was airborne for about 300 feet when the starboard wheel and landing gear struck a mound of dirt. The airplane remained airborne for about 125 feet and then both main wheels touched the ground. The airplane continued on the take-off and the port main wheel left the ground, followed by the starboard wheel about 125 feet further on. The airplane became airborne and flew about 800 feet before colliding with a stand of trees at an altitude of about 20 feet. The airplane exploded into flames as it ploughed through the trees and broke up. Killed in the crash were: 2Lt. Richard C. Englen, pilot; 2Lt. Paul E. McGranaghan, bombardier; Cpl. Linwood E. Lang, engineer; Cpl. John J. Rurack, radio operator.

3-7-45A. Gulf of Mexico, Florida. At 1005 EWT, a Curtiss P-40N crashed into the Gulf of Mexico 20 miles southwest of Pinellas Army Air Field, St. Petersburg, Florida, killing pilot Capt. Robert R. Sherbondy, 25. The airplane was part of a flight that took off from Pinellas Army Air Field on an aerial gunnery mission. The subject airplane made several passes at the towed target at about 8,000 feet. On the last firing pass the subject airplane shot the target off of the tow

ship. The subject airplane broke away after the firing pass and the port wing collided with a large pipe that was part of the target. The airplane went out of control, rolling three times and then falling into a spin at about 7,000 feet. The canopy was observed to fall away about 5,000 feet. The pilot failed to recover and the airplane continued in a tight, vertical spin to the sea. The pilot went in with the airplane and was killed instantly upon impact. Investigators noted that the airplane's guns were firing as it entered the spin. The firing ceased at 5,000 feet when the canopy fell away.

3-7-45B. Wichita Falls, Texas. At 1450 CWT, a Piper L-4H crashed five miles northeast of Sheppard Field, Wichita Falls, Texas, killing pilot 2Lt. James R. Patterson, 24. The airplane took off from Sheppard Field on a local flight. The airplane was observed in a steep diving turn. The pilot failed to recover and the airplane smashed to the ground. The airplane did not burn but was destroyed utterly. Investigators speculated that the pilot's seat pack-type parachute might have slid forward on the seat during the maneuvering, preventing the free rearward movement of the control stick.

3-7-45C. Bennington, Oklahoma. At 1530 CWT, a Vultee BT-13A performing low-altitude aerobatics crashed one mile west of Bennington, Oklahoma, killing instructor 1Lt. William J. Saxman and student A/C Robert W. Hartman. The airplane took off from Perrin Field, Sherman, Texas, on an instrument training flight. Investigation revealed that the subject airplane, by chance or by previous arrangement, rendezvoused with another BT-13 type airplane outside of the authorized flying area. The two airplanes then flew to Bennington and began flying over the town at about 100 feet agl. The airplanes made several passes over the town, and were seen diving on the railroad trestle west of town. The airplanes were then observed performing slow rolls and snap rolls at approximately 100 feet agl. The subject airplane began a slow roll and stopped momentarily in the inverted position. The airplane collided with trees as the pilot attempted to complete the roll, the port wing striking trees while in a vertical bank to the left. The port wing was severed and the airplane went completely out of control. The airplane crashed into the trees and the starboard wing was severed. The fuselage and mid-wing section smashed through the trees and burst into flames. Civilians were able to pull both fliers from the wreckage, but both were fatally injured.

3-7-45D. Wrightsville Beach, North Carolina. At 1708 EWT, a Republic P-47D crashed at Wrightsville Beach, North Carolina, killing 2Lt. Sidney V. Alley, 21. The airplane was flying in the number-three position of a three-ship flight that took off from Goldsboro Army Air Field, Goldsboro, North Carolina, on a navigation flight to Southport, North Carolina, to Whiteville, North Carolina, and return. The airplanes were seen flying to the south toward Southport at extremely low altitude over an inland waterway. The airplanes had to pull up to clear a bridge spanning the waterway. The airplanes, which were about 20 miles off of the course they were supposed to be flying, were observed in a right echelon formation flying to the northeast at approximately 15 feet above the surface along the Atlantic coast. The airplanes made three low-altitude passes over Wrightsville Beach. The formation began a turn to the right and the subject airplane collided with the surface of the sea. The airplane cartwheeled onto the beach, killing the pilot instantly. The airplane did not burn but was destroyed utterly.

3-8-45A. Coahuila, Mexico. At 0910 CWT, a Vultee BT-13B crashed 20 miles southeast of Villa Acuna, Coahuila, Mexico, killing pilot 2Lt. Dean H. Patton, 23, and seriously injuring pilot 2Lt. James T. Strickland. The airplane took off from Laughlin Field, Del Rio, Texas, on a routine instrument training flight. Lt. Strickland was flying from the front seat as the safety pilot and Lt. Patton was flying in the rear seat as the instrument pilot. Lt. Strickland climbed to 2,800 feet following the take-off. The pilot noticed that at 2,800 feet that he was too close to the base of the ceiling, so he descended to 2,000 feet (1,000 feet agl) and began flying along the Rio Grande River. Lt. Patton wiggled the stick and took control of the airplane. He had not put on his instrument hood and was flying visually. He immediately peeled off to the right and dove down into the Rio Grand Valley, flying at very low altitude in a valley on the Mexican side of the river. Lt. Strickland, fearing that the airplane was going to collide with a small hill, grabbed the control stick and pulled back. The airplane climbed away steeply, avoiding a collision with the terrain. Lt. Strickland leveled the airplane off at 300 feet agl. Lt. Strickland asked Lt. Patton if his vision forward was okay — that is, if he could see where he was going — and Lt. Patton assured him that everything was okay. Lt. Patton took control of the airplane and again peeled off, this time to the left and down into a small valley, still on the Mexican side of the river. The airplane approached rising terrain that sloped up gently in front from left to right. The starboard landing gear collided with the terrain and then the starboard wingtip. The airplane smashed into the ground and began shedding pieces as it slid across the hill. The airplane slid out of control across the hill for about 500 feet before it passed over a large depression in the terrain. The airplane passed over the depression and collided with the rise on the other side. Lt. Strickland was not wearing his safety belt and was killed when he collided with the interior of the airplane. Investigators noted that the two pilots had never flown together before and had never met before the fatal flight.

3-8-45B. Punta Gorda, Florida. At 1345 EWT, a Curtiss P-40N crashed at the AAF Bombing Range 18 miles southeast of Punta Gorda, Florida, killing 2Lt. Leo R. Tuttle, 26. The airplane was flying in the number-two position of a four-ship flight that took off at 1310 EWT from Page Field, Fort Myers, Florida, on a dive-bombing mission. The formation, flying at about 5,500 feet, arrived at the Bermont Bombing Range and the leader ordered the bombs to be armed. The leader put his flight in a line astern configuration and then entered the traffic pattern for the range. The leader climbed to 6,000 feet and then peeled off for the bomb run. The subject airplane peeled off, entered its dive and dropped its bomb. The subject airplane pulled up and then made a climbing turn to the right. The airplane then snapped over into an inverted attitude and began gaining altitude before entering an "oblique split-S," diving into the ground at an angle of approximately 50 degrees. The airplane exploded violently upon impact, scattering wreckage over an area of 600 feet.

3-8-45C. Courtland, Alabama. At 1447 CWT, a Consolidated B-24J crashed while taking off from Courtland Army Air Field, Courtland, Alabama, killing four fliers. Killed in the crash were: 1Lt. James W. Dennis, B-24 instructor; Capt. John P. Murphy, B-24 student; Capt. Harold B. Strong, B-24 student; Pfc. Francis M. Guinnaugh, engineer. The airplane took off at 1435 on Runway 5 and immediately entered traffic. The instructor radioed the control tower and requested permission to perform a touch-and-go landing. The control tower cleared the airplane for a touch-and-go landing on Runway 5. The airplane made a normal landing but the subsequent take-off roll was slow. The airplane was still rolling down the runway with only 500 feet remaining. The airplane veered to the right 15 degrees but the pilots were able to regain directional control. The airplane lifted off and cleared the end of the runway at very low altitude. The airplane was mushing along in a nose-high tail-low attitude, crossing over the field boundary at an altitude of 50 feet. The port wing fell and the airplane began a gradual turn to the left. The port wing collided with a 15-foot tree stump, severing the wing and sending the airplane out of control. The airplane crashed to the ground about 50 yards further on, smashing itself to pieces and erupting into flames as it slid for about 100 yards. Investigation revealed that the instructor had given the students a three-engine take-off scenario on the touch-and-go landing.

3-9-45. Long Beach, California. At 2005, a Lockheed P-38L crashed three miles east of the Army Air Field at Long Beach, California, killing pilot Capt. James N. Hillman, 38. The airplane took off from Long Beach at 1900 on a local flight. At 1955, the pilot returned to the field and requested landing instructions, and he was cleared to land on Runway 30. The airplane was observed flying on the downwind leg at about 600 feet agl before it went into a spin to the left and smashed to the ground, exploding into flames upon impact. Investigators speculated that the port propeller had run away, or that the port engine had stopped because of fuel exhaustion due to mismanagement of the fuel system by the pilot.

3-10-45A. Atlantic Ocean, New York. At 0910 EWT, a Consolidated B-24J crashed into the Atlantic Ocean seven miles southwest of Montauk Point, Long Island, New York, killing 12 fliers. The airplane took off at 0630 EWT from Westover Field, Chicopee, Massachusetts, on an air-to-water gunnery mission. The airplane reported to Westover Field that it was over the gunnery range south of Montauk Point at 0830. There was no further contact with the airplane. A civilian craft operating in the area found wreckage floating on the surface at about 1300 EWT. AAF crash boats, U.S. Coast Guard PBY aircraft and Coast Guard vessels were dispatched to the scene. The body of the pilot and radio operator and a small amount of airplane wreckage were recovered from the sea. Investigators noted that the horizon was indistinct and the surface of the sea was smooth. The cause of the accident was unknown, but investigators speculated that the pilot had inadvertently flown the airplane into the sea while maneuvering near the gunnery range. Killed in the crash were: 2Lt. Howard B. Tolle, pilot; F/O George F. Ruf, co-pilot; 2Lt. David H. Richey, navigator; 2Lt. Raymond G. Bushee, bombardier; Cpl. Russell L. White, gunner; Cpl. Philip W. Ayers, radio operator; Cpl. Carl E. Carlson, radio operator; Cpl. William F. Budka, radio operator; Cpl. John W. Shedlock, gunner; Cpl. Charles R. Clark, engineer; Cpl. Don J. Finger, engineer; TSgt. Harold E. Falk, engineer instructor.

3-10-45B. Padre Island, Texas. At 1720, a Republic P-47D collided with a towed target and crashed on the beach at Padre Island, Texas, killing Mexican Air Force pilot 1Lt. Javier Martinez Valle, 30. The airplane took off from Brownsville Army Air Field, Brownsville, Texas, on an aerial gunnery mission. The airplane collided with the target and cable while making a 90-degree approach to the target at an altitude of 5,000 feet msl. The airplane snap rolled to the left and then dove into the ground, exploding violently into flames upon impact. Investigators speculated that the pilot might have been incapacitated in the collision.

3-10-45C. Alexandria, Louisiana. At 2054, a Boeing B-29 crashed at Alexandria Army Air Field, Alexandria, Louisiana, killing the crew of ten. The airplane took off at 1025 CWT from Harvard Army Air Field, Harvard, Nebraska, on a long-range navigation-training mission to Whale Rock, to Freemason Island, to New Orleans, Louisiana, to Des Moines, Iowa, and return to Harvard. The pilot radioed Alexandria Army Air Field's tower and reported 20 miles out with one engine feathered and one "giving trouble." The tower cleared the aircraft to land on Runway 5, but the pilot requested a "straight-in" approach from the south. Tower personnel declared an emergency, cleared the field and alerted crash trucks and rescue personnel. The pilot was cleared to land on Runway 36. After some time had passed and the airplane failed to arrive at the field, tower personnel radioed the aircraft and asked for a position report. The pilot replied that he was north of the field. Tower personnel observed the airplane north of the field, apparently maneuvering in the traffic pattern. The tower advised the pilot that he was cleared to land on Runway 36. The pilot called in when he was on the final approach. The airplane overshot the runway on the turn from base to final, and the pilot had to turn to the left approximately 15 degrees in order to "line up" with the runway. Investigators stated, "Aircraft was observed to be overshooting when about at the intersection and at an approximate altitude of 150 feet. Power was applied with a resultant backfiring on the left side. Aircraft veered to the left in a gentle turn, which steepened until the aircraft was approximately 90 degrees to Runway 36. At this point the nose dropped abruptly, and the ship went into the ground completely out of control. The left wing and nose came in contact with the ground at the same time, and the aircraft burst into flames immediately.... The aircraft came in contact with the ground in an almost vertical position and cartwheeled, the tail coming to rest facing Runway 36 at an angle of approximately 50 degrees." Killed in the crash were: 2Lt. Valentino Tulla, pilot; 2Lt. Wayne R. Shambrook, co-pilot; 1Lt. Vincent L. Levora, navigator; 1Lt. Hubert H. Crane, radar navigator; F/O Herbert W. Callahan, bombardier; SSgt. Everett A. Neely, flight engineer; SSgt. William Cadwallander, radio operator; Cpl. William H. Adams, port gunner; Cpl. Reed D. Allison, starboard gunner; Cpl. James L. Allen, tail gunner.

3-11-45. Fort Sumner, New Mexico. At 1315 MWT, a Curtiss TP-40N crashed 23 miles southwest of the Army Air Field at Fort Sumner, New Mexico, killing pilot 2Lt. Gerald A. Frankenstein, 22. The airplane was flying in the number-two position of a two-ship flight that took off from Fort Sumner Army Air Field on a medium altitude tactical formation mission. The airplanes climbed to 10,000 feet and began maneuvering. After about 15 minutes of these maneuvers, the flight leader ordered the subject airplane in string formation and then climbed to about 13,500 feet. The flight leader performed several steep climbing turns. The flight leader then performed a half roll and a split-S type maneuver, apparently "graying" out in the pull-out at 12,000 feet. The flight leader was unable to locate the subject airplane after the maneuver. A few seconds later, the flight leader observed the subject airplane level out slightly before smashing to the ground and bursting into flames.

3-12-45A. Randolph Field, Texas. At 0950, a North American AT-6C crashed at Randolph Field, San Antonio, Texas, killing pilots 1Lt. Thomas D. Jacobs, Jr. and 1Lt. Conrad Koch. The airplane was returning to Randolph Field after completing an instrument team ride. The airplane entered the traffic pattern and stalled at about 400 feet agl while turning from the base leg to final. The airplane snapped over to the right and entered a spin at an altitude too low to allow recovery. The airplane smashed to the ground and the pilots were killed instantly. The airplane did not burn.

3-12-45B. Campbell, Texas. At 1145 CWT, Republic P-47G crashed five and one-half miles south of Campbell, Texas, killing pilot F/O Theodore P. Tuller, 20. The airplane took off from Majors Field, Greenville, Texas, on a routine training flight. The P-47 was observed performing slow rolls and other aerobatic maneuvers prior to the accident. Investigators speculated that the P-47 had probably stalled while performing an aerobatic maneuver at low altitude. Investigation revealed that the airplane struck the ground in a flat spin. The pilot fell to his death when he bailed out at an altitude too low to allow his parachute to fully deploy. The airplane did not burn but was destroyed utterly. The cause of the crash remains undetermined.

3-13-45A. George West, Texas. At 0130 CWT, a Douglas C-47 flying in a violent thunderstorm crashed 35 miles southwest of George West, Texas, killing six fliers. Killed in the crash were: F/O Walter J. Mullen, pilot; F/O Leon L. Hendren, co-pilot; F/O James D. Mitchell, co-pilot; SSgt. Milton L. Grans, engineer; Cpl. Morton N. Sax, engineer; SSgt. Paul F. Mason, passenger. The C-47 took off at 2009 CWT (3-12-45) from Greenwood Army Air Field, Greenwood, Mississippi, on a navigation flight to Dallas, Texas, to San Antonio, Texas, to Brownsville, Texas, and return to Greenwood. The airplane was cleared to fly at 6,000 feet. At 2310, the pilot reported over Waco, Texas, at 4,000 feet. The pilot was informed of a line of thunderstorms existed between Austin, Texas, and Brownsville. At 0007, while flying over the Alamo Radio Range, the pilot requested a change to 2,000 feet. The pilot was cleared to fly at 2,000 feet from Alice, Texas, to Brownsville. There was no further radio contact with the airplane and it failed to return to its station. The airplane was declared missing and a massive search effort launched. A civilian ranch hand discovered the wreckage on 3-24-45. Investigation revealed that the C-47 descended into wooded terrain at a 45-degree angle in a 15-degree bank to the right. Wreckage was scattered over an area of 185 yards. Investigators noted that the airplane was slightly off course and speculated that the pilot was attempting to "pick his way through" the thunderstorms along his flight path.

3-13-45B. Warner-Robins, Georgia. At 0651 CWT, a Curtiss RA-25A crashed just after take-off from Robins Field, Warner-Robins, Georgia, killing pilot 1Lt. Theodore R. Stellmacher, 25. The airplane took off from Runway 9 at 0650 CWT on a ferry flight to Bush Field, Georgia. The A-25 climbed to about 100 feet and began a climbing turn to the left. The airplane was observed to be trailing black smoke as it began the turn. The airplane began losing altitude and entered a steep bank to the left, but the pilot righted the airplane and entered a shallow descending glide. Small flames were observed around the engine cowl and

exhaust stack. The A-25 veered sharply to the right and then smashed to the ground, erupting into flames upon impact. Investigation revealed that the number-nine cylinder's intake valve failed and was ingested into the engine, causing the fire. Investigators noted that the airplane had been in storage.

3-13-45C. Easton, Maryland. At 1420, a Republic P-47D crashed at Easton Airport three miles north of Easton, Maryland, killing F/O Robert F. Malloy, 26. The P-47 was part of a four-ship flight that took off at 1255 EWT from Andrews Field, Maryland, on a scheduled aerobatic flight. The formation was flying at 8,000 feet over Easton Airport when the subject airplane began suffering engine trouble. The supervising instructor advised the pilot to land at Easton Airport. The pilot was cleared to land and entered the traffic pattern. The airplane appeared to lose altitude rapidly after turning from base to final. The airplane, with landing gear extended, collided with trees with the starboard wing, severing the wing. The airplane rolled left to an inverted attitude and then smashed into the terrain. The airplane did not burn but was destroyed utterly. Investigation revealed that the number-eight cylinder head failed.

3-13-45D. Santa Maria, California. At 1845, a Lockheed P-38L crashed 15 miles ENE of Santa Maria Army Air Field, Santa Maria, California, killing pilot 2Lt. Lawrence E. Nims, 23. The airplane was part of a four-ship flight that took off from Santa Maria Army Air Field on a formation training flight. The subject airplane began suffering problems with the starboard engine as the flight let down from 6,000 feet through a hole in the clouds while returning to the field. Puffs of black and white smoke were observed emitting from the starboard supercharger area. The pilot shut down the engine and feathered the propeller. The subject airplane entered a shallow descending left turn, disappearing into the scattered clouds prevalent in the area at the time. The other pilots in the flight lost sight of the subject airplane. A few minutes later the airplane crashed into a cloud-shrouded mountain approximately 40 feet from the top. Investigation revealed that the airplane smashed into the terrain in an inverted attitude.

3-14-45A. Wilmauma, Florida. At 1100 EWT, a North American P-51D crashed five miles southeast of Wilmauma, Florida, killing pilot 2Lt. Jimmie G. Martin, Jr., 21. The airplane was the lead airplane of a two-ship flight that took off from Bartow Army Air Field, Bartow, Florida, on an individual combat mission. The flight was maneuvering at 8,500 feet. During the course of the maneuvering the airplanes lost about 2,000 feet of altitude. The subject airplane was seen in a vertically banked turn to the left. The wingman followed until his airplane lost airspeed, stalled and entered a spin. The wingman recovered at about 2,500 feet. The wingman began to circle the area but was unable to locate the lead ship. The wingman then noticed a column of smoke rising and flew over to investigate, confirming that the lead ship had crashed. Investigation revealed that one of the wings had failed and separated before the airplane slammed into the ground and exploded. Further investigation revealed a hole in one of the oxygen cylinders, presumably made by a fifty-caliber bullet. Neither of the airplanes in the flight was armed and investigators could not determine when and how the bullet penetrated the oxygen tank. The bullet could not be found and several inspections of the aircraft prior to the fatal flight failed to reveal any holes in the aircraft skin, airframe or oxygen tank. The cause of the accident remains undetermined.

3-14-45B. Chester, South Carolina. At 1400, a North American B-25C crashed nine miles east of Chester, South Carolina, killing the crew of six. The airplane took off at 1300 EWT from Morris Field, Charlotte, North Carolina, on an instrument training flight. Ceiling along the proposed flight path was reported to be 5,000 feet with three miles visibility. The airplane was heard flying in the overcast in the vicinity of Chester. Ceiling at Chester was estimated to be about 3,500 feet, visibility two miles with light showers. Witnesses on the ground stated that the airplane sounded like it was in a series of dives and climbs while flying in the overcast. Moments later the airplane plunged out of the bottom of the overcast in a diving spin. The starboard wing failed and separated, smashing into the tail section and causing its progressive failure and separation. The airplane spun to ground and exploded violently into flames. The outboard wing section and pieces of the tail section were found 400 yards from the point of impact. Investigators speculated that the pilots had lost control of the airplane while flying in instrument conditions and in an attempt to regain level flight stressed the starboard wing beyond its design limitations, causing its failure and separation. Killed in the crash were: 1Lt. Charles S. Simmons, B-25 instructor/co-pilot; 2Lt. Ralph W. Zerkle, pilot; Sgt. Einar H. Hall, engineer; Pvt. Philip T. Arlotta, student engineer; Cpl. Charles W. Reeves, Jr., gunner; Cpl. Robert E. Korwitz, gunner.

3-14-45C. San Diego, California. At 1413, a Vultee BT-13A flying in instrument conditions collided with rising terrain in the Los Pinos Mountains 37 miles east of San Diego, California, killing pilot 2Lt. William E. Dryer, 34, and passenger Cpl. Joseph F. Schubert. The airplane took off at 1318 from Van Nuys Army Air Field, Van Nuys, California, on a flight to Yuma Army Air Field, Yuma, Arizona. The airplane was on the return leg of an extended navigation training flight that originated at Goodfellow Field, Texas. Investigation revealed that the airplane had crashed in a wings level attitude at cruise speed. It was speculated that the pilot was flying on instruments in instrument conditions and was not aware of his proximity to the terrain. Search planes discovered the wreckage on 3-19-45.

3-14-45D. Williamstown, West Virginia. At 1726 EWT, a Beech AT-11 crashed six miles south of Williamstown, West Virginia, killing seven fliers. The airplane took off at 1327 EWT from Eglin Field, Valparaiso, Florida, on a flight to Louisville, Kentucky. There was no further radio contact with the airplane after take-off. At 1725 EWT, the airplane appeared over Wood County Airport, which was under construction. The airplane, turning to the left, circled the field once at 500 feet. On the second circuit the landing gear was seen to extend on the downwind leg. The airplane executed a 180-degree turn and lined up with the runway, which was apparently closed and being graded with heavy equipment. The pilot retracted the landing gear as he approached the field, apparently seeing that the runway was unusable. The port engine stopped as the airplane passed over the field, followed moments later by the starboard engine. The pilot lost control of the airplane and it collided with 40-foot trees with the starboard wing. The airplane plunged through the trees and smashed into a boulder-strewn excavation near the field, bursting into flames upon impact. Investigators noted that the pilot had taken off with the tail wheel unlocked and had very little flying time in AT-11 type aircraft.

3-14-45E. Gulfport, Mississippi. At 2030, a Boeing B-17F crashed four miles north of Gulfport, Mississippi, killing five fliers and injuring five others. The airplane took off from Gulfport Army Air Field, Gulfport, Mississippi, on a medium-altitude bombing mission. The airplane was climbing out through 6,000 feet when the engineer reported that the number-one turbo-supercharger was emitting flames. The pilot attempted to blow out the fire by changing RPM and manifold pressure. He was unsuccessful so he feathered the number-one propeller. The pilot radioed Gulfport Army Air Field, informed the tower of his turbo fire and requested landing instructions. The tower officer instructed the pilot to restart the engine, unfeather the propeller and place the mixture control in the "cruise lean" position. The pilot complied but this did not put out the turbo fire. The pilot re-feathered the propeller and again asked for landing instructions. The airplane was cleared for a landing on Runway 14. When the airplane was on final approach, the pilot was alerted that the tail wheel had not extended. The pilot aborted the landing and pulled away in a three-engine go-around. The pilot ordered the co-pilot to raise the flaps all the way as the airplane climbed away. The pilot began a flat turn to the left with very little bank angle. The pilot then called for full turbos. The port wing began to drop and the pilot called for the co-pilot to help on the controls. The pilot and co-pilot were unable to raise the port wing despite full right rudder and full right aileron. The airplane entered a steep left bank and collided with a warehouse, cartwheeled into the ground and exploded into flames. Crewmembers injured in the crash were: 2Lt. Henry R. Baron, pilot;

2Lt. Robert G. Short, co-pilot; Cpl. Claude W. Hendrix, engineer; Pfc. Kenneth Weaver, gunner; Cpl. Robert B. Sublett, gunner. Killed in the crash were: 2Lt. Joseph L. Lauth, navigator; 2Lt. Charles W. Clancy, bombardier; Cpl. Boyce M. Waldo, radio operator; Pfc. Benjamin T. Goldspink, gunner; Pfc. Russell A. Wormington, gunner.

3-14-45F. Tonopah, Arizona. At 2314 MWT, a North American AT-6D crashed 20 miles WNW of Tonopah, Arizona, killing Chinese Air Force pilot Sub Lt. Shao-Chang Van, 24. The airplane had taken off at 2114 MWT from Luke Field, Phoenix, Arizona, on a night navigation flight to Yuma, Arizona, to Blythe, California, and return to Luke Field. The airplane failed to return and was declared missing. On 3-20-45, the wreck was found on Big Horn Peak, at the end of a box canyon 45 miles west of Luke Field. Investigation of the wreckage indicated that the airplane collided with the mountain in a flying attitude with the engine producing power. Wreckage was scattered over an area of 70 yards up the slope.

3-15-45A. Crown King, Arizona. At approximately 0935 MWT, a North American AT-6D crashed in rugged terrain three and one-half miles west of Crown King, Arizona, killing A/C Paul D. Summers, 24, and A/C Dalton B. Tarver, 19. The airplane took off from Luke Field, Phoenix, Arizona, on an instrument training flight to Tonopah, Arizona, to Wickenburg, Arizona, and return to Luke Field. A/C Tarver was flying in the rear cockpit as the instrument pilot; A/C Summers was flying in the front cockpit as the safety pilot. The airplane failed to return and was declared missing. Cattleman Frank Wingert found the airplane on 5-5-45. Investigation revealed that the airplane collided with the eastern slope of Bradshaw Mountain while in flying attitude with the engine producing power. The airplane impacted the 40-degree slope in a "flat" position at an elevation of about 6,000 feet msl. Investigators speculated that the pilots had attempted to pull up at the last moment. The airplane was torn to pieces as it smashed its way up the slope for almost a hundred yards. The body of A/C Tarver had been hurled from the cockpit and was found about 45 yards from the wreckage of the fuselage. A/C Summers was found in the shattered wreckage of the fuselage. Wreckage was scattered for about 100 yards up the slope. There was no fire. It was noted that there was four to six feet of snow at the site at the time of the accident.

3-15-45B. Fannin, Texas. At 1540, a North American AT-6D crashed three miles north of Fannin, Texas, killing pilots 2Lt. Edward W. Stewart and 2Lt. James D. Ditto. The airplane took off from Aloe Army Air Field, Victoria, Texas, on an instrument training flight. Lt. Ditto was flying in the rear cockpit as the instrument pilot; Lt. Stewart was flying in the front cockpit as the safety pilot. A pilot flying in the vicinity observed the subject airplane in a fast, steep spiral to the right at an altitude of 4,000 feet. The airplane continued in the spiral until it smashed to the ground and violently exploded into flames.

3-15-45C. Strauss, New Mexico. At 1825 MWT, a Douglas A-26B crashed while attempting an emergency forced landing five miles northwest of Strauss, New Mexico, killing pilot Capt. Edwin F. James, 27. The airplane took off from Palm Springs, California, on a ferry flight to Love Field, Dallas, Texas. Investigation revealed that the starboard engine accessory section was on fire, causing the pilot to attempt an off-field landing. The airplane, with 30 degrees of flaps, landing gear retracted and power off, attempted the emergency landing downwind. Neither propeller was feathered. The airplane smashed itself to pieces as slid across the terrain, coming to rest about 200 yards from the initial point of impact. The pilot was killed when he collided with the interior of the airplane. Investigation revealed that his safety belt and shoulder harness were not fastened. A small fire had erupted after the crash and it was quickly extinguished. The port battery and generator switches were found in the "off" position and the port throttle was found in the "closed" position. The starboard throttle was found "open." The starboard fuel selector valve was found in the "closed" position. The airplane's papers were found intact and indicated that required inspections of the neoprene carburetor air flex sleeve, gas pressure hoses and throat clamps had not been properly conducted in accordance with official Technical Orders. The cause of the fire in the starboard accessory section could not be determined.

3-16-45A. Enid, Oklahoma. At 0903, a North American TB-25G crashed while attempting an emergency landing at Enid Army Air Field, Enid, Oklahoma, killing pilot F/O Jimmy S. Carter, 19. Engineer-rated passenger Cpl. Raymond D. Knox was seriously injured. The starboard engine and propeller were malfunctioning so the pilot called for an emergency landing at the field. He was cleared for an emergency landing on Runway 17. The pilot made a straight-in approach and extended the landing gear. The airplane lost altitude rapidly, so the pilot ordered the engineer to retract the landing gear and bring flaps up to 30 degrees. The airplane was in a nose-high attitude as it continued the approach. Moments later the airplane veered to the right. The airspeed continued to drop and the airplane mushed toward the ground. The airplane stalled at an altitude of about 30 feet agl, falling off and smashing to the ground at a steep angle. A fire broke out in the starboard wing as the engineer scrambled to safety.

3-16-45B. Uvalde, Texas. At 1724, a Stearman PT-13D crashed seven miles north of Garner Field, Uvalde, Texas, killing navigator-rated student pilot 1Lt. Willis H. Lebo, 27. The airplane took off from Garner Field on a routine training flight. A witness in the air observed the airplane in an intentional spin at about

4,000 feet. The airplane spun about four or five turns before the spin "flattened out a bit." The airplane continued to spin and smashed to the ground, exploding into flames upon impact.

3-16-45C. Oklahoma City, Oklahoma. At 1758, a Lockheed F-5E suffering an engine failure crashed one mile north of Will Rogers Field, Oklahoma City, Oklahoma, killing Chinese Air Force pilot Sub-Lt. Ming-Ting Hsu, 25. The airplane was flying in the number-two position of a two-ship flight that took off from Will Rogers Field on a formation training flight. The airplanes successfully completed the mission and on the way back to the field the flight leader noticed a bad oil leak on the subject airplane's port engine. The flight leader ordered the subject pilot to land at Will Rogers Field immediately. The pilot alerted the control tower of his emergency and the field was cleared. The subject airplane was advised to come straight into the field, but the pilot entered the traffic pattern. The pilot was advised to use a right-hand traffic pattern to avoid turning into the bad engine. The pilot did not acknowledge and entered the normal left-hand traffic pattern. The pilot feathered the port propeller and extended the landing gear on the downwind leg. The airplane appeared to be flying satisfactorily until it was on the base leg, when it began losing airspeed. The pilot turned from base to final and the airplane entered a steep bank, stalled and then spiraled to the ground, exploding into flames upon impact.

3-16-45D. Lee Vining, California. At 1735, a Douglas C-47A flying in instrument conditions apparently broke up in mid-air and crashed ten miles southwest of Lee Vining, California, killing four fliers. The airplane took off at 1514 from March Field, Riverside, California, on a flight to Minter Field, Bakersfield, California. The C-47 failed to arrive at Minter Field and was reported missing when the fuel was calculated to be exhausted. The airplane was found in rugged terrain 69 miles southeast of the Army Air Field at Bishop, California, on 9-23-45. The port wing tip, the outer port wing panel and pieces of the port aileron were found up to a half-mile from the main wreckage. Investigation revealed that the fuselage and mid-wing section collided with a 50-degree slope at an elevation of approximately 11,500 feet msl. The tail section separated and came to rest almost upright on the steep slope. The forward fuselage and mid-wing section bounded back into the air down the mountainside and tumbled to an inverted attitude before slamming back to earth. Wreckage was widely scattered on the steep slope. The altimeter was found frozen at 11,680 feet and a wristwatch was found stopped at 1735. The landing gear was found to be in the down and locked position. Killed in the crash were: 1Lt. Raymond Moore, pilot; SSgt. Jack E. Moran, engineer; 2Lt. C.E. Beach, passenger; J.A. Rice, USN (rank unknown), passenger.

3-17-45. Sedona, Arizona. At an unknown time after 1618 MWT, a North American P-51A was abandoned and crashed in the vicinity of Mt. Mormon

17 miles east of Sedona, Arizona, killing 1Lt. Richard L. Winden, 23, who fell to his death in an unsuccessful parachute attempt. The wreckage of the airplane was found on 10-12-45; the pilot's body was found 32 miles south of Flagstaff, Arizona, on 11-2-54. The airplane took off at 1618 MWT from Kirtland Field, Albuquerque, New Mexico, on a flight to Kingman Army Air Field, Kingman, Arizona. The flight had originated at Wright Field, Dayton, Ohio, on an "urgent" flight to Los Angeles, California. Weather was reported as good at Kingman, but a front was reported near Winslow. The airplane failed to arrive at Kingman and was declared missing after the fuel supply was calculated to be exhausted. A rancher found the airplane wreckage strewn over an area of 250 yards. It was noted that the pilot had only one hour and five minutes total time in pursuit type airplanes prior to the subject flight.

3-18-45. Kingman, Arizona. At 1532 PWT, a Bell P-39Q and a Vultee BT-13A collided in mid-air and crashed 23 miles southwest of Kingman, Arizona, killing both pilots. 2Lt. Harry C. Lane, 24, was killed in the crash of the P-39Q; 2Lt. Roman R. Dryja, 28, was killed in the crash of the BT-13A. Investigators speculated that the airplanes were traveling in the same direction at the time of the collision. The P-39 smashed to earth minus the tail section and the starboard wing. The cockpit canopy was sheared off in the collision before the airplane smashed to the ground in an inverted attitude and exploded into flames. The P-39 pilot had been ejected by the collision and fell to earth without deploying his parachute, which broke open upon impact with the ground. The BT-13 burst into flames in the collision and the side panels had separated before the airplane smashed to the ground.

3-19-45A. Carrabelle, Florida. At 1139 EWT, a Curtiss P-40N crashed 15 miles east of Carrabelle, Florida, killing pilot 2Lt. Thomas G. Dennis, 22. The airplane was flying in the number-four position of a six-ship flight that took off at 1050 EWT from Dale Mabry Field, Tallahassee, Florida, on a ground gunnery mission. The airplanes flew to Alligator Point Gunnery Range and were cleared to enter the pattern. The flight changed to a string formation and entered the pattern. After completing eight firing runs, the subject pilot noticed that his coolant temperature was high. The pilot alerted the flight leader and reported that he was going to land at the gunnery range's auxiliary airstrip. The subject airplane entered the traffic pattern for the airstrip but on final approach the pilot noticed that there was a truck on the runway, so he aborted the landing. The flight leader buzzed the truck and shooed it off of the runway. The subject airplane again entered the traffic pattern but began trailing thick black smoke while flying on the base leg. The airplane overshot the turn from the base leg to final approach and began turning to line up with the runway. The airplane was about 20 feet off of the ground when the pilot was apparently overcome by the smoke and flames, causing him to lose control of the airplane. The airplane nosed down and smashed to earth about 70 yards short of the runway. The airplane smashed itself to pieces as it slid to a halt and burst into flames. Investigators noted that the truck was on the runway because the sand perimeter track had become impassable.

3-19-45B. Pacific Ocean, California. At 0840, two Lockheed P-38L airplanes collided in mid-air and crashed into the Pacific Ocean five miles south of Simmler, California, killing one pilot and seriously injuring the other. 2Lt. Robert C. Shropshire, 20, was able to parachute to safety from P-38L #44-24308, but he was not rescued and declared missing. F/O Shafeek Smith was seriously injured parachuting to safety from P-38L #44-24359. The airplanes were part of a four-ship flight that took off from Santa Maria Army Air Field, Santa Maria, California, on a routine training flight. F/O Smith was flying in the number-three position; Lt. Shropshire was flying in the number-two position. The airplanes completed an aerial gunnery exercise and were reforming their formation when the collision occurred. Investigation revealed that F/O Smith turned on both of his cockpit heaters and immediately began feeling dizzy. He reached down to shut the heaters off and felt the collision moments later. F/O Smith could not regain control of his airplane so he bailed out. Lt. Shropshire was able to regain control of his airplane, which was heavily damaged. Half of the port aileron was severed, the left wing was damaged, the port propeller feathered and the port boom bent behind the radiator. Lt. Shropshire radioed his flight leader and apprised him of the damage. Lt. Shropshire advised that he could control the airplane but that he had trouble holding altitude. The flight leader told the pilot to bail out if he had any doubt about his ability to control the airplane. Lt. Shropshire flew to the coast and bailed out. He was observed to successfully parachute into the ocean about one-quarter mile off shore. A U.S. Coast Guard Consolidated PBY circled the pilot and dropped a life raft. The pilot was observed to shed his life vest and begin swimming toward shore. The aircraft circled the area but soon lost sight of the pilot. The pilot's body was recovered on 4-20-45.

3-19-45C. Atlantic Ocean, Florida. At 1417 EWT, a Douglas C-54A crashed into the Atlantic Ocean four miles off shore of Rock Harbor, Florida Keys, Florida, killing the crew of five. The C-54 took off at 1347 from Homestead Army Air Field, Homestead, Florida, on a routine training mission. The airplane failed to return and was declared missing. Several residents of Rock Harbor stated to investigators that they heard a loud airplane noise, suggesting an airplane in a dive, followed by a loud crash. One resident exited the building he was in and ran to the beach but was unable to see anything. On 3-20-45, a U.S. Navy blimp located the wreckage in 19 feet of water about 35 miles south of Homestead Army Air Field. Weather was not considered a factor in the accident.

The cause of the crash remains undetermined. Killed in the crash were: Capt. Paul F. McMurdy, C-54 instructor; Maj. Sumner H. Reeder, pilot; Capt. Richard C. Dayton, co-pilot; Cpl. Leonard B. Schafer, engineer; TSgt. Bernard A. O'Donnell, engineer.

3-19-45D. Wheeler, Alabama. At 1430 CWT, a Consolidated RB-24E was abandoned and crashed two miles south of Wheeler, Alabama, killing pilot Capt. Robert M. Todd, who fell to his death in an unsuccessful parachute attempt. Co-pilot Maj. Abram Y. Bryson and engineer Pfc. Arthur Mendez were uninjured parachuting to safety. The airplane took off at 1400 from Courtland Army Air Field, Courtland, Alabama, on a high-altitude training mission. At 1420 the airplane was flying to the north at 3,000 feet when the number-two engine cut out and the oil pressure dropped to zero. The pilots then noticed that the number-two engine was on fire. The pilot was apparently unable to extinguish the fire using the engine fire extinguisher. The pilot ordered the crew to bail out and they complied immediately. The airplane entered slight dive and smashed to the ground about three and a half miles southeast of Courtland Army Air Field. The pilot's body was found about 100 feet from where the airplane smashed to the ground and exploded, indicating that he bailed out at an altitude too low to allow him to deploy his parachute. Investigators speculated that a faulty fuel line had caused the fire.

3-19-45E. Durham, California. At 1417, a Lockheed P-38L crashed three miles east of Durham, California, killing pilot 1Lt. William E. Hines, Jr., 28. The subject airplane was the lead ship of a four-ship flight that took off from Chico Army Air Field, Chico, California, on a cross-country flight to Tonopah Army Air Field, Tonopah, Nevada. The flight began climbing and at about 7,000 feet encountered a thin layer of clouds. The flight leader, flying in the number-four position, ordered the number-four man, who was now leading the flight, not to enter it. The airplanes, in an attempt to climb over and maneuver away from the clouds, inadvertently entered instrument conditions at about 10,500 feet. The airplanes went on instruments and the flight leader, Lt. Hines, ordered the flight to perform a 180-degree turn to return to visual conditions and head back to Chico. The number-three and number-four airplanes turned 180 degrees and descended through and popped out of the bottom of the overcast at about 7,000 feet. The flight leader and the number-two ship climbed until they popped out "on top" at about 15,000 feet. The airplanes continued climbing until the lead ship leveled off at 23,000 feet and the wingman leveled off at about 26,000 feet. The flight leader was given a fix for Chico Army Air Field and then began his let-down. The subject airplane was next observed descending out of the clouds at 7,000 feet in a steep dive. The airplane continued in the dive until it struck the ground and exploded into flames. The cause of the accident could not be determined.

3-19-45F. Columbus, Mississippi. At 2312, a Beech AT-10GL crashed one and a half miles north of the Army Air Field at Columbus, Mississippi, killing students A/C Ralph A. Mitchell and A/S Donald R. Neer. The airplane returned to the field after successfully completing a training flight and entered the traffic pattern. Investigation revealed that the airplane stalled while turning from the base leg to final approach. The airplane fell off on the starboard wing and entered a spiral to the right, remaining in this attitude until it smashed into the ground. The airplane struck the ground on the starboard wing and engine and then cartwheeled into the ground. The airplane did not burn but was destroyed utterly. The two students were flung free of the wreckage and were found still strapped to their seats.

3-19-45G. Rocky Ford, Colorado. At 2230 MWT, a North American TB-25D and a TB-25J collided in mid-air and crashed seven miles south of Rocky Ford Auxiliary Airfield, Rocky Ford, Colorado, killing five fliers. B-25 student A/C Robert O. Melton was able to parachute to safety from the B-25D, escaping injury. Instructor 2Lt. Fred Kapp, Jr. and student A/C William C. Motteran were killed aboard the B-25D. Instructor 1Lt. Robert F. Holcomb and students A/C Richard B. Johnson and A/C Hal L. Johnson were killed in the crash of the B-25J. The airplanes took off from La Junta Army Air Field, La Junta, Colorado, on separate night training flights. The airplanes flew to Rocky Ford Auxiliary Airfield to practice night landings and take-offs. The airplanes were flying in the pattern for Rocky Ford Auxiliary Airfield when the collision occurred. Investigation revealed that Lt. Kapp did not precisely follow the field's night take-off procedure directly after his last night take-off. The surviving B-25 student stated that Lt. Kapp attempted to "cut a corner" to enter the 45-degree traffic entering leg when he collided with Lt. Holcomb. The B-25D, while in a climbing turn to the left, collided with the B-25J from the bottom while the B-25J flew the 45-degree traffic entry leg to the overhead pattern. The collision ripped off the top of the B-25D's fuselage from the nose to the area aft of the cockpit, cutting off the top of the nose section and cockpit. A/C Melton, who was seated in the nose section, bailed out of the top of the shattered nose section just as the B-25D entered a steep dive. The B-25D smashed to the ground at a steep angle but did not burn. The two pilots in the B-25D were decapitated in the collision. The B-25J burst into flames as it tumbled out of control and smashed to earth, exploding into flames upon impact with the ground about three-quarters of a mile from the wreckage of the B-25D.

3-20-45A. Lake Martin, Alabama. At about 1030 EWT, a North American TB-25G flying in poor weather crashed at Lake Martin, Alabama, killing three fliers. Pilot Capt. John G. Mabry, engineer SSgt. James N. Green and passenger Capt. Charles P. Oliver were killed in the crash. The airplane took off at 0733 EWT from Bolling Field, Washington, D.C., on a flight to Maxwell Field, Montgomery, Alabama. At 0812 EWT the pilot reported over Blackstone, Virginia, and air traffic control cleared him to Spartanburg, South Carolina, at 8,000 feet. At 0847 EWT the pilot reported over Greensboro. At 0925 EWT the pilot reported over Spartanburg and requested information regarding a cold front reported on the flight path. The pilot was advised that the cold front was in the vicinity of Craig Field, Alabama. Air traffic control cleared the airplane to a point 25 miles southwest of Birmingham, Alabama, via Atlanta, Georgia. At 1008 EWT the pilot reported 40 miles northeast of Atlanta and was advised of a line of thunderstorms stretching from Atlanta to a point 50 miles west of Maxwell Field. That line of thunderstorms converged with a cold front north of Montgomery and it was reported that "moderate" turbulence existed in the system. The pilot acknowledged this weather advisory and he was cleared to Montgomery at 8,000 feet. At 1013 EWT the pilot reported over Atlanta at 8,000 feet. There was no further radio contact with the airplane. The B-25 crashed at Lake Martin five miles west of Dadeville, Alabama, at approximately 1030 EWT, killing the occupants instantly. SSgt. Green's body was recovered from the lake shortly after the crash. Navy divers were sent down to examine the wreckage. Some wreckage was recovered but investigation proved to be inconclusive. Only small amounts of human anatomy were recovered. The cause of the crash remains undetermined.

3-20-45B. Thomasville, Georgia. At 2140 EWT, a Curtiss P-40N flying in a severe thunderstorm crashed nine miles southeast of Thomasville, Georgia, killing pilot Capt. Jesse M. Bland, 27. The airplane took off at 1901 EWT from Bluethenthal Field, Wilmington, North Carolina, on a flight to Savannah, Georgia. The pilot radioed air traffic control at Savannah and requested an aerial clearance for Waycross, Georgia, which was granted. When the airplane arrived at Waycross, the pilot asked for a clearance to Thomasville, which was granted. The pilot asked for and received a homing vector from operators at Thomasville. There was a severe thunderstorm passing over the field at the time. Ceilings were reported at 100 feet agl and visibility was three-eighths of a mile in heavy rain. The airplane encountered this poor weather as it arrived in the area of Thomasville Army Air Field and crashed about one and a half miles from the runway. Investigators speculated that the airplane encountered a severe down draft that forced it into the ground.

3-21-45A. Clovis, New Mexico. At 0952 MWT, a Boeing TB-29A crashed while attempting an emergency forced landing 15 miles northeast of Clovis Army Air Field, Clovis, New Mexico, killing two fliers and injuring 13 others, four of them seriously. The airplane took off from Clovis Army Air Field at 0905 MWT on a routine training flight. Shortly after take-off power was reduced on the number-four engine because of high oil temperature. The airplane continued to climb, leveling off at 10,000 feet. It was soon discovered that the number-four oil cooler shutter would not open fully. The airplane continued on its flight and about 10 miles northeast of Hereford, Texas, the number-one engine began backfiring and the carburetor air temperature gauge reading increased to the maximum. The pilot reduced the power on the number-one engine until the carburetor air temperature went back to normal and then normal power was reapplied. The engine continued to backfire after normal power was applied. The pilot shut down the engine, feathered the propeller, and decided to return to Clovis Army Air Field. The oil temperature on the number-four engine began to rise rapidly so the power was reduced. The pilot had difficulty holding altitude and maintaining a safe airspeed. The pilot advanced the power on the number-two and number-three engines. The oil pressure on the number-four engine then dropped to zero. The pilot was unable to feather the propeller and it was wind milling at about 1,000 rpm. The airplane began to lose altitude rapidly. The pilot extended partial flaps but still was unable to arrest the rapid descent. The pilot applied emergency war power setting for the remaining engines. The airplane continued to lose altitude so the pilot ordered that the bomb bay fuel tanks be jettisoned. After some difficulty, the bomb bay fuel tanks were jettisoned and then the landing gear lever was placed in the down position. The airplane continued to lose altitude and it struck the ground before the landing gear could fully extend. The airplane smashed itself to pieces as it slid across the terrain, bursting into flames as it came to rest. Investigators stated that the airplane was too heavily loaded to fly on two engines and maintain a safe airspeed. It was speculated that the gross weight of the aircraft was about 115,000 pounds at the time of the crash landing. Navigator 2Lt. Frederick A. Vantuyl and bombardier 2Lt. Charles A. Rudauskas were killed in the crash landing. Crewmembers seriously injured in the crash were: 1Lt. John J. Naughton, B-29 instructor; 2Lt. Fred A. Lauerwasser, pilot; Cpl. Victor C. McIntyre, central fire control; Cpl. William J. Digirolamo, radio operator. Crewmembers injured in the crash were: 2Lt. Joseph D. Pijot, co-pilot; Cpl. Marshall R. Lyon, flight engineer; Cpl. Thomas F. Atkins, gunnery instructor; Cpl. Livingston R. Wilson, starboard gunner; Cpl. Kenneth A. Wyerman, port gunner; Cpl. Harrison A. Woolford, tail gunner; Cpl. Victor T. Gardner, central fire control; SSgt. Jack N. Storey, gunner; TSgt. Ralph I. Sever, gunner.

3-21-45B. Pottsboro, Texas. At 1140 CWT, a North American AT-6C crashed four and a half miles northwest of Pottsboro, Texas, killing instructor 2Lt. Vernon D. Bretz, 21, and student A/S Anthony Campo. The airplane took off from Perrin Field, Sherman, Texas, on a routine training flight. Investigation revealed that the

airplane stalled at an altitude too low to allow recovery while performing a simulated forced landing. The pilot applied throttle but the engine did not immediately respond with adequate power. The pilot attempted to stretch his glide to clear some trees but stalled in.

3-21-45C. Red Bluff, California. At 1116, a Lockheed P-38L collided with rising terrain and crashed five miles southwest of Red Bluff, California, killing pilot 2Lt. Charles J. Habig, 23. The airplane was flying in the number-two position of a two-ship flight that took off from Chico Army Air Field, Chico, California, on a low-altitude training mission and an instrument training flight. The P-38s took off, climbed to altitude, and then proceeded to fly on instruments. After about a half hour of instrument flying, the airplanes descended and entered the pattern for the low-altitude course. The airplanes — with the wing ship flying on the port wing of the leader — flew a circuit of the low-altitude course without incident. On the second circuit the flight leader ordered the subject pilot into a trail formation and then they flew down a canyon on the course. As they were about to emerge from the canyon, the flight leader ordered the subject pilot to assume the previous formation. The flight leader did not receive a reply and looked back just in time to see the subject airplane collide with a small hill on the course. The two propellers separated as the airplane burst into flames and bounded back into the air at about the same altitude as the top of the hill. The airplane then collided with a second hill in the line of flight, exploding into flames upon impact. Investigators noted that the pilot had been cautioned about flying below the level of the flight leader while flying in formation. It was noted that this was the pilot's first low altitude training mission.

3-21-45D. Goshen, Indiana. At a time between 1630 and 1710 CWT, pilot-rated passenger F/O David J. Nixon died of oxygen starvation while on board a Consolidated RB-24E that was operating at high altitude in the vicinity of Goshen, Indiana. The airplane had taken off from Newark, New Jersey, on a navigation flight back to its home station at Liberal Army Air Field, Liberal, Kansas, via Chanute Field, Rantoul, Illinois. Investigators noted that the flight had originated at Liberal several days earlier and had been delayed at Newark because of mechanical problems and weather. F/O Nixon, who flew the bomber on the inbound portion of the trip, was situated in the waist section with two other crewmembers for the take-off and climb. The airplane reached 10,000 feet and the pilot ordered everyone on oxygen. The two other crewmembers left the waist section one by one and went forward as the airplane climbed to 30,000 feet. Investigation revealed that F/O Nixon disconnected his oxygen mask from the airplane oxygen system and connected himself to a "walk-around" oxygen bottle. The walk-around bottle was empty, causing the B-24 student die of oxygen starvation. F/O Nixon was found unconscious, lying face down on the waist section floor. The B-24 descended to lower altitude and artificial respiration was performed on the stricken pilot, but he was already dead. The airplane was piloted by 2Lt. Armond Lilien and co-pilot 1Lt. Millard P. Schaaf. The pilots and six other crewmembers were uninjured.

3-22-45A. Cross Plains, Tennessee. At 1125 CWT, a Lockheed XF-5D suffering a fire in the port engine was abandoned and crashed two miles east of Cross Plains, Tennessee. Pilot 1Lt. George L. Wales, 25, was fatally injured in an unsuccessful parachute jump. The XF-5D, accompanied by a Lockheed XP-38A, took off from Wright Field, Dayton, Ohio, on a ferry flight to Key Field, Meridian, Mississippi. The airplanes were flying in a very loose formation at an altitude of 3,500 feet, with the XF-5D flying about 500 yards behind and to the left of the XP-38A. About 30 miles north of Nashville, Tennessee, a thin line of dark smoke was observed trailing the port engine of the subject airplane. Moments later, thick white smoke began trailing the port engine. The XF-5D pulled up and gained altitude. The port engine was then observed to be on fire on top of the port nacelle near the turbo-supercharger. The XF-5D crossed over to the right and fell away. The pilot bailed out and his parachute was seen to deploy. The XF-5D entered a dive and smashed to the ground, exploding into flames upon impact. The other pilot circled the parachute but could observe "no signs of life." Investigation revealed that the pilot's foot had become entangled in a wire antenna and apparently the airplane was pulling him down. The pilot deployed his parachute at high speed, creating a shock force strong enough to break the pilot's sternum and severe his left internal mammary artery, causing him to bleed to death internally. It was not known if the pilot's foot was free of the antenna at the time he deployed the parachute. Also, investigators noted that rescue personnel had cut off the pilot's parachute harness when they found him, preventing investigators from determining if the harness was properly attached.

3-22-45B. Slater, Missouri. At 1615, a Curtiss C-46A crashed 12 miles southwest of Slater, Missouri, killing pilot Maj. Thomas L. White and co-pilot Capt. Roger J. Bernard. The airplane took off from Sedalia Army Air Field, Warrensburg, Missouri, on a local instrument training flight. Investigation revealed that the starboard wing failed and struck the side of the fuselage and the tail section. The airplane broke up in flight and smashed to the ground at a speed in excess of 400 mph. Witnesses stated that the airplane smashed to the ground in a diving turn to the right. Wreckage was scattered over an area of one and a half miles wide and two miles long. Investigators could not determine with certainty why the airplane broke up. It was speculated that the pilots failed to recover from an unusual attitude and entered a high-speed dive, causing the airplane to break up.

3-23-45A. Reidsville, North Carolina. At 1845, a North American AT-6A crashed four miles east of Reidsville, North Carolina, killing pilot F/O Charles W. Stephens, Jr., and passenger Maj. Harold D. Martin. The airplane took off from Tuskegee Army Air Field, Alabama, on a cross-country flight to Winston-Salem, North Carolina. Another AT-6 flew part of the way with the subject airplane. The subject airplane was observed to drop down and begin circling the town of Rocky Mount, North Carolina. The other AT-6 turned ten degrees and flew to Roanoke, Virginia. At about 1900, the airplane was observed circling Danville, Virginia. Civilian personnel at Danville Airport turned on the runway lights and flashed lights to attract the attention of the pilot. The airplane then flew away toward Greensboro, North Carolina. The airplane arrived at Reidsville and entered the traffic pattern for the municipal airport. The airplane, flying at 300 feet agl, stalled while turning from the base leg to final approach. The airplane described one-half turn of spin before smashing to the ground about 600 feet from the airport boundary. The airplane did not burn but was completely demolished.

3-23-45B. Herington, Kansas. At 1750 CWT, tail gunner Cpl. Orville Mobley, Jr. was fatally burned when the tail gun position of a Boeing F-13A operating near Herington Army Air Field, Herington, Kansas, suffered a flash oxygen fire. Investigation revealed that the airplane was flying at 21,000 feet while on an instrument calibration mission. The crew compartments were pressurized and the crew came off of the oxygen system. The tail gunner pulled off his oxygen mask and disconnected it from the A-14 oxygen regulator. The gunner then lit a cigarette and a small oxygen fire developed. The gunner alerted the pilot of the fire in the tail gun compartment and then attempted to shut the oxygen off, but he turned the knob the wrong way and opened the valve full, causing pure oxygen to flood the compartment. There was a flash fire in the compartment. The gunner, who was severely burned, called for help over the intercom. Crewmembers came aft and pulled the gunner from the smoldering compartment. The gunner had released the emergency window hatch and was attempting to leave the airplane without a parachute, but the hatch wouldn't release cleanly. The gunner died the next day. The airplane was piloted by 1Lt. Ralph S. Learn and co-pilot 2Lt. Lloyd Dobney. The pilots and nine other crewmembers were uninjured.

3-24-45A. Old Town, Florida. At 1100 EWT, two Curtiss P-40N airplanes collided in mid-air and crashed five miles south of Old Town, Florida, killing one pilot and seriously injuring another. F/O Earl R. Goodyear fell to his death attempting to parachute to safety from P-40N #43-24362; 2Lt. Roy C. Rom was seriously injured parachuting from P-40N #44-7305. The airplanes were part of a four-ship flight that took off from Perry Army Air Field, Perry, Florida, on a low-altitude formation mission. F/O Goodyear was flying in the number-two position; Lt. Rom was flying in the number-four position. The first three airplanes took off normally, but Lt. Rom was delayed on the ground for about ten minutes. He took off and joined the formation as it flew over the field. The airplanes then flew to the Gulf of Mexico to perform the low-level mission. The airplanes were maneuvering at about 500 feet agl when the collision occurred. The second four-ship element was flying on the flight leader's port wing and attempted to cross over to the right as the flight leader turned the formation to the right. Investigation revealed that the number-four airplane's propeller collided with the tail section of the number-two ship. The number-two ship entered a spin and smashed to the ground. The number-four ship climbed slightly with the propeller windmilling. Smoke entered the cockpit and the pilot realized that the airplane was not flyable, so he bailed out. The pilot, who had been temporarily blinded by oil, collided with trees during his parachute landing.

3-24-45B. Alder Springs, California. At an unknown time between 0900 and 1430, a Vultee BT-13B crashed in a remote area eight miles west of Alder Springs, California, killing pilot Capt. Thomas W. Capps and passenger Pvt. Norbert J. Riley. The airplane had taken off from Santa Rosa Army Air Field, Santa Rosa, California, on a search flight for a missing U.S. Navy airplane. The BT-13 failed to return to Santa Rosa and was declared missing. The airplane was found 8-18-45. Investigation revealed that the pilot and passenger had survived the crash. The pilot had received head injuries and leg injuries and was initially immobile. In hopes of attracting a passing airplane, the two fliers deliberately burned the airplane the first night they were stranded. The next day, Pvt. Riley left the scene in an attempt to go for help and failed to return. His body was found in a creek bed in a ravine about three-quarters of a mile SSE of the wreck. Weather turned extremely bad and it snowed for several days following the accident. Capt. Capps realized that the poor weather was preventing search planes from finding the wreckage. The pilot, even though he was seriously injured, attempted to walk out. He was found in a creek bed in a ravine about three-quarters of a mile SSW of the wreck site. Investigators speculated that the pilot was attempting an emergency forced landing when the airplane collided with trees at low speed. The port wing was severed and the airplane smashed into the trees, coming to rest upside down. The airplane did not burn in the crash. Investigation revealed that the fuel pump had failed.

3-24-45C. El Paso, Texas. At 1038 MWT, a Republic P-47N crashed one and a half miles from Biggs Field, El Paso, Texas, killing pilot F/O Donald E. Turvey, 31. The airplane was taking off from Biggs Field while on a ferry flight from Midland Army Air Field, Midland, Texas, to Palm Springs, California. The airplane used up nearly 5,000 feet of runway on the take-off roll. The airplane lifted off and climbed slowly to 100 feet agl, and then entered a gentle turn to the left. The engine was heard "cutting out." The airplane continued in the turn and began losing altitude. The pilot alerted the tower that his engine was cutting out. The airplane continued to lose altitude and was in danger of colliding with power lines south of the control tower. The pilot pulled up and avoided the power lines. The airplane then smashed into rough terrain with landing gear retracted and flaps extended. The airplane smashed itself to pieces and burst into flames as it slid across the terrain.

3-24-45D. Marysville, California. At 1125, three Lockheed P-38L airplanes collided in mid-air approximately 9 miles east of Marysville Army Air Field, Marysville, California, killing 1Lt. John F. Knecht, 22. The two other pilots were uninjured and were able to safely land their damaged airplanes. The subject airplanes made up the fourth flight of a squadron formation consisting of four, four-ship flights that had participated in a training mission and fly-by at Camp Beale, California. The squadron was based at Chico Army Air Field, Chico, California. Lt. Knecht, flying P-38L #44-23860, was flying in the lead position of the fourth flight. F/O Clemens L. Morie, flying P-38L #44-23866, was flying in the number-three position; 2Lt. Harold L. Oster, flying P-38L #44-25622, was flying in the number-four position. The four flights were separated by about one-half mile as they flew over Camp Beale at about 1,200 feet agl. The four-ship flights were configured with the leader's wingman flying on the leader's port wing and the number-three ship flying on the leader's starboard wing. The number-four ship was positioned on the number-three ship's starboard wing, completing the standard four-ship tactical formation. As the subject airplanes conducted the fly-by, they encountered the propeller turbulence that was generated by the previous three flights. The number-four ship of the subject flight encountered the turbulence and began to drift to the left toward the number-three ship. The number-four pilot attempted to counter with right rudder but was unable to prevent a collision. The port wingtip of the number-four ship smashed into the starboard boom and nacelle of the number-three ship, sending both airplanes out of control. Both airplanes rolled to the left and the number-three ship assumed an inverted position above the flight leader, Lt. Knecht. The number-three airplane pitched downward as the pilot attempted to roll through and regain control. The number-three ship collided with the leader and severed the tail section of the lead ship. The lead ship dove straight to the ground and exploded violently into flames. Lt. Knecht made no attempt to parachute to safety and was killed upon impact. The number-three airplane's port propeller separated, but the pilot was able to regain control of the airplane and perform a successful crash landing nearby.

The number-four pilot was able to regain control of his airplane and land safely at base. The number-two ship safely rolled out of the formation when the pilot observed the collision between the number-three and number-four ships taking place. It is interesting to note that the official AAF Accident Report indicates that investigators administered thiopental sodium drugs, the so-called "truth serum," to the pilot witnesses during questioning.

3-25-45. Santa Elena, Texas. At 1626, a North American AT-6D crashed eight miles south of Santa Elena, Texas, killing student A/C William F. Leonard, 19. The airplane took off from Moore Field, Mission, Texas, on a navigation-training flight to Harlingen, Texas, to Alice, Texas, to Laredo, Texas, and return to Moore Field. The subject airplane was flying in the company of two other airplanes, but they were not flying in a formation. The subject airplane was observed to enter a descending turn to the left, rolling out about 45 degrees to the original flight path. The airplane continued to lose altitude until the pilot leveled out at low altitude above an open field. The airplane was flying toward a building so the pilot pulled up abruptly. The airplane failed to clear the building and the starboard wing collided with the structure, severing the aileron. The airplane climbed steeply, rolled over to an inverted attitude, and then plunged into a stand of mesquite trees about 250 yards beyond the building. The airplane did not burn but was totally destroyed. Investigators could not determine why the student left the approved course.

3-26-45A. Mikado, Michigan. At 0832 EWT, a Republic P-47D crashed eight miles west of Mikado, Michigan, killing French Air Force pilot Sgt. Marcel M. Oules, 25. The airplane was flying in the number-three position of a three-ship flight that took off at 0757 EWT from Oscoda Army Air Field, Oscoda, Michigan, on a scheduled aerobatic flight and dive-bombing mission. The flight made one practice run and then completed two bombing runs, each of which was started at 8,000 feet msl. After completing the last bomb run, the flight leader looked back and observed the two wing ships pulling up and noticed that no bombs had hit the target. The flight leader climbed away and radioed the subject pilot but received no response. The flight leader circled around at about 6,000 feet looking for the subject aircraft. The flight leader noticed a fire on the ground about four miles north of the AAF bombing range. The flight leader determined that the fire on the ground was an airplane crash. Investigation of the wreckage indicated that the airplane smashed into the ground at high speed and at a steep angle. The pilot left the airplane just before it hit the ground and exploded violently into flames. He slammed to earth in close proximity to the wreckage. His ripcord had not been pulled. Investigators noted that wreckage had penetrated several feet of earth.

3-26-45B. Lake Charles, Louisiana. At 1230 CWT, a Douglas A-26B crashed one and a half miles west of Lake Charles Army Air Field, Lake Charles, Louisiana, killing three fliers. Pilot 1Lt. Kenneth E. Holz, pilot-rated passenger 1Lt. Murray M. Rich and bombardier 1Lt. Roscoe T. Dwyer were killed in the crash. Investigators stated, "[The airplane] made an approach to Runway 22 but had to go around as there was another A-26 on [the runway] with a flat tire. He was instructed by the tower to make an approach to Runway 26. This was done by circling the field rather than re-entering the traffic pattern. A wide shallow turn was made into the approach, and it appeared that he overshot the runway and was more nearly in line with Runway 32. He may have confused the runways. At a point about inline with Runway 32 at an altitude of between 500 and 1,000 feet the aircraft rolled to the right and went into the ground at an angle of about 45 degrees, striking the right wing tip first." The airplane did not burn, but was completely destroyed. Investigation revealed that the starboard wing flap did not extend with the port flap, causing the pilot to lose control of the airplane. The failure of the flap to extend was caused by binding between the inboard flap trailing edge and the inboard wing trailing edge.

3-26-45C. Hinton, Oklahoma. At 1345, a Lockheed F-5E crashed eight miles east of Hinton, Oklahoma, killing pilot F/O George D. Dealey, 21. The airplane took off from Will Rogers Field, Oklahoma City, Oklahoma, on a photographic training mission to Amarillo, Texas, and return. The airplane climbed to 25,000 feet, and a short time later was observed to make several slight climbs and dives before entering a vertical dive. The pilot made no apparent effort to recover and the airplane smashed into the ground in a high-speed vertical dive. Investigators speculated that the pilot had suffered an oxygen failure and lost consciousness, causing the airplane to dive out of control to the ground. The airplane exploded violently into flames upon impact.

3-27-45A. Buffalo Gap, South Dakota. At 1145 MWT, a Bell P-63A crashed five miles east of Buffalo Gap, South Dakota, killing pilot Capt. James G. Daugherty, 27. The airplane was one of two that took off at 1016 MWT from Pueblo Army Air Base, Pueblo, Colorado, on a flight to Rapid City Army Air Base, Rapid City, South Dakota. The airplanes had encountered strong headwinds on the inbound flight and were low on fuel when they approached the area of Buffalo Gap. Sometime after 1130 MWT, the ships encountered a severe snowstorm east of Buffalo Gap. Visibility began to deteriorate and the pilot descended in an attempt to stay underneath the overcast. The pilot spotted a clear area on the ground, dropped his flaps and attempted to make a forced landing. The pilot pitched the nose downward, but apparently misjudged his altitude over the terrain and smashed into the ground at a 45-degree angle. The airplane burst into flames upon impact and the pilot was killed instantly. Investigators speculated that the pilot had trouble seeing through his windscreen. Capt. Daugherty's wingman was killed in a crash about a minute later approximately seven miles north of the crash site. See 3-27-45B.

3-27-45B. Buffalo Gap, South Dakota. At 1146 MWT, a Bell P-63A crashed four miles north of Buffalo Gap, South Dakota, killing pilot Capt. Dan W. Moore, 27. The airplane was one of a flight of two that took off at 1016 MWT from Pueblo Army Air Base, Pueblo, Colorado, on a flight to Rapid City Army Air Base, Rapid City, South Dakota. The airplanes had encountered strong headwinds on the inbound flight and were low on fuel as they approached the area of Buffalo Gap. Sometime after 1130 MWT, The airplanes encountered a severe snowstorm east of Buffalo Gap. Visibility began to deteriorate and the pilot descended in an attempt to maintain visual contact with the ground. The pilot jettisoned the starboard hatch but did not bail out, apparently because of the low altitude. The port wing collided with an earthen dam, sending the airplane out of control. The airplane smashed to the ground and burst into flames. Wreckage was scattered over an area of 100 yards. The pilot was killed upon impact. Capt. Moore's wingman was killed in a crash about a minute earlier approximately seven miles south of the crash site. See 3-26-45A.

3-27-45C. Suffolk County, New York. At 1709 EWT, two Republic P-47D airplanes collided in mid-air at Suffolk County Army Air Field, Suffolk County, New York, killing Capt. Jo J. Holstun, 25, aboard P-47D #42-22425. F/O Parker C. Nichols, 28, was seriously injured aboard P-47D #42-27824. F/O Nichols was part of a four-ship flight that was returning to Suffolk Army Air Field after completing a routine training mission. The airplanes were given permission to enter the traffic pattern for landing. Capt. Holstun arrived in the area of the field a few moments after the four-ship flight. A few minutes earlier, Capt. Holstun's wingman had bailed out and parachuted to safety and apparently he was anxious to get on the ground. Capt. Holstun was cleared to land behind the four-ship flight. Capt. Holstun flew too tight a pattern and when he turned from base to final, he cut in between the number-three and number-four ship. Capt. Holstun began to overtake F/O Nichols' airplane from above and behind on final approach. Personnel in the Mobile Control Unit observed that an unsafe situation was developing and ordered the higher ship to go around. F/O Nichols, thinking that the order was for him, applied throttle and pulled up. Capt. Holstun continued to descend and collided with the other airplane at an altitude of about 200 feet agl. The elevators and part of the rudder of F/O Nichols' airplane were severed or damaged and the P-47 went out of control, entering an extreme nose-high position before falling off on the starboard wing. The airplane then fell

squarely on top of Capt. Holstun's airplane at an altitude of about 50 feet agl. The airplanes, still stuck together, fell and smashed to the runway, collapsing the landing gear of Capt. Holstun's airplane as they slid to a halt. F/O Nichols was immediately rescued from the wreckage of his airplane, which did not burn. The starboard wing of F/O Nichols' airplane had smashed through the cockpit of Capt. Holstun's airplane, killing him instantly.

3-27-45D. Fort Myers, Florida. At 1810 EWT, a Curtiss P-40N suffering an engine failure crashed 19 miles northeast of Page Field, Fort Myers, Florida, killing F/O Neal L. Tolleson, 21. The airplane was flying in the number-three position of a four-ship flight that took off from Page Field on a low-altitude formation flight and gunnery mission. The airplanes flew to the gunnery range and began the exercise. While pulling up after the first firing run, other pilots in the flight observed the subject airplane trailing black smoke. The pilot decided to leave the gunnery range and to return to Page Field. The airplane, flying at 1,500 feet, began a turn to the left. The airplane did a three-quarter roll and dove straight down, turning slightly left just before it struck the ground and exploded into flames. Investigation revealed that a fifty-caliber bullet had ricocheted off the ground and entered the subject airplane's oil cooler, causing its failure.

3-27-45E. Yakima, Washington. At 1600, a Douglas RA-24B crashed 20 miles NNE of Yakima, Washington, killing pilot 1Lt. Phillip N. Rooney and tow reel operator TSgt. William H. Skaggs. The airplane took off from Yakima Municipal Airport on a low-altitude target-towing mission. Investigation revealed that the tow cable had snagged on the tail wheel on a target-towing mission earlier in the day. The airplane landed and the fouled target removed. The airplane returned to the exercise area for the gunnery mission. The target was shot off by fifty-caliber ground fire. The airplane left the area in order to release a new target. The operator released a new target and the cable became fouled on the arresting hook fitting. The operator was not aware the cable had been fouled and continued to let out about 1,000 feet of cable. The fouled cable formed a loop and as the operator continued to let out cable the loop became positioned aft of the airplane. The operator noticed the looped cable and attempted to recycle the cable. The looped cable became fouled on the port elevator and rudder. The elevator was forced down and the rudder was forced to the right, causing the pilot to lose control of the airplane. The airplane entered a steep diving turn to the right that progressed into a spin. The pilot stopped the spin but did not have sufficient altitude to pull out of the dive. The airplane struck the ground and burst into flames.

3-28-45A. Louisville, Kentucky. At approximately 1100 CWT, a Vultee BT-13B crashed five miles northwest of Bowman Field, Louisville, Kentucky, killing pilot 1Lt. Robert S. Hespeler and passenger Capt. Albert I. Bell. The airplane had taken off from Bowman Field on a local familiarization flight and was returning to the field when it stalled and entered a spin from which the pilot was unable to recover.

3-28-45B. Arcadia, Florida. At 1900 EWT, a North American AT-6D crashed 10 miles ESE of Arcadia, Florida, killing instructor 1Lt. John F. Kimmerle, 27. The airplane took off from Hendricks Field, Sebring, Florida, on a routine training flight. The pilot flew to Parker Auxiliary Airfield to practice take-offs and landings. The airplane was observed to take off to the east on the East-West runway and then enter a normal rectangular traffic pattern. The airplane, flying at about 300 feet agl on the base leg, was seen to climb steeply before it snapped violently to left to a near inverted attitude. The pilot was able to roll the airplane through to an upright position. The airplane was heading about 45 degrees to the left of the prior course and losing altitude rapidly. The pilot was able to regain a near level position an instant before the airplane smashed to the ground approximately one-eighth of a mile from the airport boundary.

3-28-45C. Perryville, Arizona. At 1830 MWT, a North American AT-6C crashed one mile southwest of Perryville, Arizona, killing instructor 2Lt. Daniel H. Singleton and student A/S Charles L. Parmly. The airplane took off from Luke Field, Phoenix, Arizona, and flew to Auxiliary Airfield #6 to practice landings. The airplane was observed climbing away normally after taking off to the east on the East-West Runway at Auxiliary Airfield #6. The airplane, with landing gear retracted and flaps extended about 30 degrees, made a 90-degree turn to the right and then made an immediate turn to the left. The airplane stalled and then entered a spin to the left at about 1,000 feet agl. The pilot was unable to recover and the airplane smashed to the ground in a spin to the left at about a 50-degree nose-down position.

3-29-45A. Pampa, Texas. At 0102 CWT, two North American TB-25J airplanes collided in midair and crashed five and a half miles southwest of Pampa, Texas, killing four fliers. A/C Robert D. Asher and A/C Joseph H. Anderson were killed in the crash of TB-25J #44-28869; A/C Arial D. Brown, Jr. and A/C Richard M. Browning were killed in the crash of TB-25J #43-36078. The airplanes had taken off from Pampa Army Air Field on separate student training flights. Several student airplanes flew to Reeves Auxiliary Airfield to practice landings. Landings were being made to the southeast. A student team flying TB-25J #43-35974 on the downwind leg of the upper traffic pattern at Reeves noticed TB-25J #44-28869 flying against the pattern and coming straight at them in a head-on collision course. TB-25J #44-28869 continued flying against the traffic pattern and then flashed its landing lights. The co-pilot took over control of TB-25J #43-35974 and entered a climbing turn to the

right, avoiding a collision. TB-25J #44-28869 collided head on with TB-25J 43-36078 moments later. The airplanes broke up and erupted into flames in the collision, the pieces of flaming wreckage raining to earth. The fliers made no attempts to parachute to safety and it was speculated that they were killed in the collision.

3-29-45B. Wilmington, North Carolina. At 2027, a Republic P-47N crashed three miles north of Bluethenthal Field, Wilmington, North Carolina, killing pilot 2Lt. Albert J. Seiben, 24. The airplane took off from Bluethenthal Field on a night navigation mission. The engine began "cutting out" just after take-off, so the pilot returned to the field. Ground crewmen examined the airplane but no problems could be found. The airplane took off and climbed away and the engine again began cutting out. The pilot jettisoned the auxiliary wing fuel tanks and turned back for the field. The airplane began losing altitude and collided with trees moments later, severing the starboard wingtip and aileron. The airplane rolled over to an inverted attitude and smashed to the ground, exploding into flames upon impact. Investigators could not determine the cause of the engine failure.

3-30-45A. Harrisville, Michigan. At 0938 EWT, a Republic P-47D crashed six miles west of Harrisville, Michigan, killing French Air Force pilot Lt. Serge Lazarevich, 24. The airplane was part of a four-ship flight that took off at 0800 from Oscoda Army Air Field, Oscoda, Michigan, on a high-altitude formation flight. The airplanes completed the training mission and the second element landed at about 0900 while the first element let down over open country. The subject airplane pulled up abruptly in an attempt to level out at 6,000 feet. The airplane entered a high-speed stall and fell into a spin. The pilot was able to stop the spin and recover momentarily, but the airplane entered another spin. The airplane remained in the spin until it smashed into the ground and exploded into flames.

3-30-45B. Newton Grove, North Carolina. At 1315, a Republic TP-47D crashed four miles northeast of Newton Grove, North Carolina, killing 2Lt. Louis G. Young, 20. The airplane had taken off from Seymour-Johnson Field, North Carolina, on a routine training flight. Investigation revealed that witnesses heard the subject airplane's engine "cutting out." The pilot attempted an emergency forced landing but apparently overshot the intended field. The pilot bailed out at an altitude too low to allow his parachute to fully deploy and he fell to his death. The airplane entered a nose down attitude and smashed to earth in a wooded area. The airplane exploded flames upon impact.

3-30-45C. Gulf of Mexico, Florida. At 2137 EWT, a North American P-51C crashed into the Gulf of Mexico one-half mile southwest of Venice Army Air Field, Venice, Florida, killing 2Lt. Floyd L. Derflinger, 22. The airplane was returning to the field after completing a training flight. The airplane was cleared to land on Runway 23, but an accident on that runway involving two airplanes caused control personnel to order the subject pilot to circle the field. The pilot circled the field and was cleared to land on Runway 27. The subject airplane peeled off and entered the traffic pattern for Runway 27. Control personnel on the runway gave the subject aircraft two red flares and ordered him to go around again. The pilot advanced the throttle, retracted the landing gear and climbed to about 100 feet before entering a gentle turn to the left. The P-51 continued in the turn to the left, continually losing altitude in the turn until the airplane flew into the water. There was no fire.

3-31-45A. Axis, Alabama. At 1121 CWT, a Beech AT-7C crashed three miles west of Axis, Alabama, killing five fliers. The airplane had taken off at 0951 CWT from Alexandria Army Air Field, Alexandria, Louisiana, on a navigation training flight to Morrison Field, West Palm Beach, Florida, via Tallahassee, Florida. The flight originated at the airplane's home station at Hondo Army Air Field, Hondo, Texas. The airplane was on an instrument flight plan and assigned to fly at 5,000 feet. At 1114, the airplane radioed Mobile, Alabama, Radio and gave its position as 20 miles north of Mobile at 5,000 feet. A few minutes later, witnesses on the ground observed the airplane flying underneath the overcast at about 1,000 feet agl. The airplane appeared to maneuver to the north to avoid a low-hanging black cloud to the west. The airplane passed over the witnesses and the engines were heard "sputtering." The engines "surged" as the airplane flew away and moments later it was heard to crash in the distance. Rescue personnel were unable to locate the airplane until the next morning. Investigation revealed that the pilot had switched to the auxiliary fuel tanks to test fuel consumption. It was speculated that pilot forgot to switch back to the main fuel tanks, causing the auxiliary fuel tank to run dry. The pilot was unable to switch back to the fullest tanks and restart the engines in time to avoid crashing into a wooded area. Killed in the crash were: 2Lt. Maurice M. Stimeling, pilot; 1Lt. Alfred P. Akelaitis, co-pilot; 1Lt. George W. Cenly, navigator; 1Lt. Efrem M. Adelman, navigator; 1Lt. William P. Anton, navigator.

3-31-45B. Sacramento, California. At 0931, a Lockheed P-38J crashed five miles northeast of McClellan Field, Sacramento, California, killing 1Lt. Roy B. Crane, 26. The airplane took off at 0926 from McClellan Field on a flight to Long Beach, California. The starboard engine failed as the airplane was climbing out after take-off. The pilot failed to trim the airplane and adjust power for single-engine flight and the P-38 entered a spin to the left at about 2,000 feet agl. The airplane remained in the spin to the left until it smashed to the ground and burst into flames. Investigators could not determine why the starboard engine failed.

3-31-45C. Perkinston, Mississippi. At 1610 CWT, a Consolidated RB-24E suffering an engine fire crashed into wooded terrain one and a half miles northwest of Perkinston, Mississippi, killing five fliers. Student engineer Pvt. Earl H. Kellam, 19, Calvert, Alabama, was seriously injured in the crash. Student engineers Pvt. Robert B. Coyne, 19, Pittsburgh, Pennsylvania, and Pvt. William J. Shiner, 19, Hornell, New York, parachuted to safety and escaped injury. Killed in the crash were: 2Lt. Arnold J. Grein, 26, Lakefield, Minnesota, pilot; 2Lt. Joseph E. Carr, 24, New Carlisle, Indiana, co-pilot; Sgt. Harold W. Baur, 27, Kingston, Pennsylvania, engineer instructor; Pvt. Neil V. Harrington, 21, Coudersport, Pennsylvania, student engineer. Student engineer Pvt. Thomas L. Schweitzer, 19, Crawfordsville, Indiana, died of his injuries at 0100 on 4-3-45. The airplane had taken off from Keesler Field, Biloxi, Mississippi, on a student engineer training flight. During the engine run-up, the engineer noticed that the number-four engine was vibrating more than normal. The pilots could find no problem with the engine and the flight commenced. The airplane climbed above the clouds, which topped out at about 5,000 feet, and flew above them while the students were instructed in in-flight procedures. The airplane then descended to about 1,500 feet and leveled off. The engineer turned off the number-three fuel valve to test the student engineers. The engineer sent a student to the bomb bay to check the fuel valves. The student failed to return the number-three valve to correct position. The engineer instructor went aft to return the valve to correct position when the number-four engine erupted into flames. The number-three propeller was windmilling and creating a lot of drag and the number-four engine was not producing any power. The pilots were unable to maintain altitude and the airplane began descending rapidly. The co-pilot rang the bail-out bell as flames engulfed the number-four engine nacelle. Two crewmembers bailed out and the airplane smashed into wooded terrain moments later. Investigators noted that simulated engine out procedures should not be attempted below 5,000 feet agl.

April

4-1-45A. Galt, Missouri. At 1325 CWT, a North American P-51D crashed one mile northwest of Galt, Missouri, fatally injuring pilot F/O Kenneth Marshall, 21. The airplane took off at 1300 from Rosecrans Field, St. Joseph, Missouri, on a flight to Romulus, Michigan, via South Bend, Indiana. The flight originated at Amarillo, Texas. The airplane had landed at Rosecrans Field at 1148. The pilot requested that the wing fuel tanks be filled but declined fuel for the fuselage tank. The airplane was observed flying east at about 1,500 feet agl. The airplane began to lose altitude and descended to about 50 feet agl. The airplane then entered a turn to the right, the pilot evidently attempting to get to a large open field for a forced landing. The nose went down and the starboard wing collided with trees. The P-51 continued down and the starboard wing collided with the terrain, causing the airplane to cartwheel into the ground. Wreckage was scattered over a wide area. The pilot, still strapped to his seat, was found alive. He was able to state his name and attempted to write something but was unable to remain conscious. The seriously injured pilot was taken to a civilian hospital at Trenton, Missouri, where he died at about 1900 CWT.

4-1-45B. Sweetwater, Texas. At 1504 CWT, a Republic P-47D crashed four miles west of Sweetwater, Texas, killing pilot 2Lt. Marvin P. Duncan, 21. The airplane took off at 1335 from Avenger Field on a transition training flight. The airplane returned at about 1500 and was cleared to land on Runway 31. The airplane peeled off and entered the traffic pattern. The pilot overshot the turn from base to final. The airplane, with landing gear extended and flaps up, was ordered to go around. The pilot tightened the turn in an attempt to line up with the runway and the airplane stalled. The airplane flipped to an inverted position and then entered a spin to the left. The airplane remained in the spin until it smashed to the ground, killing the pilot instantly.

4-1-45C. Alamogordo, New Mexico. At 2022 MWT, a Boeing B-29B suffering an in-flight fire crashed six miles northeast of Alamogordo, New Mexico, killing 12 fliers. Gunner Pfc. Dean F. Bittner and flight engineer 2Lt. William C. Dolton parachuted to safety, escaping injury. Gunner Pfc. John M. Macbeth fell to his death in an unsuccessful parachute jump. Killed in the crash were: 2Lt. George C. Jackson, pilot; 2Lt. Donald K. Kilgore, co-pilot; 2Lt. Vernon A. Selph, bombardier; 2Lt. Neil T. Wiley, flight engineer; F/O Richard T. Lassell, navigator; SSgt. Howard Z. Wells, engineer; Sgt. William E. Stach, radio operator instructor; Cpl. Robert E. Kramer, gunner; Cpl. James J. Panek, Jr., gunner; Pvt. Webb L. Eady, gunnery instructor; Pvt. Robert F. Mullin, radio operator. Investigators stated, "A B-29 took off at 1500 MWT [from Alamogordo Army Air Field] and flew a high altitude formation mission for approximately five hours. About 2000 MWT, the pilot called the tower for landing information and was advised to land on Runway 25 to the southwest. A short time later the pilot called in on the downwind leg and was advised to fly a pattern for Runway 15 to the south. Pilot acknowledged and turned left to a north heading for the downwind leg of Runway 15. Wheels were down and flaps coming down when a violent fire broke out in the number-four engine. The alarm bell was rung and two crewmembers bailed out of the rear door. The airplane continued straight ahead in a dive and then made a right turn of approximately 180 degrees. Just before striking the ground it appeared to flare out slightly and then dove in, right wing first, exploded and burned." It was speculated that a fuel line failure caused the fire in the number-four engine nacelle. This large fire on the starboard wing caused a great loss of lift for that wing.

4-2-45A. Dodge City, Kansas. At 0850 CWT, a Martin TB-26G crashed two miles south of Dodge City Army Air Field, Dodge City, Kansas, killing four fliers. The airplane was flying in the number-three position of a three-ship flight that took off from Dodge City Army Air Field on a routine training mission. The formation flew for a short time and the number-two ship and the subject airplane switched positions in the formation. The flight encountered freezing rain and deteriorating visibility. Ice began to form on the airplanes' leading edges and windshields. The flight leader elected to return the flight to Dodge City Army Air Field for landing. The flight leader broke up the formation and the airplanes were ordered to land separately. The airplanes arrived over the field and began to peel off for landing. The subject airplane began its peel-off to the left and entered a vertical bank. The nose dropped and the airplane entered a descending left turn, turning about 180 degrees before smashing into the ground and exploding into flames. The crew was killed instantly. Investigation revealed that both engines were producing power at the time of the crash. The cause of the accident could not be determined. Killed in the crash were: 1Lt. Warren K. Knopf, B-26 instructor; 1Lt. Lowell J. Wolfe, B-26 student; 1Lt. Donald A. Walter, B-26 student; Pfc. Russell W. Roach, engineer.

4-2-45B. Hassavampa, Arizona. At 1530 MWT, a North American AT-6D crashed ten miles northwest of Hassavampa, Arizona, killing A/C Donald H. Lodde, 22. The airplane was flying in the number-three position of a three-ship flight that took off from Luke Field, Phoenix, Arizona, on a routine training flight, which was led by an instructor. The airplanes entered a Lufbery circle at an altitude of 4,000 feet msl. The subject airplane left the circle momentarily and then rejoined the circle. The pilot rolled his airplane past a vertical bank in an apparent attempt to tighten the turn. The airplane then dove out of the circle and the pilot attempted to recover but entered a high-speed stall. The airplane snapped over to an inverted attitude and the pilot attempted to recover with a split-S type maneuver. The airplane entered a spin, describing four turns of spin before the pilot was able to stop the rotation. The airplane smashed into the ground at a 45-degree angle an instant later, exploding into flames and killing the student instantly. The elevation at the crash site exceeded 1,500 feet msl.

4-2-45C. Charlotte, North Carolina. At 1810 EWT, a Douglas A-20J crashed two miles east of Charlotte, North Carolina, killing pilot 2Lt. Budd M. Andrews, 25. The airplane was flying in the number-two position of a six-ship flight that took off at 1650 from Morris Field, Charlotte, North Carolina, on a low-level formation mission. The formation was flying about 30 miles ESE of Charlotte when other pilots noticed puffs of white smoke emitting intermittently from the starboard engine of the subject airplane. The subject pilot acknowledged and left his position in the formation. The flight leader advised the subject pilot that he would lead him back to base. The puffs of white smoke continued to issue from the engine on the way back to the field, soon turning to brownish puffs of smoke. A short time later, the starboard engine began trailing black smoke and the engine was then observed to be on fire. The pilot attempted to feather the propeller but was unsuccessful. The flames quickly spread to the entire starboard nacelle, causing the nacelle to fail and the starboard engine to separate and fall free; the outer starboard wing also failed and separated. The airplane, which had been flying at about 1,500 feet agl, went out of control and rolled to the right, entering a spin to the right. The airplane then smashed to the ground at a 60-degree angle and exploded into flames, killing the pilot instantly. The starboard engine had struck a civilian house after falling free from the airplane, and the starboard wing was found about one-half mile from the main wreckage. Investigation revealed that the starboard engine's number-one cylinder piston pin and retainer spring had failed, allowing the piston and the master rod to fail and causing the articulating rods to be misaligned and to ultimately fail. Pieces of the master rod and the piston pin were found to have been forced through the side of the number-one cylinder. The number-one cylinder was found to have broken from its mounting flange.

4-2-45D. Eagle Pass, Texas. At 2335 CWT, a North American AT-6D crashed 16 miles NNW of Eagle Pass Army Air Field, Eagle Pass, Texas, killing student A/C Luke V. Dean, Jr., 19. The airplane had just taken off from Auxiliary Airfield #3 in preparation to perform student night landing practice. The airplane failed to return to the field and was not found until daylight hours the next day. The airplane apparently flew into the ground about one mile east of Auxiliary Airfield #3. Investigators speculated that the student had set his altimeter incorrectly or had suffered from vertigo, causing him to inadvertently fly the airplane into the ground. The airplane did not burn but was destroyed utterly.

4-3-45A. Los Angeles, California. At 0947, a Vultee BT-13B struck a high-tension wire and crashed on the east bank of the Los Angeles River, Los Angeles, California, killing pilot 1Lt. Donald R. Givens, 25. The airplane took off from Grand Central Air Terminal, Glendale, California, at 0945 on a local flight. The pilot took off from Runway 30, turned left and then flew to the southeast until over northern Los Angeles, flying in the vicinity of Fletcher Drive and San Fernando Road. The airplane was observed flying to the southeast at low altitude over Edward Avenue, parallel with San Fernando Road. The airplane, still flying at very low altitude, then executed a 270-degree left turn, the radius of which being about one-half mile, in an attempt to return to the same area. The airplane, heading in a westerly direction, was still rolling out of

this turn when the port wing struck a high-tension wire. The outer port wing was torn off and it struck the starboard horizontal stabilizer, causing it to fail and separate. The airplane then rolled out of control to the left, smashing into a sewage fume disposal tower on the east bank of the Los Angeles River about one-quarter mile from the high-tension line. The airplane burst into flames in the collision with the tower and then it smashed to earth on a dry portion of the Los Angeles River bed.

4-3-45B. Elizabeth City, North Carolina. At 1337 EWT, engineer Pvt. James R. Arnold was killed when he fell from a Douglas C-47B operating two and a half miles north of Elizabeth City, North Carolina. Investigators stated, "In this accident the crew chief fell to his death when, while in flight, not wearing a parachute, he attempted to replace the emergency exit door (para-door), which had been removed to facilitate aerial photography, and was pulled thru the doorway with the door by the suction of passing air. Position was marked on map and crew landed at Weeksville Coast Guard Air Station [North Carolina], and an immediate search was instituted and continued until 6 April when body was located." The C-47 was piloted by 1Lt. Clarence E. Stark. The pilot and four other crewmembers were uninjured.

4-3-45C. Yuma, Arizona. At 1612 PWT, a Bell P-39Q crashed 14 miles northeast of Yuma Army Air Field, Yuma, Arizona, killing pilot Capt. George A. Svereika, 29. The airplane took off at 1605 PWT from Yuma Army Air Field on a routine local flight. Another pilot observed the subject airplane flying at about 5,000 feet. The subject airplane broke away to the left and entered a split-S maneuver when the other aircraft approached and came within a half mile. The subject airplane entered a spin and the pilot was able to make a partial recovery. The airplane entered a flat spin and continued in this attitude until it struck the ground and exploded into flames. The exit door was found about 50 feet from the main wreckage. The pilot was found about 70 feet from the wreckage with his parachute partially deployed. The parachute D-ring was found on the ground in between the pilot's body and the main wreckage.

4-4-45A. Bartow, Florida. At 1145 EWT, a North American P-51B crashed at Bartow Army Air Field, Bartow, Florida, killing pilot 2Lt. Norman B. Smith, 20. The airplane took off at 1115 EWT from Bartow Army Air Field on a camera-gunnery mission that was to be conducted at 23,000 feet. At approximately 1145 EWT the aircraft, flying at about 18,000 feet, was observed to enter a steep dive. The wings failed at about 8,000 feet and the airplane came apart in the air. The pilot made no effort to bail out and was killed when the airplane smashed to the ground and exploded into flames. Investigators speculated that the pilot might have suffered an oxygen problem or had been overcome by fumes in the cockpit, causing him to lose control of the airplane and enter a steep dive. It was speculated that the wings failed during an attempted recovery from the dive or that one of the main landing gear fell out of its well and caused the airplane to break up.

4-4-45B. Tucumcari, New Mexico. At 1220 CWT, a Vultee BT-13B crashed two and a half miles northeast of Tucumcari, New Mexico, killing pilots 2Lt. Alden L. Kaylor and 2Lt. Glenn W. Dexter. The subject airplane took off at 1100 CWT from Dalhart Army Air Field, Dalhart, Texas, on an instrument training flight to Tucumcari Municipal Airport and return. Lt. Dexter was flying in the rear cockpit as the instrument pilot and Lt. Kaylor was flying in the front cockpit as the safety pilot on the leg to Tucumcari. The two pilots were scheduled to switch positions upon landing at Tucumcari. Shortly after take-off from Dalhart Army Air Field, the subject airplane joined in formation with two other BT-13 airplanes that were flying on similar missions to Tucumcari. At 1150 CWT, these three airplanes landed at Tucumcari and the pilots of each airplane switched positions. At 1205 CWT, the three airplanes performed a three-ship V-formation take-off to the northwest. The airplanes then made a 360-degree turn to the left and returned over the field heading northwest at very low altitude. After buzzing the field the airplanes climbed away, leveling off at about 1,000 feet agl. The flight leader signaled a change to an echelon formation to the right and down. The number-three ship (the subject airplane) moved over from the leader's port wing and took up position on the number-two ship's starboard wing. A short time later the lead airplane pulled up into a steep wingover to the left at about 1,000 feet agl. The lead ship rolled to an inverted position and the pilot recovered by performing a split-S type maneuver, recovering level flight at about 100 feet agl. The number-two ship attempted to follow the maneuver but the pilot realized that he did not have the altitude to safely complete the maneuver, so he recovered in a very steep left turn and returned to level flight. The number-three ship, which was stacked lower than the number-two ship and the lead ship, attempted to follow the maneuver, but the pilot was unable to recover before smashing into the ground in a vertical dive. The airplane exploded into flames upon impact. The official AAF aircraft accident report indicates that Lt. Kaylor was flying the aircraft at the time of the accident.

4-4-45C. Victoria, Texas. At 1630, a Lockheed P-38L suffered a catastrophic structural failure and crashed 14 miles NNW of Foster Field, Victoria, Texas, killing 2Lt. Wayne V. Kennedy, 22. The airplane was one of two P-38s that took off from Foster Field on a local flight. The pilot of the other P-38, which was a two-seat modification, advised the subject pilot to break off from the formation because he was going to demonstrate stalls to the P-38 passenger. The subject airplane broke off from the other ship and was next seen shedding pieces

at an altitude of about 5,000 feet. The airplane then entered an outside loop until reaching an inverted 45-degree dive, remaining in this inverted attitude until it struck the ground at 45-degree angle and exploded violently into flames. Wreckage was scattered over an area of 175 yards. Investigation revealed that the elevator's lower counter balance had separated in flight and caused extreme vibration in the tail section. These vibrations caused the failure and buckling of the elevator and horizontal stabilizer. This buckling of the horizontal stabilizer caused the rudder assemblies to fail and separate. Pieces of the tail section continued to float to earth following the crash.

4-4-45D. Rockingham, North Carolina. At 1745 EWT, a Douglas A-20H crashed 14 miles NNW of Rockingham, North Carolina, killing A-20 instructor pilot Capt. Robert J. Henricks, 27. The airplane was the lead ship of a six-ship flight that took off at 1600 EWT from Morris Field, Charlotte, North Carolina, on a low-level formation mission. At about 1735, other pilots in the flight noticed white puffs of smoke emitting from the flight leader's starboard engine. The flight leader separated from the formation, feathered the starboard propeller and trimmed the airplane for single-engine operation. The flight leader was able to gain about 100 feet of altitude and then made a 90-degree turn to the left. The pilot dropped the nose and headed for a small field to attempt an emergency forced landing. The pilot dropped flaps and made an approach. The airplane was seen to make a "fishtail" maneuver in an effort to shed airspeed. The pilot advanced the throttle to clear a stand of trees on the edge of the field. The airplane collided with trees before descending into the field still under control. The field was very rough and the airplane struck an embankment, tearing off the port propeller before the airplane bounded back into the air. The ship then smashed to the ground on the starboard wing, causing the separation of the propeller and engine. The airplane smashed itself to pieces and burst into flames as it continued to slide across the terrain, coming to rest about 300 feet from the initial point of contact. Investigation revealed that the pilot had not correctly fastened his shoulder harness and was killed when he collided with the interior of the airplane.

4-4-45E. Easton, Washington. At an unknown time after 1617 PWT, a Beech F-2 crashed 20 miles NNW of Easton, Washington, killing three fliers. The airplane disappeared on a flight from Mountain Home Army Air Field, Mountain Home, Idaho, to Yakima, Washington. The pilot initially wanted a clearance to Seattle but it was denied because of poor weather. The pilot was cleared to Yakima, with instructions to land and obtain current weather and clearance before attempting to fly on to Seattle. The airplane arrived at Yakima at 1617 but did not land. The pilot requested a clearance to Seattle. The pilot was cleared to Seattle via Ellensburg at 8,000 feet. The airplane failed to arrive at

Seattle and was declared missing. Weather on the flight path included low ceilings, strong turbulence and icing conditions. The airplane wreckage was found on 7-19-46. Investigation revealed that the airplane smashed into rising terrain while in a flying attitude. Pilot 2Lt. Richard T. Payne, passengers Sgt. Fred W. Pashley and Cpl. Richard E. Tobin were killed in the crash. Investigators noted that the pilot's wife was waiting to meet him in Seattle and probably motivated the pilot to continue flight into poor weather.

4-4-45F. Oklahoma City, Oklahoma. At 1942 CWT, a Lockheed F-5E (44-24579) attempting a landing crashed onto a parked F-5E at Will Rogers Field, Oklahoma City, Oklahoma, killing Chinese Air Force pilot Sub-Lt. Siao-Nien Wang. A pilot and a mechanic were killed on the ground. The subject airplane took off from Will Rogers Field at 1826 on a navigation flight to Wichita, Kansas, and return. At 1927, the pilot radioed in to the field that his starboard coolant shutters were inoperative in the automatic position. The Control Tower Officer instructed the pilot to put the coolant switch in the "over-ride" position and to come in for a landing. The pilot was cleared to land and given landing instructions. The pilot, making no mention of an emergency situation, peeled off and entered the traffic pattern in the normal fashion, calling in on the turn from the base leg to final approach. Investigators stated, "The final approach was normal but the pilot evidently used partial power even after he had flared out for what should have been a good landing. One-third of the distance down the runway, he was advised by the tower to cut his power but the aircraft continued along the runway at approximately ten feet and at only slightly more than landing speed. With more than half of the runway behind him the pilot gained a slight amount of altitude (not more than 20 feet) and the tower then advised him to go around. Simultaneously the aircraft began to climb steeply (estimated to be 50-65 degrees) and slide to the right, and crashed directly onto another F-5E parked 200 yards to the right of the runway in use. Both aircraft burst into flames from the impact, a pilot and mechanic occupying the parked aircraft were killed, and the operator of the landing F-5E was killed." Pilot 2Lt. Richard E. Thatcher was sitting in the cockpit of F-5E #44-24588 completing paperwork after returning from a navigation mission and mechanic Sgt. Edward J. Welk was kneeling on the port inboard wing of the ship when F-5E #44-24579 smashed onto the parked airplane.

4-5-45A. Location Unknown. At an unknown time after 1025 EWT, a North American P-51D disappeared and is presumed to have crashed into the Gulf of Mexico in the vicinity of Sarasota, Florida. Pilot 2Lt. John N. Harrell, 24, was missing and presumed lost at sea. The airplane was part of a two-ship flight that took off at 1025 EWT from Sarasota Army Air Field on an instrument training flight. Lt. Harrell was scheduled to fly as the safety pilot and 2Lt. William C. Day

was scheduled to fly as the instrument pilot. The airplanes climbed to 10,000 feet and leveled off. Lt. Day then went on instruments while Lt. Harrell fell back in trail to observe. After flying on instruments for several minutes, Lt. Day looked back and noticed that Lt. Harrell was no longer behind him. Lt. Day circled the area for several minutes but was unable to locate Lt. Harrell. Lt. Day then tried to contact Lt. Harrell by radio but was unsuccessful. Control tower personnel at Sarasota Army Air Field were also unable to contact Lt. Harrell. Lt. Day continued to circle the area until he was ordered to return to Sarasota Army Air Field where he landed without incident. No trace of the airplane or pilot was ever found.

4-5-45B. Richmond, Virginia. At 1440 EWT, a Republic P-47D dove into the ground 15 miles NNW of Richmond Army Air Base, Richmond, Virginia, killing pilot 2Lt. Charles C. Hoofnagle, 23. The airplane took off from Richmond Army Air Base at 1434 EWT on a transition flight. Investigators noted that this was the pilot's first solo flight in a P-47 airplane. The pilot, who had been thoroughly checked out on the airplane by an instructor, made a normal take-off, retracted the landing gear and turned left out of the traffic pattern. Moments later the airplane was seen in a vertical dive. The pilot was able to recover, pulling out at treetop level. The airplane climbed away and was seen heading north at about 2,000 feet. A couple minutes later the airplane was seen to enter a vertical dive at about 2,500 feet. The airplane remained in the vertical dive until it smashed into wooded terrain and exploded into flames. Witnesses stated that the wings "wobbled" and "dipped" just before the airplane entered the vertical dive. The cause of the crash could not be determined.

4-5-45C. Big Spring, Texas. At 1350 CWT, a Republic P-47D crashed nine miles east of the Army Air Field at Big Spring, Texas, killing pilot 2Lt. Richard D. Riley, 23. The airplane was part of a two-ship flight that took off from Avenger Field, Sweetwater, Texas, on an instrument-training mission. The subject airplane was flying as the instrument ship and the other P-47 was flying as the safety ship. The airplanes climbed to 10,000 feet and leveled off. The subject airplane then flew on the East leg of the Big Spring Radio Range. The subject airplane arrived at the cone of silence, turned left and then flew on the South leg for five minutes before performing a procedure turn and flying back to the cone of silence. The pilot turned east and announced that he was coming off of instruments. The subject airplane then executed a chandelle to the left, climbing to 12,500 feet. The subject airplane then dove to 7,500 feet, leveled off momentarily and then entered a steep turn to the left. The airplane then suddenly snap rolled to the right through inverted and entered a spin to the left, describing one turn of spin. The pilot recovered the airplane momentarily before entering a shallow turn to the right and then diving into the ground. The airplane burst into flames upon impact and the pilot was killed instantly.

4-5-45D. Girard, Kansas. At 1410 CWT, a Lockheed F-5E crashed seven miles NNW of Girard, Kansas, killing 2Lt. Erwin M. McMillen, 20. The airplane took off from Coffeyville Army Air Field, Coffeyville, Kansas, at 1300 and returned and landed at 1335. The airplane took off again at 1346. The airplane was seen flying to the north at very low altitude near the Girard. The airplane flew out of sight and then returned a few minutes later. The airplane was observed flying over a farm field at very low altitude and at high speed when the propellers struck the ground. The airplane, which was still airborne, struck a small tree and then entered a steep climb. The airplane stalled and then fell off on the starboard wing, entering a 70-degree dive. The airplane remained in this attitude until it struck the ground and exploded violently into flames.

4-5-45E. Luke Field, Arizona. At 1527 MWT, a North American AT-6B crashed at Luke Field, Phoenix, Arizona, killing Chinese Air Force student A/C Tsih-Shiang Woo, 19. The airplane was part of a three-ship flight that took off from Luke Field on a routine training mission. The subject airplane's engine started "missing" immediately after take-off. The airplane was at about 200 feet agl as it approached the departure end of the take-off runway. The airplane turned to the left and began to lose altitude and it leveled off just before striking the ground in a semi-flying attitude. The airplane burst into flames as it bounded back into the air five feet before smashing back to the ground.

4-6-45A. Location Unknown. At an unknown time after 0800 EWT, a North American P-51K disappeared and is presumed to have crashed into the Atlantic Ocean while on a routine training mission. Pilot 2Lt. Harold C. Ham, 20, was missing and presumed killed. The airplane was flying in the number-two position of a four-ship flight that took off from Sarasota Army Air Field, Sarasota, Florida, on a combat training mission. The flight encountered poor weather and the flight leader attempted to fly above it. The flight leader became separated from the flight and instructed the subject pilot to fly west to the coast and then south to reach visual conditions. The subject pilot acknowledged the message. The other airplanes in the flight were able to find visual conditions and returned to base. The subject airplane failed to return to the field and was declared missing. A massive search was unable to find a trace of the airplane or pilot.

4-6-45B. Athens, Alabama. At 1530, a Consolidated TB-24D crashed while attempting an emergency forced landing eight miles west of Athens, Alabama, killing engineer Pfc. Billy L. Margerum. Five crewmembers escaping injury were: 1Lt. Hart Sylvester, pilot; 1Lt. Donald V. Browne, co-pilot; 2Lt. William D. Thompson, navigator; Sgt. Edward J. Lovitt, Jr., engineer; Pvt. Edward C. Nabozny, engineer. The airplane

had taken off at 1440 CWT from Courtland Army Air Field, Courtland, Alabama, on an engineering test flight. The airplane climbed to 3,000 feet and then leveled off. The pilot checked the instruments and then checked the engines at different power settings for the engineering test flight. The pilot then allowed the navigator to sit in the pilot seat and take the controls while the pilot stood behind and between the pilot seats. The navigator flew the airplane for about 20 minutes, and then the pilot took over the co-pilot position. The co-pilot remained on the flight deck. The pilot, still sitting in the co-pilot seat, moved the number-three and number-four fuel mixture controls to "idle cut-off" and feathered both propellers simultaneously. The pilot re-trimmed the airplane and flew on two engines for a short time. The pilot then unfeathered the propellers but the airplane kept veering to the right because of the differential power. The pilot was unable to get power out of the number-three and four engines and the airplane began to lose altitude rapidly. The pilot alerted the crew that a crash landing was imminent. The two crewmembers standing on the flight deck braced themselves for crash landing and the navigator remained in the pilot seat. The pilot extended the flaps as he made a long approach for a clear area. The pilot extended the landing gear just before the airplane made contact with the ground, but the landing gear did not extend fully because the hydraulic pump was not operating as a result of the lack of power in the number-three engine. The pilot moved the number-one and number-two mixture controls to idle cut-off just as the airplane touched down. The airplane slid across the terrain for 250 yards, sliding sideways for a short distance and then turning 180 degrees from the line of flight before coming to rest. The airplane did not burn but was damaged beyond repair. Investigation revealed that the pilot had failed to return the number-three and number-four fuel mixture controls to their normal positions when he was attempting to restart the number-three and number-four engines.

4-7-45. Lawrenceville, Illinois. At 1348, a Curtiss C-46A attempting a take-off crashed two miles southwest of George Field, Lawrenceville, Illinois, killing pilot 1Lt. Donald L. Ferrier and co-pilot 1Lt. Hadley R. Edmondson. The airplane took off from Runway 23 and climbed to about 400 feet before diving into the ground and exploding into flames. Investigation revealed that the port elevator lock was still in place. Investigators noted that the pilots took off before a crew chief could get out to the airplane.

4-8-45. Tellico Plains, Tennessee. At 0030 EWT, a Boeing TB-17F flew into rising terrain 27 miles northeast of Tellico Plains, Tennessee, killing ten fliers. The airplane took off from Keesler Field, Biloxi, Mississippi, on a navigation training flight. The airplane was scheduled to fly a magnetic course of 45 degrees for two and a half hours and then it was to turn around and fly the reciprocal course back to Keesler Field. The airplane failed to return to Keesler Field. The airplane was found to have collided with a 5,500-foot mountain at an elevation of 5,200 feet, exploding violently into flames upon impact. Investigation revealed that the airplane was flying straight and level on a heading of 340 degrees with all four engines producing power at cruise rpm when it flew into the side of the mountain. Killed in the crash were: 2Lt. Richard C. Henkel, pilot; 2Lt. Carl E. Lehnhardt, co-pilot; 2Lt. Walter W. Foley, navigator; 1Lt. David K. Croker, navigator instructor; SSgt. Edwin W. Bill, engineer; Cpl. Raymond S. Contino, radio operator; Cpl. John E. Stair, radio operator; Cpl. Westly J. Eaton, gunner; Cpl. John J. Fortune, gunner; Cpl. Fred Hoff, gunner.

4-9-45. Pinellas Army Air Field, Florida. At 1645 EWT, a Curtiss P-40N crashed two and a half miles northwest of Pinellas Army Air Field, St. Petersburg, Florida, killing pilot Capt. James S. Varnell, 23. The airplane had taken off from Pinellas Army Air Field on a routine gunnery mission. The airplane had completed the mission and was returning to the field when it was observed to go into a vertical dive at about 1,000 feet agl. The airplane remained in this attitude until it struck the ground and exploded. The cause of the crash could not be determined.

4-10-45A. Maywood, Nebraska. At 0400 MWT, a Boeing TB-17F flew into the ground 11 miles west of Maywood, Nebraska, killing five fliers and injuring four others, three of them seriously. Five fliers escaped injury. The B-17 had taken off from Maxwell Field, Montgomery, Alabama, on a flight to McCook Army Air Field, McCook, Nebraska, via Lambert Field, St. Louis, Missouri. The airplane had been flown to Maxwell Field to pick up a B-29 crew that had been grounded there because of mechanical problems. The airplane took off from Lambert Field at 0030 MWT on a visual clearance to McCook Army Air Field at an altitude of 6,000 feet. Because of adverse winds, the airplane was north of its intended course. The airplane approached Grand Island Army Air Field and attempted to contact the Grand Island Radio Range for an altimeter setting but was unsuccessful. The airplane contacted the Grand Island Army Air Field control tower and received an altimeter setting. The airplane approached North Platte and attempted to contact North Platte Radio Range but again was unsuccessful. The airplane then turned to the south on a heading to McCook Army Air Field, letting down to 3,900 feet. The pilots radioed McCook Army Air Field for landing instructions. The ship encountered very turbulent air as it began letting down in preparation for entering traffic at McCook Army Air Field. The pilots began to have difficulty holding the airplane straight and level in the turbulent air. Moments later, the airplane descended into the ground in a flying attitude. The airplane bounded back into the air and then smashed back to earth on its nose, crushing the compartment. The airplane bounded back into the air for

a short distance before smashing back to the ground. The airplane began sliding to the left, the fuselage breaking apart near the radio room. The airplane burst into flames as it came to rest about 500 yards from the point of initial contact. Investigators speculated that the pilots might have been given an erroneous altimeter setting by controllers at Grand Island Army Air Field. Killed in crash were: 1Lt. Earle R.G. Smith, bombardier; 1Lt. Charles P. Armstrong, passenger; 1Lt. Virgil H. Jordan, passenger; Capt. Charles L. Hynds, passenger; TSgt. Bernard P. Greeley, passenger. Crewmembers seriously injured were: 1Lt. Allan J. Nuszloch, navigator; TSgt. Francis K. Horan, engineer; Cpl. William H. Davies, gunner; Cpl. Irvin O. Mertz, passenger. Gunner Cpl. Leon S. Mieczkowski sustained minor injuries. Crewmembers escaping injury were: Lt. Col. Hadley V. Saehlenou, pilot; Maj. Gerald G. Crosson, co-pilot; Cpl. Richard A. Miesch, gunner; 2Lt. George M. Withee, passenger.

4-10-45B. Paxton, Illinois. At 0840, a Vultee BT-13A crashed eight miles WNW of Paxton, Illinois, killing pilot 2Lt. Robert Miksch, 25, in an unsuccessful parachute attempt. The airplane took off at 0737 CWT from Chanute Field, Rantoul, Illinois, on a local flight. At 0840 the airplane was observed rolling to the right before it entered a spiral to the right. The pilot was unable to recover so he bailed out. The pilot deployed his parachute but he was struck in the head by the port wing, killing him instantly. The dead pilot floated to earth in the deployed parachute.

4-10-45C. Dillon, South Carolina. At 1643, a Vultee-Stinson L-5 crashed four miles north of Dillon, South Carolina, killing glider pilots F/O James W. Schuster and F/O Edward E. Wyatt. The airplane was part of an eight-ship flight that had taken off from Laurinburg-Maxton Army Air Field, Maxton, North Carolina, on a routine training mission. The subject airplane fell out of formation and was later observed maneuvering in an attempt to make an emergency forced landing in a small field. The airplane stalled at an altitude too low to allow recovery while turning to make an approach to the field. The airplane did not burn but was completely destroyed. Investigation revealed that the pilot had run one tank dry and had failed to switch to the port tank, which contained enough fuel to allow a safe completion of the flight.

4-11-45A. Jackson, Mississippi. At 0007 CWT, two Boeing TB-17F airplanes collided in midair and crashed three miles northeast of Jackson Army Air Base, Jackson, Mississippi, killing 14 fliers. Killed in the crash of TB-17F #42-30633 were: 1Lt. George J. Canavan, pilot; 1Lt. John C. Stevens, co-pilot; 1Lt. Daniel B. Sullivan, navigator; F/O Jack C. Ewing, student navigator; F/O Harold Lisdell, Jr., student navigator; SSgt. George C. Brooks, engineer; Pvt. August N. Lilienkamp, radio operator. Killed in the crash of TB-17F #42-6010 were: 1Lt. Carl J. Shane, pilot; 1Lt. Charles J. Keathley, co-pilot; Maj. John F. Cunningham, navigator; F/O Gunter G.E. Trost, student navigator; F/O Michael B. Ellliot, student navigator; Sgt. Joseph A. Ott, engineer; SSgt. Walter L. Lark, radio operator. Both airplanes were returning to Jackson Army Air Base after completing separate eight-hour navigation training missions. The two subject airplanes and a third airplane, B-17 #42-6016, were in the traffic pattern. Investigators stated, "[B-17 #42-6016] was cleared to land and B-17 #42-6010 had reported position as on base leg and number two to land. B-17 #42-30633 had contacted the tower, and asked for landing instructions but gave no position report.... [It] is apparent that B-17 #42-30633 cut inside of B-17 #42-6010 on the base leg and struck B-17 #42-6010 while turning on the base leg and letting down from an approximate altitude of 1,500 feet on downwind and 1,000 feet on the base leg. Both witnesses stated that they were able to see only the red running light showing on B-17 #42-6010. The other running lights, passing lights, and recognition lights were apparently not operating. It is the opinion of the board that the pilot of B-17 #42-30633 did not see B-17 #42-6010, since there was obviously only one running light showing, and that his view was blanked out while turning and letting down from his downwind leg to his base leg. B-17 #42-30633 apparently struck B-17 #42-6010 amidships. Parts of radio equipment normally placed in the mid-section of a B-17 were found tangled in the landing gear of B-17 #42-30633. Both planes burst into flames and fell to the ground within 300 to 400 yards apart from each other."

4-11-45B. Gulf of Mexico, Florida. At 0922 EWT, a Curtiss P-40N collided with a towed target and crashed into the Gulf of Mexico approximately 20 miles southwest of Perry Army Air Field, Perry, Florida, killing pilot 2Lt. Stuart S. Gottleib, 20. The airplane had taken off from Thomasville Army Air Field, Thomasville, Georgia, on a camera-gunnery mission. The airplane rendezvoused with a target-towing airplane and began making "high side" passes from the right side at a sleeve target at about 5,000 feet. The subject airplane's starboard wing collided with the tow cable and it separated from the tow ship and fouled on the subject airplane. The subject airplane, with the tangled target trailing behind, went out of control and dove into the sea.

4-11-45C. Location Unknown. At an unknown time after 1115 EWT, a Consolidated B-24J disappeared and is presumed to have crashed into the Atlantic Ocean. Eleven crewmembers were missing and presumed lost at sea. The airplane was part of a five-ship flight that took off at 0709 EWT from Charleston Army Air Field, Charleston, South Carolina, on a routine training flight. At 1045 EWT, the formation broke up because of poor weather. The subject airplane was last seen 68 miles southeast of Charleston Army Air Field at about 1115 EWT. The airplane failed to return to base and was declared missing. Missing and

presumed lost at sea were: 1Lt. Edward L. Madell, B-24 instructor; F/O Charles S. Pitt, pilot; F/O Eugene S. Moore, co-pilot; 2Lt. Allen I. Klarfaen, navigator; F/O Richard A. Goff, bombardier; Sgt. James H. Ressler, engineer; Sgt. Hector A. Glotfelter, radio operator; Sgt. Jack K. Jones, gunner; Cpl. Anthony F. Souza, gunner; Cpl. Harold R. Luke, gunner; Cpl. Clifford L. Meeks, gunner.

4-11-45D. Farmington, Connecticut. At 1400 EWT, a Republic TP-47D crashed three miles west of Farmington, Connecticut, killing 2Lt. Vincent H. Core, 20. The P-47 was part of a two-ship flight that took off from Bradley Field, Windsor Locks, Connecticut, on an instrument-training mission. The other pilot in the flight was 2Lt. Ervin J. Maveal. The airplanes climbed to 6,500 feet, leveled off and then the pilots went on oxygen. Lt. Maveal began flying on instruments and the subject pilot acted as the safety observer. Lt. Maveal flew on instruments for about 15 minutes and then the pilots switched positions. The airplanes flew in this position for about 15 minutes before the subject pilot went off of instruments. The airplanes then turned to the west, climbed to about 9,000 feet and then Lt. Maveal went back on instruments. Lt. Maveal climbed to about 11,000 feet while on instruments and attempted to communicate with the subject pilot, but he received no answer. Lt. Maveal looked back and observed that Lt. Core was not in position. Witnesses on the ground observed the subject airplane crashing into a wooded area, exploding violently into flames upon impact. Investigation revealed that the airplane was in a vertical dive at the time of impact. Investigators noted that the testimony of civilian witnesses, who were unfamiliar with aviation matters, was contradictory and confused, and determined to be unreliable. The Accident Classification Committee stated that investigation of the wreckage and surrounding terrain revealed that the pilot did in no way deliberately steer the airplane to crash in the woods and away from populated areas, contrary to published eyewitness accounts of the accident. Investigators were unable to determine what caused the airplane to dive into the ground.

4-11-45E. Clearwater, Florida. At 1530 EWT, a North American P-51B crashed five miles southwest of Clearwater, Florida, killing pilot F/O John C. Farmakis, 21. The airplane was part of a flight that took off from Bartow Army Air Field, Bartow, Florida, on a ground gunnery mission. The airplane had made three passes when the range officer radioed the pilot and advised him to make level turns before heading on the approach, so that the passes would not be so low. The pilot again made a descending turn onto the approach and was flying very low when the range officer radioed him to "pull up." The pilot did not reply. The airplane was observed to pull up steeply, entering a high-speed stall. The airplane fell off and collided with trees. The airplane then "skipped" across a small lagoon, bounding back into the air for a short distance before crashing onto the range and breaking up. The airplane did not burn but was completely destroyed.

4-11-45F. Gunter Field, Alabama. At 1650 CWT, two North American AT-6C airplanes collided in mid-air one and a half miles northeast of Gunter Field, Montgomery, Alabama, killing instructor 1Lt. Joseph F. Mochen, 27, aboard AT-6C #42-44292. French Air Force student Cpl. Claude Astier, 21, was uninjured and able to safely land AT-6C #42-32747 at Gunter Field. The airplanes were part of a three-ship flight that took off from Gunter Field on a routine training flight. Cpl. Astier was attempting to assume the number-three position when he collided with the port wing of the lead aircraft, which was piloted by Lt. Mochen. Lt. Mochen bailed out but was struck on the head by the tail section, knocking him unconscious and preventing him from deploying his parachute.

4-11-45G. Douglas, Arizona. At 1704 MWT, a North American TB-25D crashed near Douglas Army Air Field, Douglas, Arizona, killing pilot 1Lt. Cady L. Daniels and co-pilot 2Lt. Robert C. West. The airplane had taken off from Douglas Army Air Field on a routine training flight. The airplane had made three take-offs and landings and had taken off a fourth time. Moments later the airplane was observed flying over the field at about 2,000 feet with the port engine trailing fire and black smoke. The airplane was observed flying south of the field when pieces were observed to fall from the airplane. Moments later, the port engine separated from the port wing and fell to earth. A parachute was seen to billow from the aircraft and then fall away. The airplane entered a flat spin to the left and remained in this position until it smashed to the ground and exploded into flames. Investigation revealed that Lt. West had deployed his parachute before clearing the airplane and was killed when he collided with the tail section.

4-12-45A. Tullahoma, Tennessee. At 0830 CWT, a Consolidated TB-24J crashed at William Northern Field, Tullahoma, Tennessee, killing the crew of three. The airplane had taken off at 0748 CWT from Smyrna Army Air Field, Smyrna, Tennessee, on a training flight to Northern Field. The airplane arrived at Northern Field at 0822 and entered the far downwind leg of the traffic pattern, which was to the right, and began a 180-degree turn to the runway heading. Investigators stated, "In the first turn to the right, the airplane was leveled out and the landing gear started to come down. The airplane was in a slightly nose high attitude and the first 180-degree turn was nearly completed when the left wing was observed to drop into a steep bank to the left, approximately 70 degrees. The airplane began to descend rapidly; when the wings were leveled, almost immediately the right wing dropped into a nearly vertical bank. The left wing dropped again and was brought nearly back to level when the airplane crashed into the trees and exploded.... It is the opinion

of this board that the airspeed was allowed to get low and attention diverted to the performance of the landing check list, turning to the runway heading, and lowering of the landing gear, resulting in a higher stalling speed. In attempting to recover from the first stall, it is believed that the stall was aggravated by over-controlling, preventing recovery." Pilot 2Lt. William A. Corley, co-pilot 2Lt. Earl W. Clift and engineer TSgt. John H. Conway were killed in the crash.

4-12-45B. Fort Sumner, New Mexico. At 0845 MWT, a Curtiss P-40N crashed 13 miles SSW of the Army Air Field at Fort Sumner, New Mexico, killing pilot 2Lt. Weldon E. Duty, 21. Investigators stated, "The pilot was on a camera-gunnery mission. While returning from the mission, Lt. Duty sighted a [Boeing] B-29 and pulled away from his wingman to make a pass at it. He came over the top of the B-29, which was flying at approximately 9,000 feet indicated, pulled up into a loop and rolled to get into position for an overhead pass. Lt. Duty made his pass but before recovery was complete the plane hit the ground."

4-12-45C. Coffeyville Army Air Field, Kansas. At 1010 CWT, a Lockheed F-5E and an F-5G collided in mid-air at Coffeyville Army Air Field, Coffeyville, Kansas, killing F-5E pilot 2Lt. Ernest E. Linville, 26. F-5G pilot 2Lt. Ronald L. Cooksey, 25, was uninjured and able to land his damaged airplane at Coffeyville Army Air Field. The airplanes were returning to the field after completing a training mission and had peeled up to enter the traffic pattern. While turning left from the downwind leg to the base leg, the F-5G, which had been flying about 100 yards to the right of the F-5E, began turning inside the F-5E as it overtook the F-5E in the turn. The F-5G's port engine collided with the starboard boom of the F-5E, severing the boom and sending the F-5E out of control. The F-5E smashed to the ground and exploded, killing the pilot instantly. The F-5G was able to make a safe landing.

4-12-45D. Columbia, South Carolina. At 1513 EWT, a Douglas A-26B crashed one mile north-west of Columbia Army Air Base, Columbia, South Carolina, killing gunner SSgt. Herbert A. Newton. Pilot 1Lt. Donald J. Hicks and navigator/bombardier 2Lt. Harry J. Goodwin were seriously injured. The airplane had taken off on a training mission and was climbing away when the navigator noticed the port engine cowling was "revolving around" the port engine. The pilot cut the port engine and feathered the propeller, and moments later the port cowling blew off. The airplane began losing altitude and the pilot increased rpm on the starboard engine. The pilot reentered the traffic pattern, turning into the dead engine. The pilot was able to line up for Runway 16 and lowered the landing gear, but the airspeed began to drop. The pilot was unable to make the field and the airplane smashed to the ground about a mile from the end of the runway. The landing gear collapsed on the rough terrain and the A-26 smashed to the ground, erupting into flames. Columbia Army Air Base rescue personnel were on the scene immediately and were able to pull two airmen to safety. Investigation revealed that the port cowling was not properly secured prior to the commencement of the flight.

4-12-45E. DeBerry, Texas. At 1715 CWT, a Boeing B-29 suffered a catastrophic structural failure of the tail section and crashed three miles east of De-Berry, Texas, killing nine fliers. Killed in the crash were: Maj. Walter H. Greer, B-29 instructor; 1Lt. Bill M. Baker, pilot; 2Lt. Ralph M. Johnson, co-pilot; 1Lt. Everette R. Lewis, navigator; 2Lt. Mearl M. Diedrich, flight engineer; TSgt. Charles W. Spriggs, gunner; TSgt. Vernon L. Michaels, gunner; TSgt. William J. Stephens, gunner; 1Lt. John Greenfield, pilot-rated passenger. Investigators stated, "Subject aircraft took off at 1355 CWT [from Barksdale Field, Shreveport, Louisiana] on a scheduled hooded instrument flight check. The aircraft was last seen under control in straight and level flight at the completion of a turn at approximately 2,000 feet altitude. The aircraft suddenly fell off to the left and dove into the ground, first striking the ground on the left wingtip and ending up on the backside of the fuselage. At the start of the dive, silver objects were observed flying from the aircraft. During the dive, a sudden surge of power was heard. The aircraft crashed in a heavily wooded area, cutting few trees on the way into the ground. This indicates the aircraft was in practically a vertical position upon striking the ground." The airplane exploded into flames upon impact. Investigation revealed that the starboard horizontal stabilizer and elevator failed and separated; the port horizontal stabilizer and elevator also failed and separated. Pieces of the tail section were found over a quarter mile from the main wreckage.

4-12-45F. Location Unknown. At an unknown time after 2103 CWT, a North American AT-6D disappeared while on a flight from Kelly Field, San Antonio, Texas, to Harlingen Army Air Field, Harlingen, Texas. Pilot 2Lt. Jack Beatty, 23, was missing and presumed killed. It was speculated that the airplane crashed into the Gulf of Mexico.

4-12-45G. Fort Green, Florida. At 2245 EWT, a North American P-51C crashed one and a half miles northwest of Fort Green, Florida, killing 2Lt. Joseph L. Olson, 21. The airplane was flying in the number-two position of a three-ship flight that took off at 2215 EWT from Venice Army Air Field, Venice, Florida, on an authorized night formation mission. The airplanes climbed to about 6,500 feet when the subject airplane fell out of formation and entered a spiral. The airplane remained in this attitude until it struck the ground and exploded.

4-13-45. Collinsville, Oklahoma. At 2000 CWT, a Beech AT-7C flying in poor weather flew into the ground two miles northeast of Collinsville, Oklahoma, killing four fliers. Passenger Sgt. William A. Reiser survived with serious injuries. The airplane took off from Chicago Municipal Airport, Chicago, Illinois, on a flight to Tulsa, Oklahoma. The airplane encountered severe thunderstorms about 150 miles from Tulsa. The pilot was able to circumnavigate around the storms but encountered very turbulent air. About 20 miles from Tulsa, the airplane again encountered a line of severe thunderstorms. These storms contained heavy rain showers and strong winds. Visibility became very poor and the pilot circled the area, attempting to identify a landmark on the ground. The AT-7 entered instrument conditions and the pilot attempted to stay underneath the very low overcast. The airplane collided with the ground at flying speed, smashing itself to pieces as it slid across the terrain. The airplane burst into flames as it came to rest 200 yards further on. Killed in the crash were: 2Lt. James R. Edwards, pilot; 1Lt. Chester F. Krecisz, navigator; 1Lt. Leonard H.J. Daigle, navigator; Sgt. Chester B. Jankowski, engineer.

4-14-45. Binger, Oklahoma. At 1230 CWT, a North American TB-25G flying in poor weather suffered a catastrophic structural failure and crashed eight miles east of Binger, Oklahoma, killing five fliers. The airplane took off from Frederick, Oklahoma, on an instrument flight to Sioux Falls, South Dakota. The airplane crashed 32 minutes after take-off. Investigation revealed that the outer port wing failed and separated. Witnesses observed the airplane "fluttering" to the ground out of control. Wreckage was scattered over an area of one mile. Killed in the crash were: 2Lt. Delbart Foster, pilot; 2Lt. Roy L. Wolfe, co-pilot; 2Lt. Gerald L. Adams, passenger; 2Lt. Mitar M. Antunovics, passenger; 2Lt. Donald C. Heyn, passenger.

4-15-45A. Eagle Pass, Texas. At 1040 CWT, a Bell P-63A suffered a catastrophic structural failure while maneuvering at Eagle Pass Army Air Field, Eagle Pass, Texas, killing pilot 2Lt. Percy Bingham, 22. The pilot had landed at Eagle Pass Army Air Field after a flight from his home station at Pyote Army Air Field, Pyote, Texas. The pilot was invited by the base commander to participate in a fly-by aerial review being conducted for Memorial Services for President Franklin D. Roosevelt and to celebrate the final graduating class at Eagle Pass Army Air Field. The pilot was instructed to fly by the reviewing stand, perform a 180-degree turn and then fly past the reviewing stand a second time. Investigators stated, "Lt. Bingham took off and circled south of the field until he received the signal to come in, and then climbed to an altitude of approximately 5,000 feet, made a diving turn toward the field from the south end of the ramp. He leveled out at about 75 feet and flew across the ramp heading north at estimated airspeed of 450 mph or more. About 200 yards north of the reviewing stand he attempted to execute a steep pull-up straight ahead, and when he had attained a nose-up attitude of about 45 degrees and an altitude of 100 to 200 feet, both wings failed almost simultaneously and became detached.... The tail section and empennage became detached immediately after the wings, due to the being struck by the right wing. The fuselage continued to rise for about 400 feet, revolving to the right, and then tumbled to the ground about 1,000 yards from the spot where the wings became detached. The engine and fuselage caught fire upon contact with the ground.... The left wing, and the right wing tank were found on Runway 8 about 50 to 100 yards north of the spot where they became detached. The right wing, the right landing wheel, and the tail section were scattered about 50 yards further north. Numerous small pieces of metal and fabric were also found in that area. The remainder of the aircraft was 600 to 700 yards further north, outside the boundary of the field." Investigators noted that the pilot had flown in the previous day and was allowed to stay overnight at his parents' home at Eagle Pass.

4-15-45B. Rock Hill, South Carolina. At 1400 EWT, a Douglas A-20G crashed four miles northwest of Rock Hill, South Carolina, killing pilot F/O Joseph S. Chandler, 20. The airplane took off from Morris Field, Charlotte, North Carolina, on a routine transition flight. The subject airplane joined up in an unauthorized formation with another A-20 airplane that was on a similar mission. The subject pilot followed the maneuvers of the other A-20, which was piloted by a squadron mate. The subject airplane stalled out of a steep left turn and snapped over to an inverted attitude. The airplane then entered a flat spin and remained in this attitude until it struck the ground and exploded. The other pilot denied that the two pilots had pre-arranged a meeting in the air. The other pilot stated that he began evasive maneuvers when he noticed the subject airplane following him and was unaware of the crash.

4-15-45C. Tuskegee, Alabama. At 1615, a North American AT-6D crashed 14 miles southeast of Tuskegee, Alabama, killing pilot Capt. Wendell O. Pruitt and passenger Pvt. Edward N. Thompson. The airplane took off from Tuskegee Institute Airfield at 1524 and began maneuvering in the area of the field. The airplane dove at the field from about 1,000 feet agl, leveled off and then began a slow roll over the field. The airplane stalled in an inverted position and the nose pitched toward the ground. The pilot was unable to recover and the airplane smashed to the ground, exploding violently into flames upon impact. The occupants were killed instantly.

4-16-45A. Fairbanks, Texas. At 0925, a Curtiss P-40N crashed four miles northwest of Fairbanks, Texas, killing pilot 2Lt. Baron D. Risling, 22. The airplane took off at 0828 from Foster Field, Victoria, Texas, on a routine local transition flight. The airplane was flying at about 5,000 feet when it observed to go into a vertical dive from which the pilot did not recover. The airplane smashed into the ground at high speed and exploded violently into flames upon impact. Investigators noted that the pilot had had a bad reaction to a round

of inoculation injections that were administered on 4-13-45, and had been running a fever until the day before the accident. The pilot was found to be in good health and was cleared for flying by the flight surgeon. Investigation revealed that the pilot had adequate sleep the night before the accident and had remarked to his roommate that he "really felt like flying." It was also noted that the crash occurred more than 100 miles from the boundary of the "local" flying area.

4-16-45B. Location Unknown. At an unknown time after 1440 EWT, a North American P-51K disappeared after taking off from Sarasota Army Air Field, Sarasota, Florida. Pilot 2Lt. Ishmael L. Abbott, 29, was missing and presumed killed. Investigators stated, "Pilot was on a scheduled high-altitude gunnery mission. He was to wait and join with the second aircraft to finish firing, he being the first to finish. Lt. Abbott called the tow ship saying that he had finished firing and peeled off. No other pilot in the formation saw him after that."

4-16-45C. Big Spring, Texas. At 2330, two Beech AT-11 airplanes collided in mid-air 20 miles south of Big Spring, Texas, killing ten fliers. The airplanes had taken off from Big Spring Army Air Field on a bombing training mission. The airplanes were flying on the AAF Bombing Range at 4,000 feet agl on a triangular pattern with illuminated targets at the three corners. The lights at Target #2 went out, so the range controller directed the airplanes to fly an oval and head back to Target #3. The two airplanes dropped bombs on Target #3 and were turning away when the collision occurred. The airplanes broke up and burst into flames in the collision. Wreckage and bodies were scattered over an area of one and a half miles. Killed in the crash of AT-11 #42-36839 were: 2Lt. Neil C. Hildebrand, pilot; 2Lt. George E. Blake, bombardier instructor; A/C Chester W. Hilgendorf, student bombardier; A/C Clyde E. Jackson, student bombardier; A/C Joseph W. Kerice, student bombardier. Killed in the crash of AT-11 #42-37635 were: 2Lt. Rowland H. Meade, pilot; F/O John Marence, Jr., bombardier instructor; A/C Leonard C. Shoberg, student bombardier; A/C Glido Smaniotto, student bombardier; A/C John W. Thompson, Jr., student bombardier.

4-17-45A. Punta Gorda, Florida. At 1005 EWT, a Boeing B-29 crashed 15 miles north of Punta Gorda, Florida, killing six fliers. Seven fliers were able to parachute to safety. The airplane took off from MacDill Field, Tampa, Florida, on a high-altitude camera bombing and gunnery mission. Investigators stated, "The take-off and climb were uneventful.... While in trail formation flying number-three of a three-aircraft flight at 30,000 feet, a high whining noise emanated from the number-one engine. Tachometer reading went to 4,500 rpm and both oil pressures dropped to 0 PSI. The pilot pushed the feathering button and ordered the engineer to use feather procedure. During this period the left scanner reported fire in the number-one engine. The engineer used his normal fire procedure, releasing both CO2 bottles, one after the other. The propeller remained out of control. The fire was not put out. The pilot was diving and slipping to try to extinguish the flames. At 25,000 feet, pilot told co-pilot to order the crew to prepare to bail out. Co-pilot did so on interphone and received an acknowledgement from the rear. F/O Mrkanich reported all in the rear were preparing to bail out. Navigator and Instructor Bombardier attempted to open pressure door of nose at 22,000 feet unsuccessfully because of pressure. Pilot was asked to depressurize which was done at 15,000 feet. At 15,000 feet pilot told co-pilot to order crew to bail out. Co-pilot states he gave bail out order on interphone. Personnel in other aircraft formation and on ground stations heard the bail out signal on VHF. Everyone in the nose bailed out except pilot, co-pilot and engineer, who elected to remain and attempt to land aircraft at Punta Gorda Army Air Field. An acknowledgement was not received from anyone in the rear. The co-pilot states he rang alarm bell once. Te aircraft was flying toward Punta Gorda, wheels down and descending at 800 feet per minute. The fire was going out and starting intermittently and at 5,000 feet the pilot ordered the remaining crewmembers on board to bail out. The engineer went first, followed by the co-pilot, who saw the engine fall off as he bailed out. The pilot left at 3,500 feet. He called on interphone once before he left, receiving no reply. He was under the impression that he was the last man out." Killed in the crash were: 2Lt. Ferdinand J. St. Angelo, navigator; F/O John G. Mrkanich, radar operator; Sgt. James E. Hall, engineer; Cpl. Richard J. Hofmann, gunner; Cpl. Lewis G. Linder, gunner; Cpl. Dewey R. Meyers, gunner. Parachuting to safety and escaping injury were: Capt. Ray F. Detwiler, pilot; 2Lt. William E. Crawford, co-pilot; 2Lt. Kenneth B. Cearley, navigator; 2Lt. Miles O. Culver, bombardier; SSgt. George C. Shuetta, engineer; Cpl. John P. Hanley, radio operator; Capt. James W. Rucker, bombardier instructor.

4-17-45B. Long Beach, California. At 1620, a Lockheed P-38L crashed three miles south of Long Beach Army Air Field, Long Beach, California, killing pilot 1Lt. Robert H. Hartung, 22. The airplane took off from Long Beach Army Air Field and climbed to about 1,000 feet. At this point the port propeller was heard to "run away." The airplane entered a descending turn to the left, remaining in this attitude until it struck the ground and exploded. Investigation revealed that the port propeller was in full low pitch. The cause of the apparent port propeller failure could not be determined.

4-18-45A. Long Island Sound, Connecticut. At 0840, a Republic P-47D crashed into Long Island Sound five miles south of Woodmount, Connecticut, killing pilot F/O Salvatore Spano, 21. The airplane had taken off from Suffolk County Army Air Field, New York, on a navigation mission. While flying over Long Island Sound on the leg from Gales Ferry, Connecticut, to a point west of Norwalk, Connecticut, the pilot descended to low altitude. Moments later, the airplane collided with the surface of Long Island Sound and then bounded back into the air. The airplane was observed in a nose high attitude as it descended back toward the water. A P-47 pilot flying nearby advised the subject pilot to get the nose down. The pilot lowered the nose and leveled off just above the surface. The airplane then pitched downward and hit the water nose first; the airplane sank immediately. When the airplane was raised, the pilot was found in the cockpit. Investigation revealed that the pilot's seat belt and shoulder harness were not fastened, causing the pilot to collide with the interior of the airplane when the airplane crashed into the water the second time.

4-18-45B. Chesterfield, South Carolina. At 1445, a Douglas A-20J crashed 10 miles south of Chesterfield, South Carolina, killing three fliers. The airplane was part of a three ship flight that took off at 1410 from Morris Field, Charlotte, North Carolina, on a navigation flight to Myrtle Beach, South Carolina, to Wilmington, North Carolina, and return to Morris Field. When the formation was in the vicinity of Pageland, South Carolina, the subject pilot radioed the flight leader and advised that he had "gas in the cockpit." The subject pilot opened his bomb doors and changed fuel tanks. The flight leader was observing and attempted to radio the subject airplane but was unsuccessful. The pilot was able to receive but unable to transmit, signaling by rocking his wings for the flight leader. The subject airplane then entered a climbing turn to the left, gaining about 800 feet through 180 degrees of turn, rolling out and leveling off at about 3,500 feet indicated altitude. The flight leader advised the subject pilot that Hartsville Army Air Field was to the right. The subject airplane then began a turn to the right and immediately fell into a spin to the right. The pilot made a partial recovery at very low altitude, the airplane entering a nose-high attitude and entering another stall just before striking the ground. The occupants were killed instantly. The airplane did not burn but was destroyed utterly.

4-18-45C. Newman, New Mexico. At 1610 MWT, a Boeing F-13A suffering an engine fire crashed 19 miles northeast of Newman, New Mexico, killing pilot 1Lt. Glen E. Erickson, 24. Tail gunner Cpl. Wayne O. Kline was seriously injured parachuting to safety. Ten crewmembers received minor injuries parachuting to safety. The airplane took off at 1330 MWT from Clovis Army Air Field, Clovis, New Mexico, on a "pinpoint and mosaic" photographic mission to El Paso, Texas, and return. The airplane made a normal take-off and climbed to 20,000 feet, pausing and leveling off twice to allow the number-two engine's cylinder head temperature to cool down. The airplane leveled off at 20,000 feet. The airplane completed the pinpoint portion of the mission and climbed to 23,000 feet for the photo mosaic mission. The crew completed the mission and turned back for base. After a short period of time, the scanners observed and reported minor oil leaks in the number-one, three and four engines. The oil leaks were not considered serious, but a short time later the number-two engine rpm dropped to 1,000 rpm. The pilot and engineer attempted to raise the rpm by manipulating the propeller governor and increasing boost from 31" hg to 55" hg, but the rpm remained steadily at 1,000 rpm and did not move up or down. A loud noise was heard from the number-two engine and it began to vibrate badly. The engineer pulled the throttle back and the vibration stopped, and the propeller slowed down considerably. Moments later, the number-two engine burst into flames. The pilot and the engineer attempted to extinguish the fire with the two engine fire extinguishers but were unsuccessful. The pilot attempted to feather the propeller but was unsuccessful. The pilot then ordered the crew to stand by to bail out. The pilot depressurized the airplane, jettisoned the bomb bay fuel tanks and lowered the landing gear. The flaming number-two engine burned furiously and fell away from the airplane, but the port wing continued to blaze. The pilot ordered the crew to bail out while he and the co-pilot remained at the controls. The crew bailed out and the pilot ordered the co-pilot to bail out. The co-pilot left his seat and began to go down the hatch. When he looked back, he saw the pilot standing in the space between the two pilot seats. As the co-pilot bailed out, the port wing burned off and fell away, causing the airplane to go out of control. The pilot was unable to escape and he went in with the airplane, which exploded into flames upon impact. Investigators speculated that a faulty tachometer (that constantly read 1,000 rpm) caused the flight crew to advance the throttle until the engine blew up. Crewmembers injured parachuting to safety were: 2Lt. Lloyd R. Juergens, co-pilot; 2Lt. William Perrier, navigator; 2Lt. Casimer G. Oksas, bombardier; SSgt. J.E. Redingfield, flight engineer; Cpl. Robert J. Boren, radar operator; Cpl. Henry Walke, gunner; Cpl. Daniel W. McClerron, starboard scanner; Cpl. Donald E. Humphrey, port scanner; Cpl. Alexander Krezel, radio operator; TSgt. Edwards S. Nano, engineer.

4-18-45D. Eloy, Arizona. At 1715 MWT, a North American AT-6C suffered a catastrophic structural failure and crashed three miles southwest of Eloy, Arizona, killing instructor pilot 2Lt. Perry C. Black, 20. Pilot 1Lt. Donald H. Erickson received minor injuries parachuting to safety. The airplane took off from Marana Army Air Field, Marana, Arizona, on an authorized aerobatic flight. Lt. Erickson performed several aerobatic maneuvers, including a slow roll, a barrel roll, a loop and an Immelmann turn. After Lt. Erickson successfully performed these maneuvers, Lt. Black took over the controls from the rear cockpit. The instructor then performed a steep dive and an abrupt pull-out, registering 8G's on the accelerometer in the front cockpit. The instructor gained some altitude and then entered another steep dive, reaching a speed of 220 mph before attempting another abrupt pull-out. The port wing failed and separated, and the airplane went out of control. Lt. Erickson opened the canopy and bailed out, pulling the ripcord immediately after clearing the airplane. Moments later, he hit the ground near where the airplane crashed.

4-18-45E. Burnet, Texas. At 2205 CWT, a Curtiss C-46D crashed eight miles northwest of Burnet, Texas, killing four fliers. Killed in the crash were: 2Lt. Willard E. Jacobson, pilot; 2Lt. Herbert D. Fleming, co-pilot; TSgt. Francis J. Pombert, engineer; Pvt. Ross A. Capparelli, radio operator. The airplane had taken off at 2031 from Bergstrom Field, Austin, Texas, on a minimum altitude navigation flight. The airplane failed to return and was not located until 4-22-45. The airplane was found to have crashed into a wooded hillside at an elevation of 1,500 feet msl. The airplane, with both engines producing power, was on a heading of 115 degrees when it collided with the hill about 60 feet from the top. The airplane exploded into flames upon impact and was completely destroyed. There were no known witnesses to the crash and investigators could not determine why the airplane flew into the hill.

4-18-45F. Bergstrom Field, Texas. At 2325 CWT, a Curtiss C-46D attempting an emergency landing crashed near Bergstrom Field, Austin, Texas, killing four fliers and seriously injuring C-46 instructor 1Lt. Eugene A. Jacobs, 25. C-46 student 2Lt. Charles Bailey and student radio operators Pvt. Robert Borge and Pfc. Gerald E. Borchard were killed in the crash. Engineer Cpl. Arthur H. Carroll, Jr. fell to his death attempting to parachute to safety. The airplane took off at 2250 CWT from Bergstrom Field on a glider-tow training mission. The Waco CG-4A glider was released and the airplane landed, the flight lasting a little over ten minutes. The airplane took off at 2311 with another CG-4A glider in tow. Just after take-off from Runway 17L, the port engine began cutting out. The instructor, sitting in the right seat, reduced power on the port engine slightly and the engine cleared up. The airplane-glider tandem climbed to about 800 feet and flew on the take-off heading for a few minutes. The pilot made right turn, heading north toward Austin, and after leveling off the instructor noticed the starboard engine was running rough and that power was falling off. When about six miles from the field, the airplane performed a turn to the east toward the field and approached for a normal glider release. The instructor notified the tower that he was coming in on a single-engine, requesting a downwind landing after the glider release. The tower officer, noticing that the starboard engine was on fire, ordered the pilots to release the glider immediately and cleared the C-46 to land on any runway. The glider pilot also noticed that the C-46's engine was on fire and he released on his own about two and a half miles from the field and made a safe landing in a plowed field. The airplane, heading east, was approaching the North-South runway at a 90-degree angle and then turned right, heading south and paralleling the runway about a mile out. The instructor pulled the starboard engine fire extinguisher but could not put out the fire. The starboard propeller would not feather and the engine began to vibrate violently. The airplane was flying at about 750 feet agl. The instructor ordered the crew to bail out and turned toward the field, barely able to maintain control because of the vibration. The blazing starboard engine fell away and the airplane continued to descend until it smashed to the ground at a shallow angle. The airplane broke up and burst into flames as it slid across the ground. One crewmember had bailed out at very low altitude and was unable to deploy his parachute. The instructor was found walking in a "dazed condition" about 150 yards from the wrecked airplane.

4-19-45A. Chester, Georgia. At 0205 EWT, a Consolidated B-24J crashed eight miles east of Chester, Georgia, killing pilot 1Lt. Peter V. Corato, 30. Seven crewmembers parachuted to safety and were uninjured. The airplane took off at 2010 EWT from Chatham Field, Savannah, Georgia, on a night navigation flight to Jonesboro, Arkansas, via Memphis, Tennessee. The airplane, flying at 6,000 feet at 180 mph, arrived near Birmingham, Alabama, let down to 4,500 feet and increased speed to 190 mph. The airplane arrived at Memphis at 2312 EWT and set a course for Jonesboro. The airplane arrived at Jonesboro and circled a couple times before setting a course back to Memphis. The airplane arrived at Memphis and turned on a course for Chatham Field. The engineer reported to the pilot that the airplane had 1050 gallons of fuel on board. The pilot descended to 3,500 feet and increased speed to 210 mph. The engineer was closely monitoring the amount of fuel left on board and the airplane's fuel consumption rate, reporting these to the pilot every few minutes. As the airplane approached Macon, Georgia, the engineer warned the pilot that the airplane would not make it to Chatham Field because of low fuel and advised that the pilot return to Macon for fuel. The pilot continued on toward Chatham Field. After a couple minutes, the pilot decided to turn around and return to Macon. The pilot completed a 180-degree turn and leveled out. A few seconds later, the number-one engine quit because of fuel exhaustion. The pilot ordered the crew to fasten parachutes and stand by. Moments later, the number-four engine quit due to fuel starvation. The pilot added power on the number-two and three engines and ordered the engineer to "set the [fuel] valves tank to engine and cross-feed." The pilot added full power to the remaining engines and moments later the number-three propeller ran away. The pilot, snapping on his parachute, ordered the crew to bail out. The engineer and co-pilot

bailed out of the open forward bomb bay. The pilot was last seen getting out of his seat. The airplane entered a diving turn to the left and smashed to the ground, bursting into flames upon impact. The pilot was unable to escape the bomber and he was killed in the crash. His body was found in the vicinity of the bomb bay. Investigators stated that the "improper" power settings increased fuel consumption to such an extent that the airplane could not safely complete the intended flight plan. Crewmembers parachuting to safety were: F/O Peter J. Porcelli, co-pilot; F/O Roger E. Purskey, bombardier; 2Lt. Harold W. Pfautz, navigator; Cpl. Henry L. Turlington, engineer; Cpl. John P. Beverly, radio operator; Pfc. Paul K. Beshore, engineer; Cpl. Samuel W. Grainger, radio operator.

4-19-45B. Wingate, Texas. At 1100 CWT, two Republic P-47D airplanes collided in mid-air three miles NNE of Wingate, Texas, killing 2Lt. Lenny S. Costley, 23, who fell to his death attempting to parachute from P-47D #42-28746. The pilot of P-47D #42-28258, 1Lt. Irving J. Cuddington, Jr., 24, was uninjured and was able to make a safe crash landing. The airplanes were part of an eight-ship flight that took off from Abilene Army Air Field, Abilene, Texas, on a strafing mission. The formation was divided into two, four-ship flights, each led by an instructor. The flights had finished their firing runs and were re-forming to fly back to base when the collision occurred. Lt. Cuddington was flying in the number-three position and Lt. Costley was flying in the number-four position of the lead flight. Investigators stated, "Witnesses in the flight stated that Lt. Costley was closing in on and overshot Lt. Cuddington's ship, putting himself directly underneath Lt. Cuddington's airplane. The propeller of Lt. Cuddington's P-47 collided with the starboard wing of Lt. Costley's ship. Lt. Costley's airplane was then observed in a steep climb before it rolled out of control. Lt. Costley bailed out at about 1,000 feet agl. The airplane dove straight to the ground and exploded. Lt. Costley's parachute failed to open and he fell to his death. Lt. Cuddington stated that he felt a jolt but did not see the airplane that he collided with. Lt. Cuddington's airplane began to vibrate violently so he picked out a field and immediately made a wheels-up landing.

4-19-45C. Shaw Field, South Carolina. At 1645 EWT, a Republic TP-47D crashed at Shaw Field, Sumter, South Carolina, killing 2Lt. Mark J. Flater, 21. The airplane had returned to the field after completing a routine training mission and entered the traffic pattern. The airplane was seen to make a descending right hand turn from the downwind leg to the base leg, recovering at about 300 feet agl. The mobile control officer ordered the pilot to go around. The airplane re-entered the traffic pattern and again fell off in a diving turn from downwind to the base leg. The mobile control officer again ordered the pilot to go around. The pilot did not reply and the airplane, flying at about

300 feet agl, stalled out of the turn from the base leg to final approach. The pilot was unable to recover and the airplane smashed to the ground and burst into flames.

4-19-45D. Palmdale, California. At 1810, a Lockheed P-38L crashed nine miles WSW of Palmdale, California, killing 2Lt. Lawrence A. Cormack, 22. The airplane was flying in the number-four position of a four-ship flight that took off at 1700 from Van Nuys Metropolitan Airport, Van Nuys, California, on a high-altitude navigation mission and aerobatic flight. The flight climbed to 21,000 feet and Lt. Cormack was allowed to lead the flight on the navigation portion to Bakersfield, California, and then to the western edge of the Mojave Desert. The flight leader took over and then ordered the pilots to perform the aerobatic portion of the mission one at a time. The subject pilot was the last one to perform his aerobatics. The flight leader ordered the subject pilot to perform a loop, a roll and then an Immelmann turn. The subject pilot peeled off to the right and performed a roll to the left as he entered a steep dive. The airplane appeared be pulling out of the dive at an altitude of 15,000 feet. Pieces were seen falling away from the airplane and the pilot was unable to recover from the dive so he bailed out. The pilot struck the horizontal stabilizer, severing his left leg at the hip and preventing him from deploying his parachute. The tail section was seen to separate from the airplane as the airplane began "flopping" out of control. The airplane described two turns of spiral before smashing into a mountain at about 5,000 feet msl, exploding into flames upon impact. The pilot's body was found about 600 feet from the main wreckage. Wreckage was scattered over an area of nearly two miles.

4-20-45A. Sweetwater, Texas. At 0640 CWT, a Douglas C-47 flying in poor weather crashed four miles southeast of Avenger Field, Sweetwater, Texas, killing three crewmembers and 22 passengers. The airplane took off at 0553 CWT from Midland Army Air Field, Midland, Texas, on a personnel flight to Nashville, Tennessee. The airplane, flying at about 5,000 feet msl, encountered a line of powerful thunderstorms near Sweetwater. At about this time the starboard propeller was apparently feathered. Investigation revealed that the airplane was heading in a northerly direction and it was speculated that the pilot was attempting to get to Avenger Field. The airplane apparently encountered very turbulent conditions or had entered an unusual attitude while flying in instrument conditions. The airplane was stressed beyond its design limitations, causing the failure and separation of the elevators and horizontal stabilizers and the progressive failure of the tail. The airplane went out of control and the tail section twisted off. The port wing failed and separated as the airplane spun crazily to earth, and the starboard wing separated just before the airplane smashed to the ground nose first. Investigation revealed that both propellers had separated from their engines as the airplane

spun to earth. Fabric from the elevators and pieces of the horizontal stabilizers were found about two miles from the scene of the main crash. Pieces of the tail section and the port wing were found scattered over an area of several hundred yards surrounding the main point of impact. Investigators noted that the airplane had logged about 4,000 hours of glider tug operations at South Plains Army Air Field, Lubbock, Texas, during the course of the war. Pilot 1Lt. James A. Bailey, co-pilot Capt. John R. Rawls and engineer Sgt. William H. Edwards were killed in the crash. Passengers killed in the crash were: 2Lt. Richard B. Arnold, 1Lt. David L. Bennell, Sgt. Robert H. Blaess, Cpl. Daniel B. Boone, SSgt. Grady O. Boyd, Pfc. Kenneth W. Carlson, Cpl. Vincent R. DiStefano, Sgt. William T. Downey, 1Lt. Leonard K. Epperly, Pvt. Robert L. Fuller, Sgt. Robert C. Goodenough, Pfc. William A. Green, Cpl. D.E. Jennelle, 1Lt. Samuel Kamrass, Pfc. James D. McLaughlin, 1Lt. Chester W. Mrozek, 1Lt. Frank A. Prete, 1Lt. Joseph A. Scieszka, Cpl. Myron V. Testament, 1Lt. James H. Wallace, Cpl. Virgle E. Walston, 1Lt. Robert L. West.

4-20-45B. Fort Myers, Florida. At 1630 EWT, a North American P-51D crashed ten miles northwest of Page Field, Fort Myers, Florida, killing 2Lt. John Molchan, 23. The airplane was flying in the number-four position of a four-ship flight that took off from Page Field on high-altitude gunnery mission. The airplanes took off at 1620 EWT from Runway 9 at Page Field and turned to the left to form up. The airplanes completed a 180-degree left turn as they formed up and were flying north of the field heading west for the aerial gunnery range. As the airplanes climbed through 2,000 feet they encountered a moderate rain shower. The flight leader was unable to contact the subject pilot over the radio and he was not in the formation when it broke into the clear. Investigation revealed that the airplane smashed into the ground in a diving turn, exploding into flames upon impact. Investigators speculated that the pilot was having radio trouble and instead of flying into the rain showers had elected to return to Page Field.

4-21-45A. Parrish, Florida. At 1030 EWT, a North American P-51C crashed four miles east of Parrish, Florida, killing 2Lt. Everett H. Greene, 21. The airplane was flying in the number-two position of a four-ship flight that took off at 1000 EWT from Venice Army Air Field, Venice, Florida, on a high-altitude formation flight. The airplanes climbed to 10,000 feet and the flight leader called the pilots on the radio for an oxygen test. Everything checked out and the flight climbed to 16,000 feet and another oxygen check was conducted. All pilots apparently checked in satisfactorily. The formation began climbing through 17,000 feet when the subject airplane was observed lagging behind. The flight turned to the right and the subject airplane caught up. The subject airplane again was lagging behind as the flight climbed through 19,000 feet. The flight leader radioed the subject pilot and asked him if he was having trouble, receiving an unintelligible answer. Moments later the airplane dropped out of the formation and passed beneath the leader as it fell away. The pilots did not see the subject airplane after that. The flight leader circled the area but was unable to locate the subject airplane. Witnesses on the ground observed the subject airplane disintegrating as it plunged to earth in a high-speed dive. The airplane smashed to earth and exploded violently into flames. Investigators speculated that the pilot had suffered an oxygen system problem, causing him to lose control of the airplane.

4-21-45B. Palmdale, California. At an unknown time after 0924, a North American AT-6C flying in instrument conditions crashed into the east slope of Mt. McDill eight miles southwest of Palmdale, California, killing pilots Capt. Fred A. Pugh and 1Lt. Thomas F. Turner. The airplane took off at 0750 PWT from Kingman Army Air Field, Kingman, Arizona, on a flight to Mines Field, Los Angeles, California, via Palmdale, California. At 0924, Capt. Pugh radioed Palmdale Army Air Field requesting an IFR clearance to Mines Field, which was granted. There was no further contact with the airplane and it failed to arrive at Mines Field. The airplane wreckage was found in the San Gabriel Mountains several months later. Investigators noted that weather in between Palmdale and Los Angeles was poor, with ceilings reported as 200 feet agl. In some areas visibility was only three-quarters of a mile.

4-21-45C. Greenwood, Mississippi. At 1637 CWT, two North American AT-6F airplanes collided in mid-air seven miles east of Greenwood Army Air Field, Greenwood, Mississippi, killing 1Lt. Robert H. Vetter, 23, aboard AT-6F #44-81824. Pilot-rated passenger 2Lt. Loran W. Macy was seriously injured. Instructor pilot 1Lt. Allen W. Hunsaker and student 1Lt. Carl A. Carter were uninjured and able to safely land AT-6F #44-81836 in a field. Investigators stated, "Lt. Hunsaker, while completing a turn onto final, was above and closer to the airport than Lt. Vetter and was making a normal approach. Lt. Vetter, approaching at a slightly lower altitude and already on a straight course for the runway, but with flaps up, arrived directly under Lt. Hunsaker's aircraft. At this particular time, Lt. Hunsaker lowered the flaps according to the prescribed curriculum in the AT-6 stage. The application of flaps required forward stick and nose-down attitude to compensate for additional drag. The sudden change in attitude of Lt. Hunsaker's aircraft with Lt. Vetter's aircraft directly under caused the propeller Lt. Hunsaker's plane to enter the left elevator [of Lt. Vetter's ship] and move forward at a small angle to the right, cutting away the left elevator and horizontal stabilizer and severing the fuselage just forward of the vertical stabilizer, causing Lt. Vetter's plane to fall out of control." Lt. Hunsaker's airplane was able to make an off-field wheels-down landing in a field following the collision.

4-21-45D. Fairfax, Washington. At 1630, co-pilot 2Lt. Carl F. Schmitt was killed attempting to bail out of a Consolidated B-24J operating over Fairfax, Washington. His body was found on 5-5-45 about five miles northeast of Fairfax. The airplane was flying in the number-two position of a six-ship flight that took off at 1515 from McChord Field, Tacoma, Washington, on a routine training flight. After the flights formed up, the flight leader led the formation through a hole in the clouds and attempted to maneuver through several cloud troughs in an effort avoid towering cumulus clouds and to climb out in contact conditions. After a short time, the flight leader realized that he was in a situation where he could not go forward or turn back without going on instruments. The airplanes were in a steep climb at 140 mph, passing through 10,000 feet. The formation continued on and a few moments later the number-three ship slid out of position, moving to the right and passing under the lead ship and ending up in front of the number-two ship. The number-two ship, the subject airplane, was unable to break off from the formation because the number-three ship was in front of it. The pilot of the subject airplane then attempted to move to the left, over the tail of the lead ship, and fall back into the number-three position. The subject airplane attempted the maneuver, but encountered propeller turbulence from the lead ship. The subject airplane went out of control, falling out of the formation and entering a spin to the right. The pilot was able to recover the airplane for a short time, during which the crew fastened their parachutes. Moments later, the airplane entered a spin to the left. The pilot recovered just as the airplane fell out of the bottom of the cloud deck. The pilot recovered the airplane and leveled out in a southwesterly direction. The crew began bailing out immediately. The co-pilot became entangled in the partition curtain while leaving the flight deck for the bomb bay. He lost his balance and fell into the bomb bay area, striking his head on the interior of the airplane. He then fell through the forward starboard bomb bay door, which was in the closed position, taking the door with him. The pilot then made a diving turn to the right and the crew bailed out. The pilot, 1Lt. Ansel A. Parrish III, was able regain control of the airplane and the tail gunner, Cpl. Jimmie W. Chiles, remained on the flight deck, acting as co-pilot. The pilot and tail gunner were able to make a safe landing at McChord Field and were uninjured. Crewmembers parachuting from the airplane were: 2Lt. Troy L. Miller, navigator; 2Lt. Robert C. Huckaby, bombardier; Sgt. Robert D. Brooking, engineer; Sgt. Robert C. Barber, radio operator; Sgt. Harold R. Hall, gunner; Cpl. Orrin K. Camenish, gunner; Cpl. Donald M. McLaughlin, gunner.

4-22-45A. Brownsville, Texas. At 1143, a Curtiss P-40N crashed while attempting to take off from Brownsville Army Air Field, Brownsville, Texas, killing pilot 2Lt. Edmund B. Overton, 25. Witnesses observed that the take-off appeared normal at first but the take-off run became exceptionally long. The airplane lifted off and climbed to about 30 feet when the engine suddenly quit. The engine caught momentarily and then quit again. The airplane, with landing gear still extended, entered a normal glide but then swerved to the right to avoid colliding with trees. The airplane descended toward the ground and the landing gear passed through a group of bushes 800 feet from the departure end of the runway. The tail struck the ground about 300 feet from the bushes and then the main landing gear contacted the terrain. The airplane rolled for a very short distance before the port landing gear collapsed. The port wing began scraping the ground and then the airplane smashed through a fence, bounding over a drainage ditch before colliding with an embankment on the other side. The port wing separated as the airplane smashed over the embankment, causing the airplane to cartwheel over anther fence before coming to rest in a field and bursting into flames.

4-22-45B. Rotan, Texas. At 1505, a Republic P-47D crashed seven miles northwest of Rotan, Texas, killing pilot 2Lt. Richard R. Branch, 25. The airplane was flying in the number-four position of a six-ship flight that took off from Avenger Field, Sweetwater, Texas, on a low-altitude formation training flight. The formation flew at very low altitude for approximately an hour before the flight leader climbed the flight to 10,000 feet to cool off because the cockpits had become very warm from flying at low altitude on a hot day. After flying at 10,000 feet for a short time the leader ordered the ships into a right echelon formation. The airplanes formed up satisfactorily and then after a short time the flight leader peeled off into a wide, gentle wingover to the left. The flight followed in a semi-string formation. When the lead ship descended to about 8,500 feet, the leader went into a gentle turn to the right. The formation followed and the subject airplane was seen to roll through to the right. The number-five pilot realized that there was something wrong and did not follow the maneuver. The number-five pilot observed the subject airplane complete "a wide roll to the right" before going down and disappearing from his view. The subject airplane continued to descend until it struck the ground and exploded into flames. The pilot made no apparent effort to recover or bail out. The cause of the accident could not be determined.

4-23-45A. Douglas, Arizona. At 1030 MWT, B-25 student A/C Junior T. Chenoweth was killed when he fell out of a North American TB-25J operating seven miles north of Douglas Army Air Field, Douglas, Arizona. B-25 instructor Capt. Arthur J. Finnell and B-25 student A/S Lester Chew were uninjured and were able to land safely at Douglas Army Air Field. The ship was on a local orientation flight and was flying at about 8,500 feet, or 4,500 feet over the terrain. A/C Chenoweth was flying in the pilot's seat

and Capt. Finnell was in the right seat. A/S Chew was in the navigator's compartment. All of the pilots were wearing B-7 "back type" parachutes. Investigators stated, "As the aircraft neared Douglas Army Air Field, Capt. Finnell started a medium turn to the right. As this turn was being completed, Capt. Finnell directed A/C Chenoweth to leave the pilot's seat, take his place in the navigator's compartment, and send A/S Chew forward to take his place in the pilot's seat. A/C Chenoweth unbuckled his safety belt and stood up. As he moved back between the pilot and co-pilot seats, facing to the rear of the aircraft, his parachute caught on the pilot's seat and was released by Capt. Finnell. At the same time, the ripcord covering on A/C Chenoweth's parachute caught on the upper hatch emergency release handle. As A/C Chenoweth struggled to release himself, he released the hatch, which immediately blew off. At the same time, it is believed that A/C Chenoweth's parachute opened, due to the ripcord and cover [being pulled out by the separating hatch] releasing the parachute, thus pulling A/C Chenoweth out of the open hatch. Evidence on the upper part of the fuselage indicates that A/C Chenoweth's path of travel was straight down the center of the top of the fuselage, striking against the bubble over the tail gunner position and then over the left horizontal stabilizer and left rudder." A/S Chew observed A/C Chenoweth being pulled out of the aircraft by the deployed parachute. Capt. Finnell, aware that the hatch blew off and concentrating on flying the airplane, motioned A/S Chew to take the pilot seat. A/C Chew assumed that Capt. Finnell knew that A/C Chenoweth had fallen out. Capt. Finnell, still unaware that the cadet had been pulled out of the aircraft, motioned to A/S Chew to look aft. A/S Chew thought that the instructor was motioning him to check for a parachute, but was motioning him to look aft because the port rudder felt unstable. Neither pilot mentioned the cadet falling out. The airplane returned to the field immediately and made a normal landing. The instructor was still unaware that the cadet had been pulled out of the aircraft until he called out the cadet's name while doing routine post-flight paperwork and was informed by A/S Chew of the incident. A/S Chew told investigators that A/S Chenoweth being sucked out of the aircraft was "instantaneous." The cadet's parachute was heavily damaged in the incident and he fell to his death. His body was found the next day.

4-23-45B. Sherman, Texas. At 1225 CWT, a North American AT-6C crashed one-quarter mile south of Perrin Field, Sherman, Texas, killing instructor 2Lt. Richard E. Mathieson and student A/C Gage T. Thurman. The airplane had taken off from Perrin Field on a routine training flight. A short time later, student flying was called off because rain showers existed in the area of the field and a large thunderstorm system was moving into the general area. As the airplanes began return to the field, low-hanging clouds were interfering with traffic east of the field. The subject airplane entered the traffic pattern and was observed turning right from the base leg to final approach. Another AT-6 airplane was turning left onto the same approach from the opposite direction. The pilot in the subject airplane apparently noticed the other airplane. The pilot in the other airplane applied full throttle and began a climbing turn, keeping the subject airplane in sight. The subject airplane tightened its turn onto the approach and stalled. The subject airplane snapped over to an inverted attitude and paused at that position momentarily before the nose dropped toward the ground and the airplane crashed in a split-S type maneuver. The airplane did not burn but was completely destroyed.

4-24-45A. Gulf of Mexico, Florida. At 0909 EWT, a Boeing B-29A suffering an engine fire was abandoned and crashed into the Gulf of Mexico about 100 miles southwest of Key West, Florida, killing 11 crewmembers. Bombardier 2Lt. John H. Matthews and navigator 2Lt. William L. Jiler parachuted into the sea and were rescued the next day. The airplane had taken off at 0753 EWT from MacDill Field, Tampa, Florida, on a navigation flight to Borinquen Field, Puerto Rico, and return. Investigators stated, "At 0845, 10,000 feet, the right scanner reported a slight oil leak in the number-three engine. The pilot ordered the engineer and scanner to maintain a close watch on that engine. At 0906, the number-three propeller ran away and reached 3,900 rpm. Simultaneously the scanner reported a bad oil leak and smoke from the number-three engine. It appears that the pilot put the nose down and was attempting to feather. The RPM was only brought down to 3,100 rpm and the command to "Prepare to bail out" was given by co-pilot on interphone at 0908 followed by a report from scanner that number-three engine was on fire. Both fire extinguishers were applied without results. The "Bail out" order was given at 0909 on interphone. Bomb bay tanks [were] salvoed and wheels put down at 7,300 feet. Airspeed was observed to be 240 indicated. Meanwhile the navigator gave coordinates to radio operator right after "Prepare to bail out" order. Radio had been giving trouble and it is doubted if transmission was effected even though key was operated. Bail out started at 7,300 feet when instructor radio operator bailed out followed by the radio operator, navigator and engineer. These went out the bomb bay. The co-pilot meanwhile went out the nose wheel well. The bombardier helped put on the pilot's parachute and bailed out. The navigator saw four parachutes in the air besides his and the bombardier saw five besides his own. The bombardier and radio operator effected contact in the water but drifted apart. The radio operator's body was recovered. The bombardier and navigator ... suffering only minor sunburn ... were rescued the following day." Crewmembers missing or drowned were: 2Lt. Daniel R. Ahern, pilot; 2Lt. Anthony Laskow, co-pilot; 1Lt. Cecil C. McKinney, B-29 instructor; 2Lt. James V. Reams, radar operator; SSgt. Erwin W. Beck, engineer; Sgt.

Thomas P. Proffitt, radio operator; Cpl. Bob L. Smith, gunner; Cpl. William L. Phillips, radio operator; Sgt. Louis T. Kapusta, engineer; TSgt. Quincy L. Zickafoose, radio operator instructor; Cpl. James C. Spears, tail gunner.

4-24-45B. Wittman, Arizona. At 1220 MWT, a Lockheed P-38J crashed 11 miles WSW of Wittman, Arizona, killing 1Lt. Frank Capel, 26, who fell to his death attempting to parachute to safety. The airplane took off 15 minutes earlier from Luke Field, Phoenix, Arizona, on a routine local transition flight. Investigation revealed that the airplane struck the ground at a 40-degree angle in a left-wing low attitude. The airplane exploded upon impact. The pilot had bailed out an altitude too low to allow his parachute to deploy. His body was found about 600 yards from the wreckage with the parachute strung out behind him.

4-24-45C. Big Spring, Texas. At 2206, a Beech AT-11 crashed seven miles southeast of Big Spring, Texas, killing four fliers and seriously injuring student bombardier A/C George B Stoolman, who miraculously survived. Killed in the accident were: 2Lt. Robert E. Mock, pilot; 2Lt. Frank P. Nemesh, bombardier instructor; A/C William C. Harris, student bombardier; A/C Clifton P. James, student bombardier. The airplane had taken off from Big Spring Army Air Field on a bombing mission. The crew had completed the mission and the ship was returning to the field when power from the engines diminished greatly. The pilot was unable to get full power from the engines and was unable to hold altitude. The pilot, apparently unaware of the low altitude, ordered the crew to bail out. The fliers had a hard time getting the exit door to jettison. One crewmember merely opened the door in the usual fashion and bailed out, the door falling back into place. The crew was able to free the door and two more airmen bailed out at an altitude too low to allow their parachutes to deploy. A/C Stoolman was standing in the doorway when the airplane descended into the ground. He was thrown clear and survived. The airplane smashed itself to pieces as it slid across the terrain, turning about 180 degrees from the original line of flight. The pilot was pulled alive from the wreckage but he died about five hours later. Investigation revealed that the port wing tank had been run dry and the starboard wing tank contained enough fuel to safely complete the flight.

4-26-45A. Centerville, Tennessee. At 0750 CWT, two Consolidated RB-24E airplanes collided in mid-air and crashed one-half mile southwest of Centerville, Tennessee, killing nine fliers. Killed in the crash of RB-24E #42-78696 were: 1Lt. Thomas G. Sanders, B-24 student; 1Lt. George M. Wenger, co-pilot; 1Lt. Golden Lang, pilot; 2Lt. Robert M. Ogburn, co-pilot; Sgt. William H. Jones, engineer. Killed in the crash of RB-24E #41-28485 were: Capt. Melvin W. Rennie, B-24 instructor; Capt. William P. Sullivan, B-24 student; 2Lt. Irvin H. Sentz, Jr., co-pilot; Pfc. Carl B.

Weigert, engineer. The two airplanes performed a formation take-off from Smyrna Army Air Field, Smyrna, Tennessee, at 0700 CWT. Investigators stated, "The two airplanes were scheduled for a routine formation flight at high altitude, normally performed at an approximate altitude of 25,000 feet. After take-off, there was no further contact with the two airplanes, and they were not observed again until seen descending from the clouds in the vicinity of Centerville, Tennessee. It is assumed that the airplanes collided above the clouds between the altitudes of 5,000 feet to 15,000 feet. When first observed, from the ground, the two airplanes appeared to be together and separated when approximately 2,000 feet above the ground. Airplane #41-28485 was observed to be missing the left wing outboard of the number-one engine and the complete tail assembly fuselage just aft of [the bottom rear hatch]. The airplane was observed to be in a steep dive, momentarily recovering and rolling into an inverted position from which the airplane dived almost vertically into the ground and burned. The wreckage showed that the airplane crashed in an inverted position and from almost a vertical angle. Airplane #42-78696 was observed, after separating, to be missing approximately 10 feet of the left outer wing panel and the horizontal and vertical stabilizers. The fuselage was next observed to separate [at the rear bomb bay bulkhead]. The airplane was then observed to go into a flat spin, crashing to the ground and burning on impact."

4-26-45B. Folsom, California. At 1615, a Vultee BT-13B crashed after it collided with a power-line strung across the north fork of the American River near Folsom, California, killing pilot Capt. Howard C. Standridge and pilot-rated passenger 2Lt. Charles M. Garmong. The airplane had taken off from McClellan Field, Sacramento, California, on an instrument training flight to Chico Army Air Field, Chico, California. The airplane's port wing was severed when it collided with the power-line, which crossed the river about 90 feet above the surface. The power-line was severed and pieces of it were found wrapped around the severed port wing, which was found about 1,200 feet from where the airplane crashed.

4-27-45A. Emison, Indiana. At 1140 CWT, a Douglas C-47A crashed one-half mile southwest of Emison Auxiliary Airfield, Emison, Indiana, killing two fliers and seriously injuring two others. Pilot F/O Paul E. Forker and engineer Sgt. Monte L. Holman were killed; co-pilot F/O Benjamin D.L. Bennett, Jr. and radio operator Cpl. William H. Herron suffered serious injuries. The airplane took off at 0730 CWT from George Field, Lawrenceville, Illinois, on a routine instrument training flight. After flying for about four hours, the pilot decided to drop a para-pack at Emison Auxiliary Airfield. The pilot descended and flew over the field at low altitude, flying from east to west. As the airplane passed approximately mid-field, the radio operator and the engineer pushed the para-pack

out of the cargo door and watched it fall. The airplane flew west to the field boundary and then was seen in a climbing turn to the left. The pilot tightened his turn, apparently in an attempt to get the deployed para-pack in sight, allowing the airplane to enter a slipping turn to the left. The pilot was able to bring the nose up but the airplane mushed into the ground. The port propeller hit the ground first, gouging a path in the terrain for about 75 feet before breaking off. The airplane veered to the right and the starboard propeller struck the ground, snapping off after a short distance. The starboard engine broke loose and was hurled 100 feet before coming to rest. The nose smashed into the ground and the airplane spun around to the right. The airplane burst into flames as it smashed itself to pieces, coming to rest heading about north. Two crewmembers were thrown clear and one crewmember was found under the port wing. He was pulled clear. A fourth crewmember was trapped in the smashed cockpit and his body was not recovered until the flames were extinguished.

4-27-45B. Claude, Texas. At 1344 CWT, a Curtiss C-46D flying in poor weather crashed 25 miles SSE of Claude, Texas, killing 11 fliers. Crewmembers killed in the accident were: 1Lt. Edward T. Kelly, pilot; 2Lt. Richard A. Gildersleeve, co-pilot; 2Lt. Forrest M. Faulkner, navigator; Cpl. Edward J. Shields, engineer; Sgt. Henry W. Curran, radio operator. The airplane took off on visual clearance at 1019 CWT from Sedalia Army Air Field, Knobnoster, Missouri, on a personnel ferry flight to Brooks Field, San Antonio, Texas, via Amarillo, Texas. The airplane encountered instrument conditions as it approached Amarillo, so the pilot went on instruments. At 1324, the pilot radioed Amarillo radio and reported that his position as 15 miles southeast of Amarillo at 6,000 feet. The pilot advised Amarillo radio that he was on an instrument clearance, which was not true, and requested weather for Abilene, Texas. The pilot radioed at 1332, reporting that he was 30 miles southeast of Amarillo at 6,000 feet. The pilot inquired about the ceiling and was advised the pilot reports indicated that the top of the overcast was about 8,000 feet. A short time later, three ranch employees heard a "sudden surge" of the airplane engines in the overcast. Moments later the airplane was heard to crash. There were no eyewitnesses to the crash. Investigators stated that the airplane had crashed into "a shelf between the rim of the river bluff and the top cap-rock of the canyon." The airplane descended into the ground at angle exceeding 45 degrees, exploding violently into flames upon impact. Pilot-rated passengers killed in the crash were: F/O Thomas G. Anderson, 2Lt. Ernest F. Crow Jr., 2Lt. Paul J. Entres, 2Lt. Robert A. Gray, Capt. Brown G. Middleton (rated-navigator), F/O Hollis K. Smith.

4-27-45C. Elmer, Louisiana. At 1500 CWT, a Curtiss P-40N suffered a catastrophic structural failure and crashed one mile northwest of Elmer, Louisiana, killing 1Lt. Jearl C. Thompson, 26. The airplane took off from Esler Field, Louisiana, on a local transition flight. The airplane was observed to emerge from the overcast in a vertical dive. Both wings separated at an altitude of 500 feet agl. The wingless fuselage smashed into the ground in a vertical dive about 90 yards from Highway 85, burying wreckage several feet into the clay and soil. The pilot made no apparent effort to leave the airplane. The wings fell to earth near where the fuselage slammed to earth.

4-27-45D. Napier Field, Alabama. At 1520 CWT, a Curtiss P-40N aborting a take-off collided with a North American AT-6C taxiing at Napier Field, Dothan, Alabama, killing the two AT-6C pilots and injuring the P-40N pilot. Pilots 1Lt. Harry L. Grainger and 1Lt. Joseph L. Green, Jr. were killed aboard the AT-6C; P-40N pilot 2Lt. William B. Cummins, Jr. sustained minor injuries. The AT-6C, which had been practicing take-offs and landings, had landed to the northeast on the grassy area about 400 feet to the left of the main Northeast Runway. The AT-6C taxied straight to the northeast, parallel with the Northeast Runway. The AT-6C taxied about 400 feet past the departure end of the runway before making a 90-degree turn to the right, taxiing about 100 feet then pausing at a point "400 feet from the end of the runway and 300 feet from the side of the runway." The P-40N was beginning a take-off on the Northeast Runway. The control tower operator advised the AT-6C pilot that he was clear to cross (from left to right) in front of the departure end of the Northeast Runway after the P-40N had cleared the field. The P-40N began its take-off roll and heavy black smoke was seen trailing the engine just as the plane became airborne. The "P-40 Mobile Control Officer" ordered the P-40N pilot to abort the take-off and land straight ahead. The P-40N landed on the runway and then ran off of the end of the runway into the grassy area. The AT-6C pilot had begun to taxi in an effort to cross in front of the runway on the grassy area when the P-40N became airborne. Seeing that the P-40N had aborted and was heading for his airplane on a collision course, the AT-6C pilot hit the brakes hard, causing the tail to rise up several feet. The AT-6C then fell back down to its normal three-point position and was immediately broadsided from the right by the P-40N. The AT-6C's starboard wing was severed and its landing gear collapsed as the airplanes smashed to a halt about 35 feet apart. The AT-6C had burst into flames in the collision. The P-40N pilot jumped from his wrecked airplane and ran over to the blazing AT-6C, helping Lt. Grainger escape from the front cockpit, which was engulfed in flames. Lt. Grainger walked about 50 feet from the wreckage with his clothing and parachute on fire. Napier Field rescue personnel immediately extinguished the clothing fire and within a couple minutes the airplane fire was extinguished. Lt. Green, in the rear cockpit, never moved and was found to be dead. Investigators stated that Lt. Grainger might have misunderstood the command given by Napier Field control tower personnel.

4-28-45A. Russell, Kansas. At 0951 CWT, a Boeing B-29 suffering an engine fire was abandoned and crashed five miles southeast of Russell, Kansas. Gunner Cpl. Roy L. Collins fell to his death in an unsuccessful parachute jump. Bombardier 2Lt. Stanley P. Hegg was seriously injured and pilot 1Lt. William D. Clifford sustained minor injuries parachuting to safety. Crewmembers uninjured parachuting to safety were: 2Lt. David H. Kaplan, co-pilot; SSgt. Snap F. Bean, flight engineer; Cpl. Ray E. Densler, gunner; Cpl. A.F. Chiarella, gunner; Cpl. Richard W. Link, radio operator; Sgt. Raymond C. Becker, gunnery instructor. The airplane took off from Great Bend Army Air Field, Kansas, on a routine training mission. Investigators stated, "Take off was normal and the pilot climbed through the first layer of overcast and emerged on top at about 4,000 feet. The lower part of the next layer began at about 6,000 feet and the pilot was about to enter [the next layer] when he saw flames coming from the number-one engine. The rpm on the number-one engine was oscillating, so he told the engineer he was feathering the number-one [propeller] and started to do so. The propeller never quite feathered so the engineer pulled the first [engine] fire extinguisher. This failed to extinguish the flames so the pilot dove the airplane to blow it out but the flames still persisted and the propeller started running away. The engineer pulled the second extinguisher but this had no effect on the fire. The landing gear was lowered and another dive was attempted but the fire was just aggravated by this and flames were extending back over the wing. At the airplane commander's order the pilot gave the bail out signal and each man evacuated normally at an altitude between 3,500 and 4,000 feet. The airplane crashed and was totally wrecked."

4-28-45B. Choctawhatchee Bay, Florida. At 1220 CWT, a Douglas A-26C was struck by its own bomb and crashed into Choctawhatchee Bay three miles northeast of Fort Walton, Florida, killing four crewmembers. The airplane had taken off from Eglin Field, Valparaiso, Florida, on a low-level bombing exercise at AAF Water Range #60. AAF photographic personnel were on hand to film the exercise and a test pilot from Wright Field, Dayton, Ohio, was piloting the aircraft. Crash Boats from Eglin Field were dispatched to the range for the exercise. Investigators stated, "[The pilot] put the aircraft into a dive to gain speed. Upon approaching the desired level the pilot pulled out and began his run on the target at a speed of approximately 400 mph. At an altitude of approximately 15 to 30 feet, the projectile was released. Upon striking the water the bomb ricocheted up, striking the aft section of the airplane, knocking off the tail section. The aircraft immediately dived into the bay and sank immediately. All personnel in the aircraft were instantly killed and the aircraft was completely demolished."

Killed in the accident were: 1Lt. Bryce L. Anderson, pilot; Sgt. Russell L. Boyer, engineer; SSgt. Dale T. Jackson, gunner; Cpl. Burton J. Naberhuis, gunner.

4-28-45C. Santa Maria, California. At 1625, a Lockheed P-38L crashed one mile southeast of Santa Maria Army Air Field, Santa Maria, California, killing pilot F/O Walter F. Taylor, 22. The airplane was flying in the number-four position of a four-ship flight that was returning to Santa Maria Army Air Field. Investigators stated, "The flight had completed their mission and were in the traffic pattern. F/O Taylor, while turning onto the base leg in a steep bank to the left in a nose high attitude and at a low airspeed, stalled out and spun in. The dive flaps were found to be in the down position. This may or may not have been a contributing factor as they could have accidentally been extended while attempting to recover from the stall." The airplane burst into flames upon impact.

4-29-45A. Victoria, Texas. At 1115, a North American P-51D suffered an in-flight break-up and crashed three miles northeast of Foster Field, Victoria, Texas, killing pilot 1Lt. William J. Leonard, 27. The airplane took off at 1052 from Foster Field on an engineering test flight. Another pilot had complained of tail section buffeting at high speed on a previous flight. The pilot climbed the airplane up to about 10,000 feet and entered a dive. Control tower personnel heard the whine of a diving aircraft and the subject airplane was observed in a 45-degree dive at about 8,000 feet, diving from northeast to southwest. The airplane began pulling out of the dive at about 5,000 feet when the port wing failed and separated. The airplane went completely out of control and entered a violent spin. The starboard wing failed and separated moments later. The fuselage smashed to the ground and exploded into flames. The pilot made no obvious effort to leave the airplane.

4-29-45B. Rapid City, South Dakota. At 2052 MWT, a Boeing B-17G suffering an engine fire was abandoned and crash landed 10 miles northeast of Rapid City, South Dakota, killing engineer Pfc. Jack A. DeVaul, who fell to his death in an unsuccessful parachute attempt. B-17 instructor pilot Capt. William K. Thomas, 29, was uninjured in the successful forced landing. The airplane had taken off from Rapid City Army Air Base at 2016 MWT on an instrument training flight. The airplane was flying on the instrument problem with landing gear down and one-third flaps when the pilots heard a loud "popping" from the number-one engine and then the airplane began yawing to the left. The instruments did not indicate that the engine was suffering a problem. Moments later the number-one engine was observed to be on fire. The instructor pilot hit the feathering button, but the propeller would not feather. The instructor then pulled the number-one engine CO2 fire extinguisher and the engineer shut the fuel valve, but the fire continued. Capt. Thomas, realizing that the airplane could not

make the field without the fire becoming a danger to the structural integrity of the port wing, ordered the crew to bail out. The instructor, flying in the co-pilot seat, held the airplane at 110 mph at an altitude of about 1,000 feet to allow the crew to safely bail out. The instructor, satisfied that the crew had bailed out, attempted to leave his seat but the airplane was very unsteady, so he decided to make a crash landing. The instructor set up the approach, forgetting that the landing gear was in the down position. The airplane made contact with the ground and started to roll. The instructor was prepared to collapse the landing gear if an obstruction materialized, but none appeared and the B-17 continued across the terrain. The fire went out as the B-17 rolled to a halt. Investigation of the number-one engine revealed that the master rod had failed, causing damage to the push rods and damage to the feathering oil line. Leaking oil was ignited by hot engine parts. Crewmembers uninjured parachuting to safety were: 2Lt. Leroy J. Allaire, B-17 student; F/O Herbert Voss, B-17 student; 2Lt. Robert M. Lang, navigator; 2Lt. George J. Rothstein, bombardier; Cpl. Charles F. Gutman, radio operator; Cpl. Alfred G. Holer, gunner.

4-29-45C. Bloomville, New York. At 2345 EWT, a North American AT-6D flying in poor weather collided with rising terrain three miles southeast of Bloomville, New York, killing Cadet Robert B. Clark, U.S. Military Academy. The airplane had taken off from Bolling Field, Washington, D.C., on a flight back to its home station at Stewart Field, Newburgh, New York. Weather conditions at the scene of the crash were reported as heavy fog with drizzle and ceiling and visibility zero. Investigation revealed that the airplane collided with the heavily wooded mountainside at an elevation of 2,400 feet msl while in a flying attitude with the engine producing power. The airplane was smashed to pieces as it collided with trees. Wreckage was scattered over an area of 100 yards.

4-30-45A. Ellington, Connecticut. At 0750 EWT, a Republic TP-47D crashed two miles northwest of Ellington, Connecticut, killing 2Lt. Edward D. Hensley, 21. The airplane was part of a two-ship flight that took off from Bradley Field, Windsor Locks, Connecticut, on a local instrument-training mission. One pilot was to act as instrument pilot and the other was to act as the safety pilot. The mission was conducted at 6,500 feet, on top of a 5,500-foot overcast. Lt. Hensley flew as the instrument pilot for a short time, and then at 0740 Lt. Hensley became the safety pilot. The other pilot switched frequencies to perform a homing exercise. The instrument pilot flew on instruments for a short time before noticing that Lt. Hensley was not in position and would not answer radio calls. Investigators stated, "[Civilians] observed a P-47 break through the bottom of the overcast 'on its side.' The airplane was flying approximately from north to south. It was then seen to roll completely over on its back, at which time an object was seen to fall from the airplane. Observers [believed] this object to have been a hat. The airplane then dove upside down into the ground and exploded."

4-30-45B. Altamont, Kansas. At approximately 0800 CWT, a Lockheed F-5E crashed seven miles ESE of Altamont, Kansas, killing 1Lt. Robert F. Hodgson, 26, who fell to his death in an unsuccessful parachute attempt. The airplane had taken off from Coffeyville Army Air Field, Coffeyville, Kansas, on a routine training flight. At 0740, the pilot radioed the tower that he had an aileron boost problem. He stated, "Every time I let go of the aileron, I go into a 60-degree bank." The tower officer directed the pilot to turn off the aileron boost switch. The pilot replied in the affirmative and nothing more was heard from him. At 0820, tower personnel were alerted to the crash. Investigation revealed that the airplane had smashed to the ground out of control. The airplane was totally destroyed and the disintegrated wreckage was scattered over a wide area. The pilot had bailed out, but he apparently deployed his parachute prematurely and his shroud lines were cut by some part of the airplane.

4-30-45C. Tinker Field, Oklahoma. At 1026 CWT, a Lockheed C-60A suffering an engine failure on take-off crashed one-half mile northwest of Tinker Field, Oklahoma City, Oklahoma, killing five fliers and seriously injuring nine others. Passengers killed in the crash were: 2Lt. Leonard Bloom; Layton Laughlin, civilian; Leo P. McCain, civilian; Arthur A. McClain, civilian; Earl Morford, civilian. Crew and passengers seriously injured in the crash were: 1Lt. John H. Burkhardt, pilot; 2Lt. Arturo C. Garcia, co-pilot; Earl R. Shanks, civilian engineer; 1Lt. Emile E. Donaldson, passenger; Arnold R. Freidkin, civilian; 2Lt. Bernice Grace, passenger; William M. Puckett, civilian; Orville M. Reeves, civilian; 2Lt. Walter S. Schuler, passenger. The C-60 was scheduled to fly a personnel ferry mission to Kelly Field, San Antonio, Texas. The C-60 taxied to Runway 3 and the pilots performed their checklist and engine run-up. The airplane commenced its take-off to the northeast and just after lift off the port engine began losing power. The airplane climbed to about 30 feet agl before it began veering about 45 degrees to the left of the take-off heading. The airplane, with landing gear still extended, passed over the parking ramp area in front of the Operations Building. The port engine fuel pressure warning light illuminated so the pilot operated the wobble pump with no effect. The engineer also tried the wobble pump but the port engine was out. The airspeed was dropping and the airplane barely cleared a large hangar and the Air Freight Loading Dock buildings. The pilot was unable to crash land straight ahead because he was passing over the residential district of Midwest City. The pilot ordered the engineer to retract the landing gear. The control tower asked the pilot if he was in trouble and he advised that he had lost the port engine. The

tower operator cleared the airplane for an emergency landing on Runway 12 and alerted rescue personnel. The pilot replied that he was not in the pattern for Runway 12 and that his direction of travel was unintentional. The pilot advised the tower that he would not be able to make it back to the field and that he intended to crash land as soon as he could find a clear spot. The tower did not receive the message, but pilots flying in the area apparently heard the transmission. The pilot made no effort to feather the port propeller because "events occurred so rapidly." Investigation revealed that the co-pilot had no previous experience in C-60 type airplanes. The co-pilot stated that he was standing by to act on the pilot's orders and at one point he had pushed the control wheel forward a bit when he noticed the airspeed beginning to drop below 80 mph. Investigators stated, "After clearing the residential section of Midwest City, the pilot attempted a belly landing in a small field. Direction of landing was to the northwest. The plane stalled into the ground in a nose high position with the left wing and tail wheel striking the ground first. The nose came down, dug into the mud, the engines jerked loose from their mounts and passed underneath the wing and at the same time the fuselage buckled just forward of the trailing edge of the wing. The wings and fuselage slid forward on in a jumbled mass and stopped approximately 100 feet past the point where the engines rested, with the wings up-side-down and the remaining portion of the fuselage upright but with the tail pointing northwest. Fire broke out immediately after impact." Two passengers were killed in the crash and three others died of their injuries a short time later. Investigation revealed that the port engine fuel pump had failed. Investigation revealed that the failure of the pilot to feather the port propeller and trim the airplane for single-engine operation, and the failure to retract the landing gear in a timely fashion, caused the airplane to lose single-engine airspeed, which it had achieved on the take-off.

4-30-45D. Rio Grande City, Texas. At 1415 CWT, two North American AT-6C airplanes collided in mid-air and crashed 10 miles northeast of Rio Grande City, Texas, killing student A/C Eugene W. Spates aboard AT-6C #42-43952. Instructor 1Lt. Howard F. Trampenau, 29, was seriously injured parachuting to safety. Student A/C Richard E. Spradling parachuted to safety from AT-6C #42-44318, suffering minor injuries. Instructor 2Lt. Warren T. Cowgill, 22, received minor injuries performing a successful forced landing in AT-6C #42-44318. The airplanes were part of a three-ship flight that took off from Moore Field, Mission, Texas, on a routine formation training flight. The instructors piloted the airplanes on the formation take-off and would fly during all change of position maneuvers. The formation climbed to 6,000 feet and leveled out. The airplanes flew in formation for a short time and the airplanes changed positions several times. The lead airplane then entered a wing-over and the other ships followed in a loose string formation. AT-6C #42-44318 was flying in the number-two position; AT-6C #42-43952 was flying in the number-three position. The lead airplane then attempted to create a Lufbery circle to the right. Lt. Cowgill stated that he could not keep the leader in view because of a frosted cockpit canopy so he peeled off of the circle. Lt. Cowgill leveled off at 4,000 feet, felt a collision and heard a "grinding sound and metal ripping." Lt. Trampenau could not control the airplane and bailed out immediately, striking some part of the airplane during the attempt. Lt. Cowgill stated that he did not see the airplane that collided with his. Investigation revealed that the starboard wing of Lt. Trampenau's airplane collided with the tail and fuselage of Lt. Cowgill's plane.

May

5-1-45. Williamsburg, Massachusetts. At 0830 EWT, a Consolidated B-24J flying in instrument conditions collided with terrain and crashed at Williamsburg, Massachusetts, killing three fliers and injuring six others, two of them seriously. B-24J #42-50995 was part of a flight that took off from Wendover Field, Chicopee, Massachusetts, on a training flight. The formation was flying under the overcast. The pilots were ordered to break up the formation, climb through the overcast and re-form on top. While the subject airplane was climbing through the overcast, the Wendover Field control tower ordered the pilot to hold at 4,000 feet. There were many airplanes operating in the area at the time and tower personnel were working several airplanes. Tower personnel directed B-24 #625, which was flying instruments on the northeast leg of the Wendover Range, to descend to contact conditions. The pilot of the subject airplane, thinking the direction was for him, began descending. The subject airplane began colliding with trees as it descended through 1,500 feet msl. The pilot attempted to pull up but was unsuccessful. The airplane passed through the treetops for a short distance before it crashed into wooded terrain and broke up. Gunners Cpl. Donald R. McKenzie, Cpl. Kenneth V. Powell and Cpl. Joseph Skwara were killed in the crash. Pilot F/O Robert B. Fair and navigator 2Lt. William Eller received serious injuries. Crewmembers injured in the crash were: F/O Hugh T. Cox, co-pilot; Sgt. Lincoln J. Harvey, Jr., gunner; Sgt. Melvin L. Irving, engineer; Sgt. Morton L. Kleven, radio operator.

5-2-45A. Goldsboro, North Carolina. At 0820 EWT, a North American AT-6F crashed five miles west of Goldsboro, North Carolina, killing passenger 2Lt. James V. Land and seriously injuring pilot 2Lt. Wesley Hjalmer, 22. The airplane took off from Seymour-Johnson Field, North Carolina, on an instrument-training mission. The airplane was observed flying at about 1,500 feet agl when the engine began making "popping" noises. The airplane made a 180-degree turn to the right and flew away losing altitude. The pilot was apparently attempting to make an emergency forced landing in a farm field. The airplane made a shallow turn to the left to make the field but collided with a tall tree. The airplane went out of control and smashed to the ground in the plowed field.

5-2-45B. Abilene, Texas. At 1207 CWT, a Republic P-47D crashed while attempting to land at Abilene Army Air Field, Abilene, Texas, killing 1Lt. William H. Goes, 23. The pilot had returned to Abilene Army Air Field after completing a routine training mission and had entered the traffic pattern. The pilot overshot on the landing, "floating" half way down Runway 13 before the tower ordered him to go around. The pilot re-entered the traffic pattern and then overshot on the turn from the base leg to final approach. The pilot attempted to S-turn to get back on the runway heading. The tower again ordered the pilot to go around, and he peeled up and re-entered the traffic pattern. The pilot again overshot on the turn from base to final. The pilot tightened the turn to a nearly vertical bank, causing the airplane to stall at about 400 feet agl. The airplane dropped suddenly and smashed to earth, bursting into flames upon impact.

5-2-45C. Pleasanton, California. At 1340, a Douglas A-20H crashed five and a half miles south of Pleasanton, California, killing pilot 2Lt. Nick Momcilonich, 26. The airplane was one of a flight of five that took off at 1310 from Hayward Army Air Field, Hayward, California, on a camera-gunnery mission. Investigators stated, "Lt. Momcilonich was making a steep diving attack and when he had dropped below the target plane, started to pull up. His plane went into a flap spin, partially recovered after several turns, then went into another spin and crashed at 1340 PWT. The plane immediately burst into flames, and was totally wrecked. The pilot was killed instantly. Verbal orders of this station prohibit all violent maneuvers in A-20 type aircraft as their high speed stall characteristics are well known to supervisory personnel at this station."

5-2-45D. Millville, New Jersey. At 1715 EWT, two Republic P-47G airplanes collided in mid-air and crashed four miles southwest of Millville Army Air Field, Millville, New Jersey, killing both pilots. 2Lt. Lee L. Pryor, 24, was killed in the crash of P-47G #42-25185; 2Lt. William D. Slater, 28, was killed in the crash of P-47G #42-25129. The airplanes were part of a three-ship flight that took off from Millville Army Air Field on a combined aerial gunnery and skip-bombing mission. Lt. Pryor was flying in the number-two position; Lt. Slater was flying in the number-three position. The flight arrived at the bombing range and entered a loose trail formation. The flight was coming on the range from the north and flying a right-hand pattern. The number-three ship, flying at about 1,000 feet agl, apparently overtook the number-two ship while both were turning from the downwind leg to the base leg on the second pass at the target. Investigators stated that the number-three ship had flown a shorter pattern on the second pass and the pilot had probably lost sight of the number-two ship.

5-2-45E. Selma, Alabama. At 1708 CWT, a Curtiss P-40N crashed five miles southwest of Craig Field, Selma, Alabama, killing pilot F/O Victor C. Karpinski, 23. The P-40N took off from Craig Field on a routine transition flight. Investigators stated, "The aircraft was heading in the general direction of west at an altitude of 3,000 to 4,000 feet. The engine backfired a number of times and the engine stopped. Smoke and some fire were seen coming from the aircraft. The aircraft turned slightly to the left then started into a dive of approximately 45 degrees, still smoking. Witnesses noticed no attempt by the pilot to pull out of the dive or to bail out. None of the witnesses saw the aircraft roll over during the dive of 45 degrees, but marks on the ground indicated that the aircraft hit at approximately a 60 to 70 degree angle on its back. The aircraft burned immediately upon impact. The pilot's wristwatch was located and was stopped at 1708 and 22 seconds." The engine fire was caused by the failure of the supercharger impeller shaft's rear bearing.

5-3-45A. San Angelo, Texas. At 0830, a North American AT-6B crashed 11 miles northeast of Goodfellow Field, San Angelo, Texas, killing instructor 2Lt. Douglas H. Duffield and student A/C Virgil R. Duncan. The airplane had taken off from Goodfellow Field on a routine training flight. The airplane was observed flying straight and level at about 4,000 feet agl. The airplane was then seen to enter a descending glide with power off. The airplane was then seen to go into a turn to the right with the engine running "very hard." The airplane entered a dive and a partial recovery was made at about 400 feet agl. Just as the airplane was leveling out, the port wing was seen to separate and strike the cockpit canopy. The airplane then rolled to the left two or three times before smashing into the ground and exploding into flames near U.S. Highway 67.

5-3-45B. Pleasantdale, Nebraska. At 1725 CWT, a North American AT-6B crashed four miles south of Pleasantdale, Nebraska, killing pilots 2Lt. Robert J. Karl and 2Lt. Warren R. Schriver. The airplane had taken off at 1430 CWT from Lincoln Army Air Field, Lincoln, Nebraska, on an instrument training flight. Lt. Karl was flying in the front cockpit as the safety pilot and Lt. Schriver was flying in the rear cockpit as the instrument pilot. The airplane returned to Lincoln Army Air Field and landed at 1630. The next take-off was at 1645. At about 1715, farmers observed the subject airplane flying in the area at low altitude. Several climbs and turns were performed before the airplane was seen to enter a steep bank. The airplane stalled out of the steep turn and spun to the ground, killing the pilots instantly. The airplane did not burn but was totally destroyed.

5-4-45A. Pinedale, California. At 1330, a Martin TB-26B crashed one and a half miles north of Pinedale, California, killing six fliers. The airplane took off at 1328 from Runway 29 at Hammer Field, Fresno, California, on a routine training flight. About a minute after take-off, the pilot radioed the control tower and advised that he was returning to the field because his port engine was "cutting out." The tower operator cleared the field for the emergency landing on Runway 11. Moments later, the airplane crashed in rough terrain about 100 feet southeast of Highway 41. The airplane careened across the highway and began to break up, bursting into flames as it came to rest about 250 yards northwest of the highway. One body had been thrown clear. Investigators speculated that the engines had been susceptible to detonation because the airplane was delayed on the ground for 15 minutes following the engine run-up prior to take-off. Killed in the crash were: 2Lt. Arthur C. Middleton, Jr., pilot; 2Lt. Bernard Prizer, co-pilot; Sgt. Joseph J. Hizney, engineer; Cpl. Paul E. Redhead, engineer; Sgt. John B. Szuce, passenger; Cpl. Paul Brown, passenger.

5-4-45B. Riverside, California. At 1607, a Consolidated B-24J crashed six miles northwest of March Field, Riverside, California, killing eight fliers. The airplane took off at 1552 from March Field on a high-altitude bombing mission. The airplane made a normal take-off and then entered a wide left turn, circling the field before climbing away. At 1605 the pilot radioed the tower and advised that he was returning to the field with the number-three propeller feathered. The pilot salvoed the bomb load but was unable to hold altitude. The airplane then made a slow left turn in a nose-high attitude. The airplane turned through about 15 degrees of turn before the airplane fell off on the starboard wing and entered a steep turn to the right. The starboard wing collided with the terrain and the airplane cartwheeled into the ground, exploding into flames and killing the crew instantly. Killed in the crash were: 1Lt. Robert H. Tieck, pilot; 2Lt. Robert P. Martine, co-pilot; 2Lt. Arthur A. Miller, navigator; 2Lt. Charles L. Yost, bombardier; Cpl. Vernon B. Behram, engineer; Cpl. Raymond E. Bengston, engineer; Cpl. Robert E. Arnold, radio operator; Cpl. Leroy M. Apple, gunner.

5-4-45C. Victorville, California. At 2315, a Consolidated B-24J crashed eight and a half miles southwest of Victorville, California, killing five fliers. Co-pilot Daniel W. Sherman, Jr. was seriously injured parachuting to safety; engineer SSgt. Giacome Tedone escaped injury parachuting to safety. The airplane took off at 2250 from Victorville Army Air Field on a radar-bombing mission. The airplane made a normal take-off and climbed out of the area. Upon reaching 13,000 feet, the number-four cylinder head temperature dropped to zero. All other instruments read normal and the airplane had not begun to yaw because of differential power. The pilots and engineer suspected that the number-four cylinder head temperature gauge was inoperative. A short time later, the number-two engine cylinder head temperature began to fall to about 40 degrees and manifold pressure had dropped to about 20 inches. The co-pilot then noticed the number-two fuel pressure drop dramatically. The co-pilot switched on the number-two fuel booster pump and almost immediately the number-two engine burst into flames. The pilot pointed the airplane toward the area of the field and the crew began to fasten their parachutes. The pilot attempted to extinguish the fire with the engine fire extinguishers but was unsuccessful. The airplane descended through 8,000 feet and the pilot ordered the crew to abandon ship. The engineer, who had been standing on the forward bomb bay catwalk, failed to shut off the fuel valve to the number-two engine and had fastened his parachute incorrectly. The bomb doors were opened and the co-pilot pushed the engineer out through the open bomb bay and then jumped himself. The airplane was observed trailing flames from the port wing as it entered a steep dive at 3,000 feet agl. The airplane remained in the dive until it struck the ground and burst into flames. Investigation revealed that a broken fuel line caused the fire in the number-two engine. Killed in the crash were: 1Lt. William K. McKain, pilot; 1Lt. John R. Wisehart, Jr., navigator; 1Lt. Willis A. Gregory, bombardier/navigator; Charles C. Griffith, Jr., navigator; Pfc. Charles C. Stowe, radio operator.

5-5-45. Russell, Kansas. At 2325 CWT, a Boeing B-29 crashed seven miles northeast of Russell, Kansas, killing nine fliers. Four crewmembers parachuted to safety. The airplane was returning to its station at Walker Army Air Field, Victoria, Kansas, after successfully completing a high-altitude formation flight. Take off was at 1754 CWT. At 2230 CWT, the pilot radioed the tower and requested a permission to make a long approach using the Walker Radio Range. Permission was granted, and weather information and landing instructions were issued. There was no further contact with the airplane. The pilot was under the hood and made his let-down, missing the field by a couple miles. The pilot increased rpm to climb away and attempt another instrument approach. Moments later, the number-one engine began to surge and then it burst into flames. Investigators stated, "Both fire extinguishers were used without success. The order to bail out was given at this time. Approximate altitude of aircraft was 1,500 feet above the terrain. [Co-pilot] states that he raised the emergency hatch over the nose wheel and as the gear went down he bailed out. He observed the airplane and could see fire streaming out all over the number-one engine and running back as far as the tail surface. The plane seemed to make a slow 360 degree turn to the left, losing altitude all the way and as it came out of the turn it dived straight into the ground, exploding upon impact." Co-pilot 2Lt. Phillip W. Stratton, navigator 2Lt. Ellis D. Parker and radio operator Cpl. Harold J. Wymer parachuted to safety and

escaped injury; engineer TSgt. John W. Zimmerman sustained minor injuries parachuting to safety. Killed in the crash were: 1Lt. Edward J. Dittus, B-29 instructor; Capt. Delbert J. Fleming, Jr., pilot; 1Lt. Sheldon S. Cole, bombardier; 1Lt. John R. Meldrum, pilot-rated passenger; Cpl. Carl G. Zugelder, Jr., gunner; Cpl. Harry S. Alexander, Jr., gunner; Cpl. Darrel A. Diehn, gunner; Cpl. John R. Albus, gunner; Pvt. John McCreary, gunner.

5-6-45A. Dalhart, Texas. At 0855, a Republic P-47G suffered a catastrophic structural failure of the tail section and crashed eight miles southwest of Dalhart, Texas, killing pilot 2Lt. James L. McCray, 20. The airplane took off from Dalhart Army Air Field on a routine local transition training flight. The airplane was observed flying at very high altitude, "making vapor trails." The airplane then entered a steep high-speed dive. Pieces of the aircraft were observed falling away. The airplane then entered a flat spin to the left, remaining in this attitude until it struck the ground level and burst into flames. The pilot's body was found about four feet to the right of the fuselage and about two feet behind the cockpit, indicating that the pilot had attempted to bail out at very low altitude. Pieces of the port elevator and port aileron were found about a half-mile from the main wreckage.

5-6-45B. Alexandria, Louisiana. At 1515 CWT, a Boeing B-17F and a B-17G collided in midair and crashed one mile southwest of Esler Field, Alexandria, Louisiana, killing 18 fliers. Killed aboard B-17F #42-5355 were: 2Lt. William Gonzalez, pilot; 2Lt. Wendell R. Stanley, co-pilot; 2Lt. Wilbur E. Lilly, navigator; 2Lt. David L. Prescott, Jr., bombardier; Sgt. Hoyt A. Young, engineer; Sgt. William R. Rooker, radio operator; Sgt. George P. Cirzan, gunner; Cpl. Joseph R. Cooper, gunner; Cpl. Robert L. Stoutburg, gunner. Killed in the crash of B-17G #42-31948 were: F/O Robert W. Lux, pilot; F/O Andrew G. Anderson, co-pilot; 2Lt. Kenneth T. Klaasen, navigator; F/O Angelo G. Gagliardi, bombardier; Sgt. Lloyd E. Battleson, engineer; Sgt. Robert H. O'Toole, gunner; Sgt. Cecil V. Vowels, gunner; Cpl. Richard O. Scott, gunner; Cpl. Carl C. Degraftenreid, gunner. The two subject airplanes were part of a 54-ship formation that was returning to Esler Field following a wing formation practice. The bombers in the wing formation were unable to drop their load of practice bombs because of low ceilings over the range. B-17F #42-5355 was the lead ship of the second element of the low squadron of the low group; B-17G #42-31948 was flying in the number-two position of the first element of the same flight. B-17 #42-5355 was too close to the lead ship of the first element; B-17G #42-31948 was above and to the right of the B-17F. The B-17G began to drift to the left and collided with the B-17F. Both airplanes broke up and burst into flames as they separated. The flaming wrecks plummeted to earth and exploded upon impact.

5-6-45C. Liberal, Kansas. At 2305 CWT, a Consolidated RB-24D and an RB-24E collided in mid-air about one mile northwest of Liberal Army Air Field, Liberal, Kansas, killing three fliers. Killed in the crash of the RB-24D were: 2Lt. Rubin Indik, pilot; 1Lt. Archie B. Caraway, co-pilot; Pfc. Aloysius H. Freigurger, engineer. The B-24D was approaching for a landing on Runway 30L at Liberal Army Air Field and was flying close in behind the B-24E, which was making a touch-and-go landing. The control tower officer observed that the interval between the airplanes was too close and that a dangerous situation was developing. The tower operator advised the B-24D pilot to either go around or land on Runway 30R. The B-24D pilot retracted the landing gear and added power for the go around but did not acknowledge the control tower instructions. The B-24D began overtaking the B-24E, which had just taken off on its touch-and-go, and the two airplanes collided. The B-24D was observed to be trailing fire from the starboard wing as it entered a spiral to the right, turning about 120 degrees off of the take-off heading before it struck the ground and exploded. Because of the lack of horizon and no moon, the B-24E pilot was flying on instruments when the B-24D flew on top of him, knocking off the B-24E's number-three propeller and causing the engine to erupt into flames. The pilot cut off the fuel, pulled the number-three extinguisher and the fire went out. The pilot continued in the traffic pattern and made a safe emergency landing. Crewmembers landing safely aboard the RB-24E were: 1Lt. Donald R. Clarke, pilot; 2Lt. William H. Clift, co-pilot; 2Lt. Lewis J. Cowart, co-pilot; Cpl. Carl E. Chuit, engineer.

5-7-45A. Tuttle, Oklahoma. At 0905 CWT, a Lockheed P-38L crashed two miles southwest of Tuttle, Oklahoma, killing pilot F/O William M. Fuller, 22. The airplane was part of a three-ship flight that took off from Will Rogers Field, Oklahoma City, Oklahoma, on a single-engine training flight. A fourth ship was unable to take off because of mechanical difficulties. The airplanes climbed to a safe altitude to practice single-engine flight and maneuvers. The instructor ordered the pilots to shut down the starboard engine and feather the propeller. The subject pilot had difficulty maintaining lateral control and fell out of position several times. The instructor ordered the pilots to extend the landing gear. The pilots complied and airspeed dropped to about 135 mph. The subject pilot did not acknowledge the flight leader's calls and it was discovered that the subject airplane was missing from the flight. The instructor assumed that the pilot had returned to the field and continued with the training mission. The airplane had apparently entered a spiral and smashed into the ground, exploding into flames upon impact. Investigators speculated that the pilot had lost control of the airplane while flying on one engine with the landing gear extended. It was also

noted that the pilot had indicated that he did not like flying P-38 type airplanes and had expressed a desire to transfer to the Troop Carrier Command.

5-7-45B. Gulf of Mexico, Mississippi. At 1125 CWT, a Consolidated OA-10A (the AAF version of the PBY-5A Catalina Flying Boat) crashed while operating on the Gulf of Mexico 12 miles southeast of Keesler Field, Biloxi, Mississippi, killing four fliers. Five crewmembers were injured, three of them seriously. The airplane had taken off from Keesler Field on a training flight to the Mississippi Sound and return. Investigators stated, "Investigation of the OA-10 aircraft wreckage reveals that the airplane apparently water-looped and then cartwheeled upon landing. It is likely that just previous to landing starboard wing was low, causing he starboard float to strike the water, creating a large amount of drag, which caused the airplane to swerve to the right and nose down. Lower part of the flight deck sheared off, following which the airplane cartwheeled and finally came to rest." Killed in the crash were: 2Lt. Buford Bell, OA-10A instructor; 1Lt. Ralph A. Weaver, pilot; TSgt. Fred F. Brown, engineer; Cpl. Paul F. Nagle, radar operator. Co-pilot 2Lt. Thomas G. Martin, navigator 1Lt. Robert B. McSweeney and radio operator Pfc. Joe R. Billingslea were seriously injured in the accident. Engineer TSgt. Donald W. Gorman and crewmember Pfc. George E. Bale sustained minor injuries in the accident.

5-8-45. Vinita, Oklahoma. At 0300 CWT, a Vultee-Stinson L-5 crashed 12 miles east of Vinita, Oklahoma, killing liaison pilot SSgt. Harry C. Hatch, Jr., 24. The airplane took off from Muskogee Army Air Field, Oklahoma, on a routine training flight. The pilot evidently became lost and low on fuel. The airplane crashed while the pilot was attempting to make an emergency forced landing. The port wing struck the terrain and the airplane went out of control and smashed to the ground. The pilot was thrown out of the aircraft through the side hatch. Investigation revealed that the port fuel tank was empty and the starboard fuel tank contained fuel.

5-11-45A. Troy, Texas. At 0045 CWT, a Beech C-45 flying in poor weather crashed three miles southwest of Troy, Texas, killing pilot 1Lt. Gerald W. Whisler and engineer-rated passenger Cpl. Russ V. Blackburn. The airplane took off from Temple Army Air Field, Temple, Texas, on a flight to Waco Army Air Field, Waco, Texas. The airplane took off at 0035 and then turned on course for Waco. A few minutes later the airplane crashed into the ground in a descending turn to the right. The airplane exploded into flames upon impact. Investigators speculated that the pilot had inadvertently flown the airplane into the ground after becoming disoriented.

5-11-45B. Frederick, Oklahoma. At 1002 CWT, a North American TB-25H flying in instrument conditions suffered a catastrophic structural failure and crashed four miles southeast of Frederick, Oklahoma, killing pilot 2Lt. Rudolph W. Kubik and co-pilot 2Lt. Donald A. Holmer. The airplane took off from Frederick Army Air Field on a navigation flight to Wichita, Kansas, to St. Joseph, Missouri, and return to Frederick. The airplane was observed descending at a steep angle with engines wide-open and shedding pieces. The airplane entered a steep dive and smashed to the ground, exploding into flames upon impact. Investigation of the wreckage revealed that the starboard wing had failed and separated, colliding with the starboard tail section and causing it to fail and separate. The airplane went out of control and dove into the ground.

5-11-45C. Florence, South Carolina. At 1720, a Douglas A-26B crashed eight miles southwest of Florence Army Air Field, Florence, South Carolina, killing three fliers. The airplane took off from Florence Army Air Field on a routine training flight, which included chandelles, peel-offs and lazy eights. The airplane was observed in a near vertical dive at about 3,000 feet agl. The airplane continued in this attitude until it struck the ground and exploded into flames. No pieces were seen to fall away from the airplane while it was in the dive and no occupant made an obvious effort to parachute to safety. Investigators could not determine what caused the airplane to enter the vertical dive.

Pilot 1Lt. Alfred E. Singer, gunner Cpl. Robert E. Wallace and passenger 2Lt. William C. Wolcott were killed in the crash.

5-11-45D. MacDill Field, Florida. At 2357 EWT, a Boeing B-29A crashed while attempting an emergency landing at MacDill Field, Tampa, Florida, killing nine crewmembers and seriously injuring two others. Killed in the crash were: Maj. William W. Roberts, pilot; 2Lt. William C. Tice, co-pilot; 2Lt. Ralph K. Hausknect, navigator; 2Lt. Jerome A. Berdy, bombardier; SSgt. John M. Petersen, engineer; Sgt. William J. Putcamp, radio operator; Cpl. Patrick J. Reynolds, gunner; Cpl. Jack G. Pardy, gunner; Cpl. Arthur M. Dantzler, tail gunner. Gunner Cpl. John R. Myers and radar operator 2Lt. Loren S. Armbruster survived but received serious injuries. Investigators stated, "Major Roberts and his combat crew took off from MacDill Field at 2348 on a long range navigation flight. Five minutes after take-off the pilot called the tower and stated that he had a runaway propeller on the number-one engine and would circle the field. Shortly thereafter he stated that the number-one propeller had been feathered and he was salvoing bomb bay tanks and was coming in to land. A 180-degree turn was made and a straight-in approach to the field was attempted. The airplane seemed to be under full control until just before the crash when the pilot stated that he was losing altitude fast and would make an emergency landing. The crew was not notified of this over the interphone. The aircraft made a very low approach and was not lined up on the runway. Marks and evidence indicate that about 12 feet of the right wing was sheared off upon impact with a tree, struck at about 40 feet from the ground. The aircraft apparently rolled to the right cutting off another tree but missing others spaced approximately 50 feet on either side. The tail section was the first to strike the ground shown by the sections of the horizontal stabilizer and rear portion of the fuselage. The aircraft crashed and burned short of the field and to the left of the runway at 2357. The rear un-pressurized portion of the fuselage was severely damaged but did not burn. Two survivors were taken from this section; all other personnel were fatalities."

5-11-45E. Tampa, Florida. At 2359 EWT, tail gunner Pfc. Eugene Eglion fell to his death when he was ejected from a Boeing B-29 that suffered a bulkhead failure while operating at high altitude five miles south of MacDill Field, Tampa, Florida. Pilot Capt. Kirkland C. Krueger, co-pilot F/O William J. Cutos and eight other crewmembers escaped injury and the bomber was landed safely at MacDill Field. The airplane had been on a night bombing mission. Investigators stated, "Pressurization had been accomplished and the plane had reached an altitude of 23,000 feet, cabin pressure being at 8,300 feet. The tail gunner was in a reclining position between the ring gunner stand and the pressure bulkhead door. He was using his parachute as a support and it was not fastened to the harness. Explosive decompression occurred at 2359 at an altitude of 23,000 feet. After the crew had adjusted their oxygen masks an oxygen check was made and the tail gunner was found to be missing. Upon investigation the forward bulkhead door in the rear pressure compartment was found to be gone and the rear bomb bay doors were open. The body of the tail gunner was located in Tampa. His parachute pack was found about 30 feet from the body and the ripcord had not been pulled. There was no parachute harness on the body, and it could not be located on the plane. Structural failure of the pressure bulkhead door is believed to be the cause of the accident."

5-12-45A. Fort Bragg, California. At an unknown time between 1750 and 2300 PWT, a Consolidated B-24M collided with rising terrain 30 miles north of Fort Bragg, California, killing 11 fliers. Killed in the crash were: 1Lt. John W. Apedaile, pilot; F/O James T. Gold, co-pilot; F/O Carrol Trowbridge, navigator; 2Lt. Adolph J. Frantz, radar operator; 2Lt. Donald S. Brown, radar operator; F/O Kenneth H. Peters, radar operator; 1Lt. Jack L. Fuller, navigator; Sgt. Robert G. Lord, engineer; Cpl. Victor W. Hamilton, engineer; Sgt. L.T. Tippets, radio operator; Sgt. Raymond B. Oppenheim, gunner. The B-24 took off at 1330 PWT from March Field, Riverside, California, on an over-water navigation flight. At 1750, after crossing the coast while flying out to sea on a northwest heading from Santa Rosa, California, the airplane transmitted a routine position report. There was no further contact with the airplane. The airplane failed to return to base and was declared missing. The airplane was found on 5-19-45. Investigation revealed that the airplane collided with large redwood trees on top of a ridge while flying in instrument conditions in a straight and level attitude with all engines producing power. The airplane broke up as it passed through the tall trees at an elevation of 1,200 feet msl. Many pieces of wreckage passed through the trees and were scattered at varying distances down the other side of the steep slope. It was not known if the airplane had completed its overwater navigation mission prior to colliding with rising terrain. Investigators speculated that the crash occurred a short time after the position report.

5-12-45B. South Bend, Indiana. At 2245 CWT, a Beech AT-7 flying in instrument conditions crashed one and a half miles north of Bendix Field, South Bend, Indiana, killing three fliers and injuring two others. Pilot 2Lt. Frederick H. Edwards, navigator 1Lt. Julius Goldstein and navigator 1Lt. Barnard M. Plaut were killed in the crash. Navigator 1Lt. Frank A. Picard and engineer Joseph M. Adragna survived with minor injuries. The airplane took off at 2217 CWT from Chicago Municipal Airport, Chicago, Illinois, on a flight to Bendix Field. The airplane was observed to fly over the field at an altitude of 400 feet agl, heading north. The airplane, which did not contact the control tower, began a descending turn to the left in an effort to enter the traffic pattern. The airplane, while still turning to the left, entered a thick fog bank northwest of the field. Moments later, personnel at the field heard the airplane crash. The surviving crewmembers believed that a faulty altimeter caused the pilot to fly into the ground.

5-12-45C. Cordelia, California. At approximately 2400 PWT, a Douglas C-47B collided with rising terrain and crashed three and a half miles southwest of Cordelia, California, killing three fliers. Pilot 1Lt. Glen K. Dorsey, co-pilot 1Lt. Francis E. Reagan and flight clerk Pvt. Harry S. Cooper were killed in the crash. The airplane had landed at 2226 PWT at McClellan Field, Sacramento, California, and Lt. Dorsey and his crew took over the airplane for the cargo flight to Hamilton Field, San Rafael, California, and Hill Field, Ogden, Utah. The airplane took off from McClelland Field at 2342. There was no further contact with the airplane and it failed to arrive at Hamilton Field. The airplane wreckage was found the next day. Investigation revealed that the airplane collided with a ridge at about 800 feet msl. The airplane collided with the ridge, which gently sloped down to the right, while in straight and level flight with both engines producing power. The port wing and fuselage smashed into the ridge, sending the airplane out of control. Both propellers separated as the airplane continued on and careened over the ridge, the starboard wing striking the ground about 120 yards from the point of initial impact. The port wing separated and fell away as the starboard wing smashed into the terrain. The starboard wing separated and the fuselage smashed to the ground

on its right side, the tail breaking off aft of the main cargo door. The shattered fuselage came to rest about 50 yards further on but the engines continued forward, coming to rest about 30 yards beyond the main wreckage.

5-13-45A. Sweetwater, Texas. At 0855 CWT, a Republic P-47D crashed 10 miles southeast of Avenger Field, Sweetwater, Texas, killing F/O Robert D. McCann, 21, who fell to his death in an unsuccessful parachute jump. The airplane was flying in the lead position of the second element of a four-ship flight that had taken off from Avenger Field on a medium-altitude unit combat mission. The subject airplane entered a high-speed stall while maneuvering at about 15,000 feet. The airplane snapped over into a half-roll, falling into a steep spin to the left. The airplane remained in this attitude until it smashed into the ground and exploded.

5-13-45B. Ocean Beach, New York. At 1302 EWT, a Beech C-45F crashed into the surf at Ocean Beach, Long Island, New York, killing 1Lt. Burton E. Goldstein, 28. The airplane took off from Albany, New York, on a flight back to its home station at Mitchel Field, New York. The airplane was observed to descend out of the overcast, flying about 200 feet over Ocean Beach. The airplane made a 180-degree turn and completed the turn without losing altitude. The ship nosed down and flew into the ground at a 10-degree angle. The airplane bounded back into the air and then crashed into Great South Bay.

5-13-45C. Ontario, California. At 1430, a Lockheed P-38L crashed at Ontario Army Air Field, Ontario, California, killing 1Lt. Willard R. Rasmusson, 26. The airplane was part of a flight that was returning to the field after completing a routine training mission. The pilot made a normal peel-off and approach for Runway 25. The airplane was a little high on the final approach, and as the airplane leveled off above the runway the pilot elected to go around. The airplane began veering to the right as it was climbing away for the go around, rolling over and smashing to the ground near the runway. Investigation revealed that a carburetor problem in the starboard engine allowed a lean mixture as the pilot added throttle for the go-around. The differential power caused the airplane to wingover to the right and into the ground at the runway intersection.

5-14-45A. Delaware Bay, New Jersey. At 1405 EWT, a Republic P-47D crashed into Delaware Bay six miles south of Greenwich, New Jersey, killing pilot 2Lt. William E. Canniff, 25. The airplane was part of a flight that took off from Millville Army Air Field, Millville, New Jersey, on a "chemical smoke mission." The airplanes formed up line abreast, but an airplane was slightly out of position and by the time the airplane regained position the airplanes were approaching the end of the range. No chemical smoke was deployed and the flight entered a 180-degree turn to the left. The subject airplane, flying to the left of the flight leader, flew into the water while attempting to maintain his relative position to the flight leader.

5-14-45B. Williams, Florida. At 1405 EWT, a North American P-51K suffered a catastrophic structural failure and crashed two miles northeast of Williams, Florida, killing pilot F/O Jerome R. Sturm, 20. The airplane was part of a two-ship flight that took off from Sarasota Army Air Field, Sarasota, Florida, on a local training flight. The airplanes climbed to about 10,000 feet. The subject airplane was observed in a steep turn to the left, trailing condensation "streamers" from the wingtips. The starboard wing failed and folded up, separating from the fuselage and smashing into the cockpit canopy, shattering it. The starboard wing, pieces of the canopy and the cockpit bulletproof glass landed about 500 yard from where the airplane slammed into the ground and exploded into flames. Investigators speculated that the pilot had overcontrolled the airplane, causing the wing to be stressed beyond its design limitations.

5-15-45A. MacDill Field, Florida. At 0509 EWT, a Boeing B-29A suffering an engine failure crashed while attempting an emergency landing at MacDill Field, Tampa, Florida, killing radio operator Pfc. Louis J. Takas, Jr., and navigator F/O James F. Kramer. Three other crewmembers were injured, two of them seriously; seven fliers escaped injury. The airplane took off from MacDill Field at 0449 EWT on a high-altitude bombing mission. Soon after take-off, the starboard scanner noticed sparks coming from the number-four engine. The engineer observed that the number-four engine was on fire and he notified the copilot. The pilot shut down the engine, feathered the propeller and the fire went out immediately. The pilot climbed to 1,000 feet agl and entered the traffic pattern for a landing. Because of hazy conditions the pilot overshot the runway while turning from the base leg, so he decided to go around. The pilot again climbed to 1,000 feet agl and entered the traffic pattern for another landing attempt. The pilot again overshot the runway on the turn to final so he went around a third time. While on the downwind leg for Runway 22, the pilot radioed the tower and requested that the field floodlights be illuminated so he could see the field. The floodlights were turned on and the pilot was able to see the general area of the runway. The pilot, satisfied that he was lined up with the runway, requested that the floodlights be turned off. The floodlights were turned off and the pilot continued his approach. The pilot had again overshot on the turn from the base leg to final approach; the airplane was to the right of the runway and had to turn to the left to get back to the runway. The airplane was approaching from the right at a 45-degree angle to the runway heading. The airplane was too low for the pilot to reacquire the runway and get the airplane realigned with the runway, so he called for the landing gear to be retracted and the flaps retracted

to five degrees. The pilot added power for the go-around. The airplane was over the hangar line apron when it started to roll to the right. The pilot reduced power on the port engines in an attempt to level the airplane. The starboard wing continued to drop and the airplane barely cleared the hangars and aircraft parked on the flight line. The airplane smashed into the ground in a skidding turn to the right, bursting into flames upon impact. Investigation revealed that the number-four engine had been written up twice for an excessive oil leak in the day preceding the accident. Repairs were made and the ship cleared for flight. Investigators speculated that excessive oil on the engine parts and airplane structure from the two previous oil leaks, and possibly oil from another oil leak, might have been ignited by hot engine parts. Gunner Pfc. Charles H. Hutchinson and radio operator Cpl. Robert E. Martens were seriously injured; bombardier 2Lt. Marvin L. Samols sustained minor injuries. Crewmembers uninjured in the crash landing were: 1Lt. Joseph P. Cullender, pilot; 2Lt. Leonard Feinstein, co-pilot; 2Lt. George M. Lethbridge, radar operator; SSgt. Leo J. Dobner, engineer; Pfc. Jack C. Dupree, gunner; Cpl. Richard W. Lockwood, engineer; 1Lt. Samuel W. Marshall, bombardier instructor.

5-15-45B. Rowe, New Mexico. At 1015 MWT, a North American F-10W (the photo-reconnaissance version of the B-25) crashed 15 miles SSW of Rowe, New Mexico, killing eight fliers. The airplane took off from Will Rogers Field, Oklahoma City, Oklahoma, on a weather-checking flight. The airplane was on an IFR clearance and was cleared to fly at 10,000 feet direct but Army Flight Service had assigned an altitude of 12,000 feet. The airplane encountered icing conditions while flying at 10,000 feet. Another airplane flying on the same airway advised the subject pilot to descend to 9,500 feet to escape the icing conditions. The subject pilot acknowledged but stated that he was going to attempt to climb above the icing conditions. There was no further radio contact with the subject airplane. The port wing failed and separated from the airplane, striking the tail section and causing its failure and separation. The airplane went out of control and smashed to earth, bursting into flames upon impact. The port wing and pieces of the tail section were found about one-half mile from the scene of the crash. Investigators speculated that the pilot had lost control of the aircraft because of severe icing conditions. It was noted that the subject airplane was not equipped with de-icing equipment. Killed in the crash were: 1Lt. Harry W. Stark, pilot; 2Lt. Robert D. Murphy, co-pilot; 1Lt. Calvin Randolph, navigator; 2Lt. Maurice Q. Paeth, crewmember; TSgt. Gilbert D. Laubinger, engineer; Sgt. Constant Diasourakis, radio operator; SSgt. E.E. Pringle, gunner; Seaman 3C C.R. Hayes, passenger.

5-15-45C. Lowell, Indiana. At 1400 CWT, a North American AT-6A flying in instrument conditions suffered a catastrophic structural failure and crashed two and a half miles north of Lowell, Indiana, killing pilot 2Lt. James W. Alexander, 27. The airplane had taken off from Tuskegee Army Air Field, Tuskegee, Alabama, on a flight to Chicago, Illinois, via Louisville, Kentucky. The airplane was observed falling out of the overcast at 700 feet agl minus the port wing. The airplane spun to the ground, killing the pilot instantly. The airplane did not burn but was totally destroyed. The severed port wing was found about 200 yards from the main wreckage.

5-15-45D. Las Vegas, Nevada. At 1231 PWT, a Martin TB-26B attempting an emergency landing crashed one mile northwest of the Army Air Field at Las Vegas, Nevada, killing three crewmembers. Pilot 1Lt. Clinton E. Barksdale, co-pilot 2Lt. Chester Ihle, Jr. and engineer SSgt. Gene E. Harkey were killed in the crash. The airplane took off at 1228 PWT from Las Vegas Army Air Field on an engineering test flight. The port engine began to emit thick black smoke a short time after take-off. The pilot turned the airplane back toward the field. The pilot was unable to maintain altitude and the airplane collided with three power lines. The airplane continued at very low altitude in a nose-high altitude for about three-quarters of a mile before encountering a reinforced ordinance bunker. The pilot pulled up, attempting to clear the building but he was unsuccessful. The airplane mushed into the top of the bunker, caroming off the top and smashing into the ground about 75 feet from the bunker. The airplane nosed into the ground and then flipped over to an inverted attitude before bursting into flames.

5-16-45. Lancaster, California. At 1413, a North American P-51D suffered a catastrophic structural failure and crashed two miles east of Lancaster, California, killing pilot 1Lt. James A. Black, 23. The airplane took off at 1228 PWT from Palmdale Army Air Field, Palmdale, California, on a routine training flight. Witnesses on the ground observed the airplane enter a steep dive at high altitude. The dive angle was estimated to be about 70 degrees. The pilot began his pull-out and the dive angle decreased to about 30 degrees. As the airplane approached about 1,000 feet agl, the port wing was seen to separate. The airplane went out of control and smashed to earth on Avenue J, exploding into flames upon impact.

5-17-45A. Bunger, Texas. At 1250 CWT, a Vultee BT-13B crashed 13 miles southeast of Bunger, Texas, killing pilots 1Lt. Joseph N. Phelan and 1Lt. Herman R. Wade. The airplane took off at 1135 CWT from Fort Worth Army Air Field, Fort Worth, Texas, on a routine pilot proficiency flight. Witnesses on the ground observed the subject airplane flying with another BT-13 type airplane at low altitude over Possum Kingdom Lake. The airplanes approached the hilly terrain near the lake at very low altitude, barely clearing a 200-foot hill. The lead airplane flew over the top and the subject airplane banked away to the right in a climbing turn, disappearing behind the hills. The subject airplane

returned to the area of the lake at low altitude and repeated the course it had flown previously. The airplane, while flying at very low altitude, collided with the terrain in a flat attitude, collapsing the landing gear and breaking the fuselage behind the rear cockpit. The engine separated and came to rest several yards ahead of the wreckage. The airplane did not burn. There were no witnesses to the actual crash. Lt. Wade died about two hours after the accident.

5-17-45B. Turlock, California. At 2022, a North American AT-6C and an AT-6D collided in mid-air at Ballico Auxiliary Airfield seven miles west of Turlock, California, killing AT-6D pilot A/C Robert E. Hutchins, 20. AT-6C pilot A/C Edward F. Hamel was uninjured and was able to safely land. The airplanes were part of a flight of student airplanes that took off from Merced Army Air Field, Merced, California, to practice take-offs and landings at Ballico Auxiliary Airfield. A/C Hutchins had flown too long of a downwind leg and was flying on a long final approach. A/C Hamel had flown a normal pattern at normal altitude and was on his final approach when A/C Hutchins' airplane climbed into his. A/C Hutchins' airplane fell out of control and crashed near the field, bursting into flames upon impact. A/C Hamel was able to make a safe landing at the auxiliary airfield.

5-18-45A. Douglas, Arizona. At 1220 MWT, a North American TB-25C crashed five miles east of the Army Air Field at Douglas, Arizona, killing A/C Francis M. Postle, who fell to his death in an unsuccessful parachute jump. A/C Odell I. Preskitt received minor injuries parachuting to safety. The airplane had taken off from Douglas Army Air Field on a routine training mission. The students had flown with an instructor during the morning hours. The airplane landed at 1045 MWT and the instructor got out. The students took off with A/C Preskitt flying in the pilot seat and A/C Postle in the co-pilot seat. The airplane landed at Auxiliary Airfield #1 and the students changed positions. The airplane flew in the area for a while and then entered the traffic pattern for Douglas Army Air Field. While flying on the final approach the students were given a red light by the mobile traffic control unit. The student added power, climbed away and left the traffic area. The airplane re-entered the traffic area east of the field. Moments later, the student radioed the field that his port engine began to surge and suffer diminished performance. The student inadvertently feathered the starboard propeller but quickly unfeathered it and feathered the port propeller. A/C Preskitt, in the co-pilot seat, radioed the field and declared an emergency. The tower operator acknowledged and the field was cleared for the emergency landing. The airplane began to yaw violently to the right and the students were having trouble controlling the airplane. A/C Postle ordered A/C Preskitt to abandon the airplane. A/C Preskitt bailed out at about 1,900 feet agl and parachuted to safety. A/C Postle bailed out at an altitude too low to allow his parachute to fully deploy. Investigators speculated that the students had mismanaged the fuel system, allowing the port engine to fail because of fuel exhaustion.

5-18-45B. Condron Field, New Mexico. At 1420 MWT, a North American TB-25H crashed four miles south of Condron Field, New Mexico, after it was damaged by shrapnel from bombs that were being dropped in a bombing demonstration at the Oro Grande Gunnery Range. Two crewmembers were killed and another was seriously injured. Four other B-25 aircraft participating in the exercise were damaged in the incident, but the pilots were able to make safe landings. No other airmen were injured. Pilot 2Lt. Calvin H. Mouser, engineer SSgt. Melvin F. Likes were killed in the crash of TB-25H #43-4520. Radio operator Pfc. Martin V. Bernardino received serious injuries. All three had been thrown clear of the aircraft during the crash. The subject airplane was flying in the lead position of the second element of a six-ship flight that had taken off from Deming Army Air Field, Deming, New Mexico, and was participating in a joint exercise with ground units. The subject airplane's element was flying close behind the lead element at about 75 feet agl when the bombs were dropped; the lead element was flying at about 100 feet agl. When the squadron's bombs detonated, shrapnel hit five of the airplanes. The subject airplane was heavily damaged and the pilot radioed ground controllers that he was heading to Condron Field for an emergency landing. Witnesses observed the subject airplane losing altitude and trailing smoke from the port engine. The airplane became uncontrollable and crashed about 40 miles north of El Paso, Texas, bursting into flames upon impact. TB-25J #43-4614 suffered damage to an engine and oil tank. The pilot feathered the starboard propeller and made a safe landing at El Paso Municipal Airport. The four remaining airplanes landed safely at Deming Army Air Field. Upon inspection, it was discovered that three of the four airplanes had been hit by shrapnel from their own bombs. Investigators stated that the pilots should be more thoroughly briefed before all missions.

5-19-45. Litchfield Park, Arizona. At an unknown time between 1500 and 1600 MWT, a North American AT-6C crashed 17 miles south of Litchfield Park, Arizona, killing pilots 2Lt. James E. Hinkle, Jr. and 2Lt. Howard M. Dillwood. The airplane took off at 1430 MWT from Luke Field on a routine training flight. Lt. Hinkle was flying in the front cockpit and Lt. Dillwood was flying in the rear cockpit. The airplane failed to return and was declared missing. The wreckage was found on 5-24-45. Investigation revealed that the port wing had failed and separated from the aircraft. The airplane went out of control and into a steep spin to the left. The airplane spun to earth, striking the ground at an 80-degree angle. The port wing was found about two and a half miles ESE of the main wreckage. It was speculated that the port wing failed due to the stress of an unknown maneuver.

5-22-45A. Foster Field, Texas. At 1600, a Martin TB-26G crashed while attempting a take-off at Foster Field, Victoria, Texas, killing one flier and injuring two others. The airplane, which was on a routine training flight, began its take-off on Runway 6. As the airplane made its take-off roll it entered an extreme nose-high attitude with the main wheels still on the surface. The airplane climbed steeply off of the runway as it passed the intersection with Runway 12. The airplane climbed to about 50 feet agl before falling off on the port wing, spinning about 180 degrees before smashing to the ground and bursting into flames. Investigation revealed that the engineer had inadvertently retracted the landing gear before the airplane was fully airborne, prompting the pilot to pull back on the control wheel in an effort to get the ship in the air and off of the landing gear. The airplane entered a steep climb and stalled above the runway. The pilot and engineer were able to escape the flaming wreckage but were unable to free the tow target operator. Pilot 1Lt. Joseph F. Cox and engineer Pfc. Willie Gavels sustained minor injuries. Tow reel operator Pfc. Ervin W. Niemyer was killed.

5-22-45B. Van Nuys, California. At 1557, a Lockheed F-5E crashed three miles east of Metropolitan Airport, Van Nuys, California, killing pilot Capt. Michael F. Dekleva, 28. The airplane took off from Biggs Field, El Paso, Texas, on a ferry flight to Burbank, California. The flight had originated at Brookley Field, Mobile, Alabama. The pilot arrived in the area of Burbank and requested an emergency landing. The pilot, thinking that he was communicating with the Burbank control tower, had actually radioed the Lockheed factory control tower. The Lockheed tower operator relayed the message and advised the pilot to tune to the Burbank frequency, which he did. The pilot radioed Burbank tower that he would have to make an off-field emergency landing immediately. The airplane, flying at approximately 1,000 feet agl, was seen to fall off into a diving turn to the right and smash to the ground. The pilot was killed upon impact. Investigation revealed that the engines failed because the pilot had failed to switch from an empty fuel tank to a tank that contained enough fuel to safely complete the flight. Investigators noted that Capt. Dekleva was an experienced P-38 pilot.

5-22-45C. Denver, Colorado. At 1750 MWT, a Consolidated RB-24E crashed one and a half miles southeast of Lowry Field, Denver, Colorado, killing engineer Pfc. Robert T. Boyle. Seven other crewmembers received minor injuries. The airplane took off from Lowry Field on an engineer training flight. The airplane made a normal take-off and flew north for about an hour and a half before performing a 180-degree turn to return to Lowry Field. The number-one engine quit a short time after the airplane completed the turn. The pilot feathered the propeller and continued on to Lowry Field. The B-24 arrived in the area of Lowry Field and entered the traffic pattern for a normal three-engine landing. The airplane began drifting to the right while on the flare out, so the pilot decided to go around. The pilot added power and apparently the airplane began to get away from him. The pilots were unable to maintain directional control because of low airspeed, nose-high position and full flaps. The pilot retarded power on the number-four engine and called for landing gear up and raise the flaps 10 degrees. The co-pilot retracted the landing gear and then the flaps. The airplane began to sink rapidly. The co-pilot retracted the flaps all the way but quickly placed the handle in the full down position. The airplane began mushing toward the surface of a small lake and the pilot added power on all operating engines. The airplane began turning to the left because of differential power. The port wing clipped the surface and the airplane cartwheeled into the lakeshore, bursting into flames upon impact. Seven of the airmen were able to escape the airplane. Pilot F/O Eugene L. Uttrich and co-pilot James E. Saxon received minor injuries. Pilot-rated student engineers injured in the crash were: 2Lt. Timothy J. Braidy, F/O David J. Martin, 2Lt. Robert J. Meyer, F/O Oliver J. Wilharber, 2Lt. Albert V. Works.

5-23-45A. Lake Charles, Louisiana. At about 1000 CWT, a Douglas A-26B crashed 20 miles south of Lake Charles, Louisiana, killing four fliers. The A-26 took off at 0822 from Lake Charles Army Air Field on a gunnery and skip-bombing mission at the Calcasieu Bombing Range. The airplane made several firing passes on the gunnery range and at 0945 the pilot radioed for clearance on the skip-bombing range. The pilot was cleared for the range and he made a dry run over the bombing range from north to south. Witnesses on the ground observed the airplane flying at very low altitude about one mile east of the bombing range heading north. The port wing was seen to drop and then the airplane pitched down about 10 degrees and dove into the ground about 100 yards from a small pond, exploding into flames upon impact. Wreckage was scattered for a half mile. Killed in the crash were: 2Lt. Jack H. Soffer, pilot; 2Lt. Robert C. Abegglen, bombardier; Sgt. William E. Haines, gunner; TSgt. Robert P. LaPaze, gunnery instructor;

5-23-45B. Knollwood Field, North Carolina. At 1448, a Douglas C-47A crashed 100 yards west of Knollwood Field, North Carolina, killing four fliers. The airplane took off from Pope Field, Fort Bragg, North Carolina, and was practicing take-offs and landings at Knollwood Field, an auxiliary airfield of Pope Field. The airplane made a landing and then a subsequent take-off at Knollwood Field. The airplane used up the entire runway and had veered to the left on the take-off roll. The airplane was pulled up abruptly and it became airborne at the extreme end of the runway. The port wing dropped and the C-47 banked sharply to the left, the port wing colliding with power lines and the tops of three tall trees. The airplane went out of

control and smashed to the ground about three miles north of Southern Pines, bursting into flames upon impact. It was speculated that the port engine might have suffered from diminished performance on the take-off roll. Killed in the crash were: 1Lt. Robert L. Hughes, pilot; 2Lt. William E. Pero, co-pilot; Cpl. Jack A. Trussell, engineer; Pfc. Richard E. Whann, radio operator.

5-23-45C. Atlantic Ocean, Virginia. At 1455 EWT, a Martin TB-26C crashed into the Atlantic Ocean near Smith Island, Virginia, killing three fliers. Tow reel operator SSgt. Michael Matyola parachuted to safety, suffering minor injuries. Pilot 1Lt. Mayo J. Reece, co-pilot 2Lt. Guy E. Abbott and engineer Sgt. Jeffress P. Craddock were killed in the crash. The airplane took off at 1310 from Norfolk Army Air Field, Norfolk, Virginia, on a target-towing mission. The airplane completed one mission and then returned to the field and dropped the target. The airplane flew back to the range at 7,000 feet and waited for another gunnery flight. While waiting for another flight to arrive, Lt. Reece began demonstrating single-engine operation for Lt. Abbott. The pilot pulled back the starboard throttle and feathered the starboard propeller. At that instant the port propeller ran away. The airplane stalled and flipped over to an inverted position, entering an inverted spin. The pilot was unable to recover and the airplane crashed into the sea near Smith Island.

5-23-45D. Kelly Field, Texas. At 1532, a North American TB-25J crashed while attempting a take-off at Kelly Field, San Antonio, Texas, killing two crewmembers and seriously injuring two others. Investigators stated, "The take-off roll appeared to be normal except that the aircraft seemed to be airborne too soon. After becoming airborne the aircraft dropped down, falling off on the right wing, then climbing back to between 50 feet and 100 feet, then stalled and fell off on the right wing, veering to the right and crashing into the ground, striking on the right wing first." Investigation revealed that the aileron controls were crossed. Pilot 2Lt. Edgar L. Brackett and civilian mechanic Frank B. Wofford were killed in the crash; co-pilot 1Lt. Gerald Daniell and civilian mechanic Paul O. Dubose were seriously injured.

5-23-45E. Seguin, Texas. At 1800, a North American AT-6C suffered a catastrophic structural failure and crashed seven miles southwest of Seguin, Texas, killing pilots F/O Floyd R. Zawistocki and F/O Carl A. Scharmer. The airplane took off at 1710 from Randolph Field, San Antonio, Texas, on an instrument training flight. The airplane, flying to the northwest at about 3,000 feet, was observed to enter a 30-degree dive. The airplane suffered a failure of the port wing as it attempted to pull out at about 1,000 feet. The wing separated and the airplane went out of control, smashing to the ground at an angle of 30 degrees. The airplane burst into flames upon impact. Investigation revealed that the port wing struck the tail section as it separated, causing its progressive failure. Wreckage was scattered over a wide area.

5-24-45A. Walkill, New York. At 1105 EWT, a North American AT-6D crashed one and a half miles southeast of Walkill, New York, killing U.S. Military Academy Cadet Arthur R. Morrison, 23. The airplane was part of a two-ship flight that took off from Stewart Field, Newburgh, New York, on a "Local Combat Practice" mission. The airplanes climbed to 10,500 feet for the exercise. The airplanes engaged in simulated combat for about ten minutes, descending to about 6,500 feet during the maneuvering. A short time later the flight climbed back to 10,500 feet to resume the exercise. The airplanes descended to about 5,500 feet during the maneuvering. The airplanes were in a steep turn to the left when the leader rolled out and entered a climbing turn to the right. He looked back and observed the subject airplane in a spin to the right. The student made a partial recovery after four turns of spin. The student pulled the nose up too soon and the airplane stalled again, entering a secondary spin to the left. The airplane described five turns of spin before the student was able to stop the rotation. The airplane straightened out momentarily before the starboard wing struck a tree. The airplane went out of control and smashed into the ground in a nose-down attitude an instant later, bursting into flames upon impact.

5-24-45B. Prattville, Alabama. At 1430 CWT, a North American AT-6D crashed at Bridge Creek eight miles northwest of Prattville, Alabama, killing 2Lt. Heinz H. Meer and seriously injuring 2Lt. Delbert L. Squires. The airplane took off from Maxwell Field, Montgomery, Alabama, on a routine proficiency flight. The airplane was observed buzzing a small boat operating on Bridge Creek. The airplane, flying at a few feet above the surface, made a steep turn to the left after passing near the boat. The airplane lost altitude in the turn and the port wing struck the surface. The airplane cartwheeled into the water. Lt. Squires was unable to remember the events leading up to the accident.

5-25-45. Charleston, Kansas. At 0905 CWT, a Martin TB-26C crashed nine miles SSW of Charleston, Kansas, killing four crewmembers. The B-26 had taken off from Dodge City Army Air Field, Dodge City, Kansas, on a routine training flight. Killed in the crash were: 2Lt. Robert A. O'Dell, pilot; Capt. Robert G. Turner, co-pilot; Cpl. Vernon B. Sommer, engineer; Pfc. Loren C. Keith, student engineer. Investigators stated, "Investigation of the wreckage indicated that the aircraft struck the ground in a near vertical dive. All control surfaces were located in the wreckage, which would indicate that no control surface was lost in flight. Both propeller power units were checked for the blade angle ... and both units indicated a blade angle of approximately 34 degrees and both units were within one-half a degree of each other. This evidence indicates that at the time of the crash both engines were developing equal power. The possibility of [flight] controls locking

was impossible to determine because of the damage to all the parts in the crash." Investigators noted that the pilot had over 1,000 hours of military twin-engine time in Douglas C-47 and C-53 type aircraft and that the co-pilot had over 2,000 hours in Vultee BT-13 type airplanes. He had been an AAF flight instructor since graduating from AAF flight school in 1942.

5-26-45A. Charleston, Kansas. At 1010 CWT, a Martin TB-26B crashed six miles south of Charleston, Kansas, killing three crewmembers. French Air Force B-26 student Aspt. Marcel Peyrouny, AAF instructor 2Lt. Charles W. Findley and AAF engineer Pfc. Royal B. Boulware were killed in the crash. The airplane took off at 0805 CWT from Dodge City Army Air Field, Dodge City, Kansas, on a routine training flight. The B-26 made two practice landings at Dodge City Army Air Field. The airplane made a third take-off at 0950 CWT and flew away from the field to the west. A short time later the airplane, flying at about 1,000 feet agl, was observed to snap over into a half-roll to the right, rolling over to a nose down attitude. The airplane rolled over again and entered a spiral to the right. The airplane remained in the spiral until it struck the ground and exploded into flames.

5-26-45B. White Oaks, Maryland. At 1315 EWT, a North American TB-25D flying in instrument conditions crashed one mile north of White Oaks, Maryland, killing four fliers. Killed in the crash were: Col. Dudley M. Outcalt, pilot; SSgt. William C. Moberly, engineer; SSgt. Frank A. Rogarelski, radio operator; Pvt. William J. Knaub, passenger. The airplane took off from Keesler Field, Biloxi, Mississippi, on a passenger flight to Bolling Field, Washington, D.C. The pilot contacted Richmond, Virginia, radio network, giving his position as five miles southeast of Richmond. A short time later, Washington, D.C., air traffic control cleared the airplane to descend to 7,000 feet and maintain 7,000 feet until further instructed. The pilot was unable to acknowledge the clearance due to poor radio contact and continued flying at about 11,000 feet. There was no further contact with the airplane. Witnesses on the ground observed the airplane flying from the northeast below the overcast at about 500 feet agl. The airplane disappeared into the overcast. Moments later it was seen to emerge from the clouds in a slight dive. The airplane disappeared behind a knoll and then collided with terrain, exploding into flames.

5-26-45C. Wright Field, Ohio. At 1559 CWT, a Curtiss XP-55 (42-78847) crashed while performing low altitude aerobatics at an air show at Wright Field, Dayton, Ohio, killing pilot Capt. William C. Glasgow, 28, Niagara Falls, New York. The experimental fighter airplane crashed to earth very close to an automobile operating near the field, spraying the vehicle with flaming gasoline. Pieces of flaming wreckage also collided with the 1941 Chevrolet 4-door sedan, which was destroyed. Three civilians riding in the automobile were fatally injured; two others in the same vehicle were seriously burned. Investigation revealed that the occupants of the vehicle, which had been headed eastbound on the Third Street extension, had just arrived near the field perimeter fence to view the air show activity. The driver asked a motorcycle policeman for directions to get into the field. The Ohio State patrol officer advised the driver to enter at Gate 24. The driver drove up to Gate 23 to turn around when the accident occurred. Investigators stated, "[Capt. Glasgow] took off from Wright Field at 1545 in the XP-55 for the purpose of flying an exhibition for the Wright Field War Bond Open House. This flight was to be the third and final exhibition of the day, and was to consist of a pass at 500 feet over a course parallel to Runway 16. This pass was to be terminated with a single slow-roll, climbing and going away. At approximately 1600 CWT the XP-55 started its pass from north to south at approximately 500 feet as specified. Just south of the tower the aircraft was pulled into a gradual climb obtaining an altitude of 1,200 and 1,500 feet. At the time this altitude was reached the aircraft was approximately three-quarters of the runway length down the field. Upon reaching this point the pilot executed a slow-roll, or barrel roll, to the right. The last phase of this roll appeared to be somewhat off, as the ship sliced out of the bottom. Upon completing this roll it appeared that the pilot was attempting another roll with the nose of the aircraft at a high angle. In this attitude, and with the aircraft at a low flying speed, it snapped suddenly to the left, the nose dropped, and the ship again snapped, this time to the right. Throughout the rest of the descent the aircraft continued to snap alternately to the left and to the right. Just prior to crashing, the ship was headed in a southeasterly direction and appeared to be in an attitude of flight similar to that of a normal approach landing. It was evident, however, that the pilot did not have flying speed and that the ship was stalling into the ground in a nose-high attitude. The aircraft finally ended up by dropping the nose and falling off on the left wing a second prior to striking the fence at the south end of the field. Upon striking the fence the aircraft carried through and immediately started to disintegrate. It is not known exactly what part of the aircraft collided with the civilian automobile just outside the fence. The automobile, however, was struck by some part and was also momentarily engulfed in flaming fuel. The five passengers of this automobile were all critically burned; three of them later died at the Patterson Field Station Hospital." Flaming wreckage also destroyed the police motorcycle. Civilians fatally burned on the ground were: Wesley B. Roehm, 23; Donna Roehm, 22 months; Kathleen Eryes, 22. Susan Roehm, 23, and infant Nina Roehm, 5 weeks, were seriously burned.

5-27-45A. Gunter Field, Alabama. At 1114 CWT, a Noorduyn C-64A crashed while attempting a landing at Gunter Field, Montgomery, Alabama, killing seven passengers. Pilot 1Lt. Donald E. Mollohan and passenger F/O Carlton E. Brown were seriously injured. The airplane took off from Napier Field, Dothan, Alabama, on a passenger flight to Gunter Field. Investigators stated, "The only witnesses to this accident were three members of the crash truck crew and the west control tower operator. The crash truck was stationed adjacent to Runway 18R and three personnel aboard the crash truck reveal that they observed the C-64 to land on Runway 18L (150' × 3,500' asphalt) and shortly after a wheels-first contact with the runway the C-64 veered to the left and rolled off the runway and struck a slight [earthen] embankment [175 feet from the edge of the runway] on the east portion of the field. Upon striking this slight rise [the airplane broke up] and burst into flames." Investigators noted that the C-64 is approved to carry only seven passengers at one time, but it was not thought that the fact that the subject airplane was carrying eight passengers at the time contributed to the accident. The pilot, who was very seriously injured, made a partial statement to investigators, stating that the port brake had "locked" during the landing roll. Passengers killed in the crash were: SSgt. Phillip W. Callaghan, Sgt. Michael Kozek, 1Lt. Joseph F. LaPlant, SSgt. Ray Osborne, Pvt. Thomas F. Taylor, Pvt. Gerald H. Thinnes, Pvt. Eugene Urban.

5-27-45B. Lapel, Indiana. At 1830 CWT, a Bell RP-63C crashed two and one-half miles northwest of Lapel, Indiana, killing 2Lt. Robert A. Barnes, 25. Investigators stated, "The pilot was on a CFR clearance from Romulus, Michigan, to Toledo, Ohio, direct, and from Toledo, to Patterson Field, Ohio, to fly airways. The pilot was approximately 70 miles west of his course when he ran low on fuel and was forced to make an emergency landing. Evidently the pilot was flying high enough to make the terrain appear level and he decided to make a wheels-up landing. After reaching a lower altitude he evidently decided that the terrain was [too uneven] to attempt a wheels up landing. Just as the pilot decided to get more altitude, the engine quit due to lack of fuel. The pilot bailed out at approximately 150 feet above the ground. His parachute opened just as he hit the ground. The pilot was killed instantly. The airplane struck the ground about 100 feet from where the pilot did, and was a major wreck."

5-27-45C. Phoenicia, New York. At 2202 EWT, a North American AT-6D flying in instrument conditions crashed into Mount Tremper two miles southeast of Phoenicia, New York, killing instructor 1Lt. Matthew M. O'Brien, 22. The airplane was part of a two-ship flight that took off at 1940 EWT from Buffalo, New York, on an instrument navigation flight back to their home station at Stewart Field, Newburgh, New York. Investigation revealed that the airplane, heading due west, collided with the east face of Mount Tremper at an elevation of 2,300 feet msl in a slightly climbing attitude with the engine producing power.

5-28-45A. Bryan, Texas. At 1201 CWT, two North American AT-6C airplanes collided in mid-air seven miles west of Bryan, Texas, killing one flier and injuring another. Pilot 1Lt. Howard W. Danks, 23, was killed in the crash of AT-6C #42-48909; 1Lt. Albert F. Leghorn sustained minor injuries parachuting to safety. Pilots 1Lt. Raymond Sculley and 1Lt. Gordon R. Hatt were uninjured and were able to make a safe landing. AT-6C #42-44180 took off from Bryan Army Air Field on a local training flight. AT-6C #44-48909 took off from Bryan Army Air Field on an instrument team ride. Lt. Danks was flying in the front cockpit as the safety pilot and Lt. Leghorn was flying in the rear cockpit as the instrument pilot. Lt. Leghorn was flying on instruments at 5,000 feet on an orientation problem. The instrument pilot determined he was outbound on the northeast leg of the radio range. The instrument pilot executed a 180-degree turn and was able to pick up the beam inbound. The airplane flew on the radio range for a short time when the collision occurred. Investigation revealed that Lt. Leghorn, flying at 5,000 feet, should have been flying an even number altitude on the inbound leg. AT-6C #44-48909 collided with AT-6C #42-44180 from underneath, damaging 42-44180's port landing gear and both of the main wheels. The pilot was able to make a safe landing at Bryan Army Air Field. Lt. Danks attempted to get to a field to land his damaged aircraft, but the airplane began to shudder violently. Lt. Leghorn bailed out at 2,400 feet and parachuted to safety. AT-6C #42-48909 went into a spiral to the left and smashed to the ground.

5-28-45B. Cherokee County, Oklahoma. At 1640 CWT, a Republic P-47D crashed at the Great Salt Plains Gunnery Range, Cherokee County, Oklahoma, killing pilot 1Lt. Gonnero Riggardo, 24. The airplane was part of a six-ship flight that took off from Strother Field, Kansas, on a routine gunnery mission. The pilot made a steep approach on his third pass and was unable to pull out in time to avoid flying into the ground. The airplane struck the terrain in a flat attitude and scraped along the ground for a few hundred yards before bounding back into the air for a hundred yards. The airplane then smashed back into the ground and burst into flames. The flaming wreckage came to rest about one mile from the point of initial contact.

5-29-45A. Mt. Meigs, Alabama. At 0950 CWT, a Douglas A-26C and a North American AT-6C collided in mid-air and crashed four miles northeast of Mt. Meigs, Alabama, killing French Air Force pilot 1Lt. Georges Maurice aboard the AT-6C; AAF instructor 1Lt. John H. Wishnick was seriously injured parachuting to safety. Pilot 2Lt. Michael P. Turok, 23, was killed in the crash of the A-26. Lt. Turok took off at 0850 EWT from Florence Army Air Field, Florence, South Carolina, on a "cruise control" mission to Hattiesburg, Mississippi, to Americus, Georgia, and return to Florence. The pilot was assigned an altitude of 6,000 feet. The AT-6C took off from Gunter Field, Montgomery, Alabama, on a local instrument training flight. Lt. Wishnick stated that AT-6C was flying at 6,000 feet with the student flying on instruments. Suddenly

the airplane "exploded" around them. The instructor was hurled from the wreckage with a broken arm. He managed to pull the ripcord and he parachuted into the Tallapoosa River. Lt. Wishnick was able to swim to the bank and make it to safety. He found one of the wings of his airplane lying in a field adjacent to river, so he sat on it until he was rescued. The A-26 was seen to spiral out of control after the collision and smash to the ground.

5-29-45B. McMinnville, Tennessee. At 1558 EWT, a Beech AT-11 flying in poor weather was abandoned and crashed two and a half miles northeast of McMinnville, Tennessee, killing two crewmembers. Three fliers escaped injury parachuting to safety. Engineer Sgt. Joseph Ralidak fell to his death when he bailed out without a parachute. Passenger Cpl. Richard G. White was also killed in an unsuccessful parachute jump. The airplane took off from Boca Raton, Florida, on an administrative flight to Evansville, Indiana, via Atlanta, Georgia. The airplane took off from Atlanta at 1445 EWT on a visual clearance. The airplane encountered deteriorating visibility, so the pilot climbed to 6,000 feet in an effort to avoid instrument conditions. The airplane entered an area of thunderstorms, encountering powerful turbulence and abundant lightning. The airplane went out of control and began dropping rapidly. A B-4 bag was observed to rise from the floor and "stick to the ceiling." The pilot was unable to regain control of the airplane so he ordered the crew to bail out. Sgt. Ralidak jettisoned the escape door and jumped first. The other three crewmembers jumped through the bomb bay. No one observed Cpl. White leaving the ship. His parachute was attached to the harness upside down and the ripcord had not been pulled.

5-29-45C. Sherman, Texas. At 1645 CWT, a North American AT-6C crashed 13 miles southwest of Perrin Field, Sherman, Texas, killing instructor 1Lt. Harrison Payne and student 1Lt. Albert F. Pjura. The airplane took off from Perrin Field on a routine training flight. The student was observed to shoot two simulated forced landings to the southwest. The airplane was next observed flying at about 400 feet agl to the northeast, entering a turn to the left. The airplane was observed to snap over to an inverted position and then fell into a spin to the left. The airplane remained in the spin until it struck the ground. The airplane did not burn but was totally destroyed. Investigators noted that the student was a rated bombardier and a combat veteran.

5-29-45D. Reno, Nevada. At 2015 PWT, a North American B-25H flying in instrument conditions collided with a mountain 11 miles southwest of Reno Army Air Base, Reno, Nevada, killing four fliers. The airplane took off from Hill Field, Ogden, Utah, at 1809 PWT on a flight to Salinas, California, via Reno. The airplane was on a visual clearance to Reno and an instrument flight plan was filed for the leg to Oakland, California (cleared to fly at 12,000 feet), and a visual clearance for Salinas at 6,000 feet. At 1958 PWT, the pilot reported that he was flying over Reno at 8,500 feet. The pilot's instrument flight plan was cleared by Oakland control and the airplane climbed to 12,000 feet. At 2010 PWT, the pilot reported that he was between cloud layers and requested a clearance for 14,000 feet. Controllers cleared the flight for 14,000 feet but the pilot never acknowledged the clearance. The airplane failed to arrive at its destination and was discovered to have crashed. The airplane, flying to the east, collided with the mountain about 40 feet from the crest. The airplane smashed into the wooded ridge in a 30-degree dive with both engines producing power. Wreckage was scattered for a quarter mile on the mountainside. It was noted that the airplane was about ten miles off course. Killed in the crash were: Capt. James F. Curtis, Jr., pilot; 2Lt. Charles R. Haas, co-pilot and rated glider pilot; Cpl. Loy E. Underwood, passenger; Pfc. Leroy J. Anderson, Jr., passenger.

5-29-45E. San Jose, California. At 2210 PWT, a Beech AT-11 flying in instrument conditions crashed into a mountain 14 miles east of San Jose, California, killing four fliers. The airplane took off at 2128 PWT from Hammer Field, Fresno, California, on an administrative flight to Mills Field, San Francisco, California. The flight had originated at Carlsbad Army Air Field, Carlsbad, New Mexico, which was the aircraft's home station. The pilot decided that if visual conditions did not exist when he arrived at Pacheco Pass that he would turn back for Hammer Field. The airplane failed to arrive at Mills Field and did not return to Hammer Field. The airplane was found on 6-4-45. The airplane had collided with Mt. Hamilton at an elevation of about 3,000 feet near the Mt. Hamilton Observatory. Investigation revealed that the airplane had collided with the wooded ridge while in a straight and level attitude with both engines producing power. Overcast in the area were reported to be at about 3,500 feet. Killed in the crash were: 2Lt. Richard C. Price, pilot; F/O John C. Gardiner, co-pilot; Pvt. Lester C. Longhurst, engineer; 1Lt. Paul S. Leung, bombardier-rated passenger.

5-30-45A. Fort Sumner, New Mexico. At 0656 MWT, a Republic P-47D crashed three miles east of Fort Sumner, New Mexico, killing pilot 2Lt. David J. Deyarmon, 20. The airplane took off from Fort Sumner Army Air Field on a high-altitude formation flight. The airplane had been airborne for about ten minutes when the pilot radioed the control tower for an emergency landing on Runway 35. The pilot advised that he was having trouble with his propeller. The control tower officer advised the subject pilot to place the propeller control in the "manual" position. The pilot complied and entered the traffic pattern. The pilot was cleared to land and he flew a normal base leg. The turn from base to final was too "shallow" and the airplane overshot the turn. The pilot tightened his turn and the

airplane entered a steep bank. The airplane stalled and flipped over to the left to a vertical bank. The pilot recovered momentarily but the airplane fell off on a wing and smashed to the ground. The airplane did not burn but was destroyed utterly.

5-30-45B. Elmendorf, Texas. At 1545 CWT, a North American AT-6C flying at low altitude crashed on a farm 10 miles southeast of Elmendorf, Texas, killing pilots F/O Alvin Westerhoff and F/O Chester A. Popek. The airplane took off from Brooks Field, San Antonio, Texas, on a local flight. Investigators stated, "The airplane came in very low over the field (in which it crashed), headed in a westerly direction and pulled up doing a roll to the left and then a roll to the right. After this he made a 180-degree turn and started a dive on the same field, again coming very close to the ground, leveling off and starting a turn to the right. At this time the right wing struck the ground, throwing the airplane into the ground with the right wing and engine striking simultaneously. The airplane then rolled to the left, knocking off the left wing and [the starboard horizontal stabilizer and elevator], and coming to rest with the fuselage inverted. Both wings, [the starboard horizontal stabilizer and elevator] and the engine were detached from the fuselage. The left wing and center section were found 20 feet from the point of impact, the engine 25 feet, left wing 100 feet, and the fuselage 120 feet."

5-30-45C. Eureka, North Carolina. At 1730, a Republic P-47D crashed one mile south of Eureka, North Carolina, killing pilot F/O Edgar E. Stockton, 21. The airplane was part of a two-ship flight that took off from Seymour-Johnson Field, North Carolina, on an instrument proficiency flight. The subject airplane was flying as the instrument ship at 8,000 feet. After flying on instruments for a short time the subject pilot began a series of diving turns until an altitude of 500 feet was achieved. The airplanes engaged in a Lufbery circle for a short time. The airplanes then buzzed a farm field a couple times. On the last pass, the subject airplane collided with trees while flying at very low altitude. The airplane went out of control and smashed to the ground, bursting into flames upon impact and killing the pilot instantly.

5-31-45A. San Angelo, Texas. At 0835 CWT, two North American AT-6D airplanes collided in mid-air and crashed 11 miles east of Goodfellow Field, San Angelo, Texas, killing four fliers. Instructor 2Lt. Alf E. Stromsborg and student A/C James L. Burnet were killed aboard AT-6D #42-86325; instructor 2Lt. Philip R. Pierson and student A/C Donald B. Ray were killed aboard AT-6D #42-86327. Both airplanes had taken off from Goodfellow Field on separate scheduled training flights. Several witnesses observed the airplanes falling after the collision, but there were no witnesses to the actual collision and no one had observed the flight paths of the subject airplanes prior to the collision. Investigation of the wreckage indicated that the starboard wings of both airplanes had made contact prior to the ships crashing to the ground.

5-31-45B. Zephyrhills, Florida. At 1150 EWT, two North American P-51D airplanes collided in mid-air and crashed 15 miles northwest of Zephyrhills Army Air Field, Zephyrhills, Florida, killing two pilots. F/O John L. Terry, 22, was killed in the crash of P-51D #44-72052; 2Lt. Robert B. Walker, 21, was killed in the crash of P-51D #44-72537. The airplanes were part of a 16-ship flight that took off from Bartow Army Air Field, Bartow, Florida, on a combined training mission with an 18-ship flight of Douglas A-26 airplanes. The P-51s made a high pass at the bombers from the right at an altitude of 9,000 feet. The fighters were breaking off their attack when the collision occurred. The two airplanes were observed breaking up and falling, but there were no witnesses to the actual collision. The two P-51s smashed to the ground, exploding into flames upon impact.

5-31-45C. March Field, California. At 1243, a Consolidated B-24J making an emergency landing collided with a truck on an inactive runway at March Field, Riverside, California, killing civilian truck driver Clovis Bingham. The airplane was part of a flight that had taken off from March Field on an air-to-air gunnery mission and a camera-bombing mission. The formation formed up over Indio, California, and then flew to Phoenix, Arizona, for the camera-bombing mission. The airplanes then returned to the area of March Field and dropped a train of bombs on Target #7 at the March Field Bombing Range. The formation then flew to Las Vegas, Nevada, for another camera-bombing mission. After completing the camera-bombing mission, the formation flew to the AAF Gunnery Range at Muroc, California, for the air-to-air gunnery exercise. During the exercise, the subject airplane was struck by fifty caliber bullets, one bullet passing through the pilot's headrest and smashing into the top turret. B-24 instructor Capt. Wallace F. McGreevy, who was positioned behind the two pilots, was struck in the head and face with flying fragments of bullets and pieces of the top turret machinery. The pilot, F/O Joseph E. Hehenstreit, who was uninjured, elected to land at Muroc Army Air Base to seek medical attention for the instructor. The instructor ordered the pilot to fly back to March Field. The airplane arrived in the area of March Field but had trouble letting down because of instrument conditions. The pilot found a hole in the clouds and descended through, entering the traffic area for March Field. The pilot was cleared for an emergency landing on Runway 30 and the airplane entered a left hand traffic pattern. The fuel booster pumps were turned on prior to landing and the number-one fuel line, which had been hit by a fifty-caliber bullet, failed and burst, causing a fuel leak. The pilot overshot the turn from the base leg to final approach and was to the right of the runway. The pilot began turning to the left to reacquire the runway. The co-pilot, F/O William

D. Blount, extended full flaps and the airplane began skidding to the left. The pilot applied full right rudder but could not stop the skid to the left. The pilot managed to line the airplane up with Runway 30L, which was under construction. The instructor directed the pilot to clear the construction equipment and workers at the approach end of the runway and then make the landing on Runway 30L. The airplane touched down and rolled for about 900 feet before the number-four propeller struck a civilian construction truck that was crossing the closed runway. The driver, who was killed instantly, was flung from the truck cab during the collision. The pilot retained control of the airplane and was able to come to a stop without further incident. Investigators noted that the engineer had failed to shut off the fuel valve for the number-one fuel booster pump. The two pilots and ten other crewmembers were uninjured.

5-31-45D. Victorville, California. At 1800, a North American B-25H crashed 27 miles southeast of Victorville, California, killing three fliers. Pilot 2Lt. August P. Marks, co-pilot F/O Charles F. Cizek and engineer Sgt. Orville V. Barton were killed in the crash. The airplane had taken off from March Field, Riverside, California, on an instrument training flight to Victorville Army Air Field, Victorville, California, and a low-level bombing mission on the Victorville Bombing Range. The airplane landed at Victorville Army Air Field and the crew was briefed on bombing range use and procedures. The airplane took off from Victorville Army Air Field at 1720 on an instrument flight to March Field and cleared to operate on the Victorville Bombing Range. The airplane failed to arrive at the bombing range and failed to return to March Field. The B-25 was found to have crashed in a box canyon that the airplane could not out climb or turn around. The airplane smashed to the ground in a turn to the left, exploding into flames upon impact and killing the occupants instantly.

June

6-1-45A. Walterboro, South Carolina. At 1510 EWT, a Curtiss P-40N crashed eight and a half miles south of the Army Air Field at Walterboro, South Carolina, killing pilot 2Lt. Willis E. Moore, 21. The airplane was flying in the number-four position of a four-ship flight that took off at 1330 from Walterboro Army Air Field on a dive-bombing and ground gunnery mission. The flight completed the ground gunnery mission and flew to the dive-bombing range, changing to an echelon right formation. The airplanes peeled off for the dive-bombing run. The number-four ship was observed to enter a dive that was steeper that normal. The first three ships pulled out of their dives at 3,000 feet. The number-four ship was observed to make a "gradual" pull-up to the left at a much lower altitude.

The airplane disappeared behind trees and a crash was heard. The pilot was unable to recover and the airplane smashed through tall trees and then dove into the ground at a 30-degree angle, exploding into flames upon impact.

6-1-45B. Cordele, Georgia. At 1450 CWT, a North American TB-25J crashed one mile north of Cordele, Georgia, killing three fliers. The airplane had taken off from Turner Field, Albany, Georgia, on a flight to Cordele Auxiliary Airfield to practice take-offs and landings. The airplane made two touch-and-go landings without incident. The airplane flew the traffic pattern and made another landing. As the pilot added power for the take-off, the airplane entered a nose-high attitude and the engines were heard mis-firing and backfiring. The airplane climbed to about 40 feet and began crabbing to the right in a nose-high attitude with the right wing down. A thin line of blue smoke was seen trailing the starboard engine. The airplane leveled out and staggered through the air for a little over a mile. The airplane, flying at very slow speed, then began a descending turn to the right and ploughed into the ground. The starboard wing fuel tank exploded into flames just as rescuers arrived. Investigators speculated that crew might have been knocked unconscious and were unable to escape the wrecked airplane before it erupted into flames. Investigators were unable to determine the cause of the apparent problem with the starboard engine. Instructor pilot 2Lt. Richard Mc-Cracken and B-25 students A/C Willis M. Matheson and A/C Jay L. Lohr were killed in the crash.

6-1-45C. Coffeyville, Oklahoma. At 1525 CWT, a Lockheed F-5E suffering an in-flight fire crashed five miles southeast of Coffeyville, Oklahoma, killing pilot 1Lt. Philip H. Kase, 27, in an unsuccessful parachute jump. The airplane took off from Coffeyville Army Air Field on a routine training flight. The starboard engine burst into flames and the fire spread to the wing. The pilot bailed out but apparently he was struck by some part of the airplane, knocking him unconscious or killing him and preventing him from pulling his ripcord. Investigators could not determine the cause of the engine fire.

6-1-45D. Spence Field, Georgia. At 2234 EWT, a North American AT-6D crashed one-half mile southwest of Spence Field, Moultrie, Georgia, killing student A/C William N. Brosnahan, 21. The airplane took off on Runway 22 from Spence Field on a routine night-training flight. The airplane climbed to about 200 feet but then began settling back toward the ground. The airplane collided with tall trees and then smashed to the ground, exploding into flames upon impact. Investigators speculated that the student put his head down in the cockpit and allowed the airplane to fly into the ground.

6-2-45A. Shadyville, Florida. At 1126 EWT, a Curtiss P-40N crashed one and a half miles east of Shadyville, Florida, killing pilot Capt. Thomas F. Reynolds, 25. The airplane took off from Dale Mabry Field, Tallahassee, Florida, on an engine "slow time" flight. The airplane was stalled while attempting to follow another P-40 in a steep wingover maneuver. The airplane was seen in a nose-high attitude when it began the wingover. The subject airplane was observed in a steep dive and the pilot made two incomplete recoveries. The airplane was stalled a third time and it fell off on the starboard wing. The airplane smashed to the ground at a 20-degree angle and burst into flames, killing the pilot instantly.

6-2-45B. Rivina, Texas. At 1150 CWT, a Beech C-45E crashed six and a half miles east of Rivina, Texas, killing pilot Maj. John E. Doran, 34, and two passengers. The airplane took off from Enid Army Air Field, Enid, Oklahoma, on a passenger flight to Kirtland Field, Albuquerque, New Mexico. The airplane took off and climbed to 4,000 feet msl. Investigators speculated that the airplane remained at 4,000 feet, which is only 400 feet above the terrain where the crash occurred. Investigators speculated that the pilot had failed to switch fuel tanks in time to prevent the engines from quitting due to fuel exhaustion. It is speculated that the pilot did not have enough time to switch to a full fuel tank because of the low altitude above the rugged, rolling terrain. The airplane smashed onto the crest of a small hill. The airplane scraped along the top of the hill for about 400 feet, becoming airborne as it left the top of the hill and then crossing a 400-foot wide and 75-foot deep gully. The airplane smashed to the ground and burst into flames on sloping terrain west of the gully about 1,500 feet west of the first point of impact. Passengers Maj. Gilbert D. Greer and Lt. (j.g.) Ruth S. Hunter, USN, were killed in the crash.

6-2-45C. Victoria, Texas. At 1440 CWT, a North American AT-6C crashed nine miles southeast of Aloe Field, Victoria, Texas, killing pilot 2Lt. John G. Bower, 26. Pilot 2Lt. Gerald R. Brady received serious injuries. The airplane took off at 1410 CWT from Aloe Field on an instrument proficiency flight. Lt. Bower was flying in the front cockpit as the safety pilot and Lt. Brady was flying in the rear cockpit as the instrument pilot. About 30 minutes after take-off, the airplane was observed flying at very low altitude in the vicinity of the Welder Ranch. The airplane buzzed a tractor operating on the ranch and passed over the Welder house. The airplane flew westward for about two miles before turning around and passing over the Welder house, clearing it by about 10 to 15 feet. The airplane was observed diving on the tractor, which was being operated by a boy who was a visitor to the ranch. The airplane passed over the tractor at very low altitude and then attempted a slow roll. As the airplane rolled inverted at 50 feet agl it struck a telephone wire. The pilot attempted to roll out to a level attitude but fell out at the bottom of the maneuver. The airplane contacted the ground in a flat attitude and bounced across a road, barely missing colliding with an automobile carrying a family before bounding back into the air. The airplane collided with another telephone wire and then the starboard wing hit the ground. The airplane cartwheeled into the ground and burst into flames as it was smashed to pieces. The shattered and wingless fuselage continued on for another 150 yards before coming to rest on its right side. The pilot in the front cockpit had been hurled from the airplane and killed instantly. The pilot in the rear cockpit was found alive strapped to his seat in the aircraft.

6-3-45A. Weiser, Idaho. At 1500 MWT, a Consolidated B-24J flying in instrument conditions suffered a structural failure and crashed ten miles northwest of Weiser, Idaho, killing eight fliers. Co-pilot F/O Raymond J. Hogan and gunner Cpl. Clarence L. Pearson miraculously survived but were seriously injured. The airplane took off at 1318 MWT from Pendleton Field, Pendleton, Oregon, on an instrument flight to Gowen Field, Boise, Idaho. The airplane climbed to 13,000 feet msl and was "500 feet on top" of the overcast. The crew made two position reports at 1425 and 1437 MWT, the pilot reporting that he was in clouds and wanted to climb on top. Controllers cleared the B-24 to climb on top of the overcast but received no acknowledgement of the message. Controllers radioed the aircraft several times but received no answer. At 1530, the local sheriff reported an airplane crash. Witnesses on the ground observed the airplane shedding pieces of the elevator as it passed over. The airplane smashed to the ground at the foot of a small knoll at 3,000 feet msl with the landing gear in the extended position. The airplane's fuselage broke in half just aft of the wing trailing edge. The after part remained intact but the forward section plowed into the terrain and smashed across the knoll, bursting into flames as it came to rest. Investigators speculated that pilot had lost control of the airplane after it entered a powerful thunderstorm system, losing 10,000 feet before control was regained. The elevators apparently failed in the recovery. Pieces of the elevators were found up to eight miles from the scene of the crash. F/O Hogan and Cpl. Pearson were unaware of the crash and the events leading up to it, and were unable to offer any information to investigators. Killed in the crash were: 1Lt. Joseph T. McCarthy, pilot; F/O Calvin C. Smith, navigator; 2Lt. Walter E. Pippert, Jr., bombardier; Cpl. Bernard R. Wood, engineer; Sgt. Carl R. Staples, radio operator; Sgt. James F. McGrath, gunner; Cpl. Harold L. Mercer, gunner; Cpl. Billie Lightfoot, gunner.

6-3-45B. Grand Prairie, Texas. At 2040, a Beech AT-11 crashed into six U.S. Navy airplanes while attempting to take-off from the airport at Grand Prairie, Texas, killing two fliers and injuring three others, two of them seriously. The airplane was taking off on the last leg of a navigation training flight back to its home station at Big Spring, Texas. The airplane was cleared to take-off on Runway 13. The airplane commenced the take-off roll and the tail came up at a speed of about 55 mph. After traveling about 650 feet down the runway the ship began veering to the left. The pilot attempted to straighten the airplane out with starboard brake and increased rpm on the port engine, but he was unsuccessful. The airplane continued to veer to the left and careened across the airfield. The pilot cut the throttles and attempted to use the brakes but he was unable to regain total control of the airplane. The airplane bounced across the field out of control, crossing Runway 22 and heading for the navy flight line. The pilot attempted to ground loop to the right but the port tire blew. The pilot observed that he would not be able to steer clear of the navy airplanes parked on the ramp. The pilot added power and attempted to fly over the navy airplanes. The airplane jumped into the air and then smashed into the second, third and fourth rows of U.S. Navy Vultee SNV-2 airplanes (similar to the BT-13 type) parked there and destroying or seriously damaging six of them before coming to rest and bursting into flames. The surviving crewmembers aboard the AT-11 had to pass through flames when they exited from the port escape hatch. Bombardier 1Lt. Edgar Green and passenger SSgt. Ray E. Youngblood were killed in the crash. Pilot 2Lt. James S. Dolin and crewmember Cpl. Almus E. Crawford were seriously injured and co-pilot 2Lt. Albert J. Collins sustained minor injuries. No one on the ground was injured or killed.

6-4-45A. Warsaw, North Carolina. At 1330 EWT, a Republic P-47D crashed three and a half miles south of Warsaw, North Carolina, killing F/O Richard C. Combs, 20. The airplane was part of a two-ship flight that took off from Seymour-Johnson Field, North Carolina, on a low-altitude formation mission. The airplanes began a series of climbing turns and crossover turns, leveling out at about 1,500 feet agl. The airplanes encountered a U.S. Navy Grumman F6F airplane and began maneuvering with it. After a few minutes the navy airplane left the area. The flight leader then turned to the left with the subject airplane following in the maneuver. The subject airplane was observed to flip over to an inverted position and plunge to the ground. The airplane exploded into flames upon impact, killing the pilot instantly.

6-4-45B. Napier Field, Alabama. At 1340 CWT, a Curtiss P-40R crashed at Napier Field, Dothan, Alabama, killing pilot 2Lt. Robert V. Parker, 20. The airplane took off at 1336 from Napier Field on a routine transition flight. The pilot radioed the Napier Field tower about four minutes later, reporting that his engine was running rough and that he was returning to the field. The airplane was observed trailing thick black smoke while flying to the southeast toward the field. The smoke changed from black to gray and then flames were seen trailing from the engine area on the starboard side. As the airplane descended between 700 and 500 feet, a large sheet of flame was observed to envelope the cockpit canopy. The airplane was seen to snap roll to the left before leveling out and entering a steep dive toward the ground. The airplane exploded into flames upon impact. Investigators could not determine the cause of the engine fire.

6-4-45C. Corona, California. At 1850, a Republic P-47D flying in instrument conditions crashed into rising terrain eight miles south of Corona, California, killing pilot 1Lt. Glen E. Tanner, 24. The airplane was part of a four-ship flight that took off from Madera, California, on a flight to Palm Springs, California. Each airplane was cleared separately. The subject pilot, who was leading the flight, changed the flight's destination to Long Beach, California. The subject airplane encountered instrument conditions and collided with terrain at an altitude of 4,200 feet, exploding into flames upon impact. Investigators noted that clouds obscured the mountains at the time of the accident.

6-5-45A. Schroeder, Texas. At 0915, two North American AT-6C airplanes collided in mid-air and crashed two miles southwest of Schroeder, Texas, killing A/C John T. Smith, Jr. aboard AT-6C #42-44019. A/S Clarence S. Rothans, 27, received minor injuries parachuting to safety from AT-6C #42-44409. The airplanes were part of a three-ship flight that had taken off from Aloe Field, Victoria, Texas, on a routine training flight. An instructor was leading the flight; A/C Smith was flying in the number-two position and A/S Rothans was flying in the number-three position. The formation climbed to 5,000 feet and began maneuvering in formation. The collision occurred during a change of position. A/C Smith was moving over to the left and climbed into the AT-6 flying on the leader's port wing. A/C Smith's propeller severed the port wing of A/S Rothans' ship. A/S Rothans airplane went out of control and he bailed out immediately. A/C Smith's airplane spun to the ground and exploded into flames.

6-5-45B. Ft. Ogden, Florida. At 1720 EWT, a North American P-51D suffered a catastrophic structural failure and crashed eight miles west of Ft. Ogden, Florida, killing pilot F/O Robert B. Smith, 22. The airplane was in the number-four position of a six-ship flight that took off from Bartow Army Air Field, Bartow, Florida, on a high-altitude formation flight. The formation climbed to 20,000 feet and the flight leader signaled the flight to form in an echelon right formation. The airplanes formed the echelon and then the leader then peeled off to the left, the flight following in the maneuver. The subject airplane snapped over to an inverted position during the peel-off and the pilot attempted to recover with a split-S maneuver. The airplane built up excessive speed during the maneuver and the pilot over-stressed the airframe with an abrupt pull-up maneuver. The port wing failed and separated during the maneuver. The airplane entered a steep dive and smashed to the ground, exploding into flames upon impact. Wreckage and the pilot's badly dismembered body were scattered over an area of 100 yards.

6-6-45A. Montgomery, Alabama. At 0006 CWT, a North American B-25J crashed three and a half miles east of Gunter Field, Montgomery, Alabama, killing seven fliers. Gunner Cpl. Selwyn Wilson survived with serious injuries. The airplane took off at 1345 CWT (6-5-45) from Godman Field, Fort Knox, Kentucky, on a navigation flight to Tuskegee, Alabama, and return. The airplane landed at Tuskegee Army Air Field at 1552 CWT. Upon landing it was discovered that there was no 100-octane gasoline available. The pilot decided to fly to Gunter Field for service, landing there at 2303 CWT. The airplane was refueled and it took off at 0005 CWT on Runway 10. About a minute later the airplane descended into a stand of trees about three and a half miles east of the runway, cutting a path through the treetops 200 yards long. The airplane burst into flames as it smashed through the trees and broke up. Investigators speculated that the pilots had retracted the flaps prematurely, allowing the airplane to settle into the terrain. Killed in the crash were: F/O Charles E. Wilson, pilot; F/O Ramon F. Noches, co-pilot; F/O Edward Glover, navigator; Cpl. Henry H. Valentine, radio operator; Cpl. Luther A. Cox, radio operator; F/O Denny C. Jefferson, bombardier-rated passenger; Pvt. James Ewing, passenger.

6-6-45B. Courtland, Alabama. At 0038 CWT, a Consolidated RB-24E crashed at Courtland Army Air Field, Courtland, Alabama, killing three fliers. Pilot 1Lt. John G. Zehren, co-pilot 2Lt. Earl O. Sizer and engineer Pfc. John T. Byrd were killed in the crash. The airplane took off from Courtland Army Air Field on a routine training flight. At approximately 0001 CWT, the airplane was called in for a full-stop landing. The pilot acknowledged and flew back to the field. The airplane arrived at the field and was cleared to land. The airplane entered the traffic pattern for Runway 5. The ship was observed to make a low final approach. Control tower personnel advised the pilot of the low approach but this message was not acknowledged. The airplane disappeared behind some hills and crashed moments later. Investigation revealed that the airplane had collided with 60-foot tall trees, plowing a path through the tree until the port wing separated. The airplane entered a bank to the left and crashed to the ground, erupting into flames upon impact and killing the crew instantly.

6-6-45C. Berlin, Georgia. At 0921 EWT, two North American AT-6D airplanes collided in mid-air and crashed four miles ESE of Berlin, Georgia, killing four fliers. Instructor 2Lt. Frederick E. Schaeffer and student A/C Jack H. Gibbs were killed in the crash of AT-6D #42-85951; instructor 2Lt. Edwin W. Goddard and student A/C Vincent W. Finewood were killed in the crash of AT-6D #41-34134. The airplanes took off individually from Spence Field, Moultrie, Georgia, on separated instrument training flights. The airplanes were observed flying near the field at an altitude of approximately 4,000 feet. The collision occurred a few minutes later. The airplanes slammed to the ground in close proximity to each other. There were no witnesses to the actual collision or the flight paths of the airplanes prior to the collision. Investigators speculated that AT-6D #42-85951 was struck from the front at an approximate 15-degree angle left of center. AT-6D #41-34134 passed through the port wing and lower fuselage of AT-6D #42-85951. A pilot flying in the area observed a large puff of black smoke at about 3,500 to 4,000 feet and then saw the airplanes breaking up as they fell to the ground.

6-6-45D. Juneau, Wisconsin. At 0845, a North American AT-6C crashed on the Nehls Farm one mile north of Juneau, Wisconsin, killing pilots 2Lt. Arthur A. Todd and F/O Charles F. Ostling. The airplane took off from Truax Field, Madison, Wisconsin, on an instrument training flight to Madison, Wisconsin, to Lone Rock, Wisconsin, to Rockford, Illinois, to Milwaukee, Wisconsin, and return to Truax Field. Lt. Todd was flying in the front cockpit as the safety pilot and F/O Ostling was flying in the rear cockpit as the instrument pilot. The airplane was observed flying at low altitude over the Nehls Farm. The AT-6C was observed circling the Nehls farmhouse at about 100 feet agl. The airplane circled to the left twice and stalled at low altitude, snapping over to an inverted position. The pilot attempted to recover but the airplane smashed to the ground just as it was approaching a wings-level attitude. Investigation revealed that the Lt. Todd was acquainted with Dorothy Nehls, a resident of the farmhouse. Lt. Todd had met with Miss Nehls on 6-4-45 and stated that he would fly over the farm in the next few days. Lt. Todd had buzzed the farm a week previous to the accident.

6-6-45E. Apalachee Bay, Florida. At 1055 EWT, a Douglas A-26B crashed into Apalachee Bay eight miles ESE of St. Marks, Florida, killing pilot F/O Charles F. Mandeville and gunner Sgt. Dominic A. Moresca. The airplane was part of a three-ship flight that had taken off from Moody Field, Valdosta, Georgia, on a low-level mission to St. Marks and return. The airplanes were flying over the surface of the bay at very low altitude when the subject airplane attempted a left turn. The pilot apparently misjudged his altitude above the water and allowed the port wing to collide with the surface. The airplane cartwheeled into the bay about one mile from the mouth of the Aucilla River.

6-7-45A. Panama City, Florida. At 0844 CWT, a Bell RP-63C crashed about a half mile west of Tyndall Field, Panama City, Florida, killing 2Lt. Richard Rollins Jr., 23. The airplane took off at 0714 CWT from Tyndall Field on a gunnery mission. The airplane rendezvoused with a Consolidated B-24 airplane for the gunnery mission. The gunnery mission was completed and the subject airplane began maneuvering with a Curtiss P-40 type airplane. The subject airplane returned to Tyndall Field and passed over the active runway at about 1,000 feet agl. After passing the halfway

point of the field the airplane peeled up to the left to enter the traffic pattern. The pilot apparently extended the landing gear just before or as he entered the peel-up maneuver rather than after, which is the correct technique. The airplane called in on the base leg and tower personnel observed that the subject airplane was in danger of overtaking a North American AT-6 airplane that was on final approach. In an effort to prevent a dangerous situation from developing, tower personnel directed the AT-6 pilot to make a touch-and-go landing and come back around for the full stop landing. The subject pilot overshot the left turn from base to final. The pilot increased the bank and the airplane stalled at an altitude of 100 feet agl. The pilot added power in an attempt to recover but the starboard wing dropped and the airplane plunged to the ground, striking on the starboard wing. The airplane burst into flames upon impact, killing the pilot instantly.

6-7-45B. Ojai, California. At 1207, a North American P-51D crashed four miles north of Ojai, California, killing 1Lt. John R. Giberson, Jr., 24. The airplane was part of a two-ship flight that took off from Oxnard, California, on a fighter instrument flight. For the first 50 minutes, the subject pilot acted as the safety pilot. The pilots then changed positions with the subject pilot flying on instruments and the other pilot acting as the safety pilot. The subject pilot was flying on instruments at about 12,000 feet for about 40 minutes when the safety pilot observed the subject airplane enter a steep climbing turn to the left. The airplane turned over to an inverted attitude and then nosed toward the ground in a split. The safety pilot radioed the subject pilot to pull out and moments later streamers were observed trailing the subject airplane's wingtips. The airplane continued in a steep dive until it entered a cloud that was covering a 4,477-foot mountain. The airplane dove into the cloud-shrouded mountaintop, exploding into flames upon impact.

6-8-45. Goodfellow Field, Texas. At 1205, a North American AT-6B and an AT-6D collided in mid-air while attempting to land at Goodfellow Field, San Angelo, Texas, killing AT-6B pilot 1Lt. Abelardo R. Mondonedo, 27. 1Lt. Julian L. Hill and 1Lt. Tomas S. Fernandez were uninjured and able to make a safe landing at Goodfellow Field. The collision occurred while both airplanes were on final approach. Investigation revealed that the AT-6D was flying directly above the AT-6B while both airplanes flew on the final approach at about 400 feet agl. The controller gave the AT-6D the red light and gave the AT-6B the green light. The controller called the upper ship to go around. Both airplanes began to go around and the AT-6B collided with the AT-6D at about 75 feet agl. Investigation revealed that the AT-6D's starboard wing collided with the vertical stabilizer of the AT-6B, apparently as the AT-6B was attempting to climb away for the go-around. The AT-6B entered a half roll and smashed to the ground. The pilot of the AT-6D was able to regain control of his airplane and make a safe landing.

6-9-45. Marble Falls, Texas. At 1645, a Stearman PT-17 crashed five miles south of Marble Falls, Texas, fatally injuring pilot F/O Kenneth Rosenfield, 22. The airplane took off from Goodfellow Field, San Angelo, Texas, on a ferry flight to Bergstrom Field, Austin, Texas. The airplane was observed flying to the north at low altitude over Highway 281. The airplane was observed to enter a roll at an altitude of about 150 feet agl. The airplane fell out of the bottom of the maneuver and collided with trees, smashing to the ground and bursting into flames. The pilot was pulled alive from wreckage by a passing serviceman but died a short time later. Before he died, the pilot stated that he was looking at his maps prior to the accident. Investigators speculated that the roll might have been unintentional and a result of sudden over-control when the pilot observed the ground coming up.

6-10-45A. Newark, Delaware. At 0830 EWT, a Beech C-45F flying in poor weather collided with a tree and crashed three miles east of Newark, Delaware, killing seven fliers. The airplane took off at 0627 EWT from Pope Field, Fort Bragg, North Carolina, on an IFR flight to Mercer Airport, Trenton, New Jersey. The pilot reported to air traffic controllers over Richmond, Virginia, at 0730 EWT, reporting that he was at 9,000 feet. At approximately 0830, the airplane was observed flying under the overcast near Newark, Delaware. Weather was reported as a ceiling of 300 feet agl and visibility less than a mile with light rain. The airplane struck a 75-foot oak tree on the farm of Judge Hugh Morris. The airplane nosed down and smashed into the terrain about 100 yards from the tree, exploding into flames upon impact. Wreckage and body parts were scattered over a wide area. Pilot Lt. Col. Douglas T. Goodale, navigator 2Lt. B. Lessin and engineer MSgt. R.Y. Lauderdale were killed in the crash. Passengers killed in the crash were: TSgt. W.E. Jenninson Jr., Pvt. J. Maragakes, Pfc. A. Ravands, Pfc. H. Sobel.

6-10-45B. Rocky, Oklahoma. At 1625 CWT, a Boeing B-29 suffering engine troubles crash-landed two miles east of Rocky, Oklahoma, killing three crewmembers and injuring eight others. The airplane was part of a four-ship flight that took off from Batista Field, Cuba, at 0545 on a bombing mission to Dalhart, Texas, and return to its station at Walker Army Air Field, Victoria, Kansas. The mission commander changed the mission while on the flight; the new mission was to fly to Galveston, Texas, and then fly a bombing mission to Albuquerque, New Mexico. The B-29s, flying at 20,000 feet, encountered instrument conditions over Texas. The airplanes broke formation and then descended through the clouds, breaking out over Apache, Oklahoma. The four B-29s circled the area at 1,000 feet agl. Because of fuel concerns, the airplanes were ordered to fly directly to Walker Army Air

Field underneath the overcast. The airplanes soon lost visual contact with each other, and the subject airplane was behind the other three. The subject airplane was flying at 200 feet agl when it began fishtailing and losing altitude. The engineer stated that the instruments were reading normally. The pilot applied right rudder and right aileron to overcome "drag" that he felt on the port side. The pilot asked for increased power but the number-one and number-two engines did not respond. The airplane continued to lose altitude so the pilot called for an emergency crash landing. The pilot repeated the warning over the intercom but had failed to ring the emergency bell. The pilot was unable to maintain altitude and the airplane descended into the ground. Investigators stated, "The belly of the aircraft skimmed the ground, knocking down wheat for approximately 100 yards, then the belly contacted the ground and continued in a straight path approximately 200 yards farther up a slight incline, and just over the crest of where it came to rest. The aircraft started burning immediately in the area of the bomb bay fuel tanks." The fliers who were able to escape were unable to rescue the crewmembers trapped in the central fire control section. Investigation revealed that the crew was unable to knock out the CFC rescue panels. The fire spread and drove them away. Investigators were unable to determine what had caused the diminished performance of the port engines. Central Fire Controller Sgt. Robert H. Bolton, gunner Cpl. Louis C. Brammer and tail gunner Sgt. William M. Clark, Jr. were killed in the crash landing. Investigators speculated that the fliers in the central fire control compartment were unaware that the airplane was making a crash landing. Crewmembers injured in the crash landing were: Capt. Roscoe C. Cutsinger, pilot; F/O Raymond E. White, co-pilot; F/O Donald Smith, navigator; 2Lt. Chester L. Cooper, bombardier; Sgt. Hubert F. Sage, Jr., flight engineer; Cpl. Morton Spaner, radio operator; Sgt. Charles R. Kramer, gunner; SSgt. Frank W. Langell, engineer.

6-11-45A. Dallas, Texas. At 0801 CWT, a Republic P-47D crashed one-quarter mile north of Love Field, Dallas, Texas, killing pilot Lt. Col. William B. Peterson, 27. The airplane was the lead ship of a three-ship flight that took off at 0800 from Love Field on a flight to Galveston, Texas; the flight had originated at Peterson Field, Colorado Springs, Colorado. The subject airplane started the take-off run at mid-field and used up most of the remaining length of the 5,200-foot runway during the take-off. The airplane staggered into the air a few hundred feet from the end of the runway. Tower personnel, anticipating that the airplane would crash, alerted rescue crews to stand by. The airplane was observed flying near stalling speed at 100 feet agl in a nose-high attitude. The airplane was trailing thick black smoke as it entered a flat turn to the left. The pilot was attempting to get back to the field, so tower personnel cleared the field for an emergency landing. The airplane completed about 180 degrees of turn to the left before the starboard wing dropped and the airplane plunged toward the ground. The airplane struck the ground in a flat attitude and exploded into flames upon impact, killing the pilot instantly. Investigators could not determine the cause of the engine failure.

6-11-45B. Rush Springs, Oklahoma. At 0820 CWT, a Boeing B-29 flying in poor weather suffered a catastrophic structural failure and crashed six miles west of Rush Springs, Oklahoma, killing ten fliers. The B-29 was part of a three-ship flight that took off at 0636 CWT from Smoky Hill Army Air Field, Salina, Kansas, on a navigation-bombing mission. The formation soon encountered a powerful thunderstorm system. Investigators stated, "Careful inspection of the wreckage, especially the portion of the shattered fuselage and engines disclosed the aircraft hit the ground in an inverted position. Inquiry revealed ... two other B-29s were flying formation with this aircraft and all aircraft decided to break formation to reduce the hazards of flying instruments [while flying] through the frontal area. The projected flight plan was to maintain 20,000 feet throughout the mission. A search of the area of the crash indicated the airplane began disintegrating before the impact as evidenced by wing and control surfaces being scattered over eight square miles of the flight path. Examination of these bits of wreckage points toward the fact that the control surfaces (i.e. ailerons, elevators, etc.) were torn from their hinges, inducing further structural failure of the aircraft." Killed in the crash were: 1Lt. James L. Groves, pilot; 2Lt. W.C. Smallwood, co-pilot; 1Lt. Robert J. Collopy, bombardier; F/O James E. Farmer, navigator; 2Lt. John H. Carson, radar operator; 2Lt. Julius Weilep, pilot-rated flight engineer; Cpl. Thomas B. Eaves, radio operator; Cpl. James A. Coveyou, central fire control; Pfc. James A. Hunt, gunner; Pfc. Matthew J. Fitzpatrick, tail gunner.

6-11-45C. Venice, Florida. At 0930 EWT, a North American P-51C suffered a catastrophic structural failure and crashed 20 miles ENE of Venice Army Air Field, Venice, Florida, killing 1Lt. Aubrey O. Hood, 28. The airplane was part of a two-ship flight that took off from Venice Army Air Field on an instrument training flight. The subject pilot was to fly as the instrument pilot and the other pilot was to act as the safety observer. The airplanes climbed to 12,000 feet to conduct the instrument flight. The subject pilot went on instruments and successfully performed several maneuvers, including steep turns in both directions. The safety pilot was able to keep the subject airplane in sight the whole time. The subject pilot had adhered to the instrument flight "checklist" that dictated his flight maneuvers. The subject airplane was attempting to roll out of a right-hand spiral at about 9,000 feet, passing beneath the safety pilot's wing. The subject airplane never re-emerged on the other side, so the safety pilot began maneuvering to re-establish visual contact with the subject airplane but he was unsuccessful. The safety pilot and ground control attempted to contact the subject pilot by radio but were also unsuccessful. The pilot apparently lost control and overstressed the airplane while attempting to re-establish level flight. Investigation of the wreckage indicated that both wings had failed and separated before the airplane had struck the ground. The pilot made no attempt to parachute to safety.

6-11-45D. Dover, Delaware. At 0940 EWT, a Republic P-47D and a P-47N collided in mid-air and crashed one and a half miles south of Dover, Delaware, killing both pilots. 2Lt. Sigmund V. Kaczmarski, 22, was killed in the crash of the P-47D; F/O James R. Trippe, 22, was killed in the crash of the P-47N. The airplanes were part of a four-ship flight that took off at 0845 from Dover Army Air Field on a gunnery mission. The airplanes completed the gunnery mission and had let down while returning to the field. The flight leader signaled the formation to switch to an echelon right formation in preparation for the peel-off for landing. Lt. Kaczmarski was flying in the number-two position of the first element; F/O Trippe, in the number-three position, was leading the second two-ship element. Lt. Kaczmarski, flying on the leader's port wing, began sliding over to take his position on the leader's starboard wing for the right echelon position and F/O Trippe began to back off to make room for Lt. Kaczmarski. Lt. Kaczmarski slid under the leader and then pulled up into F/O Trippe's airplane. The airplanes burst into flames in the collision, which occurred at about 1,000 feet agl. The airplanes remained stuck together as they entered a flat spin, striking the ground in a flat attitude and exploding into flames. Neither pilot was able to parachute to safety.

6-12-45A. Shaw Field, South Carolina. At 1032 EWT, a Republic P-47D crashed while attempting a take-off at Shaw Field, Sumter, South Carolina, killing pilot 1Lt. Lincoln R. Sherwood, 23. Two enlisted men were injured on the ground. Investigators stated, "Lt. Sherwood was given clearance by the tower to taxi to Runway 22 for take-off. He was then cleared into position and started his take-off roll at which time a steady stream of dense black smoke [was coming] from the turbo and the aircraft was gaining speed at an exceptionally slow rate. At approximately two-thirds down the runway, the aircraft was still moving far too slow for take-off, and when the pilot had used the full 4,500 feet of runway he attempted to pull the aircraft off the ground. The aircraft gained approximately five feet of altitude in a full nose-high stalled position, mushing back into the ground immediately with wheels partially retracted. The aircraft then skidded across the remaining grass area of the field and struck the concrete wall of a drain ditch about the middle of the underside of the aircraft, causing it to catch fire. The aircraft then crashed into the base trolley trailer, completely demolishing the trailer and breaking the right leg of an enlisted man, who was a passenger. Minor injuries were inflicted on an MP on duty at the time of the crash by flying debris as the aircraft continued forward and struck the right rear corner of the guardhouse. The aircraft struck the guardhouse at an angle causing [the airplane] to break in two just aft of the cockpit. The forward part of the aircraft came to rest approximately three feet beyond the left rear corner of the guardhouse and was enveloped in flames." Pfc. Frank Johnson received serious injuries and MP Pfc. William I. Reardon sustained minor injuries.

6-12-45B. Miakka, Florida. At 1135 EWT, a North American P-51K crashed eight miles northwest of Miakka, Florida, killing pilot 2Lt. James O. Clark, 22. The airplane was part of a two-ship flight that took off from Sarasota Army Air Field, Sarasota, Florida, on an instrument training flight. The subject pilot was flying as the safety observer while another pilot flew on instruments. The instrument pilot observed that the safety pilot was not in position and could not raise him on the radio. About five minutes later the instrument pilot observed a column of smoke from the crashed airplane. Investigation revealed that the subject airplane fell out of control and struck the ground at high speed, exploding violently into flames upon impact. Wreckage was scattered over an area of 100 yards. The cause of the crash could not be determined. See 6-15-45D.

6-12-45C. Barney, Georgia. At 1147 EWT, two Douglas A-26C airplanes collided in mid-air four miles southeast of Barney, Georgia, killing one flier and seriously injuring another. Pilot 2Lt. James A. Richmond was killed in the crash of A-26C #44-35515; Sgt. Louis Renzulli was seriously injured. Pilot 2Lt. Robert H. Rodeck and gunner Sgt. Edward S. Richard escaped injury and made a safe landing in A-26C #44-35198. The airplanes were part of two separate three-ship flights that took off from Moody Field, Valdosta, Georgia, on a separate low-level formation and skip-bombing missions. The two flights separately dropped bombs on the bombing range and were flying back to base at low altitude. Lt. Richmond was flying in the number-three position of a three-ship flight. He broke formation and closed in on the formation that Lt. Rodeck was leading. He apparently intended to make a swooping pass at this flight from above. Lt. Richmond dove down on and passed below the other flight, pulling up and ahead from underneath. Lt. Richmond's rudder and vertical stabilizer collided with the starboard propeller of Lt. Rodeck's airplane during the pull-up. Lt. Richmond attempted to make a crash landing but he chose a field that contained several two-foot earthen dikes that ran perpendicular to the line of flight. The airplane broke up as it slid across the rough terrain and erupted into flames. The pilot was killed and the gunner was seriously injured. Lt. Rodeck was able to make a safe landing at Moody Field.

6-12-45D. Marion, Texas. At 2230 CWT, a Boeing TB-29A crashed one-quarter mile west of Marion, Texas, killing seven crewmembers. The airplane took off at 2150 CWT from Randolph Field, San Antonio, Texas, on a night training flight. Investigators stated, "The airplane struck the ground headed east, nose down, a little past the vertical. No reason could be found for the airplane striking the ground in this attitude. The airplane was completely demolished from the impact, scattering parts of the wings, fuel tanks and tail group over an area of approximately 100-yard radius. All occupants were killed as a result of the crash." The cause of the crash could not be determined. Killed in the crash were: Capt. Robert A. Johnson, pilot; 1Lt. Paul M. Stout, co-pilot; 1Lt. Paul R. Smith, flight engineer; 2Lt. Boyda F. Miller, Jr., pilot-rated passenger; 2Lt. James D. Madigan, pilot-rated passenger; F/O William J. McCoun, pilot-rated passenger; Pfc. Kenneth T. Griffin, passenger.

6-13-45A. Lincoln, Nebraska. At 0710 CWT, pilot 2Lt. Arthur F. Savage was killed when he walked into the path of and was struck by a North American AT-6A taxiing at Lincoln Army Air Field, Lincoln, Nebraska. F/O Murray W. Riffle, who was walking with Lt. Savage, sustained minor injuries when the impact of the airplane's propeller hurled Lt. Savage's body into him. The airplane passed over the two pilots after the collision. Investigation revealed that pilot 2Lt. Royce L. Shepherd had been taxiing in a safe manner and at a slow speed. Investigators noted that there were several airplanes parked on the line with engines running.

6-13-45B. Glendo, Wyoming. At 1007 MWT, a Douglas A-20K crashed 13 miles southwest of Glendo, Wyoming, killing F/O Charles E. Lewis, 29. The airplane took off at 0828 MWT from Cheyenne, Wyoming, on a ferry flight to Great Falls, Montana. The airplane flew past Glendo and a short time later suffered a problem with the port engine. The pilot feathered the port propeller and trimmed the airplane for single-engine flight. The pilot made a 180-degree turn and headed back toward Cheyenne. A short time later the starboard engine failed, forcing the pilot to attempt an emergency forced landing. The pilot picked out the only level area and flew a small pattern for the field. Investigation revealed that the landing gear was in the extended position and the flaps were retracted while the pilot turned on the approach. The pilot attempted to retract the landing gear and extend the flaps, and then he turned to the right to avoid landing on a small ravine in his path. The starboard wing collided with the terrain and the airplane cartwheeled into the ground. The pilot, still strapped to his seat, was hurled from the airplane as it smashed itself to pieces. The A-20K, which carried Soviet markings and had a total time of 17 flying hours, did not burn but was totally destroyed.

6-13-45C. Pacific Ocean, California. At 1545, a Lockheed P-38L crashed into the Pacific Ocean about 30 miles west of Oceano, California, killing pilot Capt.

Alfonso G. Seagraves, 27. The airplane took off at 1520 from Santa Maria Army Air Field, Santa Maria, California, on an aerial gunnery mission. Capt. Seagraves was leading the four-ship flight, which had rendezvoused with a target ship at 10,000 feet about ten miles out to sea. The flight had made two firing passes at the target ship when members of the flight observed thick white smoke trailing Capt. Seagrave's port engine. Moments later flames were seen trailing Capt. Seagrave's port engine. Capt. Seagrave turned to the left and broke formation, stating over the radio that he was going to bail out. The pilot was seen to bail out on the starboard side. The pilot successfully parachuted to safety and the airplane spun into the sea. The pilot apparently drowned while waiting to be rescued. Investigators noted that the sea was rough, hampering rescue vessels and crews in their effort to find the downed pilot. The pilot's body remained missing.

6-13-45D. Albion, Oklahoma. At 2300 CWT, a Curtiss P-40N crashed six miles south of Albion, Oklahoma, killing pilot Capt. Samuel B.W. Kennedy, 24. The airplane took off from Navy Field, Norman, Oklahoma, on a flight to Stuttgart, Arkansas. At about 2250 the airplane was observed circling over Albion, flashing its landing lights. The airplane crashed into rising terrain a short time later. The airplane struck a 1,950-foot mountain about 50 feet from the peak. Investigation revealed that the airplane was about ten miles to the right of his intended course. It was noted that the ceiling was estimated to be about 2,000 feet at the time of the crash and that several peaks in excess of 1,900 feet existed in the area.

6-14-45A. Ware, Alabama. At 1020, a North American AT-6C crashed three miles southeast of Ware, Alabama, killing French Air Force student pilot Pvt. Jacques Morgeat. French Air Force instructor Aspt. Henri E. Nicolai, 21, parachuted to safety and escaped injury. The airplane took off from Gunter Field, Montgomery, Alabama, on a routine training flight. The student entered a deliberate spin to the left at 7,000 feet. After two turns of spin the student began his recovery but was unable to stop the spin, so the instructor took control of the airplane and re-established level flight. The instructor ordered the student to climb to 7,200 feet and enter another spin. As the airplane climbed, the instructor pointed out the student's errors during the previous recovery attempt. The student entered a spin to the left and was unable to recover after three turns. The student allowed the spin to become steeper so the instructor attempted to take over the controls. The student froze on the controls and the instructor was unable to overcome the student's grip on the stick. The instructor ordered the student to release the controls but the student held firm. The instructor applied throttle and rudder inputs, but the student continued to hold the stick firmly, preventing a recovery. The instructor ordered the student to bail out. The student looked back and began to open his canopy as the airplane

fell past 2,000 feet. The instructor bailed out at about 1,500 feet agl and was able to parachute to safety. The instructor stated to investigators that he had made four attempts at recovery and that he had ordered the student to bail out twice. It was also noted that the airplane spun over a dozen times before the instructor bailed out.

6-14-45B. Oakridge, Mississippi. At 1120 CWT, a Curtiss C-46D flying in poor weather crashed six miles southeast of Oakridge, Mississippi, killing 14 passengers and three crewmembers. Passenger Sgt. Raymond L. Bullock miraculously survived but received serious injuries. Passenger 1Lt. Harry L. Provasco was rescued from the wreckage, but he died about 15 minutes after reaching a hospital at Vicksburg, Mississippi. Pilot Capt. William R. Ross, co-pilot 2Lt. Robert A. Petty and flight engineer 2Lt. Don E. Hiller were killed in the crash. The airplane took off from Dallas, Texas, on a passenger flight to Jackson, Mississippi. The airplane was cleared on an instrument flight plan and was flying at about 3,000 feet. The airplane's starboard wing was apparently struck by lightning, causing an explosion in the wing. The pilot could not maintain altitude and the airplane smashed into a wooded area at a 30-degree angle. The airplane burst into flames as it smashed its way through the trees. Passengers killed in the crash were: Sgt. Henry G. Beasley, Pvt. Clifford W. Canady, Pvt. Joe Capes Jr., 1Lt. Edward P. Carnot, Pfc. Betty J. Coleman, 2Lt. John N. Ferrier, RM2C Marjorie M. Flint, Pvt. Charles H. Hines, 1Lt. Elmer F. Irwin, Ensign Everett L. Owens, SSgt. William Richards, Cpl. Willard O. Rogers, Lt. (j.g.) Robert M. Woods.

6-14-45C. Great South Bay, New York. At 1526 EWT, a Consolidated B-24J collided with the surface of Great South Bay about five miles east of Sayville, Long Island, New York, killing engineer SSgt. Michael L. Sullivan, who was manning the ball turret. The airplane made a safe landing at its home station at Westover Field, Chicopee, Massachusetts. Investigators stated, "B-24J #42-50931 after having completed low altitude gunnery firing was proceeding toward Mitchel Field at 500 feet to complete the assigned mission. Visibility according to pilot's statement was lowering to a mile and a half approximately as he approached the southern shoreline of Long Island. Pilot glanced out the window and did not notice until the aircraft had lost some 300 feet that the aircraft was in a shallow turn downward to the right. The pilot called for more power and righted the aircraft. The co-pilot did not increase power immediately but continued to adjust his throat mike equipment. Pilot again directed his attention away from flying to look at a map and in so doing inadvertently lost the remaining 200 feet of altitude. Before a recovery could be made, the ball turret struck the water and breaking apart swept the engineer, who was in the ball turret at the time, into the water. Pilot was informed of this occurrence by other crewmembers and instituted a search by circling the approximate position of the accident. When it became impossible to determine what had become of the missing man, pilot returned to base to report the incident." Pilot 1Lt. James C. Ramey, co-pilot 2Lt. Joseph E Lindsey and eight other crewmembers were uninjured.

6-14-45D. Murrieta, California. At 1628, a Lockheed P-38L crashed five miles northeast of Murrieta, California, killing pilot 1Lt. James J. O'Neil, 24, who fell to his death in an unsuccessful parachute jump. The airplane was the lead ship of a four-ship flight that took off from Ontario Army Air Field, Ontario, California, on a high-altitude aerobatic flight. Investigators stated, "Lt. O'Neil started a climbing turn to the left at approximately 11,000 feet above [Lake Elsinore]. Lt. O'Neil then dropped maneuver flaps and began a tight left turn, rolled inverted and went into a spin. After about four turns Lt. O'Neil bailed out, but hit the tail section, tearing loose his parachute pack, cutting 22 of the 24 suspension lines and receiving extensive injury to limbs and head. As a result, the parachute failed to function, resulting in death to the pilot. The aircraft crashed in a remote part of the Lake Elsinore area of the California State Forest, exploded and burned."

6-14-45E. Liberal, Kansas. At 2130 CWT, a Consolidated B-24J crashed one mile northwest of the Army Air Field at Liberal, Kansas, killing three fliers. Investigators stated, "The students took off at 2050 to circle the pattern, shoot one landing and then pick up their instructor. After take-off nothing was heard from the airplane until the tower noticed a plane turning on final approach blinking the landing lights. The tower operators immediately cleared the plane to land both over the radio and by using the green light. The airplane made a high approach, which appeared to be at a high airspeed. At about 40 to 50 feet above the ground, when the airplane was well down the field, the pilot gave it the power to go around. At this time it was noted by witnesses and later proved by investigation of the propeller that the number-one engine was feathered. It was also noted that the wheels were down and the flaps were down. The airplane started a very slight turn to the left on the go-around and appeared to be mushing. The airplane would gradually lose altitude and then the pilot would pull the nose up. This continued until the critical speed was attained, when the airplane fell off on the left wing, cartwheeling into the ground." Pilot 2Lt. Claude T. Aiken, co-pilot 2Lt. George T. Bright and engineer SSgt. Walter Martella were killed in the crash.

6-15-45A. Banner, Oklahoma. At 1125, a Lockheed F-5E crashed six miles north of Banner, Oklahoma, killing pilot 2Lt. Nairn E. Rivers, 27. The airplane took off from Will Rogers Field, Oklahoma City, Oklahoma, on a routine training flight. The airplane was observed in a spin at about 4,000 feet. The pilot apparently bailed out at low altitude and was unable

to pull his ripcord. His body was found underneath the port wing of the wrecked airplane.

6-15-45B. Abilene, Texas. At 1149 CWT, a Republic P-47D crashed 14 miles southeast of Abilene Army Air Field, Abilene, Texas, killing pilot 2Lt. Harold S. Kewish, 21. The airplane was part of a four-ship flight that took off from Abilene Army Air Field on a routine flight. The airplanes had engaged in authorized simulated combat with a supervisory pilot. The subject airplane rolled to an inverted attitude and then the pilot attempted to recover with a split-S maneuver. The pilot had entered the maneuver at too low an altitude and had built up too much speed in the dive to allow a safe recovery. The airplane smashed into the ground at about a 20-degree angle. The airplane exploded into flames upon impact and the pilot was killed instantly.

6-15-45C. Mission, Texas. At 1220 CWT, a North American AT-6C crashed five miles south of Moore Field, Mission, Texas, killing pilot 2Lt. John T. Roach, 21, and student A/C Harold A. Peterson. The airplane took off from Moore Field on a routine training flight. The airplane was seen to make a turn to the right at low altitude. The airplane stalled and fell off to the right and smashed into the ground. Investigators speculated that the student and instructor were attempting a simulated forced landing when the airplane was stalled at too low an altitude to allow recovery. The airplane did not burn but was destroyed utterly.

6-15-45D. Gulf of Mexico, Florida. At 1330 EWT, a North American P-51D crashed into the Gulf of Mexico three miles west of Long Beach, Florida, killing pilot 2Lt. Charles Wilson, 22. The airplane took off from Sarasota Army Air Field, Sarasota, Florida, on a routine training flight. The airplane was to rendezvous at 5,000 feet over Bradenton, Florida, with another P-51, but the other airplane could not take-off because of mechanical difficulties. The subject airplane was observed in a 70-degree dive at about 3,000 feet. Investigation revealed that the airplane was completely out of control when it slammed into the sea. The airplane exploded upon impact and the pilot made no apparent attempt to parachute to safety. Investigators could find no evidence of material failure and could not determine the cause of the crash. See 6-12-45B.

6-15-45E. Walla Walla, Washington. At 1752 PWT, a Consolidated B-24J crashed two miles north of the Army Air Base at Walla Walla, Washington, killing the crew of nine. The airplane had taken off from Walla Walla Army Air Base on a routine training flight. Investigators stated, "Aircraft #42-64155 came in for a landing with the number-three propeller feathered. Pilot landed long and fast and touched all wheels just before the half way mark of the runway. The airplane bounced into the air and the pilot made a three-engine go-around. The Tower Officer told the pilot he was low on the downwind leg and also told him to unfeather the number-three propeller and to use the engine if necessary; pilot acknowledged receipt of message. The downwind leg was too close to the field and pilot was forced to make a steep turn into the final approach into the dead engine. Aircraft was seen to increase bank to over 70 degrees, then go into a partial spin and crash into the ground. The airplane burst into flames and was demolished. It is believed that because the pilot had gone around on three engines once, he was too proud to go around again and was showing that he could handle a B-24 making a steep turn into a dead engine. He continued to increase the bank until he lost control." Killed in the crash were: 2Lt. Gilbert W. Milam, pilot; F/O Henry B. Eging, co-pilot; F/O Harry R. Burns, navigator; F/O Charles J. Gianini, bombardier; Cpl. Jack S. Cassidy, engineer; Cpl. Earl L. Gross, radio operator; Cpl. Raymond G. Apodoca, gunner; Pfc. Stephen Miskevich, gunner; Pfc. Wendell S. Jones, gunner.

6-16-45A. Pampa, Texas. At 1230 CWT, a North American TB-25J crashed 13 miles ESE of Pampa, Texas, killing instructor 1Lt. Anthony S. San Miguel, 24, and Brazilian Air Force B-25 student A/C Otto E. Dreher. The airplane had taken off from Pampa Army Air Field on a routine training flight. The airplane entered the overhead approach for Pampa Army Air Field at 4,500 feet. As the airplane entered the overhead pattern, controllers observed smoke trailing the port engine. A short time later the port engine was observed to be on fire. The airplane entered a spiral to the left. The instructor was heard ordering the student to bail out. The spiral became steeper and no recovery was made. The airplane smashed to the ground and exploded into flames, killing the pilots instantly. Investigators could not determine the cause of the engine fire.

6-16-45B. Willow Grove, Delaware. At 1530 EWT, a Republic P-47D crashed two miles west of Willow Grove, Delaware, killing 2Lt. Robert E. Weaver, 25. The airplane was flying in the number-two position of a four-ship flight that took off from Dover Army Air Field, Dover, Delaware, on a ground-strafing mission. The flight arrived at the ground gunnery range and was advised that there would be over an hour wait for the flight to be cleared to use the range. The flight leader acknowledged and contacted Fighter Operations requesting a change to the alternate mission. The flight climbed to 11,000 feet to wait for a new mission to be coordinated. The flight leader ordered the airplanes to assume a string formation. The flight leader then led the flight through various aerobatic maneuvers and chandelles. The flight leader performed a barrel roll and the flight followed the maneuver. The flight leader then performed a steep chandelle to the right and the subject airplane stalled while halfway through the maneuver. The airplane fell into an inverted spin. The pilot attempted a recovery, but the airplane stalled as the pilot pulled out, snapping over to the right and

plunging toward the terrain. The pilot failed to recover and the airplane smashed to the ground at a steep angle, exploding into flames upon impact.

6-16-45C. Harrah, Oklahoma. At 1435 CWT, a Douglas A-26C flying in instrument conditions crashed three miles east of Harrah, Oklahoma, killing four fliers. The airplane took off from Florence Army Air Field, Florence, South Carolina, on a flight to Tinker Field, Oklahoma City, Oklahoma. The airplane encountered deteriorating weather conditions as it approached Oklahoma City and the pilot experienced great difficulty communicating with Oklahoma City controllers. The pilot attempted to maintain visual contact with the ground, but he was eventually forced to go on instruments. A short time later, witnesses observed the airplane fall out of the overcast in a near stalling attitude. The pilot was unable to recover and the airplane smashed to the ground at a 40-degree angle, bursting into flames upon impact. Investigators speculated that the pilot had stalled the airplane at an altitude too low to allow recovery while flying in instrument conditions. It was noted that the pilot had never flown an A-26 airplane on instruments before this flight. Killed in the crash were: 2Lt. Fred A. Watson, pilot; Sgt. Burke P. Lawrence, engineer-gunner; F/O Thomas M. Barton, Jr., pilot-rated passenger; Capt. John H. Culberton, passenger.

6-17-45. Camden, Alabama. At 2045 CWT, a Vultee BT-13B crashed two miles north of Camden, Alabama, killing pilot 2Lt. Roy E. Knight and passenger 2Lt. Arthur S. Mayo. The airplane took off from Lafayette, Louisiana, on a flight to Atlanta, Georgia, via Gunter Field, Montgomery, Alabama. The airplane apparently entered a spiral and smashed to the ground, exploding violently into flames upon impact. The airplane and occupants were completely destroyed and the engine was buried deep into the hard clay. Wreckage was scattered over an area of 300 feet. Investigators speculated that the pilot had suffered vertigo and allowed the airplane to enter a descending spiral. Investigators noted that the pilot was flying the passenger to Atlanta because the passenger's mother was very ill.

6-18-45A. Bluethenthal Field, North Carolina. At 0855 EWT, a Republic P-47N crashed one-quarter mile east of Bluethenthal Field, Wilmington, North Carolina, killing pilot 1Lt. William A. Lowrance, 24. The airplane was the lead ship of a four-ship flight that was returning to Bluethenthal Field after completing a ground gunnery and skip-bombing mission. A pilot acting as the "mobile control operator" stated to investigators, "Four aircraft, led by Lt. Lowrance, broke away for a landing on Runway 23. Lt. Lowrance executed a low, tight pattern to a point about 50 degrees off the end of the runway, at an approximate altitude of 200 to 300 feet. At this point, his aircraft went into a vertical bank, and was seen to mush badly for an instant. The aircraft then snapped rolled violently to the left and crashed from a nose-down, inverted attitude, exploding upon contact with the ground. I then dispatched a crash truck."

6-18-45B. Fairchild, Montana. At 1030 MWT, a Douglas A-20H crashed 18 miles west of Fairfield, Montana, killing pilot 1Lt. James P. Layman, 29. The airplane took off at 0930 MWT from Gore Field, Great Falls, Montana, on a transition training flight. The airplane was observed to enter a vertical bank to the left while flying at very low altitude. The airplane apparently flipped over and smashed to the ground, exploding into flames upon impact. Investigators noted that this was the pilot's second solo transition flight in an A-20 type airplane.

6-18-45C. Oklahoma City, Oklahoma. At 1534 CWT, a Lockheed P-38L crashed 12 miles southwest of Oklahoma City, Oklahoma, killing pilot F/O Ben A. Peters, Jr., 22. The airplane was returning to the field following the pilot's first solo flight in a P-38 airplane. The pilot returned to the field and entered traffic for a landing. The approach was too high and the pilot was ordered to go around. The pilot retracted the landing gear and began climbing away. The airplane reached 2,000 feet and the pilot made turn to the left to re-enter the traffic circuit. The starboard wing was seen to drop suddenly and the airplane flipped to an inverted position. The airplane rolled through and entered a spin to the right. The spin flattened out as the airplane continued to fall to earth. The pilot was unable to recover and the airplane smashed to the ground in a flat attitude, bursting into flames upon impact.

6-18-45D. Gulf of Mexico, Florida. At 1635 EWT, a North American P-51K crashed into the Gulf of Mexico two miles WNW of Englewood, Florida, killing 2Lt. Marlo A. Mastrangeli, 21. The P-51 was flying in the number-three position of a four-ship flight that took off from Punta Gorda Army Air Field, Punta Gorda, Florida, on a flight to Venice Army Air Field, Venice, Florida. The flight encountered scattered clouds during the flight. The flight leader was able to maneuver around the clouds until he reached a point where he had no choice but to enter instrument conditions. The clouds were scattered and not very thick, the airplanes not spending more than a few seconds in clouds at a time. The flight broke out into the clear after flying through scattered clouds for a few miles. Once the flight was clear, the flight leader noticed that the number-three ship was missing from the formation. The number-four man stated that the subject pilot began a turn to the left just before they entered the clouds. The subject airplane was not seen again. When the subject airplane failed to return to base a search was commenced. An oil slick was found on the surface of the Gulf of Mexico. The pilot's body was recovered from the sea three days later. An autopsy determined he was killed by impact and not by drowning. The cause of the crash could not be determined.

6-19-45A. Hahira, Georgia. At 1110 EWT, a Douglas A-26C suffered a catastrophic structural failure and crashed five miles west of Hahira, Georgia, killing pilot F/O Ernest L. McLane and gunner Sgt. Raymond J. Martinolich. The airplane took off at 0835 from Moody Field, Valdosta, Georgia, on an instrument proficiency flight. Investigators speculated that the pilot dove through a hole in the clouds to get back to contact conditions and might have stressed the starboard wing beyond its design limits during a high-speed pull-up maneuver. The starboard wing failed and separated and the airplane went out of control, causing the progressive failure of the airframe. The port wing was found about 200 yards from where the fuselage smashed to earth. The starboard wing was found about 300 yards from the fuselage. Both engines, the nose section, the main wing section and assorted wreckage was scattered widely across the terrain.

6-19-45B. Crestview, Florida. At 1215 CWT, a Douglas A-26C flying in poor weather crashed ten miles northeast of Crestview, Florida, killing five fliers. The airplane took off from Eglin Field, Valparaiso, Florida, on a personnel ferry mission to Myrtle Beach Army Air Field, Myrtle Beach, South Carolina. A large thunderstorm system existed in the vicinity of Eglin Field just before take-off. The pilot was cleared to fly contact at about 1,000 feet agl, which was just underneath the ceiling, until he was clear of the storm area. This clearance was changed to 5,000 feet by Jacksonville, Florida, controllers. A short time later a farmer near Crestview observed the airplane flying in a thunderstorm at about 200 feet agl. The airplane descended into wooded terrain at flying speed moments later. The airplane burst into flames and was smashed to pieces as it passed through the trees at a 45-degree angle. Wreckage and body parts were scattered over an area of 300 yards. Killed in the crash were: 1Lt. Joseph A. McGinnis, pilot; TSgt. William J. Koger, engineer; 1Lt. Lawrence F. Schirmer, passenger; TSgt. William H. Epperson, passenger; SSgt. George L. Simmons, passenger.

6-19-45C. Wahoo, Nebraska. At 1635 CWT, a North American AT-6A crashed five miles north of Wahoo, Nebraska, killing pilots 2Lt. Kenneth E. Parsh and F/O Joe R. Kline. The airplane took off from Lincoln Army Air Field, Lincoln, Nebraska, on a local flight. The airplane was later observed in a high-speed dive. The airplane turned to the right as the pilot attempted a steep pull up maneuver. The airplane struck a tree with the wing and went out of control. The airplane contacted the ground at high speed at a shallow angle about 300 feet further on and was smashed to pieces. Wreckage was scattered over an area of 700 feet.

6-20-45A. Lake George, Colorado. At 1200 MWT, a North American AT-6A collided with rising terrain and crashed ten miles north of Lake George, Colorado, killing pilots 2Lt. Wilbur T. Horsecraft and 2Lt. Henry R. Howard, Jr. The airplane took off at

0958 MWT from Buckley Field, Denver, Colorado, on an instrument proficiency flight. The airplane was observed to fly over a mountain and then fly down the gorge leading to the Platte River at very low altitude. The airplane passed over a group of fisherman before banking steeply to the left and heading for the mouth of Longwater Gulch. The airplane flew into an area where the terrain rose up at a 50-degree angle from 300 to 1,500 feet. The airplane flew to the north to the "closed end" of the gulch and the pilot performed a "wingover" to the left and then dove back down toward the gulch. The airplane collided with rising terrain in a mushing attitude during the pull-up and was smashed to pieces as it tumbled across the rugged terrain. The airplane burst into flames upon impact.

6-20-45B. Galveston, Texas. At 1615, two Republic P-47D airplanes collided in mid-air and crashed two and a half miles northwest of Galveston Army Air Field, Galveston, Texas, killing both pilots. 2Lt. Harold H. Moore, 25, was killed in the crash of P-47D #42-8212; 2Lt. Stanley A. Roese was killed in the crash of P-47D #42-28196. Lt. Moore took off from Galveston Army Air Field at about 1600 on target towing mission. Lt. Moore made a normal take off and then entered a turn to the right. He continued to circle to the right and gain altitude for the flight to the rendezvous point. The field instructions were for target airplanes to turn right after take-off and proceed directly to the rendezvous area and if a pilot was to circle the field, he was to circle to the left. Lt. Roese was part of a flight of four that took off from Galveston Army Air Field on a gunnery mission at the west end of Galveston Island. Lt. Roese's flight was circling the field after take-off and was forming up. As Lt. Moore continued to circle, he passed under two ships of the four-ship flight and collided with the number-three man, Lt. Roese. The collision occurred at 3,500 feet agl. Lt. Moore's airplane burst into flames in the collision and then fell out of control. Lt. Roese was unable to regain control his airplane so he bailed out. Lt. Roese bailed out at an altitude too low to allow his parachute to open and he fell to his death.

6-20-45C. Tampa Bay, Florida. At 1805 EWT, a Boeing B-29A crashed into Tampa Bay seven miles southwest of MacDill Field, Tampa, Florida, killing three crewmembers. Eleven crewmembers were injured, three of them seriously. The airplane took off at 1735 EWT from MacDill Field on a bombing mission at Mullet Key, Florida. As the airplane climbed through 5,000 feet, the port scanner noticed an oil leak on the number-one engine and alerted the pilot, who decided to return to the field. The tower operator advised the pilot to make a straight-in approach to the field from the southwest because of poor visibility near the field. The airplane dropped to 1,500 feet and began its approach. The pilot made a steep and fast approach, crossing the end of the runway at about 50 feet agl at over 150 mph. The airplane floated down the runway

for a full third of its length so the pilot decided to go around. The co-pilot retracted the landing gear and the airplane began settling back toward the runway so he reversed the switch in time to avoid collapsing the landing gear. The airplane touched down and rolled for about 100 feet before becoming airborne again. The co-pilot brought up flaps very gradually and the airplane began climbing. The pilot made a right-hand away from the field and encountered a heavy rainstorm as he maneuvered to re-enter the traffic pattern, which was to the left. The pilot went on instruments and continued to maneuver for a new approach, exclaiming over the radio, "I can't see a damn thing." The co-pilot raised the final five degrees of flaps and moments later the port wing struck the surface of Tampa Bay. The airplane cartwheeled into the bay and broke up. US Navy Consolidated PBY5 flying boats spotted the wreckage and landed nearby and AAF crash boats arrived soon after. The crash boats took on the survivors. B-29 instructor 1Lt. Peter G. Weaver, radio operator instructor TSgt. Joseph E. Sleitweiler and engineer Sgt. Jack E. Crawford were killed in the crash. Bombardier 2Lt. Royce F. West, gunner Sgt. John L. McLemore and gunner Cpl. John W. Hunnicutt were seriously injured. Injured in the crash were: 1Lt. Theodore S. Will, Jr., pilot/B-29 student; 2Lt. Arnold A. Bridgewater, co-pilot/B-29 student; 1Lt. Ralph S. Johnson, navigator; TSgt. Alvin J. Gill, radio operator instructor; 2Lt. Chambers D. Adams, bombardier instructor; Pvt. Frank L. Steiger, engineer instructor; 2Lt. Glenn E. Miller, pilot-rated observer; Sgt. John R. Carter, gunner.

6-20-45D. Fontana, California. At 1719, two Lockheed P-38L airplanes collided in mid-air at Fontana, California, killing Maj. Arthur T. Williams III in the crash of P-38L #44-24324. Capt. James E. Foley was uninjured and was able to make a safe landing in P-38L #44-24746. The airplanes were part of a four-ship flight that took off at 1700 from Ontario Army Air Field, Ontario, California, on a simulated combat flight. The airplanes climbed to 11,000 feet and the flight leader split up the flight for the combat mission. The two subject airplanes were rolling out to take up positions to commence the combat mission when the collision occurred. Capt. Foley stated that he felt a "bump" while he was peeling-off. Lt. Foley's starboard wing, the underside of his starboard engine nacelle, and both propellers had collided with Maj. Williams' starboard wing and cockpit area. Maj. Williams' airplane went into a spin and he bailed out at about 600 feet agl, successfully parachuting to the ground. Maj. Williams right arm was severed and he received other serious injuries in the collision. Maj. Williams died about 30 minutes after his successful parachute jump. Maj. Williams' airplane spun into the back yard of a private residence in Fontana. No one on the ground was injured. Capt. Foley was able to fly his heavily damaged airplane back to the field and make a successful wheels-up landing on Runway 25.

6-21-45A. Bennett, Colorado. At 1030 MWT, a Consolidated TB-24J crashed one mile southeast of Bennett, Colorado, killing engineer SSgt. Richard A. Gilbert. Nine fliers were injured, one of them seriously. The airplane took off from Lowry Field, Denver, Colorado, on a flight engineer training flight. The pilot decided to demonstrate propeller feathering for the pilot-rated engineer student who was sitting in the co-pilot seat. The pilot feathered the number-four propeller, retarded the mixture control and cut the switch. After flying on three engines for a short time, the pilot attempted to restart the number-four engine. The pilot could not get the number-four propeller to re-feather. The pilot-rated flight engineer instructor, who was acting as co-pilot, took over the co-pilot seat. The pilot began flying back to Lowry Field. The co-pilot was able to get the number-four propeller to partially unfeather, but he could not get the engine to come back up to power. The co-pilot continued to operate the number-four propeller switch, but the feathering mechanism left the propeller in a position where it began to windmill backwards, creating huge drag and causing the airplane to lose altitude. The co-pilot was able to get the number-four to the full-feathered position and the pilot ordered the engineer to close the number-four fuel shut off valve. Moments later, the number-three engine began backfiring and suffering diminished performance. The airplane had lost a lot of altitude and had begun yawing to the right. The pilot realized that he could not make Lowry Field so he alerted the crew to stand by for a crash landing. The airplane was approaching a rise in the terrain and was heading for a collision with a house on top of a hill. The pilot and co-pilot turned the airplane slightly to the left and were able to avoid colliding with the house, clearing it by only a few feet. The airplane descended into the ground just after clearing the house. The airplane smashed to earth on a cultivated field and began sliding sideways to the right as it broke up and came to a halt. The airplane did not burn but was completely destroyed. Investigation revealed that the engineer had inadvertently turned off the number-three fuel shut off valve instead of the number-four. Student engineer and rated navigator F/O Glenn L. Fornes was seriously injured. Injured in the crash landing were: 2Lt. Marlowe D. Thorne, pilot; 1Lt. Harold Binder, co-pilot and flight engineer instructor; F/O Murray Siegel, navigator-rated student engineer; 2Lt. Charles D. Anderson, navigator-rated student engineer; 2Lt. Joseph F. Biddle, navigator-rated student engineer; 2Lt. Francis E. Wilke, bombardier-rated student engineer; F/O Aloysius R. Mizgalski, bombardier-rated student engineer; 2Lt. Herman Robinson, navigator-rated student engineer.

6-21-45B. Arcadia, Florida. At 1515 EWT, a Boeing TB-17F and an RB-17E collided in mid-air seven miles NNW of Arcadia, Florida, killing four fliers aboard the TB-17F. The pilot of the RB-17E was able to make a safe landing at Hendricks Field, Sebring, Florida. The airplanes were part of a six-ship flight that had taken off from Hendricks Field on a scheduled formation flight. The airplanes were maneuvering in formation, going from V-formation to trail-formation as the flights turned. The TB-17F was flying in the number-two position; the RB-17E was flying in the number-three position. The pilot of the B-17E mistakenly believed that the number-two ship was out of position and had fallen back to the number-three position while flying in trail formation. The B-17E pilot attempted to maneuver into the number-two position. The pilot of the B-17E overshot his intended position, flying over the top of the B-17F and barely missing colliding with its vertical stabilizer. In an attempt to regain the number-two position, the B-17E swung to the left and collided with the B-17F. The B-17E's port wing struck the B-17F's vertical stabilizer and the number-one propeller struck the B-17F's waist section forward of the waist windows, cutting the fuselage in half. The airplanes stuck together for a short time and then separated momentarily before slamming together again. The tail-less forward section of the B-17F fell away and dove to the ground, exploding into flames upon impact. The engineer was hurled from the broken fuselage without a parachute and he fell to his death. The severed tail section hurtled backward through the remainder of the formation, barely missing colliding with the other airplanes. The heavily damaged B-17E landed safely at Hendricks Field. Killed aboard the TB-17F were: 2Lt. Richard H. Wallis, B-17 instructor; 1Lt. Robert L. Pearson, Jr., co-pilot; 1Lt. Albert N. Roach, co-pilot; Pvt. Lee N. Tannehill, engineer. Crewmembers uninjured aboard the RB-17E were: Capt. Lawrence B. Copenhaver, B-17 instructor; 1Lt. Henry C. Rupal, co-pilot; 1Lt. George C. Rietz, co-pilot; SSgt. Harry J. Northern, engineer.

6-21-45C. Walterboro, South Carolina. At 1530 EWT, a Curtiss P-40N crashed nine miles east of Walterboro Army Air Field, Walterboro, South Carolina, killing 2Lt. Lloyd A. Carter, 20. The airplane was flying in the number-two position of a four-ship flight that took off from Walterboro Army Air Field on a gunnery mission. The gunnery range was closed due to poor weather, so the mission was changed to an escort mission. The airplanes flew escort for a North American B-25 bomber for about 45 minutes. The B-25 went in for a landing, so the P-40s broke off and headed south. The flight encountered instrument conditions and the flight leader elected to penetrate the thin clouds. When the leader emerged on the other side, none of his flight was still with him. The flight leader had observed the subject airplane, which was flying in the number-two position, still in formation about 20 seconds before breaking out in the clear. A short time later, witnesses on the ground observed the airplane flying at high speed at low altitude. The airplane was seen to enter a turn to the right before plunging into a wooded area and exploding into flames. The cause of the accident could not be determined.

6-22-45A. Denver, Colorado. At 0757 MWT, a Consolidated TB-24H crashed while attempting a take-off at Buckley Field, Denver, Colorado, killing four fliers and injuring nine others, one of them seriously. Prior to the scheduled take-off, the number-two propeller governor was found to be inoperative. The airplane was returned to the line and the propeller governor was replaced. Prior to the subsequent take-off, the pilot directed the co-pilot to watch the rpm on the number-two engine and to be ready to retard the throttle in the event of a runaway number-two propeller. Investigators stated, "The take-off run was commenced at approximately 0757 MWT, to the south, on Runway 16 with everything apparently functioning normally according to the testimony of the crew and witnesses. The airplane ran for the full length of the runway, off the end of the runway, through a fence, which bound the field, across a road, through another fence, which bounded an alfalfa field, and crashed into an irrigation ditch approximately 2,000 feet south of the field without becoming airborne.... Upon reaching flying speed, at least three-quarters of the way down the runway, a severe vibration presumably caused by the nose wheel, occurred at which point the pilot immediately closed the throttles. For reasons unknown, the brakes were either not applied or did not function. Apparently no effort was made to use the emergency brakes.... It is probable that the pilot had an apprehension regarding the function of the airplane, since he had specifically asked the co-pilot to watch the rpm on the number-two engine on take off and to catch it with the feathering button to retard the rpm in the event there was a runaway propeller on take-off, and was deliberately holding the airplane on the runway longer than he might have if he had no anticipation of trouble. Upon sudden occurrence of the excessive vibration [of the nose wheel], he instinctively chopped the throttle, without realizing that he had reached flying speed and had gone by the safe stopping portion of the runway." Killed in the crash were: 1Lt. David E. Elkin, pilot; Cpl. Joseph T. Hetter, radio operator; 1Lt. Floyd E. Lords, observer; 1Lt. James D. Gage, student engineer. Co-pilot 2Lt. Jacob W. Zansitis received serious injuries. Crewmembers injured in the crash were: Cpl. Edmund P. Woznack, engineer; 2Lt. William L. Crew, pilot-rated student engineer; F/O Charles E. Foster, student engineer; F/O John K. Hooker, navigator-rated student engineer; F/O James W. Johnson, bombardier-rated student engineer; 1Lt. Robert L. Nackmen, student engineer; 2Lt. Edward Schroeder, navigator-rated student engineer; F/O Charles H. Williams, navigator-rated student engineer; F/O Austin C. Yockey, navigator-rated student engineer.

6-22-45B. Tallahassee, Florida. At 1647 EWT, a North American P-51D crashed at Dale Mabry Field, Tallahassee, Florida, killing F/O James F. Cornish, 22. The airplane took off from Dale Mabry Field on a routine local flight. The pilot took off, retracted the landing gear, and had climbed to about 200 feet when he radioed the tower that his engine was "cutting out" and that he was returning to the field. The pilot began a turn in an effort to get to Runway 27, extending the landing gear after about 90 degrees of turn. The pilot overshot his turn onto approach for Runway 27, so he tightened the turn. The airplane stalled at about 150 feet agl and fell off on the port wing. The airplane smashed to the ground at a steep angle, killing the pilot instantly. Investigators noted that the airplane had suffered propeller damage when it was "nosed up" in a previous accident. The propeller was replaced but the engine was not changed. The airplane was test flown after the propeller change and no problems were found.

6-22-45C. Placer County, California. At 2337, a Curtiss C-46A suffered a catastrophic structural failure and crashed two miles southeast of Gold Run in Placer County, California, killing three fliers. Pilot 1Lt. James B. Soloman, co-pilot F/O Robert A. Brown and pilot F/O Hubert W. Anderson were killed in the crash. The airplane took off from Reno Army Air Field, Reno, Nevada, on a high-altitude training flight to Long Beach, California, and return. Because of deteriorating weather conditions, the pilot decided not to fly to Long Beach and elected to return to Reno early. The airplane reported in a Sacramento and was never heard from again. The airplane crashed a short time later. Investigators reached the crash site the next day. Investigation revealed that the both wings had failed and separated before the airplane slammed onto a mountain. The starboard wing was found about a mile away. Investigation of the wreckage indicated that the starboard wing had exploded prior to separating from the airplane. The cause of the in-flight explosion could not be determined.

6-23-45A. Denver, Colorado. At 0222 MWT, a Martin TB-26B crashed while attempting a take-off at Lowry Field, Denver, Colorado, killing two fliers and injuring four others, one of them seriously. Investigators stated, "The pilot held the airplane with the brakes until at least 25 inches of manifold pressure had been obtained before commencing the take-off run in order that it would not be excessive, since the instructor pilot had warned him of the high altitude (5,391 feet) at this station [Lowry Field]. There was no indication of trouble until most of the runway had been used up for the run and the airplane had not obtained sufficient airspeed for safe flying. When the ship became barely airborne with a minimum amount of airspeed, the wheels were retracted and at this time the instructor pilot realizing that the airspeed was not sufficient to enable them to clear the rise in the terrain just off the runway, took control of the airplane from the student pilot and, in an attempt to gain additional altitude with an airspeed of approximately 95 mph, he dropped full flaps. The airplane stalled, dropping the right wing, which apparently struck a boundary fence post, and the propellers started digging into the ground almost immediately afterwards. The airplane crashed and burn about 1,900 feet off the south end of the runway after having skidded for a short distance, tail first. This accident resulted from the failure to gain sufficient airspeed on the take-off run to become safely airborne and the failure to gain airspeed or altitude when the airplane did become barely airborne (possibly 10 feet altitude). The reasons for the failure to attain flying speed cannot be definitely determined, because the airplane was so completely burned." Pilot-rated passenger Capt. Melvin E. Holtman and engineer Pfc. Albert J. Dallas were killed in the crash; instructor pilot 2Lt. Allan P. Lewis received serious injuries. B-26 student 2Lt. George L. Goebel and pilot-rated passengers 1Lt. H.M. Robertson and 1Lt. Vincent I. Howard were injured in the crash.

6-23-45B. Tom Bean, Texas. At 1720 CWT, a North American AT-6A crashed one mile WNW of Tom Bean, Texas, killing students A/C Francis L. Eans and A/C Wallace L. Morgan. The airplane had taken off at 1620 CWT from Perrin Field, Sherman, Texas, on an instrument team ride. The airplane had been buzzing the area prior to the crash. The airplane was observed to enter a high-speed dive. The starboard wing failed and separated during a steep pull-out maneuver. The wing struck the tail section, causing its failure. The airplane went out of control and smashed to the ground. The airplane did not burn but was destroyed utterly.

6-24-45A. Roswell, New Mexico. At 0805 MWT, a Vultee BT-13B crashed near Roswell Army Air Field, Roswell, New Mexico, killing pilots 2Lt. Jon H. Wiggs and F/O William F. Merkle. The airplane took off from Roswell Army Air Field on a local flight. Investigators stated, "At 0802 MWT, the pilot made a normal take-off on Runway 17L and climbed to an undeterminable altitude. At a point approximately two miles from the end of the runway, the aircraft crashed.... From a study of the wreckage, apparently this airplane crashed while negotiating a turn. When the airplane crashed, the occupants were thrown clear of the airplane, which caught fire immediately and was completely destroyed." There were no witnesses to the crash.

6-24-45B. Godman Field, Kentucky. At 1112 CWT, a North American B-25J crashed while attempting to take off from Godman Field, Fort Knox, Kentucky, killing gunner Cpl. Gilmer Williams. The airplane was flying in the number-two position of a three-ship flight that was taking off on a skip bombing mission. Investigators stated, "Flight leader took off immediately, number-two aircraft began take-off after a 20 second interval with co-pilot at the controls from the right seat. Pilot stated that everything was normal on start of [take-off run].... He said aircraft cleared the ground about 3,000 feet down the runway, flew sloppily and would not climb. He took over control, decided that he would not gain sufficient altitude to clear buildings about 3,000 feet from the end of the runway, and put the aircraft back on the ground,

chopping his throttles. He lowered the nose wheel immediately and applied brakes. He and other crewmembers stated that brakes seemed to grab once, then release. Light skid marks start just east of taxiway D, about 670 feet from the end of the runway. Aircraft went off the end of the runway, crossed the road and sodded depression through perimeter fence and onto the railroad track where it stopped and burst into flames. Landing gear was sheared off when plane struck the rail bed embankment." Investigation revealed that there was a malfunction of the carburetor heat control system. Pilot 2Lt. John B. Turner and co-pilot F/O Thomas M. Flake received serious injuries.

6-24-45C. Wolcott, Indiana. At 1710 CWT, a North American AT-6D collided with a high-tension wire eight miles southeast of Wolcott, Indiana, killing pilot 2Lt. William K. Dorsett, 21. Pilot 1Lt. Kenneth H. Meyer, who was able to make a successful emergency landing, sustained minor injuries. The airplane had taken off at 1630 CWT from Chicago Municipal Airport, Chicago, Illinois, on a flight to Indianapolis, Indiana. Investigators stated, "While en route the pilot [Lt. Dorsett] was down buzzing and evidently did not see a high tension wire until it was too late to do anything except to try to fly under it. The pilot was decapitated in trying to fly under the wire.... [Lt. Meyer] took over the controls as soon as possible. He made a climbing turn to the left, [and then flew] on instruments until he was able to get the canopy open. He saw a field that he thought that he would be able to make a successful landing. Not knowing what the extent of damage to the plane was he tried to make a normal landing in an oat field. He bounced over a fence [and] landed in the adjacent field and rolled approximately 1,000 feet, striking a fence causing the plane to nose over [and come rest inverted]. The plane was a major wreck."

6-25-45. Chunky, Mississippi. At 1200 CWT, a North American P-51D suffered a catastrophic structural failure and crashed one mile south of Chunky, Mississippi, killing 2Lt. Henry E. Crist. The airplane was part of a two-ship flight that took off from Key Field, Meridian, Mississippi, on an aerobatic flight. The airplanes climbed to 18,000 feet and performed various aerobatic maneuvers. After completing the aerobatic portion of their mission, the pilots engaged in a "rat race." The flight leader performed a split-S type maneuver at about 19,000 feet, pulling out at about 12,000 feet. The flight leader leveled off at 8,000 feet and observed that the wing ship was not in position. A short time later, it was discovered that the airplane had crashed. Investigation revealed that the port wing failed and separated, striking the tail section and causing its progressive failure and separation. The airplane went out of control and slammed to the ground, but the airplane did not burn. It was speculated that the pilot had stressed the airplane beyond its design limits during a steep pull-out in an attempt to recover from the split-S maneuver.

6-26-45. Sidney, Iowa. At 2130 CWT, a Beech AT-11 flying in poor weather collided with terrain and crashed four miles northeast of Sidney, Iowa, killing two fliers and seriously injuring another. Pilot 2Lt. Harold E. Keiser and co-pilot Stephen V. Selwyn were killed in the crash. Co-pilot 2Lt. Francis J. Schwind received serious injuries. He had been hurled about 250 feet beyond the main wreckage. Investigators stated, "The weather at the time of the accident was rain, hail, wind and thunderstorms of such severity that some residents had taken refuge in storm cellars. The airplane was first sighted approaching from the southeast below the cloud ceiling. One witness watched the airplane as it slowly lost altitude and turned slightly to the right to fly just over the treetops in a wide, shallow valley. Just prior to the crash the airplane was seen to make a very steep turn to the right of approximately 180 degrees and had barely rolled out of the turn when it struck the ground, passed out of sight behind some trees and burst into flames when it stopped on top of the ridge.... The marks left on the ground indicate that the airplane struck the ground with the wings almost level and at a low angle of attack.... The airplane traveled up the slope and left the ground again approximately 75 feet beyond the crest of the rise. The amount of wreckage found in this area indicates that this was the point of greatest impact, and that the airplane cartwheeled to the top of the next ridge. The total distance traveled from the point of contact to the final resting place of the airplane was approximately 300 feet. The right wing was completely torn from the airplane, the left wing was buckled under the wreckage, and both engines were thrown a considerable distance. The fuselage was in an inverted position and the nose section was caved in. The pilot's compartment was badly burned."

6-27-45A. Dodge City, Kansas. At 1104 CWT, a Martin TB-26C crashed 15 miles south of Dodge City Army Air Field, Dodge City, Kansas, killing three fliers. Pilot 1Lt. Arthur D. Michel, co-pilot 1Lt. Davis A. Bruck and engineer Pvt. Stephen Davies were killed in the crash. The airplane took off from Dodge City Army Air Field on a routine training flight. The airplane was seen flying straight and level at about 3,000 feet when it was seen to go into a steep bank to the left. The airplane rolled through back to an upright position, staying in this position for a few seconds before the nose fell and the airplane fell off in a spin to the right. The airplane remained in the spin and smashed to the ground, exploding into flames upon impact.

6-27-45B. Brownsville, Texas. At 1400, a Bell P-63A collided with a tree and crashed ten miles northwest of Brownsville Army Air Field, Brownsville, Texas, killing 1Lt. Otto H. Schroeder, 23. The airplane took off from Laguna Madre Sub Base, Laguna Madre, Texas, on a routine pilot proficiency flight. Investigators stated, "According to the statements of the witnesses, he was observed approaching the tourist courts

in which he lived from the north at approximately 200 feet above the ground. After passing over the courts he made a 270-degree turn to the right and again headed for the courts but at a much lower altitude. As he neared the courts, he was observed to be approximately ten feet above the ground. As the pilot began to pull up, his plane struck the top of a tree and he continued upward toward some high-tension power lines, which were approximately 100 feet from where he struck the tree. The pilot failed to clear the power lines and his left wing struck a power line pole and his fuselage severed several strands of the power lines. Upon striking the pole parts of the plane were hurled through the air and the plane seemed to go out of control. After striking the pole, the plane traveled approximately 300 yards before it crashed and burned. Parts of the plane were scattered over an area of 100 yards from where it first struck the ground. The pilot received fatal injuries. The findings of the Aircraft Accident Investigative Board are that the pilot was buzzing and failed to pull up soon enough to avoid striking obstructions."

6-27-45C. Gulf of Mexico, Texas. At 1413, a Republic P-47D crashed into the Gulf of Mexico about ten miles southwest of the Army Air Field at Galveston, Texas, killing 2Lt. William A. Riley, 26. The airplane took off from Galveston Army Air Field on a gunnery mission. The subject pilot alerted the flight leader that he had propeller trouble and was heading back to base. The flight leader acknowledged and began flying back toward shore. A short time later, the subject pilot alerted the flight leader that his engine had quit. The flight leader advised the pilot to bail out. The subject pilot acknowledged and at about 1,500 feet msl he opened the canopy. A large amount of paperwork was seen to leave the airplane when the canopy was opened. The airplane then stalled and rolled over to an inverted position and then dove into the sea, exploding upon impact. The pilot was not seen to leave the airplane.

6-27-45D. Panama City, Florida. At 1434 CWT, a Martin B-26C crashed while attempting a take-off at Tyndall Field, Panama City, Florida, killing two fliers and seriously injuring another. Investigators stated, "The take-off roll was started with everything normal. When they had gone about two-thirds of the way down the runway and had gained 120 mph airspeed, the right propeller over-sped and to better than 3,000 rpm and the engine cut out and on alternately. Too much runway had been used to stop the aircraft, so the take-off was continued and the propeller was brought under control with the propeller control in automatic. The climb to 1,000 feet was made with both engines running, the right engine still missing. After reaching traffic altitude of 1,000 feet the right engine was feathered. The tower was called and told to clear the traffic pattern and runway for his landing, nothing was said of an emergency landing. Right hand traffic was being used on Runway 22 so the pilot had to turn into his feathered engine. The aircraft was still at 1,000 feet and an airspeed of 150 mph was being maintained. The turn onto the final approach was made with a skidding turn resulting from the right engine being feathered. The pilot waited until he was sure of making the field before lowering he landing gear and partial flaps. The pilot had trouble setting up his angle of descent and was jockeying the throttle on the left engine, necessitating the violent use of the rudder controls. The aircraft passed over the end of the Runway 22 at an altitude of approximately 400 feet. At this point the engineer saw that the landing would not be a normal one and went back to the navigator compartment and strapped himself in the seat. After using up about one-half the runway and still having 200 feet altitude the pilot raised the landing gear. At this point he applied full left throttle causing the nose of the aircraft to climb and roll to the right. The wings were wobbling and the aircraft turned approximately 60 degrees to the right of Runway 22 when the aircraft seemed to stall out and slide into the ground on the right wing. The right wing hit the ground first and the aircraft immediately burst into flames, cartwheeling on the right wing. The aircraft exploded and came to rest just beyond Runway 13. Both pilot and co-pilot were removed from the aircraft dead. The engineer escaped from the wreckage fully conscious but badly burned." Investigation revealed that the starboard propeller governor had failed. Pilot 2Lt. Earl S. Dennis and co-pilot 1Lt. Melvin L. Heinke were killed in the crash; engineer MSgt. Wallace M. DesJardine received serious injuries.

6-28-45A. Sullivan City, Texas. At 0930, a North American AT-6D crashed five miles northeast of Sullivan City, Texas, killing pilot 1Lt. Victor H. Shipman, 21. Pilot Capt. Wharton E. Moller received serious injuries. The airplane took off from Moore Field, Mission, Texas, on a transition flight. Capt. Moller had made several "rear seat" take-offs and landings at Auxiliary Airfield #2. Capt. Moller then performed several S-turns along a road. The overcast was breaking up and the pilots decided to get some altitude to allow some aerobatic work. Capt. Moller, who had been flying for some time, requested that Lt. Shipman take over for a while. Lt. Shipman took over the controls and at 500 feet agl entered a steep dive toward the ground. Investigators stated, "According to the student in the rear seat, Lt. Shipman circled a windmill to the left in a fairly steep turn while losing altitude. At approximately 25 feet above the ground a shallow turn to the left was started while continuing to lose altitude. The left wing struck the mesquite and after dragging in the underbrush for 70 yards, the airplane was flipped over on its back and continued to turn over and over for a distance of 65 yards before coming to a stop with both wings and the center section separated from the fuselage. The engine was thrown forward beyond the fuselage for a distance of 30 yards." Investigators noted that Lt. Shipman might have survived the crash had he been using a metal seat with shoulder harnesses

attached to the airplane structure. The airplane was equipped with a wooded seat and the safety harness was attached to the seat rather than the aircraft frame.

6-28-45B. Cheyenne, Wyoming. At 2018 MWT, a Consolidated B-24M crashed two miles east of Cheyenne, Wyoming, killing engineer Cpl. Arthur E. Deborde. Crew seriously injured in the accident were: 2Lt. Fred E. Blackmer, pilot; F/O Robert B. High, co-pilot; 2Lt. Robert E. Morrison, passenger; F/O Warren R. Anderson, pilot-rated passenger. The airplane had arrived at Cheyenne Municipal Airport after completing a flight from Scott Field, Illinois. Investigators stated, "The traffic pattern was entered under normal conditions with the airplane performing normally. Landing instructions were requested and received to land on Runway 26. The pre-landing procedure was begun and the wheels were being extended on the upwind leg. Just as the wheels were coming down all four engines stopped. When all the engines quit simultaneously the pilot immediately pushed the nose of the ship down to obtain airspeed in an effort to try to make the airport just about one-half mile to the left. The co-pilot at this point was not aware that the engines had stopped and believed the airplane was out of control and thinking that he was helping the pilot hold the nose of the airplane up and the wings level actually was pulling against him as he was trying to drop the nose for airspeed and turn left to make the field. Upon deciding that he could not stretch the glide to the airport the pilot elected to make a crash landing in an open field with safe flying speed, wheels down, flaps up and no power. The airplane hit the ground hard on the up slope side of a hill, coming to rest about 400 feet from the point of impact. [The airplane] was badly broken up and all three landing gears were broken off. The nose wheel gave way and came through the floor of the flight deck allowing the ship to slide on the nose section thereby wrecking the cockpit and breaking off all four propellers and engines."

6-29-45A. Chandler, Arizona. At an unknown time after 0808 MWT, a Consolidated B-24L crashed six miles ESE of Williams Field, Chandler, Arizona, killing seven crewmembers. The airplane suffered the failure of the number-two engine shortly after take-off from Williams Field and crashed soon afterwards. Investigators stated, "Examination of the remaining wreckage and impact marks upon the ground indicated that the ship hit with considerable force with the left wing down or while in a turn to the left. The ship crashed on a heading of approximately 330 degrees, which was almost the reciprocal of the runway [heading] upon which the ship took off. It is felt that the pilot, upon experiencing trouble, was attempting to return to the field. All four engines tore loose from the airplane, as did all four propellers. Immediately after impact the ship turned over on its back, broke in half and came to rest approximately 85 yards from the point of initial impact. The wheels were found to be in a retracted position." Investigators noted that the number-two propeller was found in the feathered position. Killed in the crash were: 1Lt. Victor S. Bruce, pilot; 2Lt. Hesler W. Ellis, co-pilot; 2Lt. Quentin P. Redden, navigator; F/O Frederick D. Gregor, student navigator; F/O E.L. Sager, student navigator; TSgt. Richard W. Hoyt, radio operator; Cpl. Arthur W. Hendrix, engineer.

6-29-45B. Frostproof, Florida. At 1030 EWT, a North American P-51D crashed two miles southwest of Frostproof, Florida, killing pilot 2Lt. Maklem W. Peake, 22. The airplane was part of a four-ship flight that took off from Sarasota Army Air Field, Sarasota, Florida, on a low-altitude formation mission. The pilot was flying at very low altitude when he collided with trees, causing the airplane to go out of control and smash into a small lake. The airplane was shedding pieces when it bounded back into the air for a short distance before smashing back into the lake.

6-29-45C. Rapid City, South Dakota. At 1120 MWT, a Boeing B-17G taxied into a building at Rapid City Army Air Base, Rapid City, South Dakota, killing navigator 2Lt. Noah S. Levine. Six other crewmembers escaped injury. Investigators stated, "In making a 90-degree right turn to park facing west between two other parked aircraft, he lost all hydraulic pressure when he had completed approximately 45 degrees of that turn. To avoid colliding with the aircraft already parked on his left, he pushed number-four throttle on and put number-four generator on. The pressure did not come up and the aircraft continued to turn left. He then pushed number-one throttle forward in an attempt to straighten out and taxi on down the ramp, but it did not overcome the turning and the aircraft taxied off the east side of the ramp, striking a maintenance building. The navigator, who was in the nose, was killed." The airplane was piloted by 1Lt. Robert H. Sattler and co-pilot 2Lt. Warren C. Looney. The pilots and four other crewmembers were uninjured.

6-29-45D. Fort Sumner, New Mexico. At 1850 MWT, a Republic P-47D crashed one and a half miles ENE of Fort Sumner, New Mexico, killing 2Lt. Harold A. Soper, 25. The airplane was part of a two-ship flight that took off from Fort Sumner Army Air Field on an instrument training flight. The airplanes completed the mission and returned to the field. The airplanes peeled off for landing and the lead ship made a normal landing. Investigators stated, "[The subject airplane] started his turn into field slightly nose high and at a presumably normal speed. The airplane's left wing dropped and Runway Control called for him to pick it up, which he did after slipping 150 to 200 feet. In recovering, pilot pulled the nose higher, killing the remaining airspeed, and stalling in from an altitude of 150 to 200 feet. Aircraft crashed into the ground nose first, killing the pilot and demolishing the aircraft."

6-29-45E. Christoval, Texas. At 2230 CWT, a Beech AT-11 crashed six and a half miles southeast of Christoval, Texas, killing three fliers. The airplane took off at 2112 CWT from San Angelo Army Air Field, San Angelo, Texas, on a routine training flight. Investigators stated, "From inspection of scene of accident, it was shown that the airplane hit the ground in a near level flight position with a slight rate of descent and at a course of approximately 345 degrees. The rate of descent was borne out by the way the airplane chopped the mesquite before it hit the ground and the way the airplane skidded along the ground after the first impact. It hit on the up-grade side of the hill, and the main part of the airplane bounced twice before reaching the crest of the hill. The first bounce was approximately 48 feet; the second bounce approximately 123 feet; the third point of impact was near the top of the hill and down the other side, a distance of approximately 420 feet from the third point of impact. The total distance the main part of the airplane traveled was approximately 660 feet. The left engine continued on for a distance of approximately 100 feet beyond the main wreckage; the right engine for a distance of approximately 300 feet. The main part of the right wing continued with the bulk of the fuselage the full distance." Pilot 2Lt. Thomas E. Nageotte and bombardiers 1Lt. Hal W. Dalton and 1Lt. J.B. Collepe were killed in the crash.

6-30-45A. Childs, Florida. At 1045 EWT, a North American P-51D crashed eight miles southeast of Childs, Florida, killing pilot 2Lt. Howard L. Nickolaisen, 29, who fell to his death in an unsuccessful parachute jump. The airplane was part of a six-ship flight that took off at 1000 EWT from Venice Army Air Field, Venice, Florida, on a low-altitude formation mission. The subject pilot, flying at 250 feet agl, began experiencing a rough engine, so he immediately climbed to about 1,400 feet agl to check it out. Another pilot observed that the airplane was on fire and advised the subject pilot to bail out. The subject pilot jettisoned the cockpit canopy at about 1,400 feet. The airplane was losing altitude gradually and when the airplane reached an altitude of about 800 feet it went into a steep dive. The airplane remained in the dive until it smashed into the ground and exploded into flames. The pilot had bailed out at an altitude of about 200 feet, too low to allow his parachute to fully deploy.

6-30-45B. Las Vegas, Nevada. At 1149 PWT, a Consolidated TB-24J crashed while attempting a landing at Las Vegas Army Air Field, Las Vegas, Nevada, killing 11 fliers and seriously injuring two others. The airplane had returned to the field and requested an emergency landing. The field was cleared and the crash crews were standing by. The airplane flew a normal pattern for Runway 2, but on the final approach the pilot aborted the landing because he had failed to extend the landing gear. The airplane re-entered the traffic pattern and the pilot reported that he was losing oil pressure in the number-one engine. Moments later white smoke was seen trailing the number-one engine. The pilot feathered the number-one propeller and continued to fly on the base leg for Runway 2. The pilot overshot the turn from base to final and was more in line with Runway 34. The pilot attempted to correct but he was approaching the intended runway at a 45-degree angle at about 75 feet agl. The pilot was unable to straighten the airplane out and was skidding toward airplanes parked on the flight line, which was in an area to the left of Runway 2. The airplane, flying at about 25 feet agl, turned to the left in a nose-high attitude as it approached the control tower. The airplane barely missed colliding with the control tower and parked airplanes as it passed over the ramp. The airplane entered a nose-high turn to the left, climbing to about 50 feet before it began sliding off to the left. The airplane fell off to the left and the port wing struck the ground. The airplane cartwheeled into the ground and burst into flames. Four airmen were found alive, but two of them died a short time later. Killed in the crash were: F/O John F. Devine, Jr., pilot; 2Lt. George W. Huntoon, co-pilot; Pfc. Charles E. Emmert, engineer; 2Lt. Charles E. Quiram, bombardier; SSgt. Ray D. Freeman, gunnery instructor; SSgt. Richard C. Hudson, gunnery instructor; SSgt. Robert F. Kozumplik, student gunner; Pvt. Glenn Duncan, student gunner; Pvt. Wayne G. Tate, student gunner; TSgt. Stanley N. Nice, student gunner; Pfc. James H. Snyder, student gunner. Bombardier 2Lt. Donald S. Smith and student gunner Pvt. Harlow E. Pendleton received serious injuries.

6-30-45C. Marshall, Wisconsin. At 1611 CWT, a North American AT-6A crashed three and a half miles north of Marshall, Wisconsin, killing pilots 2Lt. Thomas C. Bowdich and 2Lt. Jack G. Evans. The airplane took off from Truax Field, Madison, Wisconsin, on an instrument team ride. Lt. Evans was flying in the front cockpit as the safety pilot and Lt. Bowdich was flying in the rear cockpit as the instrument pilot. Investigators stated, "The plane was first sighted by witnesses at about 3,000 to 4,000 feet coming from the southwest approximately two miles northeast of Marshall, Wisconsin. It was seen to go into a slight dive for about 1,000 feet and then leveled out into a slow roll. It then started a turn to the left and completed a fairly large circle. It went into a slight dive from 1,000 to 2,000 feet and apparently completed an Immelmann. At the top of the Immelmann it apparently stalled and was seen to float as a falling leaf. It then entered into a spin. It seemed to recover once, very low, and then go into another spin and then crash.... The plane crashed almost vertically after the left wing hit a tree and stopped the plane spinning."

6-30-45D. Delaware Springs, Texas. At 2040 MWT, a Boeing TB-29A crashed into a mountain seven miles southeast of Delaware Springs, Texas, killing 12 fliers. The airplane was one of a flight of five that took off from Davis-Monthan Field, Tucson, Arizona, on a flight to Mobile, Alabama, and return. The airplane never arrived at Mobile and it was found to have crashed into the side of a mountain at an elevation of approximately 5,200 feet while flying in a straight and level attitude at cruise speed. Killed in the crash were: 1Lt. James H. Couch, pilot; F/O James A. Thomas, co-pilot; 2Lt. James Green, radar navigator; 1Lt. Joseph J. Mennen, bombardier/navigator; 2Lt. Wesley A. Waldron, bombardier/navigator; F/O Ernest E. Wilson, flight engineer; Pvt. Fayette H. Stanfield, radio operator; MSgt. Orval R. Lawless, gunner; F/O Eugene R. Chale, crewmember; Pfc. Ralph I. Martin, gunner; Cpl. Ariste Landers, gunner; Pfc. Clark J. Moore, gunner.

July

7-1-45A. Benton, Kentucky. At 0215 CWT, a Boeing B-29 broke up in mid-air and crashed at Benton, Kentucky, killing nine crewmembers. Gunner Cpl. Irving A. Elias was ejected from the wreckage as it fell and was able to parachute to safety, suffering minor injuries. Killed in the crash were: 2Lt. Ward W. Copenhaver, pilot; F/O Eugene M. Graham, co-pilot; 1Lt. Joseph F. Arone, flight engineer; F/O James R. Schetzle, bombardier; 2Lt. Richard O. Snow, navigator; Sgt. Romold A. Kryzan, radio operator; SSgt. Delmar H. Lumberg, engineer/gunner; Sgt. Arnold A. Rushton, gunner; Cpl. Roy G. Berryhill, gunner. The airplane took off at 1600 MWT from Kirtland Army Air Field, Albuquerque, New Mexico, on an instrument navigation flight to Mobile, Alabama, at 11,000 feet, to Nashville, Tennessee, at 11,000 feet, to Trinidad, Colorado, at 17,000 feet, and return to Kirtland Field at 14,000 feet. The take-off was delayed for an hour and a half because of a problem with the number-one propeller governor. The propeller governor was repaired and checked out satisfactorily. The airplane was airborne for about nine hours when it encountered an area of very poor weather, which included powerful thunderstorms, high winds and severe turbulence. The airplane broke up soon after entering these storms. The surviving gunner stated that the airplane was rocking up and down, forcing him to his knees. The airplane began shaking and vibrating and then the interior lighting failed. The airplane began breaking up and the fuselage broke open, ejecting the surviving gunner, who was wearing his parachute. Wreckage was scattered over a wide area. Investigators noted that the airplane had only 39 total engine and airframe hours since new.

7-1-45B. Spring Lake, Texas. At 1705, a Douglas C-47A flying in instrument conditions suffered a catastrophic structural failure and crashed six miles east of Spring Lake, Texas, killing five fliers. The airplane took off from Fort Sumner, New Mexico, on a flight to Majors Field, Greenville, Texas. The airplane was cleared to fly at 11,000 feet msl. The airplane was observed diving out of a cumulus cloud at a 45-degree angle with power on. The ceiling was estimated to be about 4,500 feet. The wing fillets failed and separated; simultaneously the starboard wing leading edge failed and the starboard elevator fabric failed and separated; the starboard elevator separated moments later. The airplane center section and fuselage smashed to the ground and exploded. Wreckage was scattered over a wide area. Investigators speculated that the airplane entered clouds that contained extremely powerful turbulence. The pilots might have over-controlled the airplane while attempting to re-establish level flight. Killed in the crash were: 1Lt. Frederick H. Jacoby, pilot; 1Lt. Rex M. Tharpe, co-pilot; SSgt. Joseph V. Andruskevich, engineer; SSgt. Merle W. Sullivan, assistant engineer; Pfc. Christine Davis, WAC passenger.

7-1-45C. Long Beach, California. At 1651, a Lockheed P-80A crashed into a parked Douglas A-26B while attempting to take-off from Long Beach Army Air Field, Long Beach, California, killing P-80A pilot 1Lt. Joseph Mandl. The airplane was scheduled to fly a ferry flight to Tucson, Arizona. Two other P-80 pilots assisted the pilot in his pre-flight cockpit check. The pilot's shoulder harness was found to be loose and the attending pilots had experienced trouble adjusting it. The pilot was also assisted in the engine start up by the two P-80 pilots. The airplane was then taxied to the end of Runway 30 and was prepared for take-off. Investigators noted that a maximum power run-up was not performed prior to take-off. Investigators stated, "The aircraft started its take-off roll from a stationary position at the end of the runway. The acceleration of the aircraft seemed slow as compared with other P-80 take-offs at this base, however, this was due to the other aircraft utilizing added speed of applying power and gaining some speed before entering the take-off runway. Witnesses stated the pilot made three attempts to become airborne. The first attempt resulted in a skip, and the aircraft settling to the runway; the second attempt resulted in a stall as the nose was raised very high, and witnesses indicated that the right wing dropped but did not strike the runway. Following the second attempt to become airborne the pilot either attempted to stop the aircraft, or to bring it under control by the use of the brakes. If the pilot had decided to stop the aircraft, he then changed his mind, as a third attempt to become airborne was made, which resulted in a stall and the right wing striking the ground. The aircraft swerved off of the runway to the right, skidded through a fence and into a parked A-26 on the ramp ... located at the northwest end of Long Beach Army Air Field. The P-80 struck the [port] landing gear strut of the A-26, wrecking both aircraft. A small fire was started in the left side of the cockpit of the P-80, but was extinguished by two civilians who arrived on the scene.... The pilot sustained fatal injuries. Lt. Mandl had a total of 7 hours and 30 minutes in P-80 type aircraft, however, this was his first attempt to fly a P-80 with full service." The airplane weighed 13,803 pounds prior to take-off. Investigators noted that all "full-service" flights have been discontinued until maximum gross weight take-off flight tests were completed by Lockheed Aircraft Company.

7-1-45D. Hachita, New Mexico. At 1930 MWT, a North American AT-6C crashed five miles northwest of Hachita, New Mexico, killing pilot Capt. James F. Lowery and passenger Capt. George L. Andrews. The airplane took off from El Paso Municipal Airport, El Paso, Texas, on a flight to Luke Field, Phoenix, Arizona. The pilot was issued an instrument clearance and was assigned an altitude of 10,000 feet msl. Investigation revealed that the pilot had been flying on the "Green Five Airway," apparently without clearance. The airplane evidently encountered deteriorating weather conditions. The airplane went out of control and smashed to the ground at an angle of 70 degrees, exploding into flames upon impact. The occupants were killed instantly.

7-2-45A. Barnsdall, Oklahoma. At 0810 CWT, a Lockheed F-5E crashed two miles west of Barnsdall, Oklahoma, killing 2Lt. Charles R. Schleifer, 24. The airplane took off from Coffeyville Army Air Field, Kansas, on a routine training flight. The airplane was observed flying at "high altitude," trailing an "unknown aircraft of similar type." The subject airplane then went into a half-roll and entered an inverted dive "straight toward the ground." The airplane recovered momentarily in an inverted attitude. The airplane then dove into the ground at a steep angle in an inverted attitude. The airplane exploded violently into flames upon impact, killing the pilot instantly. Investigators could not determine the identity of the pilot flying the airplane that the subject airplane was trailing.

7-2-45B. Harlingen, Texas. At 1450, a Bell TP-63A flying in instrument conditions dove into the ground ten miles northeast of Harlingen Army Air Field, Harlingen, Texas, killing pilot 1Lt. Ernest C. Oz, 26. The airplane took off from Laguna Madre Sub-Base, Texas, on a camera gunnery mission and was scheduled to rendezvous with a Consolidated B-24 near Raymondville, Texas, for the gunnery mission. The weather airplane had contacted the subject aircraft and alerted the pilots of a large thunderstorm in the rendezvous area. The weather ship pilot advised the pilots to fly to the north to avoid the storm. The advisement was acknowledged. The P-63 flew into the storm clouds and a few minutes later it was heard in what sounded like a high-speed dive. The airplane then dove out of the overcast and smashed into the ground, exploding into flames upon impact. Investigators noted that the pilot had been ordered to return to base if he could not proceed under visual conditions.

7-2-45C. Greenwich, Connecticut. At 1622, a Republic P-47D flying in poor weather crashed at Greenwich, Connecticut, killing pilot 1Lt. George S. Fitch, 24. The airplane took off at 1430 from Millville, New Jersey, on a flight to Buffalo, New York, via Newark, New

Jersey. The airplane was observed circling Greenwich at low altitude at 1620. The airplane circled in a driving rainstorm for a short time before it flew away to the northeast, crashing moments later. Investigators stated, "The first point of impact was with three trees about 75 feet high and about six to eight inches in diameter. At this point the right aileron was torn off and the right ammunition box broken open and the right wing damaged an undetermined amount. The pilot pulled the airplane up and cleared the trees immediately ahead, but apparently aileron control was lost and the plane rolled over on its back. Ammunition was falling out of the ammunition box and several rounds went through the window of the Oakley house as the airplane flew over. Approximately 1,000 feet from the first point of impact the airplane hit an apple tree. It was inverted at this point. The vertical fin was torn off, the canopy was smashed and the right wing further damaged. The airplane was in a nose-high position. Two trunks of a tree were cut off smoothly by the propeller at an angle of about 45 degrees to the ground. The tail of plane struck the ground first and pulled the nose down immediately. The remains of the right wing [separated from] the fuselage at this point. One tree about 16 inches in diameter was sheared off at the base and another about the same size was sheared off about nine feet from the ground. The plane then bounced about 30 or 40 feet into the air, the engine tore loose from the fuselage, the empennage caught in the fork of a tree about 25 feet above the ground and was torn off. The empennage remained in the tree fork. The left wing, which was still intact, and the remains of the fuselage flipped over right side up and facing in the direction from which the airplane had come and stopped about 25 feet from a road. The pilot was thrown to the side against a pine tree and the gasoline tanks exploded, burning fiercely."

7-2-45D. Hendersonville, North Carolina. At 1906, a North American TB-25H crashed 13 miles north of Hendersonville, North Carolina, killing passenger Maj. Maurice M. Boyd. Seriously injured in the crash were: 1Lt. Paul J. Kennedy, pilot; Pfc. Richard S. Court, engineer; Lt. Col. Thomas E. Howard, passenger; Maj. John R. Webb, passenger. The airplane took off from Andrews Army Air Base, Washington, D.C., on an administrative flight to Asheville-Hendersonville Airport. The airplane arrived in the area of Hendersonville at 1550, but the weather would not permit a landing at Asheville-Hendersonville Airport so the pilot diverted to Spartanburg, South Carolina, and made a safe landing. The airplane was refueled and when the weather cleared up at Asheville-Hendersonville Airport the pilot took off. The airplane arrived at Hendersonville about 1900 and the pilot entered the traffic pattern. The airplane entered the approach at about 1,000 feet agl and the pilot was starting to overshoot. The airplane touched down between one-third and half way down the runway, which was wet from recent hard

rains. The pilot applied the brakes but the airplane was traveling too fast. The airplane went off the end of the runway, hurtled over a road and then smashed into a drainage ditch about 175 yards from the end of the runway. The airplane burst into flames moments after coming to rest. The pilot was able to rescue two trapped crewmembers before flames drove him back. Investigators noted that the B-25 had been "stripped down."

7-2-45E. Carmichael, California. At 1625, a Douglas A-26C crashed three miles south of Carmichael, California, killing pilot F/O Clayton B. Sullivan. The airplane took off at 1623 from McClellan Field, Sacramento, California, on a ferry flight to Tulsa, Oklahoma, via Las Vegas, Nevada. Shortly after take-off the port engine failed. The pilot feathered the propeller and trimmed the airplane for single-engine flight. The airplane, flying slowly, entered a shallow turn to the left at an altitude of about 300 feet agl. The airplane continued to lose altitude so the pilot attempted to make an emergency forced landing in a clear area. The pilot observed that he was going to collide with power lines while on his approach for the forced landing. The pilot pulled the nose up and cleared the power lines but the ship stalled, falling off on a wing and smashing to the ground in the clear area. The airplane burst into flames upon impact, killing the pilot instantly.

7-3-45A. Climax, Georgia. At 1215 EWT, a North American P-51D broke up in flight and crashed two miles southeast of Climax, Georgia, killing 1Lt. Richard L. Hoefle. The airplane was flying in the wing position of a two-ship flight that took from Thomasville Army Air Field, Georgia, on a scheduled instrument training flight. The two airplanes climbed to 21,000 feet, the flight leader was flying on instruments and the subject pilot flying as the safety observer. The leader went off of instruments and let down to about 18,000 feet. The lead airplane made a diving turn to the left until airspeed reached 450 mph. The leader pulled out at about 10,000 feet and observed that the wing airplane was not with him. Investigators stated, "[When the flight leader] made his diving turn to the left, Lt. Hoefle lost control of his aircraft. Building up high speed, the right wheel door fairing gave way first, possibly during the pull out, causing a great pressure in the wheel well, ripping the right wing off." The airplane smashed to the ground and exploded violently into flames upon impact. The starboard wing was found some distance from the main wreckage.

7-3-45B. Montoya, New Mexico. At 1150 MWT, a Republic P-47D crashed four miles northeast of Montoya, New Mexico, killing pilot 1Lt. Millard M. White, 26. The airplane took off from Fort Sumter Army Air Field, New Mexico, on a routine training flight. After flying for about 15 minutes the subject pilot began experiencing propeller trouble. The subject pilot radioed the flight leader and the flight leader advised the subject pilot to operate the propeller in manual. The flight leader then turned the formation back toward the field. A short time later, the subject airplane began experiencing engine trouble and dropped out of the formation. The subject pilot radioed the flight leader and advised that his engine was "cutting out." The airplane lost altitude as it entered a turn to the right, turning about 90 degrees from the original heading. The airplane, flying at about 300 feet agl, made a turn to the right at low airspeed. Witnesses on the ground observed smoke and flames trailing the airplane and stated that the engine was "sputtering" as it passed over. The airplane stalled and entered a spin at an altitude too low to allow recovery, smashing into the ground at a steep angle and exploding into flames upon impact.

7-3-45C. Lake Huron, Michigan. At 1600, a Republic P-47D crashed into Lake Huron 15 miles north of Lexington, Michigan, killing French Air Force pilot Sgt. Robert R. Romand, 19. The airplane was part of a three-ship flight that took off from Selfridge Field, Mt. Clemens, Michigan, on an aerial gunnery mission. The flight made five passes at about 14,000 feet. The subject airplane made a sixth pass at the target and passed by the target and target ship at the same altitude. The subject airplane suddenly went into a dive and remained in the dive until it struck the lake and exploded. The cause of the accident could not be determined.

7-3-45D. Gulf of Mexico, Texas. At 1558, two Republic P-47D airplanes collided in mid-air and crashed into the Gulf of Mexico 22 miles southeast of Brownsville, Texas. F/O James W. Martin, 21, flying P-47D #44-32699, was apparently killed attempting to parachute to safety. 2Lt. Andrew J. Maynard, Jr., 21, was able to parachute to safety from P-47D #42-29392. The airplanes were part of a five-ship flight that took off from Brownsville Army Air Field, Brownsville, Texas, on an air-to-air gunnery mission. Lt. Martin was flying in the number-four position; Lt. Maynard was flying in the number-five position. Investigators stated, "The flight leader had three ships follow the tow ship, while he and the number-two man fired on the target. Number-three man lost sight of the tow ship and made two 180-degree turns in an endeavor to find it. Number-five man caught sight of the tow ship and firing flight, and headed toward them with three and four following. Number-three man overtook number-five and in an endeavor to lose his speed, he pulled up and over him. Number four was also coming up on number five fast but evidently he did not see him. Number four flew into number five, and both pilots bailed out. The pilot who was flying in the number-five position [Lt. Maynard] landed on the beach, uninjured, the other pilot [Lt. Martin] landed off shore approximately one-half mile. The flight leader circled the pilot in the water and observed him to be floating on his parachute slightly under water. No signs of life were seen. The pilot floated for approximately one-half hour and then sank from sight." Investigators noted that the pilot's body was never recovered.

7-4-45A. Roggen, Colorado. At 1130, a North American AT-6C crashed eight miles northwest of Roggen, Colorado, killing F/O Joseph L. Wolf, 20. The airplane was part of a three-ship flight that took off from Buckley Army Air Field, Denver, Colorado, on a routine formation training flight. The flight performed various formation maneuvers at an altitude of 2,000 feet agl. The subject airplane fell out of formation and then entered a steep dive toward the ground. The airplane pulled up slightly before smashing into the ground at an angle of approximately 70 degrees. The airplane exploded into flames upon impact, killing the pilot instantly. Investigators noted that the pilot had a hard time getting the subject airplane started and a pilot had trouble starting the airplane the day before. It was also noted that the subject airplane was a "sluggish" formation airplane.

7-4-45B. Sligo, Louisiana. At 1233 CWT, a Boeing B-29 crashed one mile south of Sligo, Louisiana, killing seven crewmembers and seriously injuring four others. The airplane was scheduled to take off at 0830 from Barksdale Field, Shreveport, Louisiana, on a long-range navigation mission. The airplane began suffering trouble in the number-one engine on the engine run up so it was taxied back to the line. The spark plugs for the front row of cylinders were changed in the number-one engine and the airplane was again run-up for take-off. The run-up checked out okay so the pilot elected to take-off. The number-one engine began cutting out as the airplane lifted off. The airplane climbed to about 500 feet agl when the number-two engine suddenly began losing power. Both engines began backfiring. The pilot struggled to hold the port wing up and attempted to set the airplane down in a large field. The airplane entered a descending left turn until the port wing struck trees and the airplane cartwheeled into wooded terrain. The B-29 broke up and burst into flames. The airplane barely missed colliding with a small house near the edge of the field. Investigation revealed that the spark plugs and their accessories for the number-one engine were badly fouled. The number-two fuel pump drive shaft failed and the reduced fuel pressure causing the number-two engine to lose power and backfire. The number-three propeller was found in the feathered position. Investigators speculated that a member of the flight crew feathered the number-three propeller inadvertently. Killed in the crash were: 1Lt. Donald R. Schultz, pilot; 2Lt. William D. Sullivan, co-pilot; 2Lt. Ludwig Novlan, crewmember; SSgt. Henry J. Tuemler, gunner; Sgt. Willis L. West, gunner; Cpl. James J. Sacco, gunner; Cpl. Robert R. MacMurray, gunner. Injured in the accident were: Sgt. Roger M. Davis, radio operator; Sgt. Rison E. Hines, Jr., gunner; 2Lt. Clair E. Lais, bombardier; 2Lt. Hal R. Stone, navigator.

7-5-45. Venice, Florida. At 1700 EWT, a North American P-51D crashed eight miles NNE of the Army Air Field at Venice, Florida, killing pilot F/O Robert

L. Haney, Jr., 21. The airplane took off from Venice Army Air Field on a "minimum power, maximum range" mission. Ground witnesses heard a loud report and observed pieces of airplane falling from the sky. Investigation revealed that the port wing failed and separated, striking the cockpit canopy and the tail section, causing its progressive failure and separation. The port wing, pieces of cockpit canopy and pieces of tail section were found about a mile from the scene of the main wreckage.

7-6-45. Los Banos, California. At 2201, a Northrop P-61B crashed six miles southeast of Los Banos, California, killing pilot 2Lt. Walter E. Fulton, Jr. and radar operator 2Lt. William W. Ailshouse. The airplane took off from Hammer Field, Fresno, California, at 2054 PWT on a low-level night navigation mission. A resident of Los Banos observed the airplane flying at about 800 feet agl. The airplane climbed slightly before it entered a 30-degree dive. The airplane remained in the dive until it struck the ground. The port wing struck the edge of the irrigation ditch and then the airplane bounded back into the air for a short distance. The starboard wing struck the edge of the irrigation ditch and then the airplane slammed into the center of the ditch, telescoping the fuselage and killing the crew instantly. The wings bridged the irrigation ditch and the fuselage was partially submerged. The airplane did not burn but was destroyed utterly. The cause of the accident could not be determined. Investigators noted that the pilot was active in golf and tennis and had suffered apparent heat sickness after participating in athletics about ten days before the accident. It was also noted that night temperatures were high at low altitudes in the area of Los Banos the night of the accident.

7-7-45A. Harlingen, Texas. At 1315 CWT, a North American AT-6D crashed 12 miles northeast of Harlingen, Texas, killing pilot 2Lt. Alton W. Zempel, 23. Pilot-rated passenger 2Lt. Clifford B. Kiah was seriously injured. The airplane took off from Harlingen Army Air Field on a pilot proficiency flight. The airplane climbed to about 8,000 feet and began to perform various aerobatic maneuvers, losing a little altitude with every maneuver. The airplane eventually reached an altitude of 4,000 feet, which was just above the cloud tops. The airplane was then seen to go into a slow roll and then enter a spin. The airplane described several turns of spin before partially recovering at an altitude of about 200 feet agl. The airplane came out of the spin in a left wing down attitude before striking the ground at a high rate of speed. The airplane cartwheeled into the ground and broke up, scattering wreckage over an area of 100 yards. The airplane did not burn but was destroyed utterly.

7-7-45B. Sterling City, Texas. At 1317 CWT, a North American P-51D flying in poor weather broke up in flight and crashed ten and a half miles northeast of Sterling City, Texas, killing pilot Capt. John D. Jackson, 24. The airplane was part of a seven-ship flight

that took off from Biggs Field, El Paso, Texas, on a cross-country navigation mission to Esler Field, Alexandria, Louisiana. The formation was flying at 22,000 feet when it began to encounter rapidly deteriorating weather. The formation attempted to climb over the clouds but could not avoid entering instrument conditions. The subject airplane, which was flying in the number-three position of the second element, became separated from the flight and entered thunderstorms that contained high winds and severe turbulence. The P-51D was apparently flying too fast for the turbulent conditions and broke up, the port wing apparently failing first. The port wing apparently folded back and struck the tail section. The starboard wing and tail section failed and separated soon after. Pieces of the aircraft were found scattered over a wide area. The shattered cockpit canopy was never located. The cockpit area completed disintegrated as the airplane broke up completely and smashed to the ground. The pilot's dismembered body was found near where the main section slammed to earth. Three other P-51Ds flying in the same formation crashed minutes later. See 7-7-45C, D, E.

7-7-45C. Robert Lee, Texas. At 1320 CWT, a North American P-51D flying in poor weather broke up in flight and crashed eight miles west of Robert Lee, Texas, killing pilot Capt. George P. Hopkins, 24. The airplane was part of a flight of seven that took off from Biggs Field, El Paso, Texas, on a cross-country navigation flight to Esler Field, Alexandria, Louisiana. The formation was flying at 22,000 feet when it began encountering rapidly deteriorating weather. The formation attempted to climb over the clouds but could not avoid entering instrument conditions. The airplane, which was flying in the number-two position of the second element, became separated from the flight and entered a thunderstorm that contained high winds and severe turbulence. The P-51D apparently was flying too fast for the turbulent conditions and broke up, the tail section apparently failing first. The port wing, the remainder of the tail section and then the starboard wing failed and separated. The airplane's main section slammed to earth and exploded into flames, killing the pilot instantly. Wreckage, which was scattered over an area of three miles, floated to earth for a few minutes after the main section had crashed to the ground. Three other P-51Ds flying in the same formation crashed within minutes of each other. See 7-7-45B, D, E.

7-7-45D. Garden City, Texas. At 1320 CWT, a North American P-51D flying in poor weather broke up in flight and crashed four miles east of Garden City, Texas, killing pilot 1Lt. William D. Martin, 28. The airplane was part of a seven-ship flight that took off from Biggs Field, El Paso, Texas, on a cross-country navigation mission to Esler Field, Alexandria, Louisiana. The formation was flying at 22,000 feet when it began to encounter deteriorating weather. The formation attempted to climb over the clouds but could not avoid entering instrument conditions. The subject airplane, which was flying in the number-two position of the first element, became separated from the fight and entered thunderstorms that contained high winds and severe turbulence. The P-51D broke away and entered a dive, and was apparently flying too fast for the turbulent conditions, causing the airplane to break up. The starboard wing apparently failed first and separated and then the airplane progressively broke up. The propeller and power plant separated in flight. The remainder of the airplane slammed to the ground and exploded into flames. Wreckage was scattered over a wide area. Investigators speculated that the first element flight leader inadvertently led the two wing airplanes into a steep left bank. The flight leader, realizing the unusual attitude, broke away and left the two wing airplanes in an unusual attitude in instrument conditions. It is likely that the two wing pilots, who were flying visually on the flight leader's wings, were unprepared for sudden instrument flight. Three other P-51Ds flying in the same formation crashed within minutes of each other. See 7-7-45B, C, E.

7-7-45E. Garden City, Texas. At 1320 CWT, a North American P-51D flying in poor weather broke up in flight and crashed four miles east of Garden City, Texas, killing pilot 1Lt. Charles W. Sinkule. The airplane was part of a seven-ship flight that took off from Biggs Field, El Paso, Texas, on a flight to Esler Field, Alexandria, Louisiana. The formation was flying 22,000 feet when it began to encounter rapidly deteriorating weather. The formation attempted to climb over the clouds but could not avoid entering instrument conditions. The subject airplane, which was flying in the number-three position of the first element, became separated from the flight and entered thunderstorms that contained high winds and severe turbulence. The P-51D broke away and entered a dive, and was apparently flying too fast for the turbulent conditions, causing the airplane to break up. It was later speculated that the port wing failed first, causing the progressive failure of the airplane. Wreckage was scattered over an area of one mile. Investigators speculated that the first element flight leader inadvertently led the two wing airplanes into a steep left bank. The flight leader, realizing the unusual attitude, broke away and left the two wing airplanes in an unusual attitude in instrument conditions. It is likely that the two wing pilots, who were flying visually on the flight leader's wings, were unprepared for sudden instrument flight. Three other P-51Ds flying in the same formation crashed within minutes of each other. See 7-7-45B, C, D.

7-7-45F. Congress Junction, Arizona. At 1220 MWT, a Lockheed P-38J crashed 19 miles WNW of Congress Junction, Arizona, killing F/O Gene R. Ott, 20. The airplane took off from Luke Field, Phoenix, Arizona, on a routine training flight. The airplane failed to return to the field and was declared missing. The pilot's body and the airplane wreckage were found the next day. The pilot's body was found about three miles from the scene of the crash. Investigation revealed that the pilot bailed out of the airplane for some unknown reason and collided with the elevator counter balance, damaging his parachute and causing him to fall to his death. The airplane continued to fly unmanned. Two civilians on the ground observed the airplane, which was flying at about 5,000 feet, enter a slight dive and then enter a moderate climb, repeating this sequence several times before finally crashing into rising terrain in a descending right turn, exploding into flames upon impact.

7-7-45G. Des Moines, New Mexico. At 1345, a Republic P-47D crashed seven miles south of Des Moines, New Mexico, killing 2Lt. William W. Amrhein. The airplane was flying in the number-four position of a four-ship flight of P-47s that took off from Fort Sumner Army Air Field, Fort Sumner, New Mexico, on a high-altitude aerobatic mission. The flight climbed to 15,000 feet and the three trainee pilots performed aerobatics while the flight leader backed off and observed. The airplanes performed loops, Immelmann turns and rolls and then the leader called for the airplanes to reform. The leader took the flight to low altitude for low-altitude formation training. The flight performed various maneuvers at an altitude of about 500 agl and then the leader climbed to about 1,500 feet and performed a 180-degree turn and set a course back for base. The airplanes descended a bit and were flying in a combat formation at about 500 feet agl on the way back to Fort Sumner. The number-four airplane fell out of formation in a descending left turn. The airplane remained in this attitude until it struck a small hill and exploded. The cause of the crash could not be determined.

7-7-45H. Dallas, Texas. At 1714, a Lockheed P-38L crashed nine miles northwest of Dallas, Texas, killing pilot 1Lt. Robert S. Young, 20. The airplane crashed about four minutes after taking off from Love Field, Dallas, on the pilot's first solo flight in a P-38 airplane. The pilot had been checked out in P-38 type airplanes in the early afternoon. The pilot had passed a cockpit check off, performed three take-offs and landings and flew around the area. The supervising instructor noted that all take-offs and landings were normal and the student was allowed to fly in the local area. While the subject pilot was taking off at 1600, a pilot witness observed that fuel was leaking from a fuel overflow vent on the underside of the subject airplane's port inboard wing. The subject pilot acknowledged this warning and continued with the take-off. A normal take-off was observed. The airplane maneuvered in the area and performed several more take-offs and landings. The airplane took off at about 1710 and climbed to about 1,500 feet on a direct line from the take-off runway. Witnesses observed that the airplane suddenly "flipped" over to the right to an inverted position before rolling upright and entering a flat spin to the right. The pilot was unable to recover and the airplane smashed to the ground at a shallow angle. The airplane burst into flames upon impact. The pilot was killed when he collided with the interior of the cockpit. Investigators speculated that the airplane suffered a failure of the starboard engine, causing the airplane to roll over to the right. Investigators noted that the fuel vent overflow was not a factor in the accident.

7-8-45A. Vevay, Kentucky. At 1059 CWT, a North American B-25J collided with the surface of the Ohio River and crashed three miles northeast of Vevay, Kentucky, killing three crewmembers and seriously injuring gunner Pvt. Napoleon B. Gibbons. Engineer Cpl. Isaiah Grice escaped injury. Pilot 1Lt. Samuel A. Black, co-pilot F/O Glenn W. Pulliam and bombardier/navigator 2Lt. Stephen Hotesse were killed in the crash. Investigators stated, "Investigation revealed that the aircraft flew from Godman Field, Kentucky, to Hayes Bombing Range, discharged its bombs and then proceeded on a low altitude cross country two-ship formation with the co-pilot taking over as operator from the co-pilot's seat. The engineer reported upon reaching the vicinity of Madison, Indiana, that the co-pilot began to let down to lower levels from approximately 1,000 feet. Let down was continued to such a low altitude that the wingman decided to pull away from the formation at which time he noted water spray from the lead ship's propellers as he pulled away. At this time the engineer reported he was making an entry on the Form 1A, and then he suddenly noted (looking out of the window) a bridge in the rear of them that apparently was passed under by his aircraft considering the low altitude at which they were flying. Approximately five minutes later, the engineer heard a splashing noise and immediately braced himself. Then looking up at the pilot and co-pilot, both were trying to recover the aircraft after having struck the water. Attempts to recover the aircraft were unsuccessful, [the aircraft] hit the water and sank. At this point, the wingman had completed his circle and was back at the scene of the accident, noting individuals in the water. He then dropped his life raft for their rescue, however, the survivors were rescued by civilians in the area.... Investigation revealed that the co-pilot was the operator at the time of the accident from the right seat (co-pilot seat). Both pilot and co-pilot attempted to recover the aircraft after it struck the water but to no avail. Investigation revealed that the pilot was cleared to fly on a low level mission at a minimum altitude of 100 feet above terrain. It was further revealed that subject aircraft was observed flying below the prescribed altitude."

7-8-45B. Pueblo, Colorado. At 1840 MWT, a Boeing B-29 flying in poor weather broke up in flight and crashed 20 miles southeast of Pueblo Army Air Base, Pueblo, Colorado, killing 13 crewmembers. Killed in the crash were: Capt. John R. Simpson, B-29 instructor; 1Lt. James R. Bettcher, pilot; 1Lt. Frank J. Fovinci,

co-pilot; Capt. William M. Dougall, bombardier; 1Lt. Edmond J. Fallon, navigator instructor; 1Lt. Clement M. Clapp, navigator; 1Lt. Richard C. Stwalley, navigator; 2Lt. August Siefried, Jr., flight engineer; 2Lt. Leo C. Marshall, bombardier; TSgt. Kenneth R. Willis, radio operator; TSgt. James K. Crook, gunner; SSgt. Eldon A. Maxey, gunner; SSgt. Leslie J. Breier, gunner. Investigators stated, "B-29 #44-86315 was scheduled for combat training mission which consisted of bombing practice to be followed by 1,600 mile navigation training mission. The ship took off [from Pueblo Army Air Base] at 1613 MWT and returned to the base for a landing at 1749, with the bombing mission completed. At this time, two passengers, the instructor bombardier and the instructor radio operator, were let out of the airplane. The ship then took off at 1803 on its navigation mission. Approximately 37 minutes later [the airplane] crashed 20 miles southeast of Pueblo Army Air Base.... The ship had evidently gone on course for some time when it encountered a severe thunderstorm. Evidently the pilot did not attempt to enter this thunderstorm and changed course to bring him back in the vicinity of where he later crashed. By one witness, the airplane was seen flying in a westerly direction with a heading of approximately 260 degrees and paralleling the front of the aforementioned thunderstorm. During the time this witness watched, the ship covered a distance of approximately five miles. The ship started into a turn to the left and at this time the only reliable witness stopped observing the airplane for a few seconds. When he looked back [at the airplane], it had completed approximately a 180-degree turn and at this time he noticed the tail was gone. The ship started into a dive of approximately 45 degrees to the ground and, on contact with the ground, exploded. At this time a huge ball of fire went into the air some 200 feet.... [The airplane suffered the] failure of both outer wing sections at approximately 20 feet in from the wing tip; failure of the vertical stabilizer and rudder, horizontal stabilizer and elevators. Failure of these parts were induced by extreme turbulence when the airplane was being flown in the vicinity of a thunderstorm.... The tail of the ship, less the vertical fin and horizontal stabilizer was found approximately 300 yards west of the final impact. The outer wing panels were found approximately 500 yards west of the impact point. At the time of the crash there was a considerable amount of lightning in this vicinity."

7-9-45A. Galveston, Texas. At 1442 CWT, a Republic P-47D crashed at Galveston Army Air Field, Galveston, Texas, killing pilot 1Lt. Frederick L. Eddy, 27. The airplane took off from Galveston on a routine target-towing mission. At 1435 the pilot advised the control tower that he had lost his target over the bayou a few miles from the field. The tower acknowledged. Investigators stated, "At 1440, a flight of five P-47s came in on an initial approach to Runway 17. The first four ships peeled off and the number-five man, 2Lt. Donald M. Hitzke, maintained his altitude, received permission to enter on initial approach by himself, and peeled off. Lt. Eddy, without having received permission from the tower to enter on initial approach, peeled off a few seconds after the number-five man, Lt. Hitzke. Lt. Hitzke lowered his landing gear and was making a wide landing pattern. Lt. Eddy pulled his ship into a very tight pattern, lowered his wheels and was cutting Lt. Hitzke out of the pattern. Noticing the tow ship was cutting the number-five man out, the Runway Control Officer called Lt. Eddy and said, 'ship turning base, go around.' Instead of acknowledging the call and rolling out of his turn, the tow ship continued tightening his turn attempting to turn on final approach. As he did so, the [subject airplane] stalled and seemed to slip straight into the ground at 1442 CWT. On impact the ship exploded and burned approximately 1,000 feet short of Runway 17."

7-9-45B. Arkansas City, Kansas. At 1850 CWT, a Republic P-47D flying in poor weather crashed two miles southwest of Arkansas City, Kansas, killing pilot 2Lt. Hugo G. Gonzalez, 20, Mexican Air Force. The airplane took off from Strother Field, Winfield, Kansas, on a navigation flight to Oklahoma City, Oklahoma, to Tulsa, Oklahoma, and return. The airplane encountered severe thunderstorms, which included heavy rain, abundant lightning and severe turbulence. The pilot apparently lost control of the airplane and was unable to recover, the airplane slamming into the ground and exploding into flames. There were no witnesses to the accident even though the airplane smashed to the ground about one-half mile from a farmhouse. Two other Mexican Air Force pilots were able to avoid the poor weather and land safely at Ponca City Municipal Airport, Ponca City, Oklahoma.

7-10-45A. Clinton County, Ohio. At 1009 CWT, a Lockheed P-38L crashed one-quarter mile ESE of Clinton County Army Air Field, Wilmington, Ohio, killing pilot 1Lt. William R. Graf, 27. The pilot had been towing a U.S. Navy Mark I pilot-less target glider on a series of stability tests. The airplane towed the 185 lb glider aloft with a 1,000-foot line to an altitude of about 2,500 feet. The glider, which had a wingspan of about 16 feet, flew in a low position behind the towing aircraft at a speed of about 250 mph. The target glider flew satisfactorily and was towed over the field five times without incident. After the last pass, the P-38 flew downwind over Runway 4 and suddenly released the glider. The P-38's port wing was seen to drop and then the airplane went into a steep spiral to the left. The pilot was unable to recover and the airplane smashed to the ground, exploding into flames and killing the pilot instantly. The glider had released smoothly and was lightly damaged when it smashed to earth near the field. Investigators stated that the glider did not cause the airplane to crash. It was speculated that the pilot was going to try to land the P-38 with the glider attached. The airplane, flying a downwind

leg with flaps extended, slowed to near stalling speed so the pilot released the glider. The airplane stalled and spiraled to the ground. Investigation revealed that several successful tests with the glider were performed with a Vultee BT-13 airplane. Although the preferred technique called for a tow pilot to release the glider while very close to the ground, the BT-13 had landed successfully with the glider in tow on several occasions. It was discovered that this was the first attempt to land the glider with a P-38 type airplane.

7-10-45B. Venice, Florida. At 1950 EWT, a Boeing B-29 exploded in mid-air and crashed five miles south of Venice, Florida, killing seven fliers. Four crewmembers were uninjured parachuting to safety. Killed in the crash were: 2Lt. John D. Vislosky, pilot; 1Lt. Hoke F. Shore, B-29 instructor; F/O James Johnston, Jr., navigator; F/O David Rappaport, bombardier; Capt. Loren W. Bell, bombardier instructor; 2Lt. William C. Farrow, flight engineer; Sgt. Willis C. Boykin, radio operator. Crewmembers uninjured parachuting to safety were: Cpl. Vincent G. Bartholome, radio operator; Sgt. Wayne D. Brodbeck, gunner; Sgt. James T. Hurst, gunner; Cpl. Carl W. Donals, gunner. The airplane took off at 1800 EWT from MacDill Field, Tampa, Florida, on a bombing mission at the AAF Venice Bombing Range. Investigators stated, "While over the range at approximately 20,000 feet the engineer reported the oil pressure on number-one engine as having dropped to zero. The scanners saw black smoke coming out of [the number-one engine] followed by a small fire, which broke out on the inboard side. This was reported to the pilot who gave the prepare to bail out signal. The fire died down momentarily then started again and began to ripple and buckle the wing skin. The men in the rear section called several times to see if the pilot was depressurizing but received no answer. The bail out signal was given followed immediately by an explosion, which blew off the port wing from just outboard of the number-one engine. The ship then went into a flat spin which broke the fuselage apart just forward of the aft bulkhead of the aft bomb bay. The four men in the rear section of the ship went out through the door into the bomb bay after unsuccessfully trying to get out of the emergency exits. The forward section of the ship was burning all over and exploding as it fell and was watched by the four men coming down in their parachutes until it hit the ground on the edge of Lemon Bay. The four men landed near the wreckage, went over to it where they were met by civilians and personnel from Venice AAB, and were taken to the station hospital overnight where they were found to be uninjured. The wreckage was spread over a wide area, indicating several large explosions in the air. The whole event was [observed] by individuals on the ground. The crew of another ship from MacDill Field, which was on the range at the same time, saw the explosion which blew off the wing off after first observing smoke coming out of the number-two engine. This crew called in the explosion to MacDill Field at 1950. From the account of this crew and from the statements of the survivors it appears the pilot attempted to feather the [number-one propeller] and to slip to put the fire out but was unsuccessful in both cases. It is the opinion of this board that the wing fire and resulting explosion were apparently caused by a fault in the number-one outboard generator line."

7-10-45C. Alanreed, Texas. At 2325, a North American TB-25J crashed five and one-half miles northeast of Alanreed, Texas, killing students A/S Ferman B. Manuel and A/S Samuel A. Manfredonia. The students were on a routine training flight that originated at Pampa Army Air Field, Pampa, Texas. The airplane took off at 2250 from Reeves Auxiliary Airfield and the students climbed to 6,000 feet and flew to an area southeast of the field for student night work. Pampa Army Air Field control tower called off student flying at 2308 and began calling in all student airplanes. The subject airplane did not reply to repeated radio calls and did not return to the field or its home station. A pilot flying in the area observed an explosion on the ground and reported it to Pampa Army Air Field. Investigation revealed that the subject airplane had dove into the ground at a steep angle with both engines producing power. Investigators speculated that the students had encountered instrument conditions and lost control of the airplane.

7-11-45A. Gage, Oklahoma. At 0755 MWT, one crewmember was killed and one was injured when they bailed out of a Boeing B-29 near Gage, Oklahoma. Navigator 1Lt. Stuart J. Komaroff fell to his death in attempting to parachute to safety. Gunner Cpl. Joseph E. Ryan was seriously injured parachuting to safety; four other crewmembers were uninjured parachuting to safety. Seven crewmembers stayed with the airplane and were able to make a safe landing at Wichita, Kansas. Investigators stated, "The Airplane Commander thought that his airspeed indicator and flight indicator had failed. However, pitot heat had not been turned on and pitot froze when airplane entered a cloudbank. Pilot ordered crew to bail out before ascertaining true nature of the trouble. Pilot was consulting the flight engineer on cruise control at the time of the first indications of trouble and was out of his seat. Airplane was flying on C-1 Autopilot, with elevator control disengaged, under the co-pilot's operation. The Airplane Commander, thinking the instruments had failed and was in a stall [in poor] weather conditions gave the bail-out signal. One man in the nose, a passenger, and five men in the tail compartment left the airplane. The Airplane Commander, after deciding the airplane was under control, after all, told the remaining crew to stay in the airplane. He proceeded on to Wichita, Kansas, and landed without further incident." Crewmembers uninjured parachuting to safety were: Sgt. Leroy M. Griffith, gunner; Sgt. Robert J. Ward, gunner; TSgt. Harold Anderson, passenger; 2Lt. Glenn

R. Prater, radar operator. Crewmembers landing safely at Wichita were: Capt. Robert F. Merrill, pilot; 2Lt. Clayton S. Jacobson, co-pilot; 2Lt. Howard C. Joseph, navigator; 1Lt. Donald C. Gow, bombardier; 1Lt. Francis D. Bajorin, flight engineer; Cpl. Floyd J. LeBlanc, radio operator; Sgt. George C. Aument, gunner.

7-11-45B. Bartow, Florida. At 1020 EWT, a North American P-51D crashed three and a half miles southwest of the Army Air Field at Bartow, Florida, killing pilot F/O Thomas R. Bassnett, 20. The airplane took off from Bartow Army Air Field on a target-towing mission. The airplane took off normally, made a right turn out of traffic and climbed to about 1,000 feet agl. The airplane flew for a short time, made a turn to the right and then the pilot jettisoned the target. The pilot jettisoned the cockpit canopy but made no further attempt to leave the airplane. The airplane then rolled over and entered a spin to the right. The pilot was unable to recover and the airplane spun to the ground. The airplane did not burn but was destroyed utterly.

7-11-45C. Camp Stewart, Georgia. At 2138 EWT, a North American B-25J collided with trees and crashed three miles SSE of the Army Air Field at Camp Stewart, Hinesville, Georgia, killing five fliers. The airplane took off from Camp Stewart on a local night transition flight. Investigators stated, "After making one take-off and landing, aircraft was taxied back to take-off position for a subsequent take-off. Tower cleared aircraft for take-off on Runway 15. Pilot acknowledged take-off instructions and executed an apparently normal take-off at 2137 EWT. At 2138 EWT, approximately two miles from the end of the runway, aircraft hit first tree, about 60 feet high, cutting through the horizontal stabilizer. Half of stabilizer and rudder were found in a cornfield about 150 feet from the tree. Aircraft continued for a half mile farther on and crashed at a very steep angle in a heavily wooded area. Wreckage was thrown forward through trees and burned. Wreckage in the area of the first impact was not burned." Investigators could not determine why the airplane was unable to gain altitude after take-off. Killed in the crash were: 2Lt. Raymond G. Spicer, pilot; F/O Wayne Schumann, co-pilot; Sgt. Walter E. Szczepaniak, engineer; Cpl. Ollie E. Highland, radio operator; Cpl. John A. Cladel, gunner.

7-12-45A. Atlantic Ocean, Georgia. At 0830, a North American P-51D crashed into the Atlantic Ocean six miles east of the south tip of Sapelo Island, Georgia, killing pilot F/O Charles F. McCarty, 20. The airplane was flying in the number-six position of a six-ship flight that took off from Waycross Army Air Field, Waycross, Georgia, on an aerial gunnery mission. The formation flew to the gunnery range area and rendezvoused with a target-towing airplane at about 7,500 feet msl. The flight leader and his wingman broke off first and entered a pattern to fire at the target aircraft while the remaining four airplanes formed an in-line type formation and began S-turning behind the target aircraft. A few minutes later, an explosion was observed on the surface of the sea. Investigators speculated that the pilot might have inadvertently entered instrument conditions that existed east of the air-to-air gunnery range area. It was speculated that the airplane stalled and entered a spiral, remaining in this attitude until it struck the sea and exploded.

7-12-45B. Grand Prairie, Texas. At 1300, a Douglas A-26C crashed two miles northwest of Grand Prairie, Texas, killing pilot Capt. Edmond J. Arbib and pilot-rated passenger 1Lt. John W. Thomas. Investigators stated, "Investigation revealed that the mission, a demonstration training flight, originated at Love Field [Dallas, Texas]. A normal take-off was reported to have been made at Love Field, and a landing was executed several minutes later at Hensley Field [Dallas, Texas]. The pilot was cleared for take-off by Hensley Field Control Tower. [Witnesses] reported that they observed black smoke emitting from both engines during the take-off run.... [It] appeared that both engines were 'sputtering, sounding like they were loading up,' and not developing full power. As the aircraft passed [the witnesses standing near the runway], the left engine was seen to be shaking violently, and acceleration seemed inadequate for a normal take-off.... [The] airplane broke the ground in a semi-stalled attitude and barely cleared obstructions at the boundary of the field. As smoke was still emitting from the engines, the left engine appeared to 'cut out.' Upon query from the tower operator, the pilot dispatched an emergency signal, however, flight was maintained at an estimated altitude of 150 feet, in a wide left turn. As the airplane disappeared from view of witnesses at the field, one of the engines is said to have stopped smoking. Witnesses assert that altitude was lost gradually during the wide turn. As the airplane was flying on a heading approximately 135 degrees from the take-off heading, the right wing of the airplane struck several trees, and then the aircraft contacted the ground, wheels up, in rough terrain. The outboard panel of the right wing was sheared off by impact with trees, and both engines were severed upon impact with an embankment. The left wing, fuselage and tail section came to rest in a mass and were enveloped in flames. The student, Lt. Thomas, was thrown clear of the cockpit, and his body was found approximately 75 feet from the wreckage, and the body of the pilot remained in the wreckage of the cockpit." Investigation revealed that the port propeller was in the full-feathered position. The spark plugs from the starboard engine were found to be heavily fouled and coated with oil. The spark plugs in the port engine were found to be normal.

7-12-45C. Darlington, South Carolina. At 1439 EWT, a Douglas A-26C and a Douglas DC-3 (NC-25647) operated by Eastern Airlines collided in mid-air seven miles southwest of Darlington, South Carolina. Pilot 1Lt. Stephen G. Jones was able to parachute from the A-26, suffering only minor injuries. Gunners Cpl. Raleigh E. Allbaugh and Cpl. Robert B. Clapp were killed aboard the A-26. DC-3 passenger Mrs. A.E. Williams received serious injuries and her infant was killed when the heavily damaged airliner made an emergency forced landing in a field. Four passengers received minor injuries and 13 others escaped injury. Three Eastern crewmembers escaped injury. The DC-3 took off from Washington, D.C. on a flight to Columbia, South Carolina. Investigators stated, "Lt. Jones ... took off from Florence Army Air Field, Florence, South Carolina, at 1355 EWT on a two-hour skip bombing and gunnery mission at McBee Bombing Range. The mission was completed successfully and everything was normal. During his return to Florence, the pilot was practicing aural nulls. The pilot remembered seeing the town of Darlington and knew he was in that general area. He stated that he was at 3,500 feet in a slight gliding angle with an indicated airspeed of 230 mph and had just rolled out of a shallow turn to the left when the collision occurred. The DC-3 was letting down from 4,000 feet and was approximately 3,100 feet with an airspeed of approximately 170 mph when the collision occurred. The airliner was on a course of 240 degrees and the A-26, which had just rolled out of a shallow, two minute left turn, made contact with the airliner at an approximate 90-degree angle. Contact was made apparently on the outboard side of the left engine. The left engine was knocked off of the transport. From there the A-26 passed under the transport and made contact with the right propeller blades of the transport. The tail surfaces of the A-26 were knocked off and the A-26 went over on its back. The pilot was unable to right the A-26 so he jettisoned his hatch, released his safety belt and parachuted to the ground. The A-26 crashed to the ground in an inverted flat spin. The gunner in the rear compartment [Cpl. Allbaugh] was cut to pieces by the propeller of the transport and evidently fell out as the rear of the airplane was torn apart. The gunner in the pilot's compartment [Cpl. Clapp] apparently was wearing a chest-type parachute and whereas his harness was on, he apparently did not have his pack in position. The pilot stated that when he abandoned the airplane the gunner was attempting to put his chest-type parachute pack on, but as he was found in the wreckage he evidently did not have sufficient time. The airliner made a considerably rough wheels-up landing in a cultivated field due to the loss of engines and the airline pilot stated that the rudder was inoperative.... The distance between the two crashed airplanes was approximately four miles. The collision occurred 13 miles WNW of Florence Army Air Field." Eastern Airlines Captain G.D. Davis received minor injuries. Eastern Airlines First Officer M.L. Martindale and stewardesses Peggy Avant and Enouch Mather escaped injury. Passengers injured in the crash-landing were: Maximo S. Asparato, Mrs. Maximo S. Asparato, Gertrude Lustig, Mrs. Annie

L.M. Walker. Passengers escaping injury were: Lillian Bullowa, Bullowa child, William I. Denning, Kenneth Duerden, Lt. Casey C. Griffith (USN), Isidore Hertzberg, Jack G. Lubelle, Carmen W. Pacheco, Lt. Albert L. Storm (USN), Doris B. Ulmer, Lt. Alfred C. Ulmer (USN), Ardell Van Horn, Van Horn child, Mrs. William H. Walker.

7-13-45A. Southport Bay, Florida. At 1030 CWT, a Consolidated B-24H crashed at Southport Bay 25 miles southwest of Tyndall Field, Panama City, Florida, killing 13 crewmembers. Co-pilot F/O Gene P. Johnson was able to parachute to safety, suffering only minor injuries. Killed in the crash were: 2Lt. Paul R. Snyder, B-24 instructor; F/O Franklin G. Osterman, B-24 student; Sgt. Ray E. Bollman, engineer instructor; SSgt. Robert Hill, gunnery instructor; Cpl. Winford Gaskins, gunner; Cpl. John S. Hindman, gunner; Cpl. Herrell C. Askew, gunner; Cpl. Eddie L. Keefe, gunner; Cpl. George N. Kamsch, gunner; Cpl. John L. Huebner, gunner; Cpl. Jack D. Howe, gunner; Cpl. Donald E. Horton, gunner; Pfc. Donald W. Nelson, engineer. The aircraft took off at 0800 CWT from Tyndall Field on a gunnery mission. Investigators stated, "After take-off, aircraft was climbed to 6,000 feet. After completion of the mission, pilot asked the co-pilot to leave the right seat. He then moved a gunnery student into the right seat, permitting him to handle the controls. This lasted for about 20 minutes. Pilot then moved another student in the right seat. The pilot then began to fly through weather that existed at 1000 CWT, giving the students instructions on using the flight instruments. Weather at this time was six to eight tenths cumulus with build-ups to 8,000 feet. Most of the clouds were rain clouds. A stratus layer was about 9,000 to 10,000 feet. At approximately 1020 CWT, pilot decided to return to Tyndall Field and by use of radio aids picked up a heading back to base. There was a line squall between aircraft and base. By making a deviation to the right of five degrees to ten degrees, pilot headed for the blackest cloud in the squall. According to survivor, a 35-degree correction to the left and weather could have been avoided. As soon as the aircraft entered the cloud, pilot, flying in the co-pilot's seat, caged the artificial horizon. Evidently to prove to the gunner that it was possible to fly actual instruments with only basic instruments. Survivor stated that turbulence was severe. The first indication of trouble was that the aircraft began to shudder. Pilot began to lose control. Evidently aircraft went into a spiral, and began to descend at a rate of 2,500 feet per minute. At this time F/O Johnson reached down, shook the co-pilot and told everybody to snap on their parachutes. There were eight men on the flight deck. F/O Johnson was standing between pilot and co-pilot's seat. The engineer was sitting in a position so that he blocked the exit to the bomb bay. F/O Johnson tried to get the engineer to open the escape door into the bomb bay. Centrifugal force was so great that

the engineer could not move. Evidently the aircraft zoomed because the centrifugal force was lessened for a moment. At this time the engineer threw open the door. The engineer, another enlisted man and F/O Johnson were thrown through the door into the bomb bay. The bomb bay doors came open but F/O Johnson does not know who opened them. He snapped the other ring of his parachute and bailed out without leg straps fastened. According to witnesses who were fishing near the scene of the accident, the aircraft emerged from a cloud in a dive then pulled up into a steep climbing turn to the left. Re-entered the cloud and when it came into view again it was in a spiral to the right. Aircraft spiraled for about 120 degrees to the right then fell off into a spin making two turns before striking the ground. Aircraft disintegrated and burned upon impact. Witnesses saw no smoke or falling parts. Engines seemed to be at full throttle."

7-13-45B. Delphos, New Mexico. At 1045 MWT, a Republic P-47D crashed five miles west of Delphos, New Mexico, killing pilot 2Lt. Stanley D. Rea, 23. The airplane was part of a flight of four that took off from Fort Sumner Army Air Field, Fort Sumner, New Mexico, on a strafing and rocket mission. The flight arrived at the Delphos Gunnery Range and the flight leader flew the formation in on a dry run. During the dry run, the subject aircraft mistakenly dropped its bomb. The flight leader advised that only rockets were to be used on the next pass, which would be the flight's first planned live fire pass. Investigation revealed that the subject pilot had pulled out late in his second and third pass on the targets, pulling out at 600 feet instead of 1,000 feet msl as briefed. On the fourth firing pass, the subject pilot began his turn for the target at a lower altitude than the rest of the flight. The subject airplane failed to recover from its diving attack and struck the ground in a flat attitude just beyond the targets. The P-47 bounded back into the air for a short distance before smashing back to the ground. The airplane burst into flames as it careened across the range and smashed itself to pieces.

7-14-45A. Sebring, Florida. At 0039 EWT, a Boeing B-17F flew into the ground three miles southeast of Hendricks Field, Sebring, Florida, killing five fliers. The airplane took off from Hendricks Field at 2136 on a navigation flight to Fort Myers, Florida, to Miami, Florida, to West Palm Beach, Florida, and return. The airplane successfully completed the navigation mission and returned to the area of Sebring Field at about 0015 EWT. The pilot asked for landing instructions and was ordered to circle to the left at 1,500 feet agl in the area southeast of the field. A short time later, officers in the control tower observed an explosion on the ground. It was discovered that the subject airplane crashed into a heavily wooded area. Investigation of the wreckage indicated that the airplane struck the ground at a 20-degree angle while in a shallow turn to the left. The airplane smashed through trees for about 100 yards, exploding into flames upon impact. Investigators speculated that the pilot lost the horizon while turning the airplane at low altitude, allowing the airplane to fly into the ground. Killed in the crash were: 1Lt. Charles L. Snowdon, pilot; F/O Willrid J. Eichert, co-pilot; F/O Earl P. Miller, pilot-rated passenger; 2Lt. William G. Medaris, Jr., pilot-rated passenger; 2Lt. Gene M. Milikan, pilot-rated flight engineer.

7-14-45B. Weldon, Texas. At 1119 CWT, a North American TB-25C suffering apparent engine trouble while flying in instrument conditions crashed one mile northwest of Weldon, Texas, killing 11 passengers and crew. The airplane took off at 1025 CWT from Esler Field, Alexandria, Louisiana, on an administrative flight to Temple Army Air Field, Temple, Texas. The Accident Classification Committee stated, "Shortly after 1100 CWT the aircraft was observed over Stumpville, Texas, a village about 25 miles northeast of Madisonville, Texas, on the highway to Austonio, Texas, coming out of a bank of clouds at an altitude of approximately 500 feet, heading in a southerly or southeasterly direction. Witnesses stated that the left engine was trailing grayish-blue smoke and emitting intermittent puffs of black smoke, and that the landing gear was down and the bomb bay doors were open. At 1119 CWT, the aircraft crashed in a previously cultivated field about one mile northwest of Weldon, Texas, nine miles southeast of where it was first observed and approximately 16 miles southeast of its course." Investigators noted that the ceiling in the area of the crash was approximately 600 feet agl with scattered thundershowers. It was speculated that the pilot was attempting an emergency forced landing and inadvertently flew the airplane into the ground. The airplane exploded into flames upon impact with the ground, scattering wreckage over an area of 400 yards. Crewmembers killed in the crash were: Capt. Harold M. Hamilton, pilot; Capt. Raymond B. Avery, copilot; TSgt. Raymond L. Tieden, engineer; Sgt. Robert W. Bagnall, radio operator. Passengers killed in the crash were: Sgt. Edward S. Craig, Cpl. Charles T. Fortenberry, Pvt. Jerome A. Harris, Sgt. Ralph S. Mann, Cpl. Robert E. Marble, Cpl. Robert E. Souders, Cpl. Carl F. Wickenhiser.

7-14-45C. Albuquerque, New Mexico. At 1630 MWT, a Boeing B-29 crashed just after take-off from Kirtland Field, Albuquerque, New Mexico, killing 13 fliers and seriously injuring B-29 instructor 1Lt. John P. Shaw. The B-29 was scheduled to fly a routine instrument navigation flight to Davis-Monthan Field, Tucson, Arizona, and return. The airplane made a normal take-off and began to climb away. Tower personnel observed that the starboard landing gear had remained in the down position after the nose wheel and port landing gear retracted. They alerted the pilot and he acknowledged. Tower personnel observed the airplane with binoculars until it disappeared into a sand storm that was approaching the area several miles distant. Minutes later the tower

was alerted that the airplane had crashed. Investigators speculated that the airplane lost power in one or more engines after entering the sand storm. Witnesses on the ground observed the airplane flying at very low altitude just before the crash. Investigation revealed that the airplane had collided with a tall tree before slamming into the ground on the port wing in a tail-low attitude about 2,300 yards further on. The starboard wing hit as the airplane smashed through several trees, the airplane striking a drainage ditch 200 feet further on. The airplane smashed into the ditch and cartwheeled, exploding into flames as it broke up. Killed in the crash were: 1Lt. Myron O. Seeder, pilot; 2Lt. Wayne E. McKinney, co-pilot; 2Lt. Richard Serensky, navigator; 2Lt. Alfred J. Johnson, bombardier; 2Lt. James F. O'Brien, flight engineer; TSgt. Loran R. Kargal, radio operator; Cpl. Robert J. Wisneske, radio operator; Cpl. Herbert H. Boger, gunner; Cpl. Martin S. Kaye, gunner; Cpl. Roy S. Dupree, gunner; F/O Jesse H. Flower, radar operator; 1Lt. Gordon, E. Cleland, navigator-rated passenger; 2Lt. Earl N. Pelton, navigator-rated passenger.

7-16-45A. Tuskegee, Alabama. At 1625, a North American AT-6A and an AT-6B collided in midair and crashed one mile south of Tuskegee Army Air Field, Tuskegee, Alabama, killing AT-6B pilot A/S Cleodis V. Todd. AT-6A pilot 2Lt. Henry T. Holland parachuted to safety and was uninjured. The two airplanes were part of two separate flights that had taken off from Tuskegee Army Air Field on routine training flights. One flight of three AT-6 airplanes took off and a short time later a six-ship flight took off. The two separate flights were flying near each other at about 9,000 feet when the instructor leading the six-ship flight ordered the three-ship flight to join on his formation, creating a nine-ship formation of three, three-ship flights. After a short time, the instructor broke the formation into four, two-ship tactical flights. The airplanes then flew a trail formation and the instructor led the airplanes through various maneuvers, which included shallow dives and climbs, slow rolls and a Lufbery circle. After the maneuvers, the instructor reformed the airplanes into four, two-ship tactical flights. The formation climbed to about 12,000 feet and resumed a trail formation. The instructor led the flight into a shallow dive and then into a turning climb. The number-six ship, piloted by Lt. Holland, overtook the number-five ship, which was piloted by A/S Todd. After the passing maneuver it was observed that A/S Todd's airplane began over-running Lt. Holland's airplane. A/C Todd's airplane struck Lt. Holland's airplane in the port stabilizer, scraped over the cockpit canopy and severed Lt. Holland's vertical fin. Lt. Holland, who heard the approaching number-five ship, ducked as he saw A/S Todd's airplane strike the tail and canopy, escaping injury. Lt. Holland's airplane entered a spin to the left and he bailed out, parachuting to safety and escaping injury. A/S Todd's airplane was observed to go into a violent spin. The airplane stopped spinning at about 2,000 feet agl and dove straight into the ground, exploding into flames upon impact and killing the student instantly. Investigators faulted both pilots for failing to maintain safe clearance and proper position while flying in formation.

7-16-45B. Warrensburg, Missouri. At 1933, a Curtiss C-46D crashed while attempting to take off from Sedalia Army Air Field, Warrensburg, Missouri, killing pilot 1Lt. Edward B. Ziegler and co-pilot F/O Thomas D. Smith. The airplane took off at 1930 on a routine local training flight. Investigators stated, "After clearing Runway 13 the ship started an exaggerated climb, becoming more exaggerated until it reached an altitude of approximately 500 feet at which point the aircraft stalled, fell off on the left wing and [began] to spin. Pilot recovered into a straight dive, striking the ground at an angle of approximately 60 degrees. The ship exploded and partially burned. A thorough investigation revealed that the crash resulted directly from the right elevator lock being left in position on the control surface.... There was no crew chief on duty with this airplane for the night flying period. It is assumed that the two pilots made their own visual and pre-flight checks for this particular flight."

7-17-45A. Salina, Kansas. At 0220 CWT, a Boeing B-17F crashed two and one-half miles south of Smoky Hill Army Air Base, Salina, Kansas, killing ten crewmembers and seriously injuring three others. Killed in the crash were: 1Lt. Lathrop P. Ellis, co-pilot; Cpl. Frank W. Burke, engineer; Pvt. Paul M. Boggs, passenger; 2Lt. Garth A. Dahinden, passenger; 2Lt. Myron W. Dawson, passenger; Pfc. DaLayne R. Denbo, passenger; Pvt. Carl C. Hackney, passenger; 2Lt. Howard P. Hamblen, passenger; F/O Richard M. Kunder, pilot-rated passenger; Pvt. Derril C. Young, passenger. Pilot Maj. John R. Millar, Jr. and passengers F/O W.D. Platt and Pvt. Stanley D. Vasquez suffered serious injuries. The airplane had taken off at about 2315 CWT (7-16-45) from Selman Field, Monroe, Louisiana, and was returning to Smoky Hill Army Air Base after having completed an administrative flight to Maxwell Field, Montgomery, Alabama. The airplane arrived in the area of Salina at about 0200 and encountered a line of severe thunderstorms, which contained heavy rain and abundant lightning. Investigators stated, "When over the city of Salina at 1,800 feet, the pilot called the Control Tower and gave his altitude and position and requested transportation for [13] men. The Tower operator answered, giving the pilot the 0130 weather at Salina. Immediately afterward the pilot called the tower and stated that he had the runway lights in sight and was making an approach to land. At this time the Tower operator contacted Weather, requesting the current weather, as it was apparent that the 0130 weather [did not reflect the] true conditions. He also requested the weather to see about closing the field. The pilot then reported that he was not

quite lined up with the runway due to heavy rain and was 'going around.' The Weather Station then contacted the Tower operator and [closed the field].... Tower operator attempted to contact the pilot to inform him that the field was closed No contact could be made. At this time the ceiling was 200 feet, visibility one-quarter mile with very heavy rain and severe lightning. A few minutes later Tower operator saw a faint glow, which appeared to be about two miles south of the field and reported it over the crash alarm system.... Upon questioning the pilot later, he stated that everything appeared normal until a blinding flash appeared, which he thought was lightning. This was the last thing the pilot remembers until after the crash. From investigation at the scene of the crash and from statements of surviving personnel, it is the opinion of the Accident Classification Committee that the pilot, after starting his go-around and while making a shallow turn to the left, either became blinded or was struck by lightning, causing him to lose control. The left wing tip hit the ground, then the number-one propeller, then number-two propeller, fairly evenly spaced, bearing out the fact that the airplane apparently hit in a fairly flat position. After this initial contact the aircraft continued in a slight left bank and disintegrated, with the main body of the wreckage coming to rest on the Missouri Pacific Railroad tracks."

7-17-45B. Pyote, Texas. At 0339 CWT, a Bell P-63A on an unauthorized flight crashed three miles NNW of Pyote Army Air Field, Pyote, Texas, killing non-pilot operator Pvt. Henry W. Suhr, 24. Investigators stated, "At approximately 0250 CWT, a P-63 was observed by the tower operator to taxi out and take off on Runway 12. No radio contact was made. The aircraft passed north of the field at about 2,000 feet, flashing a landing light. The aircraft appeared to make a normal pattern but over shot final approach and went around. Flying straight down Runway 12, the aircraft buzzed the building area at an extremely low altitude. Crash equipment was notified. On the third attempt a pull-up was made, narrowly missing the tower. Circling on the downwind leg the aircraft was seen to lose altitude, crash and burn. Crash equipment and crash alarm were set in motion. The airplane was completely destroyed." Investigators noted that the non-rated enlisted man had been relieved of his duties of airplane mechanic because of inconsistent performance. Investigation revealed that the enlisted man had received a few hours of civilian flight training before entering the army.

7-17-45C. Highland City, Florida. At about 1007 CWT, a Bell P-63A spun into the ground one and a half miles northwest of Highland City, Florida, killing U.S. Naval Aviator Lt. Commander John W. Magee, who fell to his death in an unsuccessful parachute attempt. Naval Aviator Lt. G.B. McDonald and the subject pilot were two navy pilots that were being checked off on P-63 type airplanes at Tyndall Field, Panama

City, Florida. The two pilots took off at 0910 CWT and climbed to about 7,000 feet. After maneuvering for a little while the airplanes entered into a "mild dogfight." The simulated combat was brief and the two airplanes split up at about 1000 CWT. Witnesses on the ground stated that the airplane "zoomed" down and then entered a steep climb. At the top of the maneuver the airplane appeared to "hang" in midair for a moment before falling off on the starboard wing and entering a flat spin to the right. The spin was entered at an altitude too low to allow the pilot to recover and the airplane slammed into the ground in a flat position in a vertical manner. The pilot bailed out at an altitude too low to allow his parachute to deploy and he fell to his death.

7-17-45D. Lamont, Florida. At 1715 EWT, a North American P-51D suffered a catastrophic structural failure and crashed eight miles southwest of Lamont, Florida, killing 2Lt. Leland W. Hall, 22. The airplane was part of a six-ship flight that took off from Thomasville Army Air Field, Thomasville, Georgia, on an aerial gunnery mission. The formation was flying contact under a ceiling of 2,500 feet. The flight leader was able to find a hole in the clouds and climbed the flight to about 7,000 feet, on top of the overcast. The formation encountered two groups of towering cumulus clouds and the flight leader flew in between them. Rain, which was falling from clouds at a higher altitude, was encountered in the area in between the two sets of towering clouds. The subject airplane was seen to "veer" to the left, climbing above the formation. The subject airplane then rolled to the right and leveled out above the formation. The subject airplane then entered an abrupt turn to the left, the pilot again rolling out. The airplane then entered a steep bank to the right, losing altitude as it rolled over into a diving turn to the right. The subject airplane barely missed other airplanes in the formation as it fell away into the clouds. Witnesses on the ground observed the airplane emerge from the 2,000-foot overcast in a steep dive. The pilot pulled up abruptly, causing the port wing to fail and separate. The airplane went out of control, shedding pieces as it fell to earth. The airplane slammed to the ground and exploded into flames. Wreckage was scattered over an area of 650 yards.

7-17-45E. Chino, California. At 1720, a Lockheed P-38L and a TP-38L collided in mid-air and crashed five miles southwest of Ontario Army Air Field, Chino, California, killing 1Lt. Hugh P. Hallman, 25, aboard TP-38L #44-25802. F/O Robert N. Henderson, 20, suffered burns but was able to parachute to safety from P-38L #44-23984. The airplanes were part of a four-ship flight that was scheduled to fly an aerial gunnery mission and an authorized aerobatic flight. The number-three airplane blew a main tire on the take-off roll and failed to get off. The three remaining airplanes continued on the gunnery mission. The flight leader, Lt. Hallman, led the two wing airplanes through

single-engine procedures. The airplanes then flew to the gunnery range and completed the gunnery mission. As the airplanes flew back to base, the flight leader again led the wing ships in simulated single-engine procedure. The airplanes completed the single-engine exercise and the flight leader signaled the airplanes to enter a string formation and "spread out." After the airplanes flew over the Cajon Pass they began "rat racing" and performing aerobatic maneuvers at about 9,500 feet. Lt. Hallman entered an aileron roll to left, performing a roll and a half before rolling out on a course of 185 degrees. F/O Henderson was flying straight and level, also on a course of 185 degrees. Investigators speculated that Lt. Hallman lost sight of F/O Henderson and was pulling up into a loop or Immelmann turn when he pulled up in front of and collided with F/O Henderson's airplane. Lt. Hallman's airplane was in a 70-degree climb when F/O Henderson's airplane smashed squarely into the cockpit of Lt. Hallman's ship. Lt. Hallman was killed instantly. F/O Henderson stated that he felt a sudden impact and then the cockpit filled with smoke. He bailed out and parachuted to safety, suffering burns to his back, shoulders and arms.

7-18-45. Columbus, Mississippi. At 1115, a Beech AT-10GF spun to the ground 16 miles south of Columbus Army Air Field, Columbus, Mississippi, killing pilot 2Lt. Donald W. Dungan and co-pilot F/O John N. Bailey. The pilots fell to their deaths in unsuccessful parachute jumps. The airplane had taken off from Columbus Army Air Field on a routine training flight. Witnesses on the ground observed the airplane spinning toward the ground at a 90-degree angle. The pilots were unable to recover and the airplane slammed to the ground. The airplane did not burn but was destroyed utterly.

7-19-45A. Concord, Massachusetts. At 1000, a Lockheed P-38L crashed onto a golf course one and a half miles southwest of Concord, Massachusetts, killing Capt. Daniel Kennedy, 25. The airplane took off from Bedford Army Air Field, Massachusetts, on a routine training flight. The ceiling was approximately 1,000 feet agl. Investigators stated, "At the instant before the crash, observers on the golf course saw the airplane flying low, heading south, with neither engine feathered. The pilot banked to the left around, or slightly above, a tall poplar tree. He straightened course again for an instant and then rolled left (or snapped) and crashed inverted, with a diving angle of about 40 degrees. The impact buried both engines and propeller assemblies about three feet and the wreckage bounced to a stop about 35 feet further on. Fire followed the crash in a matter of a minute or so. The accident occurred at 1000 EWT ... in the woods at the south edge of the Concord Country Club Golf Course." Investigators could not determine what caused the pilot to lose control of the airplane.

7-19-45B. Gila Bend, Arizona. At 1405 MWT, a North American AT-6D collided with a telephone pole and crashed at Gila Bend Army Air Field, Gila

Bend, Arizona, fatally injuring pilot 1Lt. Kenneth H. Wadsworth and pilot-rated passenger 1Lt. Frazer Shaw. Lt. Wadsworth took off from Luke Field, Phoenix, Arizona, on a flight to Gila Bend to pick up Lt. Shaw and bring him back to Luke Field. The airplane took off from Gila Bend (from west to east) and flew east briefly. The airplane turned around and then began to buzz the field, flying over the take-off runway from east to west at an altitude ranging from 20 to 50 feet. The airplane performed a climbing turn to the left, reaching an altitude of about 200 feet. The airplane continued in the turn to the left, turning about 280 degrees from its flight path before it began to lose altitude and collide with a telephone pole. The airplane struck the pole with the engine and propeller, severing the pole about three feet from the top. The engine cowling separated and fell to earth. The airplane went out of control and smashed on to the field about 500 yards beyond the telephone pole. The airplane broke up and burst into flames as it came to rest. Both pilots were found crawling from the smashed cockpit, alive but seriously injured. Lt. Wadsworth was taken to the AAF hospital at Davis-Monthan Field, Tucson, and Lt. Shaw was taken to the AAF hospital at Luke Field, but both died soon after. Lt. Wadsworth was able to make a statement to investigators before he died, but Lt. Shaw never regained consciousness.

7-19-45C. Wittman, Arizona. At 1810 MWT, two Curtiss P-40N airplanes collided in mid-air and crashed 11 miles southwest of Wittman, Arizona, killing 1Lt. Edward W. Linderman, who fell to his death attempting to parachute from P-40N #42-105534. Major Melvin N. Slate was uninjured parachuting to safety from P-40N #44-7742. The two airplanes took off at approximately 1735 MWT from Luke Field, Phoenix, Arizona, on personnel proficiency flights, which included tactical formation flying and authorized aerobatics. The airplanes climbed to 10,000 feet and flew formation for a short time before Lt. Linderman, who was leading the flight, signaled for a peel-off. He instructed Maj. Slate to fly in the opposite direction and then return to the area for simulated combat. The airplanes engaged in simulated combat for about 10 minutes at about 9,000 feet before the collision occurred. Lt. Linderman was in a steep bank to the right and Maj. Slate was in a steep bank to the left and slightly higher. Maj. Slate began banking to the right and climbing. Major Slate leveled the wings but remained in a climb. Moments later, Lt. Linderman's propeller collided with the tail section of Maj. Slate's airplane. Maj. Slate observed Lt. Linderman's airplane below his starboard stabilizer just before it fell away. Lt. Linderman's airplane began to bank to the left and then suddenly rolled to the right. The airplane rolled out of control until it struck the ground. Maj. Slate stated that he observed Lt. Linderman's cockpit canopy slide back about half way as the airplane rolled away from view. Maj. Slate's airplane then fell out of control so he bailed

out. Both airplanes crashed in close proximity to each other and exploded into flames upon impact. Investigation revealed that Lt. Linderman was struck by a portion of the aircraft as he bailed out. Investigators noted that Lt. Linderman was a P-40 instructor.

7-20-45A. New Castle, Delaware. At 0720, a Douglas C-54E crashed near New Castle Army Air Base, New Castle, Delaware, killing the crew of four. Killed in the crash were: Capt. Edgar L. Kinsey, C-54 instructor; Capt. Daniel E. McCafferty, pilot/C-54 student; Capt. Thomas R. Monroe, pilot/C-54 student; TSgt. Howard J. Driver, engineer. The airplane took off at 0322 from Castle Army Air Base on a local transition flight. The airplane was on a contact (visual) clearance. Investigators stated, "Patterson Field, Ohio, was to be the alternate airport as New Castle Army Air Base was forecast to close at approximately 0500. Radio contact was maintained with New Castle Radio throughout the duration of flight, which consisted of simulated instrument approaches. Upon each approach over low cone airplane was cleared across the field. The procedure is to cross the field at or above 600 feet MSL then make a climbing right turn to intercept the East Leg. At 0722, subject aircraft reported over low cone and was cleared across the field. This was the last radio contact made. Tower Operator saw [subject aircraft] cross the field. Almost immediately the residents living in Wilmington Manor, which borders this base, heard a crash and explosion. Upon investigation it was found that subject aircraft struck the ground at approximately a 30 to 40 degree angle in a Southeasterly direction, exploding on impact.... [It] appears that the subject aircraft was flying at very low altitude.... Due to conflicting reports and insufficient evidence, it was impossible for the Accident Investigation Committee to arrive at a definite cause for this accident. However, weather undoubtedly was a contributing factor as well as the low altitude of aircraft prior to the crash."

7-20-45B. Selfridge Field, Michigan. At 1000 EWT, two Republic P-47D airplanes collided in mid-air and crashed 12 miles north of Selfridge Field, Mt. Clemens, Michigan, killing both French Air Force pilots. Sgt. Henri Murat was killed in the crash of P-47D #45-49154; Sgt. Lucien Tendero was killed in the crash of P-47D #45-49372. The airplanes took off from Selfridge Field on a routine high-altitude navigation training flight. The airplanes climbed to 25,000 feet and flew for about an hour and a half before the collision occurred. Sgt. Murat was navigating for the flight and Sgt. Tendero was flying in the wing position. The AAF instructor, who was leading the flight, was flying behind the subject aircraft and observing. As the airplanes began returning to the area of the field, the flight leader/instructor ordered the students to regain formation on the leader's airplane. The flight leader then advanced his throttle and pulled up even and at the same altitude of the subject airplanes. The flight leader then began a gentle turn to the left. After completing about

60 degrees of turn, the flight leader looked back and observed the airplanes separate and fall away following the collision. Sgt. Tendero's airplane, with its tail section severed, was seen in a flat spin. The airplane continued in the spin until it struck the ground and exploded. Lt. Murat was momentarily able to gain partial control of his airplane before it fell away out of control and slammed into a small lake. The pilot made no apparent effort to bail out.

7-21-45A. Veribest, Texas. At about 0100, a North American AT-6D crashed two miles southeast of Veribest, Texas, killing rated bombardier/student pilot 1Lt. John F. Cummings, 27. The airplane had taken off from Goodfellow Field, San Angelo, Texas, on a night training flight. It was noted that this was the student's first night flying period in advanced training. The student was thoroughly briefed on the ground for night flying by his instructor. The instructor and student took off and the student was briefed in landmarks, checkpoints, traffic zones and altitudes. The instructor put the airplane in a deliberate spiral and the student was able to make a satisfactory recovery. The instructor flew out to Auxiliary Airfield #4 and the student was allowed to perform several take-offs and landings. The instructor had noticed that the student's head would "disappear in the cockpit" when he made power adjustments on climb out. The instructor cautioned the student about lowering his head just after take-off. The instructor was satisfied with the student's overall performance so he was allowed to make a night solo flight. The student performed four take-offs and landings and was to enter the traffic pattern for a fifth landing. The airplane never made its fifth landing and failed to return to Goodfellow Field. The airplane was found at approximately 0300 about one and a quarter mile east of the end of the take-off runway at Auxiliary Airfield #4. It was speculated that the student lost the horizon and flew the airplane into the ground just after take-off.

7-21-45B. Big Spring, Texas. At 1257, a North American AT-6C suffered a catastrophic structural failure and crashed 12 miles SSE of the Army Air Field at Big Spring, Texas, killing pilots 2Lt. Robert F. Wentworth and F/O John L. Schurpf. The airplane took off at 1230 from Big Spring Army Air Field on a local flight. The pilots failed to complete the proper paperwork prior to the flight and investigators discovered that the pilots were "in a hurry to get off the ground." Investigators stated, "The airplane was first seen stunting for ten or fifteen minutes about ten miles southeast of Big Spring. It was seen entering a cloud following a maneuver and was next observed coming out of or from behind this cloud in a turning dive. This may have been a spin but more likely it was a recovery from a spin or simply a diving spiral. At that point a pull out was started; the right wing separated from the airplane at a point about one-third of the way out from the wing root. The airplane whirled and fell, striking the ground

at an extreme angle not more than fifteen degrees from vertical. The right wing was found about one-quarter mile from the balance of the wreckage." The airplane burst into flames upon impact with the ground, killing the pilots instantly.

7-21-45C. Matagorda Peninsula, Texas. At 1345, a North American P-51D crashed at Matagorda Peninsula, Texas, killing pilot 1Lt. John D. Burnett, 26. The airplane was part of a flight that took off from Foster Field, Victoria, Texas, on a rocket-firing mission at AAF Rocket Range B at Matagorda Peninsula. Investigation revealed that the pilot failed to pull out of his second rocket-firing pass. The airplane was estimated to be traveling at about 400 mph when it mushed into the ground and exploded, bounding 20 feet back into the air before smashing back to earth a short distance later. Flaming wreckage was scattered over an area of 600 yards.

7-22-45A. James River, Virginia. At 1508, a Consolidated B-24M crashed into the James River three miles west of Newport News, Virginia, killing 11 crewmembers. The airplane took off from Langley Field, Hampton, Virginia, on a medium-altitude bombing mission at Mulberry Island. Investigators stated, "[The airplane] was first observed at an estimated 4,000 feet at which time the engines were heard to be intermittently sputtering and then roaring. An explosion was also heard and flames were observed coming from engine number-three and engine number-four. Pieces of the aircraft were seen floating down in flames including a large section of wing panel. The aircraft entered a cloud on a north heading and when it reappeared was in a steep dive and turning right to a heading of south. The airplane entered the water with its wings level and two distinct explosions were heard. After the aircraft hit the water gas and oil burned for an hour. Since none of the aircraft's engines or engine accessories have been recovered, it is impossible to determine definitely the cause of the accident. During the investigation there was definitive evidence of fire in the vicinity of the number-one engine." Killed in the crash were: 2Lt. Raymond L. Ferriman, pilot; 2Lt. Rosario Ferlita, co-pilot; 2Lt. Donald E. Fisher, navigator; F/O Frank C. DeMartino, bombardier; 1Lt. Charles L. Goerke, bombardier instructor; 2Lt. Hugh E. Leonard, bombardier/navigator instructor; F/O William D. Showalter, navigator; Sgt. Keith T. Hansen, engineer; SSgt. Ralph E. Forney, engineer instructor; Sgt. Walter Brown, radio operator; Cpl. Wesley Faulkner, engineer.

7-22-45B. Bryan, Texas. At 1505, a North American AT-6D crashed one mile north of the Army Air Field at Bryan, Texas, killing instructor pilot 2Lt. Preston J. Rees and seriously injuring student instructor pilot 1Lt. John M. Lile, Jr. The airplane took off from Bryan Army Air Field on an instrument cross-country flight to Fort Worth, Texas, and return. Investigators stated, "Immediately as the aircraft became airborne and the wheels were retracted, witnesses observed a streamer of gray smoke coming from the exhaust stack on the right side of the engine. As the aircraft attained an altitude of approximately 250 feet, it went into a fairly steep bank and made a turn back toward the main field. The aircraft lost altitude in the turn and just as it leveled out at approximately 50 feet altitude, the right wing dipped sharply. The aircraft was observed to come back to a level flight attitude, and then appeared to mush downward at a fairly steep angle. The aircraft struck the ground in this attitude, tearing both wings from the fuselage. Immediately upon contact with the ground, the fuselage portion broke into flames. Lt. Lile, in the rear seat, had no visible external injuries immediately after the crash, and had extricated himself from the wreckage immediately. On noting that Lt. Rees was unable to get out of the aircraft, he went back to the fuselage and removed Lt. Rees, suffering serious burns in the process." Investigation of the wreckage revealed that the impeller shaft ball bearings failed, causing the engine failure.

7-23-45A. Thomasville, Georgia. At 0944 EWT, a North American P-51D crashed into two parked P-51D airplanes and a structure while attempting to land at Thomasville Army Air Field, Thomasville, Georgia, killing a pilot on the ground. The pilot of the subject airplane (P-51D #44-84633), 2Lt. John H. Mudge, suffered only minor injuries. Pilot F/O Ollie D. Mace, Jr. was killed on the ground. Investigators stated, "Lt. Mudge was on an authorized transition flight. Completing his mission, Lt. Mudge returned to Thomasville Army Air Field and received instructions to land. Lt. Mudge made a normal landing pattern but leveled off a little high, dropping the aircraft in on the main gear. The aircraft bounced into a nose high position. Lt. Mudge applied a heavy burst of throttle, causing the left wing to go down, striking the runway, and the aircraft veered to the left. Lt. Mudge retarded the throttle and the aircraft [settled back to the ground] with the left wing down, still in a slight left turn. Lt. Mudge again applied a heavy burst of throttle, which caused the aircraft to continue to the left in a nose high attitude. Lt. Mudge retarded the throttle as the aircraft touched the ground, again on its left main gear and the left wing. By this time Lt. Mudge was to the left of the landing runway, on the grass area nearing the aircraft-parking ramp. He applied full throttle in an attempt to go around. The left wing of his aircraft struck the ground 15 feet in front of the parking ramp. The aircraft continued across the ramp, hitting two parked P-51D type aircraft and finally came to a stop in the briefing room of Training Branch II.... F/O Ollie D. Mace, Jr. was standing in the path of the aircraft when it hit the briefing room and died on 26 July 1945 from the injuries received."

7-23-45B. Sarasota, Florida. At 1558 EWT, a North American P-51D crashed at Sarasota Army Air Field, Sarasota, Florida, killing pilot 1Lt. Lawrence H. Noel II, 22. The pilot was instructed to take the airplane up on a test flight, which was to consist of a couple turns around the field and a landing. The airplane was scheduled to be assigned on the next mission. The pilot, who seemed to be in a hurry to get the flight under way, refused to sign the necessary documentation when presented by qualified enlisted personnel. The pilot taxied to the run-up area, and after checking the magnetos he immediately took off. The pilot retracted the landing gear just as the airplane became airborne and turned immediately into the pattern, lowering the landing gear just before turning to the downwind leg and requesting landing instructions. The pilot flew a "close-in pattern" at an altitude of about 300 feet agl. The pilot realized that he was going to overshoot on his turn from the base leg to final and tightened the turn. The airplane stalled and fell off. The pilot made a partial recovery but mushed into trees adjacent to the field, severing the tail section. The airplane went out of control and collided with another tree before smashing to the ground in flames. The airplane burst into flames and smashed itself to pieces as it passed under a set of telephone wires and careened across Old Bradenton Road, coming to a rest at the edge of the field. Investigators reported that the pilot died as a result of inhaling fire. Investigation revealed that the pilot, who had been stationed on the base for approximately three weeks, had completed a combat tour in P-38 type airplanes and had been reprimanded a couple of weeks earlier for making a "hot peel-off" at the subject field.

7-23-45C. Brackettville, Texas. At 1705 CWT, a Martin TB-26C made a crash landing seven miles northwest of Brackettville, Texas, killing two B-26 students who fell to their deaths attempting to parachute to safety. B-26 instructor 1Lt. Carl R. Cannop was uninjured in the crash landing. B-26 student F/O Thomas L. Rogers and engineer Pvt. William S. Eckert were uninjured parachuting to safety. B-26 students F/O Malcolm W. Sadleir and F/O Leroy R. Schlaefli were killed attempting to parachute to safety. The airplane had taken off from Laughlin Field, Del Rio, Texas, on a routine training flight. The instructor was instructing the students in single-engine procedure. The port engine had been feathered and restarted without incident. Another student took the pilot seat and the starboard engine was shut down and feathered. The instructor could not get any appreciable power from the starboard engine once it was restarted. The airplane lost altitude at an alarming rate so the instructor ordered the crew to bail out. The crew bailed out but the instructor was unable to bail out because the altitude was too low. He decided to make a forced landing straight ahead. The airplane smashed into small trees and heavy brush before coming to a rest. The airplane did not burn but was completely destroyed. It was speculated that the two pilots had deployed their parachutes at too low an altitude.

7-24-45A. Gulf of Mexico, Texas. At 1130, civilian Gillis Skeels was killed when a lead weight that was part of a target-towing rig pulled by a Republic P-47D fell on him while he was fishing on board a craft under way on the Gulf of Mexico seven miles southeast of Freeport, Texas. Pilot Capt. Melvin B. Kimball, 28, landed at Galveston Army Air Field, Galveston, Texas, after the incident. Investigators stated, "At approximately 1130 a four-ship flight of P-47s on a duly authorized low-altitude gunnery mission was flying in the vicinity of Freeport, Texas, about five to seven miles out to sea. The flight leader called that the weight that causes the target to ride vertically had been shot off and that the target was riding horizontally. All aircraft then returned to the base without incident. A fishing party of which the deceased was a member was in this area, which had for many months been a restricted zone. The deceased, Mr. Gillis Skeels, was sitting on the deck with his feet on the rail when the metal target weight struck him on the top of his head, piercing through his entire body, through his thigh and into the deck of the boat, killing him instantly."

7-24-45B. Ashland, Alabama. At 1217 CWT, a Republic P-47D crashed two and a half miles WSW of Ashland, Alabama, killing pilot 1Lt. Albert A. Walters, 24. The airplane took off from Orlando Army Air Field, Orlando, Florida, on a ferry flight to Reading, Pennsylvania, via Atlanta, Georgia. The pilot contacted Warner-Robins Army Air Field, Georgia, and requested the Atlanta weather. At approximately 1215 farmers observed the airplane flying in a west-northwest direction at about 1,000 feet agl. The airplane was then observed to enter a descending left turn. The airplane turned about 180 degrees and then flew to the southeast for about two miles before entering a spiral and smashing into a hillside. The pilot had bailed out during the final descending turn but was struck by the starboard horizontal stabilizer, killing him instantly. He evidently had pulled the ripcord immediately upon exiting the airplane. The parachute became fouled on the tail section and the pilot's body was dragged to the ground by the airplane. The airplane struck the ground at a 60-degree angle in starboard wing low attitude. The airplane did not burn but was destroyed utterly. Investigators speculated that the pilot became lost and ran out of fuel after receiving the weather report for Atlanta. Investigators described the weather in the region was "excellent."

7-24-45C. Abilene, Texas. At 1413, a Republic P-47D attempting a landing crashed at Abilene Army Air Field, Abilene, Texas, killing pilot 1Lt. Harlan W. Holden, Jr., 24. The airplane was returning to the field after completing a ground gunnery mission. The pilot made a normal peel-off for Runway 17 and entered the traffic pattern. The pilot began to overshoot on the turn from the base leg to final approach. Investigators stated, "In an attempt to keep from overshooting the runway sideways, the pilot made a very

rapid increase of his angle of bank. The airplane stalled and snapped to the left into an inverted position. The pilot recovered and rolled out right side up but only 15 or 20 feet above the ground. At this time witnesses stated that he evidently tried to pull his nose up and the airplane stalled again for almost immediately [it] snapped to the right and hit the ground inverted at approximately a 15-degree dive."

7-24-45D. Tucson, Arizona. At 1825 MWT, a Boeing B-29 suffering an in-flight fire exploded and crashed 45 miles north of Davis-Monthan Field, Tucson, Arizona, killing six fliers. Five crewmembers were able to parachute to safety, escaping injury. The airplane took off from Davis-Monthan Field on a camera-gunnery mission and cross-country navigation flight. The airplane circled the area for about 30 minutes, climbing to 15,500 feet msl. The airplane engaged in the camera-gunnery exercise with a fighter airplane. After completing the exercise the B-29 turned on course for its navigation mission and began climbing. A short time later, the starboard scanner observed a small flame trailing the number-three nacelle. Moments later the flames were observed to emit from the cowl flap area of the number-three engine. The pilot and engineer attempted to extinguish the fire with the engine fire extinguishers without success. Moments later the starboard wing erupted into flames and the number-three engine fell off. The pilot rang the bail out bell and three crewmembers bailed out of the tail section, being sprayed by small bits of molten aluminum trailing the starboard wing. The remaining two crewmembers bailed out of the aft bomb bay. No one from the forward portion of the ship was able to escape. The airplane went out of control and slammed to the ground, exploding violently into flames upon impact. Wreckage was scattered over a wide area. Portions of the starboard wing could not be found and it was later speculated that the intense fire had destroyed these portions of the wing. The number-three engine was found four and a half miles from the main wreckage. Killed in the crash were: 1Lt. Forrest R. Nichols, pilot; 2Lt. Leven L. Dorsey, co-pilot; 2Lt. Walter C. Kenney, navigator; F/O Peter Chando, bombardier; 1Lt. Arthur E. Skeats, flight engineer; Sgt. John C. Alspaugh, radio operator. Crewmembers uninjured parachuting to safety were: 2Lt. William P. Fortier, radar operator; Sgt. Harry B. Wills, gunner; Sgt. Alfred P. Marshall, gunner; Sgt. Billy G. Spikes, gunner; Cpl. George M. Miller, gunner.

7-24-45E. Hobbs, New Mexico. At 1958 MWT, a Boeing TB-17G suffering an in-flight fire crashed eight miles northwest of Hobbs, New Mexico, killing two fliers. B-17 student 1Lt. Truman D. Rose and B-17 instructor 1Lt. Charles S. Ksieniewicz were killed attempting to parachute to safety. B-17 student Capt. Dale E. Hatch and engineer Pvt. Eugene P. Poalillo parachuted to safety and were uninjured. The airplane took off at 1915 MWT from Hobbs Army

Air Field, Hobbs, New Mexico, on a routine training mission. The airplane flew around for about 15 minutes and then made a full-stop landing. The airplane taxied into position for a second take-off. The B-17 student (in the left seat) lowered the seat for an instrument take-off. The airplane made a normal instrument take-off, left the traffic area and climbed to about 2,500 feet. The pilots noticed a fire in the number-one engine and rpm began to fall off rapidly. The instructor could not get the fire under control or feather the propeller. The instructor ordered the two students to bail out and they parachuted successfully. The B-17 instructor was found fouled in his own parachute. Investigators speculated that Lt. Rose had trouble pulling his ripcord and his parachute opened an instant before he hit the ground. The airplane smashed to earth and exploded about three miles west of Runway 26.

7-25-45A. Laredo, Texas. At 0134 CWT, a Consolidated TB-24D on an unauthorized flight crashed while attempting to take off from the Army Air Field at Laredo, Texas, killing non-rated operator Cpl. Charles G. Monte and seriously injuring Pvt. Harrison M. Keller. Investigation revealed that the two enlisted men had been drinking whiskey in Laredo on the evening of 7-24-45 and continued to do so when they returned to Laredo Army Air Field. The two non-pilot enlisted men boarded the aircraft and started the engines. The airplane was taxied out to the runway and began a take-off run to the east on Runway 9. The airplane became airborne about 1,800 feet down the runway, climbing about 15 feet before the port wing dropped and began scraping the ground. The port wingtip dragged for about 600 feet as the airplane continued to turn to the left in a nose-high position. The airplane climbed abruptly and then slammed back to the ground on the nose. All four propellers contacted the surface and the airplane smashed in, turning about 180 degrees from the initial line of flight and bursting into flames as it came to rest. The crash truck and rescue crew raced to the burning wreck. Pvt. Keller was seen stumbling away from the burning airplane, collapsing about 50 feet away. The enlisted man was rescued immediately and the fire extinguished after a short time. Investigators noted that Pvt. Keller could not remember any details leading up to the crash.

7-25-45B. Tampa, Florida. At 1030 EWT, a Douglas C-47B flying in instrument conditions suffered a catastrophic structural failure and crashed 20 miles northeast of Drew Field, Tampa, Florida, killing 13 fliers. The subject airplane was part of a flight of five C-47s that took off at 0915 EWT from Sarasota Army Air Base, Sarasota, Florida, on a navigation flight to Lake Charles Army Air Field, Lake Charles, Louisiana. The last aircraft took off at 0933 EWT. The airplanes failed to join up in formation and were never in direct visual contact with each other. The airplanes were to fly north and make their turn for Lake Charles when they passed over Drew Field. The lead pilot noticed

that a line of thunderstorms was present in the area north of Tampa so he made his turn for Lake Charles south of Tampa Bay and proceeded with the flight. The lead pilot was not in radio contact with the other four airplanes and failed to alert them of the adverse weather conditions on the proposed flight path. Witnesses on the ground observed the subject airplane diving out of the low hanging overcast in a steep dive. The airplane appeared to level out momentarily before the port wing failed and separated. The tail section failed and pieces separated from the airplane as it spiraled out of control. The airplane remained in the spiral to the left and smashed to the ground in swampy terrain. Wreckage was scattered over a wide area. The port horizontal stabilizer and the starboard elevator were found about a mile from the main wreckage. The port wing and port elevator could not be found in the thick mangrove swamp, which was in a heavily flooded condition. Killed in the crash were: 1Lt. William E. Irby, pilot; F/O Kenneth J. Drauch, co-pilot; Pfc. Gerald Scruggs, engineer; Pfc. William B. Kennedy, radio operator; 1Lt. James T. Haven, navigator instructor; 1Lt. Michael J. Rossi, navigator instructor; A/C Vernon L. Larson, student navigator; A/C Leland M. Leeds, student navigator; A/C Milton Lerner, student navigator; A/C Thomas B. Lloyd, student navigator; A/C Auton C. Martin, student navigator; A/C Richard P. McClellan, student navigator; A/C Peter P. Melnik, student navigator.

7-25-45C. Wittman, Arizona. At 1845 MWT, a North American AT-6D crashed eight miles northeast of Wittman, Arizona, killing 2Lt. Stanley R. Pierce and F/O Floyd V. Penn. The airplane took off from Luke Field, Phoenix, Arizona, on an instrument team ride. Investigation revealed that the airplane stuck the ground at a shallow angle, breaking up and bursting into flames upon impact.

7-25-45D. Groom Lake, Nevada. At 1920 PWT, a North American AT-6B crashed while maneuvering at Groom Lake emergency airstrip 49 miles north of Indian Springs, Nevada, killing pilot 1Lt. Benjamin F. Garret, 22. The airplane took off from Indian Springs Army Air Field on a flight to Groom Lake to pick up an enlisted man who was guarding a Bell P-63 that had made an emergency landing there. A maintenance crew was on the strip repairing the P-63 and had witnessed the accident. The subject airplane was observed flying north up the valley, and upon arriving over the Groom Lake airstrip it was observed circling to the right. The airplane lined up with the Southeast-Northwest Runway and flew over at about 75 feet agl, heading southeast. The pilot began a slow roll to the left and the airplane began losing altitude as it approached an inverted attitude. The airplane continued to roll through and the starboard wing struck the ground, causing the AT-6 to cartwheel into the terrain. The airplane broke up and exploded into flames. The enlisted man remained at Groom Lake to guard the wreckage of the AT-6. The TP-63 was repaired and safely flown out. Groom Lake later became the site of the ultra-secret Air Force testing facility known as Area 51.

7-26-45A. Gulf of Mexico, Texas. At 1037, a Republic P-47D crashed into the Gulf of Mexico surf near Bastrop Bayou 25 miles southwest of the Army Air Field at Galveston, Texas, killing pilot 1Lt. Jack C. Fry, 26, who fell to his death attempting to parachute to safety. The airplane was part of a five-ship flight that took off from Galveston Army Air Field on an "Operational Ceiling Gunnery Mission," which was conducted at 31,000 feet. After the airplanes had expended their ammunition they were circling to the left. The flight leader ordered the airplanes to rejoin formation. The subject pilot radioed the flight leader and reported that his cockpit was filling with smoke. The flight leader advised the pilot to retard his supercharger control. The pilot complied but it did not help. The subject pilot radioed the flight leader and reported in an "excited manner" that he still had smoke in the cockpit and that he was experiencing "rudder lock." The subject airplane was then seen to enter a diving spin. The airplanes in the flight were unable to follow because of the high speed that the subject airplane built up. The airplane smashed into the surf 30 feet from shore, exploding into flames upon impact. The pilot attempted to bail out, but apparently he had pulled his ripcord while still standing in the cockpit. The parachute fouled on the aircraft and was destroyed. The pilot's body was found a short distance from the main wreckage. Examination of the parachute proved that it had been scorched by fire. The pilot's clothing also showed evidence of fire, as did his dinghy, life preserver and pilot chute. Investigators could not determine the cause of the fire.

7-26-45B. Salome, Arizona. At 1525 MWT, a North American AT-6C crashed 15 miles east of Salome, Arizona, killing instructor pilots 2Lt. Guy O. Fry and 2Lt. David B. Gardner. The airplane took off from Luke Field, Phoenix, Arizona, to supervise a student low-altitude cross-country navigation mission. Lt. Fry was in the front cockpit and Lt. Gardner was in the rear cockpit. The students were flying the mission at approximately 100 feet agl. The subject airplane was flying behind and above the four-ship student flight. Another ship carrying an instructor was flying above the others. The subject airplane was observed to pass over the student flight from left to right, clearing it by about 20 feet. The subject airplane dove to about 50 feet agl and then pulled up in a steep climbing turn to the right. The airplane apparently stalled and fell off on the right wing, turning 180 degrees from its original line of flight. The pilot leveled the wings and the airplane entered a nose-down attitude at very low altitude. The airplane remained in this attitude until it struck the ground and burst into flames.

7-26-45C. Arbuckle, California. At 1755, a Lockheed P-38L crashed four miles south of Arbuckle, California, killing pilot 2Lt. Alvin C. Hurt, 21. The airplane was flying in the number-four position of a four-ship flight that had taken off from Santa Rosa Army Air Field, Santa Rosa, California, on a low-level formation mission and aerobatic flight. The formation was flying north to south on the "deck" in the Sacramento Valley. Because of the warm weather in the valley, the formation climbed to 10,000 feet to reach cooler temperatures. After a few minutes cooling off, the formation returned to very low altitude. The formation turned around and began an authorized "treetop altitude" exercise. The airplanes flew at low altitude for about 15 minutes, climbing to 500 feet to relieve the tension of the low altitude operations. The flight, now flying in a southeastern direction, made a climbing turn to the left and then returned to the deck. The flight leader then began another climbing turn to the left. The subject airplane was then seen to snap over into a "violent roll," striking the ground on its starboard wing. The airplane burst into flames upon impact and the pilot was killed instantly.

7-27-45A. Chandler, Arizona. At 1600 MWT, a North American TB-25J exploded in mid-air and crashed while attempting an emergency landing at Williams Field, Chandler, Arizona, killing six fliers. Investigators stated, "The aircraft was in the traffic pattern, having returned prematurely from a routine radar training mission. After giving the pilot landing instructions the tower requested that the pilot give his reason for early return. The pilot replied that he had gasoline fumes in the cabin. However, the tower had difficulty in reading the ship and requested that he make another call after landing. At the time of this conversation between the pilot and the tower the ship was on the downwind leg. When on the final approach with landing gear down and everything outwardly normal the ship exploded, scattering debris over a wide area, then fell to the ground and burned.... The board agrees that the cause of the explosion was undoubtedly the accumulation of gasoline fumes within the fuselage. The board could not agree as to the source of the gasoline fumes." Killed in the crash were: F/O Everett E. Park, pilot; 2Lt. Elmer R. Miller, co-pilot; 2Lt. Richard W. Dizney, navigator; Pvt. Andrew A. Volack, radio operator; 1Lt. Roger D. Flynt, bombardier; 2Lt. Clarence A. Van Vranken, bombardier.

7-27-45B. Mobjack Bay, Virginia. At 2133 EWT, two radar operators were killed when they bailed out of a Boeing B-17G that was struck by lightning 16 miles north of Langley Field, Hampton, Virginia. F/O William B. Bishop and 2Lt. Herbert J. Gardner drowned or died of exposure before they could be rescued from the sea. The B-17G took off at 1820 EWT from Langley Field on a routine radar navigation mission. The airplane was assigned an altitude of 16,000 feet. Scattered thunderstorms were forecasted for the areas along the proposed flight plan. The subject airplane was flying over Amboy, New Jersey, when it was called back to the field. The pilot acknowledged and the airplane flew back toward Langley Field. Investigators stated, "At a point approximately 40 miles northeast of Langley Field, there was increased thunderstorm activity. The [radar operator] advised they could clear both main thunderheads on their present course. The ship hit moderate turbulence and entered light cloud. Soon afterwards, lightning was seen to strike the left wing on or near the number-two engine. The pilot rang the warning bell and a short time later the pilot could not contact the rest of the crew on the interphone. He sent the engineer back to check on them. The engineer checked and advised the pilot that two of the men, had bailed out, one of which was witnessed by the radio operator." The airplane was able to make a safe landing at Langley Field and the crew was uninjured. A search was immediately commenced but rescue personnel were unable to locate the two airmen. One body washed ashore on 7-31-45 at Diggs Wharf 20 miles northeast of Langley Field. The other body washed ashore on 8-2-45 three-quarters of a mile southeast of Diggs Wharf. The B-17 was piloted by 2Lt. Walter S. Harpool and co-pilot 2Lt. Winston E. Moore. Observer F/O Donald J. Freeburg, radio operator TSgt. Lawrence Vaughn and engineer Pfc. Robert E. Strout were uninjured.

7-27-45C. Richmond, Virginia. At 2352 EWT, a Beech AT-7C collided with trees and crashed two miles south of the Army Air Base at Richmond, Virginia, killing five fliers. The airplane took off at 2238 EWT from Greensboro, North Carolina, on an instrument clearance to Newark, New Jersey. At 2348 the pilot checked in at Richmond Radio and at 2350 gave his position as "over the Richmond range." Richmond Radio immediately advised the pilot to climb to 5,000 feet and proceed to Newark. No reply was received and a couple minutes later Richmond AAB personnel learned that the airplane had crashed. Investigation revealed that the airplane collided with trees with landing gear in the extended position and the landing lights illuminated. Light drizzle was prevalent at the time of the accident. The airplane was completely demolished by high-speed impact and fire. Investigators speculated that the pilot was attempting to get to the field for a landing. Investigators could not determine why the airplane was apparently attempting to land. Killed in the crash were: 1Lt. Harold F. Black, pilot; 2Lt. Emil J. Doletti, co-pilot; 2Lt. James D. Minor, pilot-rated passenger; 2Lt. Paul L. Duke, pilot-rated passenger; Edsel Weaver (rank unknown), U.S. Navy officer passenger.

7-28-45A. Abilene, Texas. At 0755 CWT, a Republic P-47D crashed three miles south of Abilene Army Air Base, Abilene, Texas, killing pilot 2Lt. Charles T. Streams, 20. The airplane was in the number-four position of a six-ship flight that took off from Abilene AAB on a "Theater Tactics" mission. The airplanes had dropped bombs on the range and then climbed to 13,000 msl. The flight leader led the airplanes through a split-S maneuver. The number-six ship lost a piece of its cowling during the maneuver and returned to base. The remaining five ships participated in a Lufbery circle at about 11,000 feet msl. The flight leader then led the ships in a wingover to the left, turning about 180 degrees. The flight leader and the number-two and number-three ships leveled out and then entered a steep turn to the left. The subject airplane failed to recover from the wingover and continued in a slight dive to about 5,000 feet msl. The other ship broke off and climbed back to rejoin the formation. Investigators stated, "[The Aircraft Accident Classification Committee] believe that Lt. Streams after starting a roll to the right approximately 5,000 feet above the terrain had also begun a turn to the right. It was fairly evident that in this turn he was banked beyond the vertical approximately 20 or 30 degrees.... He was gradually rolling to the left out of this inverted position at the same time decreasing his angle of dive during the turn. At the time of impact he was in a vertical bank, turning to the right, and in a dive approximately five degrees." The airplane cartwheeled to the ground and exploded, killing the pilot instantly.

7-28-45B. New York, New York. At 0952 EWT, a North American B-25D flying in instrument conditions collided with the Empire State Building, New York, New York, killing three servicemen aboard the bomber. Eleven civilians were killed and 26 were injured when the airplane slammed into the 79th floor of the 1,250-foot building on the 34th Street side and exploded into flames. The Empire State Building was the world's tallest building at the time of the collision. Killed aboard the B-25 were: Lt. Col. William F. Smith, Jr., 27, Watertown, Massachusetts, pilot; SSgt. Christopher E. Domitrovich, 31, Granite City, Illinois, engineer; Aviation Machinist Mate 2C Albert G. Perna, 20, Brooklyn, New York, passenger. The airplane took off from Bedford Army Air Field, Bedford, Massachusetts, on a flight to New York City; the flight had originated at Sioux Falls, South Dakota. Perna boarded the airplane at Bedford. The airplane was cleared on a contact clearance to LaGuardia Airport, Long Island, New York. When the airplane arrived in the area of LaGuardia Airport, the pilot requested permission from air traffic controllers to be cleared to Newark, New Jersey. The request was granted but the pilot was advised that he should return to LaGuardia if visibility fell below three miles. The pilot apparently inadvertently entered instrument conditions and lost track of his position. The airplane dropped out of the fog and turned to a southerly heading, flying directly toward the Empire State Building, the top of which was enveloped in fog. The pilot observed that a collision with the building was imminent and attempted to pull up. The B-25D, flying at over 200 mph, pulled up sharply and banked slightly to left but could not avoid a collision, smashing into the north side of the building at the 79th floor and exploding into flames. Pieces of the bomber passed clear through the building and emerged on the opposite side, striking buildings and rooftops below. Flaming gasoline and wreckage exploded through the building, smashing down people and severing elevator cables. Three elevators fell over 1,000 feet to the basement level, killing two elevator operators and a rider. One of the airplane's engines smashed into an elevator shaft, severing the cables and then falling down the shaft onto the shattered car below. The other engine smashed through the building and shot out of the south side of the structure, leaving a large hole where it had exited. The engine landed on the roof of the 12-story Waldorf Building on 33rd Street, causing minor damage. Several people who were killed on the 79th floor were employees of the National Catholic Welfare Council War Relief Services. Paul Dearing, 35, North Tarrytown, New York, a reporter for the Buffalo Courier Express (New York), jumped to his death from the 86th floor to escape the flames. His body was found on the 72nd floor parapet. Another body was found on the 65th floor parapet, apparently blown out of the building by the exploding airplane, which left a gaping 15-foot hole in the side of the structure. Several floors were damaged by fire and hundreds of windows were shattered. Airplane wreckage and building debris fell to the street below. Published accounts reported that debris fell to the street for 30 minutes following the collision. A female elevator operator opened the door of her cage on the 75th floor an instant before the bomber struck. She had stepped out of the elevator and was blown to the other side of the elevator lobby when the B-25 hit. She survived. Pieces of the airplane were imbedded in the exterior of the building and pieces of one of the propellers were found lodged in an interior wall. The blast had cleared the fog away from the building, but after several minutes the fog surrounded the top of the building again. A light drizzle was prevalent at the time of the accident. Published accounts of the accident reported that 35 people were on the 86th floor Observation Deck at the time of the collision. Louis Petley, 54, New York, an employee of the Empire State Building, guided them to safety. New York firefighters reached the 75th floor at 1035. Civilians killed in the accident were: Margaret Mullen, 33, Hoboken, New Jersey; Patricia O'Connor, 21, New York, New York; Jean Sozzi, 40, Brooklyn, New York; Mary Lou Taylor, 19, New York, New York. Joseph Fountain, 47, New York, New York, died several days after the accident. Betty Lou Oliver, 20, New York, New York, was hospitalized with severe burns. Civilians receiving minor injuries were: Ellen Lowe, Floral Park, New York; Phillip Kerby, New York; and James W. Irwin, New York.

7-28-45C. Fort Ogden, Florida. At 1115 EWT, a North American P-51D crashed six miles east of Fort Ogden, Florida, killing pilot 1Lt. Theodore R. Vanden-Houvel, 24. The airplane was part of a flight that took off from the Army Air Field at Punta Gorda, Florida,

on a training flight. The subject airplane was on the tail of the flight leader while participating in simulated combat. The flight leader performed a steep chandelle to the right in an effort to shake him off, recovering in a quarter roll to the right and achieving an inverted position. The flight leader then went into a split-S maneuver. Investigators stated, "The pilot, in attempting to stay on his flight leader's tail in unit combat, stalled out in the inverted position, recovered, and was next seen in a vertical dive. There was no apparent attempt at recovery from the dive, but the pilot did jettison the canopy upon instructions from his flight leader to "bail out." The dive of the airplane became less steep after release of the canopy and it is believed that the pilot left the airplane at this time. The airplane exploded on contact with the ground and the pilot was found about one-half mile from the airplane in line with the path of the dive with his parachute partially opened." Investigation revealed that the pilot collided with the tail section during the bail out attempt.

7-28-45D. Paducah, Kentucky. At 1120 CWT, a Waco CG-4A glider crashed during a failed take-off from Paducha-McKracken County Airport, Paducah, Kentucky, killing glider pilot Maj. George G. Branson and passenger Sgt. Maurice J. Aucoin. The glider was to be towed aloft by Curtiss C-46 #44-77724, which was piloted by Capt. William F. Carlyon and co-pilot 1Lt. Sidney C. Marlborough, Jr. The take-off was attempted to the south on a 4,000-foot runway. There was 300 feet of runway behind the glider, but the glider pilot assured the pilot that a safe take-off could be made. As the airplane-glider tandem gained speed, the glider began to swerve from side to side. At about 80 mph the glider became airborne and then began to porpoise up and down. As the tow-ship neared the end of the runway, the pilot called for one-quarter flaps. The co-pilot, anticipating that the pilot wanted to release the glider, evidently inadvertently pulled the glider release handle instead of dropping flaps. The pilot repeated the command and the co-pilot dropped one-quarter flaps. The airplane became airborne and barely cleared trees at the end of the runway. Investigators stated, "The glider pilot then, instead of pulling up as he could have done at that speed, approximately 80 mph, fell into the [propeller turbulence] of the tow ship. It is presumed that the glider was trying to make a shallow turn of approximately five degrees to land in a field just to the left of the tree obstructions. Apparently the [propeller turbulence] forced the glider's left wing into a steep bank and the pilot was unable to recover close to the ground. His left wing hit a tree and cartwheeled the nose directly into the ground." Investigators speculated that the glider passenger might have been at the controls of the glider during the initial take-off run. Experienced glider personnel who had witnessed the accident stated that they thought the glider was being "mishandled" on the take-off.

7-28-45E. Bison, Oklahoma. At 1710 CWT, two North American TB-25J airplanes collided in mid-air and crashed six miles east of Bison, Oklahoma, killing four fliers. Instructor 2Lt. Frederick Rundle and student A/C Edward F. Guiliot were killed in the crash of TB-25J #44-29212; 2Lt. Kenneth Beers and 1Lt. Robert W. Martin were killed in the crash of TB-25J #44-29420. Both airplanes had taken off from Enid Army Air Field, Oklahoma, on separate training flights. Investigators stated, "From witness it was learned that TB-25J #44-29212 was traveling from north to south and TB-25J #44-29420 was traveling from the southwest to the northeast. The left main wheel nacelle on TB-25J #44-29420 first collided with the left propeller of TB-25J #44-29212. Then the left propeller of TB-25J #44-29420 cut the left wing on TB-25J #44-29212. At the same time the left vertical stabilizer on TB-25J #44-29420 collided with the left horizontal stabilizer of TB-25J #44-29212. TB-25J #44-29212 banked sharply to the right and spun in about one mile northeast of the place of the collision. [The airplanes] exploded and burned on impact. TB-25J #44-29420 went into a steep dive and struck the ground about two miles east of the place of the collision. [TB-25J #44-29420] flipped over on its back but did not burn upon impact."

7-28-45F. Fayetteville, Tennessee. At 1945 CWT, a North American AT-6D flying in poor weather crashed four miles east of Fayetteville, Tennessee, killing pilot F/O Charles A. Waller, 26. The airplane had taken off from Napier Field, Dothan, Alabama, on a flight to Smyrna, Tennessee. The pilot evidently became lost was unable to land his airplane. The pilot was circling the area for some time before the crash. Investigation revealed that the pilot had let-down through a low-hanging overcast and immediately performed a steep turn in an effort to make an approach on a field for an emergency landing. The airplane stalled and entered a spin at an altitude too low to allow recovery.

7-29-45. Seguin, Texas. At 1205, a North American AT-6D crashed four miles south of Seguin, Texas, killing pilots 2Lt. Warren P. Moffet and 2Lt. Harold A. Weaver. The airplane took off at about 1150 from Brooks Field, San Antonio, Texas, on a routine training mission. Witnesses in the area stated that the airplane was performing slow rolls at very low altitude. The airplane was flying at high speed at low altitude when it entered a steep bank to the left. The port wing struck trees and the airplane went out of control and smashed to the ground. The pilots were killed instantly.

7-30-45. Suffolk County, New York. At 0845, a Republic P-47N crashed one and a half miles northwest of Suffolk County Army Air Field, New York, killing pilot 1Lt. Alex Datzenko, 26. The airplane was on a ground gunnery mission. Investigators stated, "The Range Officer on duty at the gunnery range noticed that Lt. Datzenko was firing at too close a range and

thus was pulling out of his pass much too low. The Range Officer warned Lt. Datzenko that he was firing and pulling out of his pass too low a total of three times. Lt. Datzenko acknowledged each message. After the third warning the pilot made two more non-firing passes on the target, which were normal passes. On his last pass at the target the pilot came down low, fired and failed to pull out with sufficient altitude. His aircraft mushed into the ground approximately 60 yards from the target and continued to bounce along the ground until it came to a stop approximately one-half mile from the point of first contact with the ground. Crash equipment and ambulance reached the scene immediately and found the pilot dead. The aircraft was a complete wreck."

7-31-45A. Blackshear, Georgia. At 1105 EWT, two North American P-51D airplanes collided in mid-air and crashed four miles northeast of Blackshear, Georgia, killing two pilots. 2Lt. Harlan L. Erwin, 24, was killed in the crash of P-51D #44-74010; 2Lt. J.A. Horn, 22, was killed aboard P-51D #44-73987. Lt. Erwin was flying in the wing position of a two-ship flight that took off at 1045 EWT from Waycross Army Air Field, Waycross, Georgia, on a routine training flight. The two airplanes joined up in formation and practiced a variety of maneuvers. Lt. Horn took off from Waycross Army Air Field at 1100, towing a target aloft for an aerial gunnery mission. Lt. Horn, with his landing gear in the extended position, was flying in easterly direction and was heading for the aerial gunnery area along the coast. Lt. Erwin, flying on the starboard wing of his leader, was climbing on a northerly heading. The flight leader noticed at the last moment that a collision with the target tow-ship was imminent. He pushed the nose of his ship down abruptly and avoided collision. The flight leader's tail section barely cleared the nose and propeller of the target tow-ship. Lt. Erwin was unable to follow the leader's maneuver and collided with the target tow-ship on the starboard side behind the cockpit and aft of the target tow-ship's starboard wing. Both airplanes broke up as they fell out of control in flames. The airplanes smashed to the ground in close proximity to each other and exploded into flames upon impact. It was speculated that Lt. Horn was killed in the collision.

7-31-45B. Midland, Texas. At 1238 CWT, a Lockheed P-38L crashed five and a half miles north of Midland, Texas, killing pilot 1Lt. Thomas B. Frederick, 25. The airplane took off from Midland Municipal Airport on a flight to Love Field, Dallas, Texas. The airplane took off on Runway 16. About two minutes later, the pilot alerted the tower that he was having trouble with the starboard engine and was returning to land. The tower cleared the airplane to land on Runway 16. The pilot entered the downwind leg and then alerted the tower that he was attempting a single-engine landing. Witnesses on the ground observed that the starboard propeller was feathered. The pilot dropped the flaps and then turned on final approach. The airplane stalled and entered a slight dive. It smashed into the ground at a ten-degree angle, bounding 20 feet back into the air before smashing back into the ground on the starboard wing 25 yards further on. The airplane burst into flames upon impact.

7-31-45C. Pacific Ocean, California. At 1232 PWT, a Vultee BT-13B crashed into the Pacific Ocean approximately five miles west of El Segundo, California, killing pilots 1Lt. Carlos A. Escardo, 30, and 1Lt. Gerald Bauer. The airplane took off at 1041 from Mines Field, Los Angeles, California, on a pilot proficiency flight. A pilot flying a North American AT-6 nearby noticed a high-speed impact on the surface as he flew over. The U.S. Navy searched the area at about 1500 and found some airplane pieces that were entangled with human flesh. There were no witnesses to the actual crash and its cause could not be determined.

7-31-45D. Foley, Florida. At 2130 EWT, a North American P-51D crashed ten miles east of Foley, Florida, killing pilot 1Lt. Aelton E. Hill, 26. The airplane was the lead ship of a three-ship flight that took off at 2130 EWT from Perry Army Air Field, Perry, Florida, on a night formation mission. The subject pilot was leading two trainee pilots when he suddenly began to maneuver erratically. Both trainee pilots broke away from the leader and lost track of the lead airplane. Unable to locate or contact the lead airplane, the trainee pilots returned to the field and landed safely. The subject pilot was heard to transmit on different frequency that he had a problem with his fuel pressure. The pilot advised that he was going to bail out. The pilot was then heard to transmit that he was too low to bail out. Two enlisted men on the ground observed the airplane fly over their quarters with the engine sputtering. The airplane was heard to crash moments later. The two enlisted men jumped into an automobile and drove to the area of the crash. They found the airplane wreckage and the pilot's body. The pilot was found next to a large tree with his parachute strung out behind him. Investigators speculated that the pilot had incorrectly set the fuel tank selector valve lever when switching fuel tanks.

August

8-1-45A. Paxton Springs, New Mexico. At about 1030 MWT, a Consolidated OA-10A (the AAF version of the PBY-5A Catalina Flying Boat) crashed ten miles southwest of Paxton Springs, New Mexico, killing seven fliers. The amphibian took off from Amarillo, Texas, on a flight to Mather Field, Sacramento, California. Investigators stated, "A [North American] P-51 circled OA-10A #44-34096 near Tucumcari, New Mexico, at 8,000 feet, because the OA-10 had one landing gear up and one down. The P-51 pilot contacted [the pilot] and asked if he was in trouble. The OA-10

pilot replied that everything was under control. Then the P-51 pilot noticed a slight smoke trail from [the OA-10's] number-one engine, and slight oil leak on top. This also was reported to [the OA-10 pilot]. The OA-10 pilot again reported that all was under control, so the P-51 left. The OA-10 passed Albuquerque, New Mexico, and crashed 80 miles west [of Kirtland Army Air Field]. He had turned approximately 90 degrees left of course and a gradual descent through the trees to the lava rock, the scene of the crash, was made.... Aircraft was heading in a southwestern direction, shearing off treetops for about 700 feet before striking the ground and burning.... The aircraft burned to an extent that investigation was impossible. It is ... believed he had one engine feathered and was turning back for Albuquerque to land. The terrain at the scene of the crash was 7,500 feet." Killed in the crash were: 1Lt. Wilson H. Parker, pilot; 1Lt. William T. Bartlett, co-pilot; 1Lt. James J. Garland, navigator; Sgt. Irwin S. Marcus, gunner; Sgt. Robert L. Crook, radio operator; Sgt. Harold L. Post, radio operator; Sgt. John M. Jackson, crewmember.

8-1-45B. Lake St. Clair, Michigan. At 1430 EWT, a Martin TB-26B crashed into Lake St. Clair five miles south of Selfridge Field, Mt. Clemens, Michigan, killing four French Air Force fliers and an AAF aerial engineer instructor. French Air Force personnel killed in the crash were: Aspt. Robert H. Comercon, pilot; Sgt. Marcel Feraud, co-pilot; Sgt. Georges Clement, radio operator; Sgt. Henri Buin, gunner; AAF aerial engineer TSgt. James Burda was also killed. The airplane was part of a three-ship flight that had taken off from Selfridge Field on a routine training flight. After take-off, the flight leader turned the formation in the traffic pattern and leveled off on the downwind leg. The flight leader looked back and noticed that the number-three ship (the subject airplane) was not in formation. He observed the subject airplane at 1,500 feet in a 15-degree bank to the left with flames streaming from the starboard engine and nacelle. The airplane entered a steep bank to the left and dove into Lake St. Clair.

8-2-45A. Big Beaver, Michigan. At 1359 EWT, a North American AT-6C crashed into a garage and house at Big Beaver, Michigan, killing pilots 2Lt. Stanley F. Perrin and F/O Louis J. Mikolajek. The airplane took off from Romulus Army Air Field, Michigan, at 1342 EWT on a routine local flight. The airplane apparently flew directly to the area of Big Beaver. The airplane was observed flying in a southerly direction. The airplane then made a 90-degree right turn to the west. The airplane then made a 180-degree turn to the left, heading east. The airplane then entered a steep turn to the right, turning about 45 degrees before it flew into the ground. The airplane careened along the terrain for about 225 feet and then became airborne again until it smashed through a small garage 100 feet further on. After leveling the garage the airplane crashed into a house. Wreckage was scattered over an area of 350 feet. No one was injured or killed on the ground. The house and the garage were completely destroyed.

8-2-45B. Brandenburg, Kentucky. At 1440 CWT, a Lockheed YP-80A broke up in flight and crashed two miles southeast of Brandenburg, Kentucky, killing Maj. Ira B. Jones, 26. The airplane took off from Wright Field, Dayton, Ohio, on a flight to Dallas, Texas, via Memphis, Tennessee. The flight was to be made at 30,000 feet at 525 mph. Investigators stated, "Majority of witnesses stated that they observed the YP-80A in a steep dive, followed by a partial pull-out to level flight. At this time there was a loud report followed immediately by rapid disintegration of the YP-80A. It appears from witnesses' statements that all heavy parts of the YP-80A immediately plummeted to earth, leaving the wings and small parts fluttering down for some time thereafter. It was also stated by a number of witnesses that after the heavier parts had struck the ground they observed what appeared to be the body of the pilot, with the parachute trailing behind, at an altitude of approximately the same as that at which the disintegration occurred." Investigation revealed that the pilot had collided with the vertical fin and possibly the horizontal fin, severing his left leg above the knee. The pilot also suffered severe head trauma; human blood and hair were found on the leading edge of the vertical fin. Blood was also found on the cockpit canopy rollers attached to the fuselage directly behind the cockpit. Investigators speculated that the pilot had deployed his parachute while standing in the cockpit. The parachute was seriously damaged when it contacted the airplane's fuselage or tail. Wreckage was scattered over an area of two miles.

8-2-45C. El Paso, Texas. At 1706 MWT, a Lockheed P-38L crashed at Biggs Field, El Paso, Texas, killing pilot 1Lt. Fred M. Garton, 23. The airplane had taken off from Coolidge, Arizona, on a flight to Biggs Field. Investigators stated, "Upon completion of last turn on the final approach this aircraft did not have sufficient airspeed and did stall and fall into an incipient spin, striking the ground at approximately a 45 degree angle to the left from the direction of flight in an inverted perpendicular position. Right wing and right engine absorbed initial contact prior to the disintegration of the aircraft. Wreckage was strewn in this 45-degree path for approximately 75 yards and came to rest approximately 25 feet within the boundary of the field 110 feet to the east from approach end of Runway 17, the runway [that] was to be used on landing. The aircraft engines were torn completely off from the nacelles. The right engine [was] approximately 40 yards from initial point of impact; the left engine [was] approximately 65 yards from the same point. The aircraft came to rest in an inverted position and was partially destroyed by fire. The pilot's body was found under the inverted right wing." Investigation revealed that

the port engine had failed while the pilot turned to final.

8-3-45A. Hutchinson, Kansas. At 1327 CWT, a North American B-25J crashed two miles south of the Naval Air Station at Hutchinson, Kansas, killing three crewmembers. The airplane took off from Hutchinson Naval Air Station on a ferry flight to Pratt, Kansas. Investigators noted that the airplane's starboard engine was changed at the station and that the necessary engineering test flights had been accomplished. The port engine was observed smoking while the airplane was on the take-off roll. The airplane became airborne and the pilot retracted the landing gear. Tower personnel alerted the pilot of the smoking port engine and the pilot advised that he was going around and coming in for a landing. The port engine was not producing any appreciable power. The airplane stalled while turning in the traffic pattern, falling off to the left and smashing to the ground on the port engine and port fuselage. The airplane burst into flames upon impact and the crew was killed instantly. Pilot Capt. James R. Hugo, engineer SSgt. Robert A. Cushman and passenger Pfc. Joseph E. Levis were killed in the crash.

8-3-45B. Ione, Nevada. At 1200 PWT, a Consolidated B-24J flying in instrument conditions collided with Arc Dome Mountain approximately 21 miles southeast of Ione, Nevada, killing three crewmembers. Pilot 1Lt. Henry G. Van Popering, co-pilot 2Lt. Nickali Thompson and engineer Pfc. Ralph A. Cote were killed in the crash. The airplane took off from Hill Field, Ogden, Utah, on a ferry flight to Tonopah Army Air Field, Tonopah, Nevada. The B-24 had departed Hill Field at 0930 PWT on a visual clearance. At 1139 PWT the pilot transmitted a routine position report while flying over Humbolt, Nevada, but he did not indicate his altitude. The crew was given weather for the Tonopah area, which was reported as ceiling 3,000 feet with light rain showers and visibility 15 miles. Investigators noted that Arc Dome Mountain is 95 miles south of Humbolt on the direct flight path to Tonopah. Investigators speculated that the B-24 encountered instrument conditions on the flight and the pilot failed to maintain a safe altitude over known obstacles on the proposed flight path. Investigators stated, "The airplane hit at an altitude of 11,200 feet and approximately 100 feet below the crest. Indications at the point of impact are such that it is believed that the aircraft was in level flight. The number-one and two propellers struck the ground first, then the nose, and last the number-three and four propellers. This was due to the slant of the hill. Pieces of the nose, bomb bay doors, and the right rudder were lying shortly beyond this point. The airplane then skidded to the crest of the mountain, leaving the outer right wing section and number-four propeller. The right vertical and horizontal stabilizer were lying on the top of the mountain. The aircraft apparently disintegrated after passing over the top. On the opposite slope the path of the airplane can be seen by bits of burned wreckage and trails of oil. At a distance of about 400 hundred yards from the top of the hill is the largest piece of wreckage that was thrown over the crest — part of the left main wing section and the number-three engine. The other engines had torn off and had been thrown to the bottom of the canyon. The outer left wing had burns."

8-3-45C. Bennettsville, South Carolina. At 1615 EWT, two Douglas A-26B airplanes collided in mid-air and crashed five miles southwest of Bennettsville, South Carolina, killing four fliers. Pilot 2Lt. William D. Napier and gunner Sgt. Robert L. MacNeil were killed in the crash of A-26B #41-39130; Pilot 2Lt. Julian A. Benson, Jr. and engineer Sgt. James J. Collins were killed in the crash of A-26B #42-22432. The airplanes were part of a 12-ship flight that took off at 1410 EWT from Florence Army Air Field on a simulated low-level combat mission. Lt. Napier was flying in the number-two position of the first element; Lt. Benson was flying in the number-four position of the first element. Investigators stated, "When approaching Bennettsville, South Carolina, on a course of 290 degrees a light plane [a civilian operated Piper Cub] was noticed approximately two miles ahead and in the path of the formation. The light plane then made a right turn and was flying a course of approximately 200 degrees. The formation made a slight turn to the right and gradually pulled up in order to be sure the formation and particularly the third element would clear the light plane. During the time the formation was pulling up and turning, Lt. Napier's airplane was seen to pull up sharply and turn to the right pulling out of formation. Lt. Benson, who was flying number-four position, had been flying a very loose formation during the whole mission and was seen to have come into his position only a few moments before the collision occurred [at 700 feet agl]. The initial point of contact was seen to be at the wingtips of both airplanes. The [starboard] wingtip of the airplane in the number-two position [Lt. Napier] was seen making contact with the [port] wingtip of the plane in the number-four position [Lt. Benson] at an angle of 45 degrees and about one and a half feet from the tip. This angle decreased and when the airplanes were last seen the wings seemed to be perfectly overlapped at about 15 feet from the wingtips. After the collision the airplanes went into a right spiral and made contact with the ground in bank of 70 degrees and an angle of approximately 45 degrees."

8-4-45A. Willis, Michigan. At 0117 EWT, a Beech C-45 crashed five miles south of Willis, Michigan, killing two fliers and seriously injuring another. Pilot 1Lt. Dorwin M. Keller was killed in the crash. Pilot 1Lt. Thomas F. Delaney fell to his death in an unsuccessful parachute attempt. Passenger Capt. Lloyd Fleischman was seriously injured parachuting to safety. The airplane took off at 2110 EWT from Romulus Army Air Field, Michigan, on a routine instrument training flight. The airplane was cleared to fly at 4,500 feet. Capt. Fleischman flew in the left seat until his practice session ended. Lt. Delaney then took the left seat. Capt. Fleischman went aft and went to sleep. He woke up a little while later when he heard a change in the engine sound and observed that the low-fuel instrument panel lights were illuminated and Lt. Keller was operating the wobble pump. The pilot was unable to get the engines started and could not find a place to land so he ordered the fliers to bail out. Capt. Fleischman bailed out first at an altitude of 900 feet agl. Lt. Delaney bailed out at about 200 feet agl and was killed when he struck the ground before his parachute could deploy. The airplane slammed into the ground a short distance away and exploded into flames upon impact.

8-4-45B. Vilas, Kansas. At 1030 CWT, a Lockheed F-5E suffering an engine fire broke up and crashed two and one-half miles northeast of Vilas, Kansas, killing pilot F/O Richard Washer. The airplane took off from Coffeyville Army Air Field, Coffeyville, Kansas, on a radio navigation flight to Claremore, Oklahoma, to Springfield, Missouri, to Chanute, Kansas, and return to Coffeyville. The airplane was letting down from 25,000 feet near Chanute. The airplane was observed to be "out of control" at an altitude of approximately 10,000 feet, trailing smoke from the starboard engine. The airplane continued to fall out of control and at very low altitude the starboard engine and wing exploded and separated from the airplane. The airplane struck the ground at an 80-degree angle and exploded into flames upon impact. The pilot had made no apparent effort to bail out. Investigation revealed that the number-three piston and piston rod had failed, causing a fire in the starboard engine. The broken piston and the broken piston rod slamming together caused so much excessive heat and deformity that the cylinder wall failed. It was speculated that the piston failure was the result of excessive detonation caused by the pilot using too lean a mixture.

8-4-45C. Wacissa, Florida. At 1355 EWT, a North American P-51B crashed three miles WNW of Wacissa, Florida, killing pilot 1Lt. George H. Pace, Jr., 27, who fell to his death in an unsuccessful parachute jump. The airplane took off at 1252 EWT from Sarasota Army Air Field, Sarasota, Florida, on routine flight to Tallahassee, Florida. Witnesses on the ground 20 miles from the scene of the crash stated that the engine was "backfiring" as the airplane flew over. The airplane was observed trailing smoke then the engine backfired and quit at an altitude of 500 feet agl. The airplane fell off on the starboard wing and smashed to the ground, exploding upon impact and killing the pilot instantly.

8-4-45D. Pyote, Texas. At 2207 CWT, a Boeing TB-29 suffering an in-flight fire was abandoned and crashed 15 miles northwest of the Army Air Field at Pyote, Texas, killing two crewmembers. Seven crewmembers received minor injuries parachuting to safety. The airplane took off from Pyote Army Air Field on a bombing mission. The airplane made two landings and was performing satisfactorily. The airplane climbed to altitude for the bombing mission. The airplane was flying at 20,000 feet and had dropped four bombs on the AAF Bombing Range near Pyote. With no warning the airplane began losing cabin pressure. The pilot ordered all hands to put on oxygen masks. The manifold pressure on the number-three engine dropped to 21" Hg. The cylinder head temperature dropped considerably. After a short time all oil pressure was lost in the number-three engine and the pilot attempted to feather the number-three propeller but was unsuccessful. Moments later, a gunner alerted the pilot that the number-three engine was on fire. The flight engineer attempted to extinguish the fire but was unsuccessful. The fire grew in intensity and appeared to spread to the number-four engine nacelle. The pilot dove the airplane in an attempt to kill the fire but was unsuccessful so he ordered the crew to stand by to bail out. The pilot attempted to jettison the bomb tanks in the forward bomb bay, but the tanks would not drop. The pilot then ordered the crew to bail out. The pilot repeated the bail out order three times and also rang the bail out bell. The airplane commander believed that the airplane had been abandoned before he bailed out. The airplane went into a steep spiral and smashed to the ground, exploding into flames upon impact. Scanner/gunner Cpl. Paul J. Price and scanner/gunner Cpl. Richard J. O'Donnell were killed in the crash. Crewmembers injured parachuting to safety were: 1Lt. James B. Nutter, pilot; F/O Douglas G. Moore, co-pilot; 2Lt. Francis W. Oliver, navigator; F/O Leroy R. Zins, bombardier; 2Lt. Joseph F. Loughran, Jr., flight engineer; Sgt. Walter E. Pinson, radio operator.

8-5-45A. Wilmington, North Carolina. At 1700 EWT, a Republic P-47D crashed 21 miles southwest of Bluethenthal Field, Wilmington, North Carolina, killing pilot Capt. Donald E. Warden, who fell to his death in an unsuccessful parachute jump. The airplane took off at 1515 EWT from the U.S. Naval Air Station at Jacksonville, Florida, on a cross-country flight to the U.S. Naval Air Station at Norfolk, Virginia. The airplane was next observed diving out of a dark cloud at a 70-degree angle. The airplane continued in this attitude until it smashed dove into the ground and exploded into flames. The pilot was found driven head-first into a pile of mud and sawdust up to his mid-chest, his partially deployed parachute strung out behind him. Civilians on the scene reported that the pilot had gasped several times before succumbing to his injuries. The airplane and pilot were stationed at Mitchel Field, Long Island, New York.

8-5-45B. Newburgh, New York. At 1856 EWT, a North American AT-6D crashed seven miles west of Stewart Field, Newburgh, New York, killing U.S. Military Academy Cadet George L. Cleere and instructor 1Lt. William F. Hall. Investigators speculated that the instructor had ordered the student to perform a simulated emergency forced landing at an altitude of about 3,000 feet. The student went through the emergency procedure, which includes switching fuel tanks. Apparently the student inadvertently moved the fuel tank selector to a space in between a fuel tank position and the off position. The student was unable to restart the engine and allowed the airplane to stall at an altitude too low to allow recovery. The airplane smashed to the ground and exploded.

8-5-45C. Ridgeland, South Carolina. At 2025 EWT, a Boeing TB-17G crashed six miles southwest of Ridgeland, South Carolina, killing five fliers. Five fliers parachuted to safety and one was seriously injured. The airplane took off at 1640 EWT from Stewart Field, Newburgh, New York, on a routine cross-country flight to Avon Park Army Air Field, Avon Park, Florida. The airplane was flying at 10,000 feet. A small oil leak was observed on the number-two engine. The number-two engine was producing power and apparently operating normally. At approximately 2015 EWT the number two engine burst into flames. The pilot feathered the number-two propeller; the co-pilot shut off the fuel to number-two and the pulled the number-two engine fire extinguisher. The fire continued so the pilot rang the bail out bell and then ordered the crew to bail out over the intercom. The pilot held the airplane at 10,000 feet for a few minutes to give the crew a chance to bail out. The pilot then dove the airplane to an altitude of 3,000 feet in an effort to extinguish the fire but was unsuccessful. The pilot leveled out, placed the airplane on auto-pilot, closed the throttles and then abandoned the airplane. Investigation revealed that the airplane smashed into the ground at high speed at a 90-degree angle. The airplane exploded into flames and was destroyed utterly. The remains of the dead crewmembers were found in the wreckage. Investigation revealed that all crewmembers had been ordered to bail out and that all crewmembers had fastened their parachutes. The pilot was found to have given the crew plenty of time to bail out. It was not known why the remaining crewmembers did not leave the airplane. Crew Chief Sgt. Leo Montenaro received serious injuries parachuting to safety. Pilot 1Lt. Dewey C. Jones, co-pilot 1Lt. Curtis L. Turner, navigator 1Lt. Lloyd H. Finnan and engineer TSgt. Matthew S. Leo parachuted to safety and were uninjured. Killed in the crash were: 1Lt. William B. Cherry, bombardier; TSgt. Edmund O. Salas, radio operator; F/O A.J. Ponessa, passenger; Cpl. Sidney G. Podherdetz, passenger; Sgt. Leo Bucaria, passenger.

8-5-45D. Hillsboro, Texas. At 2240 CWT, a North American AT-6D crashed two and a half miles northwest of Hillsboro, Texas, killing 2Lt. Glenn T. Mize. The airplane took off at 2217 from Hensley Field, Dallas, Texas, on a flight to Brooks Field, San Antonio, Texas. Witnesses on the ground observed the aircraft flying erratically with the engine "cutting in and out." The airplane was observed losing altitude before it entered a steep dive and smashed into the ground. The airplane did not burn but was destroyed utterly.

8-5-45E. Biggs Field, Texas. At 2151 MWT, a Boeing TB-29B suffering an engine fire on take-off crashed one mile east of Biggs Field, El Paso, Texas, killing seven fliers and injuring three others. The B-29 took off from Biggs Field on a bombing mission-/camera-gunnery mission and an instrument check flight. The crew completed the camera-gunnery mission but approaching darkness prevented them from beginning the bombing mission. The airplane landed at Biggs Field and picked up a B-29 instructor pilot because neither pilot had enough solo B-29 time to fly at night. The instructor boarded at 2145 MWT and the airplane took off on Runway 8 at 2150 MWT. The take-off appeared to be normal until the airplane traveled halfway down the runway when the starboard scanner observed a thick trail of white smoke trailing the number-three engine. The engineer and pilots were notified but the take-off continued. The airplane reached an altitude of about 150 feet before descending into the ground and breaking up. The airplane erupted into flames as it careened across the ground. The two scanners, who were relatively unhurt, were able to rescue the seriously injured tail gunner from the fuselage and take him to safety. Killed in the crash were: Capt. James M. Merritt, B-29 instructor pilot; 1Lt. Kenneth Payne, pilot; 2Lt. Robert W. Jones, co-pilot; F/O Jack S. McMullen, navigator; F/O Henry P. Clark, bombardier; 2Lt. Robert E. Schoo, pilot-rated flight engineer; Cpl. Roderic B. Webb, radio operator. Port scanner Cpl. William H. Graham and starboard scanner Sgt. Albert J. Engbert received minor injuries. Tail gunner Cpl. Lloyd C. Rossi was seriously injured.

8-6-45A. Micanopy, Florida. At 1500 EWT, a North American P-51D crashed three miles NNE of Micanopy, Florida, killing 2Lt. Howard L. Blackman, 26. Before taking off, the pilot had the ground crew charge the oxygen system and was heard to say that he was going to see how high the airplane could climb. The airplane was delayed on the ground because of the installation of radio equipment. The airplane took off and flew away from the area of the field. Witnesses on the ground observed the airplane flying normally at high altitude. The airplane was next observed diving straight down. The pilot failed to recover and the airplane smashed into the ground at high speed, exploding violently into flames upon impact and killing pilot instantly.

8-6-45B. Burbank, California. At 1420, a Lockheed P-80A suffered a fuel pump failure after take-off and crashed one mile southwest of the Lockheed Air Terminal, Burbank, California, killing pilot Maj. Richard I. Bong, 25, Poplar, Wisconsin, who fell to his death in an unsuccessful parachute jump. Major Bong was the top-scoring U.S. fighter ace of World War II and was the winner of the Congressional Medal of Honor. Major Bong, who was assigned to the Lockheed

Factory as a test pilot, was taking off on a routine AAF acceptance flight. The Aircraft Accident Classification Committee stated, "At 1415 P-80A Airplane Serial 44-85048 was towed to the starting line, external power was plugged in and the pilot was in the cockpit. An Army Inspector observed the start and recorded instrument readings. A 650-degree tailpipe temperature start was obtained. All instruments and engine operated normally. The tractor towed the airplane to the taxi strip and released it. Pilot taxied ship under own power to end of Runway 15 calling for a flight clearance at that time. Flight clearance was obtained and take-off was started. It appeared that full power was obtained from engine prior to pilot releasing brakes. A normal ground run was observed, [the aircraft] becoming airborne after approximately a 2,500 to 3,000-foot run. [Landing gear] retraction appeared to be normal and ship continued the average climb on out of field for that type of aircraft. Some witnesses stated fuel was seen leaving in vapor form from right wing, while others stated light smoke was seen leaving the tail cone of subject aircraft during early portion of the take-off run. At south end of field in normal take-off path aircraft appeared to be approximately 150 feet high. Approximately 500 feet beyond south end of the field power was heard to chop off and aircraft settled slightly. A large puff of white smoke was seen to leave the tail cone. An instant later [the engine sound resumed] and the ship pulled up slightly, climbing away and making a shallow turn to the right. [The airplane continued] in a general climb for approximately one-half mile forward obtaining an altitude of approximately 400 feet, at which point the [cockpit canopy] was released and seen to fall to the ground. At the time the [canopy] was released, the ship appeared to rock its wings slightly and continued on for approximately three to five seconds, nose still slightly high and then nosed over abruptly with right wing dropping simultaneously at which time it is believed the pilot attempted to abandon ship. The aircraft continued on down making approximately one-quarter turn before impact with the ground, exploding and disintegrating almost completely with instant fire starting over a radius of approximately 300 feet. From witnesses' statements near the scene of the crash, it is the opinion of the board that the pilot cleared the cockpit shortly after the aircraft began its downward dive, following the ship very close. Evidence of parachute shroud lines [coming in contact with the] leading edge portion of tail surface and rip cord being pulled indicated the parachute fouled on tail surface." Major Bong's parachute was snagged on the tail section and apparently the airplane pulled him down. He was killed when he slammed into the ground near where the airplane crashed. The airplane had made only one 45-minute production test flight previous to the accident flight. On the first flight it was found that the control stick was a bit off center and the rudder trim needed adjustment. Veteran Lockheed test pilot

Ray Meskiman conducted the aircraft's initial production test flight. Because of a shortage of fuel pumps, the original General Electric fuel governor was taken from the airplane and re-installed on another airplane. The fuel governor was eventually replaced with an earlier version of the General Electric fuel governor that had been rejected by engineers at Allison, the manufacturer of the Allison J-33 jet engine, because of its tendency "to dump too easily and had a habit of sticking in the by-pass condition." Investigators stated, "[It] was Major Bong's practice to obtain take-off rpm of 11,500 before releasing brakes for the take-off run, and that he made no throttle reduction until well airborne, thus depending fully on the [main fuel] governor. Due to ram effect, lag in governor action, and slight variation in adjustment, a considerable increase in rpm will take place, as much as 500 rpm. Then when the governor catches up it will by-pass considerable fuel. Should it then stick in the by-pass position as the [rejected fuel governor] has been known to do, it would starve the normal fuel system. Major Bong had admitted neglecting to use the I-16 emergency fuel pump on several flights." Investigators noted that Major Bong had logged only four hours on P-80 type airplanes. All P-80 type airplanes had been temporarily grounded for a short period following the accident.

8-6-45C. Seguin, Texas. At 1745 CWT, a North American AT-6C crashed eight miles south of Seguin, Texas, killing F/O Richard F. Perry, 23. The subject airplane was flying in the number-eight position of an 11-ship formation that took off from Brooks Field, San Antonio, Texas, on a routine training flight. The airplanes first flew in four-ship tactical formations and then changed to string formation. A short time later the formation changed to two, four-ship echelons and one echelon of three. The peel off signal was then given and the airplanes peeled off to the left. The leader then led the flight through a barrel roll to the right. The number-two ship got too close to the lead airplane so he pulled out of the formation and was followed by the number-three airplane. As the number-nine airplane peeled off, the pilot noticed a piece of cowling flying through the air. After the maneuvering, it was discovered that the subject airplane was missing. The airplane was then observed in a series of tight diving turns until it went out of control and smashed to the ground. Investigation revealed that the subject airplane's upper engine cowling came loose and separated, striking the windscreen and tail section. It was speculated that the pilot was incapacitated when the cowling struck the windscreen.

8-7-45A. Victoria, Texas. At 0915 CWT, a North American AT-6C crashed three miles northwest of Aloe Field, Victoria, Texas, killing student A/C Raymond R. Palmer, 19. The airplane was part of a six-ship flight that was taking off from Aloe Field on a high-altitude formation mission. The airplanes took off individually. The leader climbed out straight ahead and then made a 90-degree turn to the left. A short time later the leader made a 90-degree climbing turn to the right. The subject airplane, which was the number-five ship, was attempting to join the formation when it stalled while in a steep climbing turn to the right, entering a spin at an altitude too low to allow recovery.

8-7-45B. Stuttgart, Arkansas. At 0938 CWT, a North American F-6D (the photo-reconnaissance version of the P-51D) crashed seven miles north of the Army Air Field at Stuttgart, Arkansas, killing pilot 1Lt. Gordon R. Aitken. The airplane took off at 0928 CWT from Stuttgart Army Air Field on a routine training flight. Investigators stated, "Investigation has disclosed ... that the pilot had made four passes over a corn field at extremely low altitude. On the fourth pass he pulled up into a steep climb and suddenly began to spin down. This board is of the opinion that the pilot pulled the aircraft up into a high-speed stall and fell off into a spin. According to witnesses, the pilot apparently succeeded in making partial recovery, but fell off into a secondary spin before hitting the ground." Investigators noted that this was the pilot's first solo flight in a P-51 type airplane.

8-8-45A. Godman Field, Kentucky. At 0747 CWT, two Republic P-47D airplanes collided in mid-air three miles WSW of Godman Field, Fort Knox, Kentucky, killing pilot 2Lt. Ephrain E. Toatley, Jr., 20, who fell to his death while attempting to parachute from P-47D #42-29390. Pilot F/O Leroy Bryant, 21, was able to maintain control of his aircraft, safely landing P-47D #42-23272 at Godman Field. Investigators stated, "The mission was a four-ship flight, local gunnery. The planes were joining up shortly after take-off. Numbers one and two were in formation, turning to the left and numbers three and four were closing in. Number-three had not timed himself properly and was behind, in trail, to the right. He came in on the leader from about 5 o'clock and had too much speed and went under and in front of number-one and number-two. He slid back out to the right and pulled up in front of number-two. Number-two's propeller hit [number-three's] elevator before number-two could pull out of the way. Number-two pulled up and number-three started down in a dive to the right. [The] altitude at the time of the collision was about 1,500 feet. The flight leader called to the pilot of number-three to pull out but got no response. At approximately 700 feet, the canopy came off and the pilot left the ship at about 300 feet. The pilot struck the ground with the parachute unopened. The pilot's body bounced several yards and as it did the parachute trailed open. The ripcord had been pulled and was about five feet from the body. Death was instantaneous and the airplane completely wrecked."

8-8-45B. Moody Field, Georgia. At 1643 EWT, a Douglas A-26C crashed one mile SSW of Moody Field, Valdosta, Georgia, killing pilot 2Lt. James D. Handley and gunner Cpl. John H. Mitchelson. The pilot took off from Moody Field on his first solo flight in an A-26 airplane and landed a short time later. Investigators stated, "Pilot made second take-off 38 minutes after the first landing. Mechanical difficulty with the left engine was apparently experienced on take-off since witnesses stated left [propeller] was feathered at the time the airplane had reached a point approximately one mile off the end of the runway at an estimated altitude of 200 feet. Reliable witnesses further stated that the pilot attempted a 180-degree turn to the left at low altitude without pulling up the wheels or the take-off flaps. Pilot was apparently attempting to turn back into the field. Witnesses further stated that grayish black smoke was coming from the right engine immediately prior to the crash. Airplane apparently stalled at the completion of the 180-degree left turn, crashed and burned in a swamp, hitting the trees at a very steep angle of descent ... [in a] level attitude in a slight left bank." The airplane burst into flames upon impact, killing the occupants instantly.

8-8-45C. Kirtland Field, New Mexico. At 1510 MWT, a Bell P-63A suffered a catastrophic structural failure and crashed 15 miles northwest of Kirtland Field, Albuquerque, New Mexico, killing pilot Capt. Robert F. Touhey, 24. The subject airplane was flying in the number-three position of an eight-ship flight that took off from Kirtland Army Air Field on a dive-bombing mission. The subject airplane dove on the target, dropped its bomb and then began a steep pull-out. The subject airplane was observed coming apart in the air. The airplane went out of control and smashed into the ground, exploding into flames upon impact. Investigation revealed that the starboard door had separated from the airplane and collided with the starboard horizontal stabilizer. The starboard horizontal stabilizer failed and separated about two feet from the vertical fin. The airplane snapped over into a violent roll, causing the starboard wing to fail and separate.

8-8-45D. Milton, Florida. At 1750 EWT, a Boeing B-29 flying in poor weather suffered a catastrophic structural failure and crashed seven miles northeast of Milton, Florida, killing pilot 1Lt. Robert A. Lane, 30. Eight crewmembers received minor injuries parachuting to safety. Three other crewmembers were seriously injured parachuting to safety. The airplane took off at 0722 EWT from MacDill Field, Tampa, Florida, on a 3,000-mile round-robin navigation-training flight to San Antonio, Texas, to Brownsville, Texas, to Beaumont, Texas, and return. On the return leg, the pilot encountered several thunderstorms and they were successfully avoided. At about 1700 the weather began closing in around the airplane, which was flying at 16,000 feet. The airplane could not avoid entering instrument conditions, flying into what appeared to be a "light thunderstorm." Visibility went to zero and powerful turbulence began to pummel the airplane. As the airplane continued flying in the storm, the conditions progressively became worse and the turbulence became extremely violent. The airplane entered a ten-degree bank to the left to escape the weather. The airplane slipped into a steep spiral while in the turn. The pilot and co-pilot recovered the airplane with extreme difficulty. The airplane leveled out momentarily before going into a second spiral. The pilots again regained level flight but again entered a steep spiral. The pilots were unable to recover and the airplane entered a diving spiral, generating a speed exceeding 400 mph. The port wing failed at the aileron and separated. Moments later the starboard aileron failed and separated. The pilot rang the bail out bell and the six crewmembers in the after part of the ship bailed out immediately. The radio operator and navigator successfully bailed out through the bomb bay. Investigators stated, "Considerable difficulty was encountered by crew in the nose of the aircraft in leaving through the nose wheel well door. The bombardier and flight engineer dropped through successfully, however, as the co-pilot came down the well doors closed on his left leg around the knee and escape was made only after much violent kicking on his part and possibly with help of the pilot. As the co-pilot fell free at an altitude of 2,000 feet the well doors closed again, trapping Lt. Lane. The aircraft crashed and exploded on the bank of the Black Water River approximately seven miles northeast of Milton, Florida. The wreckage was strewn through the swampy wooded area bordering the river for a distance of some half mile." Co-pilot 2Lt. Murray Goldstein, flight engineer 2Lt. James E. Troyano and radio operator Sgt. Daniel G. Strohmeir were seriously injured parachuting to safety. Crewmembers injured parachuting to safety were: F/O Dennis L. Carr, bombardier; F/O Wallace G. Colthurst, navigator; SSgt. Elbert L. Abramson, gunner; Sgt. Monroe G. Burke, gunner; Pfc. Glenn P. Kendall, gunner; Cpl. Charles G. Gibson, gunner; 2Lt. Alexander L. Treadon, crewmember; 2Lt. Robert E. Reinhart, passenger.

8-9-45A. Smiley, Texas. At 0255, a Beech AT-7C crashed six miles north of Smiley, Texas, killing six fliers. The airplane took off from Blackland Army Air Field, Waco, Texas, on a flight to Foster Field, Victoria, Texas. Investigation revealed that the airplane descended into the ground in a slight bank to the right while on course. Both engines were producing power at the time of impact. Investigators speculated that the pilot might have fallen asleep with the auto-pilot engaged. Killed in the crash were: 1Lt. Peter R. Davis, pilot; passengers 1Lt. Harold G. Phillips, Sgt. Chester V. Seipp, Sgt. Herman D. Lawson, Cpl. Peter J. Zarrilla, Cpl. Lester O. Clotiaux.

8-9-45B. Gulf of Mexico, Florida. At an unknown time after 1005 EWT, a North American P-51D crashed into the Gulf of Mexico approximately 20 miles west of Naples, Florida. Pilot 2Lt. Robert L. Cochran, 24, was declared missing and presumed lost at sea. The airplane was flying in the number-five position of a five-ship flight that had taken off at 0915 EWT from Page Field, Fort Myers, Florida, on a high-altitude gunnery mission. The flight climbed through 10,000 feet and the flight leader called for an oxygen check. All was okay. The subject airplane dropped out of formation and did not enter the firing traffic pattern. The subject pilot called the flight leader and complained that his ears were bothering him. The subject pilot radioed the number-four ship and asked to join him on his firing run and was given permission. The subject airplane never joined the number-four ship and failed to return to base. A civilian fisherman who was underway nearby reported that he observed a single-engine fighter impact the water and explode about 18 miles due west of Naples. AAF and USN airplanes and craft rushed into the area. A P-51 pilot sighted Lt. Cochran in the water and observed that his Mae West was inflated. The crew aboard the Navy PBY could not spot the flyer and refused to land without out seeing the pilot. The AAF pilot lost sight of Lt. Cochran. Navy PT boats searched the area extensively but were unable to rescue the pilot.

8-9-45C. Nordheim, Texas. At 0950 CWT, a Curtiss TP-40N crashed seven miles south of Nordheim, Texas, killing pilot F/O Clairborne J. Wilson, 20. The airplane was flying in the number-four position of a four-ship flight that took off from Fannin Auxiliary Airfield on a flight back to the home station at Aloe Army Air Field, Victoria, Texas. The airplanes took off individually and then joined up for a climb to 10,000 feet. The flight executed a series of 90 and 180 degree turns. F/O Wilson fell out of formation during a 180-degree turn. The flight leader spotted the subject airplane in a normal glide. The subject airplane then disappeared below a thin layer of scattered clouds. The flight leader descended below the cloud layer and soon observed F/O Wilson's airplane burning on the ground. Investigation revealed that civilians had observed F/O Wilson's airplane flying at low altitude trailing smoke and fire. The airplane barely missed striking a haystack and pulled up sharply. The airplane exploded in flight and the pilot was seen falling through the air. The airplane went out of control and smashed to the ground, exploding upon impact. The pilot's body was found about one-half mile from the area of the crash. Investigators speculated that the pilot had attempted to make a wheels-up landing but pulled up because of obstructions on the ground. It was speculated that the mid-air explosion hurled the pilot from the cockpit.

8-9-45D. Fort Myers, Florida. At 1255 EWT, a Consolidated TB-24D crashed two miles southwest of Buckingham Field, Fort Myers, Florida, killing 11 fliers and seriously injuring student gunner Pvt. Wayne Rush. The airplane had taken off from Buckingham Field on a routine splash gunnery mission. Investigators stated, "At 1240, Buckingham Tower received a message from the TB-24D, stating that number-one

engine was out and aircraft was returning to the field. Tower acknowledged and instructed a southwest landing direction. Upon reaching the field, pilot again inquired about landing direction and was again informed of a southwest landing. Aircraft entered the pattern on the downwind leg and radioed in a call that he was turning on the base leg. The pilot did not make an overhead entry into the traffic pattern. Tower noticed aircraft making a turn on the approach for a west landing. Aircraft had made a poor pattern and seemed to be mixed up as to proper landing runway. The plane had to [S-turn] back to the west runway due to overshooting. All witnesses and personnel in the tower noticed the aircraft in a very steep approach and [S-turning] back to the runway. Aircraft hit the runway approximately two-thirds of the way down, bounced about 20 feet in the air and contacted the runway again 40 yards further down. Witnesses stated that the aircraft seemed under poor control. Power was applied to the three engines and aircraft immediately started a very steep climb. Soon after the application of power, wheels commenced retracting. There was not any indication of flaps retracted to 20 degrees. The aircraft reached a height of approximately 250 feet. Pilot witnesses stated that the aircraft was in a skidding attitude towards the left. Upon reaching a position 250 feet [agl] and a mile and a half southwest of the west runway the aircraft was observed falling off on the left wing and crashing into the ground. Flames immediately broke out." Killed in the crash were: 1Lt. James G. Baker, pilot; F/O Thomas Watzel, co-pilot; Cpl. Burton R. Simon, engineer; SSgt. Thomas M. Hunter, gunner; Cpl. Harry H. Davis, gunner; SSgt. Gerard J. Milton, gunner; SSgt. Robert Pelman, gunner; 2Lt. Jack D. Anderson, bombardier-rated student gunner; Pvt. Joseph E. Hemminger, student gunner; Pvt. Lawrence H. Huffman, Jr., student gunner; Cpl. Stanley R. Kalenius, student gunner.

8-9-45E. Kingfisher, Oklahoma. At 1510, a Lockheed P-38L performing low-altitude aerobatics crashed five miles southwest of Kingfisher, Oklahoma, killing pilot Capt. Walter E. Davis, 29. The airplane had taken off from Will Rogers Field, Oklahoma City, Oklahoma, on a routine transition flight. The airplane was observed flying over town from the east at about 100 agl. The airplane pulled up in a steep climbing turn toward the north. The airplane circled and returned from the north, again flying at about 100 feet agl. The airplane was observed to go into a slow roll to the left. As the pilot completed the maneuver the starboard wing struck the ground at the bottom the roll. The airplane cartwheeled into the ground in a westerly direction and exploded into flames. Wreckage was scattered over a wide area. Investigation revealed that the pilot had attended a wedding as the best man some days earlier in Kingfisher. Capt. Davis told wedding guests that he would fly over town if the opportunity presented itself.

8-9-45F. Waco, Texas. At 1735, a North American AT-6C suffered a structural failure and crashed ten miles east of the Army Air Field at Waco, Texas, killing pilots 2Lt. Donald W. Allen and 2Lt. John S. Andrews. The airplane was flying in the number-three position of a three-ship flight that had taken off from Waco Army Air Field on a routine aerobatic mission. The flight performed various aerobatic maneuvers, including an Immelmann turn. The lead airplane then pulled up sharply until close to stalling speed. The lead airplane then dove steeply and passed vertical about 15 degrees, almost beginning an outside loop. The lead airplane and number-two airplane were able to roll out of the maneuver with a 90-degree roll and steep turn to the left. Both airplanes were able to re-establish level flight. The number-three airplane also dove past vertical and also began the outside loop. The pilot attempted to roll out but quickly rolled back to the outside loop. The airplane gained excessive speed and the port wing failed and separated during the recovery. The airplane went out of control and smashed to the ground. The airplane did not burn but was destroyed utterly.

8-9-45G. Tucson, Arizona. At 2145 MWT, a Douglas A-20H flying in poor weather collided with terrain and crashed 25 miles ESE of Davis-Monthan Field, Tucson, Arizona, killing pilot 1Lt. Russell L. Mathern and pilot-rated passenger Capt. Holbrook Snyder. The airplane took off at 1957 MWT from the Army Air Field at Hobbs, New Mexico, on a flight to Davis-Monthan Field. The airplane flew into a thunderstorm and visibility went to zero and turbulence became severe. The pilot apparently began a turn to get out of the weather but apparently had lost considerable altitude in the turn. Soon after the pilot completed a right turn the airplane collided with a ridge while in a flying attitude with both engines producing power. The airplane first struck trees and then smashed into the mountain, smashing itself to pieces for about 250 yards and finally coming to rest near the peak.

8-9-45H. Greensboro, Alabama. At 2254, a Beech AT-7C collided with rising terrain five miles southeast of Greensboro, Alabama, killing six fliers. Killed in the crash were: 2Lt. James T. Quinn, pilot; 1Lt. Herman I. Seidel, navigator instructor; A/C Thomas K. Zacher, student navigator; A/C George A. Knudtson, student navigator; A/C Henry W. Herbert, Jr., student navigator. The airplane took off at 2055 CWT from Selman Field, Monroe, Louisiana, on a celestial navigation flight to Craig Field, Selma, Alabama. The airplane was assigned to fly at 7,000 feet on the outbound leg and at 8,000 feet on the return leg. The airplane apparently entered a large cloud, the only one for miles. The large cloud contained light rain showers and was about ten miles wide at its base. The pilot then made a wide turn to avoid flying through the cloud. The airplane lost altitude in the turn and collided with the south slope of a hill while in a flying

attitude with both engines producing power. The airplane exploded into flames upon impact, scattering wreckage and dismembered bodies for over 150 yards up the slope. Investigators noted that the bodies were badly dismembered and a severed hand was found clutching a closed cigarette lighter. It was also noted that the airplane was about 20 miles north of its intended course.

8-10-45. Blythe, California. At 0625, a Vultee BT-13B crashed seven miles west of Blythe, California, killing 1Lt. Warren B. Paulsell, 21. Investigators stated, "The pilot was the acting Base Operations Officer at Blythe Army Air Field at the time of the accident. Shortly after hearing a news broadcast of the reported Japanese acceptance of the Potsdam Ultimatum, he decided to make a local flight in a BT-13 assigned there. Although the airplane had not been flown the previous day nor pre-flighted for this flight, he had a mechanic stand fire guard while he started the engine inside the hangar, disregarding instructions to the contrary and taxied out of the hangar directly to the runway for take-off before calling the tower for instructions or checking the engine. He ran up the engine near the runway and was cleared by the tower for take-off. He took off at 0615 PWT. Take-off was observed to be normal. He then climbed to about 800 to 900 feet and started flying back and forth over the runways and the building area, executing steep turns, climbing turns and rolls at an altitude less than 1,000 feet. According to statements of witnesses near the scene of the crash, he was either in a steep turn or an attempted roll, or similar maneuver, when he seemed to stall or lose control of the airplane. The exact attitude of the airplane at this time could not be determined, but there is a possibility that loss of control may have been caused by momentary loss of power due to unusual attitude of the plane. The plane crashed at 0625 in an open area and burst into flames upon impact. The pilot was killed and the airplane completely destroyed."

8-11-45A. Marietta, Georgia. At approximately 1400 EWT, a Douglas C-47A crashed seven miles southwest of Marietta, Georgia, killing pilot 1Lt. Edward J. Mahoney. Virginia A. Fincher, age 12, sustained minor burns when flaming wreckage smashed into her house. Co-pilot 1Lt. Robert H. Koch, engineer TSgt. Edgar H. Wueller and WAC passenger T/4 Phoebe H. Wicks were uninjured parachuting to safety. The airplane took off at 1325 EWT from Atlanta Army Air Base, Atlanta, Georgia, on a passenger flight to Pope Field, North Carolina. Investigators stated, "Immediately upon becoming airborne, the pilot discovered that he did not have use of the aileron and elevator controls and upon investigation found both aileron control locks on but could not see whether the elevator control locks were on or not. The pilot continued on and found he could keep the airplane climbing slowly by using four degrees forward trim on the elevator trim tab and 40 inches of manifold pressure. The airplane seemed right wing heavy and continued in a slight turn to the right, this is believed to be caused by the right aileron lock slipping over and getting caught on the wing, holding the right aileron in a slight up position, causing the right wing to go down. Upon reaching 3,200 feet indicated altitude the airplane practically became uncontrollable. At this time the pilot ordered the crew and passenger to bail out. The co-pilot went back and helped the passenger and crew chief to get out safely and then went back up to the cockpit. About this time the airplane went into a violent bank to the right and the co-pilot told the pilot he was going to bail out. After the co-pilot bailed out, it is apparent that the pilot could not leave the controls long enough to put on his parachute and abandon the airplane, consequently, he crashed with the airplane and was killed. From the investigation, it is the opinion of the Aircraft Accident Board that the cause of this accident was due to the control locks not being removed before taking off. Although the crew chief stated he did remove the elevator locks, the co-pilot stated they did not have use of the elevator controls, and therefore, it is believed that the elevator locks were on as well as the aileron locks."

8-11-45B. Yucca, Arizona. At an unknown time after 1900 MWT, a North American TB-25J flying in instrument conditions crashed into a mountain 25 miles SSW of the Army Air Field at Yucca, Arizona, killing five fliers. The airplane took off at 1550 MWT from Yuma Army Air Field, Arizona, on a radar-training mission. At 1710 MWT, the airplane transmitted a position report to Las Vegas, Nevada, reporting to be 30 miles south of Needles, California, at 6,000 feet. At 1740 MWT, the airplane attempted to contact Blythe and Needles, California, and Yuma radio but was unsuccessful. At 1743 MWT, the airplane radioed Las Vegas and asked for weather information but did not give a position report. At about 1900 MWT, the airplane transmitted and asked if anyone could read them. A pilot flying 20 miles south of Kingman, Arizona, answered the subject airplane. The pilot tried to contact the subject airplane several minutes later but was unsuccessful. Investigators speculated that the airplane was flying from Las Vegas to Yuma and encountered thunderstorms while flying in the mountainous regions east of the Colorado River. Investigators stated, "Apparently the pilot ... was confused and lost. The aircraft collided with the mountain at approximately 3,500 feet on a course estimated to be 45 degrees. The aircraft cartwheeled after the initial contact with the mountain and came to rest approximately 150 feet up the mountainside. The aircraft was completely demolished and burned." Killed in the crash were: F/O Robert L. Laird, pilot; F/O Juan S. Madero, Jr., co-pilot; 2Lt. John W. Winter, navigator-rated radar student; 2Lt. William G. Winter, navigator-rated radar student; Pfc. William F. Strange, radio operator.

8-13-45. Lakeport, California. At 1645, two Lockheed P-38L airplanes collided in mid-air ten miles east of Lakeport, California, killing both pilots. 2Lt. Harold F. Mudd, 25, was killed in the crash of P-38L #44-26050; 2Lt. Albert H. Peltier, 24, was killed in the crash of P-38L #43-50241. The airplanes were part of a five-ship flight that took off from Santa Rosa Army Air Field, Santa Rosa, California, on a routine aerobatic training flight. Lt. Mudd was flying in the number-four position and Lt. Peltier was flying in the number-five position. The flight was performing aerobatics in string at an altitude of 10,000 feet in the vicinity of Clear Lake. The airplanes had performed several aerobatic maneuvers then the flight leader led the airplanes in a shallow dive, reaching 270 mph indicated airspeed. The flight leader pulled up and did a barrel roll to the left. After completing the roll the flight leader started a shallow climbing turn to the left in order to observe the individual rolls of the airplanes in the flight. The number-two and number-three airplanes successfully completed their rolls. The flight leader was maneuvering in an effort to observe the number-four and number-five airplanes when the collision occurred. The flight leader heard someone on the radio say that he was "going in." The flight leader observed Lt. Mudd's P-38 in a violent spin. Lt. Mudd's airplane spun into Clear Lake. Investigators stated, "While the number-four man was still spinning the number-five man [Peltier] was calling 'mayday' so the flight leader began trying to locate him. After locating the number-five man he observed that the left wing of Lt. Peltier's airplane was damaged and bent up at a point about four feet in from the tip at an angle of approximately 30 degrees. Lt. Peltier, who was number-five, slowed his airplane up and said that he had pretty good control with the gear either up or down at an indicated airspeed of 130 mph. The flight then told him that he would have to make the decision as to whether he was going to abandon the airplane or attempt to bring it back to the field [Santa Rosa Army Air Field]. The pilot said that he would bring the airplane back to the field so the flight started back and got clearance from the tower for an emergency landing. Lt. Peltier was on his final approach at about 4,000 feet a few miles north of the field when the flight leader saw his canopy fly off. He called [the subject pilot] asking what was wrong and the pilot replied that the airplane almost got away from him but he had it under control. He had no sooner answered when he said 'There it goes again,' and the airplane started to roll to the left. The flight leader told him to bail out but he saw no parachute before the airplane crashed [six miles east of Healdsburg, California]. The airplane went almost straight in and the pilot was found about 100 yards from the wreck with his parachute partially open as if he had bailed too late. No one in the flight saw the collision and only one witness to the accident has been found."

8-17-45. Weatherford, Texas. At 2128, two Boeing B-29 bombers collided in mid-air and crashed three miles west of Weatherford, Texas, killing 18 fliers and seriously injuring two others. B-29 #44-86276 took off from the Army Air Field at Alamogordo, New Mexico. B-29 #42-93895 took off from the Army Air Field at Clovis, New Mexico. Killed in the crash of B-29 #44-86276 were: 1Lt. Aubrey K. Stinson, pilot; 2Lt. Harold L. Swain, co-pilot; 2Lt. Gordon E. Myers, navigator; 2Lt. Benson W. Cohen, bombardier; 2Lt. Edward E. Lahniers, flight engineer; Sgt. Donald V. Lefebvre, radio operator; Sgt. Johnny A. Moseley, central fire controller; Sgt. Donald E. Reed, right gunner; Sgt. Clarence A. Jurgens, left gunner. Killed in the crash of B-29 #42-93895 were: 1Lt. Robert A. Mayer, pilot; 2Lt. Robert L. Knight, bombardier; 2Lt. John W. Burtis, navigator; F/O Robert Q. Zaleska, radar operator; SSgt. Clifford D. Longmire, engineer; Cpl. Robert H. Apirian, radio operator; Cpl. Jasper C. Wilson, Jr., gunner; Cpl. Willard A. Byerly, gunner; Cpl. Anthony J. Agliata, gunner. Co-pilot F/O Edwin F. Smith and gunner Cpl. Earl E. Wischmeier miraculously survived, parachuting to safety from B-29 #42-93895. Investigators stated, "The aircraft from Alamogordo, New Mexico [44-86276] had flown a two-hour local flight near Alamogordo, and the instructor pilot turned the aircraft over to the student airplane commander. They made out a clearance for a CFR round-robin flight from Alamogordo direct to Hudspeth to Fort Worth, Texas, [at an] altitude of 15,000 feet, IFR direct to San Antonio, Texas, at 16,000 feet, IFR to Alamogordo. The navigator's log showed that take-off from Alamogordo was made at 1808 local. They arrived at Hudspeth at 1828 on a true course of 159 degrees, arrived at Wink, Texas, 1904 on a true course of 91 degrees, arrived Big Spring, 1929 on a true course of 60 degrees, left Big Spring on a true course of 79 degrees. The last position report entered on the navigator's log was Abilene, Texas, at 1956 MWT, flying a true course of 79 degrees. Abilene is 115 miles west of Weatherford on the airways. At all times the navigator's log showed that the Alamogordo aircraft was flying at 15,000 feet indicated altitude. The indicated airspeed was 200 mph from Alamogordo, to Hudspeth and 195 mph from Hudspeth to Abilene. The ground speed was 210 mph. The aircraft from Clovis Army Air Field [42-93895] was on a routine radar-bombing mission from the home station to Fort Worth, Texas, to shoot simulated radar bombing attacks on Meacham Field from IP south of Fort Worth at 15,000 feet. The crew experienced difficulty in contacting the radar station and completed one bombing run, flying a rectangular pattern without making contact. The pilot decided to make one more run and if contact could not be established they would return to their home station. They had completed their first run and were on the downwind leg, flying a course of approximately 270 degrees at an altitude of 15,000 feet. The flight engineer noticed that his voltage regulators were not adjusted properly and while he was adjusting them the airplane commander took over the flight engineer's position. The bombardier was in the nose using a flashlight to look at his charts. The aircraft was on C-1 autopilot, with the radar operator in secondary control. The co-pilot was in the co-pilot's position watching the instruments and paying particular attention to the operation of the centering knob in order to maintain zero degrees rate of climb, altitude 15,000 feet.... The airplane commander returned to his position and shortly after that he looked up and exclaimed 'Oh my God!' The co-pilot looked up and noticed two engines and a wing of a four-engine aircraft approaching them from the right forward side at approximately a 45- degree angle. The aircraft was very close to them, and before any maneuvers could be attempted the right wing of the Alamogordo aircraft struck the right wing and number-four engine of the Clovis aircraft. The Clovis aircraft immediately caught fire. The co-pilot was thrown against the gun turret and dislocated his shoulder. The aircraft appeared to be in a spin. The co-pilot fought his way to the controls and attempted to control the aircraft, but the controls were jammed. Realizing the situation, the co-pilot opened his emergency window and climbed out of the aircraft. He was caught in the slipstream but worked his way free. The number-three propeller struck his foot and cut the top of his shoe off but did not injure him. The co-pilot pulled his ripcord and realizing that he did not give a sufficient pull attempted to open his backpack parachute by reaching behind and tearing the cover. At this time he realized that his parachute had opened and shortly after that he hit the ground. The gunner riding in the left blister of the Clovis aircraft heard the pilot state on intercom 'Oh my God!' When the crash came, not having his safety belt fastened, he was thrown free in the waist of the aircraft. He had a chest-type harness on and in the light from the fire he noticed his parachute. He picked it up and fastened it onto the harness and then attempted to break the plexiglass blister. Unable to break the plexiglass blister he succeeded in breaking the casting of the [gun] site, and being pressurized it aided in breaking the blister. He continued to kick the blister until he broke out an opening sufficient for him to get through. He freed himself from the aircraft and his parachute opened very shortly before he hit the ground. The co-pilot and the gunner from the Clovis aircraft suffered major injuries. The remainder of the crew was fatally injured. No other crewmembers of either aircraft used their parachutes. From statements of witnesses and investigation from the home station of the aircraft, it is believed that both aircraft were equipped and using blinker-type navigation lights. Both aircraft were totally wrecked."

8-18-45A. Location Unknown. At an unknown time after 1137 EWT, a North American P-51D disappeared and is presumed to have crashed. Pilot 1Lt. Francis J. Waice, 22, was declared missing and presumed killed. The pilot filed a flight plan proposing to fly from Venice Army Air Field, Venice, Florida, on a navigation flight to Philadelphia, Pennsylvania. The airplane was scheduled to refuel at Pope Field, North Carolina. After checking the weather reports, the pilot changed his flight to a CFR flight to Chatham Field, Savannah, Georgia. The airplane took off at 1137 EWT from Venice Army Air Field, but the airplane failed to arrive at Chatham Field and failed to return to Venice.

8-18-45B. Mendota, California. At 0846, a North American B-25G flying in poor weather suffered a catastrophic structural failure and crashed 22 miles south of Mendota, California, killing three fliers. The airplane took off from Porterville Army Air Field, California, on an instrument flight to Salinas Army Air Base, Salinas, California, and return. At about 0845 the pilot radioed the weather station and stated that the weather was "not too good" and he was returning to base. Moments later, witnesses on the ground observed the airplane dive out of the clouds, attempting a steep pull-up and shedding pieces. Investigation revealed that the port wing failed and separated, sending the airplane out of control. The airplane entered a violent spin, causing the starboard wing to fail and separate. The airplane smashed to the ground and exploded into flames. Investigators speculated that the pilot had lost control of the airplane while flying in instrument conditions and had stressed the airplane beyond its design limitations in an effort to re-establish level flight. Pilot F/O Arland O. Collins, co-pilot 1Lt. Gerald J. Nestor and engineer SSgt. Joseph S. Micut were killed in the crash.

8-18-45C. Gulf of Mexico, Texas. At 1430, two Republic P-47D airplanes collided in mid-air over the Gulf of Mexico about 30 miles southwest of Galveston, Texas, killing two fliers. Pilot 2Lt. Robert W. Dixon, 26, parachuted from P-47D #44-90378, but apparently drowned before he could be rescued. 2Lt. Charles A. Lane, 22, was killed in the crash of P-47D #44-32678. Both airplanes had taken off from Galveston Army Air Field on an aerial gunnery mission. Lt. Lane was towing targets for the exercise and Lt. Dixon was part of a four-ship flight that was firing on the target at 8,000 feet msl. Investigators stated, "Lt. Dixon had made several firing passes on the towed target and had [performed] in a satisfactory manner. In breaking away from his next firing pass, he evidently became engrossed in something else and rammed the towing aircraft squarely. Both aircraft burned immediately, and although Lt. Dixon was seen parachuting down, and was seen in the water apparently safe, his body remains missing.... Lt. Lane bailed out and was found in the Gulf of Mexico minus his left arm and both legs."

8-20-45. McAlester, Oklahoma. At 1038 CWT, a Boeing B-29 crashed 10 miles northwest of McAlester, Oklahoma, killing the crew of six. Killed in the crash were: Maj. Henry S. Britt, pilot; Capt. Benson H. Edwards, co-pilot; 1Lt. Carol M. Frang, flight engineer; Sgt. Melvin C. Tipton, assistant engineer; SSgt. Zander D. Bilbrey, assistant engineer; Sgt. Clifford R. Schneider, radio operator. Investigators

stated, "At 0845 [CWT] ... B-29 No. 42-24527 ... departed Eglin Field [Valparaiso, Florida]. The aircraft was on a flight test mission during which speed runs were to be made at 30,000 feet at 2,400 rpm and 43.5 inches Hg. Data was to be taken on fuel flow, airspeed, cowl flap settings and cylinder head temperature. The flight was to be conducted under Instrument Flight Rules and the plane was to complete a round-robin mission from Eglin Field to Crestview, Florida, to Oklahoma City, Oklahoma, direct at 30,000 feet, from Oklahoma City direct to Jacksonville, Florida, at 30,000 feet, then direct at 30,000 feet to Crestview for let-down and return to Eglin Field. Six hours were estimated as the time en route with nine hours of fuel aboard. The weight of the aircraft was 116,740 pounds. The aircraft was a stripped B-29 powered with special test R-3350-57A engines. The pilot was considered an excellent B-29 pilot and had flown B-29 type aircraft during all stages of its development. The engineer was a qualified flight engineer who had just returned from the 20th Air Force. At 0850 CWT, the aircraft reported over Crestview at 2,000 feet. No further reports were made by the aircraft. At approximately 1036 CWT, the aircraft disintegrated in mid-air and the wreckage fell in a four or five mile square area approximately ten miles northwest of McAlester, Oklahoma, and approximately six miles to the right of a direct course between Crestview and Oklahoma City. All crewmembers were killed. There is no evidence that there was attempted use of parachutes."

8-21-45A. Rodeo, New Mexico. At 1025 MWT, a North American B-25J and a U.S. Navy Grumman F6F-3 collided in mid-air six miles north of the Range Station at Rodeo, New Mexico, killing naval aviator Ensign Russell J. Chitwood, who fell to his death in an unsuccessful parachute jump. B-25 pilot F/O John J. Green and co-pilot F/O Patrick M. Dowling escaped injury and were able to land safely at Douglas Army Air Field, Douglas, Arizona. The B-25 took off at 0715 MWT from Douglas Army Air Field on a local proficiency flight. The airplane returned to the field at about 0815, and landed so that the pilots could change seats. F/O Green became the instrument pilot and F/O Dowling became the safety pilot. The airplane took off at 0820 and climbed to 9,500 feet. The airplane flew to the practice area between El Paso, Texas, and Tucson, Arizona, and F/O Green went under the hood to practice instrument flying. After practicing instrument flying for a couple hours, the co-pilot noticed a fighter airplane at the 2 o'clock high position, making a diving left turn toward the B-25. The co-pilot took control of the airplane and entered a shallow dive. The co-pilot observed that the fighter was continuing in the diving left turn. An instant later the co-pilot felt a sudden jar of impact. The B-25 entered a diving turn to the left. The pilots were able to regain control of the heavily damaged airplane and safely land at Douglas Army Air Field. Investigation revealed that the F6F was part of a four-ship flight that had taken off at 0936 MWT from Coolidge, Arizona, on a ferry flight to an undisclosed station on the east coast. The navy flight had originated on the west coast. Ensign Chitwood was the officer in command of the four-ship flight but had elected to allow another pilot to assume command of the formation on the leg from Coolidge to Tucson. Apparently Ensign Chitwood broke off of the navy formation and made a simulated attack on the lone B-25. The F6F struck the starboard wing of the B-25, heavily damaging the wing and aileron. The F6F went out of control and entered a high-speed diving spiral to the right. The airplane continued in this attitude until it struck the ground and exploded. The navy pilot bailed out at about 300 feet agl but he never pulled the ripcord and was killed when he struck the ground about 150 feet from the main wreckage.

8-21-45B. Pacific Ocean, California. At 1045, a North American P-51D crashed into the Pacific Ocean five miles offshore of Point Conception, California, killing 2Lt. Aubrey M. McGonigull. The subject airplane was leading a two-ship flight that took off from Santa Maria Army Air Field, Santa Maria, California, on a routine training flight. The airplanes inadvertently entered instrument conditions while attempting to climb through a hole in the overcast. The subject airplane failed to emerge when the wing ship emerged moments later. Only small pieces of wreckage and human anatomy were found on the surface. Investigators could not determine the cause of the accident.

8-21-45C. Lexington, Virginia. At 1835, a North American AT-6D crashed while attempting to take-off from a golf course at Lexington, Virginia, killing passenger Cpl. Willis A. Jetton and seriously injuring pilot 2Lt. Daniel L. Gotthilf. The airplane took off from Spartanburg, South Carolina, on a flight to Lexington. The flight originated at Tyndall Field, Panama City, Florida. Investigators stated, "At approximately [1815 EWT] ... Lt. Gotthilf ... landed without incident on the Lexington Golf Course, a distance of about one mile northeast of the Lexington Airport. Lt. Gotthilf was trying to locate the airport by circling the city and had landed on the golf course, thinking that it was the airport. Upon landing, the airplane rolled up to the sixth green, a distance of about 300 yards up a 25 to 30 degree grade from the point of initial contact with the ground. Lt. Gotthilf cut his engine and inquired of persons on the green as to the location of the airport. He was told that the airport was located about one mile southwest of his present position. Lt. Gotthilf decided to take off again and fly to the airport. The aircraft was taxied back down the hill to within 5 to 10 yards of the initial contact with the ground. The pilot went through the normal cockpit procedure, lowered flaps, ran up the engine at full throttle, held the aircraft with the brakes until full

power was obtained, then released the brakes and proceeded to take-off in the same direction he had landed, up the hill, which was a 25 to 30 degree incline and with no more than 900 feet available for his take-off run. The aircraft proceeded up the hill and, according to statements received, was airborne as it passed the sixth hole, which is located about 30 to 50 yards from a fence. Having cleared the fence, the pilot dropped the nose to gain airspeed when he noticed the cornfield in front of him. Failing to pick up airspeed, the pilot elected to pull the nose up to clear the cornfield rather than go into the field, indicated airspeed at the time being 65 mph. As a result of this sudden change in attitude of the aircraft, it dropped on the left wing. Attempts to recover from the stall were unsuccessful and the left wing dug into the ground, causing the wing to become separated from the fuselage, the engine to break loose, and the fuselage to come to rest against the base of a tree, facing 180 degrees from the take-off direction."

8-21-45D. Aguila, Arizona. At 2230 MWT, a TB-24M suffering the failure of all four engines was abandoned and crashed ten miles northwest of Aguila, Arizona, killing pilot 2Lt. Herbert Jacobi, 25. Investigators did not know why the pilot failed to leave the airplane. Crewmembers uninjured parachuting to safety were: F/O Lynn C. MacCready, co-pilot; 1Lt. William C. Bryant, navigator-rated radar operator; F/O James R. Misciagna, bombardier; F/O Theodore J. Wilson, bombardier; SSgt. James G. Crenshaw, engineer; Pfc. Charles H. Martin, engineer; Pvt. T.A. Kelly, radio operator. The airplane took off from Williams Field, Chandler, Arizona. Investigators stated, "This aircraft was approximately 45 minutes out of Williams Field [Chandler, Arizona] on a routine round-robin training flight. The aerial engineer, who has had considerable service in B-24 type aircraft, had an assistant engineer flying with him on this mission, checking him out on standard operating procedures. The engineer had asked the pilot's permission to demonstrate some emergency procedures, such as lowering the landing gear using the emergency system, etc., to the junior engineer. The pilot gave permission and stated that he would feather the number-three engine so that the electrical hydraulic pump would be the only source of hydraulic power. The engineer then went below with the junior engineer to explain procedures to him. The operator, for reasons unknown, feathered, in addition to number-three, the number-two propeller. The co-pilot stated that both engines were feathered for a very short period and that when the pilot attempted to restart the engines the AFCE failed and all four engines went out simultaneously. The co-pilot cannot remember the readings on the various engine instruments at this time.... The pilot lost about 3,000 feet in his efforts to restart the engines and at this point, being only about 3,000 feet above the terrain, ordered a bail out.... [It is the opinion of investigators] that the simultaneous failure of all four engines was in all probability caused by the inadvertent striking of the crash bar by the co-pilot." Investigators noted that B-24 co-pilots have been repeatedly warned that it was dangerous to use the crash bar as a footrest because the four ignition switches could be inadvertently turned off simultaneously by the co-pilot's foot. The co-pilot stated to investigators that he had personally fastened the pilot's parachute to his harness and thought that the pilot was right behind him when he bailed out.

8-24-45A. Eglin Field, Florida. At 0915 CWT, a North American AT-6B crashed at Eglin Field, Valparaiso, Florida, killing pilot 2Lt. John A. Sawyer and passenger Pfc. John E. Williams. The airplane took off from Eglin Field on a routine transition flight. The subject airplane encountered another AT-6 airplane and repeatedly attempted to engage it in simulated combat. The other airplane attempted to elude the subject airplane several times but was unsuccessful. The other AT-6 ended in trail of the subject airplane twice and twice tried to break away to proceed with its individual mission. The subject airplane made another pass and again attempted to engage the other AT-6. The airplanes ended up in a tight descending circle to the right. The target airplane again ended up on the tail of the subject airplane and then broke away in a roll to the right at an altitude of about 1,500 feet agl. The subject airplane continued the tight turn and stalled. The airplane fell off into a spin, describing two and a half turns of spin before the pilot effected a partial recovery. The airplane stopped spinning but smashed into a wooded area at 30-degree angle.

8-24-45B. Sells, Arizona. At approximately 1000 MWT, a North American AT-6D crashed in mountainous terrain 20 miles ENE of Sells, Arizona, killing student A/C Eugene M. Connelly. The airplane took off from Marana Army Air Field, Marana, Arizona, on a routine student solo flight. The airplane failed to return and was soon declared missing. Search planes located the airplane the next day at about 1230 MWT. Investigation revealed that the airplane suffered a loss of power while flying over mountainous terrain. The airplane stalled and entered a spin from which the student could not recover. The airplane fell into the side of a ridge at an elevation of about 5,000 feet, striking the terrain at a 40-degree angle. The airplane burst into flames upon impact.

8-24-45C. La Junta, Colorado. At 1004 MWT, a Republic P-47D crashed 17 miles northeast of the Army Air Field at La Junta, Colorado, killing pilot 1Lt. Duncan F. Matheson. The airplane had taken off from La Junta Army Air Field and had engaged a number of airplanes in simulated combat and a Lufbery circle. Investigators stated, "[The airplane] collided with the ground in a stalled position, tail first as evidenced by marks left. After the tail struck the ground, the belly section and the wing section struck in that order. After initial impact, the aircraft skidded

approximately 30 feet and apparently exploded. The empennage tore off and the cockpit section and the wings continued skidding on the ground for 150 yards and became inverted. Shortly after the aircraft exploded the supercharger and the engine were torn out and landed 50 and 75 yards respectively from the cockpit section. From witness reports it is apparent that the pilot, after the initial engagement with other aircraft, pulled away and then while flying alone pulled up in a steep climbing turn, stalled well above the minimum altitude required for the mission then fell off to the right. The recovery was almost accomplished but the aircraft was still mushing when it collided with the ground."

8-24-45D. Grenier Field, New Hampshire. At 2155 EWT, a Boeing B-17G crashed three miles southeast of Grenier Field, Manchester, New Hampshire, killing three fliers and seriously injuring three others. Killed in the crash were: F/O John E. Bafus, co-pilot; F/O Irwin J. Gingold, navigator; Sgt. Earl K. Allen, passenger. Pilot F/O Bill J. Anderson and engineer Sgt. Charles R. Jones suffered serious injuries. The airplane took off at 1852 EWT from Newcastle Army Air Base, Wilmington, Delaware, on an instrument flight to Grenier Field. The airplane arrived in the area of Grenier Field and was cleared for a standard instrument approach. The airplane missed the first approach, coming in on the right side of the runway. The B-17 was at about 700 feet agl when the pilot aborted the landing and began the missed approach turn to the right. The pilot again came in to the right on his second approach and again aborted the landing. The pilot flew the missed approach procedure and attempted a third landing. The pilot overshot the runway and aborted the landing. Instead of turning to the right for the missed approach, the pilot turned to the left in an effort to keep the field in sight. The pilot requested a 180-degree turn to the left for a visual attempt at the field. The tower denied the request and the airplane continued turning to the left. The airplane inadvertently entered instrument conditions and apparently disoriented the pilot. The airplane collided with trees and then smashed to the ground about a mile from the point of initial impact, bursting into flames upon impact. Weather at the field as reported as ceiling 1,000 feet with eight miles visibility.

8-25-45. Elgin, Oregon. At 2244, a Consolidated B-24J crashed into a mountain 17 miles NNW Elgin, Oregon, killing 15 crewmembers. Killed in the crash were: Capt. Edwin F. Zdunczyk, B-24 instructor; 1Lt. C.H. Keeler, pilot; 2Lt. R.B. Wright, co-pilot; 2Lt. G.C. Oesterreicher, navigator; 1Lt. J. Kagel, bombardier; Sgt. M.M. Pickell, radio operator; TSgt. F.G. Emmelmann, engineer; H.H. Brundahl, gunner; Sgt. B.G. Fletcher, gunner; Sgt. A.P. Lupisella, gunner; Sgt. T.R. Frazier, gunner; SSgt. R.B. Walter, gunner; Sgt. R.W. Johnson, radio operator; Sgt. Emil Eckert, engineer; Sgt. Paul E. Kleiner, engineer. The airplane took off from Sioux City Army Air Base, Sioux City, Iowa, on a "special training mission" to Walla Walla Army Air Field, Walla Walla, Washington. Investigators stated, "[The pilot] became confused as to his exact position while over LaGrande, Oregon. While flying Contact Flight Rules at night, the town of LaGrande looked similar to Pendleton [Oregon], and consequently Capt. Zdunczyk took up a heading of approximately 10 degrees in order to reach Walla Walla. Ten degrees is the heading from Pendleton to Walla Walla, and is not the heading of LaGrande to Walla Walla. He let down to an altitude of approximately 5,200 feet, and while flying under an overcast, he crashed into a mountain about 28 miles south-southeast of Walla Walla."

8-27-45. Indian Springs, Nevada. At 1430 PWT, a Bell TP-63A crashed 20 miles northwest of Indian Springs, Nevada, killing pilot 1Lt. Fay W. Maxion. The airplane had taken off at 1407 PWT from Indian Springs Army Air Field, Indian Springs, Nevada, on a routine training flight. The pilot encountered a flight of four P-63 airplanes and joined onto their formation. After flying with the formation for a couple minutes, the flight leader radioed the subject airplane and asked him to break off of the formation because he was hindering the flight's maneuvers. The subject pilot acknowledged and left the formation. Several minutes later, pilots in the flight heard the subject pilot radio that he was "going in." The flight leader reported the incident to the Las Vegas tower. Tower personnel attempted to contact the subject airplane by radio but were unsuccessful. A search was conducted immediately and the airplane wreckage was found a short time later. Investigation revealed that the airplane struck the ground while in a spin to the left. The starboard escape door was missing and the pilot's safety belt was unfastened, indicating that the pilot was contemplating a bail out.

8-29-45A. Luke Field, Arizona. At 1250 MWT, a North American P-51D crashed at Luke Field, Phoenix, Arizona, killing Chinese Air Force pilot Sub-Lt. Kuo-Zee Tien. Investigators stated, "Sub-Lt. Tien was returning from a formation flight and was landing on the concrete runway to the southwest, after making one previous attempt at landing in which he had been sent around by the Control Officer. His approach was normal in every way with the glide being broken a little high but very slowly so that the aircraft had settled to the ground gradually, assuming a three-point attitude. As the aircraft came over the end of the runway approximately five feet in the air, the aircraft settled to a solid three-point landing. The aircraft was landed straight and did not bounce. The power was off at this time. The plane rolled straight ahead for a very short distance and the Control Officer, who was approximately 100 yards away in the Mobile Control Unit, heard the engine surge and the aircraft left the ground immediately. The airplane went up in a very

nose high attitude to approximately 40 feet in the air and started to turn slightly to the left. The Control Officer instructed the pilot to 'put the nose down' and almost immediately the aircraft rolled to the left to almost an inverted position and struck the ground. The aircraft slid along the ground inverted for approximately 85 yards, stopped and burst into flame after a few seconds. The fire was quickly extinguished but the pilot had been killed upon impact."

8-29-45B. St. Andrews Sound, Georgia. At approximately 2245 EWT, a North American TB-25J crashed into St. Andrews Sound about 10 miles SSE of Brunswick, Georgia, killing three fliers. The airplane was part of a 13-ship flight that took off from Turner Field, Albany, Georgia, on a student night navigation flight to Tallahassee, Florida, to Jacksonville, Florida, to Savannah, Georgia, to Charleston, South Carolina, and return to Turner Field. At about 2245 EWT, other pilots in the flight observed the subject airplane drop out of formation, descending rapidly. The airplane failed to rejoin the formation and failed to return to the field. Wreckage from the airplane was found after dragging an area of an oil slick found on St. Andrews Sound on 9-1-45; a leg with a shoe and a sock attached was found on the surface that same day. A flier's head washed ashore at Cumberland Island on 9-2-45. Killed in the crash were: 2Lt. James K. Haag, B-25 instructor; A/C Virgil E. Kyser, student pilot; A/C Weldon L. Holsinger, student pilot.

8-31-45A. Leesburg, Georgia. At 1030 CWT, a North American TB-25J suffered a catastrophic structural failure and crashed eight miles north of Leesburg, Georgia, killing instructor pilot 2Lt. Herman J. Althoff and B-25 student A/C Joseph M. Peret. The B-25 took off at 0900 CWT from Turner Field, Albany, Georgia, on a student instrument flight. Investigation revealed that the outer panel of the starboard wing had failed and separated, striking the tail section and causing its progressive failure. The airplane fell into an inverted spin and smashed to the ground in a vertical position. The starboard wing panel and pieces of the tail section were found about a half-mile from the main wreckage. Witnesses on the ground heard the airplane in a dive until it plunged out of the overcast at a steep angle. The airplane entered a steep pull-up maneuver and rolled over on its back. The pilots evidently stressed the airplane beyond its design limitations in an effort to re-establish level flight. The starboard wing separated and struck the starboard side of the tail section, causing the entire tail section to separate. The airplane entered an inverted spin, shedding pieces as it spun to the ground. The airplane smashed to the ground and exploded into flames.

8-31-45B. Huntsville, Alabama. At 1100 CWT, a North American TB-25J crashed while attempting to take-off from the Army Air Field at Huntsville Army Arsenal, Huntsville, Alabama, killing engineer TSgt. Richard D. Welty. Nine crewmembers were uninjured. The airplane had been cleared to fly a training mission to Syracuse, New York, and return. The pilot began the take-off roll and just prior to becoming airborne had rolled on some elevator trim. The acting co-pilot, a rated bombardier, mistook the pilot's hand movement for the signal to retract the landing gear. The acting co-pilot activated the landing gear switch and the landing gear began retracting. The speed was about 65 mph. The airplane began to settle back to the runway, skidding straight ahead for about 2,000 feet, stopping on the runway and suffering major damage. A piece of propeller blade broke off and entered the fuselage, killing the engineer. Pilot 2Lt. Richard L. Sanderfur and acting co-pilot/rated bombardier 1Lt. Robert Lane were uninjured. Passengers uninjured in the crash were: 1Lt. John T. Amsden, Sgt. Morton H. Friedman, Cpl. Roger E. Lussier, Sgt. Alfred S. Eichenbaum, Sgt. John J. Smith, Sgt. Orlando A. Beltrante, Sgt. Leonard J. Leitner.

8-31-45C. San Antonio, Texas. At 1445 CWT, a North American AT-6C crashed 12 miles west of Randolph Field, San Antonio, Texas, killing pilots 2Lt. Ingram W. Varnell and 2Lt. Henry L. Steele. The airplane took off from Randolph Field on a routine training flight. A pilot flying in the vicinity of Alamo Field, San Antonio, observed the subject airplane in a spin at an altitude of about 7,000 feet agl. The subject recovered in a half roll, and then entered a vertical dive toward the ground. The airplane gained excessive speed as it fell and the pilot attempted a recovery at about 4,000 feet but was unsuccessful. The airplane entered a violent spin, describing about seven or eight turns of spin before the pilot stopped the spin and attempted a recovery. Pieces of the airplane were observed falling away. The airplane went out of control and smashed to the ground, bursting into flames upon impact.

September

9-1-45A. Hondo, Texas. At 0556, a Consolidated B-24J crashed at Hondo Army Air Field, Hondo, Texas, killing 12 passengers and crew and seriously injuring three others. Investigators stated, "[The airplane] took off on Runway 35 from Hondo Army Air Field at 0554 CWT on a Pilot-Navigator Proficiency cross-country flight, having been cleared CFR direct to Newark Airport, Newark, New Jersey. Take-off was normal, and after gaining an altitude of approximately 100 to 150 feet the plane seemed to settle in. First hitting an obstruction light 40 feet high approximately 2,500 feet from the end of Runway 35 and then a tree approximately 100 feet further on. From then on parts of the plane were found over a distance of approximately 300 feet from where the main part of the wreckage settled. No other evidence was found to bear out navigator's statement that an explosion occurred prior to the crash of the aircraft. Investigation revealed that there were two unauthorized passengers aboard." Investigators noted that no more than thirteen passengers and crew were permitted on board B-24 type airplanes. Killed in the crash were: 1Lt. Marion O. Nelson, pilot; Capt. Billy S. Warren, co-pilot; 1Lt. John A. McGrane, navigator; 1Lt. Elmer J. Murray, navigator; Pfc. Kenneth P. Palmer, engineer; TSgt. Stanley J. Rembisz, engineer; Cpl. Louis S. Ferdinand, passenger; Pvt. Barbara J. Hogan, WAC passenger; Cpl. Michael Lechus, passenger; Pfc. George A. Lowry, unauthorized passenger; 2Lt. Timothy F. Murnane, passenger; Maj. Harvey G. Wible, passenger. Seriously injured in the crash were: Cpl. Irwin L. McElliott, passenger; Pfc. Charles G. Ronkos, passenger; TSgt. Roland G. Thibault, unauthorized passenger.

9-1-45B. Richmond, Virginia. At 1100, two Republic TP-47D airplanes collided in mid-air and crashed one mile west of Richmond, Virginia, killing F/O Leon C. Kirk, 20, aboard TP-47D #44-90418. F/O George A. Patterson, 20, was seriously injured in the collision but was able to parachute from TP-47D #42-75467. The airplanes were part of a 16-ship flight that took off from Seymour-Johnson Field, Goldsboro, North Carolina. The airplanes were part of an interception/escort exercise made up of four, four-ship flights. One P-47 flight portrayed a flight of bombers; two flights were the bomber escort, and one flight portrayed enemy fighters. Both subject airplanes were the number-four ships in their respective flights. The flights maneuvered for a short time over Rocky Mount, North Carolina. The flights were badly scattered by the time they reached the Richmond area, which was the bombers' "target." The flight commander ordered the flights to reform, which they began to do while in a gentle left turn at about 8,000 feet msl. The two number-four ships were out of position. F/O Patterson was to the rear and the left of his flight and F/O Kirk was to the rear and the right of his flight and on the outside of the turn. One flight crossed underneath the other and the two subject pilots, who were attempting to rejoin on their respective wingmen, did not see the other and their airplanes collided. F/O Kirk's airplane went out of control and smashed into a grove of trees at a very steep angle, exploding into flames upon impact and killing the pilot instantly. F/O Patterson was violently hurled from his disintegrating airplane and was still strapped to the wreckage of his seat when he parachuted to the ground.

9-1-45C. Columbus, Ohio. At 1353 EWT, a North American AT-6D crashed seven miles northeast of Lockbourne Army Air Field, Columbus, Ohio, killing 1Lt. Donald M. Joy and pilot-rated passenger Capt. Walter R. Peck. Investigators stated, "The aircraft took off from Lockbourne Army Air Field at 1340 EWT on a routine training flight. The aircraft ... was seen shortly after take-off to be engaged in a series of dives and zooms, characteristic of lazy-eights or chandelles, two miles south of the city of Columbus. [The airplane] was seen by observers to start a rather steep dive from an estimated height of 5,000 feet and shortly thereafter appeared in a shallow dive, wings level, traveling at a high rate of speed at an estimated 700 feet above the ground. A few seconds later, the plane began a shallow right turn, tightened the turn, shuddered slightly, and began to descend rapidly, right wing down. As it neared the ground the engine started to roar and the low wing and nose lifted slightly, but the right wing struck the ground in the center of a cornfield and the airplane bounced and settled back to earth in a cloud of dust. The pilot, Lt. Joy, was thrown out of the wreckage; the right wing was demolished, the left wing torn off and the engine broke free from its mount and rolled across a road. The wreckage was scattered over a distance of 350 yards from the point of impact.

9-3-45. Douglas, Arizona. At 2254 MST, a North American TB-25J crashed three miles east of the Army Air Field at Douglas, Arizona, killing students A/C George W. Faust and A/C Vernon A. Witten. The airplane took off from Douglas Army Air Field on a routine training mission and immediately after becoming airborne entered a descending turn to the right. The airplane remained in this attitude until it flew into the ground and exploded. Investigators could not determine the cause of the accident.

9-6-45A. Wintersburg, Arizona. At 1245 MST, a Curtiss P-40N crashed ten miles northeast of Wintersburg, Arizona, killing Chinese Air Force pilot Sub-Lt. Kuo-Cheng Yu, 23. The airplane took off from Luke Field, Phoenix, Arizona, at 1130 MWT, flying as the leader of the second element of a four-ship flight on a scheduled oxygen mission. The airplanes joined up at about 15,000 feet. Sub-Lt. Yu signaled his wingman that he wanted him to take over the lead because of an unspecified problem. The flight leader tried to contact the subject airplane by radio but was unsuccessful. The airplanes flew in a tactical formation and performed several aerobatic maneuvers while in trail formation. The flight leader looked back and noticed that the subject airplane had dropped out of the flight and was nowhere to be seen. The flight returned to Luke Field and landed. The subject airplane failed to return and a search was commenced. Civilians found the wreckage on 2-25-46. Investigators could not determine the cause of the crash.

9-6-45B. Pearson Corners, Delaware. At 1700, a Republic P-47N suffered a catastrophic structural failure and crashed at Pearson, Corners, Delaware, killing pilot 2Lt. Benjamin M. Duff, 23. Investigation revealed that the airplane was performing aerobatics at high altitude. The airplane was seen to enter a steep dive and then two "muffled explosions" were heard. The starboard wing and horizontal stabilizer failed and separated. The airplane then entered a flat spin, remaining in this attitude until it struck the ground nine miles northwest of Dover Army Air Field,

exploding into flames upon impact. Investigators noted that the pilot was scheduled to be separated from the service in a few days and was trying to make up about eight hours of flight time in an effort to earn his flight pay.

9-7-45A. Bartow, Florida. At 1040, a North American P-51D crashed at Bartow Army Air Field, Bartow, Florida, killing pilot 2Lt. Howard A. Brandon, 25. Investigators stated, "Lt. Brandon was number-three man in a four-ship flight returning from a medium altitude formation mission in the local flying area. Landing instructions were received by the flight and they were to land on Runway 5. This is a black top, 5,000' runway with a slight rise in the center of the field such that the runway control unit is not able to view a plane in its entirety on the extreme end. The runway control unit is located 1,000' from the beginning and 150' to the left of the landing runway. The flight peeled off and set up a normal landing pattern. The leader landed and turned off at the intersection to observe the landings of his flight members. The number-two man elected to go around and when the runway control officer observed the number-three man, Lt. Brandon, he appeared to be in a slightly nose high attitude and to be holding a slightly low airspeed at about 300 feet on his final approach. He called the pilot and instructed him to lower his nose and add throttle to [get] more airspeed. The pilot made no reply and continued his approach as before. He landed very hard about 1,500' down the runway. The aircraft bounced and the pilot added a burst of throttle, landing again. He veered to the left, recovered, and then started toward the right at the intersection of Runway 9. The aircraft skidded, then held, turning approximately 40 degrees right, heading down Runway 9. He was carrying some throttle all the time. He then decided to go around and gave it full throttle. With the application of throttle, the aircraft turned approximately 45 degrees to the left, paralleling Runway 5, and approximately 200 feet to the right of the runway. His tracks showed that the aircraft was in an almost straight path. He became airborne approximately 50 yards before the end of Runway 5, climbed to about 30 or 40 feet and then the left wing stalled, causing the aircraft to roll to the left. The left wing tip struck the ground when the aircraft was in a 90-degree bank, then the roll continued with the propeller striking next. The aircraft landed inverted and remained in that position, with the tail turning slightly to the right. The propeller and engine were torn loose from the aircraft. Flames started immediately from a broken gasoline line near the cockpit. It is believed that the pilot was killed instantly when the aircraft struck the ground." Investigators noted that the pilot had flown only 60 hours from the time of his graduation in February of 1945 and was due to be separated from the service in a short time.

9-7-45B. Fort Worth, Texas. At 1405, gunner Sgt. Roy E. Rankin fell to his death at Fort Worth, Texas, when he bailed out at low altitude from of a Consolidated TB-32 that was suffering a fire in the number-three engine. The airplane took off from Fort Worth Army Air Field on an engineering test flight. The airplane suffered a fire in the number-three engine just as it became airborne. The pilot continued the take-off and climbed the airplane to about 300 feet, entering the traffic pattern so he could go around and land. The port scanner called the pilot on the interphone and asked if he should bail out. The pilot replied no. The starboard scanner pulled the emergency hatch and bailed out at about 250 feet agl without permission. The gunner's parachute did not fully deploy before he struck the ground. The pilot was able to make a safe emergency landing at Fort Worth Army Air Field. The B-32 was piloted by Capt. Winston C. Rice and co-pilot Capt. Charles E. Peretti. The pilots and four crewmembers were uninjured. The engine fire was caused by a fuel line clamp coming loose, allowing raw fuel to spill onto hot exhaust ducting in the engine nacelle.

9-7-45C. Wedowee, Alabama. At 1930, a Boeing B-29B crashed four and a half miles south of Wedowee, Alabama, killing the crew of ten. Civilian farmer John Heath was seriously injured by flaming debris when the airplane slammed into a cornfield very near where he was standing. The airplane took off from Gulfport Army Air Field, Gulfport, Mississippi, on a 13-hour navigation flight. The airplane was observed flying at about 9,000 feet when a small fire was observed on the port wing. Civilian witnesses stated that they had observed two puffs of smoke near the port wing just before observing flames. The airplane then entered a spin to the left before entering an 80-degree banked spin toward the ground. The airplane remained in this position until it struck the ground on the port wing, causing the airplane to cartwheel into a cornfield and explode in flames. All on board were killed instantly and had made no effort to bail out. Killed in the crash were: Capt. Mason L. McCormick, B-29 instructor; Capt. Robert T. Hauman, pilot; 1Lt. Thomas F. Shaughnessy, co-pilot; 2Lt. George G. Carras, navigator; 2Lt. Donald M. Coulter, bombardier; F/O Richard G. Jefferson, flight engineer; 1Lt. Robert G. Seilkop, flight engineer; Cpl. Charles F. Peck, radio operator; Cpl. Ralph F. Kellery, crewmember; Sgt. John D. April, crewmember.

9-10-45A. Ajo, Arizona. At 0635 MST, a North American AT-6B crashed at Ajo Army Air Field, Ajo, Arizona, killing pilot 2Lt. James B. Baker. The airplane was part of a six-ship flight that had taken off from Ajo Army Air Field on a scheduled formation-training mission. The six ships were to form up and then join another six-ship formation. The subject flight took off, made two climbing 90-degree turns and then began forming up at about 1,100 feet agl. The flight leader looked back and noticed that the subject airplane, which was flying in the number-six position, was missing. Moments later the flight leader noticed a cloud of dust and smoke on the ground where the airplane

crashed. Investigation revealed that the subject airplane contacted the ground on its main wheels. The airplane bounced and then bounded through the air for about 50 yards before smashing into the ground and flipping over to an inverted position. The airplane skidded along the ground on its back, killing the pilot. The aircraft was completely destroyed. Investigation of the wreckage indicated that the landing gear and flaps were in the extended position. It was speculated that the pilot had attempted an off-field emergency forced landing.

9-10-45B. Adamsville, Texas. At 1600, a North American AT-6C crashed four miles south of Adamsville, Texas, killing F/O William T. Willy. The airplane took off at 1527 from Perrin Field, Sherman, Texas, on a routine navigation flight to Aloe Field, Victoria, Texas. The pilot took off and flew directly to the area of Adamsville, which was about 80 miles off his intended course. The pilot flew over his uncle's house about three miles east of Adamsville. The pilot flew toward the uncle's house in an easterly direction and then executed a left turn and returned over the house from the north. The pilot dropped a necktie and pair of shoes, with a note attached, from the airplane as he passed over the uncle's house. The shoes and tie landed within 50 yards of the uncle's house. The pilot then flew over his father's house, which is located about three miles southwest of Adamsville. The pilot attempted a tight turn at low altitude, stalling the airplane. The airplane entered a spin at an altitude too low to allow recovery, smashing to the ground and bursting into flames about 300 yards southwest from the father's house.

9-11-45. Selma, Texas. At 1747, a North American AT-6C crashed at Davenport Auxiliary Airfield, Selma, Texas, killing pilot F/O Leander C. Stedman, Jr. The airplane took off from Randolph Field, San Antonio, Texas, on an unauthorized flight sometime after 1600 CWT. A pilot of a Beech C-45, which was flying in the area at 5,000 feet, observed the airplane climbing toward his airplane. The C-45 pilot finally had to take evasive action to avoid a collision. The C-45 pilot made a violent 90-degree bank to the right to avoid a collision, waking up a colonel sleeping in the back. Passengers on board the C-45 observed the AT-6 enter a chandelle type maneuver, leveling off at about 5,500 feet. The AT-6 then entered a steep dive, remaining in this attitude until it struck the ground at Davenport Auxiliary Airfield, exploding into flames upon impact and killing the pilot instantly. Passengers aboard the C-45 stated that the pilot of the AT-6 made no apparent effort to pull out of the dive. Investigators noted that the pilot had been suspended from flying on 7-26-45.

9-12-45. Little Rock, Arkansas. At 1552, a Boeing TB-17F crashed while attempting a landing at Adams Field, Little Rock, Arkansas, killing three civilians on the ground and injuring two others, one of them seriously. Three fliers escaped injury. The airplane had taken off from Ardmore, Oklahoma, on a ferry flight to Little Rock. The airplane arrived in the area of Little Rock and let-down through instrument conditions, breaking out under the overcast over the southeast edge of the field, heading northwest. The airplane did not contact the tower for landing instructions. The tower issued landing instructions for Runway 14 to the airplane without receiving a reply. The airplane attempted a landing on Runway 32 but was unable to line up. The pilot abandoned the landing, pulled up and retracted the landing gear, circling to the right for a go-around. Control tower personnel radioed the correct landing instructions two more times to the aircraft as it circled to the right. The subject airplane did not respond and then attempted a landing on Runway 32. Control tower personnel gave the airplane the red light signal to stop the landing attempt but the ship continued the landing, which was downwind. The red light was held on the airplane continuously during the final approach and landing instructions were again broadcast. The airplane touched down 800 feet from the approach end of the runway, running off the surface to the right. The airplane ran off the runway, knocking down a fence before leaving the airfield grounds and smashing into a civilian residence. The airplane demolished the three-room frame house, killing three occupants and injuring two others. Mother Alberta Wright, son Desmark Wright, Jr., 6 months, and stepson Harold Evans, age 12, were killed. Juveniles Irvin Lee Wright received serious injuries and Bobby Wright sustained minor injuries. Pilot Capt. Richard L. Launder, co-pilot 1Lt. Edward G. McLean and engineer Cpl. Willie W. Allen escaped injury. Investigators noted that the pilot had violated several AAF and CAA regulations during his let-down and landing attempt.

9-13-45A. Gulf of Mexico, Texas. At 1115, a North American P-51D crashed into the Gulf of Mexico five miles NNW of the Army Air Field at Matagorda Peninsula, Texas, killing pilot 1Lt. Frank A. Reid, 29. The airplane had taken off from Foster Field, Victoria, Texas, on an aerial gunnery mission. The airplane rendezvoused with the target-towing airplane at 4,000 feet about eight miles from Matagorda Army Air Field and began the gunnery mission. The pilot shot off the target, which was being towed by a Martin B-26 type airplane, and was unable to avoid colliding with the target. The pilot of the B-26 attempted to warn the P-51 pilot that the target had broken free but was unsuccessful. A steel bar that was part of the target frame smashed through the P-51 cockpit windscreen and apparently incapacitated or killed the pilot. The airplane went out of control and smashed into shallow water, bursting into flames upon impact.

9-13-45B. Atlantic Ocean, Florida. At 1415, a Vultee BT-13B crashed into the Atlantic Ocean one mile off shore of Miami Beach, Florida, killing pilot 2Lt. Charles F. Ramlow and passenger Sgt. Walter L. Smith.

The airplane was observed maneuvering over the sea at about 500 feet. The airplane went into a dive to about 200 feet and then entered a climbing turn to the left, leveling off at about 500 feet. The airplane stalled and entered a spin to the right, smashing into sea and killing the occupants instantly. The airplane had taken off from 36th Street Airport, Miami, Florida, on a local flight.

9-14-45. Glasgow, Virginia. At 2145, a North American AT-6C crashed into a mountain at Glasgow, Virginia, killing pilot 1Lt. Robert J. Kuebler and passenger Sgt. Bert Batts. The airplane was one of a flight two that had taken off at 1900 CWT from Nashville, Tennessee. The flight had originated at Aloe Field, Victoria, Texas, with Boston, Massachusetts, as the final destination. The flight was cleared to fly 500 feet "on top" of clouds on the leg to Knoxville, Tennessee. The flight was to contact Knoxville for further instructions. When the airplane arrived in the area, the pilots were unable to contact Knoxville and also unable to contact Tri-City Radio. The airplane soon entered instrument conditions. The flight attempted to get underneath but was unsuccessful. The airplanes then climbed to 6,000 feet indicated altitude and flew toward Roanoke. The flight encountered instrument conditions and very turbulent air. The airplanes became separated and the subject airplane was next seen near Glasgow. The subject airplane was observed flying at low altitude with landing lights illuminated, heading west-southwest. The airplane made a 270-degree turn and then began heading down the James River to the vicinity of the powerhouse. The valley was surrounded by 3,300-foot mountains, which were concealed by darkness. The airplane passed over the dam, then made a climbing turn to the right and smashed into the side of a mountain, which was obscured by heavy clouds. The airplane exploded into flames and the occupants were killed instantly. Investigators speculated that the pilot was lost and unaware of the rising terrain. The accompanying airplane had climbed to 7,000 feet and then made a 90-degree turn, eventually picking up the Roanoke radio range.

9-15-45. Fairfax Field, Kansas. At 0008 CWT, a Douglas C-47B attempting a take-off crashed one mile north of Fairfax Field, Kansas City, Kansas, killing 23 passengers and crew. Passenger SSgt. O.D. Delong miraculously survived, receiving serious injuries. Pilot 1Lt. Warren E. Derrickson, co-pilot 1Lt. James E. Wuest and engineer Pfc. Elbert P. Keziah were killed in the crash. The airplane was taking off on a passenger flight to Lowry Field, Denver, Colorado. The airplane took off on Runway 35, becoming airborne about halfway down the runway. The airplane cleared a 20-foot dike that was located near the end of the runway but failed to gain any significant altitude after that. The airplane then collided with tall trees on the north bank of the Missouri River. The airplane passed through the tops of a stand of willow trees for 50 feet before the starboard wing collided with a thick oak tree. The impact severed the wing outboard of the starboard engine nacelle. The airplane continued through the treetops until the port wing collided with a large cottonwood tree, shearing off the wing. The fuselage and center wing section passed through the trees and smashed to the ground on the raised roadbed of a double railroad track. The airplane broke up and burst into flames, scattering flaming wreckage and dismembered bodies over a large area. The force of the impact with the railroad embankment caused the rails to shift four inches out of line. Passengers killed in the crash were: Pfc. G.G. Brechin, 1Lt. C.J. Cody, T/5 W.H. Cory, Cpl. R.L. Dixon, SSgt. N.B. Doty, Sgt. F.C. Ebert, SSgt. P.P. Egan, T/4 J. Fleischer, Pfc. C.C. Gray, Pfc. A.O. Hanson, Pfc. C.H. Haslam Jr., TSgt. M.F. Hornor, Pfc. W.O. Lem, 1Lt. R. Martin, SSgt. H.H. Mattos, Cpl. A. Pagni, Maj. E.A. Sheridan, Pfc. J.C. Tomasini, SSgt. B.C. Tucker, T/5 W.B. Winchester.

9-16-45. Louisville, Nebraska. At 0945, a North American AT-6A crashed two and a half miles north of Louisville, Nebraska, killing pilots 2Lt. John G. Widney and 2Lt. John C. Whiteman. The airplane took off from Lincoln Army Air Field, Lincoln, Nebraska, at approximately 0730 on a routine training flight. The airplane was observed flying from the west down the Platte River at an altitude of about 1,250 feet agl. The airplane then entered a steep bank to the left, stalled, fell off on a wing and rolled over. Investigation revealed that the airplane slammed to the ground in an inverted dive, striking the highway at an angle of about 70 degrees. The airplane burst into flames upon impact, killing the occupants instantly and scattering wreckage over an area of 250 feet.

9-17-45A. Black, Missouri. At 1300, a Republic P-47N crashed at Black, Missouri, killing 1Lt. Robert E. Saunders, 21. The airplane, which was stationed at Romulus, Michigan, was part of a two-ship flight that took off at 1210 CWT from Evansville, Indiana, on a flight to Independence, Kansas. The flight climbed to 10,000 feet over the Ozark Mountains. Fifteen minutes later the subject airplane suffered engine trouble. The pilot descended in an effort to make an emergency forced landing. The pilot misjudged his glide and distance and undershot the field he was aiming for. The airplane struck a small hill about 150 yards short of the chosen field.

9-17-45B. Langford Lake, California. At 1235, a Lockheed P-38L crashed at Langford Lake Rocket Range 30 miles northeast of the municipal airport at Daggett, California, killing 1Lt. Kenneth R. Frost. Investigators stated, "[The pilot] was leading a flight of four P-38s on a Rocket Firing Mission at Langford Lake Rocket Range. Lt. Frost made one dry run on the target and was warned by the range officer that his pass was too low, the airplane clearing the ground approximately 100 feet after the pull-out. The rocket pattern is flown at 6,200 feet (which is 4,000

feet above the ground), the pull-out being effected at 1,500 feet. On Lt. Frost's second dry run, the pass was made at a 1,000-foot slant range (distance from the target) at an angle of 45 degrees, the pull-out being started at approximately a 700-foot altitude from 375 to 400 miles per hour. When attempting a rapid pull-out, Lt. Frost apparently entered into a high-speed stall, the left wing dropping slightly and striking the ground first. This cartwheeled the airplane, causing it to explode, fatally injuring the pilot."

9-18-45. Elko County, Nevada. At 0330 MWT, a Consolidated B-24J crashed in Elko County 38 miles north of Deeth, Nevada, killing seven fliers. Three fliers were able to parachute to safety, suffering only minor injuries. The airplane took off at 2025 MWT (9-17-45) from Gowen Field, Boise, Idaho, on a navigation flight to Phoenix, Arizona, and return. The airplane climbed to 13,000 feet and successfully navigated to Phoenix, arriving at 0020. On the outbound leg the airplane encountered a severe downdraft near Wheeler Peak, losing about 1,500 feet during 30 seconds of uncontrolled descent. The remainder of the flight was uneventful until the airplane reached the area near Deeth. The airplane encountered some "weather" and the pilot ordered the crew to fasten oxygen masks so that he could climb above the clouds. The airplane climbed to about 17,000 feet when the number-one engine fuel pressure dropped to zero and the engine quit. Seconds later the number-two engine fuel pressure dropped to zero and the engine also quit. The co-pilot ordered the crew to bail out over the intercom and by activating the bail-out bell. Only four men were able to bail out before the airplane went out of control and crashed to the ground, exploding into flames upon impact and killing the occupants instantly. Killed in the crash were: 1Lt. William P. Bordeker, pilot; 2Lt. William F. Carter, III, co-pilot; Sgt. David C. Boswell, gunner; Cpl. Harold W. Johnson, gunner; Pvt. Robert L. Jordan, gunner; Cpl. Bernard E. Crawford, gunner. Bombardier 2Lt. David M. Baker, who bailed out first, fell to his death in an unsuccessful parachute attempt. Engineer TSgt. Garnett E. Gayle, radio operator Sgt. Robert A. Blackburn and navigator 2Lt. Raymond A. Begley received minor injuries parachuting to safety. Investigation revealed that the port main fuel valve failed and caused the double engine failure.

9-19-45. Norfolk, Virginia. At 0930, a Republic TP-47D crashed five miles west of the Army Air Base at Norfolk, Virginia, killing pilot 1Lt. Duke A. McLeod, 21. The airplane was part of a four-ship flight that had taken off from Seymour-Johnson Field, North Carolina, on a navigation flight to Norfolk Army Air Base, to Richmond, Virginia, and return. The airplanes were flying at 3,000 feet south of Norfolk when they encountered lowering ceilings. The flight dropped down to about 1,200 feet and was flying near the Norfolk Naval Air Station when the subject airplane, in the number-two position, left the formation in a descending right turn. The flight leader began a turn in an effort to keep the subject airplane in sight. No pilot in the flight observed the subject airplane after it peeled off. A few minutes later an explosion was observed on the ground. The flight leader observed that it was the subject airplane. Investigation revealed that the subject airplane made four or five low-altitude passes over Oakdale Farms section of Norfolk. The airplane pulled up in a steep chandelle to the left and stalled out at the top. The airplane fell off on the port wing and dove toward the ground. The airplane struck the ground in a steep bank to the left. The airplane cartwheeled into the ground and exploded into flames. Investigation revealed that the pilot's mother's house and his sister's house were located in the vicinity of the crash.

9-21-45A. Inez, Texas. At 1135, two North American AT-6C airplanes collided in mid-air and crashed four miles west of Inez, Texas, killing 2Lt. Thomas J. Schroeder, 22, aboard AT-6C #42-44344. 2Lt. Arthur J. Steele, 21, received minor injuries parachuting to safety from AT-6C #42-44350. The airplanes took off from Foster Field, Victoria, Texas, on a separate transition flights. Investigators stated, "Lt. Steele stated that he and Lt. Schroeder had arranged to fly formation on this particular mission. After flying formation at 6,000 feet for approximately 15 minutes, Lt. Steele started to do trail acrobatics. After doing trail acrobatics for a short time with Lt. Steele as the leader, Lt. Steele, due to a runaway propeller, gave Lt. Schroeder the hand signal to break formation and flew in the direction of Foster Field. After a short time, the propeller of Lt. Steele's aircraft again functioned normally. He stated that after clearing the immediate area, he did acrobatics by himself. After five to ten minutes of acrobatics [the pilot] completed a barrel roll and was doing a chandelle to the right when the accident occurred. In the chandelle, Lt. Steele saw an aircraft coming up from underneath his aircraft, but at such a close range he had no chance for evasive action except to pull the control stick back. Lt. Steele stated that his aircraft was struck from below causing the engine of his aircraft to stop. Both airplanes went into a dive. Lt. Steele said that he bailed out successfully at 3,500 feet. The pilot of the other aircraft, Lt. Schroeder, failed to bail out, resulting in his death.... The two aircraft struck the ground approximately 600 feet apart and were found completely disintegrated."

9-21-45B. Palmdale, California. At 0950, a North American AT-6D crashed 15 miles WSW of the Army Air Field at Palmdale, California, killing 1Lt. Walter W. Dirris, 25. The airplane took off from Burbank, California, on a flight to Muroc Flight Test Base, Muroc, California. The airplane failed to arrive and a search was commenced. The airplane was found at 1720 on 9-22-45. Investigation revealed that the airplane smashed into rising terrain while in a flying attitude. Investigators stated, "It is believed that the pilot was attempting to fly through the pass under the clouds instead of going over the top.... It is believed that a [visual] flight could have been possible if the pilot had followed the pass the entire distance but it appears that he tried to cut across the last ridge instead of going around it [and] through the pass."

9-22-45A. Tombstone, Arizona. At 1232 MWT, a North American TB-25D crashed five miles southeast of Tombstone, Arizona, killing pilot 1Lt. Irving B. Ahrens and navigator 1Lt. Harry S. Abinanti. The airplane took off at 1130 MWT from Douglas Army Air Field, Douglas, Arizona, on a routine instrument navigation flight. The airplane was observed flying straight and level and then was seen to go into a steep dive. The airplane remained in a diving attitude until it struck the ground and exploded into flames. Both occupants were killed instantly. Investigators could not determine the cause of the accident.

9-22-45B. Freeman Field, Indiana. At 1402 CWT, a Focke Wulf FW-190D (Foreign Equipment #119) crashed near Freeman Field, Seymour, Indiana, killing pilot 1Lt. William V. Haynes, 20. The airplane had been captured from the Luftwaffe and was shipped to Newark, New Jersey, where it was assembled and test flown. Lt. Haynes flew the airplane to Freeman Field on 9-13-45. Investigators stated, "[Lt Haynes] took off in FW-190D (long nose) F.E. #119 at 1400 CWT to perform an Administration Demonstration Flight as part of a display for the Institute of Aeronautical Sciences.... The take-off was ... on Runway 13. After completing a fairly steep climb the aircraft was placed in level flight at an altitude estimated variously from 800 to 2,000 feet by statements of eyewitnesses. The aircraft maintained approximately the same altitude at which it leveled off and established a normal left-hand pattern around the field. At the end of the base leg, and in a position [that] would have been proper for a turn onto the approach leg for a landing, the airplane performed a maneuver commonly known as a wingover in a very steep dive ... and apparently with the power on. The pilot attempted a pull-out immediately but the aircraft lost altitude so fast that it was obvious that a crash was inevitable. Contrails appeared at the tips of the wings as the aircraft approached the horizontal, but the airplane pancaked into the ground at a speed estimated to be between 250 and 350 mph. The impact tore off both wings in a shower of gasoline, while the fuselage bounded into the air, across a road, and onto the flying field, throwing the body of the pilot clear. The airplane was completely wrecked, as were all of its major components and most of the individual parts. The only instrument found to be intact was the horizontal stabilizer trim indicator. It appears from the evidence available that the pilot misjudged the altitude required for a pull-out from a dive of this type. Measurements of distance from the first impact marks (that of the tail) to marks of impact of the wing indicate the aircraft was actually in a tail low altitude and 'mushing' against the pull-out."

9-22-45C. Ajo, Arizona. At 1325 MWT, a Curtiss P-40N collided with a towed target and crashed 25 miles northwest of Ajo, Arizona, killing Chinese Air Force pilot Sub-Lt. Shu-Cheng Tsao. The airplane was part of a three-ship flight that took off from Ajo Army Air Field on an aerial gunnery mission. The subject airplane made a firing pass at the target and collided with the target. The subject airplane went out of control and was observed spinning toward the ground. The airplane continued in the spin until it struck the ground, bursting into flames upon impact.

9-22-45D. Arlington, Wisconsin. At 1537, a Beech C-45F flying in instrument conditions crashed three miles southwest of Arlington, Wisconsin, killing pilot Capt. Charles P. McDonnell and six passengers. Passengers killed in the crash were: Pfc. Sanford Africk, Capt. Meyer Bruck, 1Lt. E.L. Croissant, Maj. R.C. Dench, F/O E.J. Gajnak, Maj. K.T. Grube. The airplane took off at 1338 from Sioux Falls, South Dakota, on a flight to Madison, Wisconsin. The airplane was on an instrument clearance and had been flying 500 feet on top of the clouds. Lone Rock Radio ordered the pilot to 8,000 feet. The pilot apparently was unable to maintain the assigned 8,000-foot altitude. The pilot requested permission to let down at 1334 but was told by controllers to stand by. The controller received the clearance for the let-down and attempted to contact the subject airplane at 1338 but was unsuccessful. Investigation revealed that the airplane had descended out of a 200-foot overcast and collided with the ground at a shallow angle. The airplane slid along the ground for about 100 yards, smashed through a fence and then slammed into a slight rise in the ground, bursting into flames upon impact and killing all on board instantly.

9-24-45A. Melrose, Massachusetts. At 1117, a North American TB-25J suffering an in-flight fire crashed onto Mt. Head Golf Course at Melrose, Massachusetts, killing pilot Maj. Doak A. Weston, 27, Aptos, California (see 2-23-43F). Crewmembers injured parachuting to safety were: 1Lt. William R. Ivey, navigator; 1Lt. Albert K. Owen, Jr., navigator; SSgt. Chester Miller, engineer; TSgt. Frederick J. Paquin, radio operator; SSgt. Erle T. MacDonald, engineer. The airplane took off from Grenier Field, Manchester, New Hampshire, on a flight to Boston, Massachusetts. The port engine was rough on take-off but "smoothed" out after the power was reduced. The airplane circled the field once, climbed to 4,000 feet and headed for Boston. The engineer noticed an odor of gasoline shortly after take-off. When the fumes became worse, the pilot turned the airplane back for Grenier Field. A moment later, smoke was observed pouring from the area near the navigator's table. The pilot feathered the port propeller and the engine immediately burst into flames. The pilot ordered the crew to bail out and they parachuted to safety. The fire continued and the port wing failed and separated. Moments later, the port engine fell off. The airplane smashed to the ground and exploded into flames. Investigation revealed that the airplane developed a fuel leak behind the port engine firewall and the leaking gasoline was apparently ignited by hot exhaust ducting.

9-24-45B. Ajo, Arizona. At 1014 MWT, a Curtiss P-40N crashed 28 miles northwest of the Army Air Field at Ajo, Arizona, killing Chinese Air Force pilot Sub-Lt. Chih-Yuan Cho. The airplane was part of a three-ship flight that had taken off from Ajo Army Air Field on an aerial gunnery mission. The airplane made a firing pass at the target and the port wing collided with the target, sending the subject airplane out of control. The airplane rolled out of control and entered a steep spiral. The pilot was unable to recover and bailed out at about 1,500 feet. The pilot pulled the ripcord before he was clear of the airplane. The pilot collided with some part of the airplane while bailing out, sustaining serious injuries and damaging his parachute. The parachute never fully deployed and the pilot fell to his death.

9-24-45C. Millville, New Jersey. At 1610, a Republic P-47D collided with the ground at the AAF Gunnery Range two miles southwest of Millville Army Air Field, Millville, New Jersey, killing pilot 1Lt. William S. Malone. The airplane was flying in the number-five position of a six-ship flight that had taken off from Millville Army Air Field on a ground gunnery mission. The subject airplane mushed into the ground while attempting to pull up on its third pass on the target range. The airplane bounded back into the air for a short distance before slamming back into the ground, smashing itself to pieces and bursting into flames as it slid across the range.

9-24-45D. Tacoma, Washington. At 1526, a Lockheed P-38L crashed nine miles south of Tacoma, Washington, killing pilot 1Lt. William J. Harris, 26. The airplane took off from McChord Field, Tacoma, Washington, on a navigation flight to Seattle, Washington, and return. The pilot successfully completed the navigation flight and returned and landed at McChord Field. Investigators stated, "After landing, the pilot requested permission from the tower for a local transition flight. After receiving permission pilot took off, drifting off the right side of the runway. Plane was airborne at 1524 PWT. On climb straight ahead after take-off witness stated he heard the engine misfiring. Witness was unable to distinguish which engine it was. Approximately one mile from the field the plane made a 45-degree turn to the left and flew for two miles and then made a 220-degree skidding turn to the right at approximately 1,500 feet, losing 1,000 feet in the turn. After completing the turn the pilot called the tower for permission to make a downwind landing. At approximately one and one half miles southwest of the field the plane rolled to the right and smashed into the ground from 500 feet. The P-38 crashed one mile southwest of runways on McChord Field reservation in a small cleared area. Plane exploded and burned on impact, making investigation for malfunctioning engines impossible. All controls, switches and gauges were destroyed by the crash and fire. Pilot strapped to seat was thrown clear of wreckage on impact."

9-24-45E. Bridgehouse, California. At 2040, a Boeing B-29 was abandoned and crashed three miles west of Bridgehouse, California, killing two crewmembers in unsuccessful parachute attempts. Nine fliers were injured parachuting to safety, two of them seriously. Navigator 1Lt. Clayton W. Evans and radio operator SSgt. Frederick W. Lavesque were killed attempting to parachute to safety. Co-pilot Maj. Charles B. Miller and passenger Maj. William Cluett received serious injuries parachuting to safety. Crewmembers injured parachuting to safety were: Lt. Col. Robert W. Strong, Jr., pilot; Col. John C. Pitchford, pilot-rated passenger; F/O Malcolm W. Farnum, flight engineer; SSgt. Joseph J. Dunner, gunner; SSgt. Joseph P. Miller, gunner; 1Lt. Howard M. Hayden, passenger; Pvt. J.M. Copeland, passenger. The airplane took off from Mather Field, Sacramento, California, on a flight to Washington, D.C. The flight had originated at an air base in the Pacific Theater and the airplane had landed in Hawaii before arriving at Mather Field. The airplane made a normal take-off from Mather Field. About ten minutes after take-off, the rpm on the number-two engine began to rise uncontrollably. The pilot ordered the engineer to cut off the fuel to number-two and then he attempted to feather the number-two propeller without success. The rpm on the number-two engine increased to 3,100 and remained there, defying all attempts to stop the engine and feather the propeller. Major Miller was ordered to take control of the airplane because of his extensive B-29 time and the fact that he had flown the subject airplane in combat several times. As Maj. Miller attempted to assume position in the pilot seat, the number-three and four engines began to slowly increase rpm. Maj. Miller was unable to assume the airplane commander's seat. The number-three engine surged to about 3,500 rpm and began to overspeed. Number-four engine surged to 3,000 rpm and also defied all attempts to get the engines under control. The pilot in command, Lt. Col. Strong, ordered the crew and passengers to bail out. After ascertaining that all members of the crew had abandoned ship, Lt. Col. Strong bailed out. The airplane smashed to the ground but did not burn. Investigation revealed that the number-four engine had suffered extensive fire damage before impact with the ground.

9-25-45. Gulf of Mexico, Florida. At 1445 EWT, a North American P-51D crashed into the Gulf of Mexico five miles west of Clearwater, Florida, killing 2Lt. Paul C. Hineman. The airplane was part of a two-ship flight that took off from Sarasota Army Air Field, Sarasota, Florida, on an instrument training flight. Lt. Hineman was to fly as the instrument pilot and 2Lt. Henry D. Horger was to fly as the safety pilot. Investigators

stated, "The flight took off and climbed to 8,000 feet where the mission was scheduled to be flown. After flying approximately 40 minutes at this height, Lt. Horger ordered Lt. Hineman to 10,000 feet due to clouds. After a short break, instruments were again resumed. After approximately five minutes, Lt. Hineman slowed down and Lt. Horger overtook him, slowing his own aircraft down to 120 mph to avoid a mid-air collision. Lt. Horger turned right, mushing into a cloud and losing control of his aircraft. Lt. Hineman was told to turn left but was not observed to do so. Lt. Horger was unable to contact Lt. Hineman after he [Lt. Horger] recovered and was clear of the cloud. It is believed that Lt. Hineman also hit the cloud and did not recover. A P-51 was witnessed by several men to crash about a mile off Treasure Island, Florida."

9-26-45A. Abbeville, Alabama. At 1145, a North American P-51D crashed seven miles south of Abbeville, Alabama, killing Mexican Air Force pilot 1Lt. Moreno R. Gomez. The airplane had taken off from Napier Field, Dothan, Alabama, on a routine transition flight. The airplane failed to return to the field at its assigned time and a short time later a call came in reporting the crash. Investigation revealed that the airplane crashed while in a spin. Witnesses stated that the pilot appeared to attempt to make a recovery from the spin but was unsuccessful. The pilot made no attempt to parachute to safety.

9-26-45B. Randolph Field, Texas. At 1520, a North American AT-6C crashed at Randolph Field, San Antonio, Texas, killing pilot 2Lt. James J. Duffy and seriously injuring pilot-rated passenger Capt. Marion B. Street. The pilot was entering the traffic pattern on the 45-degree entry leg when the starboard wing struck a bird. The pilot attempted his landing on the grass on the west side of the field. The aircraft bounced upon contact with the surface. The airplane began ground looping to the right. The pilot applied throttle and then the airplane entered a vertical bank to the right, turning about 45 degrees before the pilot was able to regain control of the airplane. The pilot flew near the pattern area and another AT-6 flew next to the subject airplane so that its pilot could determine if there was any damage to the wing tip or landing gear. The other pilot could not see any damage aside from the slight dent in the leading edge of the starboard wing outboard of the landing light. The subject pilot requested permission to make an emergency landing on the main runway and this request was granted. The airplane re-entered the traffic pattern and attempted to land but was given the red signal light just after the turn from base to final approach. The pilot observed the signal and made an uneventful go-around. The airplane made a normal approach with power. The airplane contacted the runway in a three-point position and rolled about 50 or 60 yards when the pilot apparently jerked the stick back to get the tail down. The airplane bounded back into the air in climbing turn to the right. The pilot added throttle and the airplane entered a vertical bank to the right. The airplane fell off to the right and plunged to the ground, cartwheeling into the terrain on the starboard wing and slamming to the ground inverted. The airplane did not burn but was totally destroyed.

9-26-45C. Valparaiso, Florida. At 2040 CWT, a North American AT-6B crashed at Auxiliary Airfield #3 ten miles north of Valparaiso, Florida, killing pilot F/O Edward V. Costello. The airplane was flying in the number-two position of an eight-ship flight that had taken off at 2035 CWT from Eglin Field, Valparaiso, Florida, on a routine night flying exercise, which included ten night landings for each pilot. The subject pilot was attempting a landing at Auxiliary Airfield #3. The number-three pilot noticed that the subject airplane was attempting to land on the area in between Runway 18 and the taxiway. The airplane appeared to level off momentarily before striking the ground with the starboard wing tip. The pilot leveled the airplane and added throttle before the airplane collided with the wind tee and a windsock pole with the starboard wing. The airplane gained about 40 feet before rolling over to the right to an inverted position and then smashing to the ground.

9-28-45A. Douglas, Arizona. At 1039 MWT, a North American TB-25J crashed three miles south of the Army Air Field at Douglas, Arizona, killing Chinese Air Force pilots 2Lt. Yen-Jui Chen and 1Lt. Chien-Lin Chin. The airplane took off from Douglas Army Air Field on a routine night flight and the students had been practicing night landings on Runway 35R. The pilots shot a successful landing and then re-entered the traffic pattern, which was to the left. The airplane turned on final approach in a normal descending turn. The pilot overshot to the right and then corrected back to the left. The airplane remained in a descending turn to the left until it struck the ground and exploded into flames.

9-28-45B. Gowen Field, Idaho. At 1114 MWT, a Consolidated B-24J crashed while attempting an emergency landing at Gowen Field, Boise, Idaho, killing two fliers and seriously injuring engineer SSgt. Edward F. Hillhouse. Passengers 1Lt. Herbert E. Pirrong and 2Lt. Russell L. Turpen escaped injury. Pilot Maj. William R. McKinley and co-pilot 1Lt. Benjamin M. Huffman were killed in the crash. The airplane took off from Gowen Field at 1054 MWT on a mission to photograph the field from the air. At 1109, the pilot reported that he was five miles northwest of the field and that he had a fire of an undetermined source in the fuselage. The airplane was cleared for a straight-in approach on Runway 10R. Investigators stated, "Approach was made with the landing gear down. Plane passed over the end of the runway at an estimated altitude of 250 feet and continued in a short close-in right hand pattern. Plane began to trail smoke just before turning on the base leg and at about the same time witnesses on

the ground heard the propellers surge or a partially runaway propeller. Pilot failed to turn early enough and had to S-turn back to the end of the runway. Contact [with the ground] was made first with left landing gear but the plane was still skidding to the right even though the left wing was slightly low. Landing was hot and no flaps were used. Left tire rim shattered on contact and tire blew out immediately thereafter. Aircraft swerved slightly to the left and off of the runway. Left landing gear sheared off completely shortly after aircraft left the runway. It came to rest about 1,500 feet from the first point of contact. Fire was observed coming from the open bomb bays when aircraft was about 30 feet in the air prior to landing. When the aircraft came to rest, the fire broke out furiously.... Pilot and co-pilot suffered fatal injuries from the fire; engineer suffered major injuries, but both crewmembers in the waist section escaped uninjured.... Limiter panel cover either came unfastened and acted as a conductor charging the trim tab cable and fuel drain line just below the panel. Small holes were burned in the fuel line, and the trim tab cable was burned in half. When the battery was turned on a second time, fuel was ignited, starting the fire."

9-30-45. Ontario, California. At 0719 PST, a Lockheed P-38L crashed one and a half miles west of the Army Air Field at Ontario, California, killing pilot Lt. Col. Ross C. Baker, Jr., 26. Investigators stated, "At 0718 PST, Lt. Col. Baker took off from Ontario Army Air Field for an extended cross-country flight, his plane was equipped with two 165-gallon belly tanks. When the aircraft had reached an altitude of approximately 150 feet over the west end of Runway 25, witnesses reported that one engine started cutting out. It has not been determined which engine it was, there was no visible smoke or apparent yawing of the aircraft. Lt. Col. Baker's plane started losing altitude, and he directed it toward an open field lying approximately one mile from Ontario Army Air Field. The aircraft passed over a group of telephone wires, cutting one, then under a second group of wires, belly landing just short of a road adjacent to the field. Fire occurred at time of impact, probably from one of the belly tanks. The aircraft skidded about 300 feet across the road and into the field, leaving a trail of fire behind it. When the plane had stopped the second belly tank blew up, enveloping the aircraft in flames. The pilot was rendered unconscious by striking his head on the gun sight, and received fatal third degree burns."

October

10-1-45. Luke Field, Arizona. At 0840 MST, two North American AT-6D airplanes collided in mid-air and crashed two miles west of Luke Field, Phoenix, Arizona, killing one flier and seriously injuring another. Instructor pilot 2Lt. Paul K. Byrne was killed in the collision and crash of AT-6D #42-85722; pilot F/O Wallace J. Christian sustained serious injuries parachuting to safety. Chinese Air Force pilot 2Lt. Kuen Chang was uninjured parachuting to safety from AT-6D #42-85380. Lt. Chang took off from Luke Field at 0745 and was returning to the field to land. Lt. Byrne and F/O Christian had taken off from Luke Field at 0800 for F/O Christian's final check ride prior to graduation from advanced school. The airplanes were flying at about 3,500 feet when the collision occurred. Lt. Chang stated that he did not see the other aircraft as he maneuvered in an effort to enter the traffic pattern. Suddenly he felt the jolt of the collision and then noticed that he could not control the airplane and it was spinning to the right. He bailed out immediately, landing in between the wreckage of the two airplanes. He was uninjured. Lt. Chang gathered his parachute and walked back to Luke Field. Lt. Byrne was flying AT-6D #42-85722 from the rear cockpit. F/O Christian felt the collision but did not see the other airplane until it was falling away on the right. Investigation revealed that the propeller of AT-6D #42-85380 severed the tail section of AT-6D #42-85722 and killed the instructor flying in the rear cockpit. F/O Christian, in the front cockpit, tried the control stick but got no response. He jumped immediately and parachuted to safety, suffering a badly broken ankle and other injuries upon landing. The severed tail section of AT-6D #42-85722 smashed to the ground near where he landed. Investigators noted that pieces of Lt. Byrne's head and anatomy were found one-half mile from the crash site. AT-6D #42-85380 slammed to the ground and exploded into flames several hundred feet from AT-6D #42-85722, which did not burn but was destroyed utterly.

10-2-45. Eglin Field, Florida. At 0900 CST, a North American AT-6C crashed 15 miles west of Eglin Field, Valparaiso, Florida, killing pilots 2Lt. Ernest E. Greenwood and F/O Marcus W. Brown. The airplane took off from Eglin Field Military Reservation Auxiliary Airfield #3 at 0835 CST on a gunnery mission. The airplane was part of a six-ship flight flying in three, two-ship elements. The airplanes entered an echelon formation as they approached the area of the range. The airplanes then peeled off to 700 feet and began their individual firing passes. The subject airplane, flying in the number-four position, was observed to enter a steep turn while turning in the firing traffic pattern, which was to the left. The subject airplane snapped over and fell into a spin to the right. The pilot attempted a recovery but was unsuccessful. The airplane was observed to enter a 45-degree dive toward the ground. The airplane dove into the ground, killing the pilots instantly. The airplane did not burn but was destroyed utterly.

10-5-45A. Upland, California. At 0115, a Curtiss C-46A collided with rising terrain and crashed north of Upland, California, killing four fliers. The airplane took off at 0100 from San Bernardino Army Air Field, San Bernardino, California, on an instrument flight plan to Reno Army Air Field, Reno, Nevada. The airplane radioed San Bernardino Army Air Field at 0108 and reported that it was "on top at 7,000 feet." There was no other communication with the airplane and it failed to arrive at its destination. At 0600, a law enforcement officer reported a fire on the summit of Mt. San Antonio. Investigation revealed that it was the subject airplane, which had smashed into 9,890-foot Devil's Backbone about 15 feet from the top. The port propeller stuck the terrain first, followed by the port wing. The airplane smashed down a small tree, skidded to the left and careened over the mountaintop for 300 feet before dropping into a canyon and coming to rest 1,000 feet below the rim of the mountain. Investigation revealed that the airplane had smashed into the mountain in a flying attitude at about cruise speed. Investigators were unable to find the two engines. Wreckage was scattered widely. Killed in the crash were: Maj. Ovid F. Pinckert, pilot; 2Lt. Edwin M. Mize, co-pilot; Cpl. Barney Cummins, engineer; 2Lt. Martha S. Betts, WAC passenger.

10-5-45B. Milton, Delaware. At 0930, a Republic P-47D crashed three miles southeast of Milton, Delaware, killing pilot Capt. Phillip G. Miller, 28. The airplane was part of a two-ship flight that took off from Dover Army Air Field, Dover, Delaware, on a supervisory flight. The two airplanes climbed to 9,000 feet as they flew to the gunnery range to observe student air work. While the other pilot began adjusting his oxygen mask, the subject airplane began performing aerobatics. Capt. Miller performed a barrel roll. A few minutes later, Capt. Miller performed a steep wingover maneuver and was observed doing aileron rolls on the way down. The other pilot lost track of Capt. Miller and was unable to establish contact with the subject airplane. Civilians on the ground observed the airplane dive out of the clouds at 4,000 feet at an angle approaching 90 degrees. The airplane failed to recover and smashed to the ground at high speed at a 90-degree angle, exploding violently into flames upon impact. The pilot had made no apparent effort to bail out.

10-5-45C. Valparaiso, Florida. At 0900 CST, a North American AT-6D crashed eight miles west of Valparaiso, Florida, killing French Air Force student pilot Sgt. Gaston Villian. The subject airplane was flying in the number-four position of a six-ship flight that took off from Eglin Field Military Reservation Auxiliary Airfield #6 at 0840 CST on a routine gunnery mission. The subject airplane was observed to pull out of its target run at about 400 feet agl. The airplane made a 180-degree turn to the left. The turn was so steep that the airplane assumed a semi-inverted attitude. The airplane stalled, snapped over to the right and then dove into the ground. The airplane exploded into flames upon impact, killing the student instantly.

10-5-45D. Gulf of Mexico, Florida. At 1002 EST, two North American P-51D airplanes collided in mid-air and crashed into the Gulf of Mexico 15 miles northwest of Key West, Florida, killing both pilots. 1Lt. Jackson A. Marshall, 22, was killed in the crash of P-51D #44-84859; 1Lt. Elmer R. Shryock, 28, was killed in the crash of P-51D #44-84937. The airplanes were part of a four-ship flight that had taken off from Venice Army Air Field, Venice, Florida, on a medium-altitude navigation flight to Key West and return. Lt. Marshall was flying in the number-one position and Lt. Shryock was flying in the number-three position. The airplanes, flying at 5,000 feet msl in standard tactical formation, had arrived in the area of Key West at 1000 EST, circled once and then set a course for the return leg. Lt. Marshall, without warning, suddenly performed a barrel roll to the left. The number-two pilot was startled and broke out and away from the formation. Investigators stated, "Lt. Shryock, in the number-three position, presumably decided to follow through on the maneuver, but the time lapse had allowed Lt. Marshall to half complete his roll before Lt. Shryock entered his. As a result, Lt. Marshall was reaching the bottom of his roll as Lt. Shryock became inverted at the top of his. At this point, the two aircraft collided, belly to belly. Lt. Shryock's aircraft was virtually demolished and fell into the gulf, while Lt. Marshall's aircraft, seemingly more or less intact, spiraled in a few seconds afterward. Both aircraft sank immediately and no wreckage was recovered. No use of parachutes was observed."

10-5-45E. Bowling Green, Florida. At 1015 EST, a North American P-51D crashed one mile south of Bowling Green, Florida, killing F/O William E. Sharpe, 22. The airplane was part of a four-ship flight that took off at 0845 EST from Venice Army Air Field, Venice, Florida, on a low-altitude navigation mission to Lakeport, Florida, to Gardner, Florida, and return to Venice. The formation was flying at an altitude of about 250 feet agl. On the leg from Lakeport to Gardner, the subject airplane dropped to very low altitude, barely clearing electric power lines. The airplane collided with a tree with the starboard wing after passing over the power lines. Part of the starboard wing was severed. The airplane rolled out of control three times and then slammed into a heavily wooded area, smashing through trees, over a railroad track, over a highway before crashing into a wooded area. The airplane burst into flames as it smashed itself to pieces.

10-5-45F. Gettysburg, Ohio. At 1145 EST, a Republic P-47N flying in poor weather crashed two miles northwest of Gettysburg, Ohio, killing 1Lt. James Dolan, 26. The airplane was part of a two-ship flight that took off from Patterson Field, Ohio, on a flight to Scott Field, Illinois, via Fort Wayne, Indiana, and Chanute Field, Illinois. The pilots were refused a contact clearance direct to Scott Field and had to fly to Fort Wayne and Chanute to avoid poor weather. The airplanes entered instrument conditions soon after taking off from Patterson Field. The airplanes descended to 2,000 feet in order to remain together and in contact conditions. A few minutes later the ceiling dropped considerably. The subject airplane descended into the ground while attempting to remain below the ceiling, exploding into flames upon impact. The pilot was killed instantly. Wreckage was scattered over an area of 200 yards.

10-5-45G. Estrella, Arizona. At 1155 MST, a North American P-51H crashed seven miles north of Estrella, Arizona, killing 2Lt. Harold C. Baker, 27. The P-51 was one of several that had taken off individually from Long Beach, California, on a ferry flight to Coolidge, Arizona. The individual ships had encountered poor weather and successfully maneuvered around it. The airplanes encountered more poor weather and many turned to the north and avoided a storm that appeared to extend down to ground level. The subject airplane apparently entered the poor weather and the pilot lost control. The pilot apparently stressed the airplane beyond its design limits in an effort to re-establish level flight. The port wing was found one-half mile northwest of the main wreckage.

10-5-45H. Amarillo, Texas. At 1620, a North American AT-6D and a Beech C-45F collided over the runway at Amarillo Army Air Field, Amarillo, Texas, killing two pilots aboard the C-45. Four passengers aboard the C-45 were uninjured. The AT-6D pilot, 2Lt. Louis A. Porter, escaped injury. Pilot Maj. John H. White and co-pilot Maj. Maxey W. McGuire were killed aboard the C-45. Passengers escaping injury were: Cpl. James A. Friedell, Sgt. J. A. Hull, Sgt. John H. Miller, Pfc. Harry J. Morgan. The AT-6D took off from Garden City Army Air Field, Garden City, Kansas, on a flight to Amarillo Army Air Field. The C-45 took off from Fort Sumner Army Air Field, Sumner, New Mexico, on a transport mission to Amarillo Army Air Field. The C-45 arrived at Amarillo and was cleared to land on Runway 21. The C-45 entered the traffic pattern, which was to the left. The C-45 pilot reported on the downwind leg, the base leg turn and the final turn and was cleared to land each time. The C-45 made an approach that was slightly lower than normal and was about to land when the tower radioed the pilot to abort the landing. The C-45 pilot pulled up and collided with the AT-6, which was flying directly overhead and descending for a landing. The airplanes locked together and both struck the runway about 100 feet from the point of collision. The airplanes skidded to the right and into the grassy area adjacent to the runway, coming to rest about 50 feet from the edge of the runway. Investigation revealed that the AT-6D pilot had mistakenly entered a right-hand traffic pattern instead of a left-hand pattern and was not properly cleared to land. No one on the ground or in the air observed the AT-6 until it was directly above the C-45.

10-5-45I. Cherokee, North Carolina. At an unknown time after 1700 CST, a Beech UC-45 flying in instrument conditions collided with rising terrain about five miles northeast of Cherokee, North Carolina, killing pilot 1Lt. Robert W. Barton and five passengers. The airplane took off at 1650 CST from Lambert Field, St. Louis, Missouri, on a direct IFR flight to Morris Field, Charlotte, North Carolina. The pilot filed his flight plan for 5,000 feet. Cincinnati Air Traffic Control changed the pilot's clearance from 5,000 feet to 7,000 feet. There was no further contact with the airplane and it failed arrive at Charlotte. After an extensive search, the airplane was discovered on 10-16-45. The airplane was found to have crashed into a 6,000-foot mountain about 900 feet from the summit approximately 20 miles southeast of Gatlinburg, Tennessee. Passengers killed in the crash were: SSgt. Raymond H. Kerkela, Cpl. Winifred R. Haines, Spec 3/C Lena E. Allred, SSgt. Hollis E. Brobrick, 1Lt. Stanley M. Lerner.

10-6-45. Enid, Oklahoma. At 2115, a North American TB-25J crashed six miles northeast of the Army Air Field at Enid, Oklahoma, killing students A/C Russell W. Blakely and A/C Gaylord G. White. The airplane took off from Enid Army Air Field on a routine student night training flight. The airplane smashed to the ground in a flying attitude. The airplane did not burn but was destroyed utterly. Investigators could not determine the cause of the accident.

10-7-45. Anton, Texas. At 2310, a North American AT-6D crashed eight miles southwest of Anton, Texas, killing pilot 1Lt. Loren R. Elsasser, 26. The airplane took off at 2002 CST from Smoky Hill Army Air Field, Salinas, Kansas, on a flight to the pilot's home station at Lubbock Army Air Field, Lubbock, Texas. The airplane arrived in the vicinity of Lubbock Army Air Field and requested landing instructions. The pilot then began to descend through a thin layer of stratus clouds, the bases of which were 100 feet agl. The airplane descended into the ground with power in a flying attitude. The pilot was killed upon impact. The airplane did not burn but was destroyed utterly.

10-8-45. Laredo, Texas. At 2059, a Consolidated TB-24D crashed three miles northeast of Laredo, Texas, killing four fliers. The airplane took off from Sheppard Field, Wichita Falls, Texas, on a flight back to its home station at Laredo Army Air Field. The pilot contacted Laredo Army Air Field at 2050 and requested landing instructions. The pilot was cleared for a straight-in approach on Runway 14L. The airplane arrived over the radio range station at 3,000 feet and the pilot requested a 360-degree overhead approach, which was granted. The airplane continued to fly toward the direction of the intended straight-in approach at an altitude of approximately 1,000 feet agl. The airplane began a turn to the left as it approached the southeast end of the field. As the airplane continued to turn, the angle of bank increased and the airplane began

to lose altitude. The airplane continued in this attitude until it was very near the ground. The pilot was unable to re-establish level flight before striking the ground. The airplane struck the ground on its belly and slid across the terrain for about 2,500 feet before coming to rest. The airplane did not burn but was destroyed utterly. Killed in the crash were: Capt. Robert C. Beck, pilot; 1Lt. Hinton B. Wright, co-pilot; Cpl. Leroy A. Lundschan, engineer; Lt. Col. Herman H. Simpson, passenger.

10-10-45. Bronson, Iowa. At 0850, a Martin TB-26C crashed three miles east of Bronson, Iowa, killing French Air Force students Aspirant Robert C. Lagier and Cpl. Yvan J. Picardat. The airplane had taken off at 0815 from Sioux City Army Air Base, Sioux City, Iowa, on a routine training mission. The airplane entered a spin to the right and began to come apart in the air as the students attempted to re-establish level flight. The port wing failed outside of the port engine nacelle, causing the wing and engine to separate. The airplane broke up, scattering wreckage over an area of over 500 yards. The forward fuselage and tail section separated from the mid-section and wings and slammed to the ground, the wings and center section exploding into flames upon impact. Investigators noted that civilian witness' statements were contradictory.

10-14-45A. Lake Michigan, Illinois. At 1215, an AAF aerial engineer was declared missing after he parachuted from an out of control Douglas C-47A over Lake Michigan approximately two miles off shore of Chicago, Illinois. After an extensive search effort, engineer TSgt. Donald Anderson was presumed lost. The pilot was able to regain control of the airplane and land safely at Glenview Naval Air Station, Glenview, Illinois. Passenger Capt. Robert R. Campbell suffered serious injuries. Pilot 1Lt. Waino K. Thompson was uninjured; passengers Pvt. Walter Herod and Pvt. Julia F. Honnicutt (WAC) suffered minor injuries. The airplane took off from Glenview Naval Air Station on a routine flight to Lunken Field, Cincinnati, Ohio. The airplane encountered rain showers over Chicago so the pilot turned about 15 degrees to the left and continued climbing. The airplane passed through a thin layer of clouds and soon encountered icing conditions at about 5,000 feet. The pilot ordered the engineer to tell the passengers to put on their parachutes. The engineer asked, "Me too?" The pilot replied in the affirmative. The pilot stated to investigators, "Immediately following his going to the back of the ship I hit severe turbulence, which I later realized to be a thunderhead. I followed the ship with my hands on the controls. At one time I seemed to be in a dive and the propellers sounded as though they were running away, and I took action to stop this by cutting the throttles, which did no good, cutting back on the propeller pitch, which did no good, and cutting back on the mixture control which did no good. I finally feathered both propellers whereupon the airplane slowed up in its reactions and

I appeared to get control but lost control immediately. Finally coming out of the base of the clouds in an upside down position, I pulled the ship out of the dive and into a normal flying position at the same time bringing both engines back. At this same time the WAC passenger informed me that that the crew chief had bailed out. After getting the engines back to normal, I immediately radioed Glenview to inform the Coast Guard that a man had bailed out over the lake about two miles east of the city of Chicago. I then put the landing gear down and informed the tower that I was going to buzz the tower and have them check my landing gear which I did and landed [at Glenview] all right."

10-14-45B. Havana, Arkansas. At 2115, a North American AT-6C flying in instrument conditions collided with rising terrain three miles northwest of Havana, Arkansas, killing pilot Capt. Edward L. Wilsey and pilot-rated passenger Capt. Norman J. Bartz. The airplane took off from Topeka, Kansas, at 1846 on a navigation mission to Coffeyville, Kansas, to Fort Smith, Arkansas, to Little Rock, Arkansas. There was no further contact with the airplane after take-off. The airplane, which was stationed at Craig Field, Selma, Alabama, failed to return to base and was declared missing. Search planes found the missing aircraft the next day. Investigation revealed that the airplane collided with the southeast slope of Mt. Magazine about 50 feet from the 2,800-foot summit. The pilots were killed instantly. The airplane did not burn but was destroyed utterly. Investigators speculated that the pilot had probably encountered instrument conditions and was probably attempting to get underneath to maintain flight in visual conditions.

10-16-45. Wheeling, West Virginia. At 1710, a Republic P-47N crashed at Wheeling, West Virginia, killing student A/C Vernon O. Enders. The airplane took off at 1323 from Bluethenthal Field, Wilmington, North Carolina, on a positioning flight to Pittsburgh, Pennsylvania. The airplane was observed circling the area of the County Airport at Wheeling. The pilot extended the landing gear and attempted an approach for the southwest runway. The engine sounded as if it were laboring on the approach. The pilot advanced the throttle but the airplane stalled and then dove straight to the ground, exploding into flames upon impact. The pilot was killed instantly.

10-18-45A. Hondo, Texas. At 0829, a Consolidated TB-24J suffering an in-flight fire crashed three and one-half miles north of the Army Air Field at Hondo, Texas, killing 13 passengers and crew. The airplane took off from Hondo Army Air Field on Runway 35 at 0826 on an engineer-training flight. Investigators stated, "The take-off was normal until the aircraft was approximately one-half mile beyond the north end of the runway when the number-three engine started smoking lightly and increasing to a heavy black smoke. The aircraft drifted slightly to the right of its

take-off heading and continued to climb normally. About one and a half miles from the field a small fire appeared on the number-three engine, and the airplane descended slightly, then started a shallow climb still heading north. At approximately three miles north of the field, the number-three engine burst into a large fire, which came from the front of the engine, then from the cowling area. The aircraft was in a medium climbing turn to the left when the right bomb bay door was blown off by an explosion in the bomb bay. A small flame was coming from the right front bomb bay for about 30 seconds before the explosion occurred. At this time the aircraft was about 1,000 feet above the terrain. It completed a turn to a westerly heading, leveled off, then started climbing abruptly for 300 to 500 feet, leveled off again and started to turn to the left. The bank increased rapidly and the aircraft nosed down abruptly, crashing in a south-southwesterly direction, hitting on the left wing and nose simultaneously. The aircraft exploded and burned upon impact with the ground. Check of the wreckage indicated that the landing gear was extended." Killed in the crash were: 2Lt. George I. Huffman, pilot; 2Lt. Kenneth W. Appleton, co-pilot; F/O Milton F. Shapiro, engineer instructor; 2Lt. Joseph D. Van Cleve, flight engineer; 2Lt. Daniel R. Zitkus, flight engineer; Sgt. John Adams, engineer; A/C Edwin Salkiewicz, student engineer; A/C Glen Schmidt, student engineer; A/C John P. Schreibe, student engineer; A/C Edward F. Rudak, student engineer; A/C Melvin R. Rowland, student engineer; A/C Albert E. Rudy, student engineer; A/C Billy J. Schaeffer, student engineer.

10-18-45B. Flat Rock, Michigan. At 1600, a Martin TB-26B crashed five miles southwest of Flat Rock, Michigan, killing six French Air Force crewmembers. The airplane took off from Selfridge Field, Mt. Clemens, Michigan, on a routine instrument-training mission. The airplane was seen flying near Flat Rock on a southwestern heading at about 3,500 feet when an explosion was heard. Witnesses on the ground then observed the airplane spinning out of control with pieces falling away. The airplane slammed to the ground and exploded, killing all on board instantly. The rudder was found over a mile and a half away from the main wreckage and apparently failed and separated first, causing the airplane to go out of control. Investigation of the wreckage revealed that the starboard wing failed out side of the engine nacelle and separated, striking the tail section and causing its progressive failure. Killed in the crash were: Sgt. Maurice M. Bouhier, pilot; Aspirant Anges B. Leca, co-pilot; Sgt. Jean R. Pietri, bombardier/navigator; Pvt. Francois Maniacci, engineer; Sgt. Georges Voilemin, radio operator; Cpl. Pierre Audonnet, gunner.

10-19-45A. Bowman Field, Kentucky. At 1915 CST, a Boeing TB-17F crashed off the end of the runway and struck a civilian automobile while attempting to land at Bowman Field, Louisville, Kentucky, killing civilian motorist Clemens P. Theisen and injuring L.G. Theisen. Two fliers sustained minor injuries and 19 others escaped injury. The airplane took off from Tinker Field, Oklahoma City, Oklahoma, on an administrative flight to Bowman Field. The airplane arrived in the vicinity of Bowman Field and the pilot requested landing instructions. The tower advised the airplane to land on Runway 32, which is 3,200 feet long. The pilot requested a longer runway and the tower approved a landing on Runway 6, which is 4,350 feet in length. The pilots stated that they executed pre-landing checks and procedures while flying in the traffic pattern. Investigators stated, "The aircraft flew an indicated altitude of 1,700 feet on the downwind leg, and made a normal base and approach. The final approach seemed to be somewhat higher than normal, according to the tower operators, and the aircraft floated for some time before finally touching the ground at approximately the intersection of Runways 1 and 6. The airplane was allowed to roll slightly and then the brakes were applied. The pilot, feeling no braking action, called for the flaps to be dumped, and put the airplane on the main gear. Both the pilot and the co-pilot rode the brakes in an unsuccessful attempt to halt the roll. The airplane rolled off the end of Runway 6, over a sod embankment, dropped to a road below the runway level, bounced from the road across the top of another sod embankment, knocking off the landing gear on the embankment, crashed through a fence, slid across a highway, striking a civilian automobile and carrying it over the embankment on the other side of the highway and onto a golf course. The airplane stopped with the nose pointed down the road embankment into the golf course and the tail pointed up over the highway. The automobile was turned over onto its side and the two occupants were pinned underneath. The crew immediately evacuated the airplane, as a small fire broke out in the grass under the number-three engine, probably caused by the hot turbo of the engine's supercharger." Rescue personnel from the field arrived immediately and extinguished the fire and rescued the trapped civilian. Investigators noted that the pilot should have gone around when it became apparent that the airplane was overshooting the runway. Co-pilot Capt. John P. Ford and passenger TSgt. John L. Mann received minor injuries. Crewmembers uninjured in the crash were: 1Lt. Guy S. Kidwell, Jr., pilot; 1Lt. Lawrence P. Brehm, navigator; 2Lt. Laurencio Nadaady, navigator; Sgt. Clair D. Johnson, engineer. Passengers uninjured in the crash were: 2Lt. Thomas Bank, 2Lt. Edward J. Brown, 2Lt. John N. Deardon, TSgt. George A. Hartfield, 1Lt. Robert T. Heil, 1Lt. Robert C. Helfrich, F/O William B. Hundley, Maj. Ned E. King, TSgt. William Madara, Jr., 2Lt. Floyd W. Martin, Capt. Ernest B. Mauck, 2Lt. Merle E. Midgley, 1Lt. Douglas C. Orr, Sgt. James S. Undercoffer, 2Lt. Norman E. Walton.

10-19-45B. Spokane, Washington. At 1802, a Beech C-45F flying in instrument conditions crashed 11 miles ENE of Spokane, Washington, killing five fliers. The airplane took off at 1716 from the Army Air Field at Walla Walla, Washington, on an instrument flight plan to Geiger Field, Spokane, Washington. The assigned altitude was 7,000 feet. Spokane Radio Range advised the pilot to descend to 6,000 feet and the pilot complied. The pilot evidently lost control of the airplane while flying in instrument conditions, the airplane entering a tight spiral to the left. The pilot was unable to recover the airplane and it smashed into terrain while in an inverted diving spiral. The airplane struck on a down-slope at an elevation of 2,400 feet msl and exploded into flames, killing all on board instantly. Killed in the crash were: Capt. George E. Matthews, pilot; SSgt. James C.R. Sheets, engineer; Brig. Gen. S.C. Godfrey, passenger; Col. J.W. Park, passenger; R.C. Hende, Geiger Field employee.

10-22-45. Danbury, North Carolina. At 1515, a Lockheed P-38K crashed three miles west of Danbury, North Carolina, killing 2Lt. Billy J. Farrington, 21. The airplane took off at 1250 from Reading, Pennsylvania, on a ferry flight to Walnut Ridge, Arkansas, via Coraopolis, Pennsylvania. The airplane was observed circling a small field near Lawsonville, North Carolina. The landing gear and flaps were in the extended position. The pilot was unable to make a landing after several attempts. The pilot retracted the landing gear and flaps and the airplane climbed away and flew in the direction of Winston-Salem, North Carolina. The airplane climbed to about 1,000 feet, which was about 300 feet above the hills in the area. The engines were observed to sputter and stop. The airplane entered a nose-down attitude and smashed to the ground. The pilot bailed out at an altitude too low to allow his parachute to fully deploy. Investigators speculated that the "electric remote indicating compass" had failed, giving the pilot erroneous readings. The pilot followed the erroneous compass, apparently making no cross reference to the magnetic compass, and became lost. The airplane ran out of fuel before the pilot was able to effect a safe landing. Investigators noted that the pilot had very little time on P-38 type airplanes.

10-25-45A. Westover Field, Massachusetts. At 1130, a Douglas C-54B was abandoned and crashed near Westover Field, Chicopee, Massachusetts, killing engineer Cpl. George K. Holloway, who was killed when he collided with the tail section while attempting to bail out. Pilot/C-54 student Maj. Richard G. Buswell escaped injury parachuting to safety; co-pilot/C-54 instructor Maj. Charles I. Longacre received minor injuries parachuting to safety. Engineer TSgt. Bernard J. Lance received serious burns when he parachuted into the exploding aircraft. Radio operator SSgt. Charles E. Walker sustained serious injuries when he collided with the aircraft while parachuting to safety. The airplane took off from LaGuardia Field, New York, on an instrument flight to Westover Field. The airplane was on a scheduled flight to check several instrument landing systems in use at various fields, Wendover Field being the first system to be flown. The airplane had been making instrument let-downs on the Westover glide path instrument system under actual instrument conditions. The pilot made five or six instrument let-downs and approaches to the point of establishing visual contact and then he would pull up and go around. The pilot flew toward Grenier Field, New Hampshire, when a massive hydraulic fluid leak erupted forward of the flight deck. The pilot turned the airplane back toward Westover Field and ordered that the landing gear be extended. The landing gear extended and locked in the normal fashion. The pilot then ordered that the flaps be lowered to landing configuration. The starboard flap extended about ten degrees and the port flap extended about five degrees and then ceased to operate. The engineer was ordered to lower the flaps using the emergency system. The engineer was unable to lower the flaps. The flap lever was returned to the neutral position and the hand pump was used to allow hydraulic pressure to build up. The flap lever was returned to the down position, but the port flap dropped only about five degrees and the starboard extended to about 25 degrees. The port flap was at about 10 degrees and the starboard flap was at about 30 degrees. The pilot then ordered the flap lever returned to the up position. The port flap retracted fully but the starboard flap continued to extend for a few more degrees. The aircraft became increasingly hard to control so the pilot ordered the crew to fasten parachutes and stand by to bail out. The co-pilot held the airplane steady while the pilot put on his parachute. The pilot then ordered the crew to bail out. The airplane entered a descending turn to the left and smashed to the ground, exploding into flames upon impact.

10-25-45B. Nashville, Tennessee. At 1455 CST, a Lockheed P-38J crashed seven miles southeast of Nashville, Tennessee, killing pilot 1Lt. Albert J. Yvon, 22. The airplane took off from Lockbourne Army Air Base, Columbus, Ohio, on a ferry flight to Nashville, Tennessee. The airplane arrived in the area of Nashville and requested landing instructions from the municipal control tower. The tower radioed landing instructions and the aircraft entered the traffic pattern. The pilot extended the landing gear but did not get a "down and locked indicator." The pilot alerted the control tower and was allowed to fly by. The tower then allowed the pilot to make a touch and go landing to ascertain if the landing gear would take the weight of the aircraft. The pilot flew the traffic pattern and was about to make an approach when he was cut off in the pattern by a U.S. Navy Douglas SBD airplane. The subject airplane went around and set up another approach. On his second approach the pilot came in high and at idle. The airplane bounced hard on the main landing gear but the nose gear did not contact. The pilot added throttle in an attempt to go around. The airplane flew straight ahead as it climbed away. Observers on the

ground noticed that the starboard engine was on fire near the supercharger. The airplane climbed to about 100 feet agl when the starboard engine lost all power. The starboard wing dropped and the airplane began a steep turn to the right, describing about 270 degrees of turn until it flew into the ground and burst into flames. The pilot was hurled from the airplane still strapped to his seat. He was rescued immediately, but died about 30 minutes after the accident.

10-25-45C. Wickenburg, Arizona. At 1425 MST, a North American AT-6C and an AT-6D collided in mid-air and crashed 20 miles SSW Wickenburg, Arizona, killing Chinese Air Force student A/C To-Chiew Sze aboard the AT-6D. Chinese Air Force student A/C Chung-Wen Ti sustained minor injuries parachuting to safety from the AT-6C. The two airplanes were part of a three-ship flight that had taken off from Luke Field, Phoenix, Arizona, on a routine student formation training flight. The formation was flying at 9,500 feet msl in a standard V-formation, which was led by their AAF instructor. The instructor signaled an echelon right formation. A/C Ti, who was flying in the number-three position, dropped down and moved over to assume his position. A/C Ti pulled up into A/C Sze's airplane, severing A/C Ti's vertical fin and rudder. A/C Ti's airplane fell away out of control so the pilot parachuted to safety. A/C Sze, who was observed to be conscious and apparently uninjured following the collision, flew straight ahead for a short time before the airplane went out of control and entered a spiral. A/C Sze made no effort to bail out and went in with the airplane.

10-25-45D. Santa Rosa, California. At 1425, a Lockheed P-38L crashed one mile south of Santa Rosa Army Air Field, Santa Rosa, California, killing pilot 1Lt. Herman A. Ladeau, 24. The airplane was flying in the number-three position of a four-ship flight that was taking off from Santa Rosa Army Air Field on a routine gunnery mission. A stream of white smoke was observed trailing the starboard engine just after the subject airplane became airborne. As the airplane climbed to about 50 feet, the starboard engine appeared to quit. The pilot retracted the landing gear and continued flying straight ahead. The airplane reached about 200 feet agl when it stopped climbing and a large burst of black smoke was seen to emit from the starboard engine. The airplane entered a glide straight ahead and appeared to be under control. The starboard propeller was not feathered and was apparently windmilling. The airplane remained on the take-off heading until it pancaked onto the slope of a small hill. The airplane bounded back into the air, slamming back to earth in a vineyard about 90 yards from the point of impact. The airplane burst into flames as it came to rest. The pilot was unable to escape the shattered cockpit and was apparently burned to death.

10-26-45. Craig Field, Alabama. At 1555, a North American AT-6D crashed at Craig Field, Selma, Alabama, killing pilot F/O Robert E. Bell, 20, and pilot-rated passenger Capt. Donald H. Wright, 24. The airplane was returning to the field after completing an instrument team ride. The airplane entered a right-hand traffic pattern for Runway 32. The tower radioed the airplane and advised the traffic pattern was to the left. The airplane was observed to enter a power-off 180-degree turn for an approach on Runway 32. The airplane continued in the turn and the bank became steeper until the airplane stalled and snapped over to the right, entering a spin and smashing to the ground. The pilots were killed instantly. The airplane did not burn but was destroyed utterly.

10-27-45A. Pennville, Indiana. At 0855 EST, a Douglas C-47B crashed three miles northwest of Pennville, Indiana, killing six fliers and seriously injuring ten others. Pilot Lt. Col. Richard E. Horner and co-pilot Maj. Emil L. Sorenson received serious injuries. Passengers killed in the crash were: Cpl. S. Feldman, Capt. F.F. Fitzgerald, 1Lt. L. Karpashewich, 1Lt. S. Liebfeld, Capt. J.S. Littrell, Maj. J.G. Magoffin. Passengers seriously injured in the crash were: Lt. (j.g.) R.E. Bowe (USN), Capt. R.W. Bratt, 2Lt. A.L. Flocken, 2Lt. D.L. Hersrud, Capt. W.A. Lien, 1Lt. J.G. Peterson, 1Lt. L.E. Schumacher, 1Lt. F.W. Zwar. The airplane took off at 0813 EST from Wright Field, Dayton, Ohio, on a transport flight to Wold-Chamberlain Field, Minneapolis, Minnesota. Investigators stated, "After take-off the aircraft climbed to 4,000 feet and leveled off with cruising power settings. The aircraft was trimmed and put on auto-pilot. Approximately 30 minutes later Col. Horner glanced down and noticed a decided decrease in airspeed. The ship was taken off of auto-pilot and at this time it was noted by both pilot and co-pilot that the manifold pressure had dropped off to 20" Hg. Both pilot and co-pilot immediately started a complete cockpit check to determine the cause of the trouble. They were unable to find the source of the difficulty and called to Capt. Lien, who was riding as passenger, to come forward to and go through the check with them again. A third cockpit check was performed without results. Maj. Sorenson indicated to Col. Horner that he would continue looking for the trouble while the Colonel picked a suitable field and got into position for landing. At this time the pilot and co-pilot estimated that they had an altitude of approximately 1,500 feet. Col. Horner selected a field and took up a heading of 270 degrees in an attempt to land. On final approach he called to Maj. Sorenson for gear down and flaps down. Maj. Sorenson responded immediately. Col Horner arrived over the field that he had picked for landing with excessive speed and slightly high. The C-47 touched down and ballooned [back into the air], hitting [the ground] approximately 100 yards further on. It was evident by now that they would be unable to execute a successful landing in this field. Col. Horner Called for gear up and flaps up and attempted to utilize the excessive speed to get the C-47 back into the air in an effort to clear five groups of trees bordering a winding creek. The C-47 became airborne again and did successfully clear the first three groups of trees, slanting down through the fourth [group of trees] and landed flat about 100 yards further on and skidded straight ahead. The C-47 skidded with considerable force into the fifth group of trees. One large tree entered the fuselage just aft of the radio compartment, causing extensive damage to the ship and personnel.... [The] loss of power still remains undetermined."

10-27-45B. Acton, Indiana. At 1715 CST, a Vultee-Stinson L-5 collided with an electric power high-tension line and crashed one-quarter mile north of Acton, Indiana, killing rated-glider pilot F/O Edward B. Watson and seriously injuring rated glider pilot F/O Chester B. Zaucha. The airplane took off from Atterbury Army Air Field, Columbus, Indiana, on a flight to Baer Field, Fort Wayne, Indiana. The airplane was observed maneuvering at low altitude prior to the accident. The pilot evidently saw the high-tension line and had pulled up abruptly but was unable to avoid the collision. The propeller and port wing struck the line, severing the wing and sending the airplane out of control. The airplane smashed to the ground in a clover field. The airplane did not burn but was destroyed utterly. Investigators noted that the airplane was 11 miles off course. Investigators speculated that the pilot was lost and attempting to ascertain his location or was deliberately buzzing the area.

10-29-45. Chandler, Arizona. At 0910 MST, a North American AT-6D crashed 17 miles southwest of Chandler, Arizona, killing instructor pilot Capt. Leslie P. Cles and Philippine Air Force student 3Lt. Andres O. Bunda. The airplane took off from Luke Field, Phoenix, Arizona, on a routine student training flight. Witnesses on the ground observed the airplane performing aerobatic maneuvers at a considerable altitude. The airplane entered a spin from which the pilots were unable to recover. The pilots bailed out at very low altitude and were killed when they struck the ground near where the airplane slammed into the ground and exploded.

10-31-45. Palm Springs, California. At 1250, a North American AT-6F crashed ten miles southeast of Palm Springs, California, killing pilot Capt. Marvin E. Moore and passenger TSgt. Ralph H. Wahlman. The airplane took off 1215 from San Bernardino Army Air Field, San Bernardino, California, on a pilot proficiency flight. The two fliers bailed out of the airplane at an altitude too low to allow their parachutes to deploy. They both fell to their deaths. Investigators were unable to determine what caused the occupants to abandon the aircraft at very low altitude. Investigators speculated that a successful forced landing could easily have been performed on the terrain surrounding the crash site.

November

11-1-45A. Ryderwood, Washington. At an unknown time after 1600 PST, a Consolidated TB-24J (42-73191) flying in instrument conditions collided with rising terrain and crashed near Ryderwood, Washington, killing the crew of four. Killed in the crash were: 2Lt. John A. Norris, pilot; F/O Robert C. McClane, co-pilot; Sgt. Lowell S. Bauer, radio operator; Sgt. John J. Sullivan. The subject airplane was flying in the number-three position of a four-ship flight that took off at 1524 from McChord Field, Tacoma, Washington, on a flight to Portland Army Air Base, Portland, Oregon. The airplanes were flying in a four-ship diamond formation. About 20 minutes after take-off, the formation encountered rain and decreasing visibility. Five minutes later the formation encountered a wall of clouds. The number-two and number-four ships went on instruments. B-24J #42-73371, flying in the number-two position, made a climbing 180-degree turn to the left and B-24J #42-109927, flying in the "slot" position, made a climbing 180-degree turn to the right. B-24J #42-109927 landed safely at Redmond, Oregon, and B-24J #42-73371 landed safely at Arcata, California. The pilot of B-24J #42-73371 reported that he observed a mountain about 100 feet off of his starboard wingtip after his climbing turn to the left. The lead aircraft (B-24J #42-64160) and the number-three ship (B-24J #42-73191) continued through the poor weather and soon both collided with rising terrain. B-24J #42-73191 collided with a mountain at an elevation of 2,100 feet. The airplane exploded violently into flames upon impact and the crew was killed instantly. The airplane was not located until 12-3-45. The lead airplane also collided with rising terrain, killing the entire crew. See 11-1-45B.

11-1-45B. Ryderwood, Washington. At an unknown time after 1600 PST, a Consolidated TB-24J flying in instrument conditions collided with rising terrain seven miles southwest of Ryderwood, Washington, killing pilot 2Lt. Dominick DiVito and pilot 2Lt. (first name unknown) Heyl. Two unknown crewmembers were also killed. The airplane was the lead aircraft of a four-ship flight that took off from McChord Field, Tacoma, Washington, at 1528 on a flight to Portland Army Air Base, Portland, Oregon. The subject airplane and the number-three ship failed to return to base and were declared missing. The subject airplane was found to have crashed into the side of a mountain at an elevation of 2,300 feet. The airplane was located on 12-10-45. The number-three ship crashed into a mountain at an elevation of 2,100 feet one mile from the subject airplane's crash site. It was found on 12-3-45. See 11-1-45A.

11-3-45. Gila Bend, Arizona. At 1023 MST, a North American AT-6D crashed two miles northeast of Gila Bend, Arizona, killing pilot 2Lt. Richard W. Windgassen and passenger Cpl. Richard L. Furgerson.

The airplane was part of a two-ship flight that took off at 1008 from the Army Air Field at Ajo, Arizona, on a routine flight. The airplanes were flying in formation at about 3,500 feet indicated altitude, the subject airplane flying on the leader's starboard wing. About 15 minutes after take-off, the subject airplane left the formation. A civilian motorist observed the subject airplane drop out of formation, diving to about 200 feet agl and then performing a barrel roll to the right. The airplane recovered in a nose-high attitude and stalled. The airplane fell off into a spin at an altitude too low to allow recovery. The airplane smashed to the ground in a nose-down attitude and burst into flames, killing the occupants instantly.

11-5-45A. St. Clair, Michigan. At 1315 EST, a Republic P-47D and a P-47N collided in mid-air and crashed three miles northwest of St. Clair, Michigan, killing French Air Force pilot 1Lt. Georges R. DeFeraudy aboard the P-47D. French Air Force pilot Sgt. Henri Tomi was able to parachute to safety from the P-47N. He was uninjured. The airplanes took off from Selfridge Field, Mt. Clemens, Michigan, on a routine training flight. Lt. Deferaudy was leading the formation and the P-47N was flying on the leader's port wing. The flight performed several turns before the leader made a climbing turn to the left. The wingman was unable to stay in position and cut across the inside of the turn, ending up in a position ahead and below the leader. The leader continued the maneuver and slammed into the P-47N from above. The P-47D's port wing severed the tail section of the P-47N, sending the P-47N into a flat spin. The P-47N pilot bailed out immediately and parachuted to safety. The P-47D spun to earth and exploded into flames.

11-5-45B. Sherman, Texas. At 1315 CST, a North American AT-6D suffered a catastrophic structural failure of the starboard wing and crashed 10 miles southwest of Perrin Field, Sherman, Texas, killing pilots 2Lt. Harold S. Friedman and F/O Philip J. Gabriel. The airplane took off from Perrin Field on a routine training flight. The airplane was observed in a spin at about 2,000 feet agl. The pilot failed to recover and the airplane began to break up. The starboard wing failed and airplane smashed to the ground out of control, striking the terrain at an angle of about 80 degrees. The tail section failed and separated as the airplane spun to earth. Wreckage was scattered over a wide area. The airplane did not burn. Investigators were unable to determine the series of events that preceded the airplane entering the spin.

11-6-45A. Atlanta, Georgia. At 1713, a North American TB-25J crashed 4 miles south of the Army Air Field at Atlanta, Georgia, killing pilot 1Lt. Davis F. Jones and engineer-rated passenger SSgt. Robert L. Smith. The airplane took off from Atlanta Army Air Field on a flight to Eglin Field, Valparaiso, Florida. The airplane took off on Runway 18 and climbed to about 300 feet when excessive black smoke was seen trailing from the port engine. The airplane began settling in a nose-high position. The airplane then snapped over into a vertical bank to the left, smashing to the ground in a wooded area and bursting into flames upon impact. The occupants were killed instantly.

11-6-45B. Chico, California. At 1909, a Boeing B-29 crashed six miles northeast of the Army Air Field at Chico, California, killing three crewmembers. Crew chief SSgt. Jerome L. Seck was seriously injured. Pilot 1Lt. Wayne B. Powell, co-pilot 1Lt. Robert M. Evans and flight engineer 1Lt. Charles S. Moffett were killed in the crash. The airplane took off from Lowry Field, Denver, Colorado, on a ferry flight to Chico Army Air Field. The airplane was cleared on an instrument flight plan via Cheyenne, Wyoming; Reno, Nevada; and Red Bluff, California. The airplane arrived in the area of Chico Army Air Field and requested landing instructions. The airplane was cleared to land and entered the traffic pattern. The pilot radioed in while turning onto the base leg. Moments later, control tower personnel observed an orange glow in the distance. The control tower operator radioed the subject airplane but received no answer. Investigation revealed that the airplane smashed to the ground in a flying attitude with power. Investigators speculated that the pilot had failed to set his altimeter to the field setting.

11-8-45A. Paxton, Illinois. At 0753, a North American AT-6D crashed three miles south of Paxton, Illinois, killing pilot Capt. Hans J. Grasshoff. Passenger Capt. Julian D. McDonald parachuted to safety and was uninjured. The airplane took off at 0732 from Chanute Field, Rantoul, Illinois, on a routine pilot proficiency flight. Capt. McDonald needed 30 minutes of flying time for the month and Capt. Grasshoff agreed to fly a 30-minute instrument training flight with Capt. McDonald. Capt. McDonald flew the airplane from the rear cockpit and climbed to about 4,000 feet indicated altitude. The pilots both smelled gasoline fumes and elected to return to the field. As the airplane flew back toward the field, Capt. Grasshoff, flying from the front cockpit, began a slow, long descent. The fuel fumes became over powering and Capt. McDonald opened his canopy to let in some air. Capt. Grasshoff pointed to droplets of gasoline spotting on the windscreen and held his nose as he motioned to Capt. McDonald. Just after descending through 3,000 feet, the front cockpit erupted into flames. The airplane rolled to the left. Capt. McDonald bailed out immediately and parachuted to safety. The airplane went out of control and smashed to the ground at a steep angle in a bank to the right. The airplane exploded into flames upon impact. The pilot in the front cockpit bailed out at low altitude but was unable to pull his ripcord, causing him to fall to his death. Examination of the pilot's body revealed that he had suffered severe burns to the legs, arms, torso, neck and face, but was killed in the fall. Investigation revealed that the cockpit fire was caused by the failure of the fuel pressure indicator line.

11-8-45B. Chemung County Airport, New York. At 1038, a North American B-25J and a Republic P-47N collided in mid-air and crashed at Chemung County Airport, New York, killing six fliers. 1Lt. Merle R. Capp was killed in the crash of the P-47N. Killed in the crash of the B-25J were: William J. Driver, pilot; 2Lt. Edwin S. Tyler, co-pilot; 2Lt. John L. Corum, navigator; Pfc. Verrel C. Shook, engineer; Cpl. Wilber G. Hackner, radio operator. The airplanes were part of a large group of ships that had participated in a Treasury Department Victory Bond Airshow at Chemung County Airport. There were ten airplanes taking off to return to Rome Army Air Field, New York. The B-25 took off third and the P-47 took off sixth. The ten airplanes took off and were circling the field at various altitudes. Investigators stated, "At approximately 1038, B-25J #44-30203 was observed flying at approximately 100 feet altitude, approaching the field from the east. The P-47N was also approaching from the east at approximately 500 feet altitude and considerably in the rear of the B-25. The B-25 was making a low-level approach over the East-West Runway, and the P-47 was making a diving approach in the same general direction. It is apparent that the pilot of the P-47 was unable to see the B-25 due to his vision being obstructed by the forward portion of the fuselage and engine. The B-25 pilot gave no indication of having seen the P-47 at any time, and as there was not radio contact between these two aircraft, it is assumed each aircraft had attempted a similar maneuver without the knowledge of the other, and the [collision was the result]. Over the boundary of the field, the P-47 struck and sheared off the right wing of the B-25 outboard of the right engine and thereupon crashed to the ground. The P-47 burst into flames immediately upon impact with the B-25. Upon being struck by the P-47, the B-25 crashed to the ground and immediately burst into flames.... Portions of both aircraft were scattered over an area approximately one-quarter mile wide by one mile long, all on the Chemung County Airport, doing slight damage to the runways and sodded portions of the field."

11-10-45A. Middlesboro, Kentucky. At 1105 CST, a North American RB-25H flying in instrument conditions collided with rising terrain and crashed seven miles west of Middlesboro, Kentucky, killing the pilot and four passengers. Killed in the crash were: Col. Norris Perry, pilot; Cpl. C.S. Anderson, passenger; SSgt. C.A. Dacus, passenger; Sgt. C.C. Marsh, passenger; Maj. F.D. Hayley, passenger. The airplane took off from Bowman Field, Louisville, Kentucky, on a flight to Knoxville, Tennessee. The airplane encountered deteriorating weather conditions shortly after take-off. Investigators stated, "Shortly before the crash, the airplane was observed flying very low along a valley in the Cumberland Mountains approximately 55 miles from Knoxville and in a southeasterly direction.

The airplane entered a cul-de-sac with a high mountain directly ahead. This mountain (named Canada) has an elevation of approximately 2,600 feet and an incline of approximately 60 degrees. The pilot apparently began pulling up as he noticed the steepening incline. Approximately 200 feet up the slope, the airplane began shearing off treetops. The pilot must have maintained a very high angle of attack as the airplane kept climbing for approximately 500 more feet before the main impact. With the main impact, the airplane exploded and the engines landed 50 feet higher. All of the occupants were killed instantly. At the time of the crash, observers stated that the top of the mountains were in the overcast and visibility was very low."

11-10-45B. Lebanon, Tennessee. At 1345 CST, a Douglas A-26C flying in instrument conditions crashed seven miles west of Lebanon, Tennessee, killing the pilot and four passengers. Killed in the crash were: Maj. Clarence E. McClaren, pilot; TSgt. John B. Lamb, engineer; 1Lt. Raymond F. Fischedick, passenger; Cpl. Charlie R. Waddell, passenger; Capt. Paul B. Young, passenger. The airplane took off from Lake Charles Army Air Field, Lake Charles, Louisiana, on a flight to Nashville, Tennessee. At 1340, the airplane arrived in the area of Nashville and the pilot was cleared to let down from 3,000 feet and land at Nashville. Investigators stated, "Letdown was to be made outbound on the northeast leg of the range, procedure turn and inbound on northeast leg of the low cone. This aircraft was seen heading north and turning back south by a farmer in the field adjoining the wooded area of the crash. This path of flight and relative geographic location indicates that the pilot was in his procedure turn on a heading approximately south back into the northeast leg when he crashed, first hitting trees approximately 70 feet high and subsequently into the ground at 1545 CST." Investigators speculated that the pilot had either misinterpreted the reading of his altimeter or unconsciously continued his let-down in the procedure turn, causing the pilot to fly the airplane into the ground.

11-11-45. Lincoln, Nebraska. At 1755 CST, a North American AT-6A on an unauthorized flight dove into the ground two miles northeast of the Army Air Field at Lincoln, Nebraska, killing pilot 1Lt. William M. Nelson, 23. The pilot, who was not authorized to fly the subject airplane, taxied the AT-6 without unfastening the rear tie-down line, which was snapped off as the airplane taxied away. The pilot taxied erratically and the airplane left the taxiway and rolled on the grass. The pilot was able to regain the pavement and then proceeded to take off to the south on the main taxiway. The control tower operator attempted to contact the airplane as it began its take-off roll and after it became airborne but was unsuccessful. Nothing was heard from the airplane and it failed to return to the field. On 11-12-45 at about 1115, a farmer discovered the airplane wreckage in his field, which was two miles

northeast of the airport from which the airplane took off. Investigation of the wreckage indicated that the airplane entered a split-S maneuver into the terrain, striking the ground in an inverted vertical dive. The airplane did not burn but was destroyed utterly. Investigation revealed that the pilot was distraught over the death of his brother and marital difficulties and had taken his own life. Investigators noted that the pilot had been drinking heavily during the afternoon before the crash and that a suicide note had been found among his personnel effects.

11-12-45. Mountain View, Arkansas. At 1745, a North American B-25J crashed three miles south of Mountain View, Arkansas, killing passengers W/O James D. Schulgen and Maj. Vincent E. Montgomery. Pilot 1Lt. Ralph M. Speck and engineer Pfc. Arthur N. Fillman parachuted to safety and were uninjured. The airplane took off from Grand Island, Nebraska, on a flight to Grenada, Mississippi. The airplane was flying at about 4,000 feet southeast of Springfield, Missouri, when the pilot noticed that the port engine was on fire. The pilot shut down the engine, feathered the propeller and trimmed the airplane for single-engine flight. A few minutes later the starboard engine erupted in flames. The pilot immediately gave the passengers and crew chief orders to abandon the airplane. The crew chief bailed out immediately and the pilot bailed out a couple minutes later at about 2,300 feet. The airplane entered a shallow turn to the left, remaining in this attitude until it struck a wooded hillside and exploded. Investigation revealed that the passengers went in with the airplane. Investigation revealed that both passengers were in the waist compartment and had parachutes fastened to their harness at the time of the crash. Investigation of the wreckage indicated that the passengers had not released the emergency escape hatch in the waist compartment.

11-14-45. Lawton, Oklahoma. At 1815, a Republic P-47N crashed six miles northeast of Lawton, Oklahoma, killing pilot 2Lt. George W. Roesler, Jr., 22. The airplane took off at 1556 MST from La Junta Army Air Field, La Junta, Colorado, on an administrative flight to Tinker Field, Oklahoma City, Oklahoma. The airplane failed to arrive at Tinker Field and was declared missing. The pilot's body was found the next day. He had apparently been struck by a part of the airplane as he bailed out and was unable to deploy his parachute. The pilot was found about one and a half miles from the wreckage of the airplane. Investigators could not determine why the pilot had to leave the airplane. It was noted that the pilot was about 65 miles off course.

11-16-45A. Oklahoma City, Oklahoma. At 1455, a Lockheed F-5G crashed at one mile northeast of Will Rogers Field, Oklahoma City, Oklahoma, killing pilot 2Lt. Audry W. Hunt, 28. The airplane took off from Will Rogers Field on a local flight. The airplane returned to the field and entered the traffic pattern. The airplane was seen to fall off on a wing and spin to the ground while turning from the base leg to final. The airplane burst into flames upon impact. Investigators speculated that one or both engines lost power while the airplane was turning at low airspeed with landing gear in the extended position.

11-16-45B. Lawton, Oklahoma. At 1545, a North American P-51D crashed at Lawton, Oklahoma, killing pilot Maj. Harold B. Williams, 28. The airplane was flying in the number-three position of a three-ship flight that took off from Will Rogers Field, Oklahoma City, Oklahoma, on a ground-strafing mission at Fort Sill, Oklahoma. The aircraft were to work with ground units in the Fort Sill area. The aircraft made several strafing runs. On the last strafing run, the first two aircraft pulled up in climbing turns. The subject airplane performed a steep pull up and then entered a violent roll to the right and then smashed into the ground in a 45-degree inverted dive. The airplane exploded violently into flames upon impact and the pilot was killed instantly. Investigators could not determine what caused the pilot to lose control of the airplane.

11-17-45. Hempstead, New York. At 1125, a Republic P-47N crashed at 126 Windsor Parkway, Hempstead, Long Island, New York, killing pilot 1Lt. Daniel A. Duncan, 23. The pilot was returning to Mitchel Field, Long Island, after completing a routine local flight. The airplane had been suffering propeller trouble just after take-off, but the pilot elected to continue with the flight. The pilot completed his mission, returned to the field and entered the traffic pattern. The pilot refused his landing attempt and applied throttle to initiate a go-around. The airplane could not climb away and the pilot attempted to crash land in an open field. The airplane struck a tree and a chimney with the port wing before mushing toward the ground. The pilot attempted to stretch the glide but the airplane smashed into a private residence, bursting into flames upon impact. The structure was completely destroyed. No one was injured or killed on the ground. It was later speculated that the propeller trouble contributed to the accident.

11-18-45. Ashland, Mississippi. At 1930, a Beech AT-7 flying in poor weather crashed five miles southwest of Ashland, Mississippi, killing five fliers. The airplane took off at 1758 from Scott Field, Belleville, Illinois, on a flight to Selman Field, Monroe, Louisiana. The airplane encountered severe thunderstorms and was observed circling the area of Ashland at low altitude. Civilians on the ground observed a red glow on the horizon a short time later. Investigation revealed that the airplane collided with trees in a 50-degree banked diving turn to the left. The airplane cartwheeled into the terrain and exploded into flames upon impact, killing the occupants instantly. Killed in the crash 1Lt. Leroy E. Reuters, pilot; 2Lt. Robert A. Sudikatus, co-pilot; 2Lt. Alfred S. Hann, navigator; F/O Robert O. Rettke, navigator; 2Lt. Joseph M. Gawal, navigator.

11-19-45. Castroville, Texas. At 1711, a North American AT-6D crashed five miles northeast of Castroville, Texas, killing instructor pilot Capt. Paul M. Wipperfurth and pilot Maj. John M. Horgan. The airplane took off at 1613 from Kelly Field, San Antonio, Texas, on a check-off flight for the major. The airplane failed to return to the field and was declared missing. The airplane was located the next day. Civilians in the area heard the crash but did not associate it with an airplane accident. Investigation of the wreckage indicated that the airplane had collided with a tree before smashing into the ground at high speed. The airplane did not burn but was destroyed utterly.

11-20-45. Guadalupe, California. At 1450, a North American P-51D crashed five miles north of Guadalupe, California, killing pilot 2Lt. John W. Rowley. The P-51D took off from Santa Maria Army Air Field, Santa Maria, California, on a routine aerobatics flight. Investigators stated, "From the testimony of witnesses, their attention was attracted by the loud noise of an airplane engine and looking up they saw a plane diving with power on. At about 4,000 feet the ship pulled out and climbed vertically. Near the top of the climb the pilot evidently did some rolls and then stalled out, going into a spin. The ship hit the ground after a good many turns of spin and the wreckage indicated a flat spin. Also, the propeller was bent in such a manner as to indicate power on. The pilot had been killed instantaneously by decapitation. It is believed by the accident board that the airplane went from a power stall directly into a flat spin from which the pilot could not recover."

11-21-45A. Bolton, Mississippi. At 1505, a North American P-51D crashed five miles southwest of Bolton, Mississippi, killing 2Lt. Edwin T. Smith, 26. The airplane was part of a four-ship flight that took off at 1450 from Jackson Army Air Base, Jackson, Mississippi, on a flight to Hensley Field, Dallas, Texas. Investigators stated, "At approximately 2,500 feet over Bolton, Mississippi, Lt. Smith received permission from the leader to return to Jackson AAB because of an oil leak. Pilot made a sharp turn to the left beneath the formation and the plane then went into a spin or tight spiral to the right. Pilot regained control momentarily, but the airplane fell into a secondary spin and crashed before he could again recover. Lt. Smith was killed instantly and the airplane was demolished. The safety belt was still fastened."

11-21-45B. Birmingham, Alabama. At 1623, a Douglas A-26C crashed while attempting to land at the Army Air Field at Birmingham, Alabama, killing crewmember 1Lt. Randolph W. Allen. Pilot 2Lt. Bradley C. Crane and co-pilot 1Lt. Walter B. Daffin suffered serious injuries. The airplane took off from Lake Charles Army Air Field, Lake Charles, Louisiana, on a flight to Birmingham Army Air Field. The airplane arrived in the area of the field while flying on instruments and was given permission from the Birmingham control tower for a standard instrument let-down. The pilot came out of the overcast and immediately went to contact. The airplane was lined up with Runway 18 and the pilot requested a landing on that runway. Permission was granted but the pilot was warned about a 30 mph crosswind component. The pilot was unable to stop the airplane on the wet runway surface. The airplane went off of the end of the runway, crashing through a chain link fence before smashing into a drainage ditch. The airplane smashed itself to pieces and burst into flames after coming to rest. Rescue personnel were on the scene immediately.

11-26-45. Douglas County, Oregon. A Curtiss C-46 flying in instrument conditions crashed at Douglas County, Oregon, killing the pilot and co-pilot. Nine crewmembers parachuted to safety. Passenger Sgt. Robert T.W. Neal, Los Angeles, California, bailed out but was not rescued and declared missing. On 12-18-59, his parachute was found tangled in the upper branches of a 200-foot fir tree that was felled in the Lake Creek area by two lumberjacks. The parachute harness was found buckled and fixed in a way that indicated that it probably contained the body of the missing flier at one time. One of the surviving crewmembers had been rescued from the upper branches of a tall fir tree a short time after the crash. The airplane had taken off from Sedalia Army Air Field, Missouri, on a flight to McChord Field, Tacoma, Washington. Published accounts of the accident reported that the airplane ran out of fuel while flying in foggy conditions.

11-27-45A. Gila Bend, Arizona. At 1030 MST, two North American P-51D airplanes collided in mid-air 27 miles northwest of Gila Bend, Arizona, killing Chinese Air Force pilot Sub-Lt. Shu-Chuang Chao aboard P-51D #44-74434. AAF pilot 1Lt. Ralph H. Donnell was uninjured and able to make a safe landing at Luke Field, Phoenix, Arizona, in P-51D #44-74691. The AAF pilot was leading a flight of four ships that had taken off from Luke Field on a formation training flight. Sub-Lt. Chao was flying in the number-two position. The formation was at 4,000 feet msl when the leader began a series of cross-over turns with the Chinese students. The number-two ship pulled up into the leader from underneath while attempting to regain position on the leader. The propeller of the lead aircraft chopped into the fuselage of the number-two ship just aft of the cockpit. The number-two ship began to break up as it fell out of control. The number-two ship smashed to earth and exploded into flames. The leader was able to regain control of his ship and make a safe landing.

11-27-45B. Missoula, Montana. At 1800 MST, a North American TB-25J flying in instrument conditions collided with rising terrain and crashed 12 miles southeast of Missoula, Montana, killing three fliers. Pilot Capt. William G. Oeder, co-pilot Capt. Robert V. Peterson and SSgt. John Brown were killed in the crash. The airplane took off from Rapid City Army Air Base, Rapid City, South Dakota, on a pilot proficiency flight to Missoula, Montana. The airplane was on an instrument clearance and assigned to fly at 9,000 feet msl. The pilot radioed in at Billings, Montana, at 1641 MST, requesting to cancel his instrument flight plan and change to contact. This request was granted. The airplane was later sighted circling the area of Deerlodge, Montana, at low altitude. The starboard wing struck a tree and the airplane rolled to the right until it struck the terrain inverted and exploded. The occupants were killed instantly.

11-28-45A. Dudley, North Carolina. At 1015, a Republic TP-47D dove into the ground four miles west of Dudley, North Carolina, killing pilot F/O Lester W. Tinnes, 22. The airplane was part of a two-ship flight that took off from Seymour-Johnson Field, North Carolina, on a navigation flight. The flight was led by Capt. Herman K. Freeman. The two pilots had agreed to execute a "compressibility dive" after completing the navigation mission. The flight leader stated that he had properly briefed the pilot on the subject of compressibility dives on the ground prior to the flight. After completing the navigation flight, the two airplanes climbed to 31,000 feet msl. The flight leader pushed over into a 65-degree dive and held it until he reached the maximum allowable speed of 490 mph. The flight leader began his pull-out and regained level flight. Instead of leveling off at 10,000 feet, the flight leader allowed his airplane to continue in a shallow dive until he leveled off at 3,000 feet indicated altitude. The flight leader could not locate the wingman and tried to contact him by radio but was unsuccessful. The flight leader flew around the area a few minutes before sighting a column of smoke rising from the ground. After further observation, the flight leader confirmed the fire to be the wreckage of the wing aircraft. Investigation revealed that the subject airplane had smashed into the ground at high speed and at an angle approaching 90 degrees, exploding violently into flames upon impact. The pilot had made no apparent effort to leave the airplane.

11-28-45B. Auburn, California. At 1910, a Douglas C-47B crashed three miles northwest of Auburn, California, killing the crew of three and five passengers. Fifteen passengers were injured, eleven of them seriously. Pilot 1Lt. Jerry Cebe, co-pilot Maj. Louie G. Martin and engineer Cpl. Paul G. Bitterle were killed in the accident. The airplane took off at 1541 PST from Palm Springs, California, on a transport flight to McClellan Field, Sacramento, California. The flight had originated at Davis-Monthan Field, Tucson, Arizona, at 1407 MST. The airplane was cleared for contact flight to Fresno, California, and then on an instrument clearance to Sacramento. The airplane was cleared to fly at 3,000 feet. The airplane flew into a hill at an elevation of approximately 1,500 feet msl while in a flying attitude with power. The airplane smashed itself to pieces as it crashed through trees, losing a wing and the tail section before coming to rest against another hill. Passengers killed in the crash were: T/4 Attilio G. DeMattei, T/3 Rolf G. Hecht, T/4 Charles S. Higa, T/4 Lawrence T. Iwamoto, T/5 Saburo Kmai. Passengers seriously injured in the crash were: T/5 Norio Akutagawa, T/5 Hiroyuki Biramoto, T/5 Robert Y. Iwamoto, T/5 Takemi Kajiwara, Cpl. Isamu Kanekuni, T/4 Howard M. Murakami, Cpl. Hideo Nakagawa, Cpl. Nick M. Shimazu, Cpl. Tomio Sunahara, Cpl. Raymond T. Tanaka, T/4 Kiyoto Yokoyama. Passengers injured in the accident were: T/5 Masaki Higa, T/5 Robert I. Ikeda, T/5 Toluo Kaneshige, Cpl. Masami O'Hara.

11-29-45A. Brandt, Ohio. At 0910, a North American P-51D crashed one and a half miles south of Brandt, Ohio, killing pilot Capt. Edmund D. Montagno, 29. The airplane took off from Patterson Field, Ohio, at 0901 on a local flight. The airplane took off from Runway 34 and crashed into the ground seven miles from the field a few minutes later. The airplane was observed to descend out of the overcast apparently under control. Witnesses on the ground stated that the wings were rocking as the airplane descended out of the overcast and into the ground. The airplane burst into flames upon impact. Civilian structures on the ground were damaged.

11-29-45B. Nanquin, Virginia. At 1330, a Republic P-47N crashed four miles northeast of Nanquin, Virginia, killing pilot 1Lt. Harold W. Cheers, 22. The airplane took off at 1224 EST from Wildwood, New Jersey, on a flight to Charleston, South Carolina. The airplane encountered severe icing conditions on the flight. The airframe picked up considerable ice and the airplane became extremely heavy and went out of control. The airplane smashed to the ground and burst into flames, killing the pilot instantly.

11-29-45C. McCook, Nebraska. At 2313 CST, a Beech C-45F crashed while in the traffic pattern for McCook Army Air Field, McCook, Nebraska, killing pilot Lt. Col. Morgan A. Giffin, 30, and co-pilot 1Lt. Graham C. Woodlaw. The airplane took off from Salina, Kansas, on an administrative flight to McCook Army Air Field. At 2303, the pilot arrived in the area of McCook and requested landing instructions. At 2308 the airplane was 15 miles east of the field and the pilot was given landing instructions. The pilot acknowledged. A few minutes later the subject airplane passed over the field and began a turn to enter the traffic pattern. The airplane passed out of the view of tower personnel, crashing a couple minutes later while flying in the pattern northeast of the field. Investigation revealed that the airplane dove into the ground at angle of 45 degrees with both engines producing power. Weather was not considered a factor in the accident.

11-30-45. Aurora, Indiana. At 2048 CST, a North American AT-6F collided with a tree and crashed two miles west of Aurora, Indiana, killing pilot 2Lt. Robert R. Frank, 20. The airplane took off from Wright Field, Dayton, Ohio, on a local flight. The airplane was not cleared to leave the Wright/Patterson local flying area. The pilot, who was soon to be separated from the AAF, had been refused a clearance to Cincinnati, Ohio. Investigators stated, "At approximately 2000 EST, the aircraft was seen over Aurora, Indiana, which is outside the local flying area and home town of the subject pilot. The pilot proceeded to buzz and fly low up and down the street on which his family resides. He was flying with landing lights on, turning them on and off intermittently. After making several passes, he flew down an adjacent valley approximately one mile, turned his landing lights off and started a turn to the right into a hill covered with trees. While in the turn of 30 or 40 degrees bank, the aircraft struck a 40-foot tree about 12 feet above the ground and severed the right wing. This caused the aircraft to roll and crash inverted about 450 feet beyond. Witnesses' statements and cuts in the earth indicate the engine was functioning properly at the time of the crash." Weather was not considered a factor in the accident.

December

12-3-45A. Tuskegee, Alabama. At 1115 CST, two North American B-25J airplanes collided in midair 15 miles SSW of the Army Air Field at Tuskegee, Alabama, killing two students. A/C Otis E. Marshall and A/C Spencer P. Isabelle were killed in the crash of TB-25J #44-30623. Instructor 1Lt. Edward J. Williams and student A/C Donehue Simmons were uninjured and able to safely land TB-25J #44-30637. The two airplanes were part of a three-ship flight that had taken off at 1045 CST from Tuskegee Army Air Field on a routine formation training flight. TB-25J #44-30623 was flying in the number-two position; TB-25J #44-30637 was flying in the number-three position. The formation had performed various maneuvers at about 4,000 feet. The flight had performed a peel-off and then attempted to rejoin. The instructor in the number-three ship noticed that a collision with the number-two ship was imminent and took evasive action. The instructor put the number-three ship into a steep bank to the left but was unable to avoid the collision with the number-two ship, which was flying toward the lead ship low and fast. The number-two ship apparently climbed into the path of the number-three ship. The number-two ship's tail was severed, sending the airplane out of control. The number-two ship plunged to the ground and exploded into flames upon impact. The number-three ship suffered serious damage to the nose and propellers but the instructor was able to make a safe landing at Tuskegee Army Air Field.

12-3-45B. Stonyford, California. At 1130, a Vultee BT-13B flying in very poor weather crashed into St. John's Mountain ten miles northwest of Stonyford, California, killing pilot F/O John F. Lanzatella and passenger Sgt. Leroy M. Craig. The airplane took off at 1020 PST from Red Bluff, California, on a ferry flight to Hamilton Field, San Rafael, California. The airplane never arrived at Hamilton Field and was declared missing. A civilian search party, working with military authorities, found the wreckage on 1-24-46 at about 1730. Investigators speculated that the pilot had attempted to stay under deteriorating weather conditions and collided with the side of the mountain. The airplane was about 12 miles off course. Civilians reported that the mountains were obscured by clouds and that heavy rain and low visibility existed in the area the day of the accident.

12-8-45. Billings, Montana. At 0213 MST, a Douglas C-47B crashed one-quarter mile south of the Billings Municipal Airport, Billings, Montana, killing Northwest Airlines pilot Capt. George Miller and First Officer Vernon W. Pfannkuck. Seventeen passengers were killed. Four passengers survived with very serious injuries. Investigators stated, "Aircraft was being used by Northwest Airlines for Army charter work of transporting veterans and was cleared [on an IFR flight plan and assigned an altitude of 6,000 feet] from Fargo, North Dakota, to Billings, Montana. Pilot left Fargo at 2311 CST [12-7-45] and checked in over Nibbe [Montana] at 0159 MST and was cleared for a straight-in letdown. Pilot reported approaching the range station at 0206 MST. He next advised that he was CFR over town at 0210 MST. The aircraft was seen south of the field, and at the same time the pilot reported in. The time was 0211. Pilot had his lights on and aircraft was seen starting a 270-degree turn to the left so as to line up with the runway. Aircraft then disappeared in the snow and two minutes later the tower operator, watching for the airplane, saw lights for an instant, which were below the level of the field.... The aircraft disappeared below the rim rock and struck the trees at the same time. All evidence seems to point to the following: The pilot crossed the field, east to west, at approximately 800 feet above the terrain. He started a 270-degree turn so as to land into the north, from which a 17 mph wind was blowing. Apparently, in attempting to line up with the runway he wished to use, pilot descended too rapidly in the turn to the left and failed to allow for the possibility of a downdraft. At any rate, the pilot wound up below the level of the field, and while still in a turning attitude, the aircraft struck trees, crashed and burned. The aircraft was in a 35-degree bank when it struck the trees left wing low. Witnesses have reported that both engines were functioning properly at the time of the crash and if there were any mechanical failure, the crew did not mention it to the tower.... The flaps were found to be slightly over halfway down and the landing gear was in the down position at the time of the crash." Passengers killed in the crash were: 1Lt. Anthony W. Alansky, Cpl. Lorrell Cassell, T/4 Fred Chapman, Maj. Ray K. Craft, Sgt. Charles Ennen, Pfc. David F. Gillett, Sgt. Don Haley,

Pfc. Mike Hobbs, T/4 Virgil E. Kinne, TSgt. Glenn G. Marr, T/5 John M. Marshall, T/5 Ned Neasham, Pfc. Walter F. Orchard, T/4 Warren L. Parrish, Pfc. Clayton Thompson, SSgt. Thomas Thomsen, T/4 Adolph Tokie. Passengers seriously injured in the crash were: Cpl. Milford A. Barnes, T/5 Raymond D. Emerson, T/5 Emil A. Hasch, Pfc. Raymond Parkins.

12-10-45. Atlanta, Georgia. At 0423, a North American AT-6D flying in poor weather crashed three miles south of the Army Air Base at Atlanta, Georgia, killing pilot 1Lt. Richard W. Hall and pilot-rated passenger 1Lt. Spugeon Ellington. The airplane took off at 0157 EST from Winston-Salem, North Carolina, on a flight to Atlanta Army Air Base. The airplane arrived in the Atlanta area and requested a clearance on the Atlanta radio range. The airplane was cleared but weather had been deteriorating steadily. Weather at the time of the accident was reported as a 300-foot ceiling. Investigators speculated that the airplane was attempting to let down when it collided with the ground. The airplane exploded into flames upon impact, scattering wreckage and damaging civilian structures.

12-11-45. Grass Valley, Oregon. At 1715, a Beech AT-11 crashed eight miles northwest of Grass Valley, Oregon, killing pilot 1Lt. Chester W. Claflin and engineer Sgt. Ralph E. Balser. The airplane took off from Klamath Falls, Oregon, at 1527 PST on an instrument flight to Redmond, Oregon, to The Dalles, Oregon, to Pasco, Washington. The airplane was cleared contact on the last two legs and cleared IFR "500 feet on top" of the overcast for the leg to Klamath Falls. The airplane arrived at The Dalles at 5,500 feet and let down to 3,500 feet. The pilot climbed to 500 feet on top of the overcast for the leg to Pasco and then asked permission for the new clearance. The pilot was granted permission but did not acknowledge the clearance. The airplane was last heard attempting to contact Pasco control but the transmission was unreadable. The airplane failed to arrive at Pasco and was declared missing. A farmer who had heard the airplane fly over found the wreckage the next day. Investigators stated, "Weather conditions at the time of the accident indicated severe icing conditions. Local inhabitants reported heavy fog. The aircraft was found by a farmer who was searching for the aircraft after hearing a news broadcast to the effect that an aircraft was missing. On arrival at the scene of the accident, it was found that the aircraft struck the ground in a left wing low attitude with a nose low attitude of not more than 10 degrees. The aircraft hit and bounced twice with the bodies of the pilot and engineer being thrown forward through the windscreen approximately 75 feet from the wreckage. The distance from the point of impact to the wreckage was approximately 375 feet. The aircraft burned."

12-12-45. Adena, Ohio. At 1215 EST, a Republic TP-47N crashed at Adena, Ohio, killing pilot Col. William D. Tipton, 52. The airplane took off from Romulus, Michigan, at 1127 EST on a flight to Bolling

Field, Washington, D.C. Investigation revealed that witnesses on the ground observed the airplane make a few shallow dives and then climb to about 2,000 feet. The airplane then dove straight into the ground with no apparent effort by the pilot to recover or bail out. Investigators could not determine the cause of the accident. Weather was clear and the engine was heard to be running smoothly.

12-15-45. El Paso, Texas. At 1440 MST, a Vultee-Stinson L-5A crashed 50 miles ENE of Biggs Field, El Paso, Texas, killing pilot SSgt. Charles T. Beaver and passenger Pfc. Gordon R. MacIssac. The airplane was part of a two-ship flight that took off from Biggs Field on a short field landing exercise and a simulated aircraft search mission. The airplanes flew to the small liaison landing strip about 50 miles ENE of Biggs Field. The two airplanes shot landings and then made a full stop landing and shut down so the pilots could discuss the landing and take-off exercise. The airplanes then took off to conduct a simulated search for a Culver PQ-14 in the area. The subject airplane took off first and circled the field as the other aircraft took off. The subject airplane did not rendezvous with the wing ship and failed to return to the landing strip. After a brief search, the subject airplane was found to have crashed near the field. Investigation revealed that the subject airplane stalled out of a steep low speed turn and entered a spin at an altitude too low to allow recovery. The airplane spun to the ground about 450 feet southeast of the landing strip. The airplane did not burn but was destroyed utterly.

12-17-45. Bay St. Louis, Mississippi. At 1545, a Boeing B-17G crashed into Bay St. Louis four miles east of the Army Air Base at Gulfport, Mississippi, killing six fliers and injuring five others, two of them seriously. Killed in the crash were: 1Lt. Edward J. Matak, instructor pilot; 1Lt. Vernon J. Hansen, pilot; 2Lt. Francis P. Barrett, co-pilot; 2Lt. Frank F. Bartosz, pilot-rated passenger; 2Lt. Jack T. Donaldson, navigator; Cpl. Edward F. Kubala, engineer. Gunner Cpl. Alfred O. Emig, radio operator Ernest E. Miller and navigator/bombardier 1Lt. George E. Webster received minor injuries in the accident. Gunner Sgt. William F. Daniels and radio operator Pfc. Linwood W. Tracy were seriously injured. Investigators stated, "Aircraft was flying at 1,100 feet altitude on a training mission on Bay St. Louis when number-two engine began vibrating and smoking. Number-two engine was feathered and smoking stopped. At this time pilot called [the control tower at Keesler Field, Biloxi, Mississippi] stating that he had an engine out and was returning to the field. Tower acknowledged this transmission. Pilot called back saying that he was unable to read the tower and that he would call back when he was nearer the field. Immediately after passing over Gulfport Army Air Base, one crewmember in the nose of the airplane and one crewmember in the rear of the airplane noticed a marked decrease in rpm on the number-one

engine. Also, at the time this was noticed, pilot gave full power to number-three and number-four engines. At this time, full flaps were put down then retracted to one-third to one-half flaps. Losing altitude at a fast rate, pilot had to make a turn into the dead engines to avoid trees on the water's edge at which time slow airspeed [and] flaps made it impossible for him to regain level attitude before striking the water. Wreckage of the aircraft being widely scattered and completely demolished makes conclusive investigation of the engines and controls impossible."

12-18-45A. Alpine Town, Utah. At an unknown time between 1100 and 1400 MST, a Vultee BT-13B crashed into rising terrain five miles northeast of Alpine Town, Utah, killing pilots 2Lt. Smith C. Coffelt and 2Lt. Jack D. Lester. The airplane took off from Salt Lake City Army Air Base, Salt Lake City, Utah, on a local flight. The airplane failed to return and was declared missing. The airplane wreckage was located a few days after the accident. Investigators stated, "Investigation at the scene of the accident indicates that the pilot was flying up a canyon; turning to the left into a branching canyon, the pilot suddenly observed it to be a dead end. Noting the canyon walls and dead end were too high to fly over he attempted a 180-degree turn to the left to leave the canyon. While in the turn the airplane was flown into a stall and mushed into the mountainside on a plane parallel to the mountainside. Upon striking mountainside the aircraft burst into flames."

12-18-45B. Cedar Keys, Florida. At 2130 EST, a Beech C-45F flying in poor weather crashed one and three-quarters mile northeast of Cedar Keys, Florida, killing four fliers. Killed in the crash was Lt. Col. Francis H. Bonham, pilot; Maj. Francis E. McKinney, passenger; Pfc. John T. Graham, passenger; Seaman 2nd Class John R. Jackson, (USCG) passenger. The airplane took off from Charleston, South Carolina, at 1616 EST on a flight to Drew Field, Tampa, Florida. The airplane never arrived at Drew Field and was declared missing. The airplane wreckage was found the next day. The airplane was found to have crashed into the surf near a narrow island at Cedar Keys.

12-22-45. Jacksonville, Florida. At 1656 EST, a Douglas C-53D crashed one mile northwest of the Army Air Base at Jacksonville, Florida, killing co-pilot F/O James J. Cook. Pilot 2Lt. Alfred G. Munford received minor injuries. The airplane took off from Bush Field, Augusta, Georgia, on a flight to Morrison Field, West Palm Beach, Florida. The airplane crashed while on approach for Jacksonville Army Air Base. Investigation revealed that the engines quit because of mismanagement of the fuel system by the flight crew. The airplane smashed into the ground just short of the field. Investigators discovered that the pilots had not been properly checked off on DC-3 type airplanes.

Appendix I: Statistics Year by Year

Statistics 1941

Collisions in full flight with other aircraft	19
Collisions in full flight with objects other than aircraft	72
Spins or stalls following engine failure	11
Spins or stalls without engine failure	46
Emergency forced landing	13
Deferred forced landing	7
Landing accident	5
Take-off accident	18
Taxiing accident	1
Fire in flight	2
Structural, mechanical, material failure	12
Miscellaneous	6
Undetermined	21
TOTAL FATAL ACCIDENTS	233
FATALITIES	417
SERIOUS INJURIES*	35
MINOR INJURIES*	51
AIRCRAFT DAMAGED OR DESTROYED*	259

Statistics 1942

Collisions in full flight with other aircraft	90
Collisions in full flight with objects other than aircraft	253
Spins or stalls following engine failure	61
Spins or stalls without engine failure	231
Emergency forced landing	53
Deferred forced landing	8
Landing accident	38
Take-off accident	67
Taxiing accident	6
Fire and flight	10
Structural, mechanical, material failure	53
Miscellaneous	45
Undetermined	193
TOTAL FATAL ACCIDENTS	1,108
FATALITIES	2,395

SERIOUS INJURIES*	135
MINOR INJURIES*	142
AIRCRAFT DAMAGED OR DESTROYED*	1,213

Statistics 1943

Collisions in full flight with other aircraft	224
Collisions in full flight with objects other than aircraft	599
Spins or stalls following engine failure	115
Spins or stalls without engine failure	459
Emergency forced landing	111
Deferred forced landing	4
Landing accident	67
Take-off accident	105
Taxiing accident	17
Fire in flight	65
Structural, mechanical, material failure	160
Miscellaneous	68
Undetermined	274
TOTAL FATAL ACCIDENTS	2,268
FATALITIES	5,632
SERIOUS INJURIES*	473
MINOR INJURIES*	574
AIRCRAFT DAMAGED OR DESTROYED*	2,536
See Note # 2.	

Statistics 1944

Collisions in full flight with other aircraft	226
Collisions in full flight with objects other than aircraft	450
Spins or stalls following engine failure	106
Spins or stalls without engine failure	363
Emergency forced landing	132
Deferred forced landing	1
Landing accident	58
Take-off accident	64

*These numbers do not include injuries and losses sustained in non-fatal accidents.

Taxiing accident	14
Fire in flight	83
Structural, mechanical, material failure	152
Miscellaneous	74
Undetermined	225

TOTAL FATAL ACCIDENTS	1,948
FATALITIES	4,984
SERIOUS INJURIES*	459
MINOR INJURIES*	557
AIRCRAFT DAMAGED OR DESTROYED*	2,213

Statistics 1945

Collisions in full flight with other aircraft	74
Collisions in full flight with objects other than aircraft	186
Spins or stalls following engine failure	48
Spins or stalls without engine failure	117
Emergency forced landing	41
Deferred forced landing	1
Landing accident	19
Take-off accident	28
Taxiing accident	2
Fire in flight	50
Structural, mechanical, material failure	83
Miscellaneous	45
Undetermined	99
TOTAL FATAL ACCIDENTS	793

FATALITIES	2,103
SERIOUS INJURIES*	256
MINOR INJURIES*	236
AIRCRAFT DAMAGED OR DESTROYED*	893

Statistics 1941—1945

Collision in full flight with other aircraft	633
Collisions in full flight with objects other than aircraft	1,560
Spins or stalls following engine failure	341
Spins or stalls without engine failure	1,216
Emergency forced landing	350
Deferred forced landing	21
Landing accident	187
Take-off accident	282
Taxiing accident	40
Fire in flight	210
Structural, mechanical, material failure	460
Miscellaneous	238
Undetermined	812

TOTAL FATAL ACCIDENTS	6,350
FATALITIES	15,531
SERIOUS INJURIES*	1,358
MINOR INJURIES*	1,560
AIRCRAFT DAMAGED OR DESTROYED*	7,114

See Note # 9.

Accidents Per Month

	1941	1942	1943	1944	1945	
Jan	8	55	168	231	111	
Feb	11	58	155	155	68	
Mar	15	72	209	183	97	
Apr	10	74	153	160	92	
May	17	93	208	205	72	
Jun	17	68	192	180	90	
Jul	17	99	205	186	91	
Aug	12	100	216	143	58	
Sept	25	121	216	130	40	
Oct	26	121	201	143	32	
Nov	29	121	173	114	31	
Dec	46	126	172	118	11	
Total	**233**	**1108**	**2268**	**1948**	**793**	**6350**

These numbers do not include injuries and losses sustained in non-fatal accidents.

Appendix II: Army Air Forces Stations in the United States, 1941–1945

Abilene Army Air Field, Abilene, Texas
Adams Field, Little Rock, Arkansas
Aiken Army Air Field, Aiken, South Carolina
Ainsworth Army Air Field, Ainsworth, Nebraska
Ajo Army Air Field, Ajo, Arizona
Alachua Army Air Field, Gainesville, Florida
Alamogordo Bombing Range, Alamogordo, New Mexico
Albany Municipal Airport, Albany, Georgia
Albuquerque Army Air Field, Albuquerque, New Mexico
Alexandria Army Air Field, Alexandria, Louisiana
Alliance Army Air Field, Alliance, Nebraska
Aloe Army Air Field, Victoria, Texas
Altus Army Air Field, Altus, Oklahoma
Amarillo Army Air Field, Amarillo, Texas
Anniston Army Air Field, Estaboga, Alabama
Apalachicola Army Air Field, Apalachicola, Florida
Ardmore Army Air Field, Ardmore, Texas
Arledge Field, Stamford, Texas
Atlanta Municipal Airport, Atlanta, Georgia
Atterbury Army Air Field, Columbus, Indiana
Avenger Field, Sweetwater, Texas
Avon Park Army Air Field, Avon Park, Florida
Avon Park Municipal Airport, Avon Park, Florida
Baer Field, Fort Wayne, Indiana
Bainbridge Army Air Field, Bainbridge, Georgia
Bakersfield Municipal Airport, Bakersfield, California
Baltimore Army Air Field, Baltimore, Maryland
Barksdale Field, Shreveport, Louisiana
Bartow Army Air Field, Bartow, Florida
Bates Field, Mobile, Alabama
Bedford Army Air Field, Bedford, Massachusetts
Bellingham Army Air Field, Bellingham, Washington
Bendix Field, South Bend, Indiana
Bennettsville Municipal Airport, Bennettsville, South Carolina
Bergstrom Field, Austin, Texas
Big Spring Army Air Field, Big Spring, Texas

Biggs Field, El Paso, Texas
Birmingham Army Air Field, Birmingham, Alabama
Bishop Army Air Field, Bishop, California
Bismarck Municipal Airport, Bismarck, North Carolina
Blackland Army Air Field, Waco, Texas
Blackstone Army Air Field, Camp Pickett, Virginia
Bluethenthal Field, Wilmington, North Carolina
Blythe Army Air Field, Blythe, California
Blytheville Army Air Field, Blytheville, Arkansas
Boca Raton Army Air Field, Boca Raton, Florida
Boeing Field, Seattle, Washington
Bolling Field, Washington, D.C.
Bonham Municipal Airport, Bonham, Texas
Bowman Field, Louisville, Kentucky
Bradley Field, Windsor Locks, Connecticut
Bowling Green Municipal Airport, Bowling Green, Kentucky
Bridgeport Army Air Field, Stratford, Connecticut
Brookley Field, Mobile, Alabama
Brooks Field, San Antonio, Texas
Brooksville Army Air Field, Brooksville, Florida
Brownsville Municipal Airport, Brownsville, Texas
Brownwood Army Air Field, Brownwood, Texas
Bruce Field, Ballinger, Texas
Bruning Army Air Field, Bruning, Nebraska
Bryan Army Air Field, Bryan, Texas
Buckingham Army Air Field, Fort Myers, Florida
Buckley Field, Denver, Colorado
Buffalo Municipal Airport, Buffalo, New York
Bush Field, Augusta, Georgia
Bushnell Army Air Field, Bushnell, Florida
Cal Aero Academy, Ontario, California
Camden Municipal Airport, Camden, Arkansas
Camden Municipal Airport, Camden, South Carolina
Campbell Army Air Field, Clarksville, Tennessee
Camp Springs Army Air Field, Washington, D.C.
Carrabelle Flight Strip, Carrabelle, Florida
Carlsbad Army Air Field, Carlsbad, New Mexico
Carlstrom Field, Arcadia, Florida

Casper Army Air Field, Casper, Wyoming
Chanute Field, Rantoul, Illinois
Charleston Army Air Field, Charleston, South Carolina
Chatham Army Air Field, Savannah, Georgia
Cheyenne Municipal Airport, Cheyenne, Wyoming
Chicago Municipal Airport, Chicago, Illinois
Chickasha Municipal Airport, Chickasha, Oklahoma
Chico Army Air Field, Chico, California
Childress Army Air Field, Childress, Texas
Cimarron Field, Oklahoma City, Oklahoma
Clarksdale Municipal Airport, Clarksdale, Mississippi
Clewiston Municipal Airport, Clewiston, Florida
Clinton County Army Air Field, Wilmington, Ohio
Clovis Army Air Field, Clovis, New Mexico
Cochran Field, Macon, Georgia
Coffeyville Army Air Field, Coffeyville, Kansas
Coleman Flying School, Coleman, Texas
Columbia Army Air Base, Columbia, South Carolina
Columbus Army Air Field, Columbus, Mississippi
Commonwealth Airport, Boston, Massachusetts
Concord Army Air Field, Concord, California
Congaree Army Air Field, Congaree, South Carolina
Connellsville Municipal Airport, Connellsville, Pennsylvania
Coolidge Army Air Field, Coolidge, Arizona
Coronaca Army Air Field, Greenwood, South Carolina
Corsicana Field, Corsicana, Texas
Corvallis Army Air Field, Corvallis, Oregon
Courtland Army Air Field, Courtland, Alabama
Cox Field, Paris, Texas
Craig Field, Selma, Alabama
Cross City Army Air Field, Cross City, Florida
Cuero Municipal Airport, Cuero, Texas
Curtis Field, Brady, Texas
Curtiss-Steinberg Field, East St. Louis, Illinois
Cut Bank Army Air Field, Cut Bank, Montana
Daggett Municipal Airport, Daggett, California
Dalhart Army Air Field, Dalhart, Texas
Daniel Field, Augusta, Georgia
Datelan Army Air Field, Datelan, Arizona
Davis, Camp, Army Air Field, Camp Davis, North Carolina
Davis-Monthan Field, Tucson, Arizona
Decatur Municipal Airport, Decatur, Alabama
Delano Army Air Field, Delano, California
Deming Army Air Field, Deming, New Mexico
Denver Municipal Airport, Denver, Colorado
DeRidder Army Air Field, DeRidder, Louisiana
Desert Center Army Air Field, Desert Center, California
Des Moines Municipal Airport, Des Moines, Iowa
Devens, Fort, Army Air Field, Fort Devens, Massachusetts
Dix, Fort, Army Air Base, Fort Dix, New Jersey
Dodge City Army Air Field, Dodge City, Kansas
Dorr Field, Arcadia, Florida
Douglas Army Air Field, Douglas, Arizona
Douglas Municipal Airport, Douglas, Georgia

Dover Army Air Field, Dover, Delaware
Dow Field, Bangor, Maine
Drane Field, Tampa, Florida
Drew Field, Tampa, Florida
Duncan Field, San Antonio, Texas
Dunnellon Army Air Field, Dunnellon, Florida
Dyersburg Army Air Field, Dyersburg, Tennessee
Eagle Pass Army Air Field, Eagle Pass, Texas
Eglin Field, Valparaiso, Florida
Ellensburg Army Air Field, Ellensburg, Washington
Ellington Field, Houston, Texas
El Paso Municipal Airport, El Paso, Texas
Enid Army Air Field, Enid, Oklahoma
Ephrata Army Air Field, Ephrata, Washington
Esler Field, Alexandria, Louisiana
Estrella Army Air Field, Paso Robles, California
Evansville Municipal Airport, Evansville, Indiana
Fairfax Field, Kansas City, Kansas
Fairfield-Suisun Army Air Field, Fairfield, California
Fairmont Army Air Field, Geneva, Nebraska
Falcon Field, Mesa, Arizona
Fargo Municipal Airport, Fargo, North Dakota
Farmingdale Army Air Field, Farmingdale, Long Island, New York
Felts Field, Spokane, Washington
Florence Army Air Field, Florence, South Carolina
Fort Sumner Army Air Field, Fort Sumner, New Mexico
Fort Worth Army Air Field, Fort Worth, Texas
Foster Field, Victoria, Texas
Franklin Field, California
Frederick Army Air Field, Frederick, Oklahoma
Freeman Field, Seymour, Indiana
Furniss Field, Selma, Alabama
Gainesville Army Air Field, Gainesville, Florida
Galveston Army Air Field, Galveston, Texas
Garden City Army Air Field, Garden City, Kansas
Gardner Field, Taft, California
Geiger Field, Spokane, Washington
George Field, Lawrenceville, Illinois
Gila Bend Army Air Field, Gila Bend, Arizona
Glasgow Army Air Field, Glasgow, Montana
Godman Field, Fort Knox, Kentucky
Goodfellow Field, San Angelo, Texas
Gore Field, Great Falls, Montana
Gowen Field, Boise, Idaho
Grand Central Air Terminal, Glendale, California
Grand Island Army Air Field, Grand Island, Nebraska
Gray Field, Fort Lewis, Washington
Grayling Army Air Field, Grayling, Michigan
Great Bend Army Air Field, Great Bend, Kansas
Great Falls Army Air Field, Great Falls, Montana
Greensboro-High Point Municipal Airport, Greensboro, North Carolina
Greenville Army Air Base, Greenville, South Carolina
Greenville Army Air Field, Greenville, Mississippi
Greenwood Army Air Field, Greenwood, Mississippi
Grenada Army Air Field, Grenada, Mississippi

Grenier Field, Manchester, New Hampshire
Grider Field, Pine Bluff, Arkansas
Groton Army Air Field, Groton, Connecticut
Gulfport Army Air Field, Gulfport, Mississippi
Gunter Field, Montgomery, Alabama
Hamilton Field, San Rafael, California
Hammer Field, Fresno, California
Hammond Army Air Field, Hammond, Louisiana
Hancock Field, Santa Maria, California
Harding Field, Baton Rouge, Louisiana
Hargrove Van DeGraaff Airport, Tuscaloosa, Alabama
Harlingen Army Air Field, Harlingen, Texas
Harris Field, Fort Valley, Georgia
Harris Neck Army Air Field, Townsend, Georgia
Harrisburg Municipal Airport, New Cumberland, Pennsylvania
Harvard Army Air Field, Harvard, Nebraska
Harvey Parks Airport, Sikeston, Missouri
Hatbox Field, Muskogee, Oklahoma
Hattiesburg Army Air Field, Hattiesburg, Mississippi
Hayward Army Air Field, Hayward, California
Hendricks Field, Sebring, Florida
Hensley Field, Dallas, Texas
Hereford Army Air Field, Hereford, Arizona
Herington Army Air Field, Herrington, Kansas
Hicks Field, Fort Worth, Texas
Hill Field, Ogden, Utah
Hillsborough Army Air Field, Tampa, Florida
Hillsgrove Army Air Field, Hillsgrove, Rhode Island
Hobbs Army Air Field, Hobbs, New Mexico
Homestead Army Air Field, Homestead, Florida
Hondo Army Air Field, Hondo, Texas
Houlton Army Air Field, Houlton, Maine
Hunter Field, Savannah, Georgia
Idlewild Airport, New York, New York
Immokalee Municipal Airport, Immokalee, Florida
Independence Army Air Field, Independence, Kansas
Indian Springs Army Air Field, Indian Springs, Nevada
Jackson Army Air Base, Jackson, Mississippi
Jackson Municipal Airport, Jackson, Tennessee
Jacksonville Army Air Field, Jacksonville, Florida
Johnson, Seymour, Goldsboro, North Carolina
Kansas City Municipal Airport, Kansas City, Missouri
Kay Field, Columbus, Mississippi
Kearney Army Air Field, Kearney, Nebraska
Keesler Field, Biloxi, Mississippi
Kellogg Field, Battle Creek, Michigan
Kelly Field, San Antonio, Texas
Key Field, Meridian, Mississippi
Keystone Army Air Field, Keystone Heights, Florida
Kingman Army Air Field, Kingman, Arizona
Kirtland Field, Albuquerque, New Mexico
Kissimmee Army Air Field, Kissimmee, Florida
Knollwood Field, North Carolina
Lafayette Municipal Airport, Lafayette, Louisiana
LaGuardia Field, Jackson Heights, Long Island, New York
Laguna Madre Auxiliary Airfield, Harlingen, Texas
La Junta Army Air Field, La Junta, Colorado

Lake Charles Army Air Field, Lake Charles, Louisiana
Lakeland Army Air Field, Lakeland, Florida
Lakeland Municipal Airport, Lakeland, Florida
Lambert Field, St. Louis, Missouri
Lamesa Municipal Airport, Lamesa, Texas
Langley Field, Hampton, Virginia
Laredo Army Air Field, Laredo, Texas
Las Animas Auxiliary Airfield, Las Animas, Colorado
Las Vegas Army Air Field, Las Vegas, Nevada
Laughlin Field, Del Rio, Texas
Laurel Army Air Field, Laurel, Mississippi
Laurinburg-Maxton Army Air Field, Maxton, North Carolina
Lawson Field, Fort Benning, Georgia
Leesburg Army Air Field, Leesburg, Florida
Lemoore Army Air Field, Lemoore, California
Lewiston Army Air Field, Lewiston, Montana
Liberal Army Air Field, Liberal, Kansas
Liberty Field, California
Lincoln Army Air Field, Lincoln, Nebraska
Lindbergh Field, San Diego, California
Lockbourne Army Air Field, Columbus, Ohio
Logan, Fort, Army Air Field, Fort Logan, Colorado
Lomita Flight Strip, Lomita, California
Long Beach Army Air Field, Long Beach, California
Louisville Municipal Airport, Louisville, Kentucky
Love Field, Dallas, Texas
Lowry Field, Denver, Colorado
Lubbock Army Air Field, Lubbock, Texas
Luke Field, Phoenix, Arizona
Luna, Camp, Army Air Field, Las Vegas, New Mexico
Luken Airport, Cincinnati, Ohio
Mabry Field, Dale, Tallahassee, Florida
MacDill Field, Tampa, Florida
Mackall, Camp, Army Air Field, Hoffman, North Carolina
Madras Army Air Field, Madras, Oregon
Majors Field, Greenville, Texas
Malden Army Air Field, Malden, Missouri
Marana Army Air Field, Marana, Arizona
Marathon Flight Strip, Marathon, Florida
March Field, Riverside, California
Marfa Army Air Field, Marfa, Texas
Marianna Army Air Field, Marianna, Florida
Marietta Army Air Field, Marietta, Georgia
Marshall Field, Fort Riley, Kansas
Marysville Army Air Field, Marysville, California
Matagorda Island Army Air Field, Texas
Matagorda Peninsula Army Air Field, Texas
Mather Field, Sacramento, California
Maxwell Field, Montgomery, Alabama
McChord Field, Tacoma, Washington
McClellan Field, Sacramento, California
McCook Army Air Field, McCook, Nebraska
Medford Army Air Field, Medford, Oregon
Memphis Municipal Airport, Memphis, Tennessee
Merced Army Air Field, Merced, California
Miami 36th Street Airport, Miami, Florida

Midland Army Air Field, Midland, Texas
Midland Municipal Airport, Midland, Texas
Mills Field, San Francisco, California
Millville Army Air Field, Millville, New Jersey
Mines Field, Los Angeles, California
Minter Field, Bakersfield, California
Mira Loma Flight Academy, Oxnard, California
Mitchel Field, Hempstead, Long Island, New York
Mitchell Field, Billy, Milwaukee, Wisconsin
Mitchell Municipal Airport, Mitchell, South Dakota
Moffett Field, Sunnyvale, California
Montbrook Army Air Field, Williston, Florida
Moody Field, Valdosta, Georgia
Moore Field, Mission, Texas
Morris Field, Charlotte, North Carolina
Morrison Field, West Palm Beach, Florida
Moses Lake Army Air Field, Moses Lake, Washington
Mountain Home Army Air Field, Mountain Home, Idaho
Muroc Army Air Field, Muroc, California
Muroc Flight Test Base, Muroc, California
Muscle Shoals Municipal Airport, Sheffield, Alabama
Muskogee Army Air Field, Muskogee, Oklahoma
Mustang Field, El Reno, Oklahoma
Myrtle Beach Army Air Field, Myrtle Beach, South Carolina
Napier Field, Dothan, Alabama
Naples Army Air Field, Naples, Florida
Nashville Army Air Center, Nashville, Tennessee
Nashville Municipal Airport, Nashville, Tennessee
National Airport, Washington, D.C.
Newark Army Air Field, Newark, New Jersey
New Albany Army Air Field, New Albany, Indiana
New Bedford Army Air Field, New Bedford, Massachusetts
New Castle Army Air Base, Wilmington, Delaware
New Haven Army Air Field, New Haven, Connecticut
New Orleans Army Air Base, New Orleans, Louisiana
Newport Army Air Field, Newport, Arkansas
Niagara Falls Municipal Airport, Niagara Falls, New York
Norfolk Army Air Field, Norfolk, Virginia
North Island Naval Air Station, San Diego, California
Northern Field, William, Tullahoma, Tennessee
Oakland Municipal Airport, Oakland, California
Ocala Municipal Airport, Ocala, Florida
Offutt Field, Fort Crook, Nebraska
Olmstead Field, Middletown, Pennsylvania
Olympia Army Air Field, Olympia, Washington
Ontario Army Air Field, Ontario, California
Orange County Army Air Field, Santa Ana, California
Orangeburg Municipal Airport, Orangeburg, South Carolina
Orchard Place Airport, Park Ridge, Illinois
Orlando Army Air Base, Orlando, Florida
Oroville Army Air Field, Oroville, California
Oscoda Army Air Field, Oscoda, Michigan
Otis Field, Camp Edwards, Massachusetts
Oxnard Flight Strip, Oxnard, California

Oxnard Municipal Airport, Oxnard, California
Page Field, Fort Myers, Florida
Paine Field, Everett, Washington
Palacios Army Air Field, Palacios, Texas
Palmdale Army Air Field, Palmdale, California
Palm Springs Army Air Field, Palm Springs, California
Pampa Army Air Field, Pampa, Texas
Patterson Field, Ohio
Parks Field, East St. Louis, Missouri
Pecos Army Air Field, Pecos, Texas
Pendleton Field, Pendleton, Oregon
Perrin Field, Sherman, Texas
Perry Army Air Field, Perry, Florida
Peterson Field, Colorado Springs, Colorado
Philadelphia Municipal Airport, Philadelphia, Pennsylvania
Phillips Field, Aberdeen, Maryland
Pierre Army Air Field, Pierre, South Dakota
Pinecastle Army Air Field, Pinecastle, Florida
Pinellas Army Air Field, St. Petersburg, Florida
Pittsburgh-Allegheny County Airport, Pennsylvania
Pittsburgh Municipal Airport, Pittsburgh, Kansas
Pocatello Army Air Field, Pocatello, Idaho
Pollock Army Air Field, Pollock, Louisiana
Pope Field, Fort Bragg, North Carolina
Port Angeles Army Air Field, Port Angeles, Washington
Portland Army Air Base, Portland, Oregon
Post Field, Fort Sill, Oklahoma
Pounds Field, Tyler, Texas
Pratt Army Air Field, Pratt, Kansas
Presque Isle Army Air Field, Presque Isle, Maine
Pueblo Army Air Base, Pueblo, Colorado
Pyote Army Air Field, Pyote, Texas
Raleigh-Durham Army Air Field, Raleigh, North Carolina
Randolph Field, San Antonio, Texas
Rankin Field, Tulare, California
Rapid City Army Air Base, Rapid City, South Dakota
Reading Army Air Field, Reading, Pennsylvania
Redding Army Air Field, Redding, California
Redmond Army Air Field, Redmond, Oregon
Reeves Field, Tulare, California
Reno Army Air Base, Reno, Nevada
Rentschler Field, Hartford, Connecticut
Rice Army Air Field, Rice, California
Richmond Army Air Base, Richmond, Virginia
Robbins Field, Madison, Mississippi
Robins Field, Warner Robins, Georgia
Rocky Ford Auxiliary Airfield, Rocky Ford, Colorado
Rogers Field, Will, Oklahoma City, Oklahoma
Rome Army Air Field, Rome, New York
Romulus Army Air Field, Romulus, Michigan
Rosecrans Field, St. Joseph, Missouri
Roswell Army Air Field, Roswell, New Mexico
Ryan Field, Apopka, Florida
Ryan Field, Tucson, Arizona
Sacramento Municipal Airport, Sacramento, California
Safford Field, Safford, Arizona

Sahuarita Flight Strip, Arizona
St. Paul Municipal Airport, St. Paul, Minnesota
Salem Army Air Field, Salem, Oregon
Salinas Army Air Base, Salinas, California
Salisbury Municipal Airport, Salisbury, Maryland
Salt Lake City Army Air Base, Salt Lake City, Utah
San Angelo Army Air Field, San Angelo, Texas
San Antonio Municipal Airport, San Antonio, Texas
San Bernardino Army Air Field, San Bernardino, California
San Diego Municipal Airport, San Diego, California
San Francisco Municipal Airport, San Bruno, California
San Marcos Army Air Field, San Marcos, Texas
Santa Ana Army Air Field, Santa Ana, California
Santa Fe Army Air Field, Santa Fe, New Mexico
Santa Maria Army Air Field, Santa Maria, California
Santa Maria Municipal Airport, Santa Maria, California
Santa Rosa Army Air Field, Santa Rosa, California
Sarasota Army Air Field, Sarasota, Florida
Scott Field, Belleville, Illinois
Scottsbluff Army Air Field, Scottsbluff, Nebraska
Scribner Army Air Field, Scribner, Nebraska
Sedalia Army Air Field, Warrensburg, Missouri
Selfridge Field, Mt. Clemens, Michigan
Selman Field, Monroe, Louisiana
Semi-Tropic Auxiliary Airfield, California
Sequoia Army Air Field, Visalia, California
Shavers Summit Field, Thermal, California
Shaw Field, Fort Sumter, South Carolina
Shelton Navy Airport, Shelton, Washington
Sheppard Field, Wichita Falls, Texas
Sherman Field, Fort Leavenworth, Kansas
Sioux City Army Air Base, Sioux City, Iowa
Sioux Falls Army Air Field, Sioux Falls, South Dakota
Sky Harbor Airport, Phoenix, Arizona
Smart Airport, Herbert, Macon, Georgia
Smoky Hill Army Air Field, Salina, Kansas
Smyrna Army Air Field, Smyrna, Tennessee
South Plains Army Air Field, Lubbock, Texas
Souther Field, Americus, Georgia
Spence Field, Moultrie, Georgia
Spokane Army Air Field, Spokane, Washington
Starkville Municipal Airport, Starkville, Mississippi
Statesboro Army Air Field, Statesboro, Georgia
Stewart, Camp, Army Air Field, Hinesville, Georgia
Stewart Field, Newburg, New York
Stinson Field, San Antonio, Texas
Stockton Field, Stockton, California
Stout Field, Indianapolis, Indiana
Strother Field, Winfield, Kansas
Sturgis Army Air Field, Sturgis, Kentucky
Stuttgart Army Air Field, Stuttgart, Arkansas
Suffolk County Army Air Field, Long Island, New York

Sweetwater Municipal Airport, Sweetwater, Texas
Syracuse Army Air Field, Syracuse, New York
Tarrant Field, Fort Worth, Texas
Thermal Army Air Field, Thermal, California
Thomasville Army Air Field, Thomasville, Georgia
Thompson-Robbins Field, Helena, Arkansas
Thunderbird Field # 1, Glendale, Arizona
Thunderbird Field # 2, Phoenix, Arizona
Tifton Army Air Field, Tifton, Georgia
Tinker Field, Oklahoma City, Oklahoma
Tonopah Army Air Field, Tonopah, Nevada
Topeka Army Air Field, Pauline, Kansas
Tri-City Army Air Field, Freeland, Michigan
Truax Field, Madison, Wisconsin
Trumbull Field, Groton, Connecticut
Tucson Municipal Airport, Tucson, Arizona
Tulsa Municipal Airport, Tulsa, Oklahoma
Turner Field, Albany, Georgia
Tuskegee Army Air Field, Tuskegee, Alabama
Tuskegee Institute Airfield, Tuskegee, Alabama
Twenty-Nine Palms Air Academy, Twenty-Nine Palms, California
Tyndall Field, Panama City, Florida
Union City Municipal Airport, Union City, Tennessee
Van Nuys Metropolitan Airport, Van Nuys, California
Venice Army Air Field, Venice, Florida
Vichy Army Air Field, Rolla, Missouri
Victorville Army Air Field, Victorville, California
Victory Field, Vernon, Texas
Visalia Army Air Field, Visalia, California
Walker Army Air Field, Victoria, Kansas
Walla Walla Army Air Field, Walla Walla, Washington
Walnut Ridge Army Air Field, Walnut Ridge, Arkansas
Walterboro Army Air Field, Walterboro, South Carolina
Watertown Army Air Field, Watertown, South Dakota
Waycross Army Air Field, Waycross, Georgia
Wendover Field, Wendover, Utah
Westover Field, Chicopee, Massachusetts
Wheeler-Sack Field, Great Bend, New York
White Field, Brady, Texas
Wichita Municipal Airport, Wichita, Kansas
Williams, Camp, Army Air Field, Camp Douglas, Wisconsin
Williams Field, Chandler, Arizona
Willow Run, Ypsilanti, Michigan
Winslow Municipal Airport, Winslow, Arizona
Winston-Salem Municipal Airport, Winston-Salem, North Carolina
Winter Garden Municipal Airport, Winter Garden, Florida
Woodward Army Air Field, Woodward, Oklahoma
Wright Field, Dayton, Ohio
Wright, Fort George, Airfield, Spokane, Washington
Yuma Army Air Field, Yuma, Arizona
Zephyrhills Army Air Field, Zephyrhills, Florida

Appendix III: Missing Aircraft

Serial #	Model	Date of Loss
36-153	Y1B-17	3-5-42E
38-254	BT-9C	7-26-43I
39-69	O-47B	2-15-42B
39-72	O-47B	2-10-43E
40-433	P-40E	12-31-42B
40-1211	BT-14	3-31-43F
40-1876	PT-17	9-2-41A
41-2052	RP-38E	5-12-43F
41-2385	P-38F	5-27-42D
41-2533	B-17E	4-15-42A
41-2597	B-17E	5-6-42E
41-5922	P-47B	2-11-43C
41-5944	P-47B	2-11-43D
41-6566	P-47C	4-27-43B
41-7519	P-38F	1-29-44D
41-9907	BT-13A	1-14-42D
41-14308	P-40F	4-25-44C
41-17002	AT-6A	5-3-43K
41-17635	B-26B	12-4-42F
41-17701	B-26B	10-4-42A
41-17966	B-26B	11-16-42G
41-18101	B-26B	12-28-42B
41-18244	B-26B	4-23-44E
41-18822	A-25	8-19-43B
41-23395	A-29	7-14-42B
41-23411	A-29	5-10-42A
41-23914	B-24D	4-21-43C
41-23935	B-24D	11-16-42C
41-24041	B-24D	10-6-43B
41-24260	B-24D	12-6-42B
41-28548	B-24E	2-26-44B
41-28723	B-24H	1-14-44A
41-29656	B-25D	3-24-43K
41-30395	B-25D	7-23-43F
41-35118	B-26C	7-24-44E
41-35805	AT-23B	5-2-44D
41-38253	P-39D	12-9-42B
42-6428	B-29A	1-9-45B
42-6765	A-24A	4-26-43A

Serial #	Model	Date of Loss
42-7225	B-24E	12-12-43C
42-7525	B-24H	4-7-44C
42-7628	B-24H	10-29-43A
42-8874	P-39N	5-5-43F
42-13386	P-38G	4-26-43G
42-20094	P-39Q	7-26-43B
42-20866	P-39Q	5-15-44I
42-20893	P-39Q	9-26-43E
42-22418	P-47D	2-25-44D
42-30555	B-17F	5-22-44I
42-35179	AT-10	8-22-44B
42-37229	AT-11	11-14-44C
42-40397	B-24D	5-29-43E
42-41108	B-24D	11-15-43D
42-48985	AT-6C	12-21-43A
42-50534	B-24J	6-20-44H
42-50946	B-24J	12-16-44A
42-58475	UC-78	1-31-44C
42-65014	B-25G	9-13-43C
42-65101	B-25G	4-16-44A
42-67953	P-38J	12-29-43A
42-89898	BT-13B	8-7-44E
42-92115	C-47A	7-21-44G
42-95611	B-24J	4-11-45C
42-100712	C-47A	7-21-44F
43-2232	P-38G	3-15-43E
43-5601	P-40M	9-24-43L
43-6247	P-51A	8-19-43I
43-9357	A-20G	1-30-44D
43-9736	P-70B	11-18-44A
43-12246	P-51B	3-30-44F
43-30664	C-47B	7-21-44H
44-7430	P-40N	8-20-44B
44-11796	P-51K	4-6-45A
44-11802	P-51K	4-16-45B
44-14909	P-51D	4-5-45A
44-15669	P-51D	10-26-44E
44-81582	AT-6D	4-12-45F
44-84922	P-51D	8-18-45A

Notes

1. No Army Air Corps accident reports found for 1-1-41.

2. No Army Air Forces accident reports could be found for 5-21-43. The AAF Form # 14 Aircraft Accident Reports for the 57 known accidents have apparently been lost.

3. The deadliest single day during the war was 1-24-43, when 51 fliers were killed in AAF aircraft accidents.

4. August 1943 was the deadliest month of the war with 598 persons being killed in AAF aircraft accidents.

5. Eighteen fatal AAF aircraft accidents occurred on 9-23-43. Ten or more fatal accidents occurred on 47 days in 1943.

6. There were 231 fatal AAF aircraft accidents during January 1944, more than any other month.

7. There is a time-zone/date anomaly for 6-13-44 and 6-14-44. The first fatal accident for 6-14-44 actually occurred *before* the final fatal accident of 6-13-44. The two accidents are chronicled by date for clarity.

8. A 2Lt. Miller (first name unknown) was killed in an airplane accident at Duncan Field, San Antonio, Texas, in December 1941. No other information available.

9. The total number of fatal stateside AAF aircraft accidents that occurred during the Second World War cannot easily be determined because of the great number of AAF Form # 14 Accident Reports missing from the microfilm record.

10. The author has been unable to determine the fate of Aviation Cadet Maurice Herzog (38428453), who disappeared after abandoning his Vultee BT-13 in Mexico on 11-3-43.

11. The following fatal accident summaries were based on newspaper articles, other periodicals or other sources: 2-19-41, 3-1-41, 3-26-41A, 6-2-41B, 8-30-41B; 1-16-42B, 2-3-42D, 3-27-42C, 11-18-42F, 11-20-42F; 11-28-42C, 12-24-42F, 1-13-43J, 3-14-43E, 4-12-43F, 4-24-43E, 5-19-43D, 6-9-43F, 8-3-43F, 11-1-43C; 5-3-44K, 5-10-44J, 5-14-44E, 5-17-44A, 7-28-45B.

12. Cessna AT-17 airplanes suffered catastrophic wing failure that resulted in a fatal accident on 25 known occasions: 5-3-42A, 7-30-42B, 8-19-42A, 11-19-42F, 12-4-42D, 12-23-42C, 12-30-42; 1-2-43D, 1-15-43C, 2-7-43E, 2-16-43A, 3-9-43K, 5-5-43E, 5-13-43I, 5-14-43D, 6-14-43F, 7-1-43B, 7-1-43D, 8-7-43D, 12-31-43D; 1-3-44D, 5-8-44C, 5-12-44B, 5-12-44L, 5-31-44C.

13. Explanation of Cessna AT-17/UC-78 wing failures can be found in two AAF "Unsatisfactory" reports found on AAF Aircraft Accident Reports: Call # 46154, 12-30-42, accident # 7; and Call # 46236, 8-7-43, accident # 11.

14. Boeing B-17 aircraft suffered six top turret fires that resulted in fatalities and loss of aircraft: 6-5-43K, 1-4-44B, 4-24-44D, 5-9-44E, 7-11-44A, 12-13-44D.

15. Martin B-26 aircraft suffered material/structural failure of the cockpit enclosure and tail section that led to fatalities and loss of aircraft on six known occasions: 11-18-41C, 5-25-42A, 8-7-42B, 9-21-42C, 9-28-42A, 8-31-43A. There was one known occurrence overseas. See AAF Aircraft Accident Reports: Call # 46208, 5-29-43, Accident # 502.

16. There were 22 known fatal compressibility accidents involving Lockheed P-38 type airplanes: 6-23-41, 11-4-41B, 11-13-41B, 4-4-42B, 4-5-42C, 5-6-42C, 6-22-42C, 8-30-42B, 9-8-42C, 9-20-42E, 11-18-42E, 11-20-42C, 11-28-42B, 12-17-42C, 2-6-43D, 7-3-43H, 10-1-43B, 10-27-43E, 12-5-43H, 9-5-44C, 9-5-44D, 9-14-44B.

17. Investigators noted that Consolidated B-24J airplanes with AAF serial numbers 42-73377 through 42-73399 suffered unusual performance problems.

18. According to an AAF memo prepared on 6-28-44 concerning the Presidential confirmation of Court Martial sentences for flying violations, 1Lt. Wesley G.W. Harju, 499th Squadron, 85th Fighter Bomber Group, was "indirectly" responsible for a fatality while performing aerobatics below 500 feet agl in a Bell P-39. The incident occurred on an unknown date before 5-30-44. Lt. Harju was sentenced to six months hard labor and dismissed from the service. The memo can be found in AAF Aircraft Accident Reports: Call # 46462, 12-28-44, Accident # 2. No other information available.

19. No official AAF Form # 14 Aircraft Accident Report could be found on AAF Accident Reports microfilm (Call # 46537) for accident 7-28-45B.

20. Martin B-26 # 41-17645 (1-5-43B) is featured on the *Flight Check — How-to-Fly* series of videos, "B-26 Marauder" episode, produced by Dastar Corp./Marathon Music and Video (Copyright 1997).

21. There were 38 fatal aircraft accidents involving Women's Air Force Service Pilots (WASP): 3-7-43D, 3-21-43E, 6-7-43F, 8-4-43G, 8-23-43C, 8-30-43F, 9-23-43I, 10-5-43C, 11-7-43B, 12-3-43D, 2-18-44E, 2-28-44L, 3-27-44B, 4-3-44B, 4-10-44B, 4-16-44B, 4-25-44E, 6-11-44B, 6-13-44A, 6-20-44C, 6-21-44B, 6-22-44D, 6-29-44B, 7-4-44C, 7-7-44F, 7-8-44A, 7-18-44D, 8-14-44E, 9-13-44E, 9-15-44C, 10-2-44C, 10-3-44G, 10-16-44F, 10-16-44G, 10-26-44E, 11-23-44C, 11-26-44C, 12-9-44B.

22. The author could find no official AAF Form # 14 Aircraft Accident Report for the AT-6 accident 12-24-42F. This summary is based on original research and interviews conducted by Chris McDoniel (Copyright 1999).

23. The memo titled "Our Safety Problem" can be found on AAF Accident Reports microfilm Call # 46104, 7-1-42, Accident # 5.

Bibliography

Anderton, David A. *The History of the U.S. Air Force.* New York: Crescent Books, 1981.

Barber, Mike. "Missing Flier Comes Out of Thin Air." Seattle: Seattle Post Intelligencer, September 1, 2003.

Baugher, Joe. *Joe Baugher's USAF Aircraft Serial Number Lists.* Online: *http://users.rcn.com/jeremy.k/serialsearch*, April 4, 2003.

Bowman, Martin W. *USAAF Handbook 1939-1945.* Mechanicsburg: Stackpole Books, 1997.

Branom, Frederick K., and Smith, Lloyd E. *The New International Atlas of the World.* Chicago: The Geographical Publishing Co., 1937.

Gero, David. *Military Aviation Disasters.* Somerset: Patrick Stephens Ltd., 1999.

Glasgow, H.W., ed. *World Atlas.* Chicago: The Geographical Publishing Co., 1937.

Hammond, C.S., ed. *Hammond's Superior Atlas and Gazetteer of the World.* New York: C.S. Hammond & Co., Inc., 1945.

The Hammond Times. Various articles. 1941–1945.

Irving, E. Eastman, ed. *The World Almanac and Book of Facts for 1945.* New York: The New York World Telegram, 1945.

LeMay, Curtis E., and Kantor, MacKinlay. *Mission with LeMay.* Garden City, IN: Doubleday, 1965.

Macha, Gary P. *Aircraft Wrecks in the Mountains and Deserts of California 1909-1996.* 2nd ed. San Clemente: Infonet Publishing, 1997.

Mason, Molloy, Jr. *The United States Air Force.* New York: Mason/Charter Publishers, 1976.

McDoniel, Chris. *P-38.com(WWII Arizona Army Air Fields...)* Online: *http://www.p-38.com*, November 22, 2000.

Thole, Lou. *Forgotten Fields of America.* Missoula, AT: Pictorial Histories Publishing Co., Inc., 1996.

United States. *Aircraft Accident Reports 1941–1945. Call Numbers 46046–46559.* Maxwell AFB: United States Air Force, 1941–1945.

United States. *The Official Pictorial History of the Army Air Forces.* Washington, D.C.: United States Army Air Forces, Historical Offices, 1947.

Watry, Charles A. *Washout. The Aviation Cadet Story.* Carlsbad: California Aero Press, 1983.

Wrynn, V. Dennis. *Forge of Freedom.* Osceola: Motorbooks International, 1995.

Yeager, Chuck, and Janos, Leo. *Yeager: An Autobiography.* New York: Bantam, 1985.

Aircraft Index (by Primary Designer/Manufacturer)

Following each designer/manufacturer name, aircraft are listed by serial number, with model number in parentheses; references are to entry numbers

Aeronca L-3

42-1930 (L-3C) 11-19-43A
42-7808 (L-3A) 1-12-42B
42-14722 (L-3B) 10-26-42B
42-14728 (L-3B) 9-12-43C
42-14785 (L-3B) 1-12-43E
42-14787 (L-3B) 4-12-44E
42-14792 (L-3B) 1-26-43F
42-14793 (L-3B) 8-17-42F
42-36082 (L-3B) 8-7-43F
42-36100 (L-3B) 8-13-42E
42-36129 (L-3B) 5-12-43C
42-36148 (L-3B) 9-4-42A
42-36170 (L-3B) 8-5-42A
42-36190 (L-3A) 7-27-42A
42-36215 (L-3B) 10-22-42A
42-36217 (L-3B) 6-28-43F
42-36258 (L-3B) 5-28-43H
43-1480 (L-3C) 9-24-43J
43-1483 (L-3C) 7-12-44G
43-1491 (L-3C) 8-6-43F
43-1501 (L-3C) 3-14-43B
43-1551 (L-3C) 5-23-44C
43-1625 (L-3C) 10-22-42A
43-1647 (L-3C) 6-24-43B
43-1648 (L-3C) 1-27-44A
43-1702 (L-3C) 7-30-44C
43-1768 (L-3C) 4-11-43F
43-1770 (L-3C) 6-22-43A
43-26800 (L-3B) 9-5-43D
43-26833 (L-3B) 6-27-43F
43-26834 (L-3B) 6-27-43F
43-26932 (L-3B) 9-12-44B
43-27002 (L-3B) 10-24-43A
43-27014 (L-3B) 3-22-44C
43-27020 (L-3B) 10-24-43A
43-27041 (L-3B) 3-22-44C
43-27236 (L-3B) 3-22-44C
43-27237 (L-3B) 3-22-44C

Aeronca TG-5

42-57378 (TG-5) 10-27-42C

Beech AT-7/F-2

40-686 (F-2) 1-6-42

40-690 (F-2) 7-8-42D
41-1146 (AT-7) 5-15-44A
41-1147 (AT-7) 5-16-41A
41-1159 (AT-7) 8-20-43C
41-1174 (AT-7) 12-18-44E
41-1198 (AT-7) 10-28-41A
41-21052 (AT-7) 4-9-42C
41-21053 (AT-7) 3-21-44B
41-21079 (AT-7) 11-18-42F
41-21087 (AT-7) 7-6-42A
41-21117 (AT-7) 4-7-43B
41-21119 (AT-7) 7-29-43E
41-21133 (AT-7) 3-13-42C
41-21140 (AT-7) 2-28-43A
41-21151 (AT-7) 1-31-43E
42-2425 (AT-7) 2-5-43I
42-2445 (AT-7) 12-24-42C
42-2463 (AT-7) 3-30-43H
42-2469 (AT-7) 6-14-43B
42-2473 (AT-7) 5-12-45B
42-2475 (AT-7) 8-24-43D
42-43505 (AT-7) 7-3-43C
42-56712 (AT-7) 4-23-43A
42-56716 (AT-7) 9-9-43H
42-56754 (AT-7) 7-10-44F
42-56772 (AT-7) 8-25-43F
42-56775 (AT-7) 7-10-44F
42-56776 (AT-7) 10-20-43D
42-56777 (AT-7) 7-6-43A
42-56788 (AT-7) 8-29-43C
42-56789 (AT-7) 11-18-45
42-56819 (AT-7) 9-30-43J
42-56821 (AT-7) 5-15-44A
42-56849 (AT-7) 1-19-45G
43-33303 (AT-7) 1-24-44C
43-33304 (AT-7) 11-24-44D
43-33322 (AT-7) 9-9-43G
43-33324 (AT-7) 8-27-43H
43-33363 (AT-7) 3-26-44H
43-33373 (AT-7) 1-24-45A
43-33381 (AT-7C) 4-18-44A
43-33400 (AT-7) 3-3-44F
43-33415 (AT-7C) 4-13-45
43-33443 (AT-7) 3-20-44E
43-33462 (AT-7C) 7-27-45C
43-33479 (AT-7C) 1-19-45G
43-33496 (AT-7C) 9-14-44D

43-33510 (AT-7C) 3-13-44F
43-33529 (AT-7C) 3-4-45A
43-33622 (AT-7C) 9-27-44F
43-49996 (AT-7C) 8-9-45A
43-50028 (AT-7) 12-18-44E
43-50061 (AT-7C) 8-9-45H
43-50099 (AT-7C) 3-31-45A
43-50100 (AT-7) 9-6-44E
44-47465 (F-2) 4-4-45E

Beech AT-10

NA (AT-10) 5-5-41
41-1721 (AT-10BH) 6-22-43D
41-1724 (AT-10) 7-5-43F
41-1734 (AT-10BH) 5-15-44E
41-1737 (AT-10BH) 10-5-44L
41-1751 (AT-10) 3-16-44F
41-1754 (AT-10) 7-21-42B
41-1756 (AT-10) 7-18-42A
41-1790 (AT-10) 1-24-43H
41-1799 (AT-10) 1-24-43H
41-1828 (AT-10) 3-28-43A
41-1832 (AT-10BH) 12-29-42A
41-1851 (AT-10) 7-24-43C
41-2372 (AT-10) 11-7-44B
41-9265 (AT-10BH) 8-6-43A
41-9294 (AT-10) 5-3-44H
41-9297 (AT-10BH) 1-11-43C
41-9300 (AT-10BH) 9-12-42A
41-9301 (AT-10BH) 7-20-43E
41-9309 (AT-10BH) 10-28-42F
41-9326 (AT-10) 5-15-44F
41-9327 (AT-10) 11-23-42B
41-9330 (AT-10BH) 9-25-42B
41-9339 (AT-10BH) 1-23-43I
41-9365 (AT-10BH) 9-19-42A
41-9381 (AT-10BH) 1-25-44G
41-9384 (AT-10BH) 6-1-43A
41-9395 (AT-10BH) 6-29-43D
41-9408 (AT-10BH) 1-3-43D
41-9413 (AT-10BH) 3-22-44D
41-9414 (AT-10) 1-27-43F
41-9417 (AT-10) 11-7-44B
41-9419 (AT-10BH) 6-6-44A
41-9434 (AT-10BH) 10-1-42A
41-9435 (AT-10BH) 9-26-44A

41-26260 (AT-10BH) 1-3-43D
41-26280 (AT-10BH) 9-23-43N
41-26291 (AT-10) 3-31-43C
41-26293 (AT-10BH) 10-27-42F
41-26317 (AT-10BH) 12-10-42E
41-26323 (AT-10BH) 11-3-42C
41-26325 (AT-10) 3-9-43A
41-26330 (AT-10BH) 11-10-42C
41-26331 (AT-10) 3-7-44B
41-26347 (AT-10) 6-10-44D
41-26368 (AT-10) 2-27-43C
41-26370 (AT-10) 2-22-43A
41-26376 (AT-10) 11-10-42A
41-26379 (AT-10BH) 1-9-45A
41-26412 (AT-10) 5-31-44H
41-26422 (AT-10) 7-20-43E
41-26430 (AT-10) 2-23-43J
41-26463 (AT-10) 3-2-44A
41-26469 (AT-10) 3-31-43C
41-26475 (AT-10) 3-16-43B
41-26501 (AT-10) 7-29-43G
41-26638 (AT-10) 1-7-44A
41-26655 (AT-10) 4-11-44E
41-26704 (AT-10BH) 2-27-43A
41-26707 (AT-10BH) 2-5-43B
41-26722 (AT-10BH) 5-3-43M
41-26747 (AT-10BH) 3-27-43G
41-26760 (AT-10) 5-31-44G
41-26772 (AT-10) 5-5-44D
41-26774 (AT-10) 8-9-43B
41-26813 (AT-10BH) 2-10-44A
41-26814 (AT-10) 1-18-44L
41-26821 (AT-10) 6-5-43B
41-26827 (AT-10) 3-7-44B
41-26837 (AT-10BH) 5-27-43B
41-26839 (AT-10BH) 3-5-43C
41-26862 (AT-10BH) 12-14-44E
41-26871 (AT-10) 10-20-44F
41-26874 (AT-10BH) 1-13-44F
41-26887 (AT-10) 12-14-43C
41-26902 (AT-10BH) 8-5-44E
41-26992 (AT-10BH) 2-22-44A
41-26995 (AT-10BH) 2-22-44B
41-26999 (AT-10BH) 12-4-44A
41-27004 (AT-10B) 5-25-43H
41-27023 (AT-10) 3-29-43L
41-27024 (AT-10) 10-30-43H

1195

Bell P-59

Bell P-63

Boeing B-17

42-50534 (B-24J) 6-20-44H
42-50550 (B-24J) 12-11-44B
42-50552 (B-24J) 11-5-44A
42-50559 (B-24J) 10-17-44A
42-50564 (B-24J) 8-25-45
42-50571 (B-24J) 8-3-45B
42-50691 (B-24J) 1-16-45C
42-50778 (B-24J) 11-18-44G
42-50780 (B-24J) 7-2-44B
42-50817 (B-24J) 10-23-44F
42-50834 (B-24J) 8-19-44D
42-50850 (B-24J) 8-23-44F
42-50866 (TB-24J) 4-12-45A
42-50869 (TB-24J) 6-21-45A
42-50871 (B-24J) 7-15-44B
42-50877 (B-24J) 3-8-45C
42-50890 (B-24J) 9-15-44A
42-50931 (B-24J) 6-14-45C
42-50946 (B-24J) 12-16-44A
42-50968 (B-24J) 4-19-45A
42-50969 (B-24J) 3-3-45D
42-50973 (B-24J) 10-28-44B
42-50975 (B-24J) 3-10-45A
42-50983 (TB-24J) 6-30-45B
42-50985 (B-24J) 9-17-44A
42-50992 (B-24J) 12-15-44D
42-50995 (B-24J) 5-1-45
42-51001 (B-24J) 11-8-44E
42-51022 (B-24J) 7-4-44F
42-51034 (B-24J) 12-27-44D
42-51037 (B-24J) 10-6-44A
42-51044 (B-24J) 6-28-44A
42-51045 (B-24J) 10-29-44B
42-51067 (B-24J) 10-16-44A
42-51111 (B-24H) 4-24-44C
42-51220 (B-24H) 6-12-44D
42-51313 (B-24J) 7-25-44A
42-51435 (B-24J) 7-4-44F
42-51444 (B-24J) 10-20-44C
42-51453 (B-24J) 8-19-44B
42-51469 (B-24J) 7-10-44A
42-51478 (B-24J) 12-14-44G
42-51482 (B-24J) 8-18-44G
42-51515 (B-24J) 9-9-44A
42-51517 (B-24J) 8-21-44A
42-51617 (B-24J) 7-19-44B
42-51619 (B-24J) 2-7-45A
42-52386 (B-24H) 1-26-44E
42-52395 (B-24H) 1-8-44F
42-52493 (B-24H) 1-8-44D
42-52626 (B-24H) 1-2-44F
42-63754 (RB-24D) 5-6-45C
42-52756 (B-24H) 2-18-44D
42-63757 (B-24D) 5-25-43E
42-63767 (B-24D) 10-4-43C
42-63772 (B-24D) 10-23-44A
42-63777 (TB-24D) 10-22-44B
42-63895 (B-24D) 2-12-45C
42-63907 (B-24D) 4-17-44G
42-63910 (B-24D) 11-9-44A
42-64013 (B-24D) 10-31-43E
42-64037 (B-24D) 1-16-45A
42-64038 (B-24D) 9-14-44J
42-64039 (B-24D) 12-25-43C
42-64056 (F-7A) 3-30-44A
42-64108 (B-24J) 5-25-44C
42-64111 (B-24J) 2-14-44A
42-64127 (B-24J) 12-8-43B
42-64154 (B-24J) 1-29-44M
42-64155 (B-24J) 6-15-45E
42-64160 (TB-24J) 11-1-45B
42-64163 (B-24J) 10-22-44C
42-64238 (F-7A) 5-25-44E
42-64395 (B-24E) 8-19-44F
42-64397 (B-24E) 10-15-43D
42-64398 (B-24E) 3-20-44A
42-64403 (B-24E) 7-12-44E

42-64412 (B-24E) 10-31-43A
42-64417 (RB-24E) 10-1-44B
42-64425 (RB-24E) 7-7-44A
42-72859 (B-24D) 4-12-44G
42-72862 (B-24D) 8-1-44A
42-72867 (B-24D) 11-14-43E
42-72880 (B-24D) 10-11-43A
42-72882 (B-24D) 9-17-43G
42-72887 (B-24D) 1-12-44E
42-72894 (B-24D) 4-7-44F
42-72929 (B-24D) 5-1-44D
42-72931 (B-24D) 1-24-44A
42-72944 (B-24D) 9-29-43C
42-73012 (B-24J) 9-16-43F
42-73128 (B-24J) 10-30-43F
42-73161 (B-24J) 9-29-44E
42-73184 (B-24J) 5-4-45B
42-73190 (B-24J) 11-24-43I
42-73191 (TB-24J) 11-1-45A
42-73198 (B-24J) 2-24-44E
42-73228 (B-24J) 3-23-44I
42-73229 (B-24J) 11-3-43A
42-73234 (B-24J) 9-1-44A
42-73235 (B-24J) 10-23-43F
42-73239 (B-24J) 3-21-44H
42-73291 (B-24J) 6-14-45E
42-73344 (B-24J) 11-30-44A
42-73347 (B-24J) 1-12-44F
42-73349 (B-24J) 11-15-43B
42-73352 (B-24J) 3-5-45F
42-73356 (B-24J) 11-18-43F
42-73357 (B-24J) 11-30-44A
42-73358 (B-24J) 1-3-44B
42-73359 (B-24J) 9-14-44G
42-73362 (B-24J) 7-20-44G
42-73365 (B-24J) 1-8-44G
42-73367 (B-24J) 1-18-45B
42-73369 (B-24J) 7-13-44D
42-73375 (B-24J) 4-5-44D
42-73377 (TB-24J) 4-21-45D
42-73378 (B-24J) 6-3-44C
42-73382 (B-24J) 9-8-44A
42-73386 (B-24J) 4-27-44F
42-73387 (B-24J) 5-16-44G
42-73390 (B-24J) 2-1-44C
42-73411 (B-24J) 9-24-44B
42-73514 (B-24J) 11-16-44F
42-78522 (B-24J) 8-1-44A
42-78535 (B-24J) 8-9-44F
42-78537 (B-24J) 9-13-44A
42-78538 (B-24J) 6-3-45A
42-78549 (B-24J) 9-18-45
42-78550 (B-24J) 11-18-44D
42-78551 (B-24J) 9-28-45B
42-78553 (B-24J) 5-31-45C
42-78574 (B-24J) 1-29-45C
42-78579 (B-24J) 9-30-44A
42-78581 (B-24J) 2-5-45
42-78582 (B-24J) 11-2-44C
42-78621 (B-24J) 1-20-45A
42-78632 (B-24J) 10-25-44H
42-78696 (RB-24E) 4-26-45A
42-78701 (B-24J) 2-21-45C
42-94731 (B-24H) 3-8-44D
42-94956 (B-24H) 7-13-45A
42-95069 (TB-24H) 6-22-45A
42-95389 (B-24H) 10-9-44B
42-95392 (B-24H) 10-20-44A
42-95403 (B-24E) 3-4-45B
42-95405 (B-24H) 11-1-44A
42-95467 (B-24H) 4-25-44F
42-95499 (B-24H) 9-14-44F
42-95523 (B-24J) 12-5-44A
42-95525 (B-24J) 3-6-45B
42-95545 (B-24J) 10-17-44D
42-95559 (B-24J) 1-1-45B
42-95568 (B-24J) 1-2-45C

42-95580 (B-24J) 10-4-44C
42-95591 (B-24J) 2-28-45A
42-95605 (B-24J) 4-27-44A
42-95606 (B-24J) 11-16-44G
42-95611 (B-24J) 4-11-45C
42-95614 (B-24J) 6-30-44C
42-95627 (B-24J) 6-16-44C
42-99902 (B-24J) 3-14-44G
42-99946 (B-24J) 8-21-44A
42-99962 (B-24J) 4-14-44A
42-99969 (B-24J) 1-21-44G
42-99998 (B-24J) 12-29-44D
42-100002 (B-24J) 5-20-44G
42-100010 (B-24J) 12-14-43D
42-100014 (B-24J) 12-30-43E
42-100019 (B-24J) 6-29-44F
42-100023 (B-24J) 6-17-44G
42-100024 (B-24J) 5-27-44D
42-100032 (B-24J) 2-2-44C
42-100043 (B-24J) 6-24-44C
42-100050 (B-24J) 2-9-45C
42-100056 (B-24J) 11-30-44C
42-100058 (B-24J) 1-18-45H
42-100063 (B-24J) 7-14-44A
42-100065 (B-24J) 7-7-44B
42-100069 (B-24J) 10-8-44B
42-100070 (B-24J) 12-19-43E
42-100082 (B-24J) 4-26-44D
42-100088 (B-24J) 1-18-44E
42-100092 (B-24J) 7-31-44G
42-100104 (B-24J) 9-19-44B
42-100106 (B-24J) 2-15-44D
42-100114 (B-24J) 11-20-44C
42-100124 (B-24J) 5-5-44E
42-100137 (B-24J) 9-5-44E
42-1001 (B-24J) 7-20-44G
42-100155 (B-24J) 11-24-44A
42-100171 (B-24J) 10-4-44E
42-100183 (B-24J) 1-19-44B
42-109834 (B-24J) 1-26-45A
42-109868 (B-24J) 12-20-44C
42-109877 (B-24J) 7-23-44A
42-109899 (B-24J) 7-26-44B
42-109902 (B-24J) 2-19-44A
42-109920 (B-24J) 3-2-45C
42-109930 (B-24J) 6-22-44J
43-30565 (C-87) 3-2-45A
44-10605 (B-24J) 7-26-44C
44-28208 (B-24J) 10-24-44D
44-40523 (B-24J) 12-22-44C
44-40552 (B-24J) 9-17-44D
44-40608 (B-24J) 4-25-44F
44-40758 (B-24J) 6-7-44A
44-41015 (B-24J) 8-19-44I
44-41995 (B-24M) 5-12-45A
44-42211 (TB-24M) 8-21-45D
44-48772 (B-24J) 5-4-45C
44-48800 (B-24J) 8-19-44I
44-48865 (TB-24J) 10-18-45A
44-48900 (B-24J) 10-3-44A
44-48934 (TB-24J) 9-1-45A
44-49061 (B-24L) 2-21-45B
44-49180 (B-24L) 1-30-45F
44-49669 (B-24L) 11-29-44A
44-49873 (B-24L) 1-22-45B
44-49908 (B-24L) 12-15-44A
44-50103 (B-24L) 6-29-45A
44-50190 (B-24L) 2-19-45A
44-50307 (B-24M) 6-28-45B
44-50343 (B-24M) 12-20-44B
44-50788 (B-24M) 2-2-45E
44-51522 (B-24M) 7-22-45A

Consolidated B-32

41-141 (B-32) 5-10-43I
42-108473 (B-32) 3-5-45C
42-108516 (B-32) 9-7-45B

Consolidated LB-30

41-527 (LB-30) 7-1-42A
AL-601 (LB-30) 6-4-42B
AL-607 (LB-30) 1-8-42C
NA (LB-30) 6-2-41B

Consolidated OA-10

44-33919 (OA-10A) 5-7-45B
44-34096 (OA-10A) 8-1-45A

Culver Aircraft

Culver PQ-8

42-96931 (PQ-8A) 11-5-43C

Culver PQ-14

43-44284 (PQ-14A) 7-31-44B
44-21782 (PQ-14B) 11-22-44D

Culver Cadet

NC-32470 (Cadet) 1-22-43E

Curtiss A-25

41-18781 (A-25A) 5-17-43C
41-18822 (A-25A) 8-19-43B
42-20169 (RA-25A) 9-23-44C
42-54885 (RA-25B) 7-15-44D
42-79670 (A-25A) 9-2-43A
42-79678 (RA-25A) 10-28-43E
42-79682 (RA-25A) 10-26-43C
42-79695 (RA-25A) 12-24-43H
42-79715 (RA-25A) 11-26-44A
42-79804 (RA-25A) 7-23-44C
42-79823 (RA-25A) 12-29-43B
42-79826 (RA-25A) 7-23-44C
42-79862 (RA-25A) 3-13-45B
42-79895 (RA-25A) 12-20-43D
42-79901 (RA-25A) 6-13-44B
42-79928 (RA-25A) 12-15-44F
42-79939 (RA-25A) 7-24-44F
42-79960 (A-25A) 3-11-44C
42-80118 (A-25A) 6-13-44D
42-80121 (RA-25A) 7-10-44C
42-80146 (RA-25A) 3-18-44F
42-80148 (A-25A) 6-13-44D

Curtiss AT-9

41-5757 (AT-9) 4-2-42F
41-5771 (AT-9) 10-8-43A
41-5779 (AT-9) 5-9-44A
41-5786 (AT-9) 1-22-43B
41-5789 (AT-9) 10-2-42G
41-5790 (AT-9) 7-14-42A
41-5793 (AT-9) 8-19-42D
41-5803 (AT-9) 2-19-42B
41-5808 (AT-9) 1-17-44A
41-5819 (AT-9) 9-11-42D
41-5828 (AT-9) 12-30-42J
41-5831 (AT-9) 7-9-43A
41-5835 (AT-9) 10-26-42E
41-5846 (AT-9) 1-10-44C
41-5849 (AT-9) 3-5-42A
41-5852 (AT-9) 3-12-43B
41-5857 (AT-9) 2-19-42B
41-5864 (AT-9) 2-11-43E
41-5865 (AT-9) 9-25-43H
41-5867 (AT-9) 6-3-42D
41-5871 (AT-9) 7-9-44
41-5873 (AT-9) 1-17-44A
41-11943 (AT-9) 11-4-42B

P-40R refers to 300 Merlin-equipped P-40F and P-40L airplanes that were re-fitted with Allison V-1710-81 engines.

Douglas A-24/A-33

Douglas A-26

Douglas B-18

Douglas B-23/C-67

Douglas C-54

Douglas DB-7

Douglas O-38

Douglas O-46

Douglas Transport Series

C-39, C-47, C-48, C-49, C-53, DC-3

*R37 refers to an AAF version of the Ventura II Model 37-27-01 modified to USAAF standards and equipped with Pratt & Whitney R-2800-31 engines, also known as R-Model 37 or RM-37. The AAF retained 264 Ventura II/Model 37s from an RAF order.

North American
B-25

Ryan Trainers

PT-16, PT-20, PT-22, PT-25

Schweizer TG-2

Seversky P-35

Sikorsky Aircraft

Sikorsky R-5

Sikorsky S-43

Stearman Trainers

PT-13, PT-17, PT-18, PT-27

Location Index (by State)

References are to entry numbers

Name Index

References are to entry numbers

Aaron, Joseph E. 9-24-44A
Abbate, Paul F. 11-18-44D
Abby, Kenneth 2-17-43D
Abbott, Guy E. 5-23-45C
Abbott, Ishmael L. 4-16-45B
Abbott, Raymond C. 5-2-44G
Abegglen, Robert C. 5-23-45A
Abel, Donald 8-10-43A
Abel, George L. 10-4-44B
Abelein, Jay D. 7-26-43G
Abeles, Anson S. 8-1-43D
Abell, Stuart S. 8-16-43J
Abercrombie, Aaron R. 4-13-44A
Abernathy, W. E. 10-8-43C
Abinant, Harry S. 9-22-45A
Ables, Robert E. 11-30-42A
Abmyer, Richard R. 12-29-43C
Abney, James A. 12-30-42A
Abraham, Jack W., Jr. 10-9-43J
Abraham, Spero 8-3-44B
Abrahamson, Paul M. 4-17-44G
Abramezyk, Stanley E. 8-16-43J
Abrams, Lawrence R. 5-27-42H
Abramson, Elbert L. 8-8-45D
Abramson, Shelly L. 8-15-43A
Abramson, Sydney 1-17-45G
Abrell, John D. 4-26-43E
Abromowitz, Elliot A. 12-11-43B
Abruzzo, Thomas S. 8-25-43E
Abts, Thomas J. 7-9-43A
Achberger, Eugene G. 1-4-44B
Achterberg, Robert W. 7-11-44E
Acker, Charles P. 9-4-43D
Acker, Edward J. 4-5-44E
Acker, Hoyt, Jr. 7-30-42D
Ackerman, Ellis J. 11-5-42A
Ackerman, William P. 4-28-43C
Acklin, Evan C. 5-12-44E
Acree, William F. 8-3-44F
Adair, Ellis H. 10-5-44L
Adair, Frank G. 11-4-43D
Adair, William G. 5-5-44F
Adams, Aaron R. 1-1-43B
Adams, Alvin P. 1-23-43H
Adams, Billy G. 5-2-43
Adams, Chambers D. 6-20-45C
Adams, Clyde E. 8-8-44A
Adams, Clyde M. 5-2-43
Adams, Frank W., Jr. 8-21-44A
Adams, George 5-31-43H
Adams, George L. 9-2-43C

Adams, George R. 1-3-43C
Adams, Gerald L. 4-14-45
Adams, Herman 10-23-42A
Adams, James R. 7-14-42C
Adams, Jesse P. 11-16-43G
Adams, John 10-18-45A
Adams, Kenneth 7-29-43A
Adams, P. C. 4-21-42B
Adams, Peter H. 1-12-44E
Adams, Peter Henry 5-10-44J
Adams, Postal A. 9-21-42
Adams, Richard V. 7-10-44G
Adams, Robert 10-6-44A, 12-16-44A
Adams, Robert E. 7-7-42C
Adams, Robert F. 1-14-42B
Adams, Robert H. 1-11-45B
Adams, Robert S. 12-4-42E
Adams, Stephen E. 10-29-43A
Adams, Thiel E. 7-16-43F
Adams, William F. 7-9-43H
Adams, William H. 3-10-45C
Adamson, Lindel M. 5-13-43C
Adamson, Russell W. 2-2-44D
Adcock, Luther E. 4-25-44H
Addy, Noel D. 12-16-42G
Adelman, Efrem M. 3-31-45A
Adelson, Horace J. 7-9-43I
Aden, James K. 7-2-43D
Aderholt, Warren F. 3-5-44B
Ades, Gilbert L. 11-8-43
Adin, William M. 2-21-45A
Adkins, Frank E. 2-23-45C
Adler, Richard L. 7-18-42B
Addys, William R. 9-5-42B
Adell, Bruce N. 4-17-44G
Adell, Robert F. 5-8-44C
Adelsky, Jacob 2-9-43D
Adkins, Ralph C. 1-17-44F
Adler, John P. 7-7-44G
Adley, Albert S. 11-2-44C
Admeyers, Glen S. 8-18-44G
Adragna, Joseph M. 5-12-45B
Adriance, Elmer R. 7-15-43D
Affeldt, Martin F. 8-9-44F
Affinito, Theodore V. 3-5-44B
Affrine, Milton B. 1-16-42B
Africk, Sanford 9-22-45D
Agee, Lewis W. 2-23-45A
Agin, Lambert 10-14-44E
Agliata, Anthony J. 8-17-45
Agnew, Casper 3-30-43B

Agnew, Eugene I. 10-13-44A
Agnew, Francis M. 2-9-45A
Agnew, Robert E. 11-2-41A
Agostini, Amerigo L. 12-13-43B
Agriesti, Joseph L. 3-8-44E
Ahart, Charles W. 7-24-43B
Ahern, Daniel R. 4-24-45A
Ahlbeck, Torsten W. 1-18-44E
Ahlers, Robert C. 2-20-43
Ahlert, Joseph F. 1-3-44B
Ahlmann, Richard A. 7-2-44
Ahr, William A. 7-9-43G
Ahrens, Irving B. 9-22-45A
Aiello, Ralph J. 2-2-44C
Aiken, Claude T. 6-14-45E
Aikey, Royal H. 9-1-43I
Ailshouse, William W. 7-6-45
Aitchison, John C. 1-5-43B
Aitken, Ellsworth W. 3-5-42E
Aitken, Gordon R. 8-7-45B
Aitken, Richard L. 11-16-44C
Akelaitis, Alfred P. 3-31-45A
Akers, William C. 2-8-44B
Akielaszek, Stefan J. 12-14-43D
Akin, Ira C. 8-11-44E
Akins, James H. 12-4-42E
Akins, Wallace O. 2-5-43A
Akridge, James M. 3-25-44H
Akutagawa, Norio 11-28-45B
Alabaster, J. H. 12-16-42G
Alaniz, Alfonso M. 11-4-43C
Alansky, Anthony W. 12-8-45
Albanese, Frederick T. 1-3-44A
Alber, John E. 3-4-43D
Alber, Wayne R. 3-15-44I
Albers, Robert J. 7-14-42B
Albert, John F. 8-3-44F
Albert, Lawrence 7-16-42F
Albert, Sidney J. 7-25-42B
Albert, William L. 4-22-44B
Alberty, Ernest H., Jr. 9-27-44G
Alberty, Vic 2-28-42D
Albin, Samuel J. 8-20-44C
Albondy, Anthony D. 2-5-45
Albrecht, George R. 12-11-43B
Albright, Jack C. 6-21-43E
Albright, P. J. 7-14-44D
Albring, William C. 7-25-43C
Albro, William C. 12-23-43C
Albus, John R. 5-5-45
Alchorn, George C. 11-8-42D
Alcoa Prospector 2-27-43B

Alden, Frank R. 6-9-42B
Alderman, Frederick J. 1-20-44C
Alderman, Walter 1-15-43A
Alderson, John W. 7-9-43K
Aldridge, Dean C, 7-17-44G
Aldridge, Robert P. 3-8-44A
Alessi, Vincent A. 8-26-44B
Alevy, Victor I. 5-15-44A
Alexandar, Robert L. 6-8-43A
Alexander, Andrew J. 6-15-43A
Alexander, Aubrey H. 2-18-44A
Alexander, Ernest R. 10-3-44C
Alexander, Harry S., Jr. 5-5-45
Alexander, J. V. 1-23-43A
Alexander, James D. 1-20-44F
Alexander, James W. 1-23-43H
Alexander, James W. 8-1-43D
Alexander, James W. 5-15-45C
Alexander, Kenneth M. 2-23-45A
Alexander, Raymond C. 5-8-42D
Alexander, Robert L. 7-7-44E
Alford, George G. 1-25-44H
Alford, John T., Jr. 8-16-43L
Algeria, Alfred 12-15-44D
Algosi, Albert 5-17-43A
Allain, Jules B. 6-22-42C
Allaire, LeRoy J. 4-29-45B
Allard, William T. 6-11-43L
Allbaugh, Raleigh E., Jr. 7-12-45C
Allee, Edward S. 5-31-44C
Alleman, Alden A. 1-6-43F
Allen, Annie R. 7-11-44G
Allen, Charles J. 5-14-44A
Allen, Charles M. 1-14-43A
Allen, Charles P., Jr. 3-25-44G
Allen, Don 11-5-42B
Allen, Donald W. 8-9-45F
Allen, Earl K. 8-24-45D
Allen, Edmund T. 2-18-43
Allen, Edwin W. 2-7-45A
Allen, Elwood J. 10-1-43F
Allen, Francis E. 2-11-44D
Allen, Fulton T. 7-11-43B
Allen, Gail F. 6-13-44D
Allen, George B. 5-7-44B
Allen, Henry M. 3-17-44B
Allen, James A. 7-19-43E
Allen, James L. 3-10-45C

1235

Arnett, Walter B. 3-16-44D
Arney, Mafalda E. 5-22-45H
Arney, Rector 5-3-43A
Arnold, Claude 10-12-42H
Arnold, Donald T. 5-5-43L
Arnold, Edward W. 12-29-43C
Arnold, Herbert C. 6-20-43G
Arnold, James R. 4-3-45B
Arnold, John W. 8-31-43A
Arnold, Joseph R., Jr. 10-18-43J
Arnold, Norris V. 7-22-44A
Arnold, Patrick A. 3-7-44B
Arnold, Paul H. 4-23-44E
Arnold, Reuben E. 5-11-44H
Arnold, Richard B. 4-20-45A
Arnold, Robert E. 5-4-45B
Aro, Arthur A. 1-2-44F
Arone, Joseph F. 7-1-45A
Arp, Harry C. 12-30-43A
Arrillaga, John F. 9-10-43G
Arrington, Thomas N. 5-27-43 G
Arseneault, Clayton M. 11-16-43A
Artland, P. 9-20-43A
Arthur, Charles E. 3-3-43A
Arthur, William G. 5-15-44F
Artman, Karl A. 11-12-44A
Asay, Gene R. 9-17-44A
Aschenbrenner, Robert R. 3-9-44C
Ash, Arnold R. 10-5-43B
Ashba, John W. 2-12-44E
Ashby, Hugh R. 1-3-44H
Ashcraft, Blain 11-15-43B
Ashcraft, Miller G. 1-13-43D
Asher, Robert D. 3-29-45A
Ashley, John C. 6-13-44E
Ashley, William M., Jr. 5-27-44D
Ashner, Herman 8-27-42C
Ashpole, Donald J. 6-16-43G
Ashurst, Roger D. 9-1-43B
Askew, Herrell C. 7-13-45A
Asmer, Lewis J. 5-10-43G
Asparato, Maximo S. 7-12-45C
Asparato, Maximo S., (Wife) 7-12-45C
Asper, Donald R. 10-25-44F
Aspinwall, Randolph H. 3-18-44D
Asplund, Robert H. 12-22-44C
Assam, Moneer 12-25-44A
Assell, Lawrence B. 12-23-43B
Ast, Zygmunt A. 8-21-43H
Astier, Claude 4-11-45F
Aten, Arley F. 2-5-45
Atherton, Louis E. 2-16-44A
Athey, Milton W. 7-25-44E
Atkin, Thomas W. 4-10-42D
Atkins, Frank A. 9-11-42C
Atkins, James P. 10-29-43H
Atkins, Thomas F. 3-21-45A
Atkins, William L. 12-25-43B
Atkinson, Charles E. 12-12-44B
Atkinson, Earl G. 9-2-43G
Atkinson, Edward N. 6-12-44D
Atkinson, Philip E. 2-27-43B
Atkinson, Ralph J., Jr. 10-27-43G
Atkinson, Richard L. 5-5-43L
Attaberry, James L. 5-9-44A
Attison, L. R. 4-5-42A
Attridge, James F., Jr. 9-13-42A
Atwood, Calvin G. 8-30-43F
Atwood, Ralph D. 10-22-43F
Atwood, Robert W. 1-18-44E

Atwood, Simeon H. 7-27-43B
Atz, John F. 1-8-44F
Aubert, William B. 3-18-43A
Aubin, Martin R. 1-19-44L
Auborn, Seymour E. 5-29-43B
Aubrey, Henry S. 6-7-43F
Aubry, Jacques 1-18-44L
Aucoin, Maurice J. 7-28-45D
Audino, Emanuel P. 1-18-45B
Audio, Joseph B. 6-23-43A
Audius, John W. 12-25-44A
Audonnet, Pierre 10-18-45B
Aufiery, John D. 3-5-45C
Augelli, Archille P. 10-25-43C
Auld, Dewitt C. 1-23-44G
Aulsbury, Robert J. 7-18-42D
Aulwes, William W. 7-24-43A
Aument, George C. 7-11-45A
Auranen, Allen H. 9-26-43F
Aus-Lander, Pincus H. 8-16-43L
Ausburn, Jess J. 1-1-45A
Austenfeld, Eugene J. 11-16-43I
Austin, Daniel V. 11-30-42D
Austin, Donald W. 6-27-43J
Austin, Howard C. 10-4-42B
Austin, Joe, Jr. 5-9-44A
Austin, John K. 5-8-42A
Austin, Lewis G. 1-12-44H
Austin, Robert E. 8-14-44H
Avansino, Harold V. 4-24-43C
Avant, Peggy 7-12-45C
Avedisan, Souron S. 1-16-43B
Aveni, Anthony C. 4-12-44A
Averill, Paul C. 8-12-42E
Averitt, William T. 5-13-43F
Avery, R. B. 3-4-45B
Avery, R. E. 1-8-42C
Avery, Raymond B. 7-14-45B
Averings, Byron C. 10-20-44C
Axen, Laurice G. 10-28-43C
Aycock, Julian C. 12-7-42A
Ayers, Clarence F. 3-5-42C
Ayers, Francis N. 1-15-43I
Ayers, Philip W. 3-10-45A
Ayling, Albert E., Jr. 8-11-42A
Aylor, Meredith M. 3-19-43A
Azevedo, Joseph P. 1-19-44-K
Azzinnaro, Gerald 9-29-44D

Babb, Eugene W. 5-27-42D
Babb, Frederick 6-5-43K
Babb, John J. 6-21-43A
Babbitt, Richard K. 7-16-42C
Babcock, Durward A. 3-27-44G
Babcock, Franklin H. 8-3-44B
Babcock, Warren E. 8-6-43D
Baber, James H. 6-9-44B
Babinetz, Marjory L. 7-27-44D
Babkewicz, John W. 1-21-44G
Bacca, George P. 3-27-43E
Bachelder, Donald R. 1-18-45H
Bachhuber, Andrew J. 11-15-43A
Bachrach, William H. 7-10-44E
Backstrom, C. J. 3-29-43C
Bacon, Dwight G. 6-21-43F
Bacon, Joseph J. 9-20-43A
Bacon, Philip H. 4-16-43C
Bacon, Thomas P. 11-18-41C
Bacon, Warren O. 6-11-43L
Bacot, Henry P. 6-12-41B
Bacsik, John, Jr. 8-31-43E
Bade, Robert O. 10-20-43B
Bader, Peter 2-28-43E
Badgett, Charles R. 1-15-43E
Badura, J. C. 2-12-42A
Baebler, Bernard G. 10-13-43G

Baehr, Albert J. 6-10-43B
Baer, Alvin L. 4-4-43B
Baer, Howard D. 4-21-43D
Baer, James W. 5-14-42C
Bafford, Thomas W. 11-1-41
Bafus, John E. 8-24-45D
Bagby, J. S. 2-28-42E
Bagby, W. W. 8-28-42D
Bagdley, Benjamin 10-4-43H
Baggett, J. 1-6-43G
Baggett, James E., Jr. 7-16-41
Bagley, A. L. 7-10-43D
Bagley, Edward N. 7-13-44D
Bagley, W. M. 8-13-42D
Bagnai, A. W. 10-15-42A
Bagnall, Robert W. 7-14-45B
Bagwell, Edward A. 2-26-43C
Bahling, Donald M. 5-10-44D
Baie, Wendell M. 11-4-42B
Bailey, Allen R. 9-5-41B
Bailey, Arnold J. 2-4-43C
Bailey, Charles 4-18-45F
Bailey, Eugene R. 4-13-44G
Bailey, Holcomb D. 1-24-44A
Bailey, James A. 4-20-45A
Bailey, John G. 5-13-44F
Bailey, John J. 4-9-44B
Bailey, John L. 3-15-42B
Bailey, John N. 7-18-45
Bailey, Ralph W. 1-26-43E
Bailey, Randolph E. 4-23-43E
Bailey, Richard A. 2-26-44B
Bailey, Robert C. 6-20-43D
Bailey, Robert E. 3-21-44A
Bailey, Robert W. 7-21-43E
Bailie, Edward E. 9-27-44F
Bailiff, Edward C. 11-16-44E
Baima, A. L. 1-6-43G
Bainbridge, William G. 6-16-44A
Bains, Lios C. 2-25-44F
Baiocchi, Joseph 8-14-43J
Bair, Floyd V. 10-31-43D
Bair, Henry L. 7-5-43D
Baird, Calvin C. 12-30-43C
Baird, Dale M. 4-5-44D
Baird, Harry W. 2-18-43
Baird, Robert E. 3-21-44A
Bajorin, Francis D. 7-11-45A
Bakas, Vytant J. 4-18-43B
Baker, A. J., Jr. 7-4-44F
Baker, Albert B., Jr. 11-4-43C
Baker, Alfo L. 9-19-42D
Baker, Bill M. 4-12-45E
Baker, Clande G. 5-13-43J
Baker, Clinton 10-23-42D
Baker, David M. 9-18-45
Baker, David R. 9-29-44E
Baker, Donald P. 1-28-42A
Baker, Donald T. 9-6-42
Baker, Elwood w. 2-24-44E
Baker, Frank J. 1-2-44C
Baker, Gerald W. 2-17-43D
Baker, Gloria 2-27-43A
Baker, Harold C. 10-5-45G
Baker, Harold M. 10-9-43H
Baker, James B. 9-10-45A
Baker, James G. 8-9-45D
Baker, James O. 10-1-43F
Baker, John D. 9-28-43I
Baker, John M. 5-6-43B
Baker, Lowell R. 5-25-44E
Baker, Lyle R. 7-11-43D
Baker, Myrtle 6-6-43C
Baker, Robert A. 4-9-42C
Baker, Robert B., Jr. 10-17-44A
Baker, Ross C., Jr. 9-30-45

Baker, Sam M. 9-30-42D
Baker, Thomas W. 10-3-43E
Baker, William A. 9-28-43G
Baker, William C. 2-25-44H
Baker, Woolford B. 1-31-41
Bakker, Marvin L. 1-16-44C
Bakula, Robert G. 10-19-44B
Balagia, Jack M. 2-19-44A
Balch, F. 2-17-43F
Baldsiefen, Richard E. 3-4-42A
Baldwin, Allen E. 3-2-45A
Baldwin, Elbert E. 3-16-43A
Baldwin, Ida Beatrice 7-30-42B
Baldwin, John S. 11-10-44C
Baldwin, Joseph 7-7-44J
Baldwin, Louis B., Jr. 6-24-44C
Baldwin, Ralph R. 9-17-43G
Baldwin, Raymond M. 9-19-41D
Baldwin, Robert L. 8-16-43E
Baldwin, Thomas S. 6-15-44D
Bale, George E. 5-7-45B
Bales, Marvin E. 5-10-43A
Baliski, Walter P. 3-22-43C
Balkcum, Cleve C. 4-22-44E
Balkin, F. 9-20-43A
Balkovich, William 3-15-43D
Ball, Carroll B. 8-1-44A
Ball, Charles C. 11-19-41B
Ball, John R. 4-14-44A
Ball, Kyle E. 1-11-43A
Ball, Lewis S. 6-20-44H
Ball, Lloyd L. 2-11-43F
Ball, Phillip H. 10-23-44F
Ballard, John H. 8-4-43I
Ballard, Harold D. 11-30-44A
Ballard, Howard W. 7-18-43C
Ballard, Marshall W. 9-28-44D
Ballard, Norman E. 7-10-44E
Ballard, William J. 1-18-44G
Ballas, Jack 3-21-44G
Ballas, John W. 6-9-43D
Ballenger, E. M. 9-20-43A
Ballew, Herbert H. 6-7-42B
Ballif, Leonard H., Jr. 4-9-42C
Ballman, William R. 2-7-44H
Ballos, Emmanuel J. 1-11-45B
Balmer, David 9-28-43J
Balser, Ralph E. 12-11-45
Balser, Robert A. 7-7-44K
Balser, Clifford A. 5-15-44A
Balter, R. D. 10-17-44A
Baman, Walter R. 3-14-43D
Bamberger, Alan J. 3-24-43I
Banaszek, Joseph J. 12-22-43I
Bancom, Joseph P. 2-26-43B
Bancroft, John W. 8-27-43I
Bancroft, Richard F. 11-1-44A
Bandenelli, Vito W. 6-16-43A
Bane, Glenn O. 9-24-43K
Bank, Thomas 10-19-45A
Banks, Paul W. 7-12-43F
Banks, William E. 10-21-43C
Bankston, Buell A. 1-27-43C
Bankston, Howard D. 2-10-43F
Banner, Armon B. 4-14-44D
Bannerman, Frank W. 10-29-44C
Barag, Herbert M. 6-22-42A
Baran, Michael, Jr. 10-21-43B
Baran, Walter F. 1-25-43C
Barar, Harold H. 10-29-44C
Barbalat, John 2-28-44D
Barbanti, Robert A. 8-29-43C
Barber, Donald A. 2-29-44A
Barber, Elliot R. 5-25-43D
Barber, Frank L. 4-2-42B

Bucklin, Fred R. 5-5-44A
Buckner, Kermit C. 1-31-42
Buckner, Seymour K. 10-13-44A
Buckner, Warren J., Jr. 2-16-43B
Budenholzer, Frank E. 3-21-42
Budimirovich, Bernard 2-19-43A
Budka, William F. 3-10-45A
Budovsky, George A. 9-8-44C
Budraski, George 7-25-42D
Budryk, Adolph A. 10-6-42C
Buecker, John F. 6-9-43A
Buege, Walter W. 6-5-43D
Buela, Ramon G. 2-28-43B
Buess, Louis J. 10-15-43D
Buffington, George P. 9-17-44D
Bugner, Leo S. 9-22-43A
Bugra, John 10-9-42D
Buhrmann, Deanec 12-30-43F
Buick, Warren H. 11-16-43I
Buie, Jacob M. 5-12-44C
Buin, Henri 8-1-45B
Bulcavage, Edward A. 1-11-45B
Bulha, James H. 3-8-44D
Bulleigh, Donald L. 7-10-43F
Bullington, Rudolph L. 9-3-44A
Bullis, Clay U. 1-25-42A
Bulloch, Robert N. 2-12-44E
Bullock, Frederick E. 12-25-42A
Bullock, Gerald D. 1-15-43K
Bullock, Harry 3-5-42C
Bullock, Raymond L. 6-14-45B
Bullough, Alma D. 3-21-44A
Bullowa, (Child) 7-12-45-C
Bullowa, Lillian 7-12-45C
Bullukian, David 9-5-43G
Bumberg, Irving 1-31-44C
Bumpass, Robert A. 11-9-44A
Bumgarner, Carl T. 9-1-43F
Bunce, Harold W. 3-29-43B
Bunch, Paul W. 3-8-41B
Bunch, Shirley R. 4-19-44A
Bund, Durward F. 12-2-43D
Bunda, Andres O. 10-29-45
Bundy, Carl H. 2-16-45
Bundy, Charles W. 12-12-41
Bundy, Edwins S. 1-10-45A
Bunker, Ray M. 3-19-44E
Bunn, henry 11-30-42A
Bunn, Robert D. 10-31-42A
Bunn, Roy M. 1-18-44M
Bunting, Billy H. 8-9-43A
Burall, Jesse E. 6-20-42A
Buran, Curtis L. 2-16-44A
Burbank, Albert M. 10-5-44K
Burbank, Lawrence 5-22-43G
Burch, Francis M. 1-15-45E
Burch, Thomas N. 1-17-45H
Burcham, Milo G. 10-20-44G
Burchell, Richard E. 8-28-43G
Burchfield, Harvey C. 1-9-45C
Burd, G. E. 10-23-43D
Burd, Raymond E. 10-12-42H
Burda, James 8-1-45B
Burdette, Harry L. 3-18-43C
Burdick, Richard M. 6-20-43F
Burey, Joseph 1-16-45C
Burfeind, Henry G. 1-27-45
Buford, Bernard P. 9-29-44E
Burford, John T. 1-21-44G
Burford, William J. 12-11-42F
Burgan, Donald E. 6-27-43E
Burdorfer, Kenneth G. 8-20-42A
Burge, Theodore J. 10-27-44B
Burge, William B. 10-12-42F

Burgess, Clifford J. 1-14-44A
Burgess, Eugene V. 3-24-43I
Burgett, Orville N. 6-21-44C
Burham, Howard D. 6-22-44D
Burk, Theora B. 1-26-43A
Burke, A. J. 2-26-43B
Burke, Carr 7-19-43B
Burke, Edward J. 8-17-42A
Burke, Frank W. 7-17-45A
Burke, James C., Jr. 8-30-44F
Burke, Jules C. 2-10-43F
Burke, Louis R. 5-6-42A
Burke, Monroe G. 8-8-45D
Burke, Raymond D. 2-11-43B
Burke, Thomas J. 10-22-42E
Burke, Thomas J. 9-11-43F
Burke, Vance G. 5-23-43C
Burke, Walter J., Jr. 9-4-43C
Burke, William J. 5-23-43E
Burkett, Ray G. 5-19-43A
Burkhall, Walter H. 7-26-43E
Burkhardt, John H. 4-30-45C
Burks, John Adams 1-11-43B
Burkus, W. A. 2-15-45B
Burlow, Marshall J. 6-19-43G
Burleson, Clark L. 5-11-44H
Burman, Alfred W. 10-3-43E
Burner, James W. 11-3-44B
Burnet, James L. 5-31-45A
Burnett, Billie 12-11-44B
Burnett, Carl A. 9-16-43E
Burnett, Charles E. 1-31-43A
Burnett, Charles F. 8-24-43E
Burnett, Donald H. 9-28-43E
Burnett, Donald R. 9-1-43C
Burnett, Jack M. 2-26-44J
Burnett, John D. 7-21-45C
Burnett, Lyda, A. 10-20-44A
Burnison, Harmon E. 2-18-43
Burns, Charles J. 1-24-43J
Burns, Clifford F. 10-2-44A
Burns, Eliam B. 4-14-41B
Burns, Harry R. 6-15-45E
Burns, Henry C. 12-23-43A
Burns, Homer C. 12-12-41
Burns, Hugh R. 7-17-43D
Burns, Robert J. 6-5-43D
Burns, Robert W. 6-7-43A
Burns, Warren V. 3-30-44A
Burnside, M. B. 12-11-42F
Burnside, Walter R. 2-12-45C
Burr, Cecil C. 10-10-43A
Burr, William Glenn 6-9-42B
Burrell, Frank J. 3-21-44A
Burrell, Sammy S. 4-17-42B
Burris, Charles 9-21-44A
Burris, Lloyd S. 8-19-44A
Burris, William H. 1-2-44G
Burroughs, Raymond W. 2-12-43C
Burrows, Daniel W. 10-26-42B
Bursaw, Frank L. 5-12-44C
Bursey, Robert C. 12-5-43A
Burson, Ralph 12-21-43D
Burson, Vance 2-19-45D
Burtis, John W. 8-17-45
Burton, Donald D. 8-19-43F
Burton, James J. 2-16-44A
Burton, Joseph H. 2-19-45A
Burton, Moseley 1-3-44B
Burton, Richard C. 9-13-42C
Burton, Thomas L. 7-24-43E
Burton, W. J. 1-2-43E
Burton, William A. 6-18-43D
Burwell, Edward L. 6-13-44D
Burwell, Harold H. 10-10-42A
Bury, Kermit M. 1-20-44F

Busbee, William S., Jr. 6-8-44C
Busby, Jack D. 5-13-44B
Busby, Robert E. 7-24-44F
Busby, Robert E. 9-19-43C
Busby, Roland H., Jr. 8-5-43A
Busch, Keith N. 3-3-43A
Busch, Walter A. 10-9-44B
Bush, Charles J. 10-13-43D
Bush, D. C. 7-10-41
Bush, Elery L. 12-31-44B
Bush, Harold D. 3-3-45E
Bush, Herbert H. 12-8-43B
Bush, Kelton G. 1-8-44F
Bush, Richard C. 8-28-43G
Bushee, Raymond G. 3-10-45A
Bushey, Orin J. 9-27-44F
Bushnell, Jordan L. 5-3-43H
Busick, Brazel P. 2-23-45A
Busiek, John D. 8-12-42C
Buskey, Douglas H. 12-19-43F
Buss, Elmer 6-21-43D
Bussey, Carver T. 10-27-44B
Bussey, John William 9-22-42A
Busta, Claude B. 11-3-44B
Buswell, Richard G. 10-25-45A
Butcher, Albert E. 6-7-43C
Butcher, Morris A. 10-19-43H
Butcher, Paul E. 8-26-44B
Butela, William F. 9-18-42B
Butler, Cyril B. 8-17-43A
Butler, Don J. 10-9-42D
Butler, Edward W., Jr. 1-23-43A
Butler, Gravin S., Jr. 1-11-45B
Butler, George A. 5-7-42B
Butler, Jack H. 11-6-41
Butler, James H., Jr. 11-29-41
Butler, James Herring 6-12-42A
Butler, Parks M. 9-2-43E
Butler, T. R. 9-17-42D
Butner, J. A. 5-27-42G
Butorac, Melvin A. 4-5-44E
Butt, William R. 7-10-43F
Butterbaugh, Jack W. 1-16-43E
Butterfield, William H. 1-17-44A
Butterwick, Charles R. 2-22-43B
Buttler, Richard C. 3-12-44A
Buttner, Sol 12-19-43C
Button, Ervine J. 5-3-43H
Buturia, Joseph 6-12-41B
Buzzanca, Frank A. 6-5-44G
Buzzard, Ernest M. 7-13-44A
Bybee, Max 10-15-42B
Bybee, Walter E. 2-19-43A
Byerly, Willard A. 8-17-45
Byers, Charles W., Jr. 3-27-44G
Byers, Ralph T. 12-9-43A
Byers, William H. 1-24-44D
Byington, Robert M. 1-16-44N
Bynum, E. V. 11-17-41A
Bynum, Franklin C. 3-25-43D
Byran, Harlan L. 10-23-44F
Byrd, Charles E. 12-1-41A
Byrd, Charles W. 11-4-42A
Byrd, Homer L. 6-8-43K
Byrd, Jimmie C. 7-25-43D
Byrd, John T. 6-6-45B
Byrd, Nordi 6-20-44A
Byrd, William H. 6-13-44C
Byrne, Charles E., Jr. 1-20-44C
Byrne, Frank T. 12-25-44C
Byrne, John F. 4-24-44F
Byrne, Paul K. 10-1-45
Byrne, Richard L. 11-29-42B
Byrnes, George W. 5-30-43A

Byrnes, James E. 1-6-44A
Byrnes, John P. 11-11-44A
Byron, Lee Dwell 3-15-43B
Bystrom, Leroy 7-28-43D
Bywaters, William 12-5-42C

Cabe, Henry W. 7-30-43E
Cabe, Jarvis B. 12-8-43B
Caccavale, Joseph F. 8-4-43I
Caceves, Radames E. 12-22-43A
Caddell, J.W. 4-5-42A
Cadle, William E. 6-23-43B
Cadlolo, Charles, Jr. 8-9-43G
Cadwallader, Donald J. 1-12-44H
Cadwallader, William M. 3-10-45C
Cadwell, Harold L., Jr. 12-28-44A
Cadwell, N.L. 1-11-43D
Cady, Robert P. 10-24-42D
Caffney, Bernard G. 2-24-44D
Cagle, Clyde J. 4-28-44D
Cagney, Joseph J. 4-27-43A
Cahaly, Dimitri M. 3-26-44B
Cahill, Andrew J. 3-21-44C
Cahill, Burr T. 12-29-44D
Cahill, Edward E. 5-15-42C
Cahill, Marion N. 11-14-41
Cahoon, Wells L. 11-28-42A
Caillier, James E. 12-22-42A
Cain, James A. 1-26-43B
Cain, Joseph C. 2-5-43B
Cain, Kenneth D. 3-12-43D
Cain, Nigel H. 4-5-42D
Cain, Robert A. 2-16-45
Cain, Theodore J. 11-19-43K
Caines, Wootz W. 7-4-44F
Cairney, Alexander 2-14-45B
Cairns, A.J. 1-6-43D
Caissie, Wilfred J. 5-15-42A
Calame, Harry M. 6-29-44D
Calder, D.J. 10-3-42E
Calderwood, William E. 5-20-43H
Caldwell, James R., Jr. 12-26-42A
Caldwell, James W., Jr. 7-15-44B
Caldwell, Jay H. 2-27-43D
Caldwell, Leon G. 7-22-44F
Caldwell, Maxwell W. 12-26-42A
Calhoon, Donald J. 10-23-43D
Calhoon, Ralph M. 1-13-44A
Calhoun, Benjamin P. 10-8-44A
Calhoun, Hal A. 2-5-45
Calkins, Ralph T. 12-19-43B
Calkins, Wilbert S. 11-13-42A
Call, Ronald L. 11-4-43D
Callaghan, Phillip W. 5-27-45A
Callahan, Alvin F. 10-7-42C
Callahan, Edward M. 5-26-44B
Callahan, H.R. 3-29-43C
Callahan, Herbert W. 3-10-45C
Callahan, Jeremiah 1-30-45A
Callahan, John L. 2-18-44G
Callahan, Noel E. 7-27-43A
Callahan, Richard A. 5-25-44E
Callahan, Robert A. 3-14-43E
Callahan, Stanley E. 7-5-44G
Callaway, James B. 12-29-42B
Callaway, Pete 7-16-43F
Callender, Bernhard M. 4-27-44F
Callender, Donald A. 6-23-43B

Edwards, Harold E. 7-26-44C
Edwards, Hoover 6-29-42A
Edwards, Howard L. 11-17-41D
Edwards, Jack N. 6-21-44C
Edwards, James H. 10-11-42A
Edwards, James H. 6-28-43B
Edwards, James R. 4-13-45
Edwards, Jerome Thompson 5-7-43F
Edwards, John 8-24-44A
Edwards, Lucius D. 1-12-42B
Edwards, Marjorie D. 6-13-44A
Edwards, Milton O. 5-31-43I
Edwards, Ralph C. 11-11-43E
Edwards, Ralph M. 6-13-42A
Edwards, Richard Francis 3-6-43C
Edwards, Robert K. 5-23-43A
Edwards, Rubio 12-16-44A
Edwards, Sidney A. 7-1-43F
Edwards, Thomas L. 12-24-43D
Edwards, William H. 4-20-45A
Edwards, William M. 1-8-42C
Edwards, Wilson W. 10-28-42F
Edvean, John H. 4-7-43b
Effrige, Frederick 5-25-43A
Egan, Carl J. 8-19-43F
Egan, James M. 3-21-43A
Egan, Michael F. 11-12-44B
Egan, P.P. 9-15-45
Egan, Vincent W. 11-3-43A
Egar, John K. 7-20-44A
Egelhoff, C.H. 11-13-42D
Eggeling, G.W. 7-6-42A
Egger, Louis P. 10-7-42C
Eggers, Emsley M. 7-18-43C
Eggleston, Maxon W. 9-19-44A
Eging, Henry B. 6-15-45E
Eglian, Eugene 5-11-45E
Ehlers, Walter H. 3-5-43A
Ehret, Roland C. 12-15-42D
Ehrman, Melvin C. 11-2-44A
Ehrhardt, Lawrence W. 3-21-44G
Eichamer, George 3-11-44D
Eichelberger, Ernest C. 1-29-44M
Eichenbaum, Alfred S. 8-31-45B
Eichenberger, Kent E. 8-1-43C
Eichenlaub, George J. 4-25-44G
Eicher, Howard R. 9-22-43E
Eichert, Willrid J. 7-14-45A
Eichholzer, Joseph P. 12-8-44B
Eickhoff, Jack K. 8-29-44A
Eickmeyer, Alva 11-11-44B
Eickstadt, Robert 12-6-44F
Eierstein, Irving 11-20-43B
Eifert, Fredie C. 3-11-43K
Eilers, John M. 10-3-43B
Eisele, Harold E. 6-13-44F
Eiseman, Willliam H. 5-28-43A
Eisenbarth, Walter 7-5-43D
Eisenhart, D.H. 9-6-43G
Eisenhart, Frank O. 11-6-44C
Eisler, Carl J. 10-11-44B
Eitzen, Orville H. 8-14-44E
Ekberg, Vernon C. 11-19-43E
Ekdahl, Kenneth L. 11-22-43E
Ekelund, Roy L. 10-20-43D
Eklof, William A. 7-28-43C
Ekstrom, Donald I. 9-25-42B

Ekstrom, Manfred A. 5-14-43E
Ekstrom, Rudolph E. 12-12-44B
Elbert, Gaylord W. 3-26-43D
Elder, Charles S. 4-13-43D
Elder, Douglas M. 2-18-44B
Elder, Herman 10-29-43A
Elder, W.F. 6-22-44J
Elder, William K. 6-28-42C
Eldon, David A. 11-16-43H
Eldredge, Wynn W. 1-24-44A
Eldridge, George W. 1-24-45C
Eley, Vernon C. 5-30-44D
Eley, William B. 11-14-43E
Elhai, Salvatore J. 10-9-43J
Elias, Irving A. 7-1-45A
Elinsky, Morris 10-18-43C
Elisha, Charles J. 7-20-44C
Elivian, Ernest E. 2-19-44A
Elkin, David E. 6-22-45A
Elkin, Ernest M., Jr. 5-3-43F
Elkins, Darrill P. 2-15-45C
Elkins, Quintin R. 8-7-43A
Eller, William 5-1-45
Ellett, George M. 2-9-45C
Elli, Milton F. 11-16-43H
Ellingson, Lyle C. 1-29-44C
Ellington, Hal 2-17-43F
Ellington, John K. 2-9-44
Ellington, Spugeon N. 12-10-45
Elliot, A.N. 3-29-43C
Elliot, Arnold R. 3-1-43C
Elliot, Billy G. 3-16-43E
Elliot, Donald G. 5-26-44D
Elliot, Everett G. 2-3-43C
Elliot, Glen N. 6-12-43D
Elliot, James 6-5-44F
Elliot, Lawrence H. 4-15-44D
Elliot, Michael B. 4-11-45A
Elliot, Milton A. 4-6-43J
Elliot, Robert A. 12-16-44G
Elliot, W.D. 5-30-44F
Elliott, Billy Y. 11-2-42F
Elliott, Clifford 1-2-43C
Elliott, George S. 12-29-42E
Elliott, Jack C. 12-9-43A
Elliott, Kenneth E. 10-8-41
Elliott, Otis L. 8-21-41B
Elliott, Paul O. 8-5-43B
Elliott, Rae 9-14-44F
Elliott, Stewart P., Jr. 12-22-42A
Ellis, Bruce G. 6-15-43A
Ellis, E.L. 2-18-44G
Ellis, Earl E. 9-7-43C
Ellis, Glenn M. 10-28-43A
Ellis, Harold A. 11-16-44G
Ellis, Harry W. 5-26-44D
Ellis, Hesler W. 6-29-45A
Ellis, Hugh B. 2-17-44C
Ellis, James T. III 11-12-44B
Ellis, John E. 3-25-44C
Ellis, Lathrop P. 7-17-45A
Ellis, Louis B. 5-22-44I
Ellis, Melvyn T. 9-25-43C
Ellis, Raymond F. 7-19-43D
Ellis, Richard G. 1-22-44A
Ellis, Robert E. 7-12-42A
Ellis, Ralph Russell 7-9-43J
Ellis, Solon E. 9-11-43F
Ellison, Alden V. 2-28-45B
Ellison, Richard T. 6-27-43K
Elliston, Milton O. 11-24-43F
Ells, B.W. 5-22-41C
Ellwood, Albert J. 3-19-43E
Ellwood, August A. 5-4-43E
Elmer, James Harris 9-29-42A
Elmore, Wayne K. 8-16-43J

Elnitsky, Walter 6-29-44A
Elrod, James L. 7-11-42A
Elrod, Robert A. 1-10-44D
Elrod, W.R. 6-29-43D
Elsasser, Loren R. 10-7-45
Elsie, Hubert J. 1-29-44I
Elson, John F. 2-25-43D
Eltzroth, James M. 4-18-43B
Elwell, Joseph L. 10-27-43D
Elvin, Arthur E. 1-22-45C
Ely, Howard C. 7-24-43I
Ely, Robert G. 4-17-44G
Ely, Theodore B., Jr. 9-10-42B
Emberson, Channing B. 1-31-42
Embry, Allison 10-20-41A
Emerich, James F. 7-19-43E
Emerson, Arthur R. 11-26-43A
Emerson, Clifford L. 6-7-44I
Emerson, James P. 6-28-43A
Emerson, Raymond D. 12-8-45
Emerson, William F. 2-14-44A
Emery, John P. 9-8-43C
Emery, Valmer C., Jr. 4-17-44C
Emig, Alfred O. 12-17-45
Emigh, Elmer E. 8-29-44F
Emmelmann, F.G. 8-25-45
Emmert, Charles E. 6-30-45B
Emminger, J,W, 11-8-42F
Emrie, Aress 8-18-44H
Ence, Albert G. 2-10-44A
Endemano, Ernest H. 8-17-43A
Enders, Vernon O. 10-16-45
Endres, Earl G. 5-26-44D
Endres, Meinrad J., Jr. 8-8-42B
Endsley, Harry B. 12-19-43F
Endy, Gerald D. 9-5-43A
Enegbrepson, Charles C. 5-7-43H
Eng, James C. 8-2-43B
Engbert, Albert J. 8-5-45E
Engel, Leibel I. 6-5-43G
Engeman, Charles T. 3-25-41
Enghadl, Clarence E. 10-23-44C
England, Daniel C., Jr. 7-10-44A
Englebrecht, Vernon H. 12-30-41B
Englebrook, Donald E. 2-12-43B
Englen, Richard C. 3-6-45E
Englerth, Lonnie W. 12-1-43D
Engles, Richard J. 10-22-44A
English, Arthur J. 9-6-44E
English, Darrow E. 9-24-43K
English, Joseph D. 12-6-42A
English, Oliver R. 7-23-43C
English, Richard J. 8-14-43A
English, William H. 9-21-44G
Ening, P.K. 5-30-44F
Enloe, Joseph R., Jr. 11-5-44A
Enlos, J.W. 11-21-43C
Ennen, Charles 12-8-45
Ennis, Robert F. 10-13-44A
Ennis, W.R. 3-29-43C
Ennis, William T. 11-4-43D
Enoch, Charles H. 8-19-44D
Enos, Martin B., Jr. 5-25-43H
Ensiminger, William D., Jr. 2-2-43F
Enswiler, Homer R. 7-7-42D
Entres, Paul J. 4-27-45B
Entricau, William A. 10-13-43D
Enyeart, Robert N. 12-30-41B
Epler, Robin B. 1-28-44D
Epperly, Leonard K. 4-20-45A
Epperson, William H. 6-19-45B
Eppinger, Thomas G. 4-24-44D

Epstein, Elwyn E. 2-12-44G
Epstein, H.L. 6-22-44J
Erard, Andre C. 8-2-44A
Erb, John W. 8-16-43J
Erbe, Raymond L. 4-3-44A
Erck, William G., Jr. 2-19-42B
Erdoesy, Everett W. 1-31-44B
Ericksen, Dick L. 8-15-44D
Erickson, Arthur W. 1-7-44E
Erickson, Charles W. 11-13-43B
Erickson, Dan M. 1-1-43B
Erickson, Donald H. 4-18-45D
Erickson, Elizabeth 4-16-44B
Erickson, Glen E. 5-20-43C, 4-18-45C
Erickson, K.G. 2-23-43B
Erickson, Lester S. 7-1-42B
Eirckson, Melville A. 11-5-42B
Erickson, R.V. 10-11-43A
Erickson, Roy W. 10-22-42B
Erler, Wallace E. 4-4-43C
Ernest, William E. 2-28-43C
Ernst, Herman E. 12-24-43D
Ernst, Raymond C. 10-29-44A
Ertam, Zaim 7-5-44B
Ertz, Jeffery C., Jr. 1-7-43F
Ertzbischoff, F.P. 1-21-44D
Erwin, Frank W. 5-3-42D
Erwin, George P. 1-26-45D
Erwin, Harlan L. 7-31-45A
Erwin, Woodrow W. 2-26-43B
Escardo, Carlos A. 7-31-45C
Esckilsen, Gordon L. 12-7-43D
Eshbach, James R. 7-20-43B
Esinhart, Hartley L. 7-20-44C
Esler, Wilmer 4-11-41
Eslinger, Lawrence H. 9-17-44E
Espenship, Clark 6-5-43B
Espinosa, Marcus H. 7-20-44C
Esser, Donald F. 12-11-44B
Esser, Jerome J. 9-29-44D
Esser, Joseph O. 5-31-43C
Esser, William F. 5-1-44C
Estabrook, Norman B. 8-14-43H
Estaver, George W., Jr. 8-3-43D
Estep, Howard 6-17-43C
Estes, Benjamin R. 1-13-44A
Estes, Billy J. 11-3-43D
Estes, D.C. 12-5-44A
Estes, Mark E. 7-5-43C
Etheridge, H.R. 9-5-43A
Etheridge, Vernon K. 2-17-43D
Etherton, Raymond A., Jr. 1-15-43L
Etkinson, Frank L., Jr. 5-6-44I
Etler, Harold L. 2-12-43H
Ettelt, Richard J. 9-24-43H
Ettinger, Cecil R. 7-16-44B
Etz, Herbert R. 9-28-44C
Eubank, Rollen H. 1-15-43L
Eubanks, Marvin W. 7-10-44F
Eucinas, William 3-24-44A
Eustice, John Clifford 6-3-42D
Euston, Walter H. 9-5-44E
Euzen, Norris B. 5-18-44A
Evans, Aaron A. 10-3-44A
Evans, Billy J. 5-16-43A
Evans, Charlie F. 9-24-44F
Evans, Clayton W. 9-24-44E
Evans, Donald W. 11-15-42C
Evans, Elmer 9-5-42B
Evans, F.D. 8-4-42C
Evans, Frederick B. 6-14-44H
Evans, George D. 6-13-43
Evans, George V., Jr. 10-8-43D

Miller, Elmer R. 7-27-45A
Miller, Ernest E. 12-17-45
Miller, Frank P. 4-23-43E
Miller, Fred M. 8-3-44B
Miller, Geneva Eliza 1-6-42
Miller, George 12-8-45
Miller, George H. 1-15-45D
Miller, George J. 11-21-43A
Miller, George M. 7-27-45D
Miller, Glenn E. 6-20-45C
Miller, Glenn W. 8-5-43C
Miller, Gordan C. 6-12-43G
Miller, Harold E. 5-18-43F
Miller, Harry D. 2-2-43B
Miller, Harvey E. 11-14-43E
Miller, Henry 12-25-44A
Miller, Howard Herman 11-19-42A
Miller, Howard S. 10-6-43G
Miller, James 4-5-43F
Miller, James E.L. 1-14-44A
Miller, James H. 3-5-42E
Miller, James H. 4-2-43A
Miller, James T. 10-4-43E
Miller, John A. 7-10-43D
Miller, John C. 11-7-44A
Miller, John E. 1-17-44A
Miller, John H. 10 5 45H
Miller, Johnson S. 10-8-44B
Miller, Joseph M. 6-19-43A
Miller, Joseph M. 3-26-44I
Miller, Joseph P. 9-24-45E
Miller, Lee V. 6-29-43A
Miller, Lewis H. 1-8-42B
Miller, Lorin D. 10-1-44C
Miller, Louis E. 7-9-43I
Miller, M. A. 11-13-41A
Miller, Martin J. 9-21-42A
Miller, Max E. 3-29-43A
Miller, Paul D. 1-12-43A
Miller, Phelp 2-16-44B
Miller, Phillip G. 10-5-45B
Miller, R. W. 12-5-44A
Miller, Raymond Edward, Jr. 4-28-42A
Miller, Raymond E., Jr. 5-9-42A
Miller, Richard 6-22-44E
Miller, Richard A. 12-28-44C
Miller, Richard L. 1-19-44B
Miller, Richard R. 1-20-43C
Miller, Robert J. 8-10-43B
Miller, Robert J. 1-20-44F
Miller, Robert L. 11-3-43D
Miller, Robert L. 4-1-44F
Miller, Robert M. 10-29-43C
Miller, Robert M. 7-15-44E
Miller, Ross K. 10-2-44A
Miller, Sam 10-11-41
Miller, Seldon T. 11-17-41D
Miller, Troy L. 4-21-45D
Miller, Vernon 1-2-43E
Miller, Vernon E. 1-19-44B
Miller, W. F. 7-8-42C
Miller, Wendell C. 8-22-42B
Miller, Wilber E. 11-10-44C
Miller, Wilbert 8-28-43G
Miller, Wilfred E. 10-16-42A
Miller, William 3-29-43A
Miller, William 7-2-44C
Miller, William A. 6-24-44B
Miller, William E. 1-19-44F
Miller, William L. 4-2-44C
Miller, William R. 8-26-42A
Miller, William R. 9-26-43G
Miller, William W., Jr. 1-14-45B

Millet, Sidney M. 8-28-43F
Millich, Joseph, Jr. 5-24-43D
Milligan, Douglas F. 11-25-43F
Milligan, Harold C. 8-27-43C
Milligan, James D. 1-17-43D
Milligan, Mather P. 5-22-43D
Millikan, George R. 3-14-43D
Millikan, James W. 4-21-43C
Milliron, E. C. 12-25-44C
Millis, Hugh W. 5-6-42E
Millman, Alex 7-27-43A
Millman, Odean R. 7-4-43F
Millowitz, Mendall 10-22-43E
Mills, Arthur D. 12-14-44F
Mills, Carl L. 10-3-43D
Mills, David L. 9-4-43I
Mills, H. E. 4-23-43D
Mills, Harold L. 10-22-42D
Mills, Hawley P. 3-9-43E
Mills, Irving J. 10-20-43F
Mills, James S. 7-19-43A
Mills, John F. 12-11-42D
Millsap, John C. 9-6-43C
Milner, Roland F. 12-3-43B
Milner, Victor, Jr. 12-1-41B
Milton, Arthur 9-26-43G
Milton, Gerard J. 8-9-45D
Milton, James R. 1 22 45A
Miltz, Raymond E. 5-15-44A
Mimmack, Russell F. 10-21-44
Mims, W. T. 2-17-43F
Minarcin, Milton W. 3-8-42
Minden, Herman 5-16-44G
Miner, Denver E. 7-27-42A
Miner, Richard H. 2-25-44J
Minervini, Michael S. 10-29-44B
Mingle, Henry W., Jr. 9-28-44D
Mings, Benjamin S. 9-4-43I
Minichiello, Ralph F. 10-29-43D
Minichillo, Thomas J. 2-24-43C
Mink, Wesley F. 4-22-42A
Minnich, H. A. 8-28-42C
Minnich, LuVerne C. 10-15-44D
Minnick, Franklin L., Jr. 7-20-44E
Minnick, Thomas J. 5-9-44E
Minor, Francis J. 7-17-42D
Minor, James A. 6-1-43B
Minor, James D. 7-27-45C
Minski, Thadday 11-19-43I
Minton, Glenn R. 9-4-43I
Mirata, Charles 3-18-44F
Miro, Rudolph M. 8-18-43D
Mirro, Anthony J. 10-27-42F
Misak, Otto R., Jr. 11-10-43D
Misciagna, James R. 8-21-45D
Mishey, George C. 11-16-43H
Miskell, Robert J. 7-21-44H
Miskevich, Stephen 6-15-45E
Miskopf, Richard F., Jr. 12-6-42B
Missana, Charles 2-17-43D
Mitcham, Oriel 9-5-42D
Mitchel, Mancil L. 4-27-44F
Mitchell Anthony B. 3-25-43A
Mitchell, Artie U. 5-12-43D
Mitchell, Clarence B. 8-30-43D
Mitchell, Dan D. 12-15-44A
Mitchell, Douglas T. 11-14-42A
Mitchell, Edward B. 10-31-43A
Mitchell, George 11-22-44C
Mitchell, Hewitt F. 8-8-43C

Mitchell, Irving F. 8-9-43A
Mitchell, James B. 5-12-44M
Mitchell, James D. 3-13-45A
Mitchell, James E. 1-23-44F
Mitchell, James H. 6-15-42A
Mitchell, James I. 2-21-43B
Mitchell, James W. 12-23-44B
Mitchell, John J. 8-29-41
Mitchell, Leon D. 6-6-43C
Mitchell, Lyle S. 10-9-43H
Mitchell, M. T. 9-6-43B
Mitchell, Ralph A. 3-19-45F
Mitchell, Richard B. 3-3-45A
Mitchell, Rufus E. 4-13-43C
Mitchell, Samuel C. 5-13-43K
Mitchell, Samuel L. 11-6-42B
Mitchell, Thomas G. 12-16-42D
Mitchell, William D. 5-6-44G
Mitchell, William E. 1-15-43J
Mitchell, William L. 3-4-45C
Mitchelson, John H. 8-8-45B
Mitchler, James T. 3-15-42B
Mitterando, Louis T. 3-27-44G
Mittleman, Adolf R. 5-3-44E
Mittuch, Emery M. 6-17-43C
Mitzka, Michael 2-17-43D
Miullo, M. A. 12 9 41A
Mixon, Raymond L. 7-10-44E
Mize, Edwin M. 10-5-45A
Mize, Elzie L. 4-15-42C
Mize, Glenn T. 8-5-45D
Mizell, William D., Jr. 11-14-43E
Mizerski, Edward J. 1-17-44H
Mizgalski, Aloysius R. 6-21-45A
Mizner, Robert L. 8-3-43D
Mlady, Wilford C. 1-7-44F
Moberly, William C. 5-26-45B
Mobley, Orville, Jr. 3-23-45B
Mobley, Alfred L. 6-16-44E
Mobley, Rufus 12-4-43F
Mochen, Joseph F. 4-11-45F
Mock, Claire L. 10-22-43F
Mock, Orion L. 7-14-43D
Mock, Robert E. 4-24-45C
Mockler, Robert J. 7-6-43A
Mode, John R. 11-2-41B
Modes, Harry G., Jr. 1-18-45A
Modine, Elton C. 12-2-44E
Modjeski, Carl A. 3-21-44D
Moduno, Dominic T. 9-2-42E
Modzewski, Walter 11-2-44C
Moeller, C. E. 2-12-42A
Moeller, Fredrick W. 6-11-43C
Moen, Alton H. 10-3-43B
Moen, Hanford J. 2-15-44E
Moen, Merland E. 5-15-44I
Moeller, Berthold A., Jr. 1-19-44M
Moffatt, Virginia C. 10-5-43C
Moffet, Warren P. 7-29-45
Moffett, Charles S. 11-6-45B
Moffett, John H. 12-29-43C
Moffitt, Richard I. 1-24-45C
Moffitt, Thomas L. 9-6-44F
Mogan, Vincent P. 8-14-43K
Mogan, William M. 10-25-44H
Mogavero, Albert R. 10-25-43C
Mogenson, Walter E. 2-19-45B
Mohn, Fritz 2-18-43
Moholt, Arnold 5-18-44B
Mokol, Emanuel 11-3-44A
Mokwa, Ralph 3-12-44C
Molander, Stanley V. 12-1-42D
Molchan, John 4-20-45B

Moldenhauer, Don P. 8-13-44A
Moldovan, Louis R. 7-15-44B
Mole, Anthony G. 1-13-43F
Moler, Charles R. 9-24-43G
Moles, George L. 8-9-43C
Molique, Elmer J., Jr. 1-12-43A
Molitor, William G. 9-18-44C
Moll, Andrew J. 4-18-43A
Moller, Wharton E. 6-28-45A
Mollicut, Frank 1-1-42B
Mollohan, Donald E. 5-27-45A
Molloy, Joseph D. 5-21-42H
Molyneaux, Howard R., Jr. 1-8-44B
Momcilonich, Nick 5-2-45C
Monahan, Thomas F., Jr. 5-25-43E
Monan, Daniel A. 7-30-44C
Monceaux, Oren 3-17-43E
Mondahl, Leonard N. 10-3-44F
Mondino, Renato, Andre 2-29-44E
Mondonedo, Abelardo R. 6-8-45
Monesi, Louis P. 5-20-44G
Monetta, James J. 11-22-44E
Monger, Ernest M. 12-25-44A
Monhof, A. 4-21-42B
Monk, George S. 1-7-44A
Monk, William B. 6-8-43K
Monoham, Harold J. 3-8-43D
Monohan, William J. 9-14-44F
Monroe, Bart S. 3-11-43A
Monroe, Frank R. 6-1-43B
Monroe, Thomas R. 7-20-45A
Monson, Milton W. 5-26-43B
Monsulick, Charles 4-14-44A
Montagno, Edmund D. 11-29-45A
Montastier, Jean 7-20-44F
Monte, Charles G. 7-25-45A
Montemayor, Gustav A. 9-16-43A
Montenaro, Leo 8-5-45C
Montero, Jesus Gomez 8-20-43G
Montgomery, A. 9-20-43A
Montgomery, Alvin G. 5-5-44B
Montgomery, Archie W. 5-5-44C
Montgomery, Daniel 10-18-44F
Montgomery, Milton R. 3-11-43L
Montgomery, Otis A. 8-19-43F
Montgomery, Pat L. 9-6-44D
Montgomery, Russell L. 2-21-41
Montgomery, Vincent E. 11-12-45
Montmeat, John K. 1-11-43C
Montmorency, John D. 6-16-43D
Montoney, Edward H. 9-28-44D
Montville, Frank 8-15-44F
Mood, John L. 6-5-43B
Mood, Lester M. 7-20-44C
Moody, Calvin C. 12-16-44H
Moody, Frank H. 4-11-44A
Moody, George P. 5-5-41
Moody, Timothy M. 4-16-44A
Moody, Ward R. 7-5-43F
Moody, William T. 9-20-44A
Moon, James J., Jr. 11-3-44F
Moon, Robert L. 12-18-44B
Moon, Robert S. 10-27-44B

Roberts, Finis E., Jr. 11-15-42D
Roberts, Frank L. 2-26-44J
Roberts, Frank W. 1-29-43D
Roberts, Gerald F. 9-4-42C
Roberts, Gleanna 6-20-44C
Roberts, Harry S. 2-20-42D
Roberts, Herbert W. 8-14-43J
Roberts, James A. 2-9-43C
Roberts, James E. 4-28-43F
Roberts, James G. 11-30-43D
Roberts, Joseph L. 12-7-44C
Roberts, Leslie B. 8-3-44F
Roberts, Lewis 6-19-42C
Roberts, M.A. 9-14-44D
Roberts, Madison E. 12-23-42D
Roberts, Pat N., Jr. 8-3-44F
Roberts, Paul T. 7-27-43E
Roberts, Rayford C. 9-10-43F
Roberts, Robert M. 6-23-44A
Roberts, Thomas C. 2-18-44A
Roberts, Thomas T. 9-2-42E
Roberts, Wayne 2-28-45B
Roberts, Wendell E., Jr. 8-31-43H
Roberts, William 9-20-43A
Roberts, William E. 9-9-42C
Roberts, William S. 2-2-43F
Roberts, William W. 4-16-44D
Roberts, William W. 10-8-44A
Roberts, William W. 5-11-45D
Roberts, Zail E. 6-7-43A
Robertson, Ann 7-11-44G
Robertson, Bertrand H. 6-26-43A
Robertson, Clarence J. 5-3-44D
Robertson, Edward B. 3-3-45D
Robertson, Edward R. 11-19-42C
Robertson, Ernest W. 4-29-42B
Robertson, F.C. 2-1-43E
Robertson, George J. 7-11-44G
Robertson, H.M. 6-23-45A
Robertson, Henry H. 4-2-42A
Robertson, Hulan 1-31-44B
Robertson, J.M. 10-2-42D
Robertson, John E. 10-22-43F
Robertson, John M. 2-25-44C
Robertson, Orville H. 1-18-45E
Robertson, Paul W. 9-24-43B
Robertson, Rita 7-11-44G
Robertson, Roy 12-27-41C
Robertson, Vincent D. 6-7-44I
Robertson, William B. 8-1-43B
Robertson, William J. 8-21-43C
Robertson, William J. 1-7-44F
Robey, Clarence V. 2-15-45A
Robillard, Harvey J. 8-23-43C
Robin, Chester F. 4-7-43E
Robinett, Harold N. 10-17-42B
Robinette, Alexander S. 1-19-45C
Robinson, Comyn F. 12-21-42A
Robinson, Decatur M. 1-16-44G
Robinson, Donald T. 2-4-43C
Robinson, Everette F. 11-27-44A
Robinson, Herman 6-21-45A
Robinson, Hugh W. 12-29-43C
Robinson, Jack C. 10-23-43F
Robinson, James R. 7-10-44E
Robinson, John L. 5-18-43H
Robinson, Julius G. 6-15-43A
Robinson, L.D. 7-25-41
Robinson, Lawrence A. 9-22-42C
Robinson, Lyle W. 7-28-43F
Robinson, Raymond L. 2-26-44B

Robinson, Richard T. 1-23-44F
Robinson, Robert L. 11-19-43E
Robinson, Robert W. 3-4-44F
Robinson, Thomas B. 2-25-43A
Robinson, W. 2-17-43F
Robinson, W.C. 9-5-43F
Robinson, Walter P. 9-26-43B
Robinson, William E. 8-27-43H
Robinson, William H., Jr. 5-12-42C
Robison, Donald V. 3-20-44A
Robison, Edward O. 8-21-43F
Robison, George S. 5-16-43C
Robison, Henry W. 10-13-44C
Robison, Merle F. 3-12-43E
Robson, William M., Jr. 10-18-43C
Roby, Roy A. 9-20-43A
Rocheleau, Norman N. 8-27-44B
Rochelle, Liston L. 2-17-45B
Rochester, Eugene E. 8-23-42A
Rock, Lawrence A. 12-6-43A
Rock, Walter E. 10-8-44A
Rocke, Billy B. 12-4-42F
Rockett, Milburn W. 7-23-43F
Rockney, Arlo D. 12-15-43G
Rockwell, Ernest B. 7-20-44C
Rockwell, Harry T. 3-18-43B
Rockwell, Stephen G. 1-4-44B
Roddy, Charles N. 8-26-42B
Rodeck, Robert H. 6-12-45C
Roden, Perry L. 2-24-44J
Roderick, Charles H. 9-29-44E
Rodge, George C. 1-2-42A
Rodger, John R. 2-14-43G
Rodger, William G. 7-25-44A
Rodgers, Adolph I. 4-14-44F
Rodgers, Edwin P. 2-15-44D
Rodgers, Henry J. 6-16-42C
Rodgers, J.T. 1-18-44C
Rodgers, Michael F. 8-24-43E
Rodgers, R.D. 1-10-45C
Rodgers, Robert H. 1-5-44F
Rodgers, Willis M. 5-7-43H
Rodriguez, Buster, 7-20-44G
Rodriguez, James E. 8-26-44B
Rodriguez, Louis A. 4-26-44D
Rodriguez, Luis R. 2-22-44I
Rody, George 1-22-44A
Roe, Alfred W. 11-4-43C
Roe, Charles E. 10-11-44C
Roehm, Donna 5-26-45C
Roehm, Nina 5-26-45C
Roehm, Susan 5-26-45C
Roehm, Wesley B. 5-26-45C
Roelle, Raymond L. 8-4-44B
Roemlein, George R. 9-16-44C
Roese, Stanley A. 6-20-44B
Roesler, George W. , Jr. 11-14-45
Rogarelski, Frank A. 5-26-45B
Rogers, Allen K. 5-11-44G
Rogers, Clyde 7-18-43A
Rogers, Dolan J. 2-3-45D
Rogers, Donald S. 7-4-44F
Rogers, Edgar D. 7-19-43A
Rogers, Edward C. 7-29-44C
Rogers, Frank 5-28-42A
Rogers, Gale W. 9-13-43A
Rogers, George B. 10-9-43B
Rogers, George C. 10-4-43A
Rogers, Henry T. 10-15-44H
Rogers, James P. 9-25-43C
Rogers, James W. 11-3-43C
Rogers, John D. 12-29-43C

Rogers, Lynn E. 3-10-41
Rogers, Phillip J. 8-16-43D
Rogers, Richard G. 7-21-44B
Rogers, Robert E. 6-7-44A
Rogers, Robert L. 12-23-44D
Rogers, Rubert E. 10-23-44A
Rogers, Stephen A. 12-24-43H
Rogers, Thomas L. 7-23-45C
Rogers, Willard O. 6-14-45B
Rogers, William S., 1-28-44D
Rogers, William W. 6-16-43H
Rogo, Clevio R. 6-20-42B
Rogove, Herbert J. 8-21-44A
Rohlfs, Leo F. 4-23-43G
Rohozka, Frank 2-3-43B
Rohr, Edmund 11-1-42
Rohrbach, Alan I. 11-2-44A
Rohrbach, William H. 6-3-44C
Rohrbaugh, James B. 1-3-44E
Rohrer, Richard 7-27-43A
Rohrs, Albert C. 1-25-44E
Roiland, Ralph G. 9-20-43A
Rojas, Rodney G. 2-3-44B
Roland-Gosselin, Pierre 4-24-44B
Rolfe, Benjamin E. 2-3-44B
Rolfe, James A. 3-12-43C
Rollins, Fred F. 12-6-42D
Rollins, James E. 8-27-43A
Rollins, James K. 1-31-43A
Rollins, Richard, Jr. 6-7-45A
Rollow, George V. 11-17-44A
Rollston, Frederick P. 4-24-43C
Rom, Roy C. 3-24-45A
Roman, Clarence 12-17-41A
Roman, George T. 2-14-43B
Roman, Reinhard W. 11-19-44
Roman, Sam 5-22-44I
Romand, Robert R. 7-3-45C
Romanin, Aldo L. 4-10-44A
Rome, Emery J. 5-11-44A
Rome, Evo J. 6-28-44F
Romo, Peter E. 10-15-43A
Ronoghan, William R. 6-29-43A
Ronkos, Charles G. 9-1-45A
Ronne, Richard D. 1-13-44D
Ronnfeldt, Leroy L. 4-8-43A
Ronning, John G. 10-1-44B
Ronning, O.E. 3-4-45B
Rood, Robert C. 8-10-44C
Rook, Cecil E., 12-30-42F
Rooker, Jack D. 5-8-43D
Rooker, William R. 5-6-45B
Rookstool, George L. 2-9-45A
Rooney, John H. 5-1-44A
Rooney, Phillip N. 3-27-45E
Root, Bradford K. 1-25-43D
Root, Gordon H. 11-26-43E
Roper, R.A. 10-15-42B
Roper, Thomas D. 5-15-44K
Rosado, George D. 10-2-44C
Rosalez, John S. 2-4-44A
Rosato, Antonio 7-30-42E
Rosblatt, Saul Peter 10-4-43G
Rose, Charles A. 2-21-45B
Rose, George L. 1-15-43L
Rose, James A. 1-4-45A
Rose, James H. 11-21-43D
Rose, Orville J. 3-22-43E
Rose, Paul E. 6-7-44A
Rose, Robert V. 1-14-43C
Rose, Robert W. 5-15-44A
Rose, Samuel D. 9-19-42B
Rose, T.D. 3-14-41
Rose, Truman D. 7-24-45E
Rose, Vincent J. 1-7-45B

Rose, Wayland W. 9-24-44F
Rose, Wilbur C. 7-30-42C
Rose, Willie F. 1-12-43A
Roseberry, Rufus M. 11-20-42B
Rosebush, Richard G. 4-18-42B
Rosell, Woodrow O. 4-13-43D
Rosella, Mary 2-18-43
Rosen, Benjamin N. 12-31-44A
Rosen, Harold 1-16-44G
Rosen, Harry C. 7-7-43C
Rosen, Stanley L. 12-15-42E
Rosen, Herbert Perry 3-23-44F
Rosenbaum, Alvin L. 8-2-43F
Rosenberg, David L. 12-6-42B
Rosenberg, Edwin R. 3-2-44C
Rosenberg, Gilbert 9-19-42E
Rosenberg, Melvin E. 3-26-44G
Rosenberg, Millard F. 6-5-44D
Rosenberg, Murray 11-16-44C
Rosenblatt, George 12-25-44A
Rosenburg, Charles W. 5-17-43E
Rosenchein, Robert 7-3-43C
Rosendahl, Carl 3-8-44E
Rosendorf, Frank C. 5-2-42D
Rosene, Lavern 11-13-42C
Rosenfield, Kenneth 6-9-45
Rosengren, Jess A. 2-19-43C
Rosenstock, Gordon F. 3-10-41
Rosenthal, Benjamin J. 11-19-43K
Rosenthal, Jules E. 7-29-41A
Rosoff, Ernest 11-24-43I
Ross, Beaman D. 7-17-43C
Ross, James A. 2-12-45C
Ross, James W. 11-19-43K
Ross, Lee G. 1-19-45B
Ross, Lloyd L. 10-6-43C
Ross, M. 2-17-43F
Ross, Richard H. 7-19-43C
Ross, Robert T. 9-24-42B
Ross, Russell R. 5-15-44C
Ross, Sam 7-19-43A
Ross, William R. 6-14-45B
Rosser, Joseph R. 10-23-42D
Rosser, William 7-1-42A
Rossi, Lloyd C. 8-5-45E
Rossi, Michael J. 7-25-45B
Rossler, Eldon F. 6-15-43A
Rostankowski, Leonard R. 4-26-43E
Rote, Bernard 8-10-43A
Rotegard, Glenn M. 8-3-43A
Rotert, William E. 1-2-45C
Roth, Adolph H. 9-28-43M
Roth, August C. 5-12-44B
Roth, Edmund L. 12-18-44E
Roth, Harold D. 9-10-43E
Roth, James A. 7-16-43D
Rothans, Clarence S. 6-5-45A
Rothblatt, Meyer 3-1-43B
Rothermel, Lee E. 11-15-41
Rothstein, George J. 4-29-45B
Rothstein, Simon 1-17-43B
Rottier, Dale R. 10-15-42C
Rotundo, Mitchel F. 3-23-43F
Rouhier, Robert N. 4-10-43C
Rouintree, Jesse G. 5-14-44C
Rouner, Floyd H., Jr. 1-13-44D
Rounsaville, Marvin M. 2-19-43B
Rounsaville, John C. 8-1-41
Rountree, Thomas B. 1-19-44B
Rouse, Barney L. 5-15-43B
Rousek, Jay P., 9-3-44B
Roush, George W. 1-28-44C
Roush, Orville E. 10-23-44A